Carbohydrate Chemistry

Volume 12

A Specialist Periodical Report

Carbohydrate Chemistry

Volume 12

A Review of the Literature Published
during 1978

Senior Reporters
J. F. Kennedy, *University of Birmingham*
and
N. R. Williams, *Birkbeck College, University of London*

Reporters
B. E. Davison, *Polytechnic of North London*
I. M. Morrison, *University of Glasgow*
R. J. Ferrier, *Victoria University of Wellington, New Zealand*
C. M. Sturgeon, *University of Edinburgh*
R. J. Sturgeon, *Heriot-Watt University, Edinburgh*

The Royal Society of Chemistry
Burlington House, London W1V 0BN

British Library Cataloguing in Publication Data

Carbohydrate chemistry. – Vol. 12. –
(A Specialist periodical report)
1. Carbohydrates
I. Royal Society of Chemistry
547.7'8 QD321

ISBN 0-85186-940-8
ISSN 0576-7172

Organic formulae composed by Wright's Symbolset method

Typeset and printed by John Wright & Sons (Printing) Ltd.
at The Stonebridge Press, Bristol

Preface

This report, the twelfth in the series, covers the new literature available to us between January and December 1978. We have attempted to produce a comprehensive survey of publications on carbohydrate chemistry in as concise yet readable form as possible; the ever-expanding volume of literature makes this an increasingly formidable endeavour, and we apologise for the inevitable errors and omissions that accrue. As has been the policy in previous years, the patent literature and *Abstracts of the American Chemical Society Meetings* have not been abstracted. We would be grateful to receive any constructive comments on the scope and methods of our coverage and presentation which may serve to improve the value of the Report.

The publication of this volume of *Carbohydrate Chemistry* marks another milestone in the progress of the series. Professor J. S. Brimacombe has now relinquished his position as Senior Reporter, and we should like to acknowledge here our debt of gratitude to him for his monumental labours in editing these reports over the past eight years.

The high standards he has set by his own painstaking scrutiny throughout have given many guidelines to help us, and challenge us to preserve in future volumes a clear reflection of his good work. The value of his productions has been repeatedly acknowledged in reviews of the various volumes (*e.g.* ref. 1).

In view of the ever enlarging degree of specialization even within the subject of carbohydrates and the increased size of the volume, and in order to facilitate production of the two parts of the volume concurrently, the Report is now being organized by two persons: Dr. N. R. Williams is acting as Senior Reporter of Part I (mono-, di-, and tri-saccharides and their derivatives), and Dr. J. F. Kennedy is acting as Senior Reporter for Part II (macromolecules).

It is a pleasure to welcome Dr. B. Davison as a reporter of Part I, and Dr. C. M. Sturgeon and Dr. I. M. Morrison as reporters for Part II who succeed Dr. J. M. Williams and Dr. B. L. Catley, respectively, whose valuable contributions we hereby acknowledge. It is also a pleasure to acknowledge the willing secretarial and administrative assistance of Miss Vivienne J. Hunt.

Finally, the invaluable services provided by Dr. P. G. Gardam (Manager, Books) and his staff, particularly Mr. K. Wilkinson at The Royal Society of Chemistry in the production of this Report have been much appreciated.

J. F. K.
N. R. W.

[1] A. H. Haines, *Carbohydrate Res.*, 1978, **67**, C31.

Contents

Part I Mono-, Di-, and Tri-saccharides and their Derivatives

Part II Macromolecules

Abbreviations

The following abbreviations have been used:

ADP	adenosine diphosphate
ATP	adenosine triphosphate
c.d.	circular dichroism
CDP	cytidine diphosphate
CMP	cytidine monophosphate
DBU	1,5-diazobicyclo[5,4,0]undec-5-ene
DCC	dicyclohexylcarbodi-imide
DEAE	diethylaminoethyl
DMF	*N,N*-dimethylformamide
DMSO	dimethyl sulphoxide
DNA	deoxyribonucleic acid
dpm	dipivaloylmethanato
e.s.r.	electron spin resonance
fod	2,2-dimethyl-6,6,7,7,8,8,8-heptafluoro-3,5-octanedionato
g.l.c.	gas–liquid chromatography
HMPT	hexamethylphosphotriamide
i.r.	infrared
NBS	*N*-bromosuccinimide
n.m.r.	nuclear magnetic resonance
o.r.d.	optical rotatory dispersion
py	pyridine
RNA	ribonucleic acid
THF	tetrahydrofuran
Thp	tetrahydropyranyl
TMS	trimethylsilyl
UDP	uridine diphosphate

Part I

MONO-, DI-, AND TRI-SACCHARIDES AND THEIR DERIVATIVES

By

B. E. Davison

R. J. Ferrier

N. R. Williams

1

Introduction

The general terms of reference remain the same as for previous volumes in this series. In order to minimize the time required for editorial collation, the previous system of numbering diagrams and references sequentially throughout Part I has been discarded in favour of each chapter carrying its own set of diagram and reference numbers as in Part II. We hope this alteration will not detract too seriously from the usefulness of the report. We have attempted to maintain the previous policy of providing a generous system of cross-references between chapters.

More than one thousand references quoted in Part I clearly reflect the continued high level of interest in the chemistry of carbohydrates, and the report demonstrates the particular interest now being shown in the synthesis of natural and analogous glycosides and disaccharides (Chapter 3), carbohydrate antibiotics (Chapter 18), and of course nucleosides (Chapter 19), which together account for over one-third of all the references cited.

Texts published this year of general interest to carbohydrate chemists have included issues in the Advances in Carbohydrate Chemistry and Biochemistry series, which are prefaced by appreciations of the late Professors E. J. Bourne [1] and E. L. Hirst;[2] these texts include review articles on 1,6-anhydro sugars,[3] cyclic acetals of aldoses and aldosides,[4] the Koenigs–Knorr reaction,[5] carbohydrate boronates,[6] and the biosynthesis of sugars in antibiotics.[7] Extensive surveys of carbohydrate derivatives are also included in several books devoted to antibiotics.[8, 9]

1 H. Weigel, *Adv. Carbohydrate Chem. Biochem.*, 1977, **34**, 1.
2 M. Stacey and D. J. Manners, *Adv. Carbohydrate Chem. Biochem.*, 1978, **35**, 1.
3 M. Černy and J. Staněk, *Adv. Carbohydrate Chem. Biochem.*, 1977, **34**, 24.
4 A. N. de Belder, *Adv. Carbohydrate Chem. Biochem.*, 1977, **34**, 179.
5 K. Igarashi, *Adv. Carbohydrate Chem. Biochem.*, 1977, **34**, 243.
6 R. J. Ferrier, *Adv. Carbohydrate Chem. Biochem.*, 1978, **35**, 31.
7 H. Grisebach, *Adv. Carbohydrate Chem. Biochem.*, 1978, **35**, 81.
8 'Kirk-Othmer Encyclopaedia of Chemical Technology', 3rd edn., Vol. 2, ed. M. Grayson and D. Eckroth, Wiley, New York and London, 1978.
9 'Topics in Antibiotic Chemistry', Vols. 1 and 2, ed. P. G. Sammes, Wiley, New York and London, 1977 (Vol. 1) and 1978 (Vol. 2).

2

Free Sugars

A review of aspects of the structure and reactivity of carbohydrates has appeared.[1] The free-radical telomerization of vinylene carbonate (1,3-dioxal-2-one) with polyhalomethanes has been dealt with in a review; telomers so derived may be converted to aldoses containing from three to eight carbon atoms[2] (see below). *Ab initio* SCF calculations have been used to predict the electron distribution, the electrostatic molecular potential around the oxygens, and the hydration and cation binding schemes of $C_{3'}$-*endo*-gg-ribose. The latter were correlated with recent experimental results.[3]

1 Isolation and Synthesis

Melibiose, raffinose, glucose, and fructose were detected in developing rice (strain IR29), whereas in strain IR28 sucrose, glucose, fructose, raffinose, maltose, melibiose, glucodifructose, maltotriose, and higher oligosaccharides were found.[4] More free sugars were present in the developing grain than when it was mature. Bran contains a higher proportion of raffinose and fructose than does milled rice. The free sugars, water-soluble gums, and hemicelluloses present in barley grains were compared before and after malting.[5] Sugars detected were arabinose, xylose, galactose, glucose, fructose, maltose, and sucrose, levels of which increase on malting. The hydrolysate of the extracellular polysaccharides of *Rhizobium* (strain CB756) was shown to contain 6-deoxy-L-talose, 3-*O*-methyl-D-glucose, and 6-*O*-methyl-D-galactose.[6] D-Galactofuranose was found to be present in the capsular polysaccharide of *Klebsiella* serotype K41.[7]

Per-*C*-deuteriated D-glucose has been synthesized[8] using the sequence shown in Scheme 1. A simple synthesis of L-idose from (1) by sequential treatment with sodium borohydride, sodium methoxide, and acid has been reported.[9] Also prepared were L-idose di(ethylthio)acetal and 1-deoxy-L-iditol.

[1] L. Hough and E. Tarelli, *Sci. Prog. (Oxford)*, 1977, **64**, 569 (*Chem. Abs.*, 1978, **88**, 38 082h).
[2] T. Kunieda and T. Takizawa, *Heterocycles*, 1977, **8**, 661 (*Chem. Abs.*, 1978, **88**, 89 938z).
[3] H. Berthod and A. Pullman, *Theor. Chim. Acta*, 1978, **47**, 59 (*Chem. Abs.*, 1978, **88**, 191 266r).
[4] C. G. Pascual, R. Singh, and B. O. Juliano, *Carbohydrate Res.*, 1978, **62**, 381.
[5] G. R. Wollard, E. B. Rathbone, and L. Novellie, *J. Sci. Food Agric.*, 1977, **28**, 469 (*Chem. Abs.*, 1978, **88**, 73 198x).
[6] L. D. Kennedy, *Carbohydrate Res.*, 1978, **61**, 217.
[7] G. Chambat, J. P. Joseleau, M. Lapeyre, and A. Lefebvre, *Carbohydrate Res.*, 1978, **63**, 323.
[8] H. J. Koch and R. S. Stuart, *Carbohydrate Res.*, 1978, **64**, 127.
[9] M. Blanc-Muesser and J. DeFaye, *Synthesis*, 1977, 568 (*Chem. Abs.*, 1978, **88**, 23 280x).

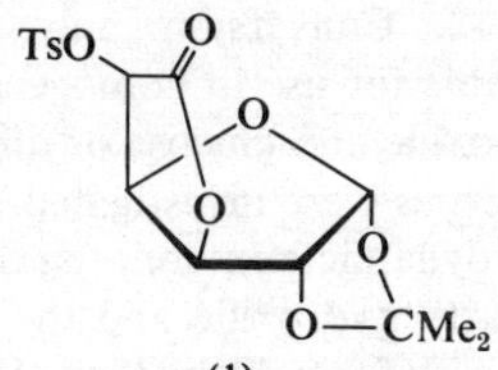

(1)

The polymerization of formaldehyde has been studied using various catalysts and complexing co-catalysts.[10, 11, 12] When a bed of NaX-zeolite spheres was used, aqueous formaldehyde at 95 °C gave an initial conversion to formose sugars of 50% at pH 5—7. Rapid irreversible catalyst deactivation then occurred due to the presence of formic acid produced in the undesired Cannizaro reaction. The problem was overcome by incorporation of 0.86 cm^3 sodium hydroxide per cm^3 formaldehyde into the combined feed to the reactor

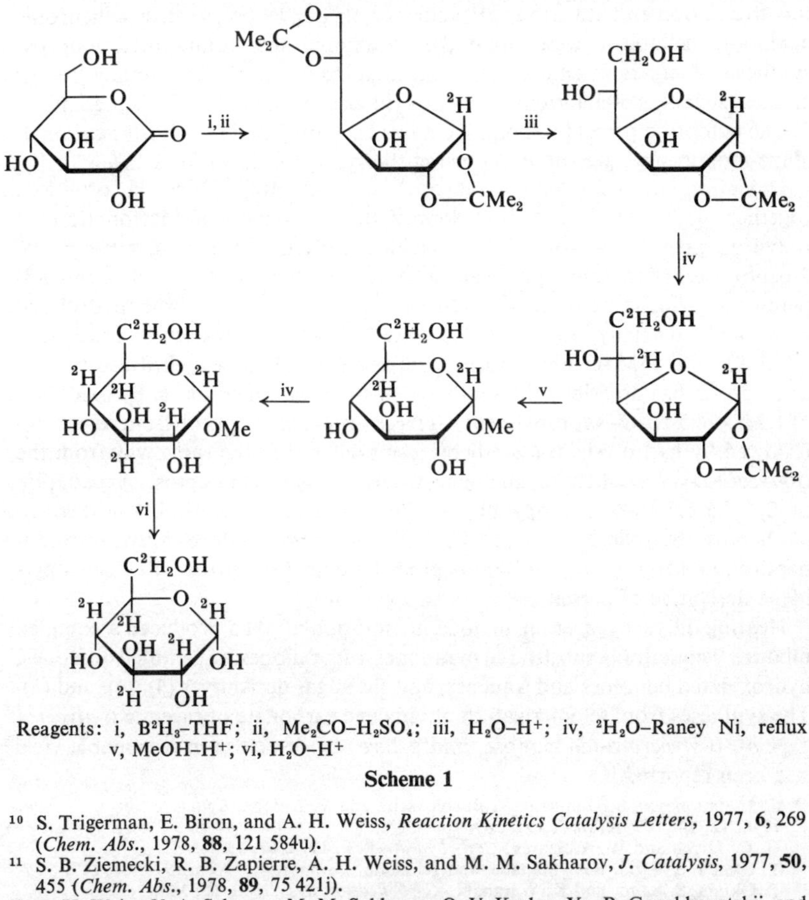

Reagents: i, B^2H_3–THF; ii, Me_2CO–H_2SO_4; iii, H_2O–H^+; iv, ^2H_2O–Raney Ni, reflux v, MeOH–H^+; vi, H_2O–H^+

Scheme 1

10 S. Trigerman, E. Biron, and A. H. Weiss, *Reaction Kinetics Catalysis Letters*, 1977, **6**, 269 (*Chem. Abs.*, 1978, **88**, 121 584u).

11 S. B. Ziemecki, R. B. Zapierre, A. H. Weiss, and M. M. Sakharov, *J. Catalysis*, 1977, **50**, 455 (*Chem. Abs.*, 1978, **89**, 75 421j).

12 A. H. Weiss, V. A. Seleznev, M. M. Sakharov, O. V. Krylov, Ya. B. Gorokhovatskii, and N. P. Evmenenko, *J. Catalysis*, 1977, **48**, 354 (*Chem. Abs.*, 1978, **88**, 121 585v).

thus maintaining pH 10—12. Conversion only fell from 95% to 92% during 3 h.[10] Catalysts suitable for use in homogeneous polymerization were prepared from calcium hydroxide and glucose or dihydroxyacetone, and their physical and chemical properties were investigated.[11] ESCA and i.r. analysis suggest a loosely structured dynamic mixture of species derived by co-ordination between the alkaline earth hydroxide and the hydroxylated compound. When the catalyst was used with formaldehyde in ^2H_2O no C—^{2}H bonds were formed, showing that water does not participate in the formose reaction and that the latter occurs within the complex. When glucose was used as the complexing co-catalyst with calcium hydroxide in a tank reaction, the Cannizaro reaction was reduced to ~ 2% with near-complete conversion to formose sugars.[12] Although calcium hydroxide is a good catalyst for the Cannizaro reaction, the calcium hydroxide–glucose complex is not.

Cold-plasma decomposition of methane–water mixtures produced simple organic compounds containing formaldehyde; on apatite surfaces under u.v. irradiation and at various pH values erythrose, ribose, glucose, glucuronic acid, and cellobiose were obtained. Ammonia was found to favour the synthesis of sugars when used to create alkaline media.[13] (The authors report all these sugars as D-enantiomers but no evidence for optical purity is presented.)

The octodiose present in apramycin has been prepared [14] as its di-isopropylidene-monobenzyl derivative (2), using the synthesis shown in Scheme 2.

Oxidation of 1,2:4,5-di-*O*-isopropylidene-D-xylitol, which is produced together with the 2,3:4,5-di-*O*-isopropylidene isomer on acetonation of D-xylitol, gave the 3-oxo-derivative, acid-hydrolysis of which gave D-*erythro*-3-pentulose. The same publication describes the synthesis of L-*threo*-3-pentulose: 2,4,5-tri-*O*-acetyl-1,3-*O*-benzylidene-L-arabinitol was hydrolysed with acid, partially acetylated to give 1,2,4,5-tetra-*O*-acetyl-L-arabinitol, oxidized with DMSO–phosphorus pentoxide and finally hydrolysed.[15] The same group has described the syntheses of D-*manno*-3-heptulose, by oxidation of 1,2:3,4:6,7-tri-*O*-isopropylidene D-*glycero*-D-*manno*-heptitol (from volemitol) followed by hydrolysis, D-*ido*-3-heptulose prepared in the same way from the D-*glycero*-L-*galacto*-heptitol analogue, D-*glycero*-L-*galacto*-heptose by oxidation of 2,3:4,5:6,7-tri-*O*-isopropylidene-D-*glycero*-L-*galacto* heptitol, D-*glycero*-D-*ido*-heptose by oxidation of 2,3:4,5:6,7-tri-*O*-isopropylidene-D-*glycero*-D-ido-heptitol, and D-*glycero*-D-*galacto*-heptose by oxidation of the tri-*O*-isopropylidene derivative of perseitol.[16]

Heating dihydroxyacetone in aqueous solution at pH 5 produced a complex mixture [17] containing substituted pyranones, dipyranones, substituted quinones, hydroxylated benzenes and toluenes, and the sugar derivatives (3), (4), and (5). The syntheses from telomerization of vinylene carbonate of racemic D-*glycero*-L-*gulo*-, D-*glycero*-D-*ido*-heptose, and D-*threo*-D-*ido*-octose in reasonable yield has been reported.[18]

[13] C. I. Simionescu, S. Dumitriu, V. Bulacovschi, and V. I. Popa, *Cellulose Chem. Technol.*, 1978, **12**, 143 (*Chem. Abs.*, 1978, **89**, 163 855j).
[14] H. C. Jarrell and W. A. Szarek, *Carbohydrate Res.*, 1978, **67**, 43.
[15] T. Okuda, S. Satio, K. Watanabe, and H. Isono, *Carbohydrate Res.*, 1978, **67**, 117.
[16] T. Okuda, S. Saito, and K. Watanabe, *Carbohydrate Res.*, 1978, **65**, 183.
[17] T. Popoff, O. Theander, and E. Westerlund, *Acta Chem. Scand.*, 1978, **32B**, 1.
[18] Y. Nii, T. Kunieda, and T. Takizawa, *Chem. and Pharm. Bull. (Japan)*, 1978, **26**, 1999.

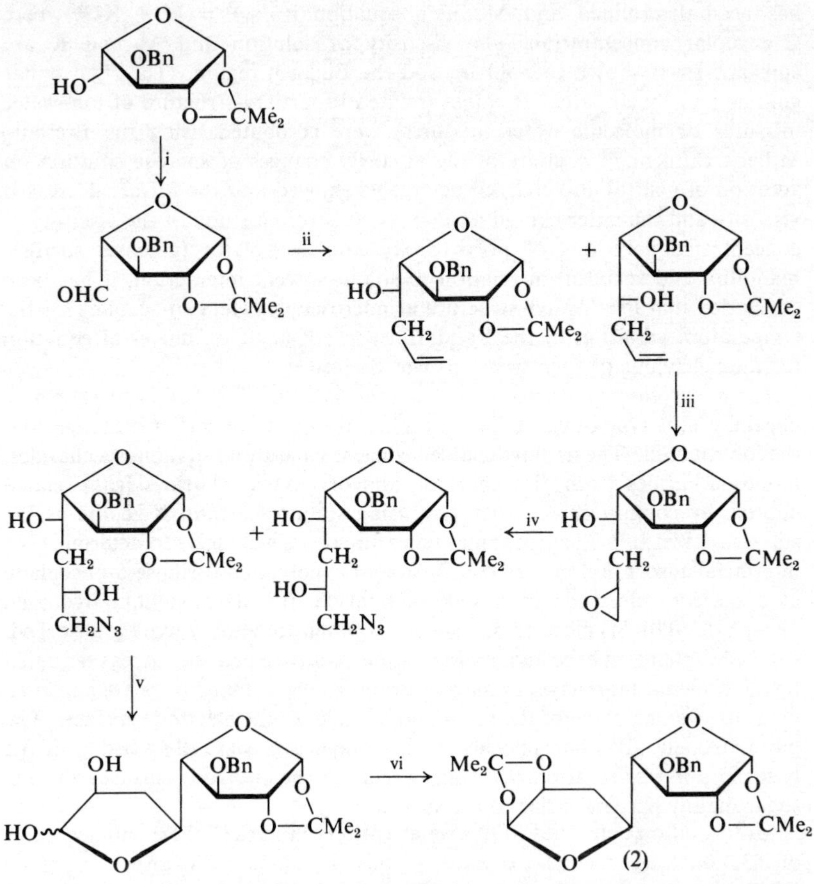

Reagents: i, CrO_3–dipy; ii, $CH_2{=}CH{-}CH_2MgBr$; iii, *m*–$ClC_6H_4CO_3H$; iv, NaN_3–NH_4Cl v, *hν*; vi, dimethoxypropane–TsOH

Scheme 2

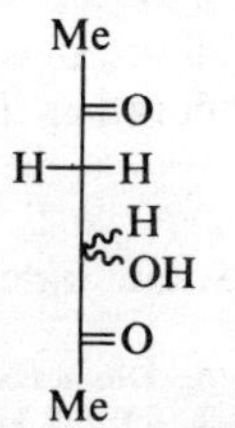

(3)

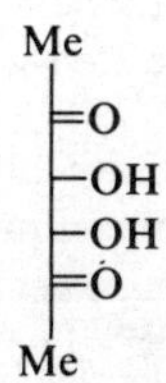

(4)

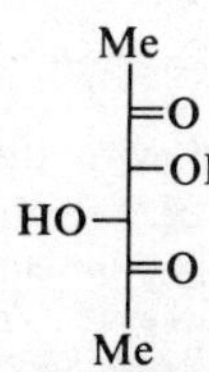

(5)

2 Physical Measurements

The viscosity behaviour of the ternary systems of sodium or potassium halides (0.125–3 M) in aqueous solutions of D-xylose (0.4 M) at 25, 30, 35, and 40 °C has been determined and Monlik's equation [$(\eta/\eta_0)^2 = M + KC^2$ where C = molar concentration, η = viscosity of solution and M and K are constants] was shown to hold beyond the Einstein region. The rigid molar volume and the apparent B values (related to tertiary structure of ion–water mixtures or molecule–water mixtures) were computed using the Breslau–Miller treatment.[19] A study of the viscosity changes of sucrose solution on addition of neutral and charged detergents showed that the former decreased viscosity and the latter caused an increase.[20] From a study of the viscosity of concentrated solutions of polyhydroxy non-electrolytes (glucose, sucrose, mannitol, and sorbitol) in relation to solute–solvent interaction, it has been concluded that local water structure at microscopic level can account for the temperature variation of the Vand–Einstein constant. A universal equation for the behaviour of such solutions was derived.[21]

The adsorption of fructose and glucose on X-zeolites has been shown to depend on the type of cation in the zeolite, the extent of cation exchange, and the solvent used. The findings enabled efficient separations of monosaccharides, monosaccharides from disaccharides, and of monosaccharides from related alditols by column chromatography using sodium calcium X-zeolite as the stationary phase.[22] Electrocapillary measurements were used to determine the interfacial adsorption of D-ribose, 2-deoxy-D-ribose, and D-ribose-5-phosphate at a mercury electrode in a sodium fluoride (0.5 M)–di-sodium hydrogen phosphate (0.01 M) electrolyte. Each was found to be only weakly absorbed, the 5-phosphate, as expected, mainly at the positive electrode. Areas occupied by molecules at monolayer surface saturation suggest that sugars are absorbed with the average plane of the sugar ring parallel to the electrode surface. The more strongly absorbed nucleosides in comparison were absorbed with the bases in a flat orientation (*i.e.* ring parallel to the electrode surface) and the sugar mainly perpendicular to the surface.[23]

Mutorotation continues to receive attention. Studies [24] of the mutarotation of 2,3,4,6-tetra-*O*-methyl-α-D-glucopyranose in aqueous dioxan and aqueous DMSO by optical rotation over a wide range of water concentration, with and without catalysts, have shown that the uncatalysed reaction transition-state contains one mole of sugar and two moles of water in an acyclic bonded structure (6). In the pyridine-catalysed reaction one mole of water may be replaced by one mole of pyridine (Scheme 3). The evidence suggests that the reaction takes place by an intimate step-wise mechanism rather than synchronously.

Agreement with the above findings on water dependence is reported by Sørenson *et al.*[25] for high water concentrations only; at low water concentra-

[19] V. Misra and V. P. Misra, *Carbohydrate Res.*, 1978, **60**, 29.
[20] N. S. D'Yackenko, Ya. G. Ropotenko, A. A. Ivanyuk, I. A. Dumanskii, and N. U. Fishchuk, *Ukrain. khim. Zhur.*, 1978, **44**, 536 (*Chem. Abs.*, 1978, **89**, 110 178j).
[21] S. P. Moulik and D. P. Khan, *Indian J. Chem., Sect. A*, 1977, **15A**, 267 (*Chem. Abs.*, 1978, **88**, 7218c).
[22] T. M. Wortel and H. Van Bekkum, *Rec. Trav. chim.*, 1978, **97**, 156 (*Chem. Abs.*, 1978, **89**, 75 422k).
[23] V. Brabec, S. O. Christian, and G. Dryhurst, *J. Electrochem. Soc.*, 1978, **125**, 1236 (*Chem. Abs.*, 1978, **89**, 170 868u).
[24] C. S. Chin and H. H. Huang, *J.C.S. Perkin II*, 1978, 474.
[25] A. M. Kjaer, P. E. Sørensen, and J. Ulstrup, *J.C.S. Perkin II*, 1978, 51.

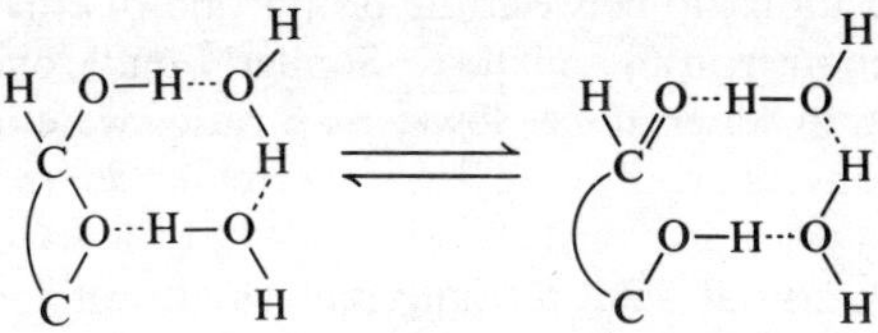

(6)

tions (⩽11 M) the order with respect to water falls to one for dioxan and acetonitrile but increases to 3.7 in DMSO. This behaviour parallels a decrease and increase respectively in both activation energy and entropy of activation for the two groups of solvents. The data were shown to be compatible with concerted proton transfer *via* water catalyst molecules on the basis of known properties of mixed solvents and modern quantum theories of proton transfer

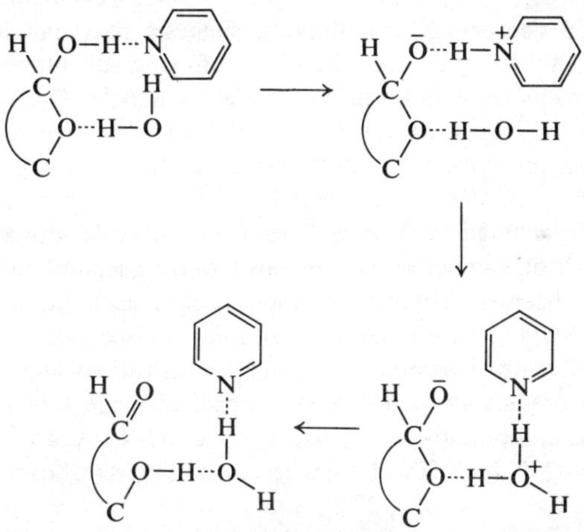

Scheme 3

reactions in polar media. An investigation of the isotope effect in acid-catalysed mutarotation of D-glucose using polarography has shown that the effect may be used to establish the existence of general or special catalysis by acid, or whether there is a pre-equilibrium involving the participation of protons. The rate constants, and hence isotope effects, were calculated for the systems $NaClO_4$–$HClO_4$ at pH 3—3.5 in water, heavy water, water–acetone, and heavy water–acetone.[26] The effect of ethanol on optical rotation, velocity of mutarotation, and equilibrium constant of lactose has been reported.[27] The specific rotation is less in ethanol than in water and there is a linear relationship between the percentage ethanol in aqueous ethanol and the final specific rotation. Equilibrium studies on the $\alpha : \beta$ anomer ratio have shown that

[26] C. Dragulescu, A. Calusaru, A. Nimara, A. Balint, and E. Princz, *Internat. Symp. Specific Interact. Mol. Ions, Proc.*, 3rd, 1976, **1**, 119 (*Chem. Abs.*, 1978, **89**, 24 640x).

[27] F. Majd and T. A. Nickerson, *J. Agric. Food Chem.*, 1978, **26**, 207 (*Chem. Abs.*, 1978, **88**, 74 501c).

there is a direct relationship between the proportion of ethanol in water and the amount of α-anomer in the mixture. Similarly mutarotation is slower in ethanol–water than in water alone; lower α : β ratios were obtained showing that the rate of conversion of α to β was decreased more than the reverse reaction.

The effect of the crystal habit-forming impurities, raffinose, dextrose, and potassium chloride on the growth of sucrose crystals from various seeds has been studied.[28] A mechanism for the oxidation of D-galactose by Nessler's reagent in alkaline media *via* the enediol has been proposed on the basis of kinetic measurements.[29] The reaction is zero-order with respect to Hg^{II} and first-order with respect to galactose. The rate is inversely proportional to the concentration of iodine ion.

The presence of 1.6% septanose form in solutions of D-idose has been detected [30] using ^{13}C n.m.r. in ^2H_2O solution at 37 °C. A report on the diffusion of glucose, maltose, and maltotriose in cross-linked gels at different gel concentrations and degrees of cross-linking suggests that diffusion velocity is restricted not only by interaction between diffusing substances and gel components but also by steric hindrance of the gel matrix.[31] The lyoluminescence of irradiated solid carbohydrates in ice has been suggested to be due to the formation of singlet oxygen species.[32] The difficulties in invoking such species in light emission studies are discussed. The rise in conductivity on flash photolysis of oxygenated aqueous hydrogen peroxide solutions at pH 5 of *meso*-erythritol, D-xylitol, D-glucitol, *myo*-inositol, D-glucose, and methyl α-D-glucopyranoside has been measured as a function of time.[33] All showed transient conductivity decaying within several hundred milliseconds, which was attributed to formation of hydroperoxy radicals which dissociate to give protons and $O_2^{\dot{-}}$. The precursors of hydroperoxy radicals were thought to be sugar α-hydroperoxy compounds. Polyols were found to show two first-order processes, $k \simeq 200$ and $3000\ s^{-1}$, whereas D-glucose shows three parallel first-order processes, $k \simeq 400$, 2700, and $>70\,000\ s^{-1}$; the separate reactions were assumed to be due to the different points of attachment of the hydroperoxy radical with the fastest rate for D-glucose being due to the C-1 adduct and the slowest due to that at C-6. Dehydration of β-D-glucose and β-D-xylose in the presence of Cr^{3+}, Al^{3+}, Ti^{3+}, Ga^{3+}, and In^{3+} to yield 5-hydroxymethylfurfural and furfural respectively has been shown to proceed by a unimolecular mechanism.[34] As part of the continuing studies on molybdic acid transformations of free sugars the kinetics of the isomerization of DL-glyceraldehyde to dihydroxyacetone at pH 3.1 and 70 °C have been shown to be first order with

[28] G. Mantovani, C. A. Accorsi, and G. Vaccari, *Ind. Sacc. Ital.*, 1977, **70**, 137 (*Chem. Abs.*, 1978, **89**, 24 656g).
[29] M. P. Singh, A. K. Singh, and M. Kumar, *Z. Naturforsch.*, 1978, **33B**, 657 (*Chem. Abs.*, 1978, **89**, 75 424n).
[30] T. Bruce Grindley and V. Gulasekharam, *J.C.S. Chem. Comm.*, 1978, 1073.
[31] K. Nakanishi, S. Adachi, S. Yamamoto, R. Matsuno, A. Tanaka, and T. Kamikubo, *Agric. and Biol. Chem.* (*Japan*), 1977, **41**, 2455.
[32] P. J. Baugh, M. G. Mahjani, S. C. Ellis, and D. R. Evans, *Radiat. Phys. Chem.*, 1977, **10**, 21 (*Chem. Abs.*, 1978, **88**, 7221y).
[33] E. Bothe, D. Schulte-Frohlinde, and C. Von Sonntag, *J.C.S. Perkin II*, 1978, 416.
[34] V. I. Krupenskii, *Khim. Drev.*, 1978, 72 (*Chem. Abs.*, 1978, **88**, 170 405v).

respect to the trioses, but the results did not distinguish between an enediol mechanism or an intramolecular hydride transfer.[35] A case in which an intramolecular C-1 to C-2 hydride transfer did occur has been reported for the conversion of D-*threo*-pentulose to D-xylose under acid-catalysis.[36] This is in line with other aldose–ketose conversions (see Vol. 9, p.8). The thermogravimetric analysis of α-lactose hydrate has shown that α-lactose first loses its water of crystallization, followed by a small exotherm, then by the melting endotherm. An anhydride which was characterized by an endotherm at 180 °C and an exotherm at 190 °C was produced by heating in methanol.[37]

3 Reactions

The alkaline degradation of D-glucose produced coloured polymeric and acidic compounds which were isolated and characterized by gel filtration, gel electrophoresis, ion-exchange, and spectral analysis.[38] Another group of workers have studied the stoicheiomctry of the alkaline degradation of D-glucose and D-fructose in the temperature range 30—70 °C, and shown that almost two moles of alkali are consumed per mole of the carbohydrate. Although the degradation is partly controlled by the dielectric constant of the medium, such additives as acetone and urea have specific effects, presumably due to complex formation characterized by presence of a dark red colour (λ_{max} 320 nm).[39] The reaction of urea with lactose produced lactosyl urea in maximal yield (40%) at pH 2 during 5 days at 50 °C or 4 h at reflux. Some hydrolysis of lactose occurred in the process, giving rise to minor urea–glucose and urea–galactose compounds. Increasing the alkalinity resulted in a reduction of the yield of lactosyl urea and an increase in production of lactulose.[40]

The sodium aluminate-catalysed isomerization of D-glucose to D-fructose has been shown to proceed *via* a β-D-glucopyranose–1,3-aluminate complex which is converted into an α-D-fructofuranose–1,3,6-aluminate complex by way of an enol complex (Scheme 4).[41] The latter prevents formation of a D-mannose aluminate complex. The D-fructose complex reacts further to form a D-psicose aluminate complex in substantial yield. In another paper[42] the same authors report the product distribution and separate rate constants for the isomerization of D-glucose by sodium aluminate and have shown that lower temperature favours peak yields of D-fructose (~60% at 0 °C) whereas higher temperatures favour formation of D-psicose. The acid-catalysed isomerization and dehydration of DL-glyceraldehyde and 1,3-dihydroxy-2-propanone have been studied by performing the reaction in 0.5 M sulphuric acid

[35] J. Königstein, *Coll. Czech. Chem. Comm.*, 1978, **43**, 1152.
[36] S. Ralapati and M. S. Feather, *J. Carbohydrates, Nucleosides, Nucleotides*, 1978, **5**, 297.
[37] T. Itoh, M. Satoh, and S. Adachi, *J. Dairy Sci.*, 1977, **60**, 1230 (*Chem. Abs.*, 1978, **88**, 121 564n).
[38] N. Martinez and S. G. Kara-Murza, *Rev. CENIC, Cienc. Fis.*, 1976, **7**, 141 (*Chem. Abs.*, 1978, **89**, 60 015f).
[39] S. P. Moulik, D. Basu, and P. K. Bhattacharya, *Carbohydrate Res.*, 1978, **63**, 165.
[40] J. Cerbulis, P. E. Pfeffer, and H. M. Farrell, *Carbohydrate Res.*, 1978, **65**, 311.
[41] A. J. Shaw and G. T. Tsao, *Carbohydrate Res.*, 1978, **60**, 327.
[42] A. J. Shaw and G. T. Tsao, *Carbohydrate Res.*, 1978, **60**, 376.

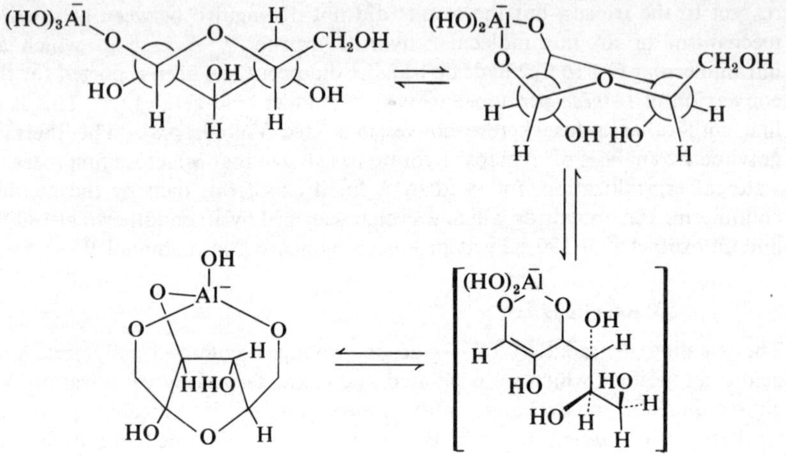

Scheme 4

in ^3H_2O at 100 °C. The pattern of incorporation of tritium in the pyruvaldehyde showed that little interconversion of starting materials occurred under these conditions. In a separate experiment using [2-^{3}H]-2,3-dihydroxypropanol an intramolecular hydride transfer from C-2 to C-3 occurred.[43]

The electrochemical reduction of D-ribono-γ-lactone to D-ribose in aqueous or propylene carbonate solution has been effected using lithium–lead or sodium–lead intermetallic compounds.[44] The use of vanadium pentoxide in M sulphuric acid as an oxidant for carbohydrates has been reported.[45] Aldoses give the next lower aldose and formic acid whereas aldonic acids give carbon dioxide and the next lower aldose. Complexation and epimerization of aldoses by molybdic acid have been shown[46] to be maximized at pH 2.5—3.5, whereas the specific rotation reaches extremum values between pH 2.6 and 6.8. All aldopentoses undergo molybdate ion catalysis to furnish the same equilibrium of xylose (34%), arabinose (29%), lyxose (24%), and ribose (13%) having the same D- or L-configuration as the starting material. The inversion was found to occur by mutual replacement of the vicinal hydrogen in the *vic*-diol-molybdate complex.[47] 3-Deoxy-D-*arabino*-hexose and 3-deoxy-D-*ribo*-hexose are transformed under molybdic acid catalysis in a non-reversible reaction to 3-deoxy-D-*erythro*-hexulose without formation of the epimeric 3-deoxy-aldohexose. A similar finding was made when calcium hydroxide catalysis was used.[48] Molybdate has been used to prepare[49] some ω-deoxyaldoses of L-ribose, L-talose, D-talose, and L-*glycero*-L-*talo*-heptose in low yield (9—19%).

[43] G. L. Lookhart and M. S. Feather, *Carbohydrate Res.*, 1978, **60**, 259.
[44] I. A. Avrutskaya, I. G. Kiseleva, N. N. Tomashova, A. V. Chekavtsev, T. G. Tsar'kova, N. A. Begun, B. N. Kabanov, and M. Ya. Fioshin, *Elektrokhimiya*, 1978, **14**, 1048 (*Chem. Abs.*, 1978, **89**, 187 864n).
[45] K. M. Haldorsen, *Carbohydrate Res.*, 1978, **63**, 61.
[46] V. Bilik, L. Petrus, and J. Zemek, *Chem. Zvesti*, 1978, **32**, 242.
[47] V. Bilik, L. Petrus, and V. Farkas, *Coll. Czech. Chem. Comm.*, 1979, **43**, 1163.
[48] V. Bilik and L. Petrus, *Coll. Czech. Chem. Comm.*, 1978, **43**, 1159.
[49] V. Bilik, L. Petrus, L. Stankovic, and K. Linek, *Chem. Zvesti*, 1978, **32**, 372.

This paper also reports the synthesis from L-rhamnal of 7-deoxy-L-*glycero*-L-*galacto*-and -L-*talo*-heptose by the nitromethane route. Degradation of D-galactose phenylhydrazone by alkaline hydrogen peroxide–molybdate gave a 50% yield of D-lyxose, whereas the corresponding 2,5-dichlorophenylhydrazine gave a 4 : 1 mixture of D-lyxose and D-galactose and the 2,4-dinitrophenylhydrazone gave D-lyxose and D-galactose in a 1 : 9 mixture. The pentose was not formed in the absence of the molybdate.[50]

The formation of oligosaccharides with DP 2—10 on heating D-galactose at 150 °C for 100 min has been reported.[51] Pulse radiolysis and ^{60}Co γ-radiolysis have been used to elucidate the mechanism of formation of malonaldehyde from irradiated carbohydrates; attack by hydroxy free-radicals at C-5 and C-6 of D-glucose may initiate the reaction, while substitution at C-2, C-3, or C-6 decreases the yield which is unaffected by substitution at C-1 by phosphate. The mechanisms for D-glucose, L-arabinose, and maltose are given.[52]

Condensation of formaldehyde with D-[^{14}C]glucose catalysed by 3—5 M calcium hydroxide at 100 °C gave D-fructose, D-mannose, D-arabinose, dihydroxyacetone, glyceraldehyde, and glycolaldehyde. Isotope studies confirmed that the rate of retroaldol cleavage of the aldose was considerably less than the rate of carbohydrate synthesis.[53] The pH changes and the complex formation of the same reaction with either calcium hydroxide or lead hydroxide have been investigated.[54]

The *C*-deuteriation of carbohydrates by hot ^2H_2O containing Raney nickel has been shown to proceed readily with retention of configuration, although prolonged exposure to the catalyst causes isomerization (Scheme 5). The

$$R_2CHO^2H + Ni \rightleftharpoons R_2C{=}O + Ni(H^2H)$$

$$R_2C{=}O + Ni(^2H)_2 \rightleftharpoons R_2C^2HO^2H + Ni$$

Scheme 5

presence of an isopropylidene group, as in 1,2:5,6-di-*O*-isopropylidene-α-D-glucofuranose and 1,2:3,4-di-*O*-isopropylidene-β-D-fructopyranose, inhibits interaction with the catalyst and prevents deuterium exchange.[55]

Conditions for formation of reductones by pyrolytic degradation of cellulose, dextrin, maltose, and glucose have been reported.[56] The photoreactions of (7) with R,R′CO (R,R′ = Ph or Me) in deoxygenated solution under nitrogen

[50] L. Petrus, V. Bilik, K. Linek, and M. Misikova, *Chem. Zvesti*, 1978, **32**, 701.

[51] P. Barbetti and I. Chiappini, *Ind. Aliment.* (*Pinerolo, Italy*), 1977, **16**, 61 (*Chem. Abs.*, 1978, **88**, 150 936r).

[52] T. Bucknall, H. E. Edwards, K. G. Kemsley, J. S. Moore, and G. O. Phillips, *Carbohydrate Res.*, 1978, **62**, 49.

[53] T. I. Khomenko, O. A. Golovina, A. A. Berlin, M. M. Sakharov, Yu. E. Sinyak, and V. L Masenko, *Izvest. Akad. Nauk S.S.S.R., Ser. khim.*, 1978, 195 (*Chem. Abs.*, 1978, **88** 152 881t).

[54] A. Kh. Vais, V. A. Seleznev, M. M. Sakharov, B. Ya. Gorokhovatskii, and N. P Evmenenko, *Izvest. Akad. Nauk S.S.S.R., Ser. khim.*, 1978, 1614 (*Chem. Abs.*, 1978, **89** 180 245c).

[55] H. J. Koch and R. S. Stuart, *Carbohydrate Res.*, 1978, **67**, 341.

[56] G. Matsui, M. Obata, K. Shikata, T. Tokuyama, and T. Ueno, *Nippon Nogei Kagaku Kaishi*, 1978, **52**, 69 (*Chem. Abs.*, 1978, **89**, 6478w).

produce pinacols, cross-adducts, and a dimer of (7), whereas in the presence of air, (8) and (9) were obtained. Also reported were rates of similar photoreactions of monosaccharides and disaccharides.[57]

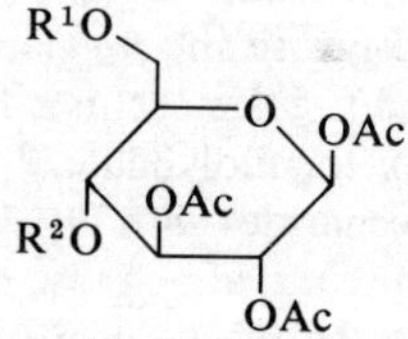

(7) R^1, R^2 = PhCH<

(8) R^1 = H, R^2 = Bz
(9) R^1 = Bz, R^2 = H

A study of the formation of dimers and trimers of 2,4-*O*-ethylidene-D-erythrose has appeared.[58]

The reaction between glucose and phenol leading to a mixture of products is mentioned in Chapter 3. Reference to the reaction of glucose with vanillyl alcohol is made in Chapter 4.

[57] M. Suzuki, R. Matsushima, T. Inai, and S. Tsujimoto, *Asahi Garasu Kogyo Gijutsu Shoreikai Kenkyu Hokoku*, 1977, **30**, 235 (*Chem. Abs.*, 1978, **89**, 180 240x).
[58] R. Anderson, O. Theander, and E. Westerland, *Carbohydrate Res.*, 1978, **61**, 501.

3
Glycosides

1 *O*-Glycosides

A review, containing 57 references, on aspects of the chemistry of disaccharides has appeared.[1] Readers are referred to Chapter 20 for references to several n.m.r. (especially ^{13}C) studies of glycosides and di- and higher saccharides.

Synthesis of Monosaccharide Derivatives.—A further study has been carried out on the methanolysis of D-fructose, and rates of the initial glycosidation, the anomerization of the furanosides, and the furanoside–pyranoside conversion were measured. The results were compared with those obtained using arabinose, and the differences observed considered in relationship to the steric and polar effects of the hydroxymethyl group bonded to the anomeric centre of the ketose.[2] As expected, methyl α-D-glucopyranoside obtained from D-[5-^{3}H]glucose by Fischer glycosidation was produced without loss of the radiolabel. Alternatively, D-[5-^{3}H]glucuronic acid lost 85% of its activity when converted into the same glycoside by way of the methyl ester methyl glycoside followed by reduction with sodium borohydride, which signifies that prior to reduction the ester group had undergone enolization.[3] Reaction of enantiomeric sugars with chiral alcohols is referred to in Chapter 22.

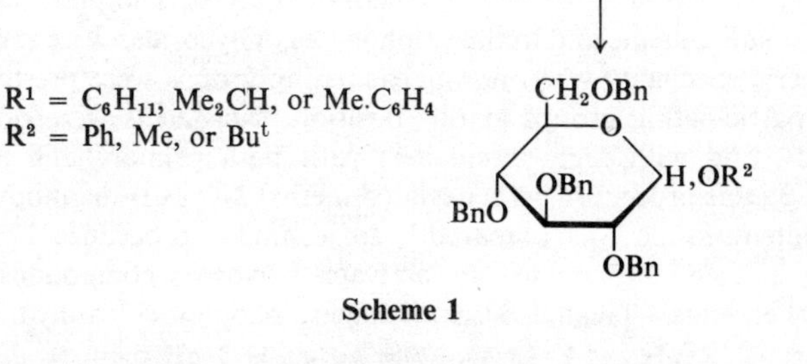

$R^1 = C_6H_{11}$, Me_2CH, or $Me.C_6H_4$
R^2 = Ph, Me, or Bu^t

Scheme 1

[1] L. Hough and A. C. Richardson, *Pure Appl. Chem.*, 1977, **49**, 1069.
[2] W. Szeja and T. Kiersznicki, *Polish J. Chem.*, 1978, **51**, 2097.
[3] H. S. Prihar, R. Maganathan, and D. S. Feingold, *Carbohydrate Res.*, 1978, **67**, 271.

A new approach to glycoside synthesis involves the use of glycosylisoureas, prepared from 1-hydroxysugars and carbodi-imides in the presence of copper chloride (Scheme 1). Alkyl and aryl glycosides were made in good yields, the proportion of α-products varying from low to high according to the nucleophile used. Theophylline gave mainly an α-nucleoside derivative when used in this way with 2,3-*O*-isopropylidene-5-*O*-triphenylmethyl-D-ribose.[4] A somewhat related procedure involves the preparation of mixed sulphites followed by thermal extrusion of sulphur dioxide, but these and related reactions of sulphamido analogues provided glycosides in modest yields only (Scheme 2). In the latter case yields were in the 10% range, free sugars and

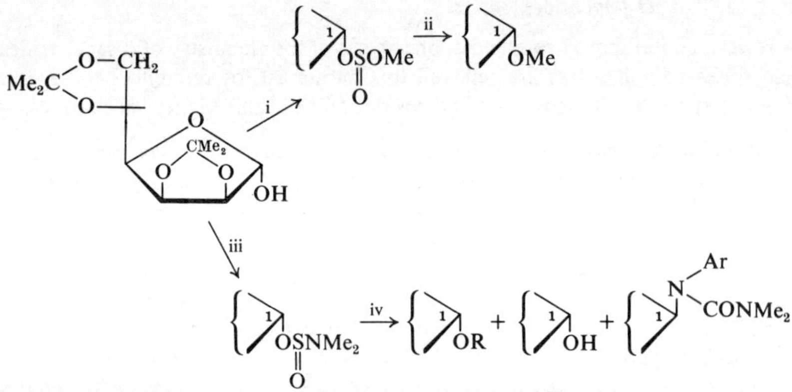

Reagents: i, MeOSOCl; ii, Δ; iii, $OS(NMe_2)_2$; iv, ArNCO, ROH

Scheme 2

glycosylamine derivatives accounted for the majority of the products and, in addition, some disaccharides were formed.[5] A similar approach, but one using glycosyl sulphonate esters, has yielded far more encouraging results in the D-galactopyranose series. In general, methanesulphonates and toluene-*p*-sulphonates, in solvents like ether, gave α-products with high stereoselectivity when a variety of alcohols, including sugar derivatives, were used. β-Glycosides were favoured with trifluoromethanesulphonates. Reactions were carried out with a variety of compounds having different substituent groups at C-2, -3, -4, and -6.[6] An extension of this work in the D-mannopyranose series led to an investigation of a range of substituted glycosyl sulphonates varying in the sugar substituents and in the sulphonates. Glycosides were formed with low anomeric specificity when non-participating groups were present at C-2, but with participating groups at this position, α-products were obtained in high yields, and with high specificities with both primary and secondary alcohols. By this procedure perbenzylated methyl 2-*O*-(α-D-mannopyranosyl)-α-D-mannopyranoside was prepared.[7] In a similar procedure Leroux and Perlin have used sulphonates to activate 1-hydroxy-compounds without isolating the esters. Treated with methanesulphonic acid anhydride or its

[4] H. Tsitsumi, Y. Kawai, and Y. Ishido, *Chem. Letters*, 1978, 629.
[5] J. R. Dormoy and B. Castro, *J. Chem. Research*, (S), 1978, 21; (M), 1978, 0321.
[6] V. Marousek, T. J. Lucas, P. E. Wheat, and C. Schuerch, *Carbohydrate Res.*, 1978, **60**, 85.
[7] E. S. Rachaman, R. Eby, and C. Schuerch, *Carbohydrate Res.*, 1978, **67**, 147.

hexafluoro analogue in the presence of halide ion and *S*-collidine, the hydroxy-compounds were converted into glycosyl halides which then gave glycosides when treated with alcohols in 'one-pot' reactions. Several α-D-glucopyranosides, *e.g.* the cholesteryl and 1,2,3,4-tetra-*O*-acetyl-β-D-glucopyranos-6-yl compounds, were prepared in high yields by use of 2,3,4,6-tetra-*O*-benzyl-D-glucose *via* the glycosyl bromide. The use of 2,3,4,6-tetra-*O*-acetyl-D-glucose is complicated by orthoester formation, and various aspects of this process were investigated. 2,3-*O*-Isopropylidene-5-*O*-triphenylmethyl-D-ribofuranose and 2,3:5,6-di-*O*-isopropylidene-D-mannofuranose gave glycosides without high stereoselectivity.[8]

A range of aromatic glycosides have been described following acid-catalysed condensations between phenols and glycosyl acetates. β-Glucosides having a variety of substitutes at C-6 and at the *p*-position on the aromatic ring were studied as anti-viral agents, the 6-deoxy-6-halogeno compounds and those having *p*-alkyl groups being found to be the most potent.[9] In a study of the preparation of glucosides and galactosides of 4-methylumbelliferone, penta-*O*-acetyl-β-D-glucopyranose and -galactopyranose were condensed with the compound in the presence of zinc chloride; mixed products were obtained, the galactosides containing some β-furanoside. This is believed to be the first example of ring contraction occurring in such a reaction. Use of the trimethylsilyl ether of the phenol increased slightly the proportion of α-pyranosides, but a more effective method of obtaining these products involved condensation of the sodium phenate with tetra-*O*-acetyl-β-D-glucopyranosyl and -galactopyranosyl chloride in HMPT. Mannopyranosides were also prepared, and a number of the products were oxidized catalytically to the corresponding uronosides.[10] α-*p*-Nitrophenyl glycosides were obtained from the peracetates of β-D-mannopyranose and β-L-rhamnopyranose by use of tin(IV) chloride as catalyst,[11] and the same glycosides of α-D-talopyranose and β-D-ribopyranose were prepared from the α-penta-acetate[12] and β-tetra-acetate,[13] respectively. Alternatively, the β-D-ribopyranoside of 7-hydroxy-4-methylcoumarin was obtained by use of 2-*O*-methanesulphonyl-D-arabinose (which reacted *via* the 1,2-anhydride).[13] In the dideoxyhexose series, compounds (1)—(3) were used

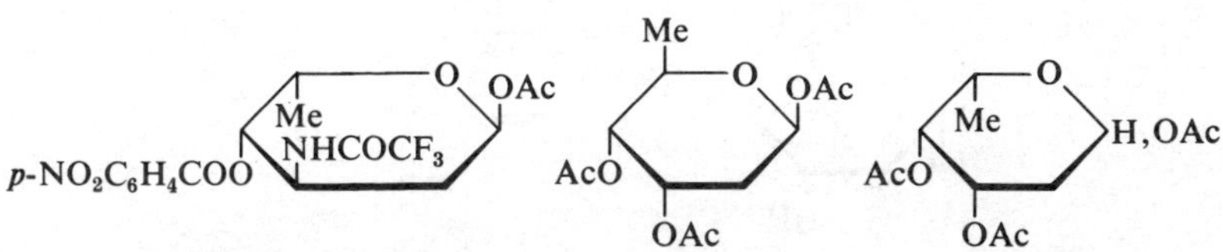

(1) (2) (3)

[8] J. Leroux and A. S. Perlin, *Carbohydrate Res.*, 1978, **67**, 163.

[9] H. Arita, K. Sugita, A. Nomura, K. Sato, and J. Kawanami, *Carbohydrate Res.*, 1978, **62**, 143.

[10] M.-C. Courtin-Duchateau and A. Veyrières, *Carbohydrate Res.*, 1978, **65**, 23.

[11] J. Garegg, H. Hultberg, and T. Iversen, *Carbohydrate Res.*, 1978, **62**, 173.

[12] R. H. Shah and O. P. Bahl, *Carbohydrate Res.*, 1978, **65**, 47.

[13] M. Claeyssens, E. Saman, C. K. de Bruyne, and A. de Bruyn, *J. Carbohydrates, Nucleosides, Nucleotides*, 1978, **5**, 33.

in benzene in the presence of toluene-p-sulphonic acid to give phenyl and methyl glycosides.[14]

Aryl glycosides to have been synthesized by use of glycosyl halides are: p-nitrophenyl 2,3,4,6-tetra-O-acetyl-β-D-allopyranoside (formed together with a methyl 1,2-orthoester),[15] p-nitrophenyl β-D-mannopyranoside (4,6-di-O-acetyl-α-D-mannopyranosyl bromide 2,3-carbonate used),[16] 2-O-β-D-glucopyranosyl-4-O-methylphloroacetophenone,[17] and the D- and L-β-glucosides of the dihydrochalcones (4). These last compounds were prepared together with a variety of analogues containing different disaccharide units as well as different patterns of substitution in the aromatic rings as part of a study of sweetness–structure correlations. Some were tasteless, but several were sweeter than saccharin, compounds (5) and (6) being 4 and 4.5 times as sweet,

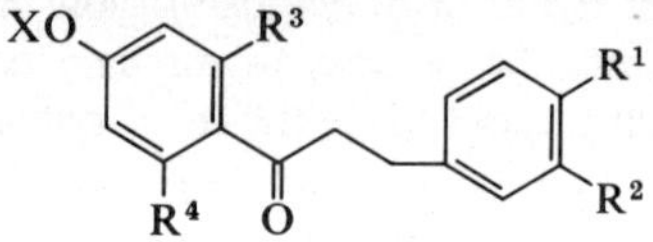

(4) R^1, R^2, R^3, R^4 = OMe, OH, or H; X = H
(5) R^1 = OMe, $R^2 = R^3 = R^4$ = OH;
X = 6-O-methyl-β-D-glucopyranosyl
(6) R^1 = OMe; $R^2 = R^3$ = OH; R^4 = H;
X = 2-O-(α-L-rhamnopyranosyl)-β-D-galactopyranosyl

respectively.[18] Appropriate glycosyl halide derivatives were also used in the preparation of the 3-O-α-L-rhamnopyranoside and 3-O-β-rutinoside of 7-O-methylquercitin, the 3-O-α-L-rhamnoside of 3′,4′,7-tri-O-methylquercitin[19] and the 7-O-β-D-glucopyranoside of 7-hydroxy-3′,4′-(methylenedioxy)isoflavone.[20] Glucosylation of the corresponding substituted aminophenols gave compounds (7)—(9) required as agents for photoaffinity labelling,[21] and tri(4-glycosyloxyphenylazo)phloroglucinol compounds were prepared using diazotized p-aminophenyl β-D-glucopyranoside, β-D-mannopyranoside, and α- and β-D-galactopyranoside. Aggregation of these compounds was studied by various physical methods and they were used as artificial antigens.[22]

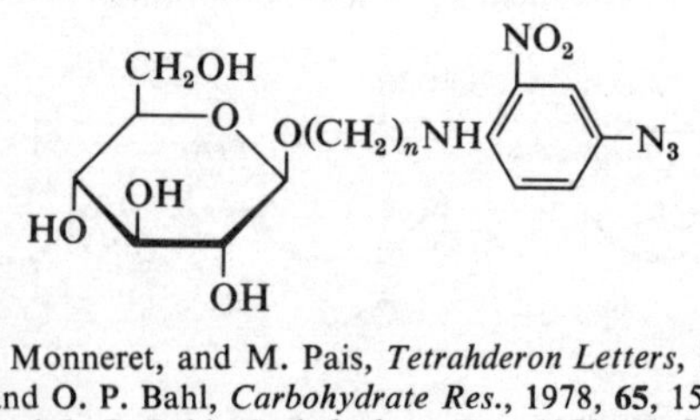

	n
(7)	10
(8)	6
(9)	3

[14] J. Boivin, C. Monneret, and M. Pais, *Tetrahderon Letters*, 1978, 1111.
[15] R. H. Shah and O. P. Bahl, *Carbohydrate Res.*, 1978, **65**, 153.
[15a] R. H. Shah and O. P. Bahl, *Carbohydrate Res.*, 1978, **62**, 261.
[15b] M. Blanc-Muesser, J. Defaye, and H. Drignez, *Carbohydrate Res.*, 1978, **67**, 305.
[16] F. Maley, *Carbohydrate Res.*, 1978, **64**, 279.
[17] G. R. Nagarajan and V. S. Parmar, *Indian J. Chem., Sect. B*, 1977, **15B**, 955.
[18] S. Kamna, F. Konishi, and S. Esaki, *Agric. and Biol. Chem. (Japan)*, 1978, **42**, 941.
[19] K. Székely, B. Vermes, L. Parkas, and M. Nógradi, *Acta Chim. Acad. Sci. Hung.*, 1978, **97**, 241.
[20] A. Levai and R. Bognar, *Pharmazie*, 1978, **33**, 235 (*Chem. Abs.*, 1978, **89**, 110 264j).
[21] M. Hagedorn, R. R. Sauers, and A. Eicholz, *J. Org. Chem.*, 1978, **43**, 2070.
[22] E. F. Woods, G. G. Lilley, and M. A. Jermyn, *Austral. J. Chem.*, 1978, **31**, 2225.

Other glycosides of biochemical significance to have been produced are the glucosides of cholesterol, β-sitosterol, and 28-*O*-acetylbetulin (acetobromoglucose and mercury(II) acetate used in benzene; poor stereospecificity – α : β, 40 : 60),[23] and the 3- and 16-glucosides of estriol.[24] Reaction of the iridoid aglycone (10) with tetra-*O*-acetyl-α-D-glucopyranosyl bromide and silver

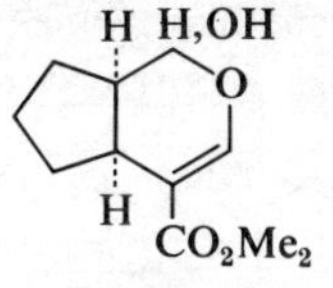

(10)

perchlorate, and tetra-*O*-acetyl-α-D-glucopyranose and boron trifluoride gave access to α- and β-linked products but, in addition, 2-substituted carbohydrate derivatives were obtained following reaction at C-2 of 1,2-orthoester intermediates.[25] When the same hydroxy-compound was treated with tri-*O*-acetyl-1,2-anhydro-α-D-glucopyranose in the presence of boron trifluoride, three α-glycosides were obtained having different stereochemistry within the aglycone.[26] Standard glycosylation of 1,2-di-*O*-benzyl-*sn*-glycerol with tetra-*O*-acetyl-α-D-mannopyranosyl bromide gave 75% of products with an α : β ratio of 3 : 2, whereas use of the corresponding t-butyl orthoacetate gave 87% of a 7 : 1 mixture.[27] Related compounds (11) are cerebroside analogues, and were prepared from D,L-*threo*- and *erythro*-3-phenylserines.[28] Condensation of tetra-*O*-acetyl-α-D-glucopyranosyl bromide with acetone cyanohydrin led to the cyanogenic glucoside linamarin,[29] and a series of β-D-glucosides, -2-acetamido-2-deoxyglucosides, and -galactosides, related to the 'cluster glycoside' (12), were prepared for attachment to proteins or solid matrices as

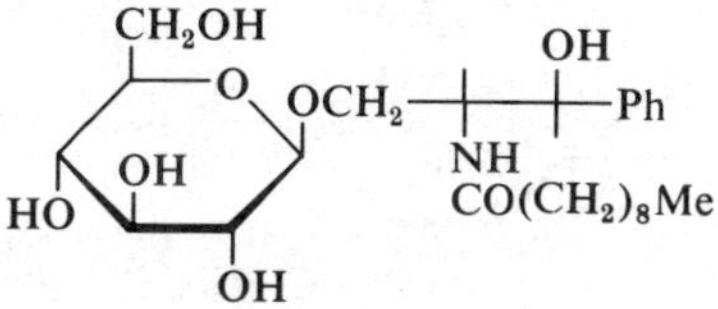

(11)

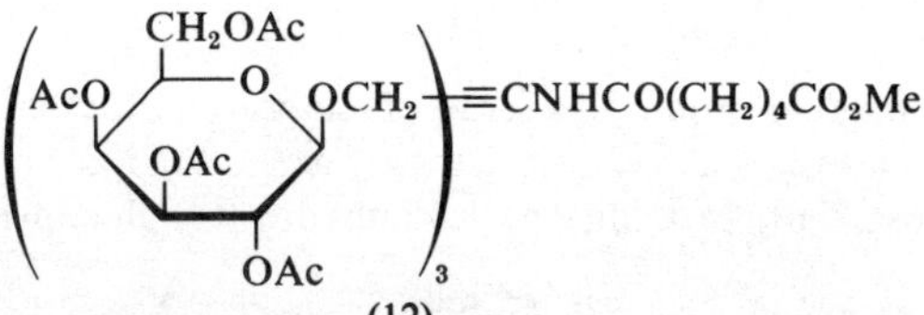

(12)

[23] N. I. Uvarova, L. N. Atopkina, N. F. Samoshina, and G. Elyakov, *Bioorg. Khim.*, 1977, **3**, 1493, (*Chem. Abs.*, 1978, **88**, 191 321e).
[24] T. Nambara, J. Goto, H. Furuyama, and H. Kato, *Chem. and Pharm. Bull.* (*Japan*), 1978, **26**, 591.
[25] L.-F. Tietze and U. Niemeyer, *Chem. Ber.*, 1978, **111**, 2423.
[26] L.-F. Tietze and P. Marx, *Chem. Ber.*, 1978, **111**, 2441.
[27] A. P. Kaplun, V. I. Shvets, and R. P. Evstigneeva, *Zhur. org. Khim.*, 1977, **13**, 1606 (*Chem. Abs.*, 1978, **88**, 23 315n).
[28] B. Weiss, *Chem. Phys. Lipids*, 1977, **19**, 347 (*Chem. Abs.*, 1978, **88**, 23 323p).
[29] B. K. Malik, *Indian J. Chem.*, 1978, **16B**, 76 (*Chem. Abs.*, 1978, **88**, 191 329p).

part of a biochemical study.[30] Transglycosylation occurred from maltose to esculin when they were treated with an acetone extract of pig liver, and 3′- and 4′-α-glucosides were formed.[31] Glycosylation of L-ascorbic acid is referred to in Chapter 15.

In the uronic acid series, glucuronides of phenolphthalein,[32] *N*-hydroxy-4-aminobiphenyl,[33] and the methaqualone metabolite (13)[34] have been synthesized by standard Köenigs–Knorr procedures.

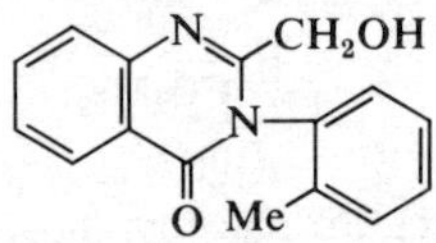

(13)

Synthesis of Disaccharides and their Derivatives.—A method has been devised for the preparation of alkyl α-glycosides of reducing disaccharides. Conversion of the peracetylglycosyl bromides to the β-*N*,*N*-dialkyldithiocarbamates was followed by deacetylation and treatment with the alcohol (methanol and benzyl alcohol were used) in the presence of mercury(II) chloride and cadmium carbonate. Good yields of products containing mainly α-glycosides were obtained from cellobiose, lactose, and maltose.[35]

Condensation between 1,2,3,4-tetra-*O*-acetyl-6-*O*-triphenylmethyl-β-D-glucose and the pyruvic orthonitrile derivative (14) gave gentiobiose β-octa-acetate, and self polycondensation of the 6-*O*-triphenylmethyl analogue of compound (14) gave a β-1,6-linked glucan.[36] Action of a fungal glucosyl

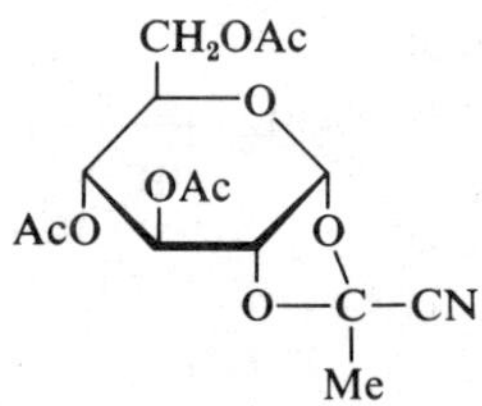

(14)

transferase on maltose produced a series of oligosaccharides including nigerose, kojibiose, isomaltose, maltotriose, panose, isomaltotriose, and 6-*O*-D-glucopyranosylpanose, but, in addition, 1,6-anhydro-β-D-glucopyranose was also

[30] Y. C. Lee, *Carbohydrate Res.*, 1978, **67**, 509.
[31] K. Uchida and Y. Suzuki, *Nippon Nogei Kagaku Kaishi*, 1978, **52**, 53 (*Chem. Abs.*, 1978, **88**, 191 338r).
[32] T. Takizawa, *Ogata Igaku Kagaku Kenkyusho Kenkyu Hokoku*, 1974, 59 (*Chem. Abs.*, 1978, **88**, 136 900t).
[33] H. R. Moreno and J. L. Radomski, *Cancer Letters*, 1978, **4**, 85 (*Chem. Abs.*, 1978, **89**, 44 043h).
[34] O. Ericsson, *Acta Pharm. Suec.*, 1978, **15**, 81 (*Chem. Abs.*, 1979, **89**, 180 294t).
[35] C. T. Gi, H. Ishihara, and S. Tejima, *Chem. and Pharm. Bull. (Japan)*, 1978, **26**, 1570.
[36] A. F. Bochkov, I. V. Obruchnikov, and V. M. Kalinevich, *Ref. Doklady Soobshch.–Mendelcevsk. S'ezd. obshchei priklad. Khim., 11th*, 1975, **6**, 95 (*Chem. Abs.*, 1978, **88**, 191 278w).

produced.[37] In related work, a transferase from *Klebsiella pneumoniae* gave, from maltose and maltotriose, the series of malto-oligosaccharides up to the eleven-unit compound, and also cyclo-hexa-, -hepta-, and -octa-amyloses. Transfers were then effected onto *p*-nitrophenyl α- and β-glucopyranoside and other glycosides from cyclohexa-amylose to give compounds containing up to 12 residues. These were separated by h.p.l.c.[38] Reaction of 1,2-*O*-cyclopentylidene-α-D-glucofuranurono-6,3-lactone with tetra-*O*-acetyl-α-D-glucopyranosyl bromide gave 45% and 38% of the α- and β-linked disaccharide derivatives, respectively. Reduction with lithium aluminium hydride followed by removal of the acetal rings afforded 5-*O*-α- and -β-D-glucopyranosyl-D-glucofuranose. Base-catalysed isomerization of the α-compound then gave crystalline 5-*O*-α-D-glucopyranosyl-D-fructopyranose (leucrose) in 53% yield.[39] Selective acetalation of α,α-trehalose with ethyl or methyl isopropenyl ether in DMF with toluene-*p*-sulphonic acid as catalyst gave the mono-4,6-*O*-isopropylidene derivative which was isolated as its hexa-acetate in 38% yield. Hydrolysis of the acetal ring and a sulphonate displacement at the liberated secondary site gave access to the D-*gluco*-D-*galacto*-disaccharide, whereas standard reactions at the primary site gave 6-deoxy-6-fluoro- and 6-deoxy-6-iodo-trehalose.[40] Galactosylation of methyl 2,3,6-tri-*O*-benzyl-α-D-galactopyranoside with tetra-*O*-benzyl-α,β-D-galactopyranosyl bromide, prepared *via* the 1-*O*-(*N*-phenylcarbamate), afforded a route to 4-*O*-α-D-galactopyranosyl-D-galactose,[41] and the corresponding 4-*O*-α-D-mannopyranosyl-D-mannose was obtained by condensation between tetra-*O*-acetyl-α-D-mannopyranosyl bromide and 1,6-anhydro-2,3-*O*-isopropylidene-β-D-mannopyranose.[42] Also in the mannobiose series the 3-*O*-α-linked dimer has been reported following glycosylation of 1,2,4,6-tetra-*O*-acetyl-α-D-mannose, and in this work a range of 3-substituted mannoses were synthesized,[43] Xylosylation of methyl and benzyl 4,6-*O*-benzylidene-α-D-mannoside gave the 3-*O*-β-linked products; and analogous reaction of the 3-acetates led to the 2-bonded disaccharide derivatives together with minor proportions of α-linked isomers.[44] Studies of the synthesis of 4-*O*-β-D-galactosyl-D-mannose (epilactose) have shown that it can be made with 9 : 1 stereospecificity by the hydroxylation with hydrogen peroxide and sodium molybdate of lactal, whereas the specificity is much lower (3 : 2) when produced by the nitromethane extension reaction applied to 3-*O*-β-D-galactopyranosyl-D-arabinose. Maltal underwent similar hydroxylation to give the D-*manno*- and -D-*gluco*-products also in the ratio of 9 : 1, and lactal gave 2-deoxylactose in 84% yield when treated with aqueous trichloroacetic acid. These 1,4-linked disaccharides did not undergo epimerization

[37] J. H. Pazur, Y. Tominaga, C. W. De Brosse, and L. M. Jackman, *Carbohydrate Res.*, 1978, **61**, 279.
[38] K. Wallenfels, P. Földi, H. Niermann, H. Bender, and D. Linder, *Carbohydrate Res.*, 1978, **61**, 359.
[39] W. A. R. van Heeswijk, F. R. Wassenburg, and J. F. G. Vliegenthart, *Carbohydrate Res.*, 1978, **62**, 281.
[40] J. Defaye, H. Driguez, B. Henrissnt, J. Gelas, and E. Bar-Guilloux, *Carbohydrate Res.*, 1978, **63**, 41.
[41] D. D. Cox, E. K. Metznar, and E. J. Reist, *Carbohydrate Res.*, 1978, **62**, 245.
[42] P. L. Durette and T. Y. Shen, *Carbohydrate Res.*, 1978, **65**, 314.
[43] M. M. Ponpipom, *Carbohydrate Res.*, 1977, **59**, 311.
[44] R. T. Lee and Y. C. Lee, *Carbohydrate Res.*, 1978, **67**, 389.

under the influence of molybdate under conditions which caused isomerization of monosaccharides.[45]

In the L-rhamnose series, benzyl 2,3-*O*-isopropylidene-α-L-rhamnopyranoside has been used in a synthesis of muricatin B (15), a component of resins of the Convolvulaceae, establishing the nature of the anomeric bonding in the

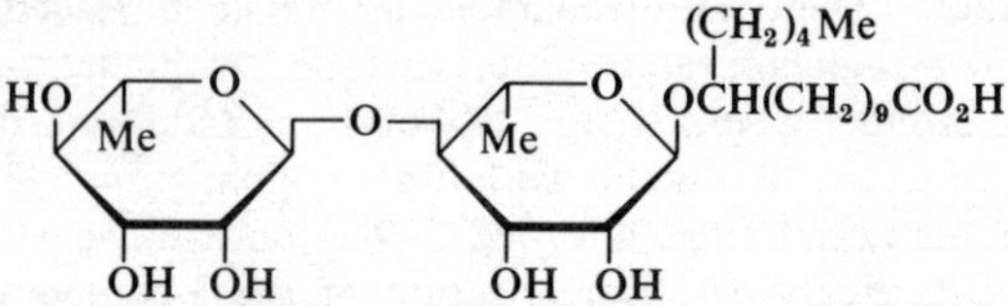

(15)

compound.[46] Glycosylation with tri-*O*-acetyl and -*O*-benzoyl-α-L-rhamnosyl bromide has been used to obtain rutinose, 3-*O*-α-L-rhamnopyranosyl-L-rhamnose, and 6-*O*-α-L-rhamnopyranosyl-D-galactose and some trisaccharides (see later). With this range of compounds it is difficult to determine anomeric configurations by standard n.m.r. methods, and it was therefore shown that these configurations and positions of substitutions can be assigned from chemical shifts of the protons at C-5 and C-6.[47] 2-*O*-α-L-Rhamnopyranosyl-L-rhamnose, part of the immunodeterminant of a streptococcal polysaccharide, was obtained by glycosylation of 1,4-di-*O*-benzoyl-β-L-rhamnose, which is somewhat surprising since substitution at position-2 is known to be relatively difficult – especially when an axial hydroxy-group is involved.[48]

A high yield of 1,4-β-linked product was obtained when methyl 2,3-anhydro-β-D-ribopyranoside was condensed with 3,4-di-*O*-acetyl-2-*O*-benzyl-α-D-xylopyranosyl bromide,[49] and use of tri-*O*-acetyl-α-D-xylopyranosyl bromide afforded the related product which, on base-catalysed hydrolysis, gave the methyl β-xylobioside.[50] Several 2-*O*-glycosyl-D-arabinoses have been prepared by glycosylation of benzyl 3,4-*O*-isopropylidene-β-D-arabinopyranoside, and from benzyl 2-*O*-β-D-glucopyranosyl-β-D-arabinoside the aldobiuronic acid was produced.[51]

In the field of dideoxyhexosylmannosides the syntheses of the disaccharides corresponding to the repeating units of salmonella O-antigens, 3-*O*-α-abequosyl-, 3-*O*-tyvelosyl-, and 3-*O*-α-paratosyl-α-D-mannopyranosides, have been reviewed,[52] and *p*-aminophenyl glycosides of this type were synthesized as outlined in Scheme 3.[53] The tetradeoxydisaccharide glycosides, methyl 4-*O*-α-D-digitoxosyl-α-D-digitoxoside[54] and methyl 3-*O*-α-D-olivosyl-α-D-olivo-

[45] V. Bílik, E. Jurčová, and V. Sutoris, *Chem. Zvesti*, 1978, **32**, 252.
[46] A. Liptak, V. M. Chari, B. Kreil, and H. Wagner, *Phytochemistry*, 1978, **17**, 997 (*Chem. Abs.*, 1978, **89**, 197 879k).
[47] C. Laffite, A. M. Nguyen Phnoc Du, F. Winternitz, R. Wylde, and F. Pratviol-Sosa, *Carbohydrate Res.*, 1978, **67**, 91.
[48] W. Schalch, W. Hochstrasser, and D. G. Braun, *Tetrahedron Letters*, 1978, 4153.
[49] J. Hirsch and P. Kováč, *J. Carbohydrates, Nucleosides, Nucleotides*, 1978, **5**, 373.
[50] P. Kováč, J. Hirsch, and V. Kováčik, *Chem. Zvesti*, 1978, **32**, 514.
[51] R. S. Sarfati and L. Szabó, *Carbohydrate Res.*, 1978, **75**, 11.
[52] B. Gotthammar, *Chem. Comm., Univ. Stockholm*, 1977, 1 (*Chem. Abs.*, 1978, **88**, 191 256n).
[53] G. Ekborg, P. J. Garegg, and S. Josephson, *Carbohydrate Res.*, 1978, **65**, 301.
[54] J. Thiem and J. Schwentner, *Tetrahedron Letters*, 1978, 459.

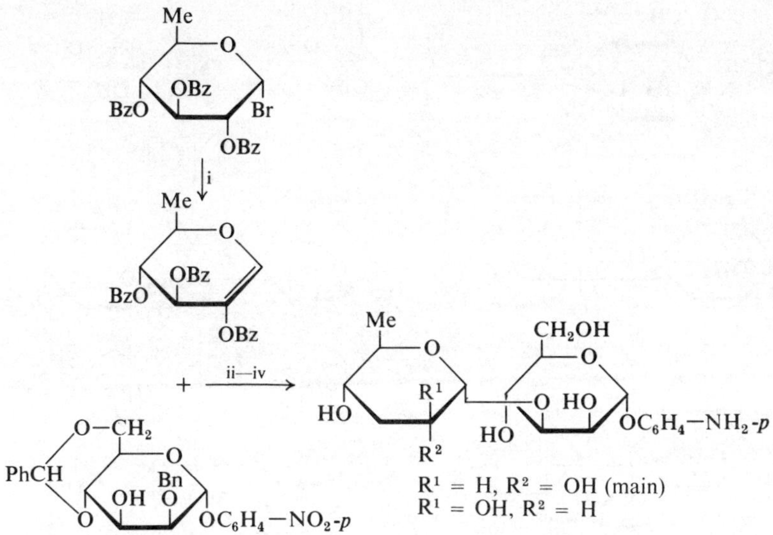

Reagents: i, NaI–Et_2NH; ii, H_2–PtO_2; iii, H_2–Pd; iv, MeOH–NH_3

Scheme 3

side,[55] were obtained by related use of glycal derivatives as shown in Schemes 4 and 5, respectively. 3-*O*-α-D-Galactopyranosyl-D-galactose has been made by

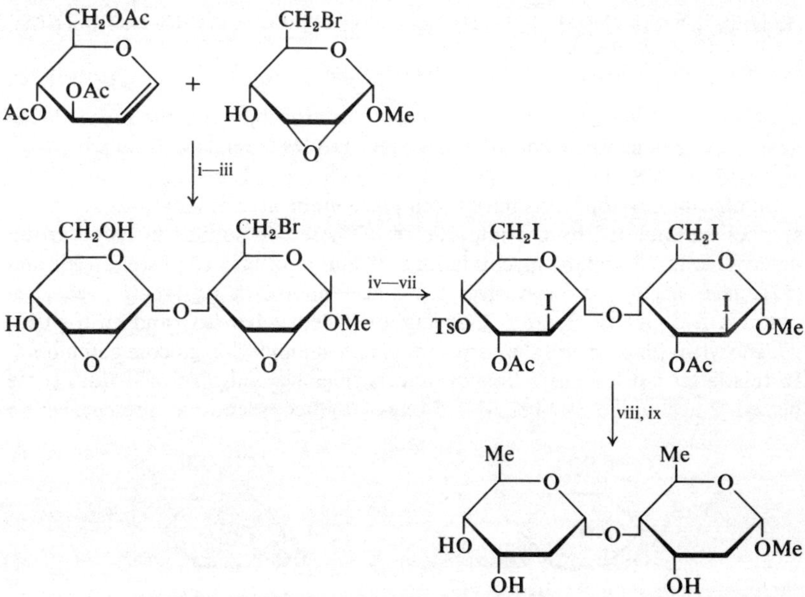

Reagents: i, BF_3,Et_2O; ii, NaOMe; iii, H_2O_2–MeCN; iv, TsCl–py; v, LiI; vi, Ac_2O–py; vii, NaI–MeCOEt; viii, $NiCl_2$–$NaBH_4$; ix, Na–Hg

Scheme 4

[55] J. Thiem and H. Karl, *Tetrahedron Letters*, 1978, 4999.

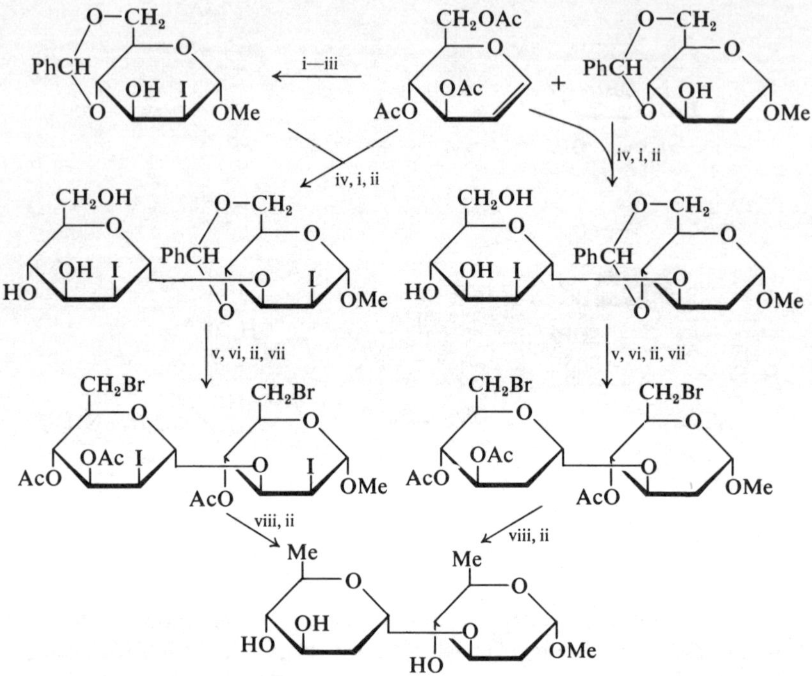

Reagents: i, NIS; ii, NaOMe; iii, $PhCH(OMe)_2-H^+$; iv, BF_3Et_2O; v, $PhCHO-ZnCl_2$; vi, NBS; vii, Ac_2O–py; viii, NiB_2

Scheme 5

application of the glycal-nitrosyl chloride approach,[56] and the use of lactal hexa-acetate in syntheses of di- and tri-saccharides is referred to in Chapter 8 and later in this Chapter.

Significant developments have taken place in the area of amino-disaccharide synthesis stimulated by the importance of these compounds in blood group determinants. Standard glycosylations of 2-aminosugar derivatives (16) and (17) gave access to derivatives of 2-acetamido-2-deoxy-3-*O*-(β-D-galactopyranosyl)-D-glucose[57] and 2-acetamido-2-deoxy-4-*O*-(α- and β-D-glucopyranosyl)-D-glucose and the corresponding α-mannosyl-D-glucose compound. In this latter work some 6-linked disaccharides were also formed from some benzyl 2-acetamido-3-*O*-benzyl-2-deoxy-α-D-glucopyranoside present as an

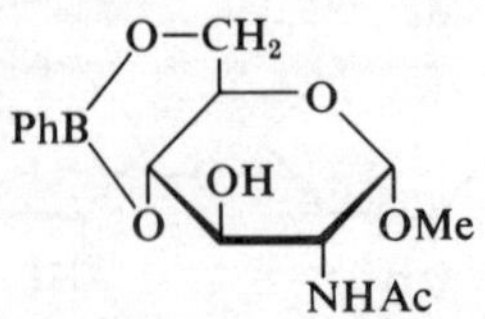

(16)

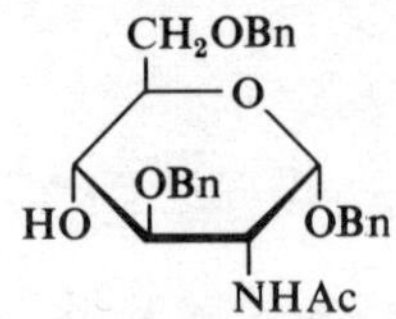

(17)

[56] R. U. Lemieux and R. V. Stick, *Austral. J. Chem.*, 1978, **31**, 901.
[57] C. J. Griffiths and H. Weigel, *Carbohydrate Res.*, 1978, **62**, 169.

impurity in compound (17).[58] Compounds with amino-groups in the non-reducing portions have also been described, 3-*O*-(2-acetamido-2-deoxy-α-D-galactopyranosyl)-D-galactose having been synthesized by the glycal-nitrosyl chloride procedure,[56] and the oxazoline (18), condensed with allyl 2-acetamido-3, 4-di-*O*-benzyl-2-deoxy-α-D-glucopyranoside gave, after deprotection, the 1,6-linked isomer of di-*N*-acetyl chitobiose. A similar attempt to prepare the 1,4-isomer was less successful.[59] Paulsen and his group have made major contributions in this area, in particular through their use of the glycosylating agent (19) which was prepared from 1,6:2,3-dianhydro-4-*O*-benzyl-β-D-

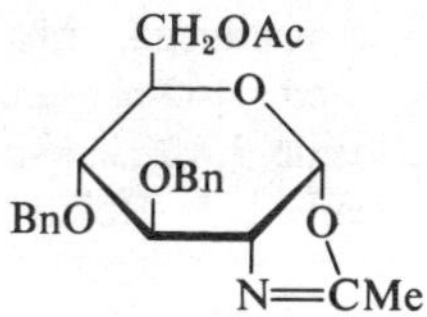

(18)

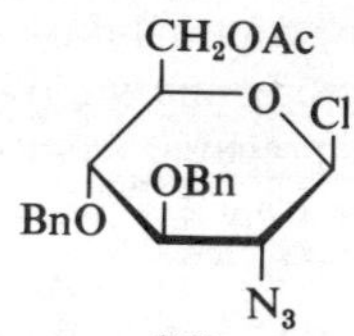

(19)

mannopyranose *via* the α-glycosyl bromide. Disaccharide derivatives (20)—(22) were described,[60, 61] and in the D-galactose series the analogue of compound (19) was used to prepare the α-linked disaccharides (23) and (24).[62] (See below for trisaccharides produced in developments of this work.)

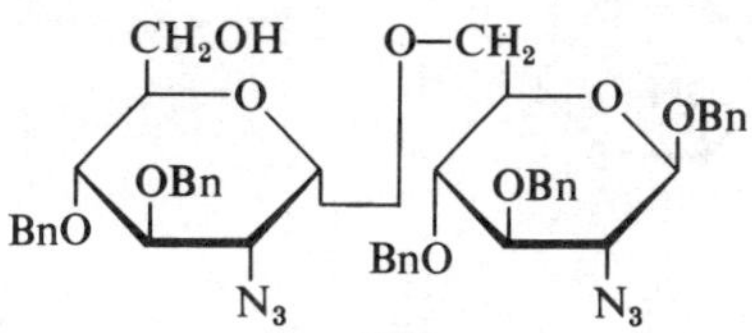

(20)

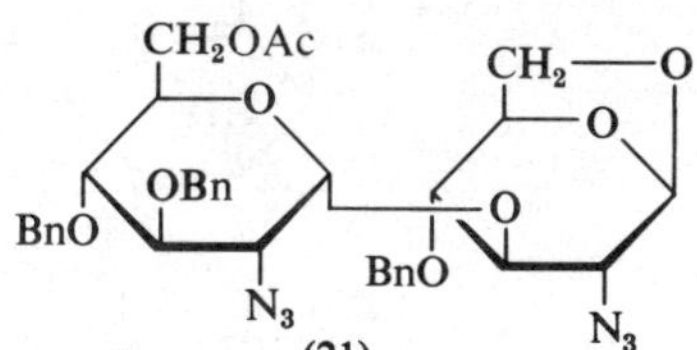

(21)

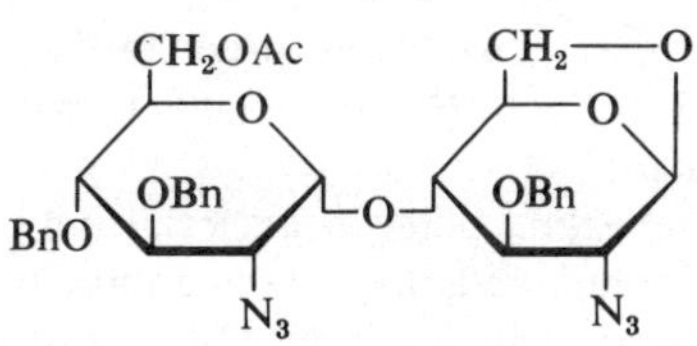

(22)

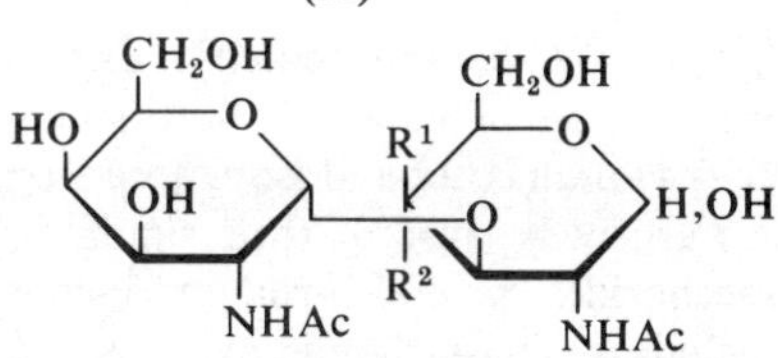

	R^1	R^2
(23)	H	OH
(24)	OH	H

In the area of aldobiouronic acids methyl 2-*O*-(methyl 4-*O*-methyl-α- and β-D-glucopyranosyluronate)-β-D-xylopyranoside were synthesized using methyl 3,4-di-*O*-acetyl-β-D-xylopyranoside and the glycosylating agents (25) and (26). From the bromide the α- and β-linked compounds were produced in the ratio 1.4 : 1 [mercury(II) cyanide in benzene, nitromethane], whereas this ratio was 18 : 1 when the chloride was used (silver perchlorate in ether).[63]

[58] E. F. Hounsell, M. B. Jones, and J. A. Wright, *Carbohydrate Res.*, 1978, **65**, 201.
[59] C. D. Warren, M. A. E. Shaban, and R. W. Jeanloz, *Carbohydrate Res.*, 1977, **59**, 427.
[60] H. Paulsen and W. Stenzel, *Chem. Ber.*, 1978, **111**, 2334.
[61] H. Paulsen and W. Stenzel, *Chem. Ber.*, 1978, **111**, 2348.
[62] H. Paulsen, Č. Kolář, and W. Stenzel, *Chem. Ber.*, 1978, **111**, 2358.
[63] P. Kováč and R. Palovčík, *Chem. Zvesti*, 1978, **32**, 501.

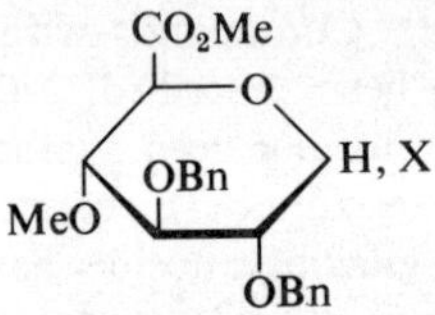

(25) X = Br (α-isomer)
(26) X = Cl

Potentially valuable selective acylations of disaccharides are mentioned in Chapter 6 and, from methyl β-cellobioside 2,6,2′,3′,4′,6′-hexa-acetate, methyl 4-*O*-β-D-glucopyranosyl-β-D-allopyranoside was produced. Likewise, methyl 4-*O*-α-D-galactopyranosyl-β-D-allopyranoside and 4-*O*-α-D-glucopyranosyl-D-allose were prepared from methyl β-D-maltoside 2,6,2′,3′,6′-pentabenzoate and β-maltose 1,3,6,2′,4′,6′-hepta-acetate, respectively.[64] Related methods led to *galacto*-sucrose.[65]

Synthesis of Tri- and Tetra-saccharides.—Two specific trisaccharides have been produced by galactosylation of lactose. The 6-β-linked compound (27) was made from the 2,2′,3,3′-tetra-*O*-acetyl-1,6-anhydro derivative,[66] and the 4-α-linked isomer (28), as its methyl β-glycoside, from the methyl glycoside

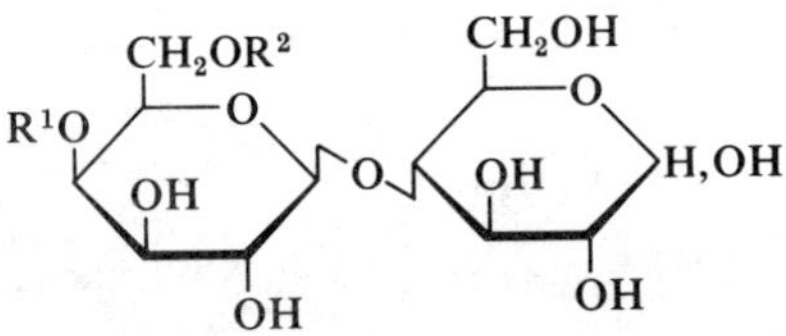

(27) R^1 = H, R^2 = β-D-galactopyranosyl
(28) R^1 = α-D-galactopyranosyl, R^2 = H

2,2′,3,3′,6,6′-hexabenzoate.[67] Enzymic methods have been used for the preparation of panose [*O*-α-D-glucopyranosyl-(1 → 6)-*O*-α-D-glucopyranosyl-(1 → 4)-D-glucose] and isopanose from pullulan.[68]

Largely because of their significance as repeating units in bacterial polysaccharides, several L-rhamnose-containing compounds have been reported, including *O*-α-L-rhamnopyranosyl-(1 → 3)-*O*-α-L-rhamnopyranosyl-(1 → 6)-D-galactose and -D-glucose[47] and *O*-β-D-mannopyranosyl-(1 → 4)-*O*-α-L-rhamnopyranosyl-(1 → 3)-D-galactose.[69] *p*-Aminophenyl *O*-β-D-glucopyranosyl-(1 → 4)-*O*-α-L-rhamnopyranosyl-(1 → 3)-β-D-galactopyranoside, after synthesis employing the glucopyranosyl rhamnopyranosyl bromide and *p*-nitrophenyl 2,6-di-*O*-acetyl-β-D-galactopyranoside, was coupled to protein and to sepharose for biological work.[70] Previously developed methods which use unsaturated

64 K. Takeo and S. Okano, *Carbohydrate Res.*, 1977, **59**, 379.
65 L. Hough, M. S. Chowdhary, and A. C. Richardson, *J. C. S. Chem. Comm.*, 1978, 664.
66 T. G. Chung, H. Ishihara, and S. Tejima, *Chem. and Pharm. Bull.* (*Japan*), 1978, **26**, 2147.
67 D. D. Cox, E. K. Metzner, and E. J. Reist, *Carbohydrate Res.*, 1978, **63**, 139.
68 Y. Sakano, M. Kogure, T. Kobayashi, M. Tamura, and M. Suekane, *Carbohydrate Res.*, 1978, **61**, 175.
69 N. K. Kochetkov, B. A. Dmitriev, and A. V. Nikolaev, *Izvest. Akad. Nauk S.S.S.R., Ser. khim.*, 1977, 2578 (*Chem. Abs.*, 1978, **88**, 152 884w).
70 N. K. Kochetkov, B. A. Dmitriev, and A. Va. Chernyak, *Bioorg. Khim.*, 1977, **3**, 752 (*Chem. Abs.*, 1978, **88**, 23 281y).

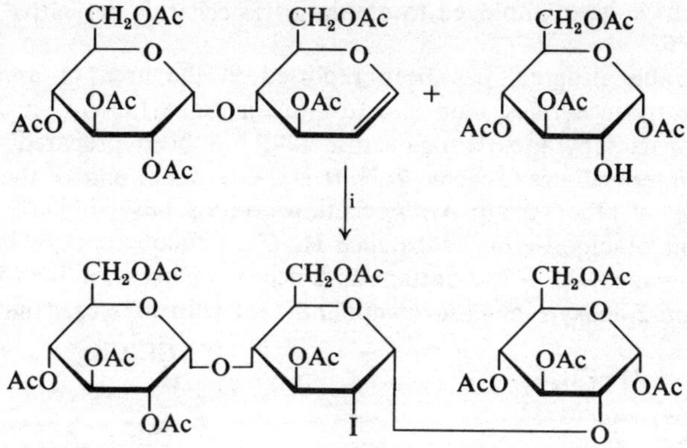

Reagents: AgOAc–I_2–MeCN

Scheme 6

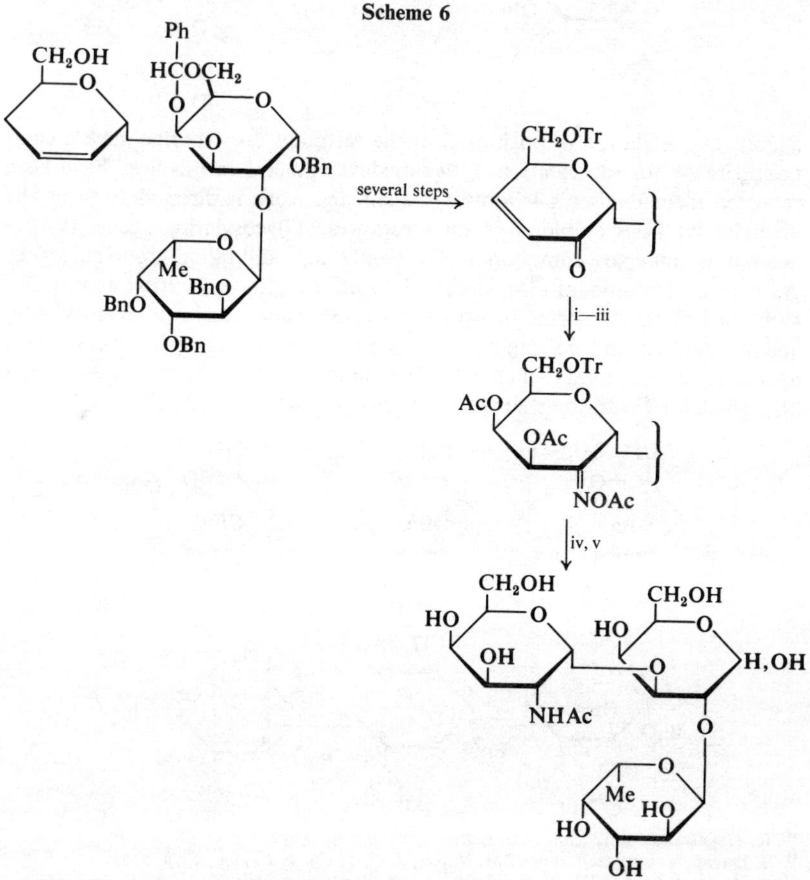

(29)

Reagents: i, OsO_4; ii, NH_2OH; iii, Ac_2O–py; iv, LAH; v, H_2–Pd

Scheme 7

structures have been employed to obtain a trisaccharide derivative as shown in Scheme 6.[71]

Considerable progress has been reported in the area of amino-sugar-containing trisaccharides, and 2-*O*-(α-L-fucopyranosyl)-3-*O*-(2-acetamido-2-deoxy-α-D-galactopyranosyl)-D-galactose (29) has been prepared using unsaturated intermediates (Scheme 7).[72] It is the terminal unit of the antigenic determinant of blood-group A. Kochetkov's group have similarly made the determinant of blood-group substance H, *O*-α-L-fucopyranosyl-(1 → 2)-*O*-β-D-galactopyranosyl-(1 → 3)-2-acetamido-2-deoxy-D-glucose, in high yield from the 2-amino-2-deoxy-D-glucose compound (30) with (31) as galactosylating

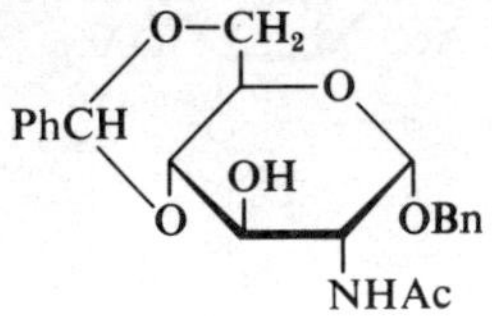

(30)

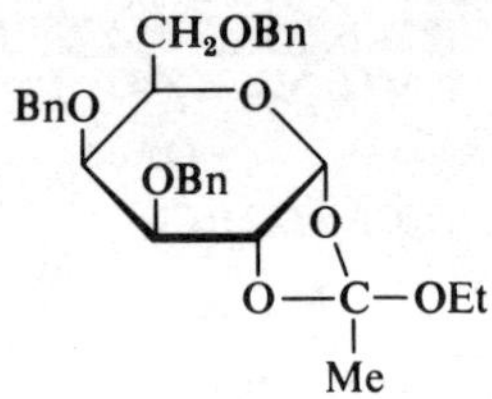

(31)

agent. Deacetylation at position-2 of the resultant disaccharide gave a compound ready for fucosylation.[73] Major developments in this area have been reported from Paulsen's laboratory; again the work is directed towards the artificial synthesis of blood-group substances. Glycosylating agent (19) has been used to prepare compounds (32)[60] and (33),[61] and blood-group substance A, Type 1, determinant (34) was synthesized using the galactosyl analogue.[74] While all of these studies involve the use of monosaccharide glycosylating agents, Paulsen and colleagues have shown that disaccharide glycosylating agents can also be used (Scheme 8). The mannosylgalactosyl bromide (35) was also condensed with the rhamnoside to give trisaccharide (36).[75]

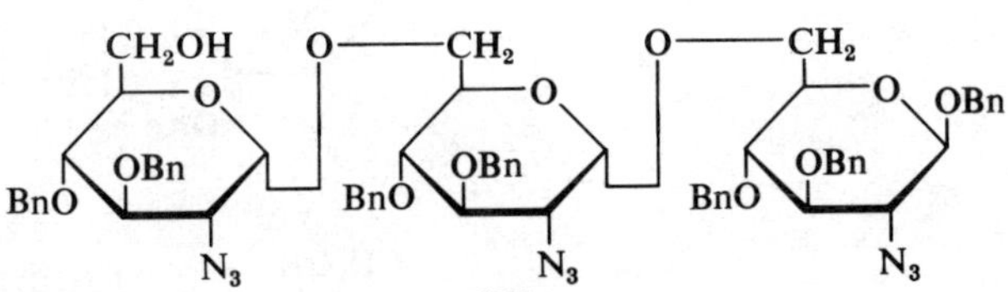

(32)

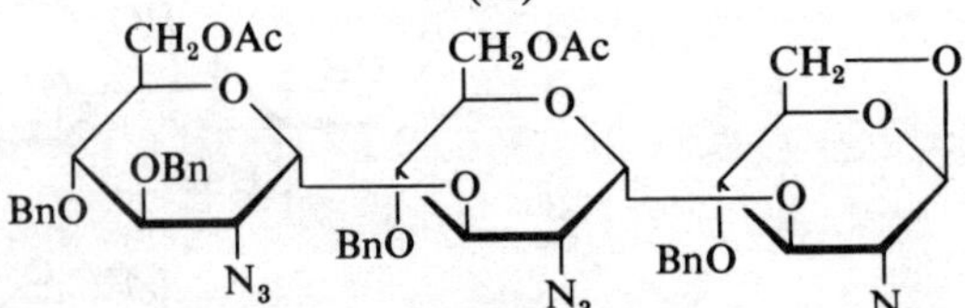

(33)

[71] K. Heyns and M. T. Lim, *Tetrahedron Letters*, 1978, 891.
[72] S. David, A. Lubineau, and J.-M. Vatèle, *J. C. S. Chem. Comm.*, 1978, 535.
[73] V. A. Derevitskaya, O. S. Novikova, A. Yu. Evstigneev, and N. K. Kochetkov, *Izvest. Akad. Nauk S.S.S.R., Ser. khim.*, 1978, 450 (*Chem. Abs.*, 1978, **88**, 191 277v).
[74] H. Paulsen, Č. Kolář, and W. Stenzel, *Chem. Ber.*, 1978, **111**, 2370.
[75] H. Paulsen and O. Lockhoff, *Tetrahedron Letters*, 1978, 4027.

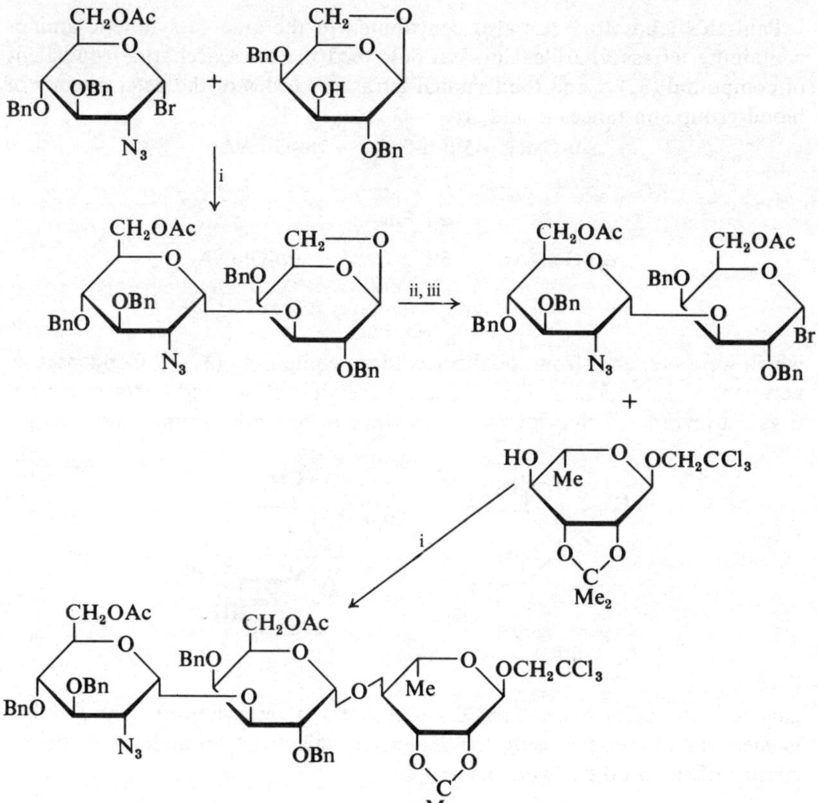

Reagents: i, $HgBr_2$, $Hg(CN)_2$; ii, Ac_2O, CF_3CO_2H; iii, Me_3SiBr

Scheme 8

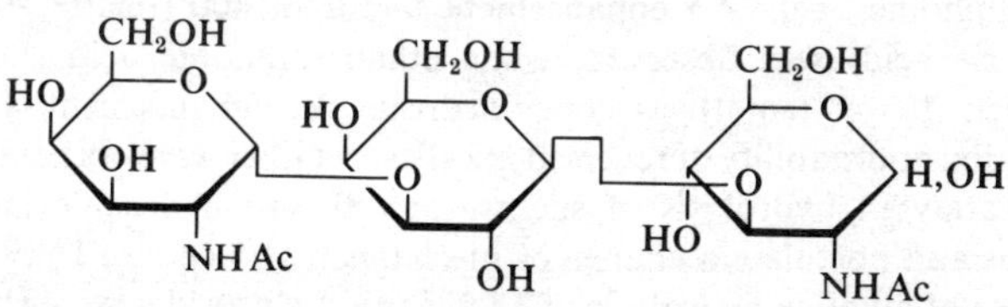

(34)

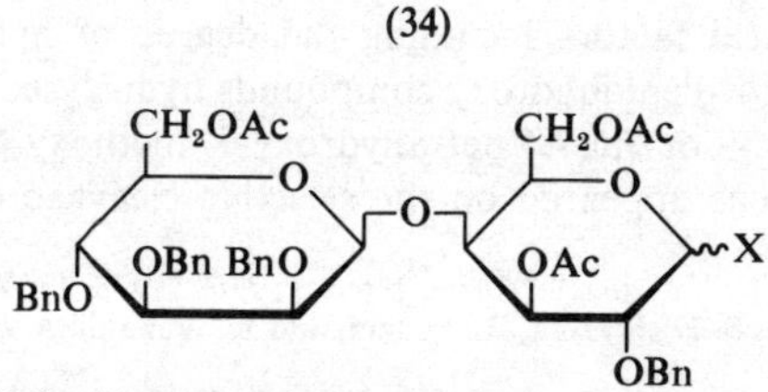

(35) X = Br, α-isomer

(36) X = trichloroethyl 2,3-*O*-isopropylidene-α-L-rhamnosid-4-yl (α-isomer)

Paulsen's laboratory has also contributed to the area of synthetic amino-containing tetrasaccharides and has described the tetrasaccharide equivalent of compound (32).[60] and the terminal tetrasaccharides of the determinants of blood-group substances A and B:

α-D-Gal(1 → 3)β-D-Gal(1 → 3)D-Glu*N*Ac
2
↑
α-L-Fuc

α-D-Gal*N*Ac(1 → 3)β-D-Gal(1 → 3)D-Glu*N*Ac
2
↑
α-L-Fuc

which were prepared from the dibenzylidene compound (37).[76] Condensation between 1″,2,2″,3,3′,3″,4,4′,4″,6′-deca-*O*-acetylraffinose and tetra-*O*-benzyl-D-galactopyranosyl chloride in the presence of tetraethylammonium chloride

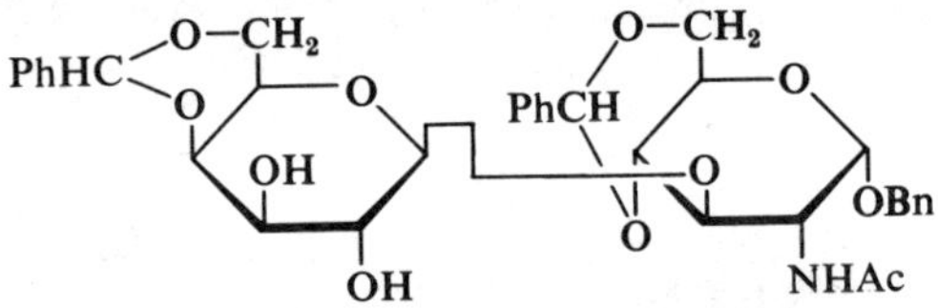

(37)

gave a route to stachyose tetradeca-acetate (α-galactosylation). The β-linked isomer was obtainable using the acetylated galactosyl bromide, and the ^{13}C spectra of the products were discussed.[77]

Hydrolysis and Other Reactions.—Continuing earlier studies (*Bull. Chem. Soc. Japan*, 1976, **49**, 1059), Japanese workers have now examined the hydrolysis of sucrose solubilized in benzene by a reversed micelle formed from dodecyl-benzenesulphonic acid. An enhancement factor of 400 (relative to aqueous hydrochloric acid) was observed, while ethanesulphonic acid showed only slight effect. It was tentatively concluded that the enhancement is related to the formation and stability of reversed micelles.[78] Other workers have examined the acid-catalysed hydrolysis of sucrose at different hydrochloric acid concentrations and postulate a change of mechanism at 5.68 mol l^{-1}.[79]

The rates of alkaline hydrolysis of 13 flavanoid glycosides were shown to be dependent on several factors including the degree of hydroxylation of the aglycones. 3,5,7,3′,4′-Pentahydroxy compounds hydrolysed more rapidly than 3,5,7,4′-tetrahydroxy- or 3,5,7,4′-tetrahydroxy-3′-methoxy analogues.[80]

A short review has appeared on the selective cleavage of the glucuronide

76 H. Paulsen and Č. Kolář, *Angew. Chem. Internat. Edn.*, 1978, **17**, 771.
77 T. Suami, T. Otake, S. Nishiyama, R. Adachi, and K. Nakamura, *Bull. Chem. Soc. Japan*, 1978, **51**, 1826.
78 K. Arai and Y. Ogiwara, *Bull. Chem. Soc. Japan*, 1978, **51**, 182.
79 G. I. Gorokhov, T. Ya. Lugina, V. V. Derevenko, T. N. Platonenko, and O. B. Krayanskii, *Trans. Krasnodar, Politekh. Inst.*, 1976, **70**, 75 (*Chem. Abs.*, 1978, **89**, 6484v).
80 M. D. Alaniya, *Khim. prirod. Soedinenii*, 1977, 646. (*Chem. Abs.*, 1978, **88**, 105 710b).

linkages in oligosaccharide saponins by photolysis, lead tetra-acetate, acetic anhydride–pyridine degradation, and anodic oxidation.[81]

Detailed information has appeared on the thermal degradation of sucrose. Above melting point, transglycosylation occurs to give oligomers and polymers as well as 1,6-anhydro-β-D-glucopyranose and volatile products including 5-hydroxymethylfurfural. At the melting point initial reaction is believed to be dissociation into glucose and a fructose derivative, conceivably an anhydro compound which would be an efficient glycosylating agent, which, for example, converts sucrose into difructosyl-glucoses.[82] Heating galactose at 150 °C gave rise to oligosaccharides. After 100 min, 79% of material was present as oligomers higher than disaccharides with the average degree of polymerization being 9.6.[83]

γ-Radiolysis of deoxygenated, nitrous oxide-saturated, aqueous solutions of maltose, lactose, gentiobiose, trehalose, and sucrose has been found to give products originating from scission of the glycosidic bonds and O-values were found to vary only between 1.9 and 3.5 with the nature of these bonds.[84]

Interestingly, anomerization of 1,2-*trans*-related glycofuranosides into the thermodynamically less favoured *cis*-isomers can be brought about by the use of Grignard reagents. Methyl 5-*O*-benzyl-β-D-ribofuranoside can be converted in 95% efficiency to the α-anomer by treatment with t-butylmagnesium bromide, the intermediate, it is presumed, being a species like (38). With methylmagnesium iodide, some of the 1-deoxyhexitol derivative (39) was formed together with the α-glycoside. When methyl 5,6-*O*-cyclohexylidene-3-deoxy-2-*C*-methyl-β-D-*ribo*-hexofuranoside was then treated with methylmagnesium iodide, anomerization occurred to give the α-isomer (40), but products (41) and (42) of nucleophilic opening of the cyclohexylidene ring were then formed.[85]

Investigations have been made of pyridine as a reaction solvent for the

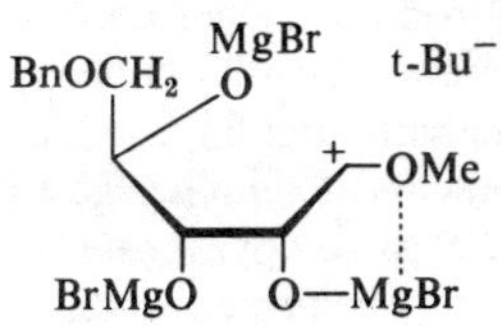

(38)

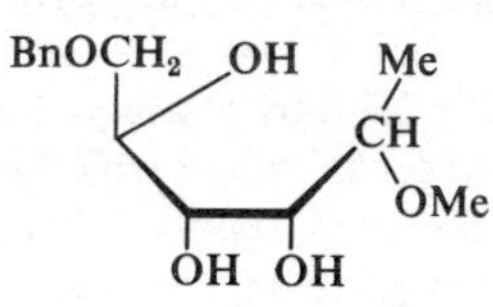

(39)

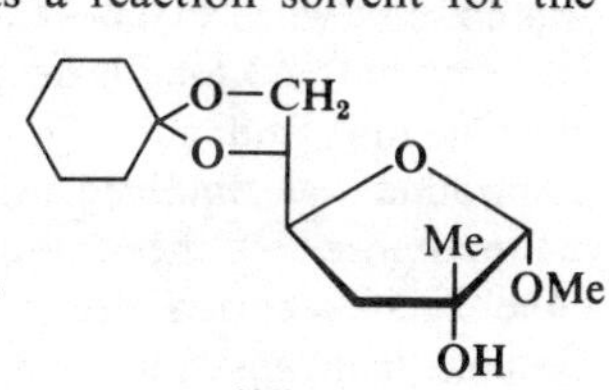

(40)

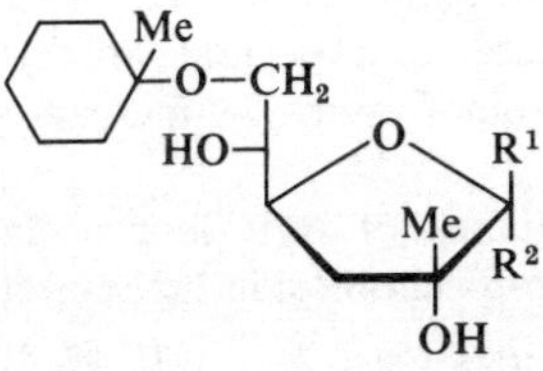

(41) R¹ = H, R² = OMe
(42) R¹ = OMe, R² = H

[81] I. Kitagawa and M. Yoshikawa, *Heterocycles*, 1977, **8**, 783 (*Chem. Abs.*, 1978, **88**, 170 394).
[82] G. N. Richards and F. Shafizadeh, *Austral. J. Chem.*, 1978, **31**, 1825.
[83] P. Barbetti and I. Chiappini, *Ind. Aliment* (*Pinvolo, Italy*), 1977, **16**, 61 (*Chem. Abs.*, 1978, **88**, 150 936r).
[84] H. Zegota and C. von Sonntag, *Zeit. Naturforsch., B: Anorg. Chem., Org. Chem.*, 1977, 1977, **32B**, 1060 (*Chem. Abs.*, 1978, **88**, 7250g).
[85] M. Kawana and S. Emoto, *Tetrahedron Letters*, 1978, 1561.

periodate oxidation of glycosides, and the structures of saponins and glycolipids were examined by the method. Dialdehydes initially produced by diol cleavage were not further oxidized.[86]

2 *S*-Glycosides

Interest in thioglycosides is enhanced by their potential as glycosidase inhibitors and for other exhibited biological activity. 2,3,4-Tri-*O*-benzoyl-1-thio-β-D-ribopyranose, prepared from the isothiouronium salt, gave the *p*-nitrophenyl thioglycoside ester on treatment with *p*-nitrobromobenzene,[13] and *p*-aminophenyl 1-thio-β-D-allopyranoside[15] and *p*-aminophenyl 1-thio-α-D-talopyranoside[12] and the anomeric D-idopyranose analogues[15a] have been made using *p*-nitrobenzenethiol. Likewise, *p*-nitrophenyl 1-thio-α- and β-D-glucopyranosides have been synthesized, as have several non-reducing and 4-linked reducing disaccharide derivatives with sulphur as the inter-unit atom.[15b] The structure–activity relationships of several aminoalkyl and aminoaryl glycosides (linked through oxygen, sulphur, and nitrogen) with insulin-like activity have been reported. Highly specific activity was found, it being greatest with 6-aminohexyl-1-thio-β-D-mannopyranoside (43), and it was

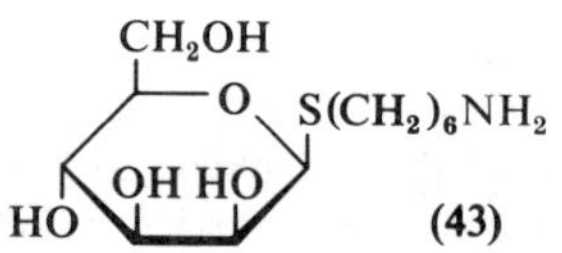

(43)

concluded that both the sugar and aglycone are involved in determining the biological activity.[87] Some thioglycosides involving cholesteryl units in the aglycones were synthesized using the 6-iodohexyl ether of cholesterol which was coupled with 2,3,4,6-tetra-*O*-acetyl-1-thio-α-D-mannose and -β-D-galactose and methyl 2,3,6-tri-*O*-acetyl-1-thio-β-D-glucopyranuronate. Deprotection gave neutral and acidic glycolipids and a corresponding 6-amino-6-deoxy compound was obtained using 3,4,6-tri-*O*-acetyl-1-methanesulphonyl-1-thio-α-D-mannose.[88] Other thioglycosides of possible biochemical significance to have been reported are β-*S*-glucopyranosyl and β-*S*-galactopyranosyl L-cysteine (condensation of acetylglycosyl bromides and *N*-protected cysteine methyl ester), the hydrolyses of which were studied at different pH values and temperatures[89] (a related synthesis of thioglucosides used β-D-glucopyranosylisothiouronium bromide tetra-acetate and several methyl-substituted L-alanine derivatives[89a]), and α-L-arabinofuranosyl-6-mercaptopurine and some of its esters.[90]

In the disaccharide series the syntheses of 6-(trifluoroacetamido)hexyl, heptyl, benzyl, and *p*-nitrobenzyl 1-thio-β-cellobioside hepta-acetates as well

[86] S. Honda, Y. Ohkaru, and K. Kakehi, *Analyst. Chim. Acta*, 1978, **98**, 85 (*Chem. Abs.*, 1978, **89**, 110 247f).
[87] P. L. Durette, R. L. Bugianesi, M. M. Ponpipom, T. Y. Shen, M. A. Casieri, M. S. Glitzer, and H. M. Katzen, *J. Medicin. Chem.*, 1978, **21**, 854.
[88] J. C. Chabala and T. Y. Shen, *Carbohydrate Res.*, 1978, **67**, 55.
[89] E. Baran and S. Drabarek, *Polish J. Chem.*, 1978, **52**, 941.
[89a] M. L. P. Monsigny, D. Delay, and M. Vaculik, *Carbohydrate Res.*, 1977, **59**, 589.
[90] D. Arnot and A. Graffi, *Z. Chem.*, 1977, **17**, 224 (*Chem. Abs.*, 1978, **88**, 7268n).

as methylene bis(hepta-*O*-acetyl-1-thio-β-cellobioside) have been effected using hepta-*O*-acetyl-1-thio-β-cellobiose and an appropriate halide. *p*-Aminophenyl and *p*-chlorophenyl 1-thio-β-cellobioside hepta-acetate were obtained, alternatively, using the acetylated glycosyl bromide and the thiophenols.[91] The α-*S*-benzyl glycoside of 4-*O*-α-D-mannopyranosyl-D-mannopyranose was reported following the preparation of the glycosyl bromide which was derived from 1,6-anhydro-4-*O*-(tetra-*O*-acetyl-α-D-mannopyranosyl)-β-D-mannopyranose.[42]

3 *C*-Glycosides

D-Glucose and phenol treated in an acidic medium gave a mixture of products from which the following were isolated: phenyl α- and β-D-glucopyranoside, 1-deoxy-1,1-di-(*p*-hydroxyphenyl)-D-glucitol, tri-(*p*-hydroxyphenyl)methane, some polymeric material, and the *C*-glycoside (44). Pyrogallol condensed with glucose in aqueous medium at pH 4.6 to give the analogue (45).[92]

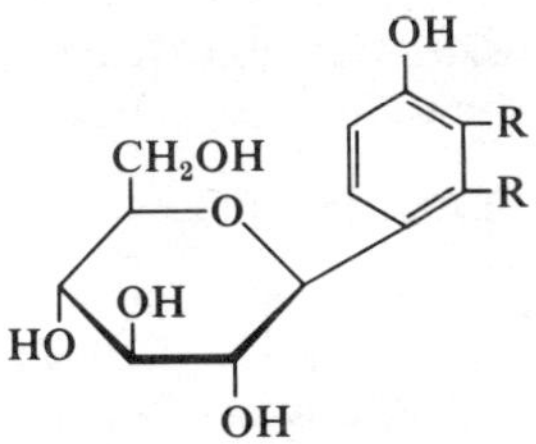

(44) R = H
(45) R = OH

In addition to the bicyclic products formed by photochemical (2 + 2) cycloadditions between tri-*O*-acetyl-D-glucal and an acetone-propan-2-ol mixture (*Carbohydrate Res.*, 1973, **29**, 459), compound (46) and minor proportions of (47) and (48) have now been shown to be produced [(46) presumably arises by hydrolysis of the main bicyclic adduct]. On acetylation of compound (46) the secondary hydroxy-group was substituted specifically. The INDOR n.m.r. technique was used in the product analysis.[93]

1,4-Addition took place when the enone (49) was treated with the thiolane anion (50) to give the β-product (51) as well as an isomer with inverted stereochemistry at C-4 and C-1. The *X*-ray analysis of this latter product was carried out, and from the two products four isomeric oct-2-ulose derivatives, *e.g.* (52), were obtained.[94]

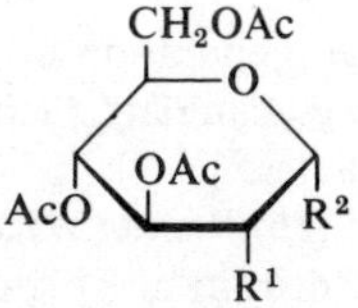

(46) R^1 = HO, R^2 = CMe_2OH
(47) R^1 = CMe_2OH, R^2 = H

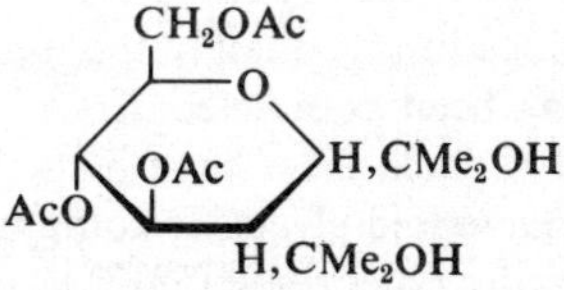

(48)

[91] P. L. Durette and T. Y. Shen, *Carbohydrate Res.*, 1978, **67**, 484.
[92] P. A. Pernemalm, *Acta Chem. Scand.*, 1978, **32B**, 72.
[93] Y. Araki, K. Senna, K. Matsuura, and Y. Ishido, *Carbohydrate Res.*, 1978, **60**, 389.
[94] H. Paulsen and H. Bünsch, *Chem. Ber.*, 1978, **111**, 3484.

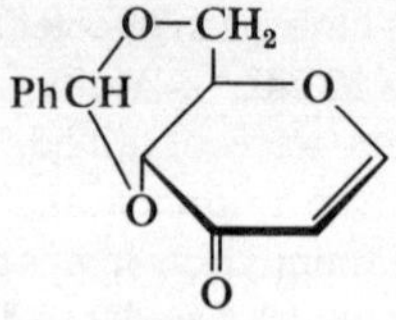

(49)

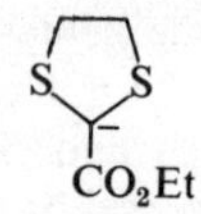

(50)

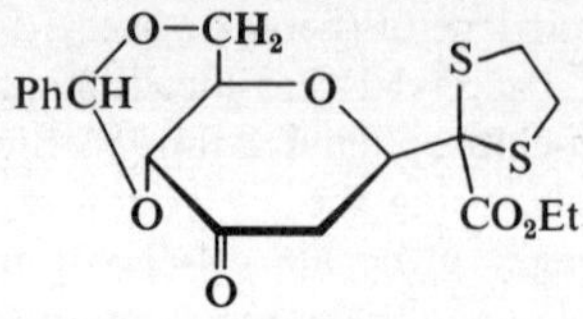

(51)

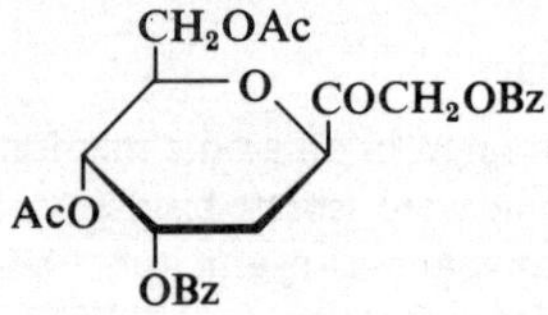

(52)

In the area of *C*-nucleosides (see also Chapter 19), the homonucleosides (53) has been prepared as shown in Scheme 9; isomerization occurred in basic

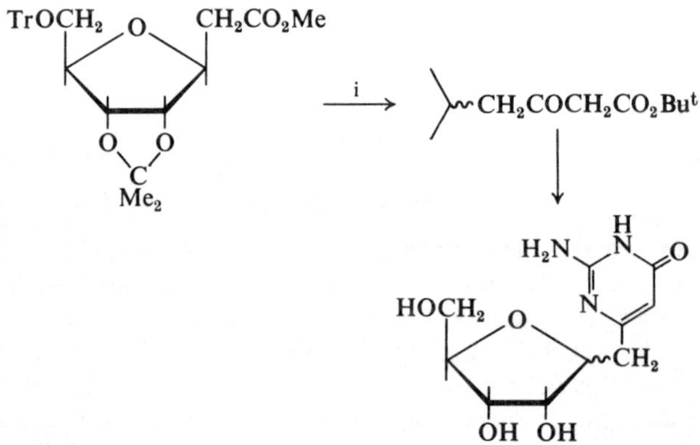

(53)

Reagents: i, $LiCH_2CO_2Bu^t$; ii, $NH_2C(NH_2)=NH$; iii, H^+

Scheme 9

conditions to give mainly the α-anomer.[95] The preparation of pseudouridine (54) has been completed from achiral starting materials (Scheme 10), and related *C*-nucleosides were made by parallel routes.[96]

The flavonoid glycoside compounds (55) and (56),[97] and (57) and (58),[98] and the *p*-hydroxybenzoate (59) [99] have been isolated from plant sources.

[95] J. A. Secrist, *J. Org. Chem.*, 1978, **43**, 2925.
[96] R. Noyori, T. Sato, and Y. Hayakawa, *J. Amer. Chem. Soc.*, 1978, **100**, 2561.
[97] B.-G. Österdahl, *Acta Chem. Scand.*, 1978, **32B**, 93.
[98] G. G. Zapesochnaya and N. A. Laman, *Khim. prirod. Soedinenii*, 1977, 862 (*Chem. Abs.*, 1978, **88**, 136 902v).
[99] M. Goetz and A. Jacot-Guillarmod, *Helv. Chim. Acta*, 1978, **61**, 1373.

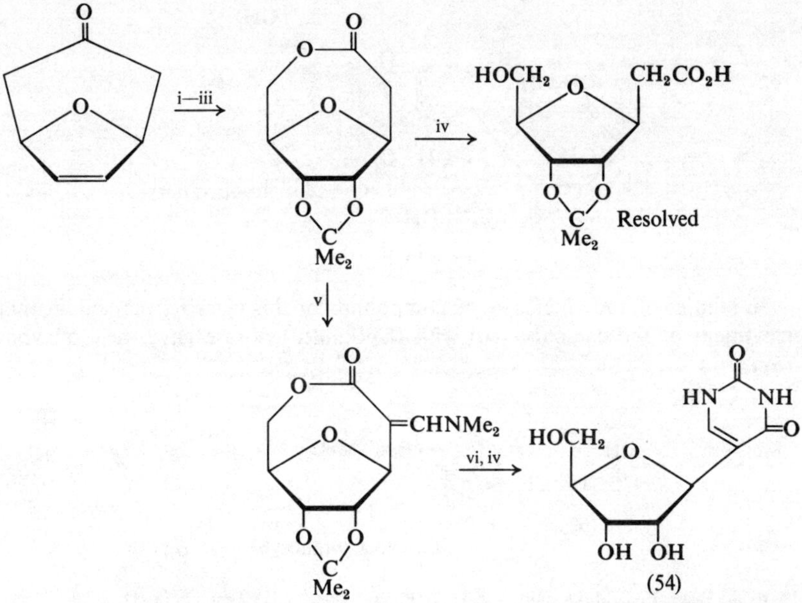

Reagents: i, H_2O_2–OsO_4–$HClO_4$; ii, Me_2CO–H^+; iii, CF_3CO_2OH; iv, H^+; v, $(Me_2N)_2CHOBu^t$; vi, NH_2CONH_2–EtO^-

Scheme 10

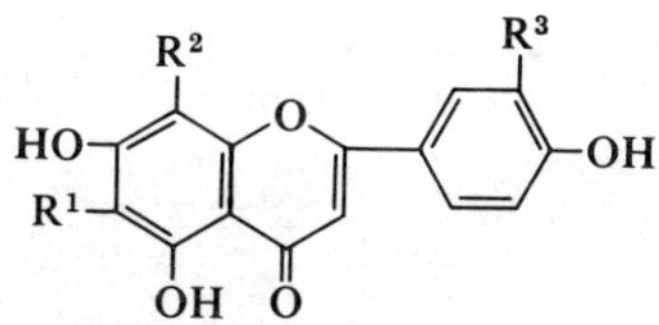

	R^1	R^2	R^3
(55)	G	G	H
(56)	G	G	OH

G = β-D-glucopyranosyl

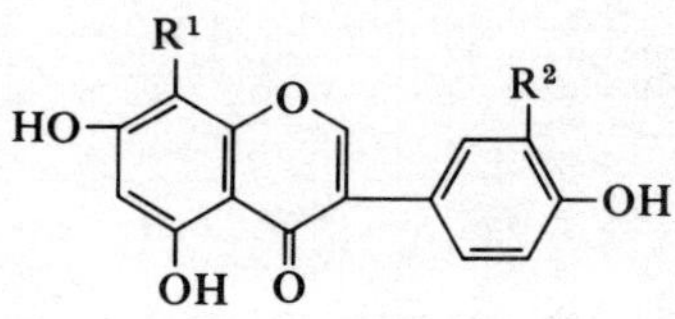

	R^1	R^2
(57)	G	H
(58)	G	OH

G = β-D-glucopyranosyl

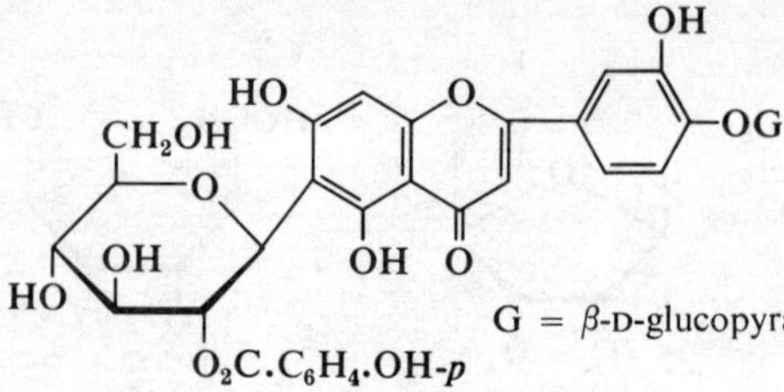

G = β-D-glucopyranosyl

(59)

In studies of the chemistry of compounds of this class it was shown that treatment of the chalcone (60) with thallium(III) nitrate gave the isoflavone (61).[100]

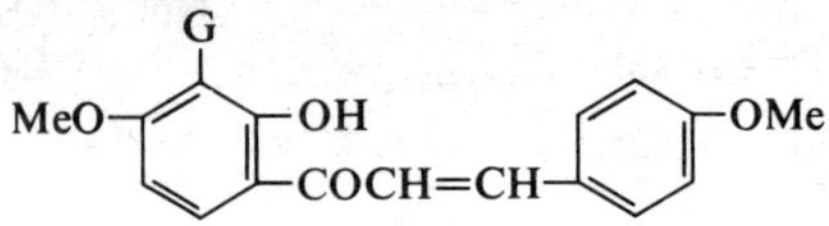

(60)

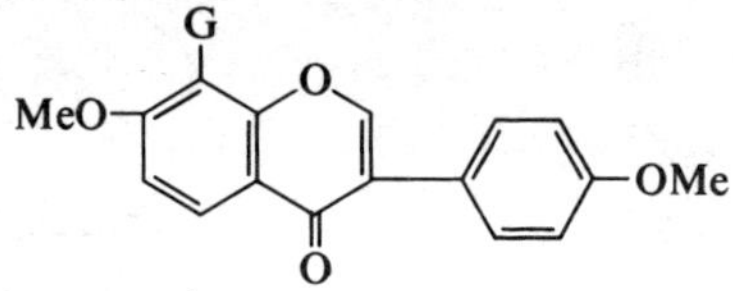

G = β-D-glucopyranosyl

(61)

[100] R. A. Eade, F. J. McDonald, and H.-P. Pham, *Austral. J. Chem.*, 1978, **31**, 2699.

4

Ethers and Anhydro-sugars

1 Ethers

Methyl Ethers.—A new methylation procedure developed in Angyal's laboratory involves the treatment of primary, secondary, or tertiary alcohols with acetic anhydride, acetic acid, and DMSO to give good yields of methylthiomethyl ethers which can be reductively desulphurized by use of 'nickel boride'. Methyl 4,6-*O*-benzylidene-β-D-glucopyranoside gave the diether in 47% yield together with the ketones derived by oxidation of both monoethers. Methylation of acetyl derivatives occurs without ester migration.[1] A procedure has been developed for calculating the rate constants for alkylation (methyl iodide in acetonitrile containing sodium hydroxide used) and acylation of the hydroxy-groups of carbohydrate diols. Methyl 2,6-dideoxy-α-D-*ribo*-hexopyranoside was used in the study.[2] Another theoretical paper relates to the calculation of the number of allowable conformational states of various partially methylated hexopyranosides.[3] A further investigation of the partial methylation of methyl β-L-arabinopyranoside was undertaken using the Haworth, Kuhn, and Hakamori procedures and, in each case, the initial relative reactivities of the hydroxy-groups decreased in the order 2, 3, 4, but as the di- and tri-ethers are produced, the relative reactivities change. Despite this, the Hakamori method tends to give mainly the triether from the beginning of the reaction, while the Kuhn procedure can be used to obtain the 2,3-dimethyl derivative.[4]

3,5-Di-*O*-methyl-D-glucose and thence the corresponding dimethyl-D-fructose was produced from 1,2-*O*-isopropylidene-6-*O*-triphenylmethyl-α-D-glucofuranose and purified by way of its α-1,2,6-triacetate, catalytic treatment of which, with sodium methoxide, gave the 6-acetate. Reaction with a stronger solution of this reagent afforded the fructose dimethyl ether. 1,6-Anhydro-3,5-di-*O*-methyl-β-D-glucofuranose, produced together with the dimethylglucose, is mentioned in the later part of this Chapter.[5]

In the D-mannose series, selective methods for obtaining the 2- and 3-methyl ethers have been developed by use of the dibutylstannyl derivative of methyl 4,6-*O*-benzylidene-α-D-mannopyranoside. Direct methylation (methyl iodide in DMF) gave the 3-ether, and the 3-benzyl and 3-allyl analogues were em-

[1] P. M. Pojer and S. J. Angyal, *Austral. J. Chem.*, 1978, **31**, 1031.
[2] M. Marek, P. Chuchvalec, K. Kefurt, and J. Jarý, *Coll. Czech. Chem. Comm.*, 1978, **43**, 115.
[3] V. P. Panov, V. N. Gritsan, and R. G. Zhbankov, *Doklady Akad. Nauk S.S.S.R.*, 1978, **22**, 238 (*Chem. Abs.*, 1978, **89**, 44 093z).
[4] E. V. Evtushenko and Yu. S. Ovodov, *Khim. prirod. Soedinenii*, 1977, 753 (*Chem. Abs.*, 1978, **88**, 152 916h).
[5] J. Kuszmann, P. Sohár, and L. Kiss, *Carbohydrate Res.*, 1978, **63**, 115.

ployed in obtaining the 2-methyl derivative by standard methods.[6] Benzyl 3-*O*-benzyl-4,6-*O*-benzylidene-α-D-mannoside can be made by selective, reductive opening of the 2,3:4,6-diacetal with $LiAlH_4$–$AlCl_3$, and it served as starting material for the preparation of the 2-ether, the 2,6-diether, and the 2,4,6-triether.[7] 5,6-Di-*O*-methyl-D-galactose was obtained from methyl β-D-galactofuranoside *via* the 5,6-*O*-isopropylidene compound,[8] and standard methods have been used to obtain the 2-, 3-, and 4-methyl ethers of phenyl β-D-quinovoside.[9] 2,5,6- and 3,5,6-Tri-*O*-methyl-D-galactose have been prepared from methyl β-D-galactofuranoside and converted into their alditol acetates.[9a] The 2,3:6,7- and 2,3:4,7-di-*O*-cyclohexylidene acetals of methyl α-D-*glycero*-L-*manno*-heptopyranoside have been used to prepare the 4,6,7- and 2,3,6-trimethyl ethers of the heptose by a partial hydrolysis and methylation sequence.[9b]

Methylation (methyl iodide in DMF) of 1,5-anhydro-4,6-*O*-benzylidene-D-glucitol with limited reagent gave the 3-*O*-methyl,2-*O*-methyl and 2,3-di-*O*-methyl ethers in yields of 50%, 14%, and 18% respectively.[9c]

In the disaccharide series the 3,3′-dimethyl ether of α,α-trehalose was obtained by complete methylation of 4,6:4′,6′-dibenzylidene 2-esters or by selective substitution using the Kuhn procedure with the bisacetal. From the 2,3,6,2′,3′,6′-hexabenzoate and the 2,3,4,2′,3′,4′-hexa-acetate the 4,4′- and 6,6′-dimethyl compounds were prepared.[10] Unselective methylation of cellobiose, cellobi-itol (as well as of 2-acetamido-2-deoxy-D-glucose and -D-glucitol) using methyl iodide and potassium t-butoxide in DMSO (dimsyl potassium) under nitrogen gave high yields of methylated products.[11]

Other Alkyl and Aryl Ethers.—Methylthiomethyl ethers are referred to at the beginning of this Chapter. Various alkoxymethyl ethers of D-glucose have been separated by t.l.c. methods,[12] and the preparations of similar derivatives of sucrose have been studied. The surface activities of the products with alkyl residues containing 6—18 carbon atoms and various degrees of substitution were studied. Their wetting properties were good, but foaming and emulsifying characteristics were not as good as those of other non-ionic surface-active agents.[13] Several 'pseudopolysaccharides' involving sugar residues linked by ether bonding to synthetic polymers were prepared as shown in Scheme 1. Various oxygenated allyl ethers were synthesized in the course of the work.[14]

[6] M. A. Nashed, *Carbohydrate Res.*, 1978, **60**, 200.
[7] A. Lipták, A. Bobak, and P. Nánási, *Acta Chim. Acad. Sci. Hung.*, 1977, **94**, 261 (*Chem. Abs.*, 1978, **88**, 191 268t).
[8] G. Chambat, J.-P. Joseleau, M. Lapeyre, and A. Lefebvre, *Carbohydrate Res.*, 1978, **63**, 323.
[9] L. Kiss, A. Lipták, and P. Nánási, *Acta Chim. Acad. Sci. Hung.*, 1978, **98**, 353.
[9a] A. S. Rao and N. Roy, *Carbohydrate Res.*, 1977, **59**, 393.
[9b] P. Szabo, *Carbohydrate Res.*, 1977, **59**, 580.
[9c] Y. Kondo, K. Yabunchi, H. Ono, and S. Hirano, *Agric. and Biol. Chem.* (*Japan*), 1978, **42**, 673.
[10] C. K. Lee and M. G. Lindley, *Carbohydrate Res.*, 1978, **63**, 277.
[11] N. V. Bovin, V. A. Nesmeyanov, and A. Ya. Khorlin, *Izvest. Akad. Nauk S.S.S.R., Ser. khim.*, 1977, 2834 (*Chem. Abs.*, 1978, **88**, 121 595y).
[12] D. Biniakiewicz and J. Szymanowski, *Chem. analit.* (*Warsaw*), 1977, **22**, 767 (*Chem. Abs.*, 1978, **88**, 105 661m).
[13] J. Szymanowski, *Abh. Akad. Wiss. D.D.R.*, 1976. (Publ. 1977) 1 (Originalheitr. – Int. Tag. Grenzflaechenakt. Stoffe, 4th, Tech 1), 317 (*Chem. Abs.*, 1978, **88**, 191 273r).
[14] B. Kraska and L. Mester, *Tetrahedron Letters*, 1978, 4583.

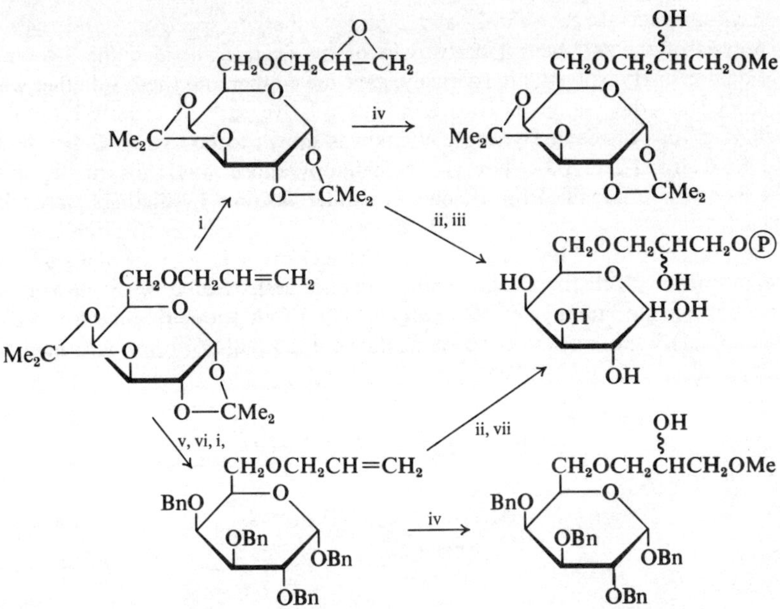

Reagents: i, m-$ClC_6H_4CO_3H$; ii, Ⓟ-OH-DMSO–KOH; iii, H^+; iv, MeO^-; v, BnOH–H^+; vi, BnBr–NaH; vii, H_2–Pd

Scheme 1

The 6-, 6′-, and 6,6′-trityl ethers of α- and β-cellobiose have been produced by standard methods,[15] and boron trifluoride etherate in methanol has been used as a detritylation procedure for primary ethers. A range of common ether, ester, and acetal groups were stable to the reagent.[16] Removal of trityl ethers in the nucleoside series has been effected using trifluoroacetic acid, sometimes diluted with butanol.[16a]

The enhancement of regioselectivity provided by cyclic stannyl derivatives has been employed to obtain a series of benzyl and allyl ethers of hexopyranosides. Methyl α-D-glucopyranoside treated with bis(tributylstannyl)oxide (1 .5 mol equiv.) followed by benzyl bromide gave methyl 2,6-di-*O*-benzyl-α-D-glucopyranoside (31%) together with the 6-ether (49%). Similar results were obtained when allyl bromide was used. In parallel work with methyl β-D-galactopyranoside and methyl α-D-mannopyranoside, 3-allyl, 3,6-diallyl, 3,6-diallyl-2,4-dibenzyl, and 2,4-dibenzyl ethers were synthesized in both series.[17] The diastereoisomeric benzyl 3,4-*O*-benzylidene β-D-arabinopyranosides and their 2-benzyl ethers were separately reduced with $AlCl_3$ – $LiAlH_4$ to give synthetic routes to the 3-, 4-, 2,3-, and 3,4-benzyl ethers of the glycoside.[18] Similar work with benzyl α-L-rhamnopyranoside *exo*- and *endo*-2,3-*O*-

[15] K. Koizumi and T. Utamura, *Yakugaku Zasshi*, 1978, **98**, 327 (*Chem. Abs.*, 1978, **89**, 44 014z).
[16] K. Dax, W. Wolflehner, and H. Weidmann, *Carbohydrate Res.*, 1978, **65**, 132.
[16a] M. MacCoss and D. J. Cameron, *Carbohydrate Res.*, 1978, **60**, 206.
[17] T. Ogawa and M. Matsui, *Carbohydrate Res.*, 1978, **62**, C1.
[18] A. Lipták, *Carbohydrate Res.*, 1978, **63**, 69.

benzylidene acetals gave the 3- and 2-benzyl ethers, respectively, and hydrogenolysis of the 4-*O*-benzyl derivatives of the acetals afforded the 3,4- and 2,4-diethers. Hydrolysis, alternatively, gave the 4-ether and the 2,3-diether was obtained *via* the 4-allyl derivative.[19] The relative ease of acetolysis of the benzyl groups of mannopyranose ethers was shown to be C-6 > C-4 > C-3. 1,2,4,6-Tetra-*O*-acetyl-3-*O*-benzyl-α-D-mannopyranose was thus made, and the 3-methyl analogue from it, and in similar fashion 3,4-diethers were obtained.[20]

Several substituted benzyl ethers have been examined. As part of a study of lignin–carbohydrate bonding, vanillyl alcohol was treated with glucose in aqueous acetate buffer at 60 °C and gave 48% of an anomeric mixture of the 6-ethers (1).[21] In work on the same theme 1,2:5,6-di-*O*-isopropylidene-α-D-

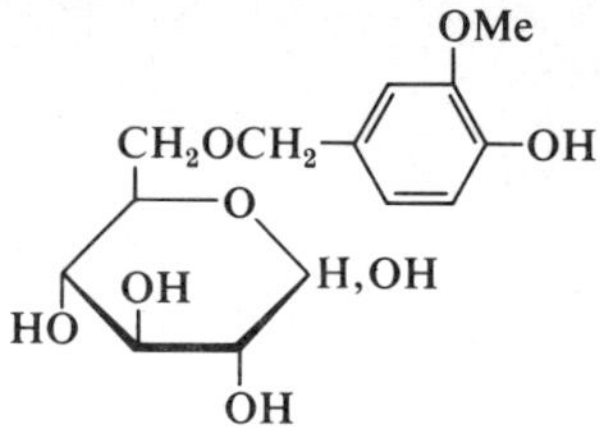

(1)

glucose was treated with *p*-methoxy- and 3,4-dimethoxy-benzyl halides to give the 3-ethers, the ether linkages of which were of comparable stability to acid to that of the 1,2-isopropylidene group.[22] Benzyl ethers can be cleaved by anodic oxidation, and those with electron-releasing groups are most readily removed. The *p*-methoxy compound mentioned in the last reference was converted completely into the 3-hydroxy analogue by treatment in aqueous acetonitrile with lithium perchlorate as electrolyte at 1.85 V using a current of 0.7 A.[23] 2,3:4,5-Di-*O*-isopropylidene-β-D-fructopyranose treated with vinylbenzyl chloride in the presence of sodium hydride gave the 1-vinylbenzyl ether which then gave a range of polymers on co-polymerization with styrene. The acetal groups were removed and the substituted and unsubstituted mixed polymers were studied by a variety of physical methods.[24]

D-Mannitol 1,2,5,6-tetraoctadecyl ether has been made from the corresponding di-*O*-isopropylidene derivative by way of the 3,4-di-*p*-methylbenzyl ether using a phase-transfer procedure.[25] and dodecoxymethyl derivatives of glucose were prepared directly by use of dodecoxymethyl chloride. The isomeric mono-substituted and the di- and tri-substituted products were separated.[26]

19 A. Lipták, P. Fügedi, and P. Nánási, *Carbohydrate Res.*, 1978, **65**, 209.
20 M. M. Ponpipom, *Carbohydrate Res.*, 1977, **59**, 311.
21 J. A. Hemmingson, G. J. Leary, I. J. Miller, W. A. Thomas, and A. D. Woodhouse, *J. C. S. Chem. Comm.*, 1978, 92.
22 L. Kosáková, B. Kosíková, and D. Joniak, *Chem. Zvesti*, 1978, **32**, 420.
23 S. Czernecki and G. Gorson, *Tetrahedron Letters*, 1978, 4113.
24 W. K. Busfield, F. P. Franke, and R. D. Guthrie, *Austral. J. Chem.*, 1978, **31**, 2559.
25 A. F. Rosenthal, L. A. Vargas, and J. F. Dixon, *Chem. Phys. Lipids*, 1977, **20**, 205 (*Chem. Abs.*, 1978, **88**, 121 599c).
26 D. Biniakiewicz, J. Szymanowski, and S. Owsiana, *Stärke*, 1977, **29**, 353 (*Chem. Abs.*, 1978, **88**, 136 853e).

Stereospecific 4-*O*-lactyl and (1-hydroxymethyl)ethyl ethers of glucose were obtained as outlined in Scheme 2, the products derived by use of (*R*)-2-chloropropanoic acid having (*S*) stereochemistry at the ether centre and *vice versa.*[27] [4-*O*-(2-Acetamido-2-deoxy-β-D-glucopyranosyl)-*N*-acetylmuramyl]-L-alanyl-D-isoglutamine in which a dipeptide and disaccharide are bonded through an ether linkage was prepared as shown in Scheme 3.[28]

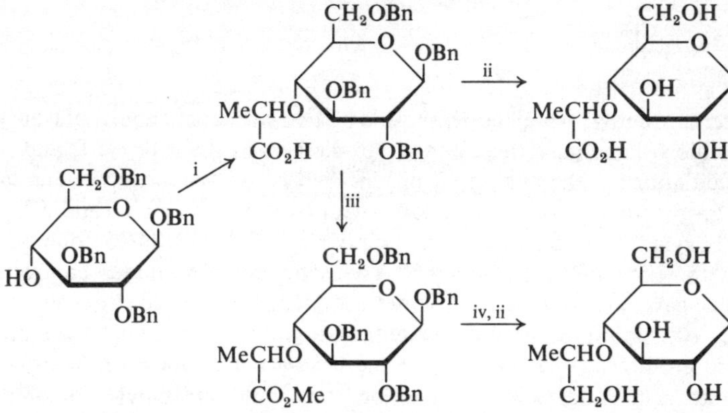

Reagents: (*R*)-2-chloropropanoic acid–DMF; ii, H_2–Pd; iii, CH_2N_2; iv, $LiAlH_4$

Scheme 2

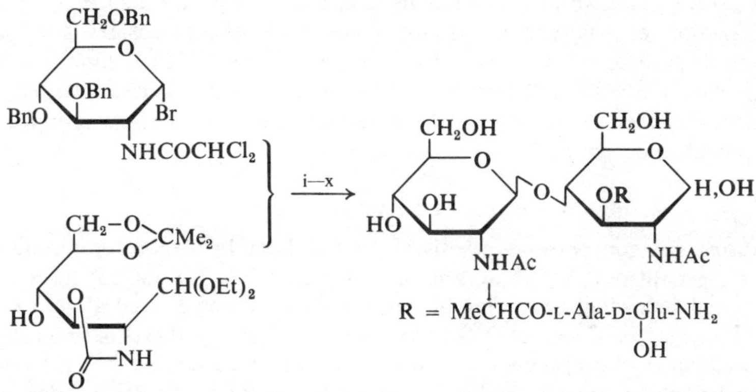

Reagents: i, $Hg(CN)_2$; ii, $Ba(OH)_2$; iii, Ac_2O–py; iv, NaOMe; v, (R)–MeCHClCO$_2$H; vi, CPh_2N_2; vii, ^{-}OH; viii, L-alanyl-D-isoglutamine benzyl ester–DCC; ix, H^+; x, H_2

Scheme 3

Methyl 4,6-*O*-benzylidene-α-D-glucopyranoside treated with bis(2-chloro-, 2-iodo-, or 2-tosyloxy-ethyl)ether gave the crown ethers (2) and (3) which were obtained in about 15% yield (by a phase-transfer procedure).[29] Isomeric crown ethers are also mentioned in Chapter 23.

[27] J.-M. Petit and P. Sinaÿ, *Carbohydrate Res.*, 1978, **64**, 9.
[28] S. Kusumato, K. Yamamoto, and T. Shiba, *Tetrahedron Letters*, 1978, 4407.
[29] L. Toke, L. Fenichel, P. Bakó, and J. Szejtli, *Acta Chim. Acad. Sci. Hung.*, 1978, **98**, 357.

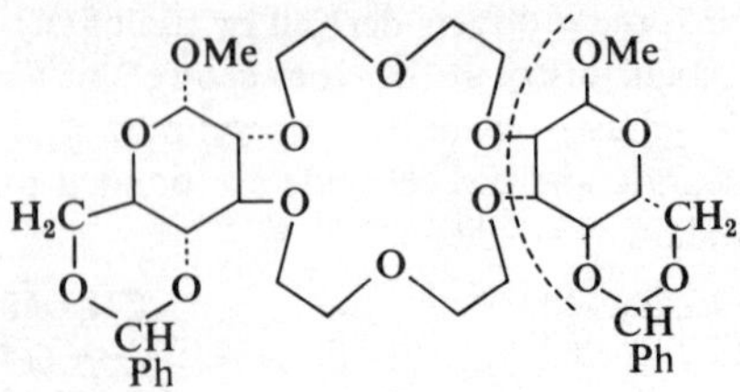

(2)

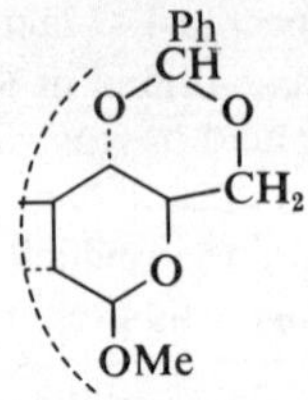

(3)

Silyl Ethers.—Methyl α-D-glucopyranoside treated with 2 mol equiv. of t-butyldimethylchlorosilane gave the 2,6- and 3,6-diethers in the ratio 4 : 1, and the β-glycoside afforded the analogous products in equal proportions. The last product (β-glycoside 3,6-diether), treated with triphenylphosphine and diethyl azodicarboxylate, isomerized to give mainly the 4,6-disubstituted compound but also to some of the 2,6-ether.[30] The same reagent, under controlled conditions, gave a crystalline 6,6′-diether, isolated in 36% yield, from sucrose. A further 10% could be obtained by chromatographic treatment of the non-crystalline products. Acetylation gave the hexa-acetate which, on hydrolysis with dilute acid, underwent desilylation, but with tetrabutyl ammonium fluoride gave 2,3,6,1′,3′,4′-hexa-*O*-acetylsucrose following 4 → 6 acetyl migration. When the work was repeated with benzoates similar migration again took place with the ammonium salt, but not with the acid.[31]

Reaction of t-butyldimethylchlorosilane with ribonucleosides and their 5′-ethers gave access to mono-, di-, and triethers, and optimal conditions for each were developed. Selectivity can be high, so that, for example, 65% of the 2′,5′-diether of uridine can be obtained. The use of these products in oligonucleotide synthesis was discussed.[32]

2 Intramolecular Ethers (Anhydro-sugars)

Oxirans.—A convenient synthesis of methyl 2,3-anhydro-α-D-ribofuranoside uses the path outlined in Scheme 4.[33] The epoxides (4) and (5) have been prepared following the finding of methods of obtaining 3′- and 4′- and 3′,4′-sulphonyl esters of sucrose.[34] Scheme 5 illustrates new syntheses of furanosides of D-epihamamelose; the C-3 epimer of the starting material led to the D-*xylo*-branched-chain sugar. Application of these reactions to methyl 2-deoxy-2-methylene-β-D-*erythro*-pentopyranoside gave the *ribo*-epoxide and hence methyl β-D-hamamelopyranoside; epoxidation of the α-D-*threo*-alkene failed.[35]

Other Anhydrides.—Hungarian workers have shown that treatment of 1,2,3,4-tetra-*O*-acetyl-β-D-glucopyranose with stannic chloride affords tri-*O*-acetyl-1,6-anhydro-β-D-glucopyranose by way of an isolable complex containing the

[30] H. H. Brandstetter and E. Zbiral, *Helv. Chim. Acta*, 1978, **61**, 1832.
[31] F. Franke and R. D. Guthrie, *Austral. J. Chem.*, 1978, **31**, 1285.
[32] K. K. Ogilvie, S. L. Beaucage, A. L. Schifman, N. Y. Theriault, and K. L. Sadana, *Canad. J. Chem.*, 1978, **56**, 2768.
[33] F. M. Unger, R. Christian, and P. Waldstätten, *Carbohydrate Res.*, 1978, **67**, 257.
[34] R. Khan, M. R. Jenner, and H. Lindseth, *Carbohydrate Res.*, 1978, **65**, 99.
[35] J. C. Depezay and A. Duréaut, *Tetrahedron Letters*, 1978, 2869.

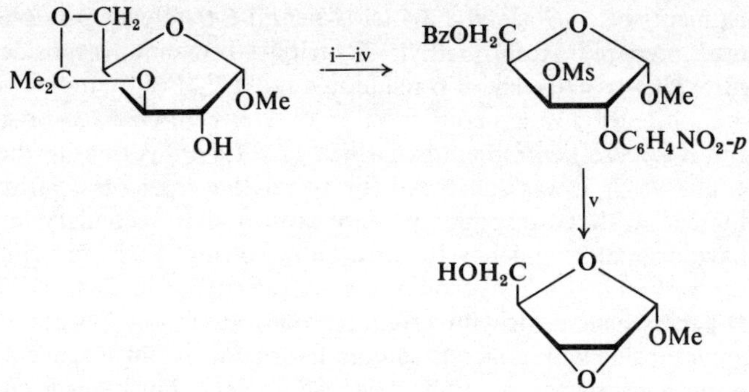

Reagents: i, $O_2NC_6H_4Cl$; ii, H^+; iii, BzCl–py; iv, MsCl–py; v, NaOMe

Scheme 4

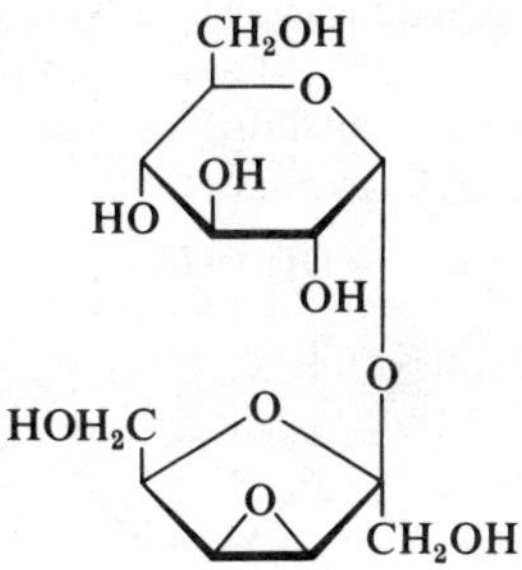

(4)

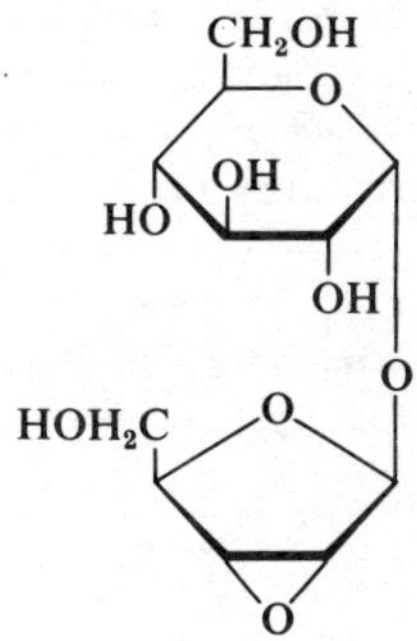

(5)

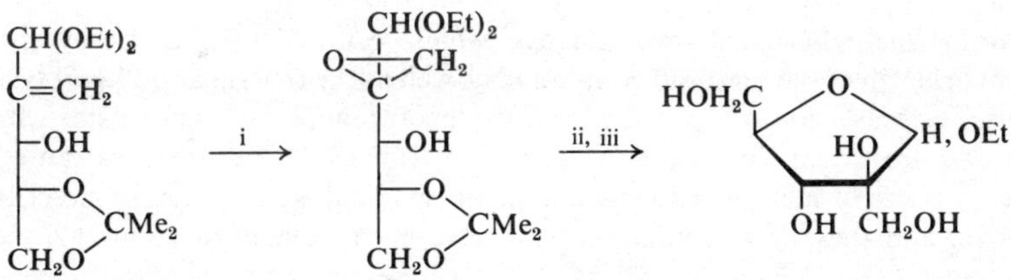

Reagents: i, Bu^tO_2H–VO(Acac)$_2$; ii, NaOH; iii, H^+

Scheme 5

catalyst and acetic acid,[36] and that the same product can be obtained from the trityl ether of the above starting material by use of titanium tetrachloride in chloroform.[37] Action of a fungal glucosyl transferase on maltose produced some 1,6-anhydro-β-D-glucopyranose as well as a series of oligosaccharides.[38]

[36] E. Zára-Kaczián and G. Deák, *Acta Chim. Acad. Sci. Hung.*, 1977, **93**, 201 (*Chem. Abs.*, 1978, **88**, 89 979p).

[37] E. Zára-Kazcián and G. Deák, *Acta Chim. Acad. Sci. Hung.*, 1978, **96**, 311.

[38] J. H. Pazur, Y. Tominaga, C. W. DeBrosse, and L. M. Jackman, *Carbohydrate Res.*, 1978, **61**, 279.

Base treatment of 1-*O*-acetyl-2,3,4-tri-*O*-benzyl-6-*O*-toluene-*p*-sulphonyl-α-D-mannose, prepared from methyl 6-*O*-trityl-α-D-mannopyranoside, gave 1,6-anhydro-2,3,4-tri-*O*-benzyl-β-D-mannopyranose.[39] From studies of acid-catalysed equilibration in aqueous solution of D-*glycero*-L-*manno*-, D-*glycero*-L-*allo*-, and D-*glycero*-L-*altro*-heptose (which gave 1,6-anhydrides to the extent of 15, 54, and 98%), it was concluded (by correlating these observations with those obtained with corresponding aldohexoses) that secondary hydroxy-groups have a greater tendency to form anhydro-rings than do primaries.[40] Several derivatives of 1,6-anhydrohexoses are referred to in Chapter 7.

Several papers have dealt with 1,6-anhydrohexofuranoses. Heating phenyl α-D-glucopyranoside with aqueous sodium hydroxide at 170 °C gave as much as 70% of both levoglucosan and its furanose isomer,[41] but it has been shown that the furanose derivatives can be obtained, sometimes in high yield, directly from free sugars by heating in DMF solution with toluene-*p*-sulphonic acid, with removal of the water formed. Galactose gave 87% of 1,6-anhydro-α-D-galactofuranose (33% isolated by direct crystallization). Likewise, allose, talose, mannose, and glucose gave 78, 86, 22, and 35% of analogous products.[42] 1-*O*-Acetyl-2,3-*O*-isopropylidene-β-D-allose was converted into both possible anhydro-derivatives (6) and (7) by controlled treatment with acid,[43] and

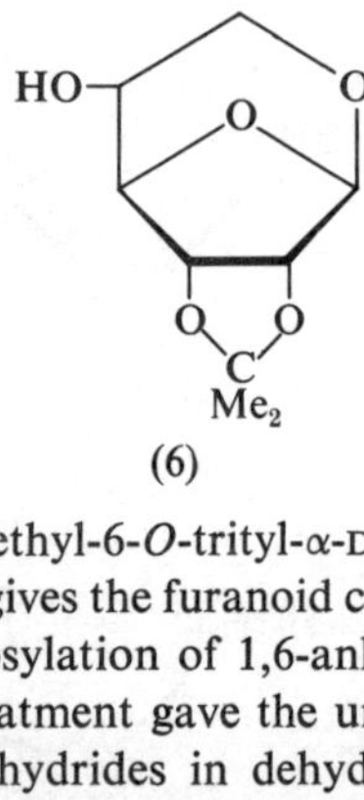

(6)

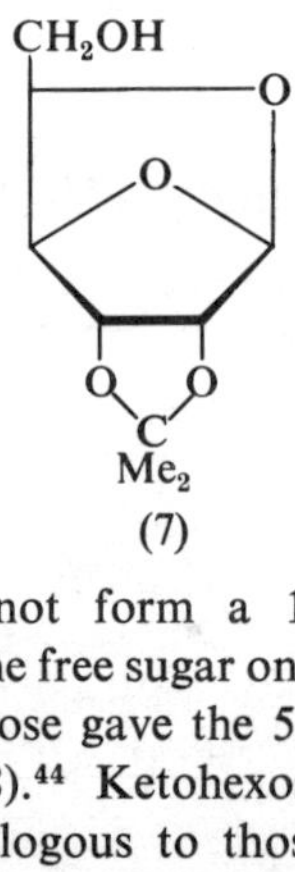

(7)

3,5-di-*O*-methyl-6-*O*-trityl-α-D-glucose which cannot form a 1,6-pyranoid anhydride gives the furanoid compound as well as the free sugar on hydrolysis.[5] Selective tosylation of 1,6-anhydro-β-D-glucofuranose gave the 5-ester which on base-treatment gave the unusual dianhydride (8).[44] Ketohexoses can only give 2,6-anhydrides in dehydration processes analogous to those described above, and these [*e.g.* compound (9)] have been isolated in about 1% yield

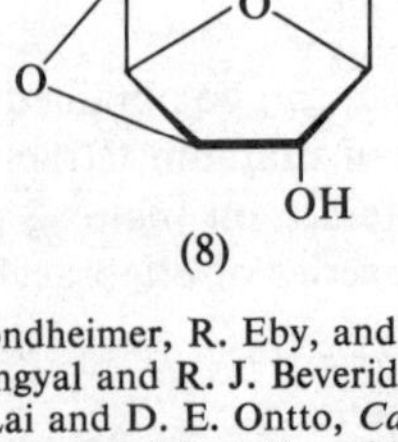

(8)

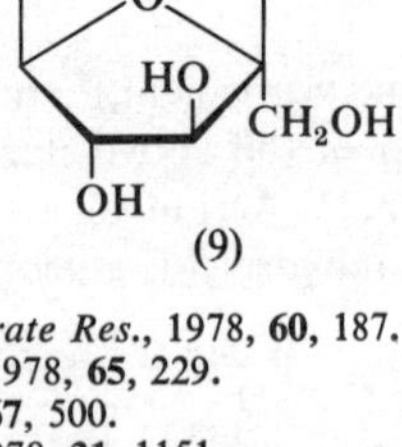

(9)

[39] S. J. Sondheimer, R. Eby, and C. Schuerch, *Carbohydrate Res.*, 1978, **60**, 187.
[40] S. J. Angyal and R. J. Beveridge, *Carbohydrate Res.*, 1978, **65**, 229.
[41] Y.-Z. Lai and D. E. Ontto, *Carbohydrate Res.*, 1978, **67**, 500.
[42] S. J. Angyal and R. J. Beveridge, *Austral. J. Chem.*, 1978, **31**, 1151.
[43] P. Köll, *Tetrahedron Letters*, 1978, 51.
[44] P. Köll and J. Schulz, *Tetrahedron Letters*, 1978, 49.

following vacuum pyrolysis of D-fructose, D-tagatose, D-psicose, and L-sorbose.[45]

Application of the Corey–Winter reaction to compound (10) did not give

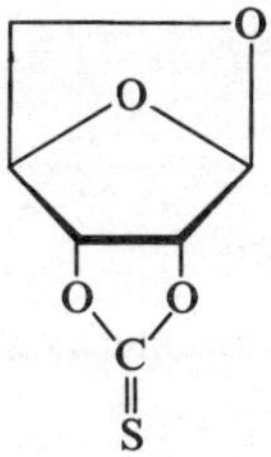

(10)

the anticipated alkene; in consequence the bicyclo parent compound (11) of the 1,5-anydroaldofuranoses was prepared as shown in Scheme 6.[46]

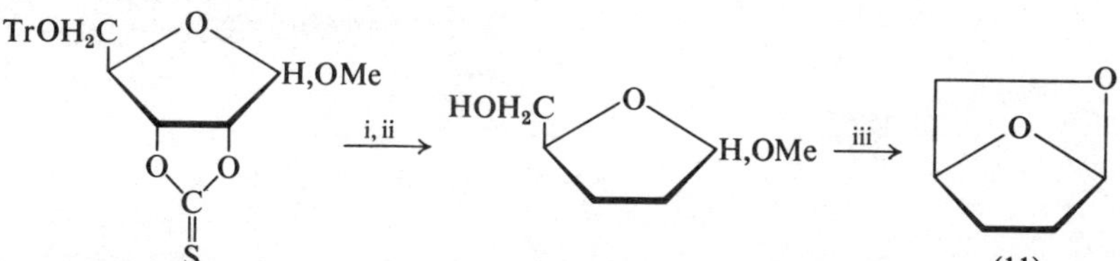

(11)

Reagents: i, $(MeO)_3P$; ii, H_2–Ni; iii, $(CO_2H)_2$–Δ

Scheme 6

Compounds (12)—(14) are reported to isomerize to give 1,4-anhydro products on treatment with mercury(II) bromide. The relative rates decreased in the order (12) > (13) > (14).[47] 1,4:3,6-Dianhydro-α-D-glucose (15) was produced together with levoglucosan and 1,6-anhydro-3,4-dideoxy-β-D-*glycero*-hex-3-enos-2-ulose (16) on acid-catalysed pyrolysis of glucose, amylopectin, and cellulose, and the dianhydride could be transformed into the enone

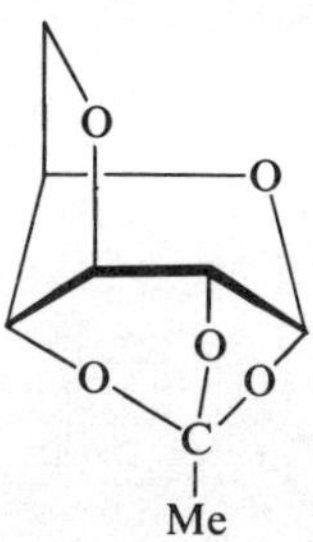

(12)

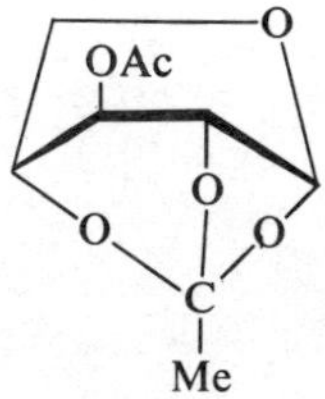

(13)

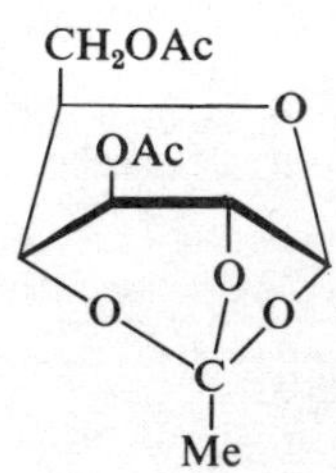

(14)

[45] P. Köll, S. Deyhim, and K. Heyns, *Chem. Ber.*, 1978, **111**, 2909.
[46] P. Köll and S. Deyhim, *Chem. Ber.*, 1978, **111**, 2913.
[47] Ya. V. Voznyi and A. F. Bochkov, *Tezisy Doklady – Vses. Konf. 'Stereokhim. Konform. Anal. Org. Neftekhim. Sint', 3rd*, 1976, 75 (*Chem. Abs.*, 1978, **88**, 170 408y).

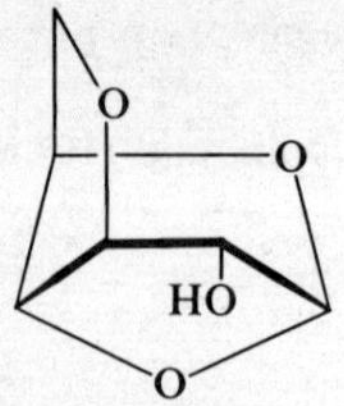

(15)

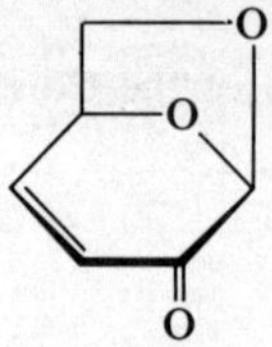

(16)

showing that it could be a reaction intermediate. Possible routes for this last conversion were proposed.[48]

[48] F. Shafizadeh, R. H. Furneaux, T. T. Stevenson, and T. G. Cochran, *Carbohydrate Res.*, 1978, **61**, 519.

5

Acetals

1 Synthesis

Phase transfer catalysis has been used in the methylenation of carbohydrates: methyl 4,6-*O*-benzylidene-glucosides and galactosides gave the *trans*-fused 2,3-acetals in about 60% yield with methylene dibromide and alkali. Methyl β-D-ribofuranoside gave the fully substituted dimeric compound (1), and its 5-trityl derivative afforded the monomeric 2,3-acetal as anticipated.[1]

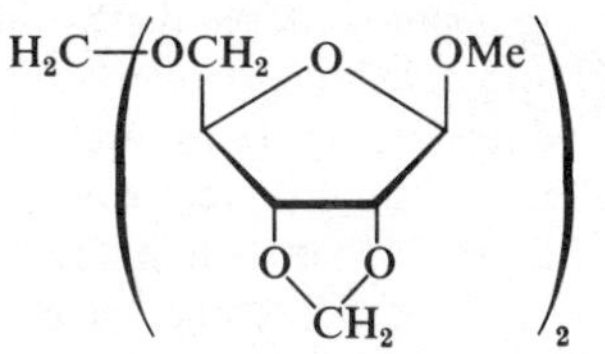

(1)

Isopropenyl methyl ether in DMF in the presence of catalytic amounts of toluene-*p*-sulphonic acid gave 4,6-*O*-isopropylidene derivatives of methyl α-D-mannopyranoside, D-mannose, D-glucose, and D-galactose in nearly quantitative yields. D-Gulono- and D-galactono-1,4-lactones afforded 5,6-acetals, the latter leading to 5,6-*O*-isopropylidene-D-galactose, but this gave the 1,2:3,4-diacetal on further reaction rather than the 1,2:5,6-substituted isomer.[2] Gelas and Horton have also reported on the reaction between D-mannose and isopropenyl methyl ether, and have shown that the 4,6-acetal is the kinetic product formed prior to the 2,3:4,6-diacetal. From these, α-D-mannopyranose 1,2,3-triacetate and 1-*O*-acetyl-2,3-*O*-isopropylidene-α-D-mannopyranose were obtained.[3] D-Galactose treated with acetone, acetic acid, and concentrated sulphuric acid gave 6-*O*-acetyl-1,2:3,4-di-*O*-isopropylidene-α-D-galactose directly and in 40% yield.[4]

Isopropenyl acetate in acetone has been used in the acetalation of 7-β-D-ribofuranosyltheophylline, methyl β-D-ribofuranoside, adenosine, D-ribose, and D-glucose, and vinyl acetate has been employed to produce ethylidene acetals from ribofuranosyl derivatives.[5]

[1] K. S. Kim and W. A. Szarek, *Synthesis*, 1978, 48.
[2] C. Copeland and R. V. Stick, *Austral. J. Chem.*, 1978, **31**, 1371.
[3] J. Gelas and D. Horton, *Carbohydrate Res.*, 1978, **67**, 371.
[4] N. Martinez, *Rev. C.E.N.I.C., ciesic. Fis.*, 1976, **7**, 135 (*Chem. Abs.*, 1978, **89**, 110 151v).
[5] Y. Araki, Y. Hijioka, Y. Ishido, and T. Sato, *Carbohydrate Res.*, 1978, **64**, 309.

In the amino-sugar series, 2-acetamido-2-deoxy-D-glucose and 2,2-dimethoxypropane in DMF with toluene-*p*-sulphonic acid gave, initially, mainly the 4,6-isopropylidene acetal together with small amounts of the 5,6-acetal and the unsaturated compound (2).[6] Subsequently, methyl 2-aceta-

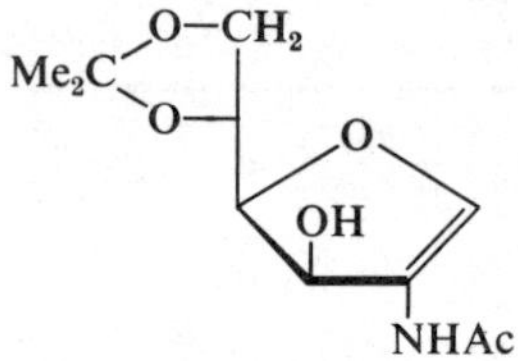

(2)

mido-2-deoxy-5,6-*O*-isopropylidene-β-D-glucofuranoside was formed together with smaller proportions of methyl 2-acetamido-2-deoxy-4,6-*O*-isopropylidene-β-D-glucopyranoside and 2-acetamido-2-deoxy-3,4:5,6-di-*O*-isopropylidene-*aldehydro*-D-glucose dimethylacetal.[6] Treatment of 3-acetamido-3-deoxy-D-*glycero*-D-*galacto*-heptose diethyldithioacetal with acetone in the presence of sulphuric acid and copper sulphate gave the crystalline 6,7-acetal in 85% yield. More forcing conditions afforded the 4,5:6,7-diacetal from which the terminal group could be removed selectively by hydrolysis.[7]

Selective acetalation of α,α-trehalose with methyl isopropenyl ether gave the 4,6-acetal as major product, and this could be isolated as its hexa-acetate. Removal of the acetal ring with trifluoroacetic acid, mesylation of the diol, benzoate displacement, and saponification of the per-ester gave the *gluco-galacto*-disaccharide.[8] Acetalation of sucrose with dimethoxypropane at room temperature gave the 1′,2:4,6-acetal (15%), and the 4,6-acetal (55%); with cyclohexanone dimethylacetal the 4,6-cyclohexylidene derivative, and with α,α-dibromotoluene in toluene the 4,6-benzylidene acetal were obtained.[9]

Acid-catalysed monobutylidenation of 2-deoxy-D-*arabino*-hexitol, 2-deoxy-D-*lyxo*-hexitol, and 2-deoxy-D-*erythro*-pentitol yielded the 1,3-acetals, as kinetic products, in each case. The thermodynamic products were 4,6-acetals from the first two alditols and a 3,5-acetal from the deoxypentitol. 2-Deoxy-D-*lyxo*-hexitol also yielded diastereoisomeric 4,5-monoacetals.[10] Reaction of 1,2,3-triols with chloroacetone gives 3,6,8-trioxabicyclo[3.2.1] octanes, *e.g.* (3) and Scheme 1.[11]

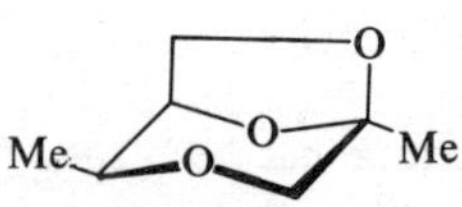

(3)

[6] A. Hasegawa and M. Kiso, *Carbohydrate Res.*, 1978, **63**, 91.
[7] J. M. Beau and P. Sinaÿ, *Carbohydrate Res.*, 1978, **65**, 1.
[8] J. Defaye, H. Driguez, B. Henrissat, J. Gelas, and E. Bar-Guilloux, *Carbohydrate Res.*, 1978, **63**, 41.
[9] R. Khan, K. S. Mufti, and M. R. Jenner, *Carbohydrate Res.*, 1978, **65**, 109.
[10] T. G. Bonner, D. Gibson, and D. Lewis, *Carbohydrate Res.*, 1978, **66**, 265.
[11] P. Calinaud and J. Gelas, *Canad. J. Chem.*, 1978, **56**, 2292.

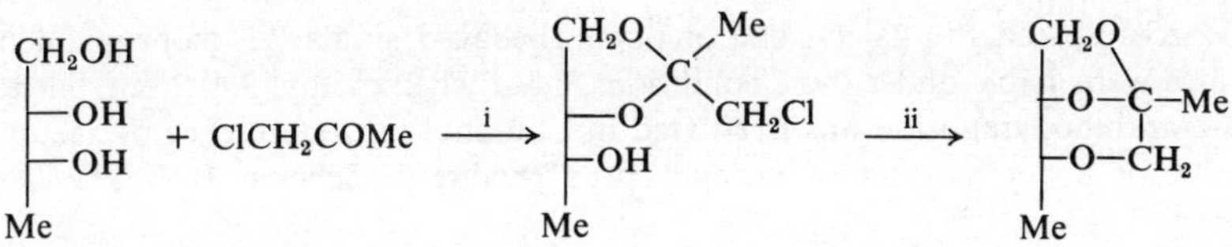

Reagents: i, H^+; ii, ^-OH

Scheme 1

Benzylidenation of 1-deoxy-D-galactitol gave a mixture of 2,3:4,5- and 2,3:4,6-diacetals with undetermined stereochemistry at the acetal centres. Similar reaction of 1-deoxy-L-mannitol afforded the 2,5:4,6-diacetal with (*S*) stereochemistry at both acetal sites.[12] A range of substituted benzylidene compounds was found using methyl α-D-glucopyranoside with the aldehyde dimethylacetals and toluene-*p*-sulphonic acid as catalyst. Yields near 80% were reported for the anisylidene, vanillylidene, syringylidene, and veratylidene compounds.[13]

2 Reactions

Wet silica gel has been used for the smooth deacetalation of simple acetals and ketals; added oxalic acid catalysed the reactions and yields were much higher than those achieved using aqueous acid.[14]

Reductive cleavage of the *p*-methoxybenzylidene acetal (4) gave the two products (Scheme 2) in similar proportions, but the dimethyl ether (5) afforded

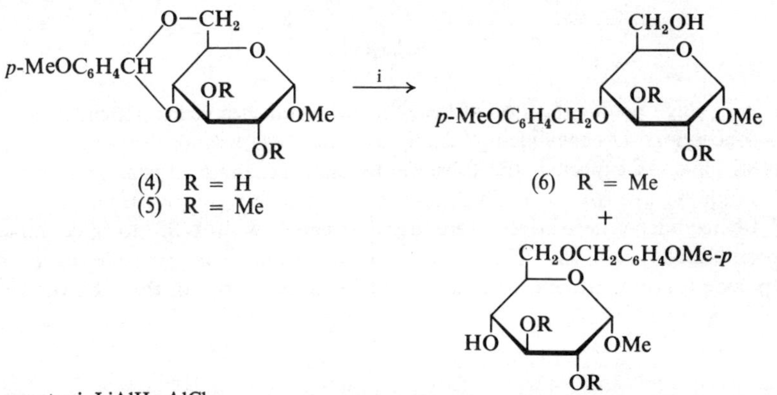

(4) R = H
(5) R = Me

(6) R = Me

Reagents: i, $LiAlH_4$–$AlCl_3$

Scheme 2

the 4-ether (6) almost exclusively. As expected, the *p*-methoxybenzyl products were very readily hydrolysed by acid.[15] Related photochemical oxidative ring opening of the acetal of methyl 2,3-di-*O*-benzoyl-4,6-*O*-benzylidene-α-D-glucopyranoside is reported to give the corresponding 2,3,6- and 2,3,4-tri-

[12] P.-E. Jansson and B. Lindberg, *Carbohydrate Res.*, 1978, **65**, 291.
[13] D. Joniak, B. Kosíková, and L. Kosáková, *Chem. Zvesti*, 1977, **31**, 106 (*Chem. Abs.*, 1978, **88**, 191 326k).
[14] F. Huet, A. Lechevallier, M. Pellet, and J. M. Conia, *Synthesis*, 1978, 63.
[15] D. Joniak, B. Kosíková and L. Kosáková, *Coll. Czech. Chem. Comm.*, 1978, **43**, 769.

benzoates, and the galactoside analogue behaved similarly. Isopropylidene rings were stable under these conditions,[16] but when methyl 3,4-*O*-ethylidene-β-L-arabinopyranoside was irradiated in t-butanol in the presence of acetone initiator a complex set of products was produced (Scheme 3).[17] The best

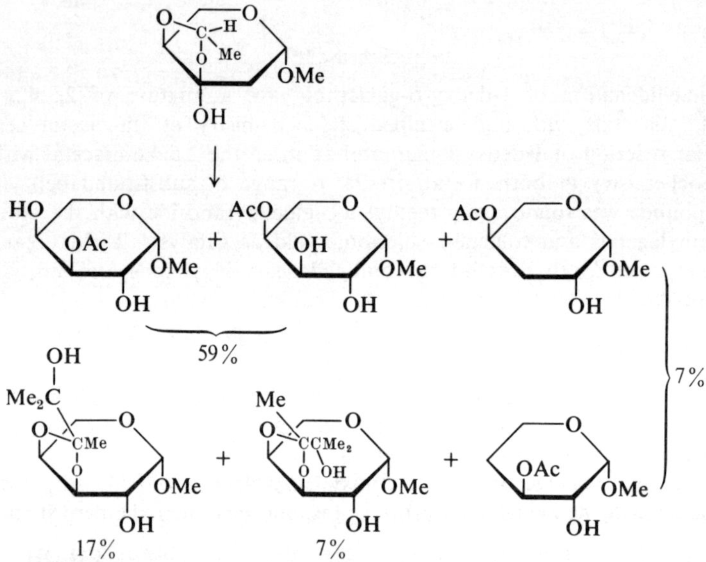

Scheme 3

known ring opening reaction of benzylidene acetals has been applied to methyl 3-*O*-acetyl-4,6-*O*-benzylidene-2-deoxy-α-D-*ribo*-hexapyranoside which, with NBS gave, as expected, the 6-bromo-6-deoxy benzoate. From this, methyl 2,6-dideoxy-α-D-*ribo*- and β-L-*lyxo*-hexopyranoside were prepared, and their 3,4-benzylidene derivatives were again treated with NBS to give mixed brominated products. Those obtained from the *ribo*-glycoside gave the epoxide (7) on treatment with base.[18] Other applications of the NBS reaction

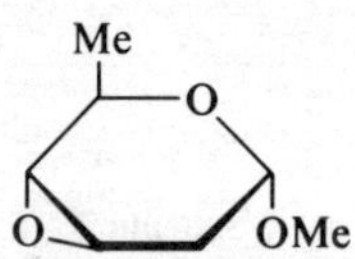

(7)

are reported in Chapter 7 and further ring openings of acetals are referred to in Chapters 4, 11, and 12.

[16] W. Szeja and M. Lapkowski, *Polish J. Chem.*, 1978, **52**, 673 (*Chem. Abs.*, 1978, **89**, 44 198h).
[17] W. A. Szarek, R. J. Beveridge, and K. S. Kim, *J. Carbohydrates, Nucleosides, Nucleotides*, 1978, **5**, 273.
[18] J. Boivin, M. Pais, and Co Monneret, *Compt. rend., C*, 1978, **286**, 51.

6
Esters

1 Carboxylic Esters

Using *ab initio* LCAO–MO calculations on alkyl formates as a basis, the electronic structures and conformations of polysaccharide esters have been interpreted. Acetate units were calculated to be 30° to the eclipsed position due to non-bonded interactions.[1]

A major bile pigment has been shown [2] to have the structure (1). In post-obstruction bile the acyl group may be found bonded to O-2, O-3, or O-4. A

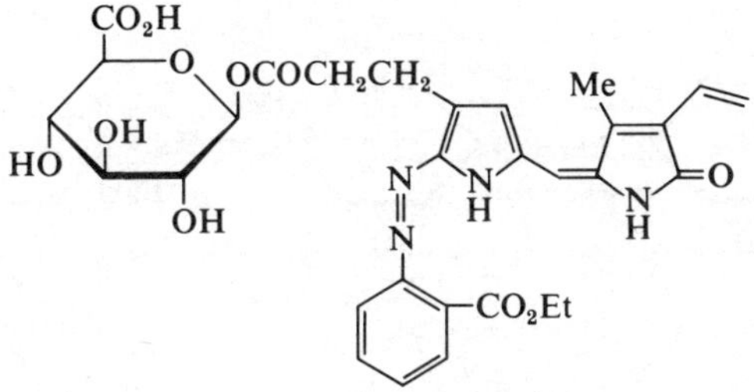

(1)

β-glucoside with a 3-methylbutynyl ester group at C-2 has been isolated from green coffee-beans,[3] and the seeds of *Banakadali* (*Ensete superba*) have been found to contain a partially acetylated sucrose ester of *p*-coumaric acid which has antifertility activity.[4] Chemical degradation and spectral evidence suggested that the fully acetylated compound was a 2-, 3-, or 4-(hepta-*O*-acetyl-α-D-glucopyranosyl-*β*-D-fructofuranoside ester of 4-*O*-acetylcoumaric acid.

The product from periodate-oxidation of 4,6-*O*-ethylidene-D-glucose after acetylation was shown to be the erythrose dimer (2).[5] Acetylation of an

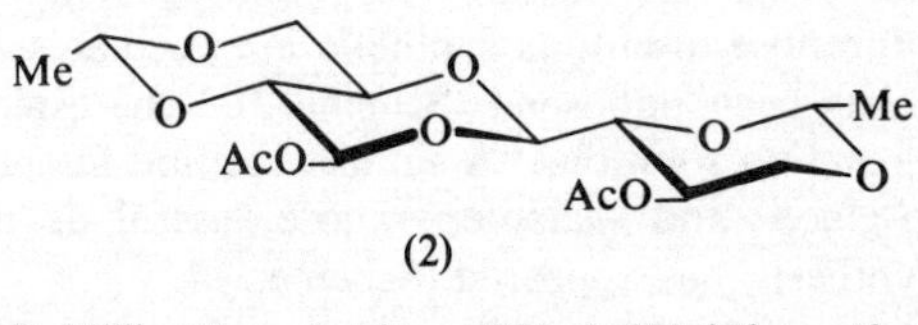

(2)

[1] A. Meunier and P. Millie, *Nouv. J. Chim.*, 1978, **2**, 151 (*Chem. Abs.*, 1978, **89**, 44 023b).
[2] F. Compernolle, G. P. Van Hees, N. Blanckaert, and K. P. M. Heirwegh, *Biochem. J.*, 1978, **171**, 185.
[3] H. Richter and G. Spiteller, *Chem. Ber.*, 1978, **111**, 3506.
[4] A. J. Amonkar, G. K. Trivedi, and S. C. Bhattacharyya, *Indian J. Chem., Sect. B*, 1978, **168**, 12 (*Chem. Abs.*, 1978, **88**, 191 276u).
[5] R. Andersson, O. Theander, and E. Westerlund, *Carbohydrate Res.*, 1978, **61**, 501.

equilibrated solution of D-erythrose in pyridine yields a mixture containing mainly the expected triacetates (3), whereas if D-erythrose is dissolved in pyridine and immediately treated with acetic anhydride, acetylated dimers (4) and (5) are formed in addition to (3),[5] Methyl 2,3-di-*O*-acetyl-β-D-xylopyranoside has been synthesized [6] by the route shown in Scheme 1, while the

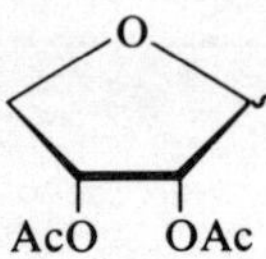

(3)

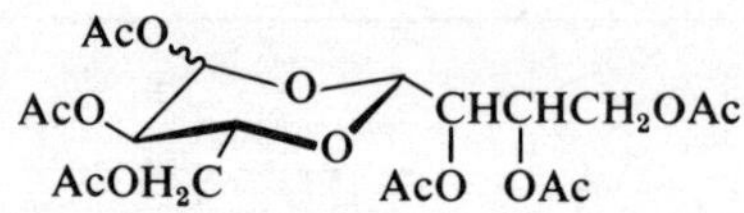

(4)

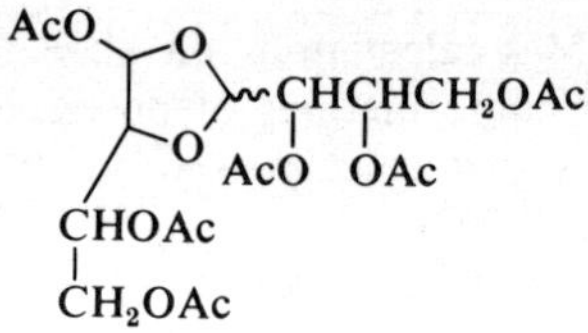

(5)

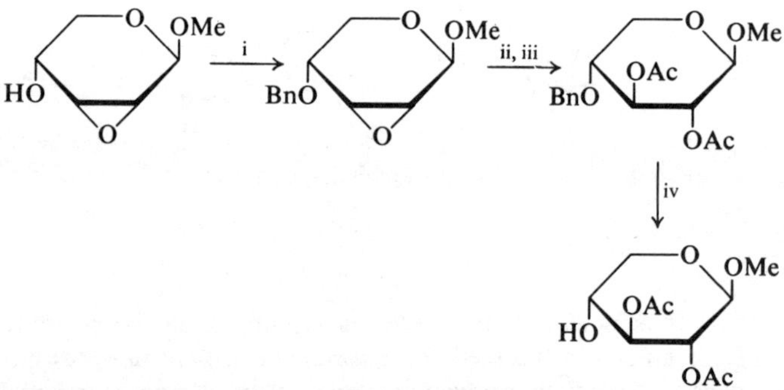

Reagents: i, BnBr–NaH; ii, KOH; iii, Ac_2O; iv, H_2–Pd

Scheme 1

corresponding 3,4-di-*O*-acetyl compound has been prepared conventionally from 2-*O*-benzyl-D-xylose.[7] A method for preparing 1,3,5-tri-*O*-acetyl- and -benzoyl-α-D-ribofuranose, useful as synthetic intermediates for 2′- modified nucleosides, *etc.*, has been published (Scheme 2). The esters (6) crystallize directly from the reaction solution.[8] A simpler method for preparing 1,3,4,6-tetra-*O*-acetyl-β-D-gluco- and -galactopyranose, useful as intermediates in oligosaccharide synthesis, has appeared (Scheme 3).[9]

[6] P. Kovac and J. Alfoldi, *Chem. Zvesti*, 1978, **32**, 519.
[7] P. Kovac and R. Palovcik, *Chem. Zvesti*, 1977, **31**, 98 (*Chem. Abs.*, 1978, **88**, 191 325j).
[8] C. Chavis, F. Dumont, and J.-L. Imbach, *J. Carbohydrates, Nucleosides, Nucleotides*, 1978, **5**, 133.
[9] V. M. Chari, M. Jordan, and H. Wagner, *Acta Chim. Acad. Sci. Hung.*, 1977, **93**, 99 (*Chem. Abs.*, 1978, **88**, 51 103k).

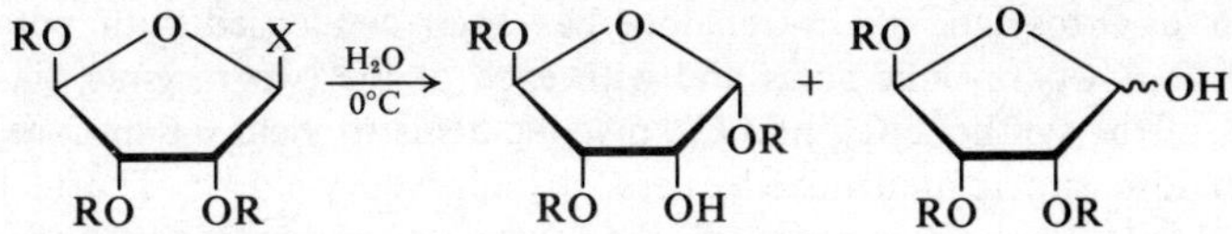

X = Br, R = Bz
X = Cl, R = Ac

(6)

Scheme 2

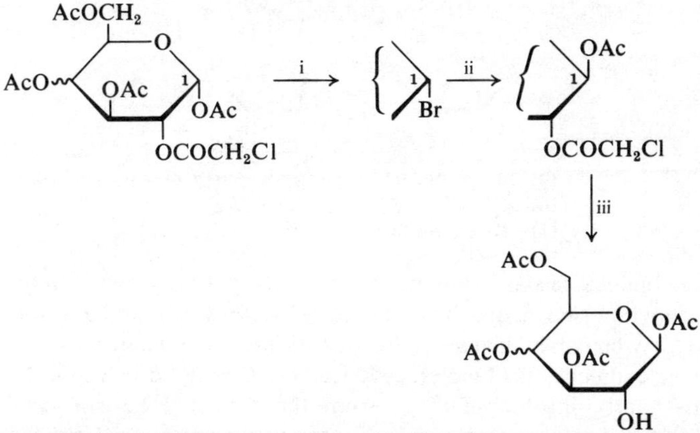

Reagents: i, HBr–HOAc; ii, $Hg(OAc)_2$–HOAc; iii, $CS(NH_2)_2$(1 mol equiv.)–MeOH–$CaCO_3$

Scheme 3

Acetyl migration from O-4 to O-6 occurred when peracetylated 6,6′-di-*O*-t-butyldimethylsilyl-sucrose was desilylated with tetrabutyl-ammonium fluoride but not when dilute acid was used. Similar results were obtained with the silylated perbenzoyl derivative.[10]

Standard methods have been used to synthesize dodecyl β-D-glucopyranoside, 3-*O*-dodecyl-D-glucose, and 6-*O*-dodecyl-D-galactose.[11] Reaction of sucrate ion (from sucrose and sodium–potassium alloy or alkali metal hydride) with methyl esters gave sucrose esters with 6—8 acyl groups per mole in greater than 90% yield.[12] 6,6′-Di-*O*-corynomycolyl-α,α-trehalose, a cord-factor analogue, was synthesized from (7) by reaction with (8) in HMPA.[13]

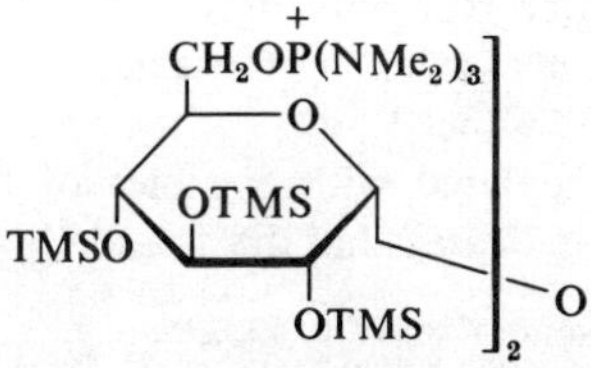

(7)

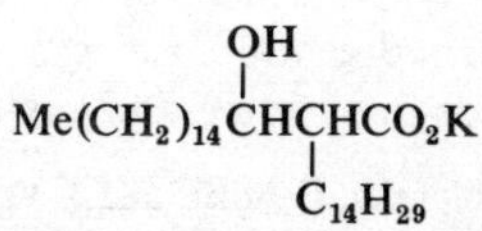

(8)

[10] F. Franke and R. D. Guthrie, *Austral. J. Chem.*, 1978, **31**, 1285.

[11] B. Havlinova, M. Kosik, P. Kovac, and A. Blazej, *Tenside Deterg.*, 1978, **15**, 72 (*Chem. Abs.*, 1978, **89**, 113 059g).

[12] G. P. Rizzi and H. M. Taylor, *J. Amer. Oil Chem. Soc.*, 1978, **55**, 398 (*Chem. Abs.*, 1978, **88**, 154 756c).

[13] R. Toubiana, C. Pizza, Y. Chapleur, and B. Castro, *J. Carbohydrates, Nucleosides, Nucleotides*, 1978, **5**, 127.

The 6,6′-di-*O*-tosylate of αα-trehalose has been condensed with potassium salts of C_{80}—C_{90} mycolic acids and with each of the two racemic diastereoisomers of the synthetic C_{44} and C_{32} mycolic acids to yield 6,6-di-*O*-mycolyl-α,α-trehalose and related mixed esters in improved yield.[13a] The maltoside esters (9) to (11) have been prepared; the deuteriated compound (9) was used to assign the n.m.r. signal due to the 4′-*O*-acetyl group unambiguously.[14]

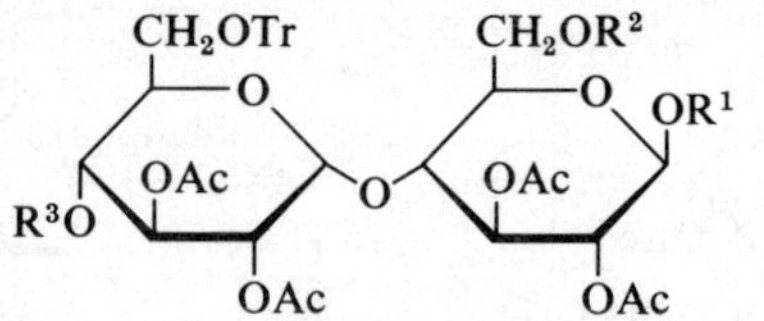

(9) R^1 = Bn, R^2 = Ac, R^3 = C^2H_3CO
(10) R^1 = Me, R^2 = Ac, R^3 = Ac
(11) R^1 = Me, R^2 = Tr, R^3 = Ac

1-Acetylimidazole showed no selectivity in the partial acetylation of methyl α-D-glucopyranoside, giving 30% 2-*O*-, 15% 3-*O*, 20% 4-*O*, and 35% 6-*O*-acetyl derivatives when used in equimolar quantities.[15] D-Fucose with acetic anhydride–sodium acetate (3 moles) gave 1,2,3-tri-*O*-acetyl-β-D-fucose (18%) and 1,2,4-tri-*O*-acetyl-β-D-fucose (3%); from the former 1,2,3-tri-*O*-acetyl-4-*O*-methyl-β-D-fucose was prepared. L-Fucose was reported to yield the corresponding 1,2,3- and 1,2,4-tri-*O*-acetyl derivatives in 9.5 and 12% yield respectively.[16] The partial benzoylation of the 4,6-*O*-benzylidene-1,5-anhydro-D-glucitol has been re-examined. Using 1 mole benzoyl chloride in pyridine at 0 °C for 3 h gave the 2-*O*- and 3-*O*-benzoyl derivatives in 27.5% and 39.9% yield respectively; after 24 h, the respective amounts were 27.6 and 42.3%. Thus, in contrast to results reported by Newth (*J. Chem. Soc.*, 1959, 2717) the 3-OH was found to be more reactive.[17, 18] Benzoylation of benzyl 4,6-*O*-benzylidene-β-D-glucopyranoside using 1.1 mole *N*-benzoylimidazole gave 3-*O*- and 2-*O*-benzoyl derivatives in 18.4 and 15% isolated yields respectively, whereas the α-anomer gave 2-*O*- and 3-*O*-benzoyl derivatives in 40 and 12% isolated yields respectively. Migration of the benzoyl group from O-3 to O-2 in the β-anomer and from O-2 to O-3 in the α-anomer occurs in acetone–water–0.5M-sodium hydroxide. The greater reactivity of the 2-OH in the α-anomer is explained by hydrogen-bond activation by the axial benzyloxy-group at C-1 which cannot occur in the β-anomer.[19]

When β-gentiobiose was partially benzoylated with 8 moles of benzoyl chloride in pyridine, the mixture obtained indicated that the 4- and 4′-hydroxy

[13a] J. Polonsky, E. Soler, and J. Varenne, *Carbohydrate Res.*, 1978, **65**, 295.
[14] K. Koizumi and T. Utamara, *Carbohydrate Res.*, 1978, **63**, 283.
[15] J. Gelas and D. Horton, *Carbohydrate Res.*, 1978, **64**, 323.
[16] E. Lee, A. Bruzzi, and G. Keaveney, *Proc. R. Irish Acad., Sect. B*, 1977, **77B**, 495 (*Chem. Abs.*, 1978, **89**, 110 150u).
[17] Y. Kondo, *Agric. and Biol. Chem.* (*Japan*), 1977, **41**, 2481.
[18] Y. Kondo, K. Yabuuchi, H. Ono, and S. Hirano, *Agric. and Biol. Chem.* (*Japan*), 1978, **42**, 673.
[19] A. Liptak, E. Csik, and P. Nanasi, *Acta Chim. Acad. Sci. Hung.*, 1977, **94**, 267 (*Chem. Abs.*, 1978, **88**, 191 328n).

groups were the least reactive.[20] A study of the partial benzoylation of methyl β-maltoside monohydrate has shown that the relative reactivity of the hydroxy-groups is 2, 6 > 2′, 3′ > 3. The same paper reports the preparation of 1,2,2′,3′,4′,6,6′-hepta-*O*-benzoyl-β-D-maltose in 87% yield by treating maltose with 10 mole equivalents of benzoyl chloride in pyridine without the need for column chromatography. The partial benzoylation of methyl β-cellobioside with 7.9 molar equivalents of benzoyl chloride gave a mixture of the per-*O*-benzoyl and 2,2′,3′,4′,6,6′-hexa-*O*-benzoyl derivatives in 56 and 27% yield respectively. The latter was epimerized at C-3 to give 4-*O*-D-glucopyranosyl-D-allopyranose derivatives by a conventional sulphonate ester displacement sequence.[21]

The constitution of the half-ester derivatives formed by treating sucrose with 3 moles of phthalic or succinic anhydride in DMF has been determined using mass spectrometry and substitution was found to occur mainly at positions 4, 6, and 6′. Organotin, organolead, and organogermanium complexes of the half-ester products were prepared.[22] Treatment of sucrose with pivaloyl chloride under varying conditions has led to a variety of partially esterified derivatives; in particular, the 1′,3,3′,4′,6,6′-hexa-ester (12) could be isolated by direct crystallization, and the hepta-ester 4-ol derivative (13), obtained in

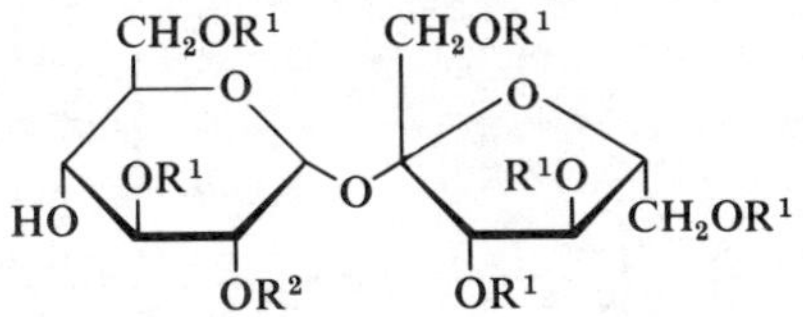

(12) $R^1 = Bu^tCO, R^2 = H$
(13) $R^1 = R^2 = Bu^tCO$

40% yield after column chromatography, was converted to *galacto*-sucrose, 4-chloro-4-deoxy-sucrose, and 4-deoxy-sucrose by standard procedures.[23]

A new method for *N*-deacetylation of 2-acetamido-2-deoxy-sugars has appeared. A mixture of trifluoroacetic anhydride–trifluoroacetic acid was used to transacylate the *N*-acetylamido compound and trifluoroacetylate all free hydroxy-groups; treatment with either methanolic ammonia or ethanolic sodium borohydride removed all the trifluoroacetyl groups. When applied to capsular polysaccharides of *Diplococcus pneumoniae* minimal degradation was observed.[24]

Selective deacetylation of the sucrose derivative (14) with methanolic ammonia gave the 3,6′-diacetate (15) and the 3-acetate (16).[25] β-D-Glucosyl esters have been prepared with high stereospecificity using tributylstannyl derivatives in two ways as outlined in Scheme 4. The method was used to prepare β-1-*O*-acetyl and *p*-nitrobenzyol derivatives from 2,3,4,6-tetra-*O*-benzyl-α-D-glucose and the plant β-glucosyl ester, stevioside, from 2,3,4,6-

[20] K. Takeo, *Carbohydrate Res.*, 1977, **59**, 584.
[21] K. Takeo and S. Okano, *Carbohydrate Res.*, 1977, **59**, 379.
[22] A. Parkin and R. C. Poller, *Carbohydrate Res.*, 1978, **62**, 83.
[23] L. Hough, M. S. Chowdhary, and A. C. Richardson, *J.C.S. Chem. Comm.*, 1978, 664.
[24] B. Nilsson and S. Svensson, *Carbohydrate Res.*, 1978, **62**, 377.
[25] R. Khan, M. R. Jenner, and H. Lindseth, *Carbohydrate Res.*, 1978, **65**, 99.

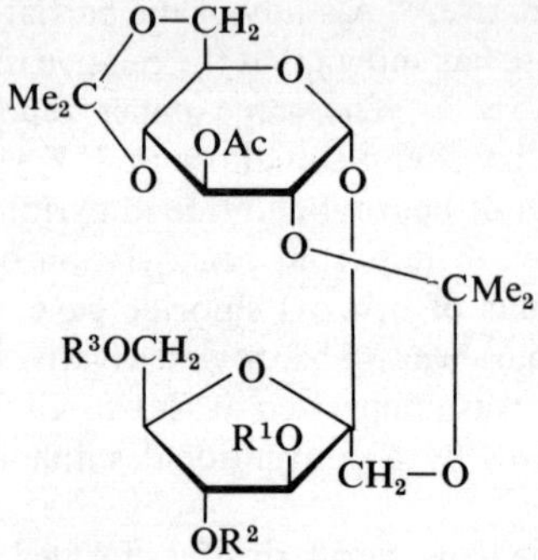

(14) $R^1 = R^2 = R^3 = Ac$
(15) $R^1 = R^2 = H, R^3 = Ac$
(16) $R^1 = R^2 = R^3 = H$

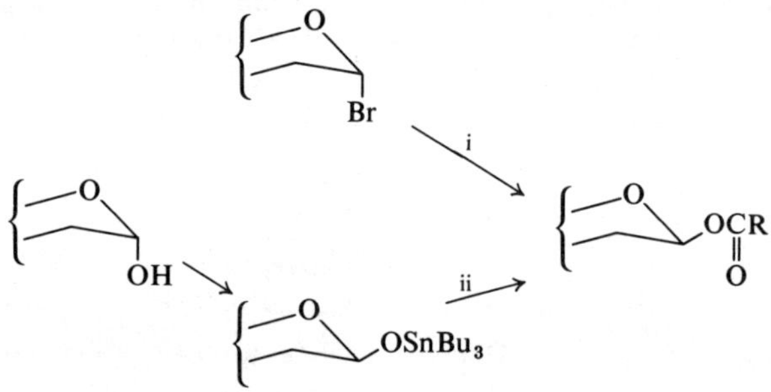

Reagents: i, RCO_2SnBu_3–NEt_4Br; ii, RCOCl

Scheme 4

tetra-*O*-acetyl-α-D-glucopyranosyl bromide and tributylstannyl steviobioside hepta-acetate.[26]

A study on the stability of reducing sugars towards trifluoroacetolysis has shown that the ratio of trifluoroacetic acid to trifluoroacetic anhydride is critical. When the former was in excess considerable degradation of the sugars occurred, particularly of pentoses and 6-deoxy-hexoses. 2-Deoxy-sugars such as 2-deoxy-D-*erythro*-pentose and 2-deoxy-D-*lyxo*-hexose were rapidly destroyed under all conditions, whereas 2-amino-2-deoxy-D-galactose was stable when the anhydride was in fifty-fold excess. The products of non-degraded sugars were the pertrifluoroacetates in each case. The method allowed selective cleavage of reducing terminal 2-acetamido-sugars from oligosaccharides.[27]

Ring and open-chain derivatives of L-erythritol 1,4-di-cinnamate (17) and (18) have been irradiated to give truxinic acids by intramolecular [2 + 2] cycloaddition. The stereochemistry of the products was determined.[28]

[26] T. Ogawa, M. Nozaki, and M. Matsui, *Carbohydrate Res.*, 1978, **60**, C7.
[27] B. Nilsson and S. Svensson, *Carbohydrate Res.*, 1978, **65**, 169.
[28] B. S. Green, A. T. Hagler, Y. Rabinsohn, and M. Rejto, *Israel J. Chem.*, 1977, **15**, 124 (*Chem. Abs.*, 1978, **88**, 23 288f).

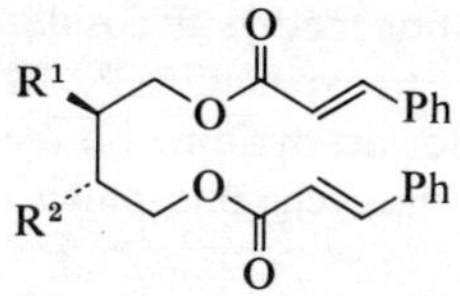

(17) R¹ = R² = OMe
(18) R¹, R² = $-O\diagdown CMe_2 \diagup -O$

Benzoates and acetates derived by oxidative photolysis of benzylidene and ethylidene acetals, respectively, are noted in Chapter 5.

2 Phosphates and Related Esters

Methods for the synthesis of phosphate monoesters and phosphoramidates of carbohydrates and nucleic acids have been reviewed.[29]

A simple enzymic synthesis of barium glucose 6-phosphate from starch has appeared.[30] Ths use of 2-methylthio-4*H*-1,3,2-benzodioxaphosphorin-2-oxide (MTBO) (see M. Eto *et al.*, *Tetrahedron Letters*, 1971, 4263) continues to receive attention. A study of twenty amines as catalysts has shown that primary n-alkyl and alicyclic amines are the most effective, followed by secondary and tertiary amines. Increasing the branching reduced the catalytic efficacy while cyclic imines gave low yields.[31] MTBO has been used to synthesize adenosine 2′,3′-cyclic phosphate in 50% yield from 5′-*O*-acetyladenosine (Scheme 5).[32] D-Arabinose 5-phosphate and D-xylose 5-phosphate have been

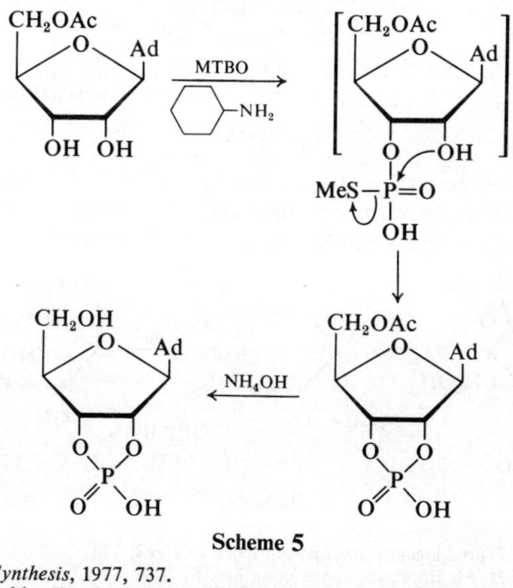

Scheme 5

[29] L. A. Slotin, *Synthesis*, 1977, 737.
[30] H. Susumu, *Hakko Kogaku Kaishi*, 1978, **56**, 31 (*Chem. Abs.*, 1978, **88**, 121 581r).
[31] M. Eto, M. Iio, H. Maeda, and H. Omura, *Agric. and Biol. Chem. (Japan)*, 1978, **42**, 299.
[32] M. Eto, M. Iio, H. Omura, and M. Eto, *Agric. and Biol. Chem. (Japan)*, 1978, **42**, 199.

prepared from the corresponding methyl glycosides using diphenyl or dibenzyl phosphochloridate followed by reduction.[33] The action of phosphorus trichloride in carbon tetrachloride–pyridine on methyl α- or β-D-ribopyranoside has been shown to yield the cyclic phosphites (19) and (20) respectively.[34]

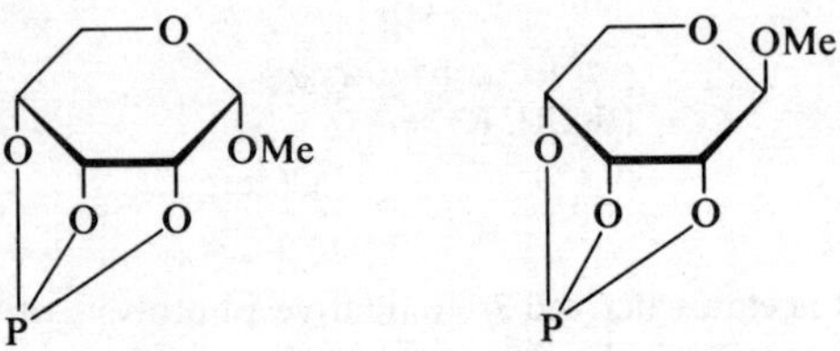

(19) (20)

L-Fucopyranose tetracetate as the α and β anomeric mixture has been converted to β-L-fucopyranosyl phosphate *via* the orthoester (21) (Scheme 6). The

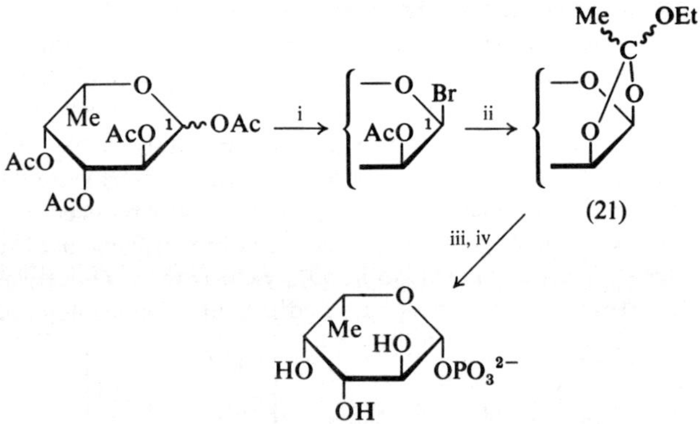

Reagents: i, HBr–HOAc; ii, EtOH–*sym*-collidine; iii, H_3PO_4; iv, NaOMe–MeOH

Scheme 6

exo- and *endo*-isomers of (21) were not separated.[35] 1-Deoxy-1-dihydroxyphosphinylmethyl-D-fructose, an analogue of D-fructose-1-phosphate, has been synthesized and shown to be a growth inhibitor for *E. coli* (Scheme 7).[36]

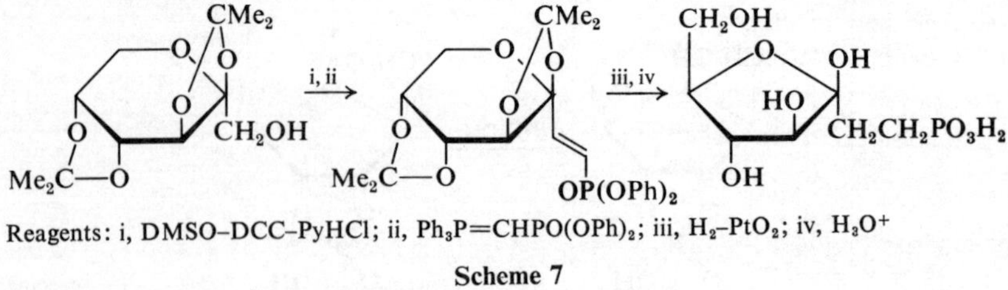

Reagents: i, DMSO–DCC–PyHCl; ii, $Ph_3P{=}CHPO(OPh)_2$; iii, H_2–PtO_2; iv, H_3O^+

Scheme 7

33 H. Maehr and J. Smallheer, *Carbohydrate Res.*, 1978, **62**, 178.
34 A. C. Bellaart, H. M. Buck, P. A. Leclercq, and L. J. M. Van De Ven, *Rec. Trav. Chim.*, 1977, **96**, 293 (*Chem. Abs.*, 1978, **88**, 89 975j).
35 J.-H. Tsai and E. J. Behrman, *Carbohydrate Res.*, 1978, **64**, 297.
36 J.-C. Tang, B. E. Tropp, and R. Engel, *Tetrahedron Letters*, 1978, 723.

α-Glycosyl bromides react with di-*O*-alkylphosphorothiolate anions to give almost quantitative amounts of either β-*S*- or β-*O*-glycosyl thiolates (22) and (23) respectively; when tetraethylammonium salts of the anion were used

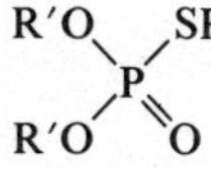

(22)

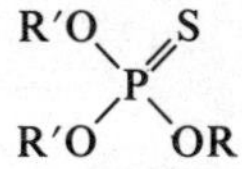

(23)

R = β-Glycosyl, R^1 = alkyl

the *S*-derivative predominated, whereas the presence of Ag^+ favoured the formation of the *O*-derivatives. The *O*-glucosyl derivative was found to isomerize thermally to the *S*-glucosyl derivative with retention of configuration, attributed to participation by the neighbouring acetyl group. The products were characterized by i.r. and ^{31}P n.m.r. and by the syntheses outlined in Scheme 8.[37]

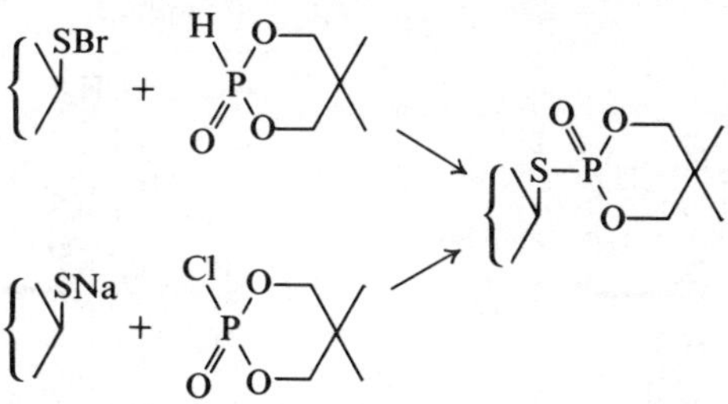

Scheme 8

β-D-Glucopyranosyl citronellyl phosphate has been synthesized from the corresponding t-butyl α-D-glucopyranosyl ortho-acetate derivative and citronellyl benzyl phosphate.[38] Several phosphoric esters of L-ascorbic acid have been prepared by the reaction of phosphoryl chloride at 0—5 °C on 5,6-*O*-isopropylidene-L-ascorbate. The product at pH 12—13 in high concentration of pyridine was quantitatively the 2-phosphate. Hydrolysis of the acetal gave a 70% yield of L-ascorbate 2-phosphate. In the absence of pyridine the products were the 2-phosphate and the diphosphate (24).[39] 5,6-Dideoxy-6-phosphono-D-*arabino*-hexose (25), an isosteric analogue of D-arabinose 5-phosphate, has been synthesized.[40] 1,2-*O*-Isopropylidene-6-thio-α-D-glucofuranose reacted with phosphorus trichloride–triethylamine in absolute ether to give the 3,5,6-bicyclothiophosphite (26).[41]

α-D-Glucopyranosyl derivatives of methyl 2-acetamido-2-deoxy-α-D-glucosyl-1-phosphate (27) and its 3-*O*-methyl ether (28) have been prepared by

[37] M. Michalska, J. Michalski, and I. Orlich, *Tetrahedron*, 1978, **34**, 617.

[38] L. L. Danilov, L. V. Volkova, and R. P. Evstigneeva, *Zhur. obshchei Khim*,. 1977, **47**, 2137 (*Chem. Abs.*, 1978, **88**, 7236g).

[39] C. H. Lee, Paul A. Seib, Y. T. Liang, R. C. Hoseney, and C. W. Deyoe, *Carbohydrate Res.*, 1978, **67**, 127.

[40] F. M. Unger, D. Stix, E. Moderndorfer, and F. Hammerschmid, *Carbohydrate Res.*, 1978, **67**, 349.

[41] E. E. Nifant'ev, M. P. Koroteev, S. A. Ermishkina, S. A. Lysenko, and N. K. Kochetkov, *Zhur. obshchei Khim.*, 1978, **48**, 1904 (*Chem. Abs.*, 1978, **89**, 180 249g).

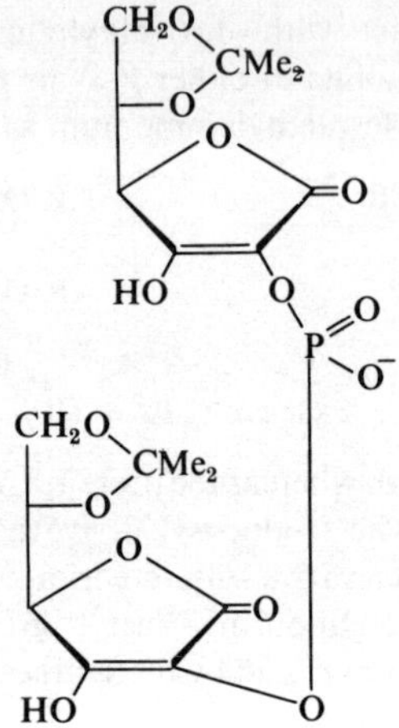

(24)

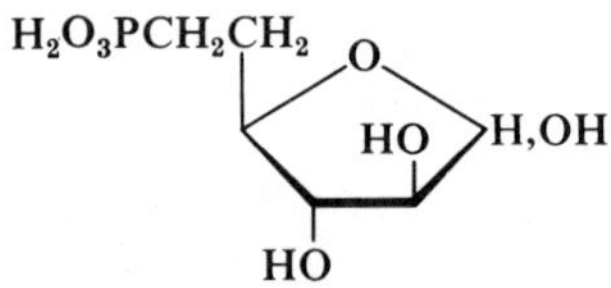

(25)

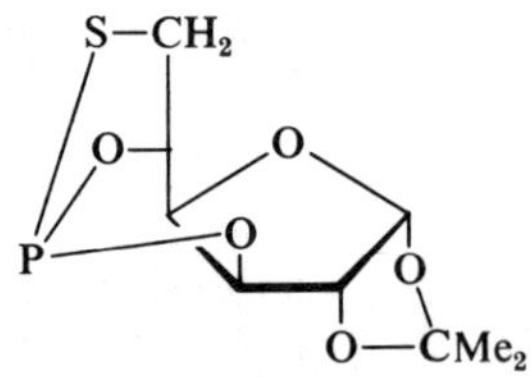

(26)

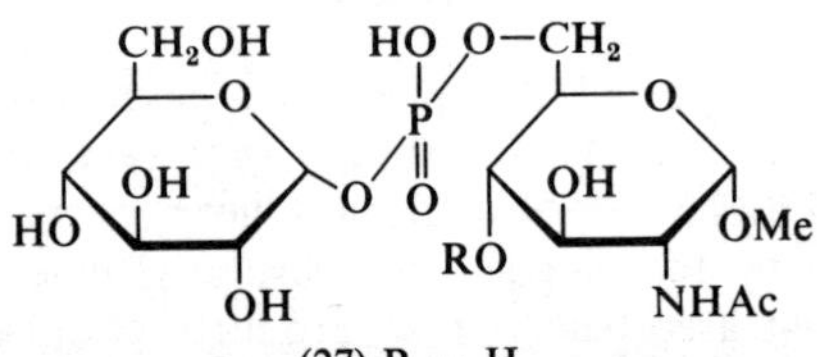

(27) R = H
(28) R = Me

coupling appropriate derivatives of α-D-glucose-1-phosphate and methyl 2-acetamido-2-deoxy-α-D-glucopyranoside, followed by deblocking, and the compounds were then used to study the phosphoric ester linkage in the cell-wall of *Micrococcus lysodeikticus*.[42]

The phosphates (29) and (30) have been prepared by reaction of the corresponding peracetylated sugar 1,2-oxazoline derivative with dibenzyl hydrogen

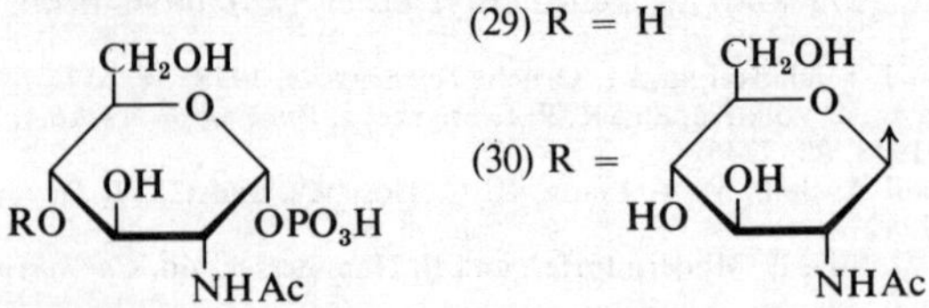

(29) R = H
(30) R =

[42] C. D. Warren, Nasir-ud-Din, and R. W. Jeanloz, *Carbohydrate Res.*, 1978, **64**, 43.

phosphate followed by *O*-deacetylation and hydrogenolysis; the peracetate of compound (30) was then condensed with diphenyl dolichyl pyrophosphate to give the amino-disaccharide dolichyl pyrophosphate (31) after deacetylation.[43]

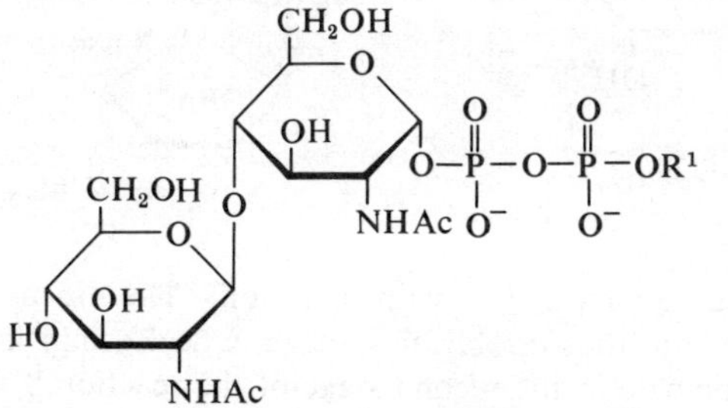

(31) R^1 = $-CH_2CH_2CH(Me)CH_2[CH_2CH{=}C(Me)CH_2]_{18}H$

The peracetylated 1-phosphates of 3-deoxy-D-*arabino*-hexose, 4-deoxy-D-*lyxo*-hexose, and 6-deoxy-D-mannose have been prepared.[44]

The effect of pH and temperature on the product distribution and mode of opening in the hydrolysis of α-D-glucose 1,2-cyclic phosphate has been investigated. At pH 1, α-D-glucopyranosyl phosphate was the main product below 60 °C but at higher temperatures complete hydrolysis to D-glucose and inorganic phosphate occurred. In the pH range 5—13 at 80 °C hydrolysis of the cyclic phosphate was fast relative to further hydrolysis of the monoesters produced and substantial amounts of the 2-phosphate were obtained. Carrying out the reaction in $H_2{}^{18}O$ gave results consistent with formation of the 2-phosphate by phosphorus–oxygen cleavage in both 0.1M-NaOH and 0.1M-HCl at 100 °C. These findings are in contradistinction to the proposal that ring strain governs the rate of hydrolysis of five-membered cyclic phosphates by accelerating cleavage of C-1-*O* or C-1-*P* bonds.[45]

3 Sulphonates

Selective mesylation of methyl α-D-glucopyranoside using 3.1 moles of methane sulphonyl chloride gave predominantly the 2,4,6-tri-*O*-mesylate whereas under the same conditions toluene-*p*-sulphonyl chloride gave mainly the 2,6-di-*O*-tosyl derivative together with approximately equal amounts of the 2,3,6- and 2,4,6-tri-*O*-tosyl derivatives. The reaction with *p*-bromobenzene sulphonyl chloride gave an unresolved complex mixture of products.[46]

The reactivity of the secondary hydroxy-groups in the lactose derivative (32) towards tosylation has been explored. Analysis of the mixture of partially sulphonylated derivatives obtained under varying conditions suggested the order of reactivity: $2 > 3' \gg 3 > 2'$.[47]

The stereochemistry of nucleophilic displacement reactions of the di-*O*-sulphonate (33) by sodium benzoate in DMF during 6 h at reflux showed that

43 C. D. Warren, A. Herscovics, and R. W. Jeanloz, *Carbohydrate Res.*, 1978, **61**, 181.
44 S. Kucar, J. Zamocky, J. Zemek, and S. Bauer, *Chem. Zvesti*, 1978, **32**, 414.
45 F. C. Kokesh, D. A. Cameron. Y. Kakuda, and P. V. Kuras, *Carbohydrate Res.*, 1978, **62**, 289.
46 Y. Ali and M. Hussain, *J. Iraqi Chem. Soc.*, 1977, **2**, 43 (*Chem. Abs.*, 1978, **88**, 152 920e).
47 T. Takamura and S. Tejima, *Chem. and Pharm. Bull. (Japan)*, 1978, **26**, 1117.

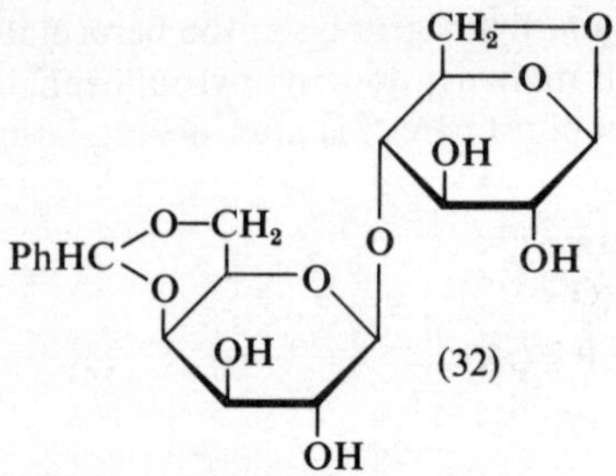

(32)

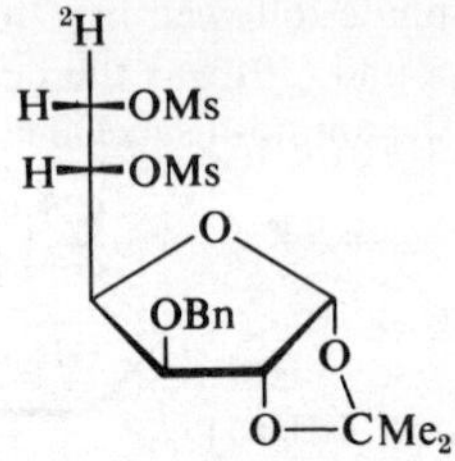

(33)

the preferential displacement occurs at C-6 with inversion. The dibenzoate was also formed with inversion, thus establishing that the acyl group introduced at C-6 does not participate in the second stage of the reaction.[48] The selectivity of 5-(dimethylamino)-1-naphthalenesulphonyl chloride (dansyl chloride) for primary hydroxy-groups has found use in the synthesis of 6-*O*-dansyl derivatives of D-galactose and 6-*O*-α-D-galactopyranosyl-D-galactose;[49] these were then used to investigate the competitive binding of galactosyl galactose to immunoglobulins by means of fluorescence changes.[50] Selective tosylation of 1,5-anhydro-4,6-*O*-benzylidene-D-glucitol using one molar equivalent of reagent gave the 2-*O*-, 3-*O*-, and 2,3-di-*O*-tosyl derivatives in 40—45, 6—10, and 8—9% yields respectively.[18] The 3′-hydroxy-group of sucrose has been shown to be more reactive than the 4′-hydroxy-group towards tosylation by treatment of (15) with limited amounts of tosyl chloride in pyridine.[24]

The i.r. spectra of sulphonic esters of carbohydrates are treated in Chapter 21.

The use of triflates for introducing iodide and fluoride into carbohydrates is mentioned in Chapter 7. Glycosyl synthesis using glycosyl sulphonates is described in Chapter 3.

4 Other Esters

A brief review of glucosinolates and other naturally occurring *O*-sulphates has appeared.[51]

D-Glucose reacted with dodecyl chloroformate in DMF in the presence of sodium carbonate to produce a mixture of carbonates which was separated into mono-, di-, and tri-carbonates by t.l.c. on silica gel.[52] Sucrose monoesters have been prepared in 44—85% yield from sucrose and 1-alkoxyethyl isocyanates in DMF containing triethylamine.[53] The syntheses and properties of sucrose *N*-(aryloxyalkaneacyl)-carbamates from sucrose and the corresponding isocyanate have been reported.[54] As part of a study of the behaviour of the

48 K. Kakinurna, *Tetrahedron Letters*, 1978, 765.
49 E. Zissis and C. P. J. Glaudemans, *Carbohydrate Res.*, 1978, **63**, 99.
50 A. K. Bhattacharjee and C. P. J. Glaudemans, *Carbohydrate Res.*, 1978, **67**, 536.
51 A. Kjaer, *A.C.S. Symp. Ser.*, 1978, **77** (Carbohydrate Sulphates), 19.
52 W. Meissner and R. Zielinski, *Zesz. Nauk – Akad. Ekon. Poznaniu, Ser.* **1**, 1976, **69**, 122 (*Chem. Abs.*, 1978, **88**, 7233d).
53 T. Lesiak and J. Iwaszkiewicz-Nowak, *J. prakt. Chem.*, 1977, **319**, 727 (*Chem. Abs.*, 1978, **88**, 191 285w).
54 J. Arct and Z. Eckstein, *Pol. J. Chem.*, 1978, **52**, 11 (*Chem Abs.*, 1978, **89**, 44 030b).

carbamoyl group in bleomycin, methyl 3-*O*- and 2-*O*-carbamoyl-α-D-mannopyranoside have been prepared either from the corresponding 2,3-cyclic carbonate with ammonia or by the method shown in Scheme 9. Under

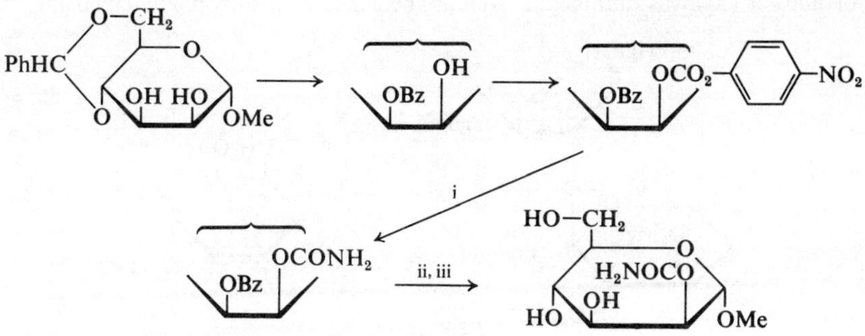

Reagents: i, NH_3–MeOH; ii, H_2–Pd; iii, NaOMe

Scheme 9

alkaline conditions carbamoyl migration between *O*-2 and *O*-3 occurs to give an equilibrium mixture of the two.[55] A study of the reactivity of furanosyloxyphosphonium salts of D-mannose, D-allose, and D-ribose, prepared from suitably blocked free anomers and tri(dimethylamino)phosphite, has been reported (Scheme 10).[56] Sugar azidosulphates have been prepared from sugar chlorosulphates by treatment with potassium azide in the presence of 18-crown-6-ether.[57] Treatment of sodium heparinate with methanol–0.06 M-HCl

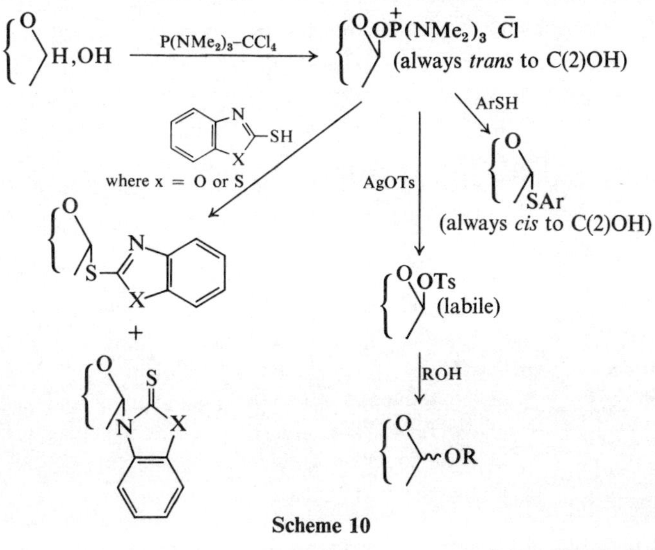

Scheme 10

[55] M. M. A. Abdel Rahman, *Indian J. Chem., Sect. B*, 1977, **15B**, 1032 (*Chem. Abs.*, 1978, **88**, 152 918k).
[56] R.-A. Boigegrain, F. Chretien, B. Castro, and B. Gross, *J. Chem. Res.* (S), 1978, 85; (M), 1978, 1229.
[57] N. T. Naidoo and H. Parolis, *Carbohydrate Res.*, 1978, **62**, C5.

followed by nitrous acid gave 10% monosulphated disaccharide which was presumed to be the sodium salt (34).[58] When tetra-*O*-acetyl-α-D-allopyranosyl bromide was treated with alkaline *p*-nitrophenol in chloroform–methanol the orthoester (35) was obtained as well as the expected β-nitrophenylglycoside.[59]

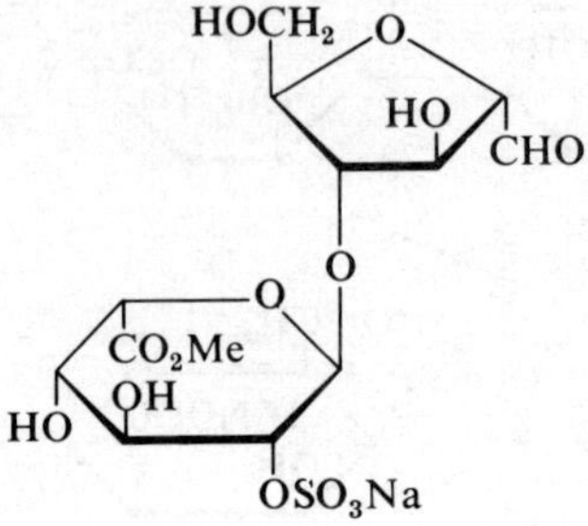

(34)

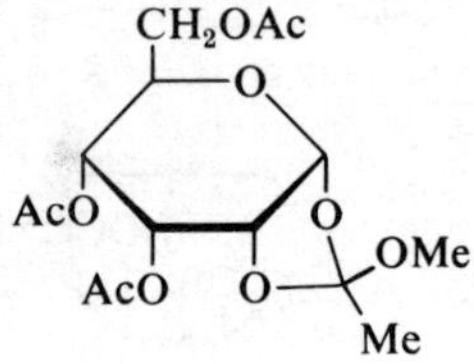

(35)

The use of unsymmetrical sulphites in glycoside synthesis is mentioned in Chapter 3. Reduction of dimethylsulphamoyl esters to deoxy compounds is noted in Chapter 11. Reference to ascorbic acid sulphates and phosphates is made in Chapter 15. Phenylboronates used for disaccharide synthesis are mentioned in Chapter 3. Chapter 4 includes a mention of an alkene synthesis *via* a thionocarbonate.

[58] L. C. Ginsberg, D. T. DiFerrante, and N. DiFerrante, *Carbohydrate Res.*, 1978, **64**, 225.
[59] R. H. Shah and O. P. Bahl, *Carbohydrate Res.*, 1978, **65**, 153.

7
Halogenated Sugars

1 Glycosyl Halides

The use, by Paulsen and his colleagues, of 2-azido-2-deoxy-D-glucopyranosyl and -D-galactopyranosyl chlorides in the preparation of α-bonded amino-sugar oligosaccharides is referred to in Chapter 3. An independent paper details the syntheses of a range of such compounds. See Scheme 1 for an example. The

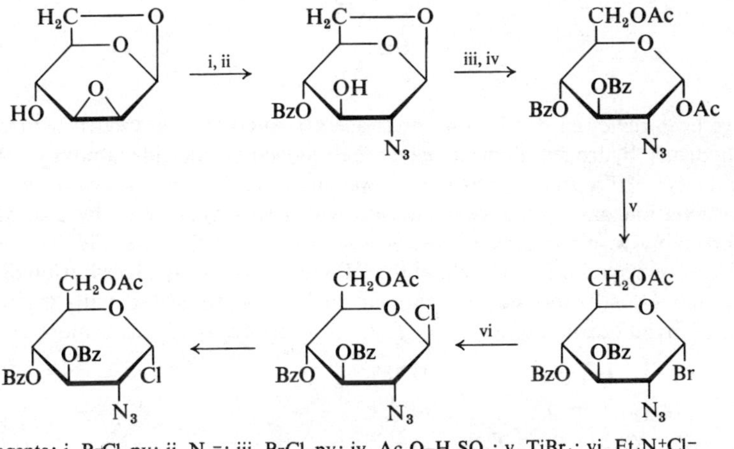

Reagents: i, BzCl–py; ii, N_3^-; iii, BzCl–py; iv, Ac_2O–H_2SO_4; v, $TiBr_4$; vi, $Et_4N^+Cl^-$

Scheme 1

paper contains data, including ^1H and ^{13}C n.m.r. data, on more than 60 compounds, and the kinetics of some of the displacement reactions at the anomeric centres were reported.[1] Related work afforded the anomerically pure 3-*O*-acetyl-2-azido-5-*O*-benzoyl-2-deoxy-α- and β-D-arabinofuranosyl chlorides (1), (2), (Scheme 2) used in condensations with silylated uracil. From the β-linked product the aminonucleoside (3) was obtained.[2]

Peracetates of mono- and di-saccharides (*e.g.* α-D-glucopyranose and β-lactose) were treated with dichloromethyl methyl ether and tin(IV) chloride at room temperature to give the more stable acetylated glycosyl chlorides in high yields,[3] and this and other uses of this reagent in carbohydrate chemistry

[1] H. Paulsen, A. Richter, V. Sinnwell, and W. Stenzel, *Carbohydrate Res.*, 1978, **64**, 339.
[2] M. Bobek and V. Martin, *Tetrahedron Letters*, 1978, 1919.
[3] I. Farkas, I. F.-Szabó, M. M. Menyhárt, E. R.-Dávid, and R. Bognár, *Acta Chim. Acad. Sci. Hung.*, 1977, **94**, 153 (*Chem. Abs.*, 1978, **88**, 136 867n).

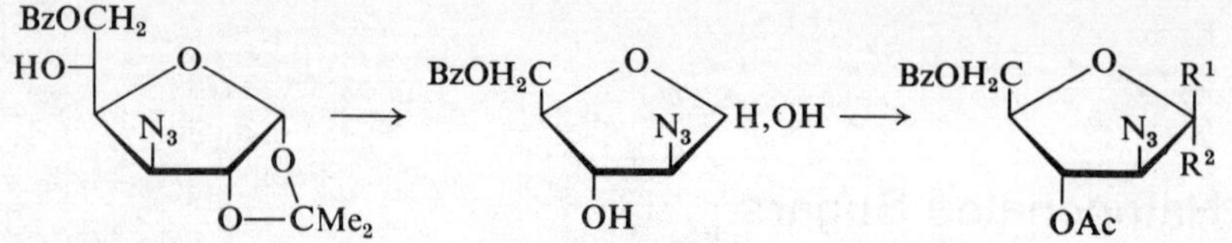

	R^1	R^2
(1)	H	Cl
(2)	Cl	H

Reagents: i, H^+; ii, $^-IO_4$; iii, Ac_2O–py; iv, $TiCl_4$

Scheme 2

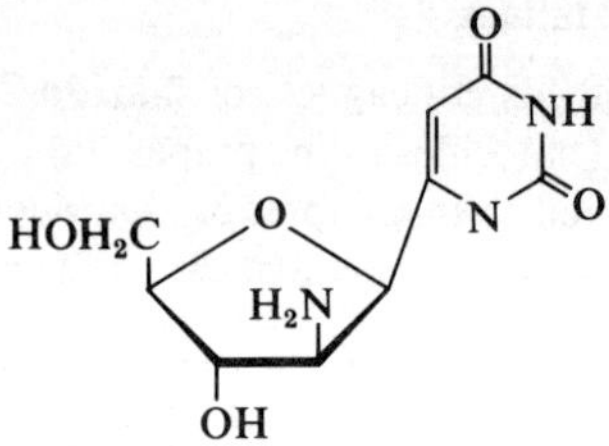

(3)

have been reviewed.[3a] Maltose hepta-acetate, alternatively, treated with cold, anhydrous hydrogen fluoride gave the α-glycosyl fluoride amongst other products. ^{19}F n.m.r. spectroscopy was used in the characterization, and deacetylation gave α-maltosyl fluoride which was hydrolysed by β-amylase appreciably faster than was maltotriose.[4] Lactosyl fluoride and its hepta-acetate were synthesized to help in the ^{19}F spectroscopic characterization of the products (4)—(7) formed by treatment of lactal hexa-acetate with trifluoromethyl hypofluorite (Scheme 3). The additions occurred much more slowly

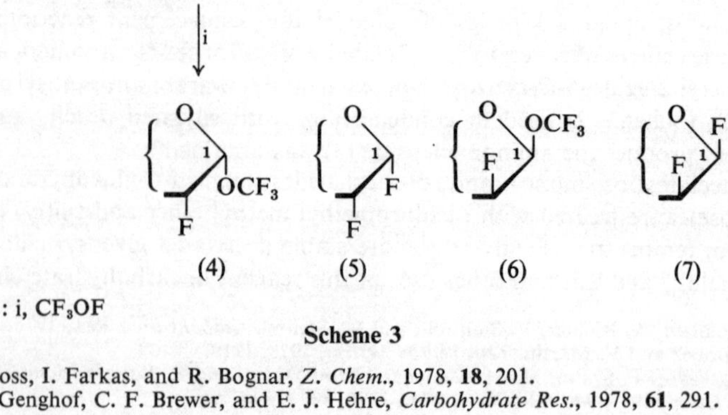

(4) (5) (6) (7)

Reagents: i, CF_3OF

Scheme 3

[3a] H. Gross, I. Farkas, and R. Bognar, *Z. Chem.*, 1978, **18**, 201.
[4] D. S. Genghof, C. F. Brewer, and E. J. Hehre, *Carbohydrate Res.*, 1978, **61**, 291.

than in the monosaccharide series, and this observation and the product ratios were discussed in terms of the influence of the saturated residue on the glycal.[5]

Bromination with bromine in carbon tetrachloride under bright light causes substitution of H-5 of tetra-*O*-acetyl-*β*-D-xylopyranose and penta-*O*-acetyl- and *O*-benzoyl-*β*-D-glucopyranose and gives products, *e.g.* (8), which can be considered to be 5-acyloxyglycopyranosyl bromide peresters. Substitution of the bromine atom of (8) led to compounds (9)—(11), and elimina-

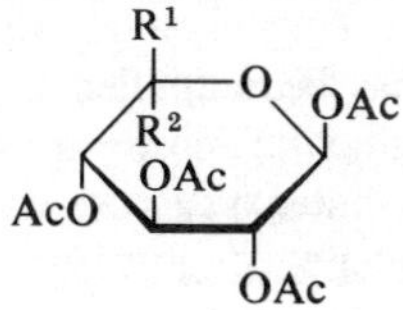

	R^1	R^2
(8)	H	Br
(9)	OMe	H
(10)	OAc	H
(11)	SAc	H

tions in this and in the hexose series gave a set of 4 and 5-enose derivatives (see Chapter 12).[5a]

2 Other Halogenated Derivatives

Fluorination of tri-*O*-acetyl-D-glucal with $^{18}F_2$ gave the *α-gluco-* and *β-manno*-1,2-difluorides in the ratio 4 : 1, and in the course of the work fluorine and trifluoromethyl hypofluorite were used with D-[*U*-^{14}C]glucose to give labelled 2-deoxy-2-fluoro-D-glucose.[6] An efficient new synthesis of 3-deoxy-3-fluoro-D-glucose involves treatment of 1,2:5,6-di-*O*-isopropylidene-α-D-allose with diethylaminosulphur trifluoride. The diacetal formed by inversion at C-3 was produced in 90% yield, and this can also be obtained almost quantitatively from the triflate of the same D-*allo*-starting material using caesium fluoride in DMF. Di-*O*-isopropylidene-D-glucose treated with diethylaminosulphur trifluoride in the presence of pyridine gave the alkene (12) in good yield.[7] 6-Deoxy-6-

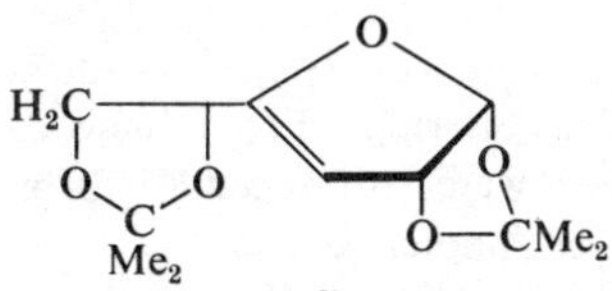

(12)

fluoro-α-D-galactopyranose-^{18}F was obtained from 1,2:3,4-di-*O*-isopropylidene-6-*O*-toluene-*p*-sulphonyl-α-D-galactose by use of labelled tetraethylammonium fluoride.[8]

[5] P. W. Kent and S. D. Dimitrijevich, *J. Fluorine Chem.*, 1977, **10**, 455 (*Chem. Abs.*, 1978, **88**, 136 864j).
[5a] R. J. Ferrier and P. C. Tyler, *J. C. S. Chem. Comm.*, 1978, 1019.
[6] T. Ido, C. N. Wan, V. Casella, J. S. Fowler, A. P. Wolf, M. Reivich, and D. E. Kuhl, *J. Labelled Compounds Radiopharm.*, 1978, **14**, 175 (*Chem. Abs.*, 1978, **89**, 110 142t).
[7] T. J. Tewson and M. J. Welch, *J. Org. Chem.*, 1978, **43**, 1090.
[8] D. R. Christman, Z. Orhanovic, W. W. Shreeve, and A. P. Wolf, *J. Labelled Compounds Radiopharm.*, 1977, **13**, 555 (*Chem. Abs.*, 1978, **88**, 105 665r).

Selective acid-catalysed methanolysis of 2,3,2′,3′-tetra-*O*-benzyl-4,6:4′,6′-di-*O*-benzylidene-α,α-trehalose gave the monoacetal which was converted into the dimesylate and from this 6-deoxy-6-fluoro-α,α-trehalose was obtained. In the course of the work 4-chloro-4,6-dideoxy-6-fluoro-α-D-galactopyranosyl α-D-glucopyranoside was produced (as a derivative), and 6-iodo, 6,6′-di-iodo, 6-azido-, and 6,6′-diazido-trehalose compounds were also prepared.[9] 6-Deoxy-6-fluoro- and 6-deoxy-6-iodo-trehalose were also obtained by way of the 4,6-isopropylidene acetal.[10]

The use of triphenylphosphine and carbon tetrahalide for selectively halogenating primary hydroxy-groups has been applied to methyl α-D-glucopyranoside, 1,2-*O*-isopropylidene-α-D-glucofuranose, inosine, and uridine with good results, and 6,6′-dichloro-6,6′-dideoxysucrose was obtained by this method. 5,6-Anhydro-1,2-*O*-isopropylidene-α-D-glucofuranose gave the 6-bromo- and 6-chloro-6-deoxy-D-glucose acetals.[11]

Reference to the application of the ring opening of benzylidene acetals to give bromo-monobenzoates is given in Chapters 5 and 11, and an interesting study of diacetals has appeared. Methyl 2,3:4,6-di-*O*-benzylidene-α-D-mannopyranoside treated with NBS gave the 3,6-dibromide (13), but also some of the C-3 epimer and, more surprisingly, some of the 4,6-dibromo-4,6-dideoxy-D-idose compound (14) formed by a benzoate migration. Methyl 2,3:4,6-di-*O*-benzylidene-α-D-allopyranoside, with the same reagent, gave the anticipated 2,6- and 3,6-dibrominated products (15) and (16), whereas the strained

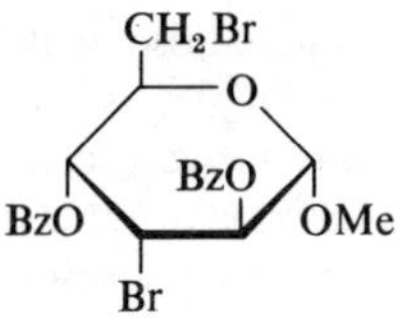

(13)

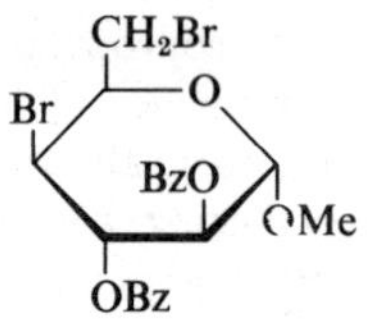

(14)

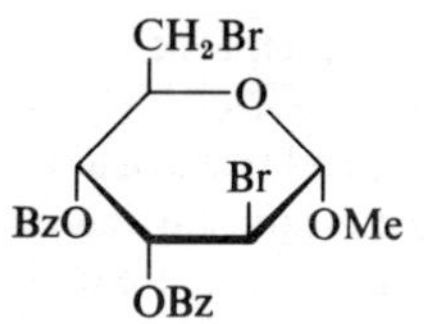

(15)

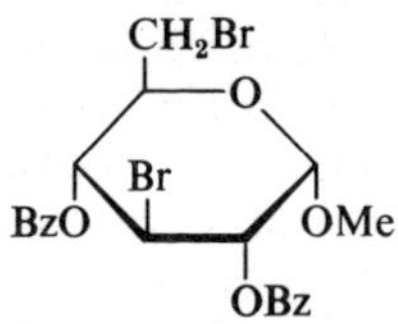

(16)

α-D-glucose analogue gave 6-bromo-hydroxyesters formed by direct oxidative opening of the *trans*-fused acetal ring. From these dibrominated products (13)—(16) several epoxides were produced.[12] Brominated disaccharide compounds are referred to in Chapter 3.

A very efficient method of introducing iodide (and presumably other nucleophiles) into carbohydrates involves the use of trifluoromethanesulphonate esters and their displacement with tetrabutylammonium iodide in refluxing benzene. Several monodeoxymonoiodo compounds were made in this way including compounds (17) and (18) which are difficult to obtain by more usual methods.[13]

[9] A. F. Hadfield, L. Hough, and A. C. Richardson, *Carbohydrate Res.*, 1978, **63**, 51.
[10] J. Defaye, H. Driguez, B. Henrissat, J. Gelas, and E. Bar-Guilloux, *Carbohydrate Res.*, 1978, **63**, 41.
[11] A. K. M. Anisuzzaman and R. L. Whistler, *Carbohydrate Res.*, 1978, **61**, 511.
[12] J. Thiem and J. Elvers, *Carbohydrate Res.*, 1978, **60**, 63.
[13] R. W. Binkley and D. G. Hehemann, *J. Org. Chem.*, 1978, **43**, 3244.

5′-Bromo-, chloro-, and iodo-derivatives of showdomycin are referred to in Chapter 18

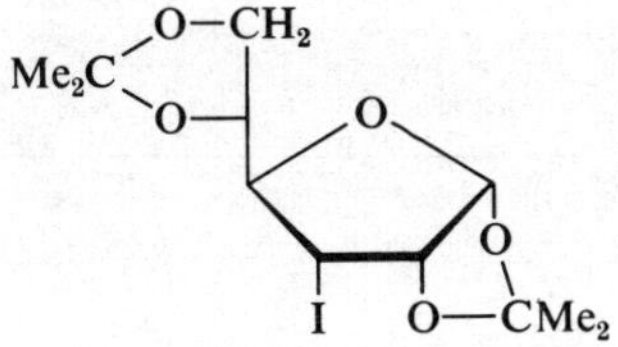

(17)

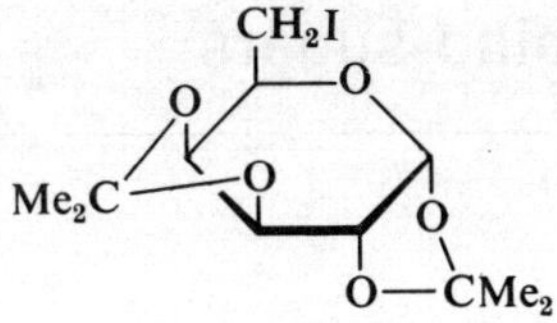

(18)

8
Amino-sugars

The characterization of megosamine, a constituent of megamycin, as 2,3,6-trideoxy-3-dimethylamino-L-*ribo*-hexopyranose, and not the corresponding D-*lyxo* C-5 epimer D-rhodosamine previously claimed, is referred to in Chapter 18.

1 Synthesis

A synthesis of *N*-acetyl-L-acosamine(3-amino-2,3,6-trideoxy-L-*arabino*-hexose) (1) has been described, starting from a derivative of sorbic acid (2) (Scheme 1).[1]

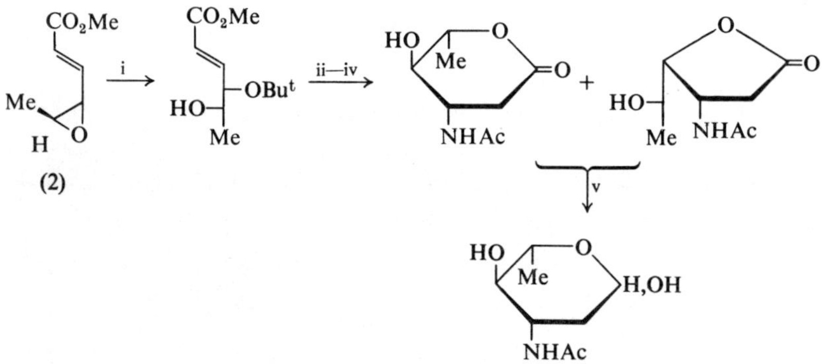

(2)

(1)

Reagents: i, $Bu^tOH–BF_3$; ii,  , H^+; iii, NH_3–MeOH; iv, Ac_2O; v, $Bu^i{}_2AlH$

Scheme 1

The full paper on the synthesis of D-forosamine(4-dimethylamino-2,3,4,6-tetradeoxy-D-*erythro*-hexose) from (2) has also appeared.[2] Standard transformations have been used in a synthesis of derivatives of L-ristosamine (3-amino-2,3,6-trideoxy-L-*arabino*-hexose), and its C-4 epimer, from L-rhamnal.[3] Non-carbohydrate precursors have been utilized for the synthesis of the derivative (3) of racemic daunosamine (3-amino-2,3,6-trideoxy-*lyxo*-hexose), in a sequence involving allylic amination of a hexenal derivative using a selenium–chloramine T reagent, followed by *cis*-hydroxylation which led to a crystalline product with the required *lyxo* configuration (Scheme 2).[4]

[1] I. Dyong and H. Bendlin, *Chem. Ber.*, 1978, **111**, 1677.
[2] I. Dyong, R. Knollmann, N. Jersch, and H. Luftmann, *Chem. Ber.*, 1978, **111**, 559.
[3] F. Sztaricskai, I. Pelyvas, L. Szilagyi, R. Bognar, J. Tamas, and A. Neszmelyi, *Magyar Kem. Folyoirat*, 1978, **84**, 320 (*Chem. Abs.*, 1978, **89**, 180 262f).
[4] I. Dyong and R. Wiemann, *Angew. Chem. Internat. Edn.*, 1978, **17**, 682.

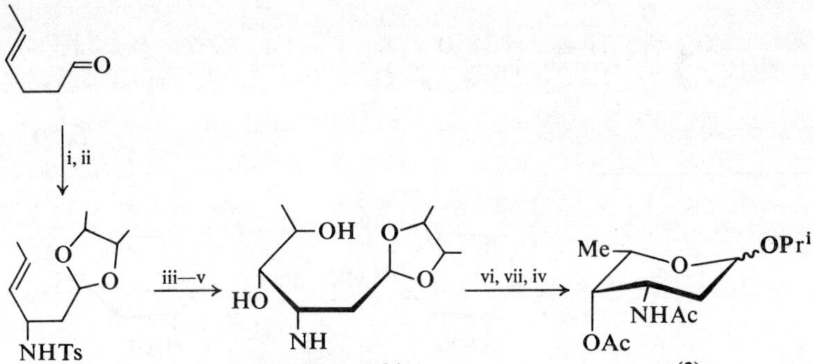

(DL-*lyxo*, *xylo*)
mixture

(3)

Reagents: i, MeCH(OH)CH(OH)Me–TsOH; ii, $Se(NTs)_2$; iii, $Na–NH_3$; iv, Ac_2O;

v, OsO_4— 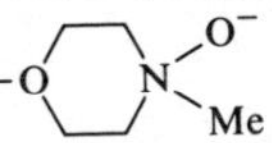

; vi, HCl; vii, Pr^nOH

Scheme 2

The dihydropyran derivative (4) obtained from the arylselenyl compound (5) (see also Chapter 11) reacts with chloramine T to give the isomeric 4-amino-4-deoxy sugar derivatives (6) and (7) (Scheme 3).[5] The 4-deoxy-DL-daunosamine

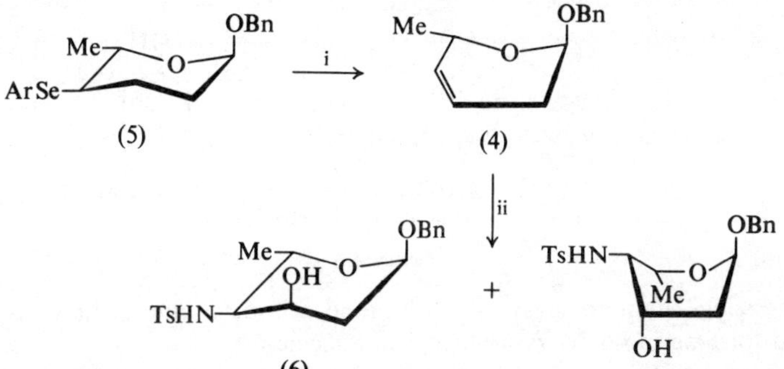

(5) (4) (6) (7)

Reagents: i, H_2O_2; ii, Chloramine T–OsO_4

Scheme 3

(8) has been prepared from an oxazole derivative by a sequence outlined in Scheme 4.[6]

Singlet oxygen adducts of 1-acyl-1,2-dihydropyridines have been utilized for the stereoselective synthesis of 5-amino-5-deoxy-DL-lyxo- and ribo-pyranosides ('piperidinoses') by the route outlined in Scheme 5.[7] A new procedure for

[5] S. Current and K. B. Sharpless, *Tetrahedron Letters*, 1978, 5075.
[6] H.-K. Hung, H.-Y. Lam, W. Niemczura, M.-C. Wang, and C.-M. Wong, *Canad. J. Chem.*, 1978, **56**, 638.
[7] M. Naksume, Y. Sekine, and H. Soyagimi, *Chem. and Pharm. Bull.* (*Japan*), 1978, **26**, 2188.

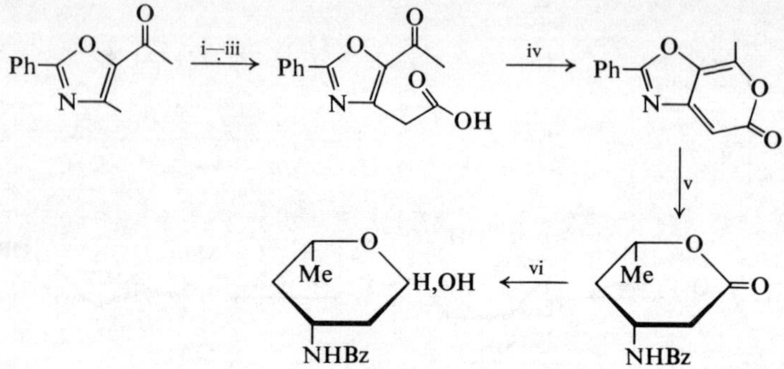

(8)

Reagents: i, NBS–$h\nu$; ii, KCN–HCN; iii, H^+–H_2O; iv, $SOCl_2$; v, H_2–Pt; vi, 'Red-Al' reagent

Scheme 4

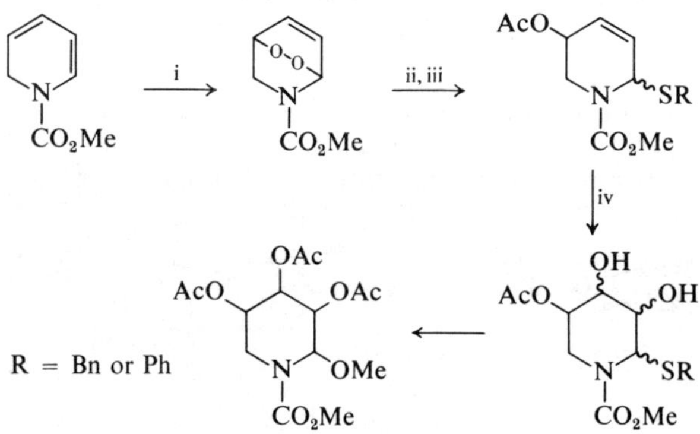

Reagents: i, O_2–$h\nu$; ii, RSH; iii, Ac_2O; iv, OsO_4; v, MeOH–Hg^{2+}

Scheme 5

synthesizing ω-amino-sugars *via* 1,2-iminocarbonate intermediates is illustrated for 6-amino-6-deoxy-D-glucose in Scheme 6.[8]

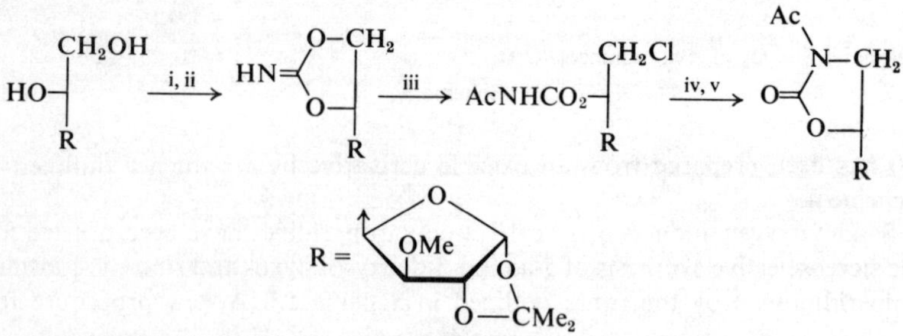

Reagents: i, BuLi; ii, CNBr; iii, AcCl; iv, NaH; v, Ac_2O–Py

Scheme 6

[8] D. H. R. Barton and W. B. Motherwell, *Nouv. J. Chim.*, 1978, **2**, 301 (*Chem. Abs.*, 1978, **89**, 197 850u).

7-Acetamido-6,7,8-trideoxy-1,2:3,4-di-*O*-isopropylidene-α-D- and β-L-*glycero*-D-*galacto*-octapyranoses have been prepared from D-galactose *via* the unsaturated nitro-sugar (9) by standard reactions.[9]

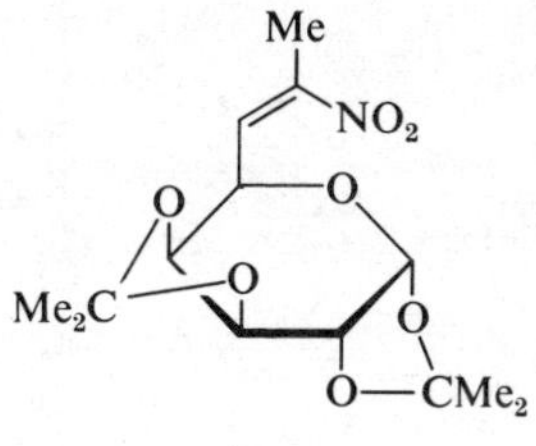

(9)

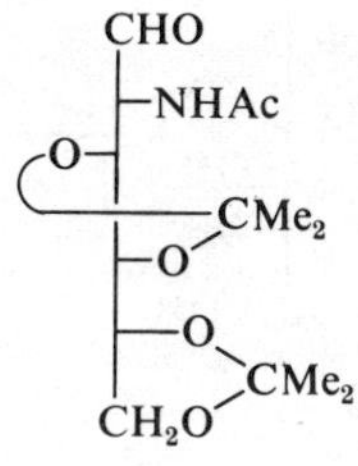

(10)

Condensation of pentodialdo-1,4-furanose derivatives with ethyl isocyanoacetate has been used to prepare 6-amino-6-deoxyhepturonic acid derivatives, illustrated in Scheme 7; corresponding 5,6-unsaturated and 6-deoxy sugars were also synthesized conventionally.[9a]

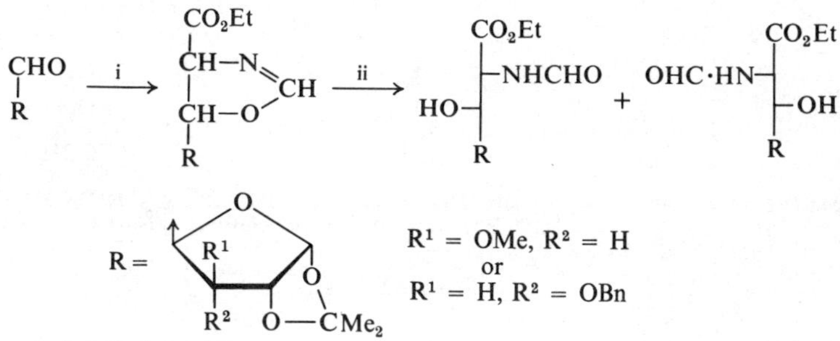

Reagents: i, $EtO_2C{\cdot}CH_2NC$–NaCN

Scheme 7

Improved yields of 3-aminoheptoses were obtained from 2-acetamido-2-deoxy-hexoses by the nitromethane route when the *aldehydo*-sugar (10) was used instead of the free-sugar for the condensation with nitromethane.[10] The same authors have developed a multi-step procedure for the sequential build-up to *N*-acetylneuraminic acid from 2-acetamido-2-deoxy-D-mannose which can be applied to the preparation of specifically labelled compounds and stereoisomers (Scheme 8).[11]

4-Amino-1,6-anhydro-4-deoxy-β-D-mannopyranose (11) and the corresponding D-altrose derivative (12) have been prepared by the sequence outlined in Scheme 9.[12] Multi-stage syntheses using conventional procedures have been reported for derivatives of L-holantosamine (4-amino-2,4,6-trideoxy-3-*O*-methyl-L-*arabino*-hexose) and of L-epiholantosamine, the corresponding

[9] J. C. A. Boeyens, E. B. Rathbone, and G. R. Woolard, *Carbohydrate Res.*, 1978, **62**, 39.
[9a] K. Bischofberger, A. J. Brink, O. G. De Villiers, R. H. Hall, and A. Jordaan, *Carbohydrate Res.*, 1978, **64**, 33.
[10] L. Benzing and M. B. Perry, *Canad. J. Chem.*, 1978, **56**, 691.
[11] L. Benzing-Nguyen and M. B. Perry, *J. Org. Chem.*, 1978, **43**, 551.
[12] M. Cerny, I. Cerny, and T. Trnka, *Carbohydrate Res.*, 1978, **67**, 33.

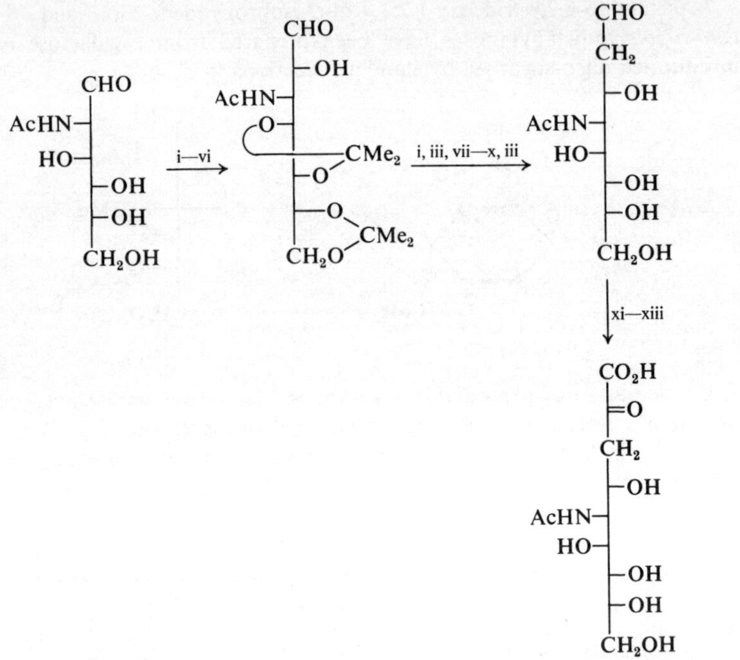

Reagents: i, $MeNO_2$–NaOMe; ii, $Ba(OH)_2$; iii, H^+; iv, MeSH–H^+; v, $(MeO)_2CMe_2$–H^+; vi, HgO–$HgCl_2$; vii, Ac_2O–H^+; viii, $NaHCO_3$; ix, $NaBH_4$; x, NaOH; xi, NaCN; xii, H_2O; xiii, $KClO_3$–V_2O_5

Scheme 8

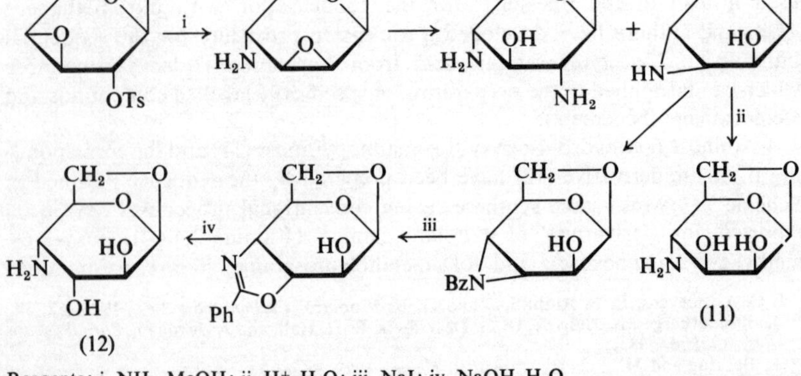

(11)

(12)

Reagents: i, NH_3–MeOH; ii, H^+–H_2O; iii, NaI; iv, NaOH–H_2O

Scheme 9

L-*lyxo*-hexose 4-epimer,[13] and also for methyl 3-amino-2,3,6-trideoxy-β-L-*xylo*-hexopyranoside by two closely related sequences involving the unsaturated azido sugar (13).[14, 15]

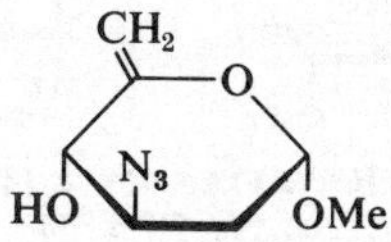

(13)

2-Amino-2-deoxy-D-[1-^{13}C]glucose and the corresponding D-mannose isomer have been prepared from *N*-benzyl-D-arabinosylamine with [^{13}C] cyanide.[16] The glyculose-oxime route has been used to convert methyl 2,3-di-*O*-benzyl-6-deoxy-α-D-galactopyranoside to the corresponding 4-amino-4,6-dideoxy-D-galactoside.[17] Methyl 2-acetamido-2-deoxy-α-D-galactopyranoside has been converted to its 6-deoxy analogue (*N*-acetyl-D-fucosamine) by standard procedures,[18] and this sugar together with its benzyl 3-*O*-benzyl α- and β-glycosides has also been prepared from 2-acetamido-2-deoxy-D-glucose.[19] Epimerization of 2-acetamido-2-deoxy-D-glucose or the corresponding mannose derivative in basic tritium oxide gave a mixture of 2-acetamido-2-deoxy-D-[2-^{3}H]glucose and 2-acetamido-2-deoxy-D-[2-^{3}H]mannose in which the former predominated.[20]

1-*N*-Alkylamino-1-deoxy-D-*threo*-pentuloses have been prepared by Amadori rearrangement of the corresponding xylosylamines derived from long-chain amines (C_8—C_{18}).[21] Further references to Amadori products are given in Chapter 9.

α,α-Trehalose has been converted to a number of 3,3′-diaminodideoxy non-reducing disaccharide analogues containing glucose or mannose units by sequential periodate oxidation and condensation of the bis-dialdehyde product with nitromethane to give dideoxydinitro-disaccharides which were subsequently reduced.[22] Disaccharide glycals have been used to prepare 3-azido-deoxy derivatives, and hence 3-amino-deoxy compounds, by reaction with sodium azide–boron trifluoride, *e.g.* (14) → (15) and (16). Di- and tri-saccharides were also prepared from 3-azido-deoxy glycals by iodine-catalysed addition of suitably protected mono- or di-saccharides.[23] Koenigs–Knorr condensation of 3-azido-3-deoxy-2,4,6-tri-*O*-benzyl-α-D-glucopyranosyl chloride with a di-*O*-isopropylidene derivative of D-pinitol led to the amino-

[13] C. Monneret, C. Conreur, and Q. Khuong-Huu, *Carbohydrate Res.*, 1978, **65**, 35.
[14] T. M. Cheung, D. Horton, R. J. Sorenson, and W. Weckerle, *Carbohydrate Res.*, 1978, **63**, 77.
[15] J. Boivin, M. Pais, and C. Monneret, *Carbohydrate Res.*, 1978, **64**, 271.
[16] T. E. Walker and R. Barker, *Carbohydrate Res.*, 1978, **64**, 266.
[17] C. L. Stevens and S. H. Czernecki, *Carbohydrate Res.*, 1978, **63**, 307.
[18] D. Horton, G. Rodemeyer, and H. Saeki, *Carbohydrate Res.*, 1977, **59**, 607.
[19] D. Horton and H. Saeki, *Carbohydrate Res.*, 1978, **63**, 270.
[20] M. A. E. Sallam, *Carbohydrate Res.*, 1978, **63**, 127.
[21] J. Fernandez-Bolanos and M. P. Gimenez Gracia, *Anales de Quim.*, 1978, **74**, 298 (*Chem. Abs.*, 1978, **89**, 148 491u).
[22] H. H. Baer and A. J. Bell, *Canad. J. Chem.*, 1978, **56**, 2872.
[23] K. Heyns and M. T. Lim, *Tetrahedron Letters*, 1978, 891.

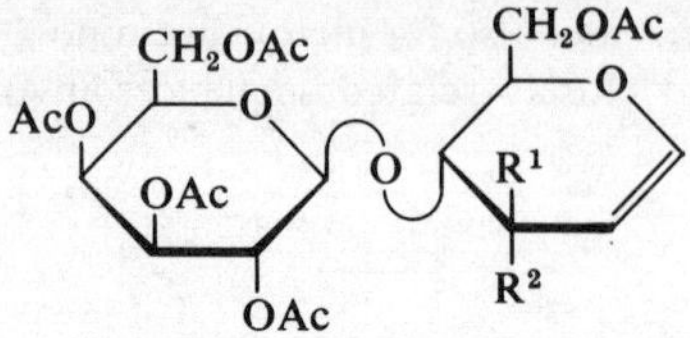

(14) R^1 = OAc, R^2 = H
(15) R^1 = N_3, R^2 = H
(16) R^1 = H, R^2 = N_3

glycoside (17) and its 3′-acetamidino-3′-deoxy- analogue (18). The 3,6-dideoxy-3,6-epimino derivative (19) was also prepared, but none of these compounds

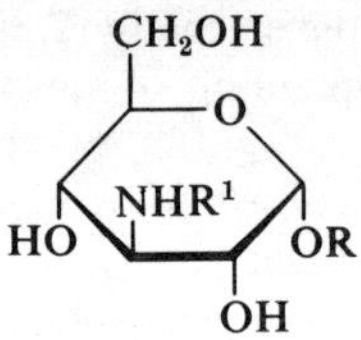

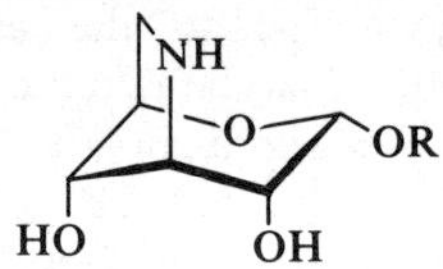

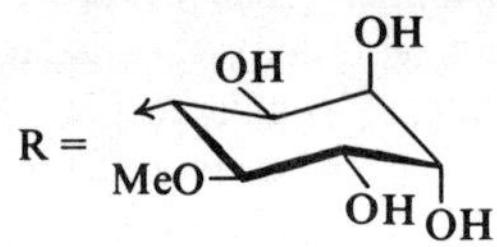

(17) R^1 = H
(18) R^1 = $MeC{=}\overset{+}{N}H_2\overset{-}{Cl}$

(19)

showed any antibiotic activity.[24] Standard procedures have been used to prepare 6-acetamido, 6′-acetamido-, and 6,6′-diacetamido-deoxy derivatives of lactose from lactose or glucose precursors.[25]

The synthesis of D-evernitrose utilizing branched-chain amino-sugar intermediates is mentioned in Chapter 13. Some amine-containing acid compounds are reported in Chapter 15.

2 Reactions

The γ-irradiation of 2-amino-2-deoxy-D-glucose as a free base or as its hydrochloride has been studied; mechanisms were proposed to account for the formation of ammonia and eighteen nitrogen-free products which involve radical-induced deamination initiated through hydrogen abstraction by hydrogen or hydroxy radicals; in the presence of oxygen, peroxy radicals were considered to be the reactive agents.[26] E.s.r. measurements on 2-amino-2-deoxy-D-glucose hydrochloride at 77—337 K confirmed the presence of three different temperature-dependent radicals formed from a primary radical. At low temperature 2-acetamido-2-deoxy-D-glucose gave signals for five separate radicals.[27] The nitrous acid deamination of methylated amino-oligosaccharide glycosides has been explored as a potential method for analysing amino-sugar oligo- and -poly-saccharides, the procedure giving 2,5-anhydro or 2-*C*-formyl-sugars which can be analysed by g.c.–m.s.[28]

[24] J.-Y. Tang and L. Anderson, *Carbohydrate Res.*, 1977, **59**, 319.
[25] I. Ježo, *Chem. Zvesti.*, 1978, **32**, 493.
[26] A. G. W. Bradbury and C. von Sonntag, *Carbohydrate Res.*, 1978, **62**, 223.
[27] M. A. Nadzhafova and V. A. Sharpatyi, *Doklady Akad. Nauk, Azerb. S.S.R.*, 1977, **33**, 41 (*Chem. Abs.*, 1978, **88**, 121 607d).
[28] G. O. Aspinall, E. Przybylski, R. G. S. Ritchie, and C. O. Wong, *Carbohydrate Res.*, 1978, **66**, 225.

2-Amino-2-deoxy-D-[1-^{14}C]glucose with sodium borohydride yielded 2-amino-2-deoxy-D-[1-^{14}C]glucitol.[29]

Treatment of benzyl 2-acetamido-4,6-*O*-benzylidene-2-deoxy-D-glucopyranoside and related muramyl-peptide derivatives with sodium in liquid ammonia removed both the benzyl and benzylidene groups simultaneously, providing a useful alternative to catalytic hydrogenation for glyco-peptides which are catalyst poisons.[30] Benzyl 2-acetamido-2-deoxy-α-D-glucopyranoside has been converted to the muramic acid derivative (20) which was used to

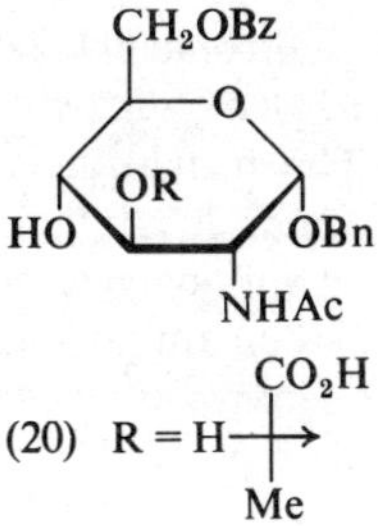

(20)

prepare a β-(1-4)linked disaccharide with 2,3,4,6-tetra-*O*-benzoyl-α-D-glucopyranosyl bromide.[31]

The synthesis of amino-sugar disaccharides incorporating muramic acid is mentioned in Chapter 4, and reference to several other amino-sugar disaccharides and higher saccharides is made in Chapter 3.

The *N*-bromo-, chloro-, and fluoro-acetyl derivatives of 2-amino-2-deoxy-D-glucose and -D-galactose have been prepared and tested for cytotoxic activity; the per-*O*-acetylated *N*-bromoacetyl derivatives in either the glucose or galactose series were most effective against Friend erythroleukemia cells.[32] *N*-nictotinoyl derivatives of 2-amino-2-deoxy-D-glucose and its ethyl α- and β-pyranoside have been prepared.[33]

Di- and tri-*O*-benzylated oxazoline derivatives derived from 2-acetamido-2-deoxy-D-glucose have been prepared *via* allyl and prop-1-enyl glycoside intermediates, and these were then used to prepare corresponding α-1-phosphate derivatives and a 1 → 6 linked analogue of di-*N*-acetyl-chitobiose.[34]

p-Nitrophenyl 2-deoxy-2-(thioacetamido)-β-D-glucopyranoside, the galactose isomer, and their corresponding 1-thio analogues, of interest as inhibitors of 2-acetamido-2-deoxy-β-D-glucosidase, have been prepared from the corresponding acetamido sugars by sulphur–oxygen exchange using phosphorus pentasulphide in pyridine; 1,2-thiazoline derivatives were also obtained.[34a]

Treatment of 2-acetamido- and 2-benzamido-sugars with methyl perchlorate in presence of silver oxide yielded both *N*- and *O*-alkylated products,

[29] M. A. E. Sallam, *Carbohydrate Res.*, 1977, **59**, 612.
[30] M. Zaoral, J. Ježek, R. Straka, and K. Masek, *Coll. Czech. Chem. Comm.*, 1978, **43**, 1797.
[31] A. Hasegawa, K. Bito, and M. Kiso, *Gifu Daigaku Nogakubu Kenkyu Hokoku*, 1977, **40**, 95 (*Chem. Abs.*, 1978, **88**, 191 340k).
[32] T. P. Fondy, S. B. Roberts, A. S. Tsiftsoglou, and A. C. Sartorelli, *J. Medicin. Chem.*, 1978, **21**, 1222.
[33] S. Hirano, Y. Ohe, and N. Matsuda, *Carbohydrate Res.*, 1978, **60**, 404.
[34] C. D. Warren, M. A. E. Shaban, and R. W. Jeanloz, *Carbohydrate Res.*, 1977, **59**, 427.
[34a] G. S. Bedi, R. H. Shah, and O. P. Bahl, *Carbohydrate Res.*, 1978, **62**, 253.

whereas in benzene–chloroform in presence of silver perchlorate–silver oxide only *O*-alkylated products formed; in the absence of silver oxide, benzamido-sugars also yielded oxazoline derivatives. 2-Methyl-oxazolines derived from 2-acetamido sugars rearranged to 1-*O*-acetyl-2-amino-2-deoxy sugars in perchloric acid–aqueous dioxan, whereas corresponding 2-phenyl-oxazolines were unaffected under similar conditions.[35] Partially *O*-methylated *N*-acetyl-*N*-methyl-β-D-neuraminate methyl glycosides have been prepared as references for methylation analysis, and their g.c.–m.s. data detailed.[36] 2-Acetamido-2-deoxy-D-mannono-1,5-lactone has been prepared and shown to act as a competitive inhibitor against *p*-nitrophenyl 2-acetamido-2-deoxy-β-D-glucopyranoside with 2-acetamido-2-deoxy-β-D-glucosidase; the isomeric 1,4-lactone behaved similarly.[37] The formation of benzylidene derivatives involving amino and hydroxy-groups in amino-sugars and the partial de-*N*-formylation of per-*N*-formylated aminoglycoside antibiotics are referred to in Chapter 18, and [13]C n.m.r. studies on amino-sugars are covered in Chapter 20.

Syntheses of *O*-(2-amino-2-deoxy-β-D-glucopyranosyl) derivatives of L-serine and L-threonine have been reported.[38]

3 Diamino-sugars

A full paper on the synthesis of prumycin [4-(D-alanylamino)-2-amino-2,4-dideoxy-L-arabinose] from D-xylose has appeared[39] (see Vol. 10, p. 137); an intermediate in that synthesis, benzyl 2,3-anhydro-4-azido-4-deoxy-β-L-ribopyranoside, was also used to prepare the isomeric diamino sugar, 4-(D-alanylamino)-3-amino-3,4-dideoxy-L-xylose. A closely related synthesis of the D-xylose isomer (21) of prumycin, from the same laboratory, is outlined in Scheme 10.[40] A further report on the synthesis of prumycin from benzyl

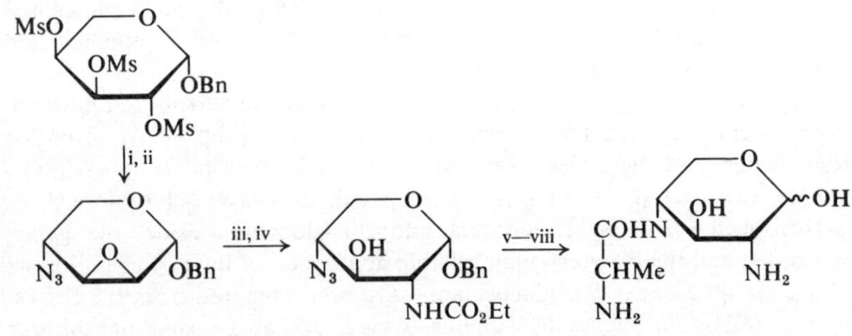

(21)

Reagents: i, N_3^-; ii, NaOMe; iii, NH_3; iv, $N_3CO_2Bu^t$; v, H_2–Pt; vi, $BnOCONHCHMeCO_2C_6H_4NO_2$; vii, HCO_2H; viii, H_2–Pd

Scheme 10

[35] J.-R. Pougny, U. Kraska, and P. Sinaÿ, *Carbohydrate Res.*, 1978, **60**, 383.
[36] H. van Halbeck, J. Haverkamp, J. P. Kamerling, J. F. G. Vliegenthart, C. Verslius, and R. Schauer, *Carbohydrate Res.*, 1978, **60**, 51.
[37] M. Pokorny and N. Pravdic, *Croat. Chem. Acta*, 1977, **49**, 797 (*Chem. Abs.*, 1978, **89**, 110 207t).
[38] M. G. Vafina, A. Kim, and N. V. Molodtsov, *Carbohydrate Res.*, 1978, **64**, 334.
[39] H. Hashimoto, T. Nishide, F. Chiba, and J. Yoshimura, *Carbohydrate Res.*, 1978, **60**, 75.
[40] H. A. Khan, H. Hashimoto, and J. Yoshimura, *Bull. Chem. Soc. Japan*, 1978, **51**, 951.

2-(benzyloxycarbonyl)amino-2-deoxy-β-D-glucofuranoside has been published; mild acid hydrolysis of the di-*N*-benzyloxycarbonyl methyl α-pyranoside of prumycin (22) led to the furanoid prumycin dihydrochloride (23) (Scheme 11), whereas the β-anomer only gave uncharacterized products.[41]

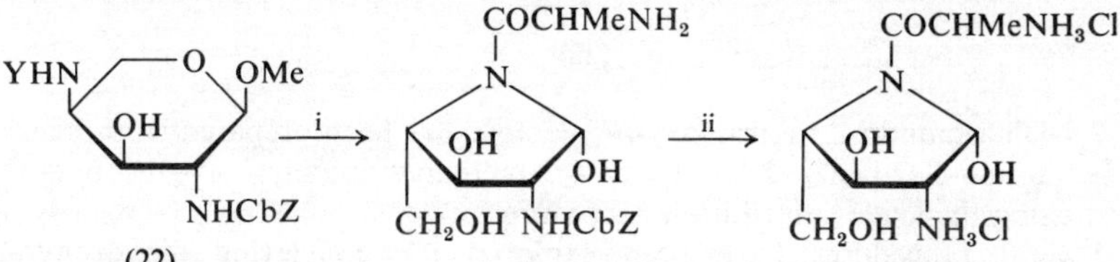

(22)

(23)

Y = COCHMeNHCbZ

Reagents: i, HOAc—2M-HCl, 40 : 1; ii, H_2–Pd

Scheme 11

Benzyl 2,4-diamino-2,3,4,6-tetradeoxy-α and β-*arabino*-hexopyranoside have been prepared by direct azide displacement reactions on the corresponding benzyl 3,6-dideoxy-2,4-di-*O*-methanesulphonyl α and β-*xylo*-hexopyranosides.[42] A new synthesis of the tobrosamine derivative (24) from methyl α-D-manno-pyranoside is outlined in Scheme 12. Direct reaction of (25) with

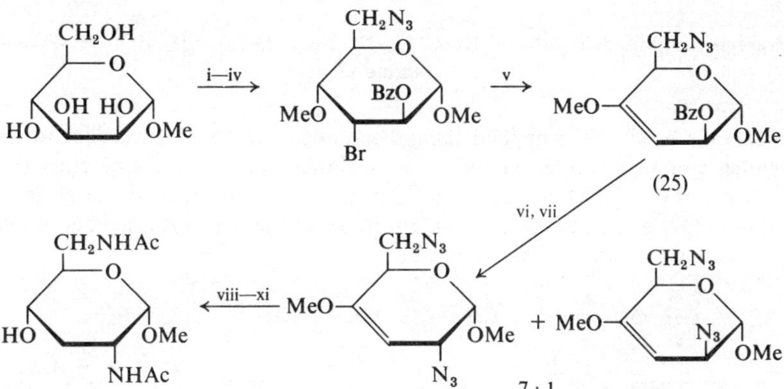

(25)

(24) (26) (27)

Reagents: i, Ph_3P–CCl_4–NaN_3; ii, $PhCH(OMe)_2$–H^+; iii, Me_2SO_4–OH^-; iv, NBS; v, AgF–Py; vi, $Ba(OH_2$; vii, Ph_3P–HN_3–$EtO_2CN{=}NCO_2Et$; viii, H_2–Pd; ix, Ac_2O; x, HCl; xi, $NaBH_4$

Scheme 12

sodium azide gave the opposite ratio of the unsaturated diazides (26) and (27) (1 : 5).[43] An efficient synthesis of 6-*N*-methyl purpurosamine C (28) has been described, which uses a sequence of standard transformations from 2-amino-2-deoxy-D-glucose.[44]

[41] A. Hasegawa, T. Sakurai, and M. Kiso, *Agric. and Biol. Chem.* (*Japan*), 1978, **42**, 153.
[42] J. Yoshimura, K. Sato, H. Hashimoto, and K. Shimizu, *Bull. Chem. Soc. Japan*, 1977, **50**, 3305.
[43] J.-C. Florent, C. Monneret, and Q. Khnong-Huu, *Tetrahedron*, 1978, **34**, 909.
[44] Y. Ohashi, S. Okuno, K. Takeda, and Y. Ito, *Carbohydrate Res.*, 1978, **67**, 503.

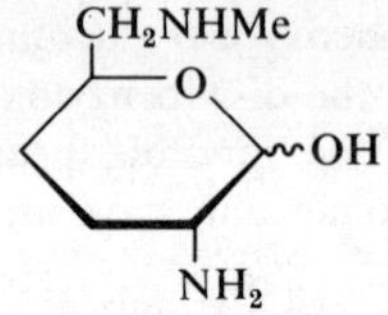

(28)

2,4-Diacetamido-2,4,6-trideoxy-D-galactose has been prepared from benzyl 2-acetamido-3-*O*-benzyl-2,6-dideoxy-4-*O*-methanesulphonyl-α-D-glucopyranoside using the azide substitution procedure.

Several procedures have been explored for converting dihydropyran derivatives to the racemic 3,6-diamino-2,3,4,6-tetradeoxy-DL-*threo*-hexopyranose derivative (29) required as an intermediate for the synthesis of DL-negamycin (see Vol. 6, p. 135); the most efficient entailed the sequence shown in Scheme 13.[46]

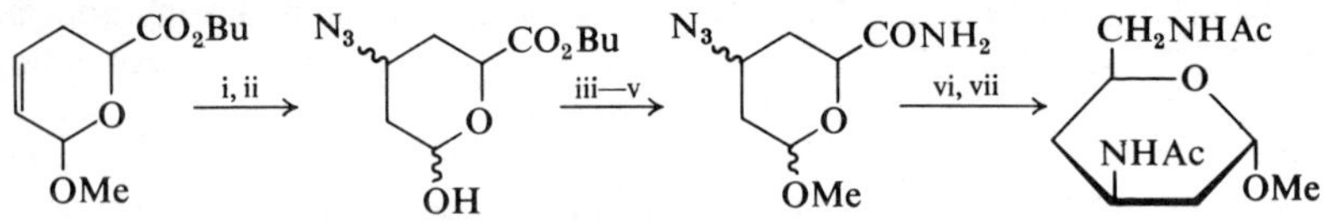

(29)

Reagents: i, H_2SO_4aq; ii, NaN_3; iii, $HC(OMe)_3$–H^+; iv, NaOMe; v, NH_3; vi, LAH; vii, Ac_2O

Scheme 13

David's group have applied their diene approach to disaccharides to the synthesis of kasuganobiosamine (30). Cycloaddition of the dienyl ether (31) with menthyl glyoxylate gave a glycal which was converted to the glycenosidulose (32) which, in turn, led to (30) by sequential hydrazoic acid addition, oximation, reduction, and deblocking.[47]

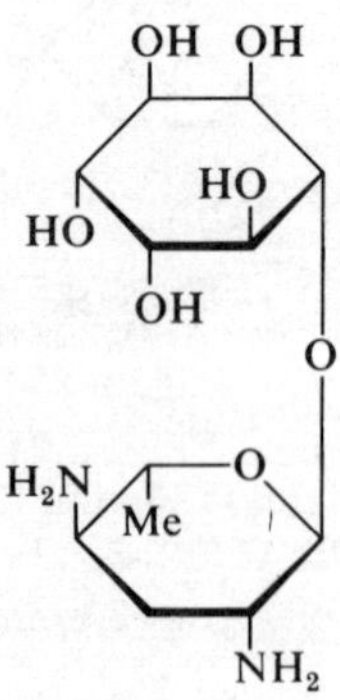

(30)

$CH_2{=}CH{-}CH{=}CHOR$

(31)

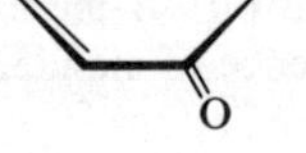

(32)

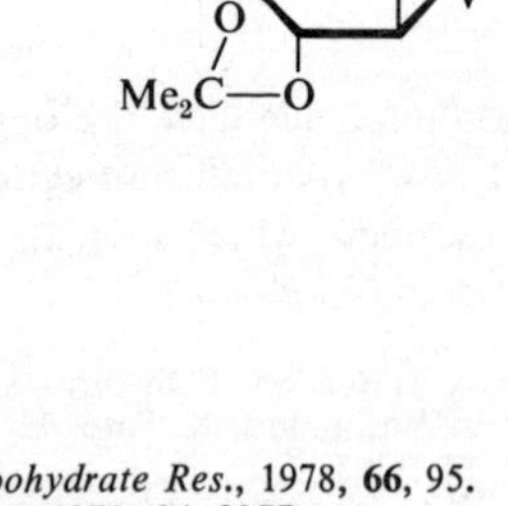

[45] A. Liav, I. Jacobson, M. Sheinblatt, and N. Sharon, *Carbohydrate Res.*, 1978, **66**, 95.
[46] M. Chmielewski, J. Jurczak, and A. Zamojski, *Tetrahedron*, 1978, **34**, 2977.
[47] S. David and A. Lubineau, *Nouv. J. Chim.*, 1977, **1**, 375.

4,6-Diamino-4,6-dideoxy-D-galactose as a free base forms the 1,6-anhydro-furanose analogue (33), and its di-*N*-acetyl derivative exhibited the expected rotamers about the *N*-acyl bond.[48]

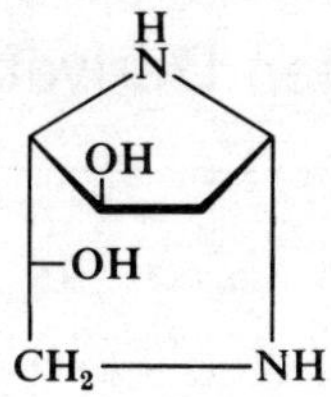

(33)

[48] H. Paulsen, G. Landsky, and M. Kebernick, *Chem. Ber.*, 1978, **111**, 3699.

9

Miscellaneous Nitrogen Derivatives

1 Glycosylamines

A number of *N*-phenylglycosylamine derivatives of 2,3,5-tri-*O*-benzoyl-[1] and -*p*-nitrobenzoyl-α- and -β-D-ribofuranose[2] have been prepared and studied spectroscopically, extending previous work on L-arabinose (see Vol. 10, p. 71). *N*-Arylglycosylamines have also been prepared from *o*-aminoacetophenone[3] and *o*- and *m*-aminoacetophenone oxime.[4]

2-Acetamido-3,4,6-tri-*O*-acetyl-2-deoxy-β-D-glucopyranosylamine has been prepared from the corresponding α-glycosyl chloride with lithium azide, and its dimerization has been studied as a model for its biological activity; spectroscopic evidence suggested that an acyclic immonium ion analogue of the aldehydo sugar was involved in the formation of the β,β and α,β-linked dimers.[5]

Periodate oxidation of *N*-acetyl-*N*-β-D-glycopyranosyl-*p*-bromoaniline or its *p*-anisidine analogue followed by condensation with nitromethane yielded the corresponding *N*-acetyl-(3-deoxy-3-nitro-*N*-β-D-glucopyranosyl) aniline as the major product in both cases.

A series of 1-*N*-acyl derivatives of 2-acetamido-3,4,6-tri-*O*-acetyl-2-deoxy-β-D-glucopyranosylamine have been prepared from the appropriate acid anhydride of benzoic acid, nicotinic acid, and a sequence of fatty acids (C_2—C_{18}).[7]

The α,β-configuration of octa-*O*-acetyl-'α'-di-D-glucosylamine (see Vol. 11, p. 76) and the β,β-configuration of an *N*-acetyl analogue have been established by ^{1}H and ^{13}C n.m.r. spectroscopy.[8] The stability of *N*-acyl-glycosylamines in methanolic ammonia has been investigated; β-anomers and mannose derivatives were less stable than other pyranose forms, and although the *N*-acetyl and *N*-toluene-*p*-sulphonyl derivatives of 1-amino-1-deoxy-D-glucitol were considerably decomposed in alkali, their *N,N*-bis-substituted analogues were relatively stable.[9]

1 R. Walczyna and J. Sokolowski, *Polish J. Chem.*, 1978, **52**, 1933.
2 R. Walczyna and J. Sokolowski, *Polish J. Chem.*, 1978, **52**, 2139.
3 J. Sykulski, *Acta Pol. Pharm.*, 1977, **34**, 471 (*Chem. Abs.*, 1978, **88**, 191 336p).
4 J. Sykulski and T. Czyzewska, *Acta Pol. Pharm.*, 1977, **34**, 595 (*Chem. Abs.*, 1978, **89**, 129 877h).
5 B. Paul and W. Korytnyk, *Carbohydrate Res.*, 1978, **67**, 457.
6 Z. Fialkiewiczowa, *Polish J. Chem.*, 1978, **52**, 1425.
7 S. Hirano, H. Iwaki, and Y. Kondo, *Carbohydrate Res.*, 1978, **65**, 307.
8 G. Tóth, I. Pinter, J. Kovács, A. Messmer, and W. Dietrich, *J. Carbohydrates, Nucleosides, Nucleotides*, 1978, **5**, 225.
9 S. Delpy and A. S. Cerezo, *Anales Asoc. Quím. Argentina*, 1976, **64**, 277.

A number of 1-pentopyranosyl-isatins (1) have been prepared either from corresponding *O*-acylated-*N*-glycosylphenylamines by the Stolle reaction or by chromic acid oxidation of *O*-acylated-1-pentopyranosyl-indoles.[10]

The reaction of D-glucuronic acid with proteins has been studied using an L-lysine-containing tripeptide as a model system; under physiological conditions, the dimeric glycopeptide (2) was obtained as the major product.[11]

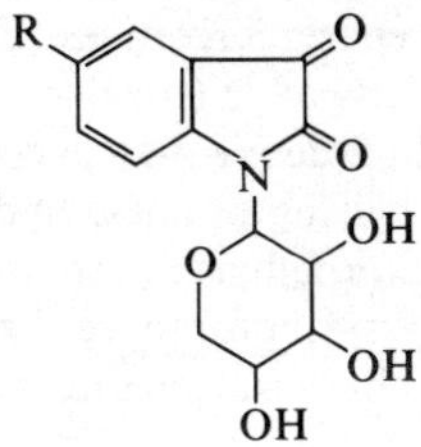

(1) R = H, Me, NO_2

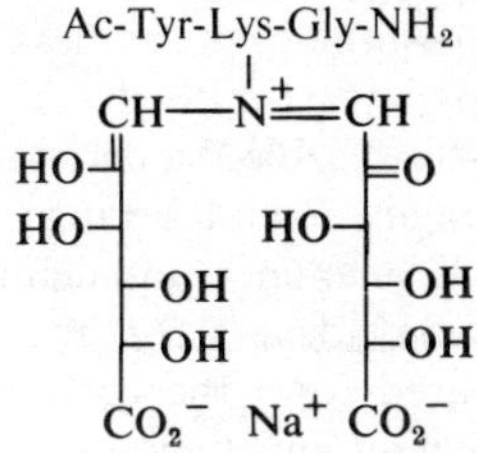

(2)

The influence of divalent metal ions on the Maillard reaction between glucose and glycine has been studied; copper ions strongly accelerate colour formation at pH 3, whereas calcium, zinc, or manganese ions showed no effect or were weakly inhibitory at pH 6.2.[12] In another study, the effect of radical initiators or inhibitors on the formation of 5-hydroxymethylfuraldehyde was investigated, and a free-radical mechanism in deaerated aqueous solution discussed.[13] The reaction of glucose with methionine and tryptophan has also been studied.[14] The chemical structures and routes of formation of compounds derived from sugar–amine reactions which contribute to food and tobacco flavours have also been investigated; about 70 products were identified in pyrolyses either of hexoses with amino-acids or cyclic amines, or of the related Amadori compounds, and some α-pyrrolidinone and 5,6,7,8-tetrahydro-indolizine derivatives were detected as new components of such reactions.[15]

The reaction of *N*-*p*-tolyl-D-glucosylamine with *p*-toluidine yielded 3-deoxy-D-glucosone and reductone derivatives, and D-arabinose, D-erythrose, and D-glyceraldehyde together with their corresponding glyconic acids were also detected in the browning of D-glucose with *p*-toluidine.[16]

Glycosyl sulphamides derived from D-glucose and D-ribose have been prepared conventionally; further reaction with diketen led to corresponding glycosyl *N*-sulphamoyl-pyrone carboxamide derivatives.[17]

[10] I. V. Yartseva, L. V. Ektova, V. I. Sakharova, Ya. V. Dobrynin, N. P. Yavovskaya, T. G. Nikolaeva, Z. P. Sof'ina, and M. N. Preobrazhenskaya, *Zhur. org. Khim.*, 1977, **13**, 1743 (*Chem. Abs.*, 1978, **88**, 7227e).

[11] Y. Takeda, Y. Kyogoku, and M. Ishidate, *Carbohydrate Res.*, 1977, **59**, 363.

[12] S.-H. Wang, F. O. Bobbio, and P. A. Bobbio, *Anais Acad. brasil. Cienc.*, 1978, **50**, 55 (*Chem. Abs.*, 1978, **89**, 106 184r).

[13] P. Barbetti, *Ind. Ailment.* (*Pinerolo, Italy*), 1977, **16**, 151 (*Chem. Abs.*, 1978, **88**, 61 276h).

[14] F. Orsi and E. Dworschak, *Acta Aliment. Acad. Sci. Hung.*, 1978, **7**, 41 (*Chem. Abs.*, 1978, **88**, 150 942q).

[15] H. Shigematsu, *Nippon Sembai Kosha Chuo Kenkyusho Kenkyu Hokoku*, 1976, **118**, 119 (*Chem. Abs.*, 1978, **88**, 7217b).

[16] N. Kinae, Y. Nakamura, M. Abe, and T. Ozawa, *Yakugaku Zasshi*, 1978, **98**, 153 (*Chem. Abs.*, 1978, **89**, 129 839x).

[17] P. Goya and M. Stud, *J. Heterocyclic Chem.*, 1978, **15**, 477.

2,3,4,6-Tetra-*O*-acetyl-α-, and -β-D-gluco- and -D-galactopyranosyl isocyanides, have been prepared from the corresponding α-glycosyl bromide with silver cyanide.[18]

The anomeric 2,3,4,6-tetra-*O*-benzoyl-D-glucopyranosyl isocyanides, prepared from the corresponding β-D-glucosylamine by formylation and dehydration with phosphorus oxychloride, have been polymerized with nickel chloride to give two helical polymers with opposite screw orientations.[19]

Glycosyl isothiocyanates have been used to prepare derivatives of D- and L-arabinopyranosyl-thiosemicarbazide,[20] derivatives of 2-(D-galactopyranosylamino)-oxadiazole,[21] a derivative of 2-acetamido-2-deoxy-β-D-glucopyranosyl-L-asparagine,[22] and a range of glycosyl ureas using amines, hydrazines, amino-acids, enamines, and diamines.[23] 1-[3-(2-Chloroethyl)-3-nitrosoureido]-2,3-*O*-isopropylidene-5-*O*-*p*-nitrobenzoyl-α and β-D-ribofuranose have also been prepared from the corresponding ribofuranosyl isocyanate, and both anomers exhibit anti-tumour activity.[24]

A number of *N*-(2,3,4,6-tetra-*O*-acetyl-β-D-glycopyranosyl)-5-aralkylidene-rhodanines, which were prepared conventionally from the appropriate rhodanine and glucosyl chloride, have been shown to possess antibacterial activity.[25]

The anomeric *N*-xylofuranosyl-pyridinium salts derived from nicotinamide have been prepared as nucleoside analogues, and equilibration of their dihydro derivatives was shown to favour the β-form.[26]

^{1}H-N.m.r. evidence indicates that α-D-mannopyranosylisocyanuric acid (3) and its derivatives adopt a skew conformation, whereas *X*-ray data suggest a $^4C_1$ conformation for its *N*,*N*-dimethyl derivative.[27]

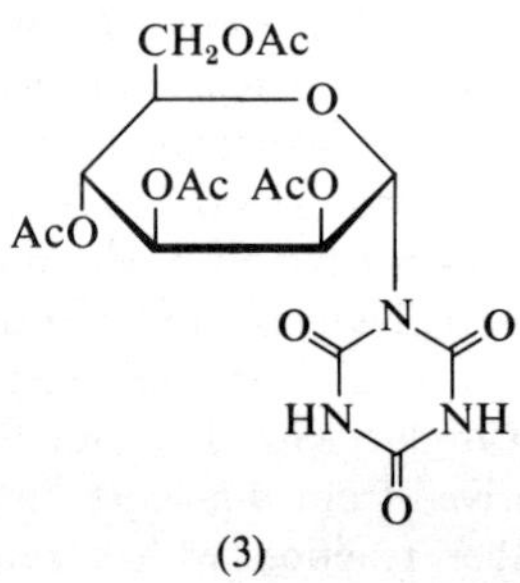

(3)

18 M. Martin-Lomas and M. E. Chacon-Fuertes, *Carbohydrate Res.*, 1977, **59**, 604.
19 R. J. M. Nolte, J. A. J. van Zomeren, and J. W. Zwikker, *J. Org. Chem.*, 1978, **43**, 1972.
20 E. Lada and W. Wieniawski, *Acta Pol. Pharm.*, 1977, **34**, 29 (*Chem. Abs.*, 1978, **88**, 51 111m).
21 C. Gmernicka-Haftek and W. Wieniawski, *Acta Pol. Pharm.*, 1977, **34**, 23 (*Chem. Abs.*, 1978, **88**, 51 124t).
22 S. E. Zurabyan, R. G. Macharadze, and A. Ya. Khorlin, *Bioorg. Khim.*, 1978, **4**, 1135 (*Chem. Abs.*, 1978, **89**, 180 263g).
23 H. Ogura and H. Takahashi, *Heterocycles*, 1977, **8**, 125.
24 J. L. Montero, M. Rodriguez, and J. L. Imbach, *European J. Medicin. Chem.-Chim. Ther.*, 1977, **12**, 408 (*Chem. Abs.*, 1978, **88**, 51 099p).
25 W. O. Foye and P. Tovivich, *J. Pharm. Sci.*, 1977, **66**, 1607 (*Chem. Abs.*, 1978, **88**, 121 604a).
26 N. J. Oppenheimer, *J. Carbohydrates, Nucleosides, Nucleotides*, 1978, **5**, 251.
27 W. Depmeier, H. von Voithenberg, and J. C. Jochims, *Chem. Ber.*, 1978, **111**, 2010.

2 Azido-sugars

The photolysis of a series of glycosylazides either yielded the next lower aldose or gave an intermediate which reverted to the starting azide on storage; a mechanism was proposed, illustrated for β-D-glucopyranosylazide which gave D-arabinose (Scheme 1); 2,3,4,6-tetra-*O*-acetyl-β-D-glucopyranosyl azide,

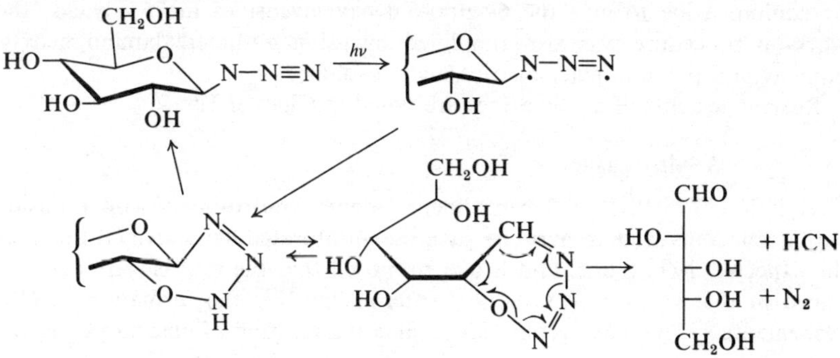

Scheme 1

with the *C*-2 hydroxy-group blocked, gave the corresponding gluconic acid instead, probably *via* the gluconimine (4).[28] 1,2-Alkylidene derivatives of 5-azido-5-deoxy-L-idofuranurono-6,3-lactone (5) together with the D-gluco analogues (6) have been prepared from the corresponding glucuronolactone (7) using the triphenylphosphine-azodicarboxylate procedure with hydrazoic acid.[29] Similarly methyl 2,6-di-*O*-t-butyldimethylsilyl-α-D-glucopyranoside with this reagent gave the corresponding 4-azido-4-deoxy-D-galactoside (8) whereas the β-anomer gave instead the 3-azido-3-deoxy-D-allopyranoside analogue (9).[30] A better yield of the 1,2-*O*-isopropylidene derivative (5) was

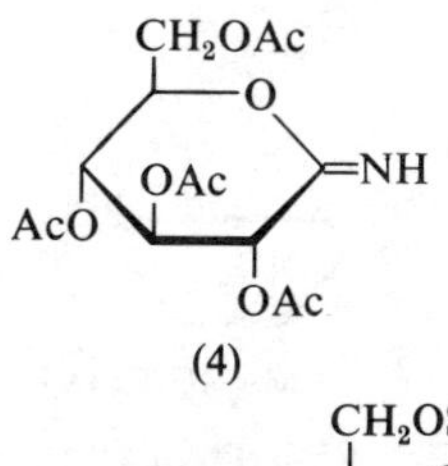

(4)

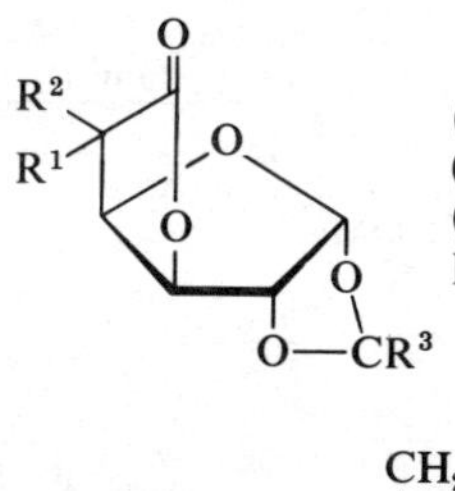

(5) R^1 = H, R^2 = N_3
(6) R^1 = N_3, R^2 = H
(7) R^1 = OH, R^2 = H
R^3 = Me_2 or $(CH_2)_5$

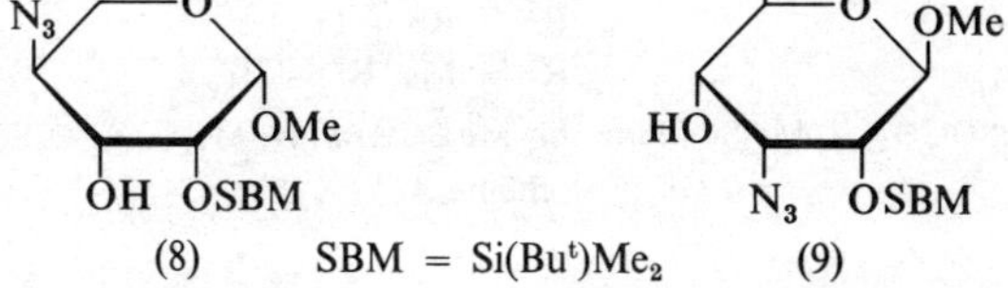

(8) SBM = Si(Bu^t)Me_2 (9)

[28] W. A. Szarek, O. Achmatowicz, J. Plenkiewicz, and B. K. Radatus, *Tetrahedron*, 1978, **34**, 1427.
[29] J. Schweng and E. Zbiral, *Tetrahedron Letters*, 1978, 119.
[30] H. M. Brandstetter and E. Zbiral, *Helv. Chim. Acta*, 1978, **61**, 1832.

obtained from the 5-*O*-chlorosulphate of the glucuronolactone (7) by treatment with potassium azide in the presence of 18-crown-6-ether.[31] (See also Chapter 6 for the synthesis of sugar azido sulphates by this procedure.)

A one-flask procedure for the synthesis of methyl 6-azido-6-deoxy-α-D-hexopyranosides has been described, in which the corresponding glycoside is treated with triphenylphosphine and *N*-bromosuccinimide in DMF followed by sodium azide to give the 6-azido-6-deoxy glycosides in 75% yield; the work-up procedure was also simplified by using *p*-(dimethylaminophenyl)-diphenylphosphine in place of triphenylphosphine.[32]

Rearrangements of allylic azides are noted in Chapter 12.

3 Nitro-sugars

Acetylation of methyl 4,6-*O*-benzylidene-2-deoxy-2-nitro-β-D-glucopyranoside with acetic anhydride in pyridine gave the unsaturated derivative (10) besides the expected 3-*O*-acetate, and it was proposed that the nitro-sugar (11) produced in the work-up was the key intermediate for the formation of (10) (Scheme 2).[33] This same group has studied the reaction of unsaturated nitro-sugars with sulphur ylides. The β-anomer (12) gave the cyclopropyl or branched-chain sugar derivatives shown in Scheme 3.[34] The corresponding α-anomer gave analogous products with the methylene and ethoxycarbonyl-

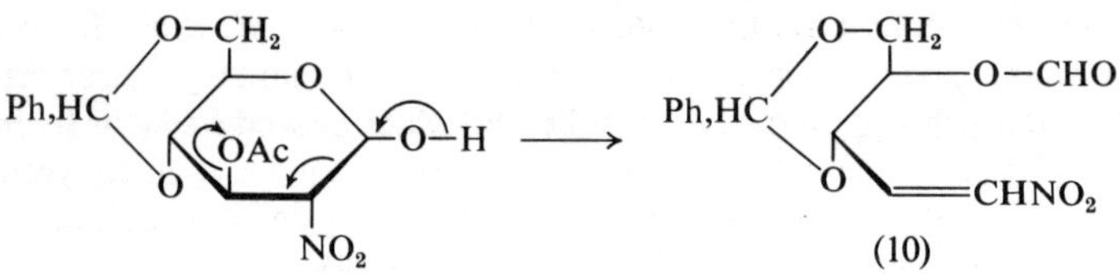

(11) (10)

Scheme 2

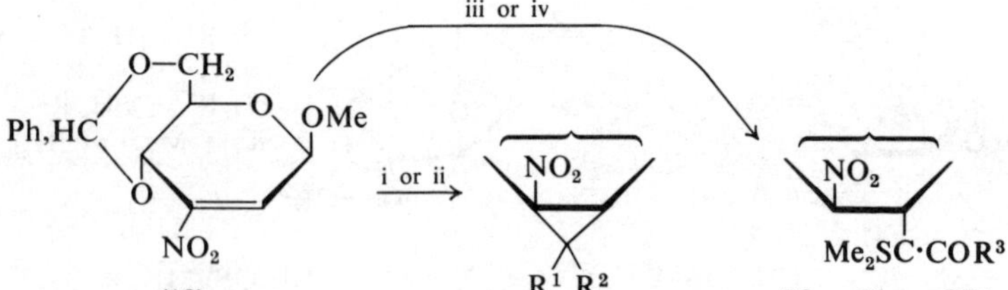

(12)

$R^1 = R^2 = H$

$R^1 = Me, R^2 = Bz$

$R^3 = Ph$ or OEt

Reagents: i, $Me\overset{+}{S}O\overset{-}{C}H_2$; ii, $Me_2\overset{+}{S}\overset{-}{C}MeBz$; iii, $Me_2\overset{+}{S}\overset{-}{C}HBz$; iv, $Me_2\overset{+}{S}\overset{-}{C}HCO_2Et$

Scheme 3

[31] N. T. Naidoo and H. Parolis, *Carbohydrate Res.*, 1978, **62**, C5.
[32] S. Hanessian, D. Ducharme, R. Masse, and M. L. Capman, *Carbohydrate Res.*, 1978, **63**, 263.
[33] T. Sakakibara and R. Sudoh, *Bull. Chem. Soc. Japan*, 1978, **51**, 1263.
[34] T. Sakakibara, R. Sudoh, and T. Nakagawa, *Bull. Chem. Soc. Japan*, 1978, **51**, 1189.

methylene reagents, but in addition the heterocyclic derivatives (13) were also obtained.[35] The formation of anhydro-deoxy-nitroalditols is mentioned in Chapter 17. The reduction of α-nitro-epoxide to deoxy compounds is mentioned in Chapter 11.

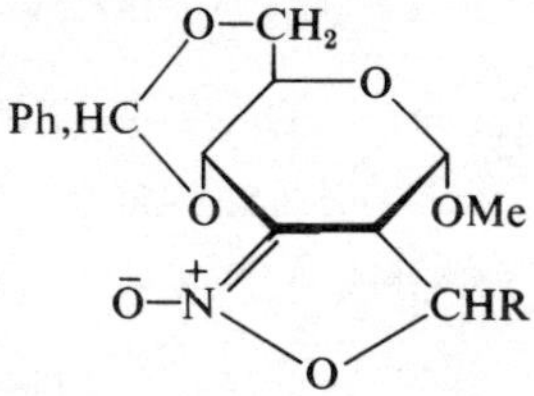

(13) R = H or CO_2Et

4 Phenyl Hydrazones, Osazones, and Formazans

Treatment of brominated 3,4-di-*O*-acetyl-D-xylal with *p*-nitrophenylhydrazine gave a mixture of the 3,4-di-*O*-acetyl-2,5-anhydro-D-xylose and -D-xylose *p*-nitrophenylhydrazones; both these compounds and their deacetylated derivatives exhibited a preference for the $^3T_4$ conformation.[37] Acetylation of L-rhamnose benzoylhydrazone with acetic anhydride and pyridine has been shown to give the α-pyranose derivative (14) rather than the acyclic hydrazone (15) previously claimed (see Vol. 10, p. 74). With acetic anhydride and zinc chloride, compound (14) gave 1,2,3,4-tetra-*O*-acetyl-α-L-rhamnose, whereas (15) gave the oxadiazoline (16).[37]

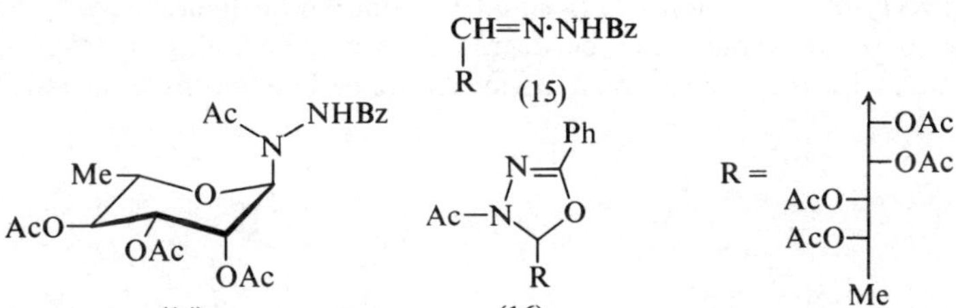

(14) (16)

The reduction of the bis-tolyl-*p*-sulphonylhydrazone derivative of periodate-oxidized methyl 4,6-*O*-benzylidene-α-D-glucopyranoside with sodium borohydride yielded an 2,3-epimino-mannose derivative, whereas sodium cyanoborohydride yielded either a mixture of 2-deoxyhexose derivatives or, in the presence of an acid resin, a 2,3-bis-tolyl-*p*-sulphonylhydrazine allose derivative (Scheme 4).[38]

El Ashry's group have continued studies on the chemistry of hydrazone derivatives of L-ascorbic acid and its D-*erythro* epimer. 2-Mono and 2,3-bis-hydrazones of dehydro-L-ascorbic acid can be prepared, which may be cyclized

[35] T. Sakakibara and R. Sudoh, *Bull. Chem. Soc. Japan*, 1978, **51**, 1193.
[36] A. Rockenbauer, J. Hemela, G. Toth, and A. Gerecs, *Magyar Kóm. Folyóirat*, 1977, **83**, 442 (*Chem. Abs.*, 1978, **88**, 170 402s).
[37] L. Somogyi, *Carbohydrate Res.*, 1978, **64**, 289.
[38] S. Honda, K. Kakehi, and S. Oguri, *Carbohydrate Res.*, 1978, **64**, 101.

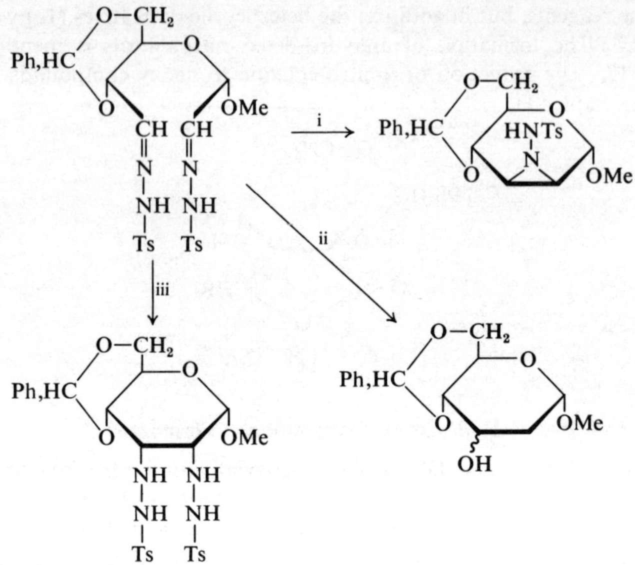

Reagents: i, $NaBH_4$–EtOH; ii, $NaBH_3CN$–EtOH–H_2O; iii, $NaBH_3CN$–IR 120 (H^+)

Scheme 4

to pyrazoledione or pyrazoline-pyrazoledione derivatives (see Vol. 11, p. 88). The glyceryl-pyrazolediones (17) obtained from the bis-hydrazones by alkali-catalysed rearrangement can be degraded to corresponding hydroxymethyl derivatives by the conventional periodate oxidation–borohydride reduction

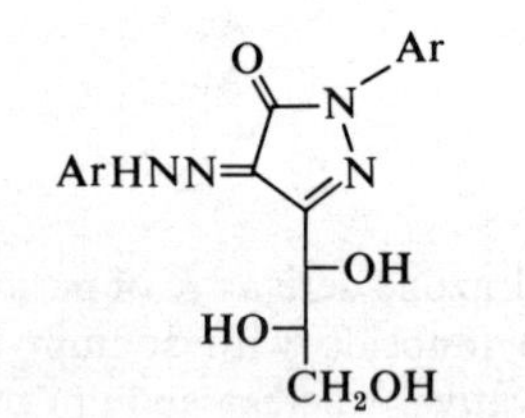

(17) Ar = Ph or p-C_6H_4R

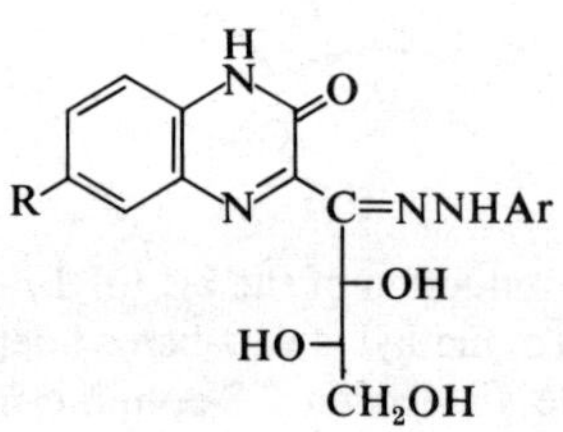

(18) R = H or Cl

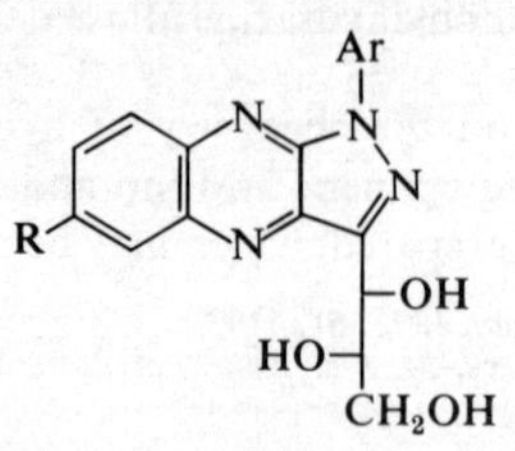

(19)

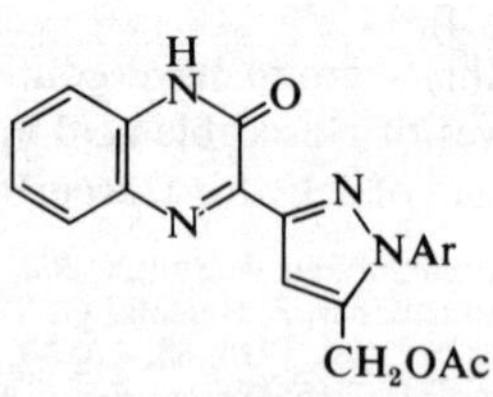

(20)

sequence.[39, 40] Reaction of L-ascorbic acid with *o*-phenylenediamine and arylhydrazine gave substituted quinoxalin-2-one derivatives (18), which with alkali rearranged to flavazole compounds (19) or in boiling acetic anhydride gave the pyrazole-quinoxalinone derivatives (20).[41] The glyceryl side-chain of (18) was also degraded to give corresponding formyl derivatives by periodate oxidation.[42] Related 1-methylquinoxalin-2-one derivatives analogous to (18) and (20) were also prepared.[43] 4-Chlorophenylenediamine led to 6-chloro-quinoxalin-2-one derivatives by analogous reactions.[44] Dehydro-D-erythrobic acid likewise gave the D-*erythro* epimers of (18) and (19), and the same pyrazole derivative (20).[45] The use of *p*-(*N*-2-pyrimidinylaminosulphonyl)phenyl-hydrazines with L-ascorbic acid and its D-*erythro* epimer leading to corresponding pyrazoledione and flavazole derivatives related to sulphanilamide drugs was also reported.[46]

Arylhydrazone and bis-arylhydrazone derivatives of lactose have been prepared, and their periodate oxidation studied; the formation of an osone derivative by hydrazinolysis of the bisbenzoylhydrazone is noted in Chapter 14.[47]

Zemplen deacetylation of D-galactose diphenyl formazan required more than catalytic amounts of sodium methoxide, and led to 2,5-anhydro-D-talose diphenyl formazan; a mechanism was proposed involving neighbouring group participation of the hydrazone group leading to the intermediate (21) which undergoes S_N2 substitution at C-2 to give the 2,5-anhydro ring.[48]

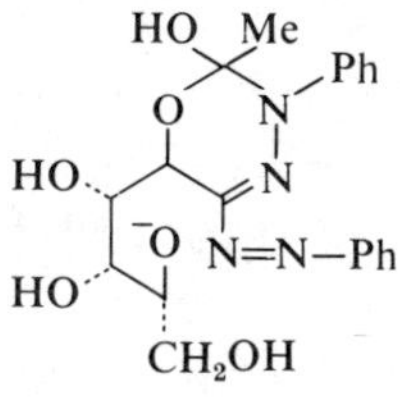

(21)

5 Heterocyclic Derivatives

In a further study of dipolar addition reactions of carbohydrate acetylenic-ketone derivatives leading to heterocyclic compounds (see Vol. 11, p. 91),

[39] E. S. H. El Ashry, I. E. El Kholy, and Y. El Kilany, *Carbohydrate Res.*, 1977, **59**, 417.
[40] E. S. H. El Ashry, Y. El Kilany, and F. Singab, *Carbohydrate Res.*, 1978, **67**, 415.
[41] E. S. H. El Ashry, I. E. El Kholy, and Y. El Kilany, *Carbohydrate Res.*, 1978, **60**, 303.
[42] E. S. H. El Ashry, I. E. El Kholy, and Y. El Kilany, *Carbohydrate Res.*, 1978, **60**, 396.
[43] E. S. H. El Ashry, I. E. Kholy, and Y. El Kilany, *Carbohydrate Res.*, 1978, **64**, 81.
[44] E. S. H. El Ashry, M. M. A. Rahman, M. A. Nassr, and A. Amer, *Carbohydrate Res.*, 1978, **67**, 403.
[45] E. S. H. El Ashry, M. M. A. Rahman, N. Rashed, and A. Amer, *Carbohydrate Res.*, 1978, **67**, 423.
[46] R. Soliman, E. S. H. El Ashry, I. E. Kholy, and Y. El Kilany, *Carbohydrate Res.*, 1978, **67**, 179.
[47] M. M. A. Rahman, E. S. H. El Ashry, and N. Rashed, *Carbohydrate Res.*, 1978, **64**, 89.
[48] V. Zsoldos, A. Messmer, I. Pintér, and A. Neszmélyi, *Carbohydrate Res.*, 1978, **62**, 105.

González's group have prepared a series of isoxazoline derivatives (22) using hydroxylamine with the corresponding acetylenic keto-sugar (23).[49]

Cyclocondensation of 1-alkylamino-1-deoxy-D-*threo*-pentuloses with ethyl acetoacetate led to *N*-alkyl-pyrrole derivatives (24).[50] Aldonic acid lactones react with ethyl isocyanoacetate in presence of DBN to yield oxazole derivatives, *e.g.* 2,3:5,6-di-*O*-isopropylidene-D-mannono-1,4-lactone yields the oxazole (25),[51] whereas with potassium hydride as base octenonate derivatives

$R-C\equiv C-COR'$

(23)

R =

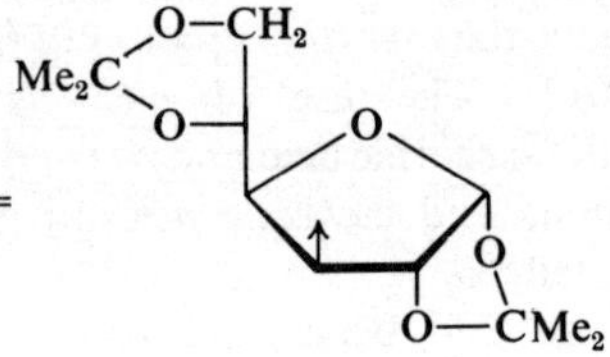

or

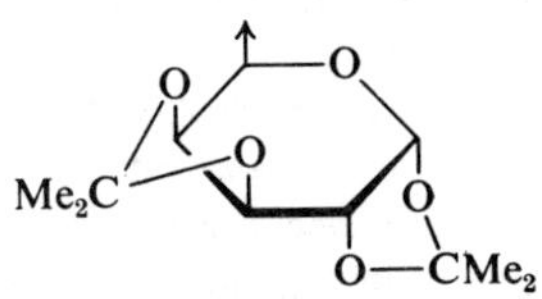

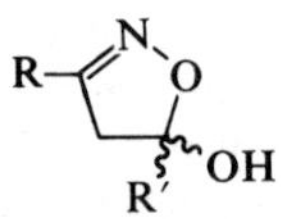

(22)

$R' = H, Bu, CHEt_2, Ph,$ or $p\text{-}C_6H_4Ph$

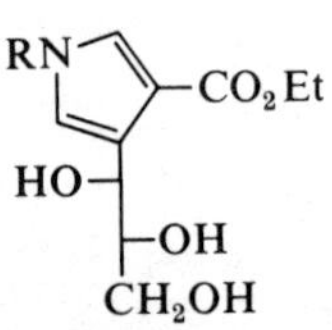

(24)

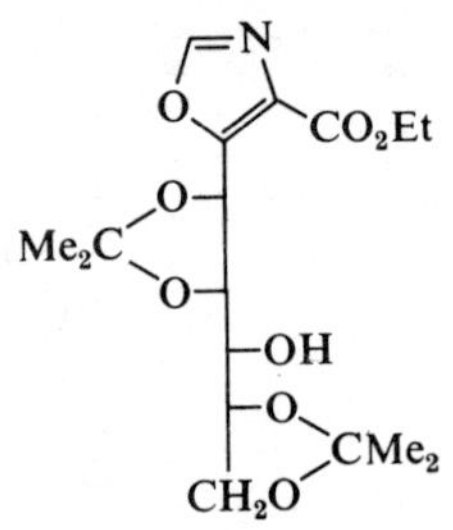

(25)

were previously reported (see Vol. 11, p. 71); 2,3,4,6-tetra-*O*-benzyl-D-glucono-1,5-lactone has now been found to give an oxazole derivative with potassium hydride as well. Both products were considered to arise from a spiro-oxazolide anion intermediate (Scheme 5).

The reaction of L-rhamnose with aqueous ammonia leading to heterocyclic compounds has been studied; sixty-five components of the complex product mixture were identified, which were principally pyrazine, pyrrole, and imidazole derivatives, and possible mechanisms for their formation were discussed.[52]

[49] A. Llamas, A. Gonzalez, and E. Martinez, *Carbohydrate Res.*, 1978, **67**, 515.

[50] F. G. Gonzalez, J. Fernandez-Bolanos, S. M. Guerra, and G. M. Jimenez de la Plata, *Anales de Quím.*, 1978, **74**, 315 (*Chem. Abs.*, 1978, **89**, 180 244b).

[51] R. H. Hall, K. Bischofberger, S. J. Eitelman, and A. Jordaan, *J.C.S. Perkin I*, 1977, 2236.

[52] T. Shibamoto and R. A. Bernhard, *J. Agric. Food Chem.*, 1978, **26**, 183 (*Chem. Abs.*, 1978, **88**, 74 495d).

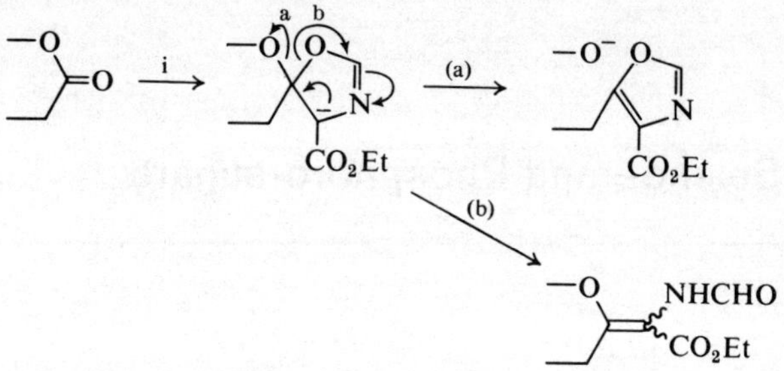

Reagents: i, $OCNCH_2CO_2Et$

Scheme 5

The elongation of sugar chains *via* heterocyclic derivatives is mentioned in Chapter 8. Thiazoline derivatives are referred to in Chapter 10.

10
Thio-, Seleno-, and Phosphoro-sugars

1-Thiosugars and thioglycosides are dealt with in Chapter 3.

A new method for converting amino-sugars to thioacetamido-sugars *via* a thiazoline intermediate has appeared, and the reaction has been applied to the 2-amino-2-deoxy-3-*O*-tosyl-altroside (1), enabling synthesis of the 2-amino-2-deoxy-3-thio-mannopyranose derivative (2) (Scheme 1).[1]

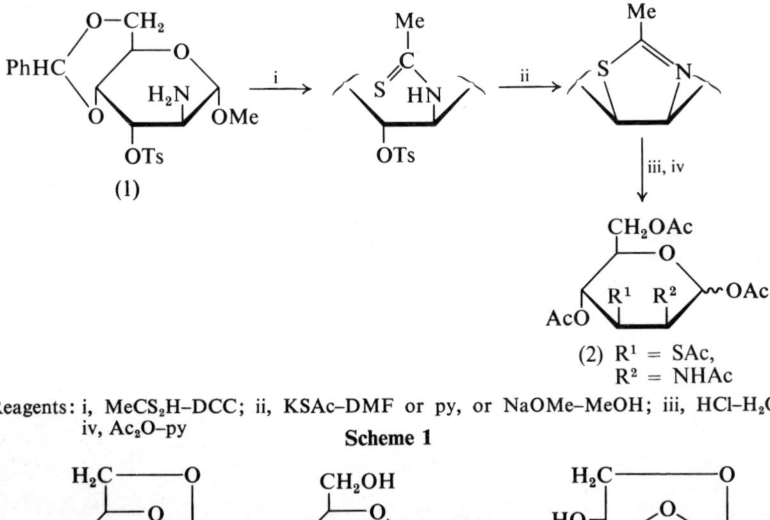

(2) R^1 = SAc, R^2 = NHAc

Reagents: i, $MeCS_2H$–DCC; ii, KSAc–DMF or py, or NaOMe–MeOH; iii, HCl–H_2O; iv, Ac_2O–py

Scheme 1

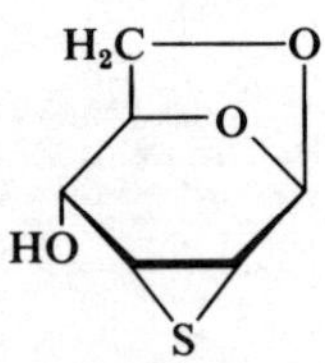

(3)

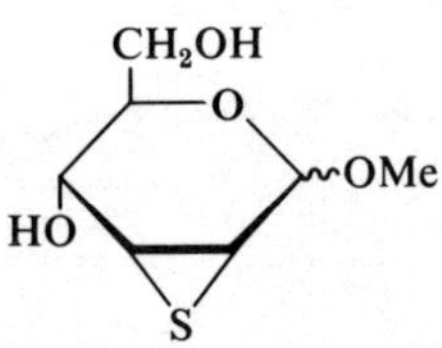

(4)

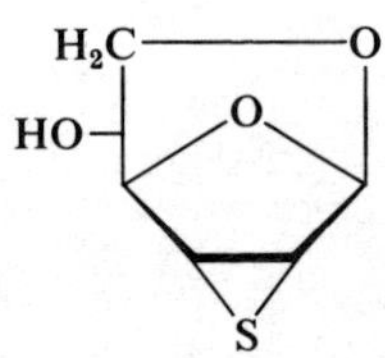

(5)

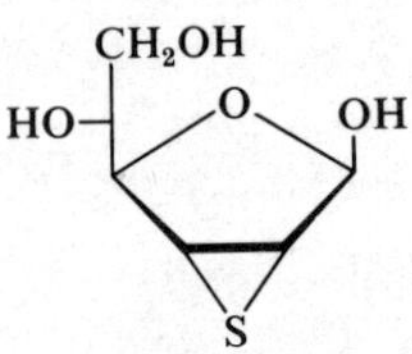

(6)

[1] P. Herczegh and R. Bognar, *Acta Chim. Acad. Sci. Hung.*, 1978, **98**, 321.

The mass-spectrometry of the 2,5-epithio-D-allose derivatives (3)—(6) has been studied.[2] (See Chapter 21 for detailed analysis.)

In a preliminary communication the conversion of oxo-sugars to thiosugars *via* thiones has been reported.[3] Thus the 3-thio-allose derivative (8) was synthesized from (7) as shown in Scheme 2.

The 1-*O*-acetyl-4-thio-tetrofuranose derivative (9) reacts with alcohols and thiols in the presence of toluene-*p*-sulphonic acid to give *O*- and *S*-glycosides. The former were produced with high stereoselectivity giving the β-anomer whereas the latter gave approximately equal quantities of α- and β-thioglycosides. The thioglycosyl bromide (10) was prepared from (9) and found to be less reactive in glycoside synthesis than (9) itself (Scheme 3).[4] A synthesis

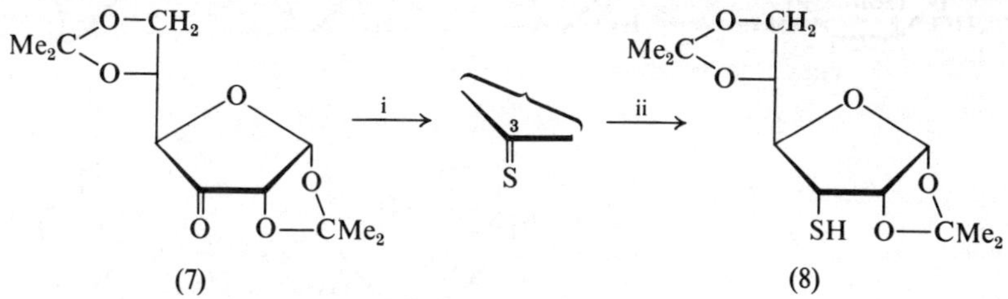

Reagents: i, H_2S–DMF–Et_3N; ii, Na–NH_3(liq.)

Scheme 2

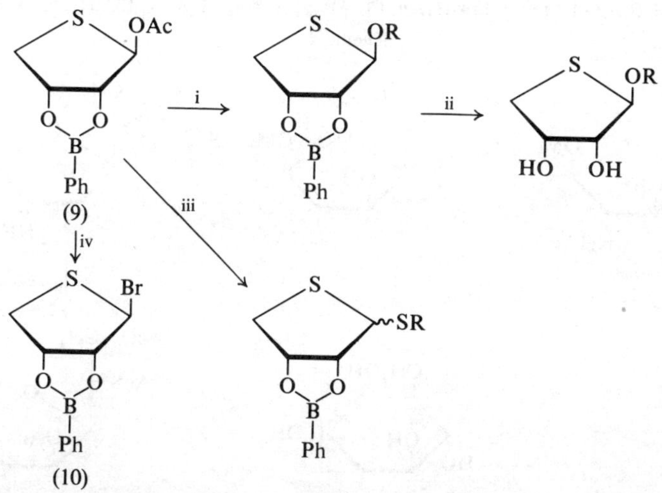

Reagents: i, ROH–$CHCl_3$–TsOH; ii, $HOCH_2CH_2OH$; iii, RSH–$CHCl_3$–TsOH; iv, HBr–HOAc

Scheme 3

[2] T. Yamaguchi and M. Kojima, *Org. Mass Spec.*, 1978, **13**, 113.
[3] M. M. A. Abdel-Rahman, *Acta Chim. Acad. Sci. Hung.*, 1977, **93**, 425 (*Chem. Abs.*, 1978, **88**, 121 568s).
[4] J. E. McCormick and R. S. McElhinney, *J.C.S. Perkin I*, 1978, 64.

of the 6-deoxy-4-thio-D-galactose derivative (12) *via* the thioketone (11) as shown in Scheme 4 has been published.[5]

5-Thio-D-glucose has been demonstrated to be a substrate for lactose synthetase which converts it into 5-thio-lactose.[6] The reaction of *vic*-epoxides with thiourea to yield *vic*-epithio-compounds has been used[7] to synthesize 2-acetamido-2-deoxy-5-thio-D-glucopyranose as portrayed in Scheme 5. Two further papers on the reaction of epimines with sulphur-containing compounds have appeared[8, 9] (see Vol. 11, p. 97). Reactions leading to

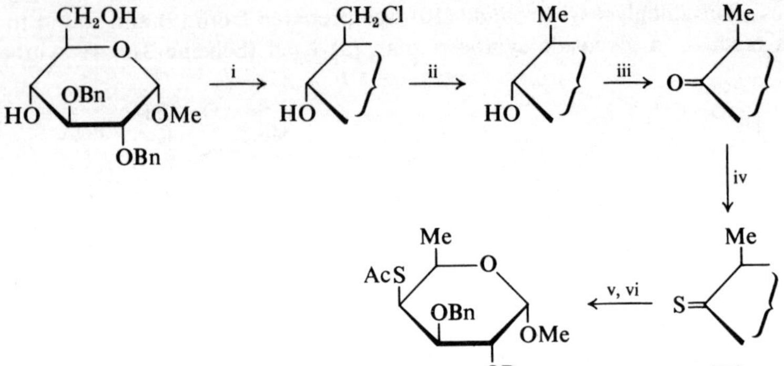

(11)

(12)

Reagents: i, $SOCl_2$–DMF; ii, $LiAlH_4$; iii, Cl_2–Me_2S; iv, Na_2S–P_2S_5; v, $LiAlH_4$; vi, Ac_2O–py

Scheme 4

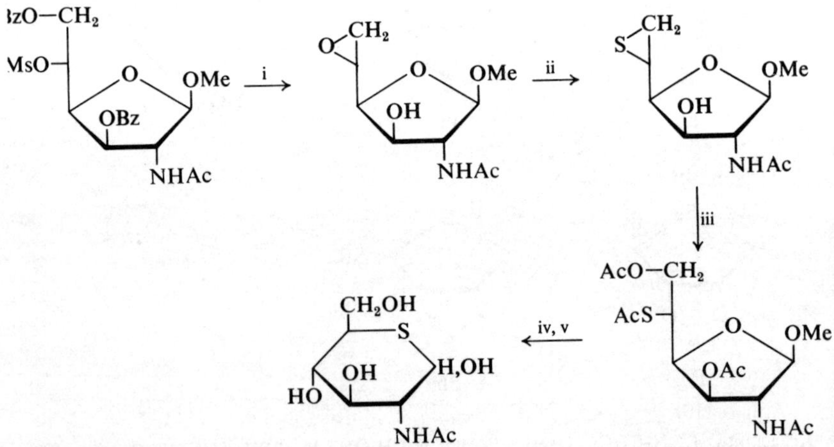

Reagents: i, NaOMe–$CHCl_3$; ii, H_2NCSNH_2; iii, KOAc; iv, H^+; v, NaOMe–MeOH

Scheme 5

[5] P. Simon, J. C. Ziegler, and B. Gross, *Carbohydrate Res.*, 1978, **64**, 257.
[6] A. A. Bushway and T. W. Keenan, *Biochem. Biophys. Res. Comm.*, 1978, **81**, 305.
[7] A. Hasegawa, Y. Kawai, H. Kasugai, and M. Kiso, *Carbohydrate Res.*, 1978, **63**, 131.
[8] B. Bannister, *J.C.S. Perkin I*, 1978, 274.
[9] B. Bannister, *J.C.S. Perkin I*, 1978, 967.

methyl 6-acetamido-8-deoxy-7-thio-D-*galacto*-L-*threo*-thio-octosides, which are analogues of the active portion of lincomycin, are represented in Scheme 6 [reactions with cyclic thioethers and α,ω-di(methylthio)alkanes [8]], and Scheme 7 [reaction with ω-(alkylthio)alkanethiols [9]]. The epimines (13) and (14) do not react with simple thiols. It was shown that compound (15) was formed by direct alkylation of the reagent thiol sulphur atom with anchimeric assistance by the other thioether sulphur atom, and not *via* methyl transfer at the sulphonium salt stage.[9] Sulphide-thiols of the general formula (16) are also formed in the reaction of (13) with ω-(alkylthio)alkanethiols.[8]

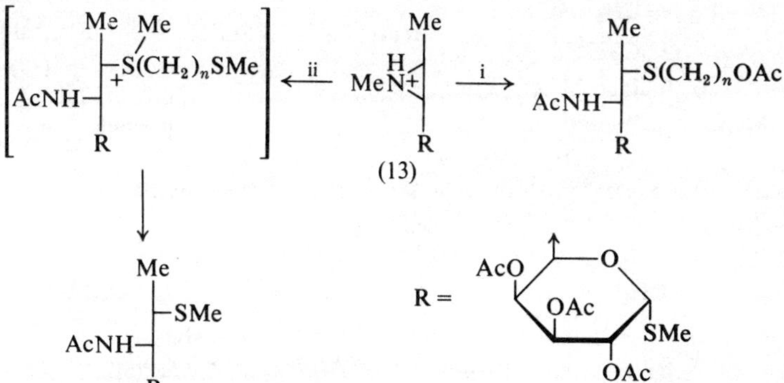

Reagents: i, $(CH_2)_nS$; ii, $(MeS)_2(CH_2)_n$

Scheme 6

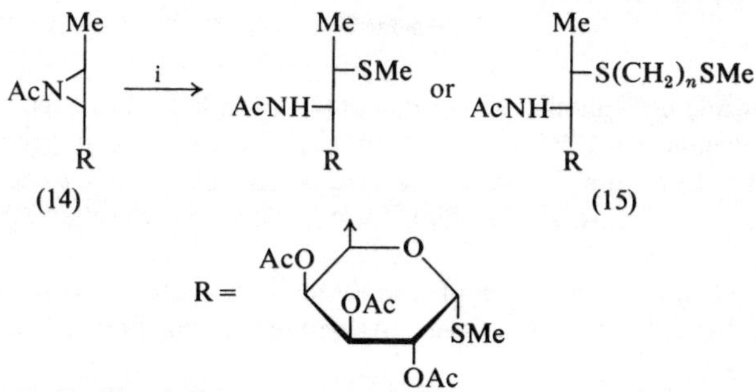

Reagents: i, $MeS(CH_2)_nSH$–HOAc

Scheme 7

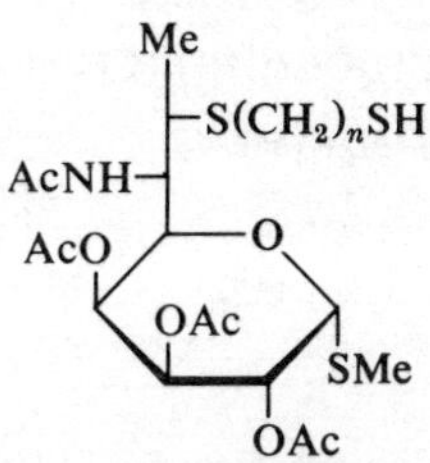

(16)

6-Seleno- and 6-thio-D-galactose esters of dimethylarsinous acid have been synthesized by reaction of the appropriate 6-selenol or 6-thiol with dimethylarsinous chloride (Schemes 8 and 9). Removal of the isopropylidene groups in the ester (17) by means of Dowex-1 (H^+) in aqueous methanol also removed the dimethylarsenyl group and the parent galactose ester was not synthesized. Compound (18) was shown to be carcinostatic against mouse lymphocytic leukemia in *in vivo* tests. Mass spectral and u.v. data of the products were also reported.[10]

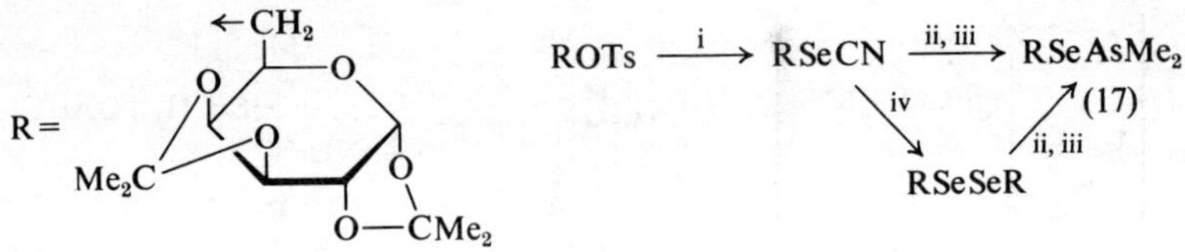

Reagents: i, KSeCN–DMF; ii, $NaBH_4$; iii, Me_2AsCl; iv, NaOMe–MeOH

Scheme 8

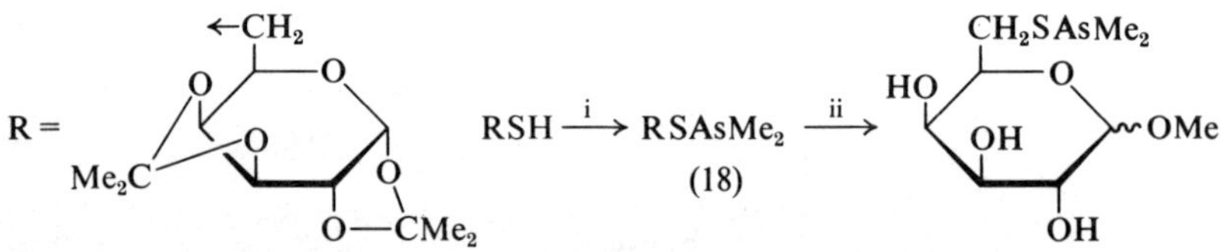

Reagents: i, Me_2AsCl; ii, $MeOH–H_2O–H^+$

Scheme 9

The glucoside 6-sulphonic acid (19), a chloroplast lipid, has been studied in different solvents using ^{13}C n.m.r. A monomeric structure is proposed for methanol solutions but in water the results are interpreted in terms of a bilayer vesicular structure. In chloroform an inverted micellar structure is postulated.[11]

Various selenium derivatives of L-arabinose, D-ribose, and D-xylose, *e.g.* (20)—(24), have been synthesized. Attempted cyclization of 2,3,4-tri-*O*-

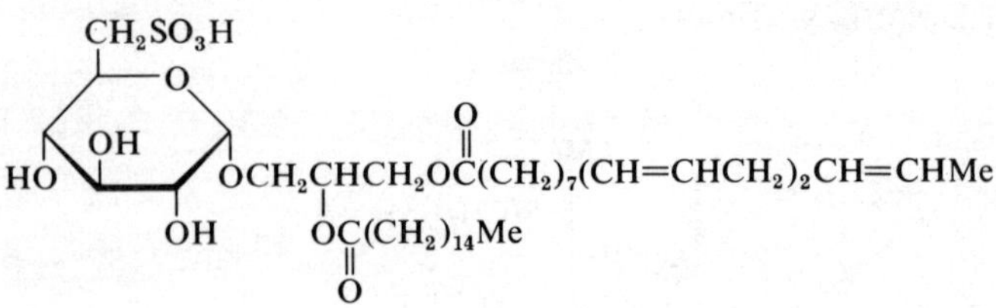

(19)

[10] J. R. Daniel and R. A. Zingaro, *Carbohydrate Res.*, 1978, **64**, 69.
[11] S. R. Johns, D. R. Leslie, R. I. Willing, and D. G. Bishop, *Austral. J. Chem.*, 1978, **31**, 65.

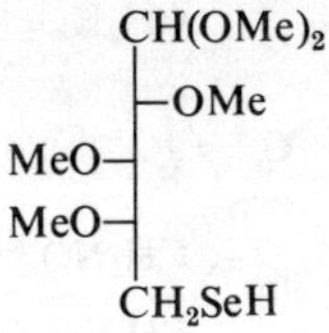

(20)

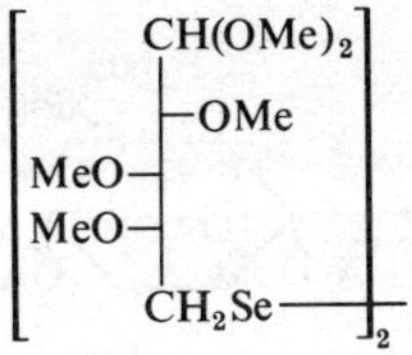

(21)

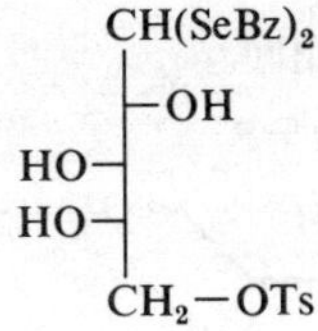

(22)

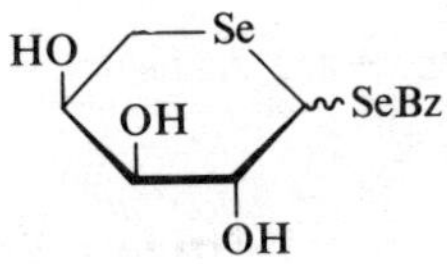

(23)

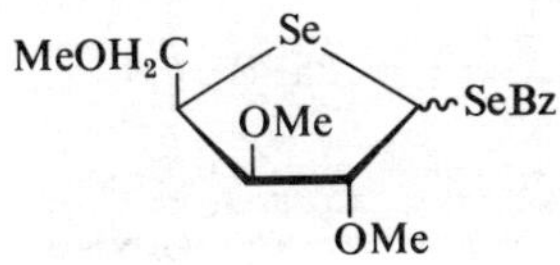

(24)

methyl-5-seleno-L-arabinose dimethylacetal (20) in acid solution gave the diselenide (21).[12]

Carbohydrates containing *C*-phosphorus linkages including those with phosphorus in the ring have been synthesized *via* addition of phosphorus nucleophiles to 6-nitro-*xylo*-hex-5-enofuranose derivatives (Schemes 10 and 11). The acetyl migration of (25) to yield (26) occurred spontaneously.[13]

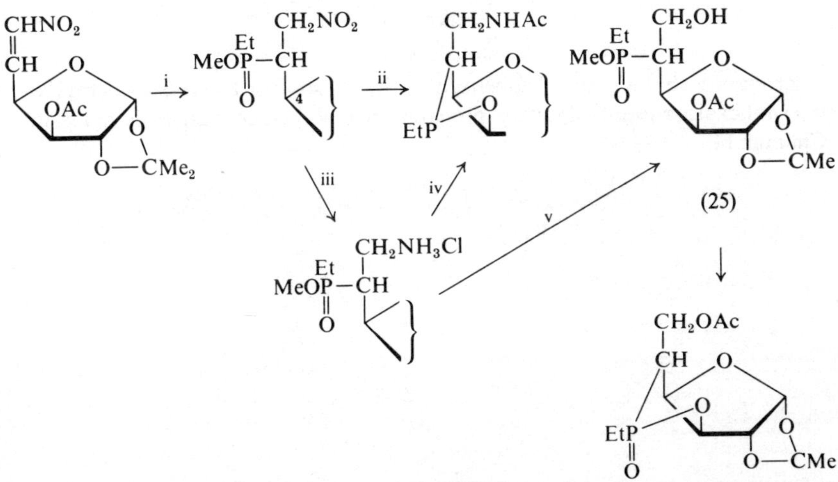

(25)

(26)

Reagents: i, MeOP(O)HEt; ii, H_2–Ni–MeOH; iii, H_2–PtO_2–HCl; iv, resin(HO^-); v, HNO_2

Scheme 10

[12] K. Blumberg, A. Fuccello, and T. van Es, *Carbohydrate Res.*, 1977, **59**, 351.

[13] H. Takayanagi, K. Seo, M. Yamashita, H. Yoshida, T. Ogata, and S. Inokawa, *Carbohydrate Res.*, 1978, **63**, 105.

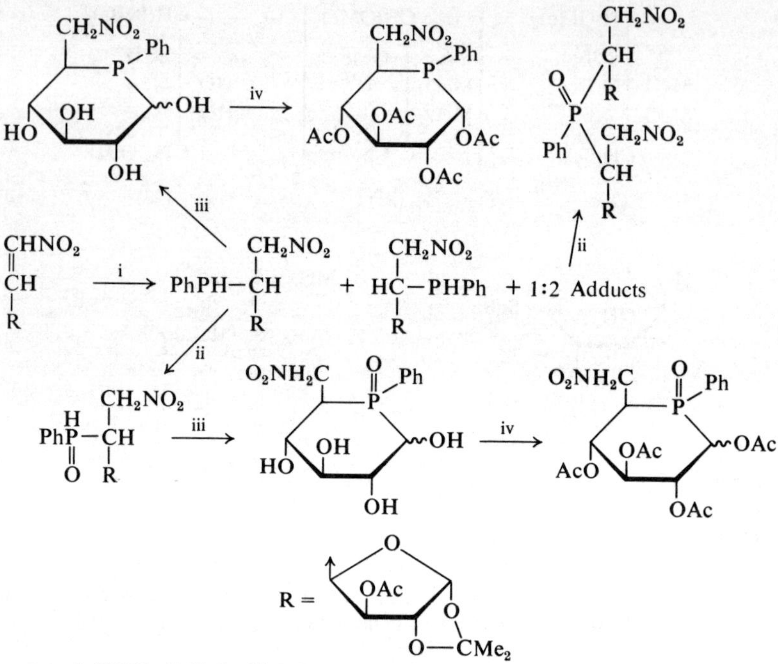

Reagents: i, $PhPH_2$; ii, H_2O_2; iii, HCl–H_2O; iv, Ac_2O–TsOH

Scheme 11

Reference to the use of 6-benzylthio-1,2-*O*-isopropylidene-3-*O*-methyl-α-D-*xylo*-hexofuranose-5-ulose for preparation of other derivatives is made in Chapter 14.

11

Deoxy-sugars

Several references to deoxy derivatives of amino-sugars are given in Chapter 8.

A new method of preparing deoxy-sugar derivatives involves treatment of α-nitro-epoxides with sodium borohydride; in this way, for example, compound (1) gives the 5-deoxy derivative (2), and (3) gives (4). However,

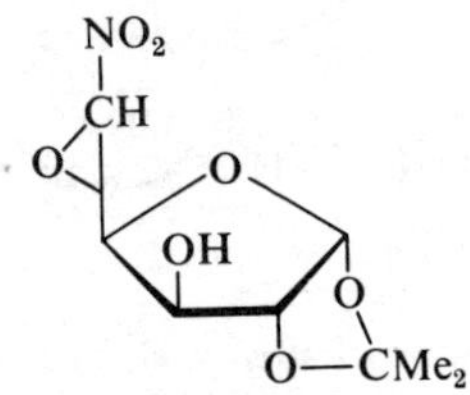

(1)

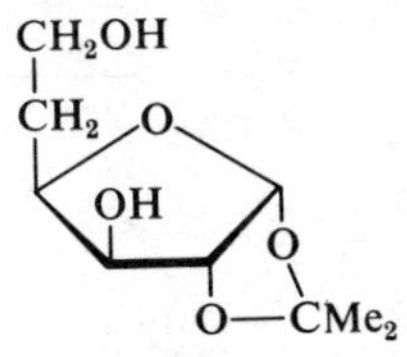

(2)

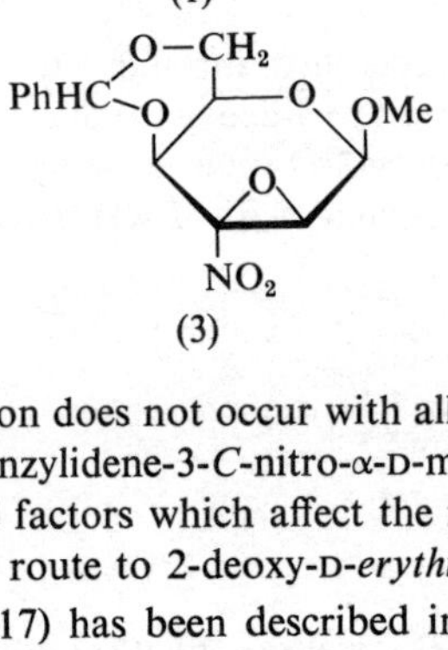

(3)

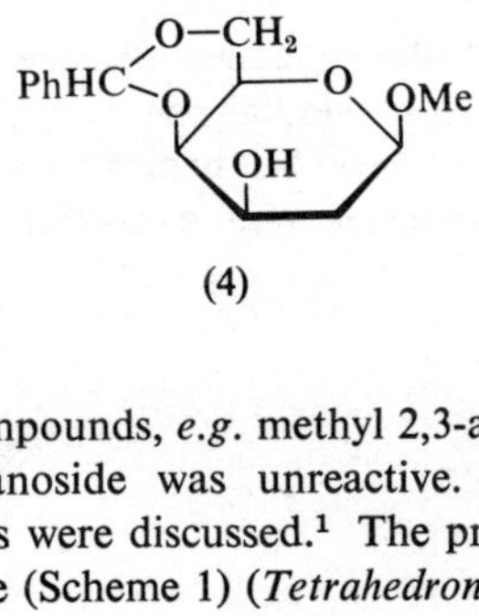

(4)

denitration does not occur with all such compounds, *e.g.* methyl 2,3-anhydro-4,6-*O*-benzylidene-3-*C*-nitro-α-D-mannopyranoside was unreactive. Stereochemical factors which affect the reductions were discussed.[1] The previously reported route to 2-deoxy-D-*erythro*-pentose (Scheme 1) (*Tetrahedron Letters*, 1977, 1617) has been described in detail, and the analogous reactions from D-xylose gave 2-deoxy-D-*threo*-pentose. Deuteriation studies showed that the

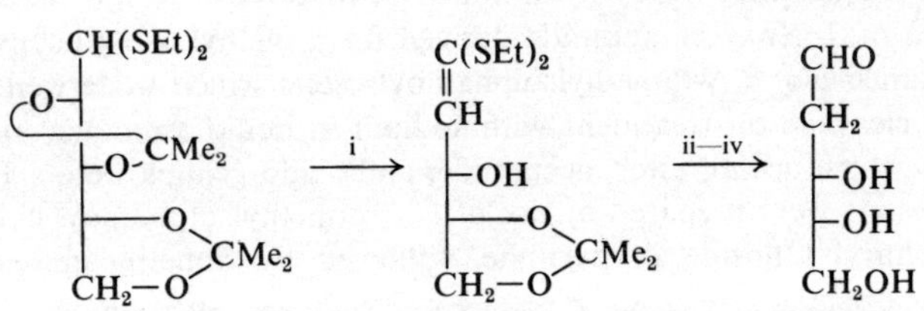

Reagents: i, $KOBu^t$; ii, LAH; iii, MeOH–HgO–$HgCl_2$; iv, H^+

Scheme 1

[1] H. H. Baer and C. B. Madumelu, *Canad. J. Chem.*, 1978, **56**, 1177.

hydride was delivered exclusively *trans*- (Fischer projections) to the hydroxy-groups at C-3.[2]

Various reductions leading to 2-deoxyhexoside derivatives were applied to the enone (5) as illustrated in Scheme 2; the isomerization of the 4-ulose (6)

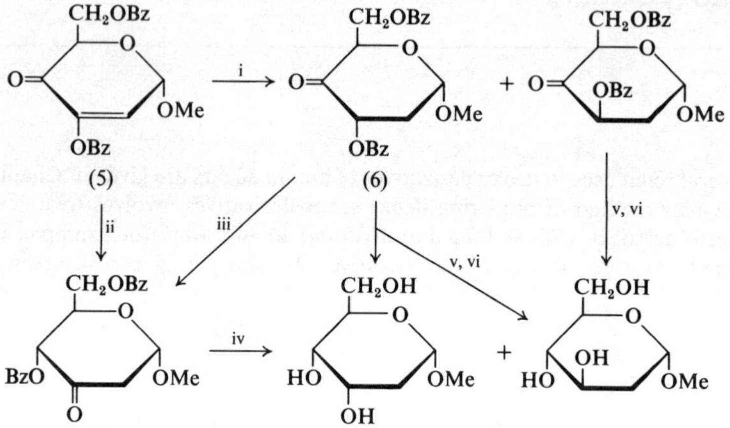

(7)

Reagents: i, H_2–Pd; ii, $Zn(BH_4)_2$; iii, contact with SiO_2; iv, $NaBH_4$; v, H_2–Pt_2; vi, NaOMe

Scheme 2

to the 3-ulose (7) was deemed to have proceeded through an ortho-acid intermediate. Similar reactions of enone (8) led to 4-deoxy products.[3] Reduction of phenyl 3-*O*-benzoyl-4,6-*O*-benzylidene-2-*O*-toluene-*p*-sulphonyl-β-D-glucopyranoside and a corresponding 3-tosylate with LAH gave 2- and

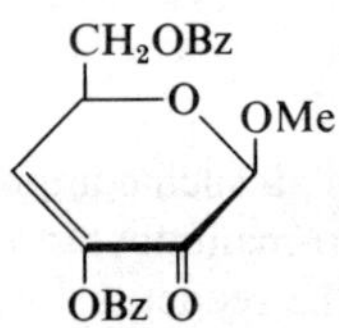

(8)

3-deoxy-phenyl glycosides by direct C—O cleavage of the sulphonates. Epoxide intermediates were shown not to be involved.[4]

A series of 3-deoxy compounds derived from methyl α-D-glucopyranoside were obtained *via* *N*,*N*-dimethylsulphamoyl esters which underwent efficient reductive cleavage on treatment with sodium in liquid ammonia under conditions in which acetal, ester, acetamide, and azido groups were stable. The required esters were prepared by use of *N*,*N*-dimethylsulphamoyl chloride, or with sulphuryl chloride in pyridine, followed by dimethylamine.[5] Other

[2] M. Y. H. Wong and G. R. Gray, *J. Amer. Chem. Soc.*, 1978, **100**, 3548.
[3] F. W. Lichtenthaler, V. Kraska, and S. Ogawa, *Tetrahedron Letters*, 1978, 1323.
[4] L. Kiss, *Acta Chim. Acad. Sci. Hung.*, 1978, **97**, 345.
[5] T. Tsuchiya, I. Watanabe, M. Yoshida, F. Nakamura, T. Usui, M. Kitamura, and S. Umezawa, *Tetrahedron Letters*, 1978, 3365.

methods of deoxygenation at position 3 of hexopyranose units of antibiotics are referred to in Chapter 18.

An improved route to 4-deoxy-D-*xylo*-hexose (4-deoxy-D-glucose) is outlined in Scheme 3,[6] and its phenyl β-glucoside was prepared as shown in Scheme 4.[7] 5-Deoxy-D-*xylo*-hexose and its L-*arabino*-isomer were synthesized,

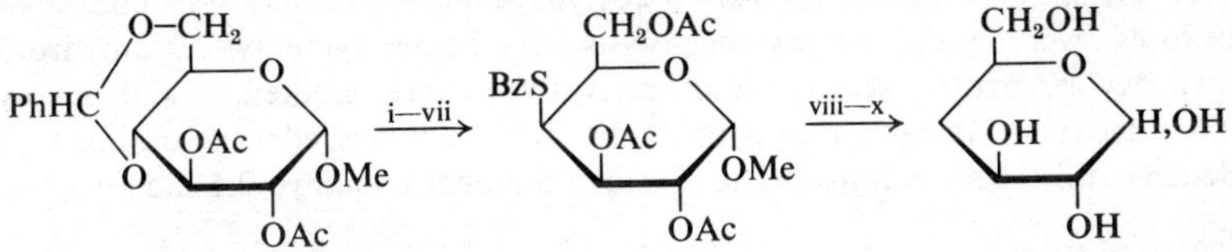

Reagents: i, H^+; ii, Me_3SiCl–py; iii, H_2O; iv, Ac_2O–py; v, MeOH–AcOH; vi, MsCl–py; vii, BzSK–DMF; viii, Ni; ix, MeO^-; x, H^+

Scheme 3

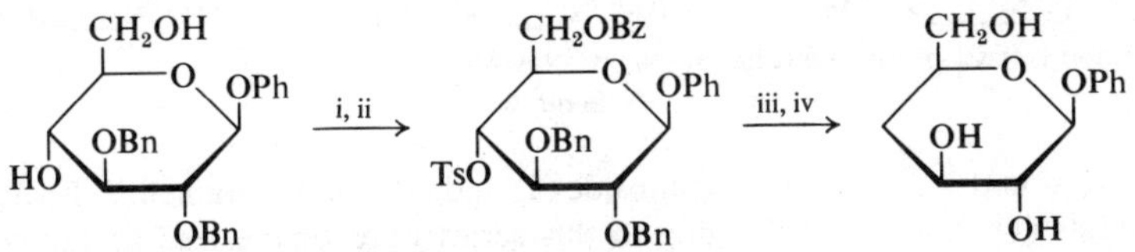

Reagents: BzCl–py; ii, TsCl–py; iii, LAH; iv, H_2–Pd

Scheme 4

respectively, from 5,6-dideoxy-1,2-*O*-isopropylidene-3-*O*-toluene-*p*-sulphonyl-α-D-*xylo*-hex-5-enofuranose and methyl 5,6-dideoxy-2,3-di-*O*-toluene-*p*-sulphonyl-α-L-*arabino*-hex-5-enofuranoside by the addition of iodine trifluoroacetate followed by treatment with Raney nickel in ethanol in the presence of triethylamine. Protecting groups were removed photolytically and by acid hydrolysis. From the products, 9-(5-deoxy-β-D-*xylo*-hexofuranosyl)-adenine and the α-L-*arabino*-isomer were obtained.[8]

Access to 6-deoxy-D-glucose is obtainable through 1,2:3,5-di-*O*-methylene-6-*O*-toluene-*p*-sulphonyl-α-D-glucofuranose which was reduced in high yield to the 6-deoxy compound with lithium triethyl borohydride,[9] and 6-deoxy-D-*altro*-heptose was made as outlined in Scheme 5.[10] The catalytic hydrogenation

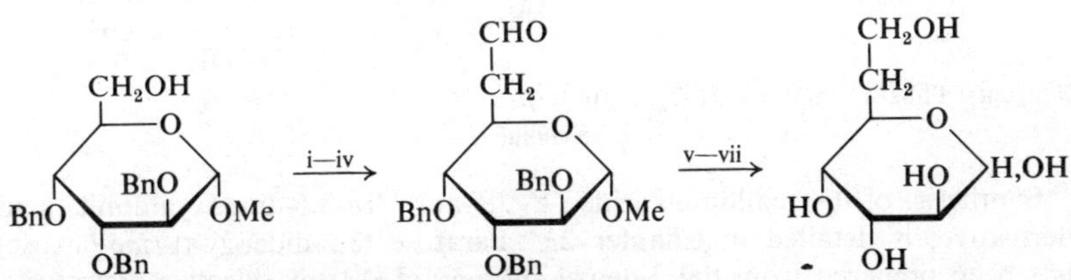

Reagents: i, TsCl–py; ii, Bu^n_4NCN; iii, $Al(Pr^i)_2H$; iv, H^+; v, BH_4^-; vi, H_2–Pd; vii, H^+

Scheme 5

[6] M. Brockhaus, E.-F. Fuchs, and J. Lehmann, *Chem. Ber.*, 1978, **111**, 811.
[7] L. Kiss and P. Nánási, *Acta Chim. Acad. Sci. Hung.*, 1978, **98**, 349.
[8] W. A. Szarek, R. G. S. Ritchie, and D. M. Vyas, *Carbohydrate Res.*, 1978, **62**, 89.
[9] V. K. Srivastava and L. M. Lerner, *Carbohydrate Res.*, 1978, **64**, 263.
[10] P. J. Garegg, J. Hoffman, B. Lindberg, and B. Samuelsson, *Carbohydrate Res.*, 1978, **67**, 263.

of 6-deoxy-maltose, -lactose, and -cellobiose-5-ene hepta-acetates usually gave predominantly 6-deoxy-4-*O*-glycosyl-L-idose products, with the β-linked disaccharides affording greater selectivity. From the β-maltose compound, however, palladium in methanol gave products with a *gluco-:ido* ratio of 1 : 0.6.[11]

In the dideoxy series, 1,3,6-tri-*O*-acetyl-2,4-dideoxy-DL-*erythro*- and *threo*-hexoses were prepared by acetolysis of 2-acetoxymethyl-2,3-dihydro-6-methoxy-2*H*-pyran, and various derivatives were studied,[12] and methyl 2,6-dideoxy-α-D-*arabino*-hexopyranoside was synthesized as outlined in Scheme 6.[13] From appropriate 6-deoxy-5-alkenes methyl 2,6-dideoxy-3-*O*-

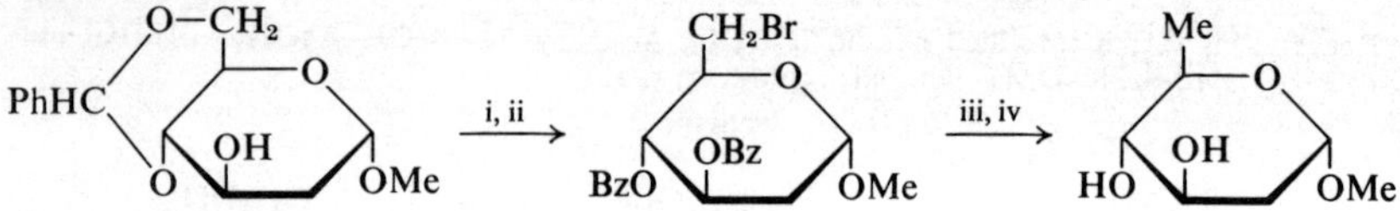

Reagents: i, BzCl–py; ii, NBS; iii, H_2–Ni; iv, NaOMe

Scheme 6

methyl-α- and β-L-*lyxo*-hexopyranoside (9) and (10) were obtained.[14] Racemic products (11) and (12) belonging to this series were synthesized as shown in Scheme 7.[15]

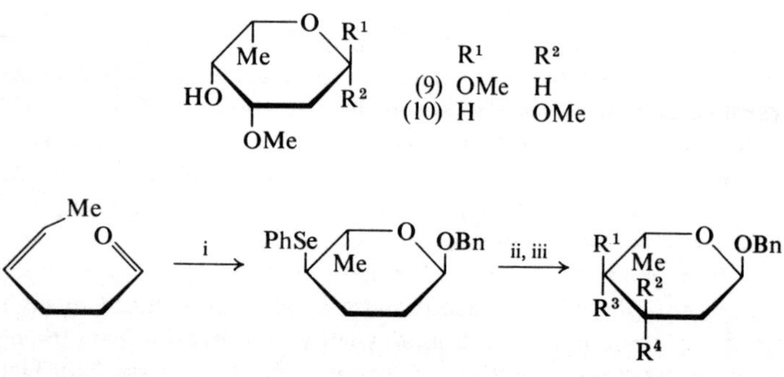

	R^1	R^2
(9)	OMe	H
(10)	H	OMe

	R^1, R^2	R^3, R^4
(11)	H	OH
(12)	OH	H

Reagents: PhSeBr–BnOH; ii, H_2O_2; iii, Bu^tO_2H

Scheme 7

Synthesis of D-epiallomuscarine, a 2,5-anhydro-3,6-dideoxyaldonic acid derivative, is detailed in Chapter 23. Paratose (3,6-dideoxy-D-*ribo*-hexose) has been prepared from the 3-deoxy compound (13) by selective hydrolysis, tosylation at the primary position, and reduction using lithium aluminium

[11] H. Goto, M. Mori, and S. Tejima, *Chem. and Pharm. Bull. (Japan)*, 1978, **26**, 1926.
[12] J. Jurczak, M. Chimielewski, and A. Zamojski, *Pol. J. Chem.*, 1978, **52**, 743 (*Chem. Abs.*, 1978, **89**, 110 159d).
[13] J. Staněk, M. Marek, and J. Jarý, *Carbohydrate Res.*, 1978, **64**, 315.
[14] C. Monneret, C. Conreur, and Q. Khuong-Huu, *Carbohydrate Res.*, 1978, **65**, 35.
[15] S. Current and K. B. Sharpless, *Tetrahedron Letters*, 1978, 5075.

hydride followed by acid-catalysed hydrolysis.[16] Colitose (3,6-dideoxy-L-*xylo*-hexose), on the other hand, was obtained from the acetal (14) (diastereoisomers were separated) which, on treatment with NBS, gave mainly methyl 4-*O*-benzoyl-3-bromo-3,6-dideoxy-α-L-gulopyranoside (15). In similar fashion,

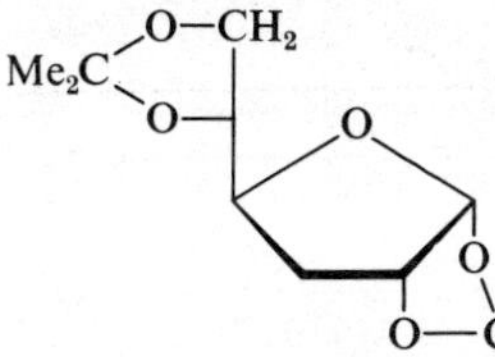

(13)

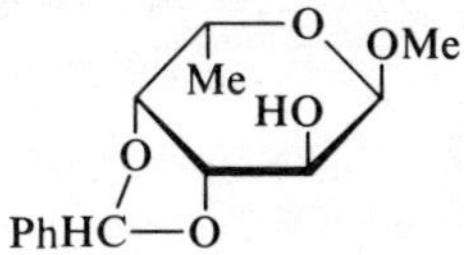

(14)

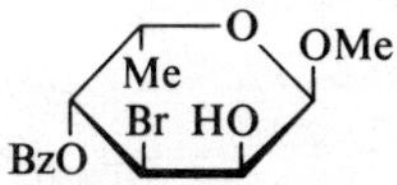

(15)

methyl α-L-rhamnopyranoside provides a route to ascarylose (3,6-dideoxy-L-*arabino*-hexose).[17]

L-Amicetose (2,3,6-trideoxy-L-*erythro*-hexose) has been obtained from the styryl derivative (16) by hydrogenation, ozonolysis to cleave the aromatic ring,

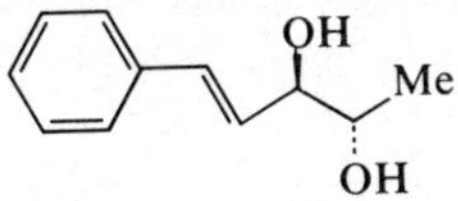

(16)

and partial reduction of the lactone produced.[18] A 2,6,7-trideoxyheptose branched-chain sugar derivative is described in Chapter 13. Other trideoxyhexose compounds have been produced by hydrogenation of unsaturated derivatives (Chapter 12).

[16] J. J. Patroni and R. V. Stick, *Austral. J. Chem.*, 1978, **31**, 445.
[17] D. R. Bundle and S. Josephson, *Canad. J. Chem.*, 1978, **56**, 2686.
[18] C. Fuganti and P. Grasselli, *J. C. S. Chem. Comm.*, 1978, 299.

12
Unsaturated Derivatives

1 Glycals

Two reports have appeared on methods applicable to the synthesis of both pyranoid and furanoid glycal derivatives; both employ strongly basic reagents and can be used, therefore, only with compounds having base-stable protecting groups. Ireland and co-workers employed lithium in liquid ammonia and obtained the L-rhamnal ether (1) in 90% yield,[1] whereas a South African group prepared this compound (46% yield) using sodium naphthalide in THF [2] (Scheme 1). The former group also applied their procedure to obtain

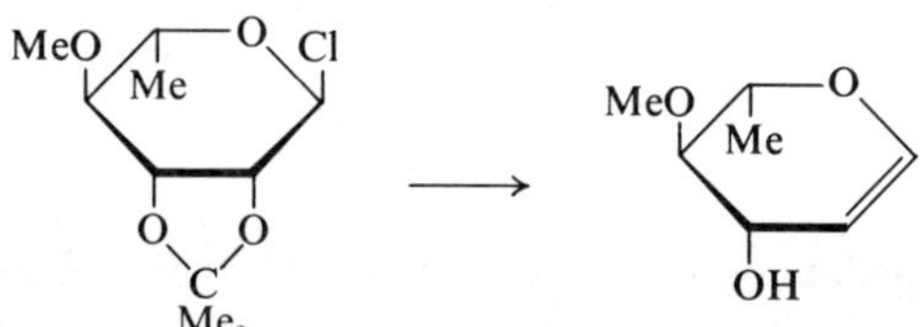

(1)

Scheme 1

the furanoid compound (2) from 2,3-*O*-isopropylidene-5-*O*-methyl-β-D-ribofuranosyl chloride (small proportions of the 1-deoxy compound formed by reductive cleavage of the carbon–chlorine bond were also produced), whereas the latter produced the analogous compound (3) from 2,3:5,6-di-*O*-iso-

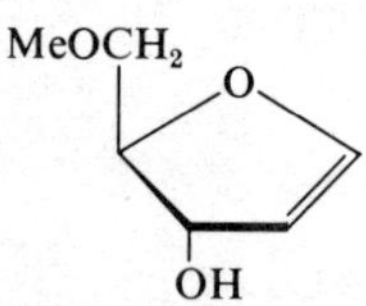

(2)

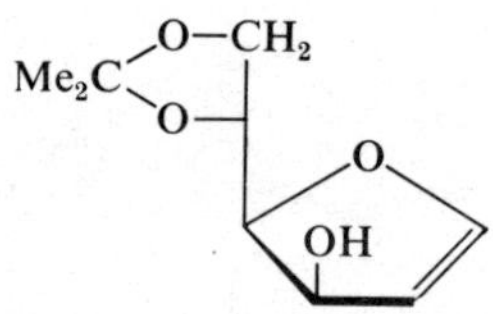

(3)

propylidene-α-D-mannofuranosyl chloride. When they treated this compound with sodium or potassium in THF the oxyanion of the product reacted with the starting material to give a 3-*O*-mannofuranosyl glycal derivative. (This same group of workers have pointed out an error in an earlier publication of

[1] R. E. Ireland, C. S. Wilcox, and S. Thaisrivongs, *J. Org. Chem.*, 1978, **43**, 786.
[2] S. J. Eitelman, R. H. Hall, and A. Jordaan, *J.C.S. Perkin I*, 1978, 595.

theirs on furanoid glycal preparation.[3]) Treated with basic resin, compounds (4) and (5) give the glycal derivative (6), which with alcohols in the presence of acid reacts to afford 2,3-unsaturated glycosides (7) and with acid alone dimerizes to give (8) (Scheme 2).[4]

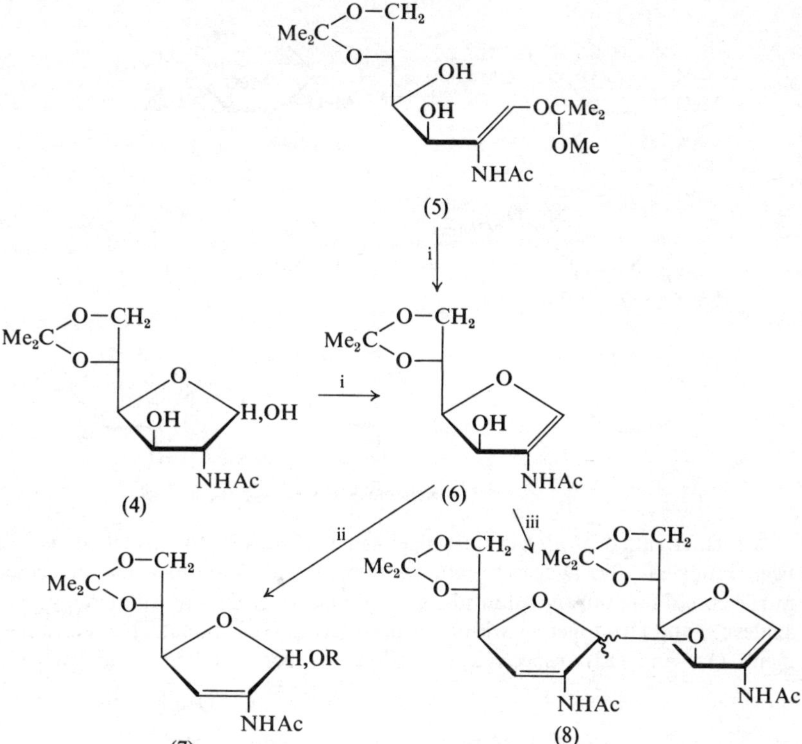

Reagents: i, Resin(HO⁻); ii, ROH–H⁺; iii, H⁺

Scheme 2

A wide range of glucal and alkyl 2,3-dideoxy-D-*erythro*-hex-2-enopyranoside derivatives have been reported in a study of the preparation of monosaccharide compounds containing two double bonds. 6-Sulphonyl esters of D-glucal could be converted into analogues with 6-deoxyalk-5-ene functionality by way of 6-deoxy-6-iodo intermediates, but corresponding methods did not lead to 2,3:5,6-di-unsaturated products.[5]

Further unsaturated compounds have been described as main products of the treatment of fully substituted pyranoid acetals with n-butyl-lithium; yields are only moderate (Scheme 3).[6]

[3] S. J. Eitelman and A. Jordaan, *J.C.S. Chem. Comm.*, 1977, 552; 1978, 32.
[4] A. Hasegawa and M. Kiso, *Carbohydrate Res.*, 1977, **59**, C17.
[5] G. Descotes, J.-C. Martin, and Tachi-Dung, *Carbohydrate Res.*, 1978, **62**, 61.
[6] A. Klemer and D. Balkau, *Chem. Ber.*, 1978, **111**, 1514.

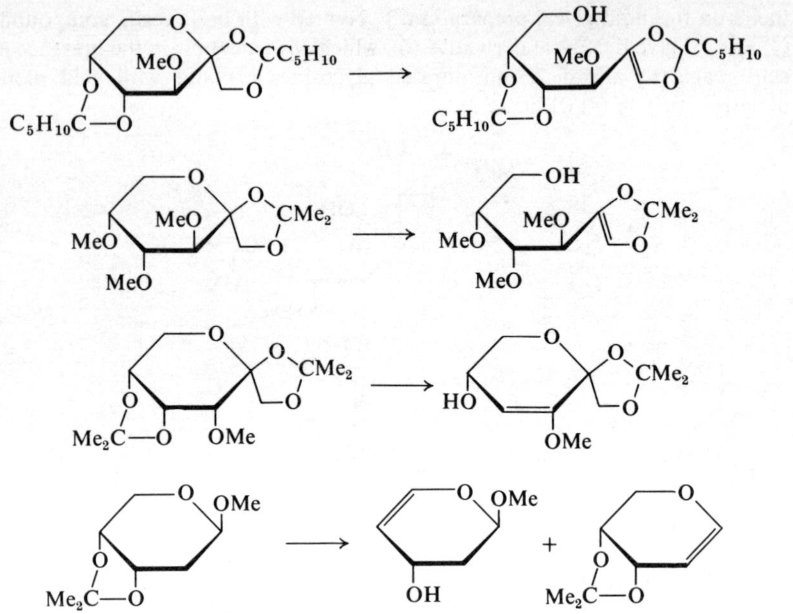

Scheme 3

A further range of allylic rearrangement reactions in the glycal series has been reported. Tri-*O*-acetyl-D-glucal, treated with sodium azide in acetonitrile containing boron trifluoride, gave a mixture of the unsaturated glycosyl azides (9) and (10) together with their products of [3,3]-sigmatropic rearrangement, (11) and (12) respectively. Methyl and benzyl ethers of glucal gave

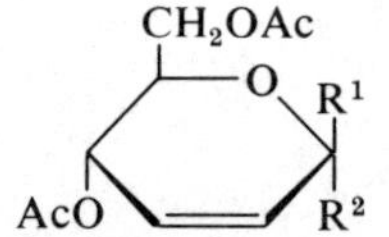

	R^1	R^2
(9)	H	N_3
(10)	N_3	H

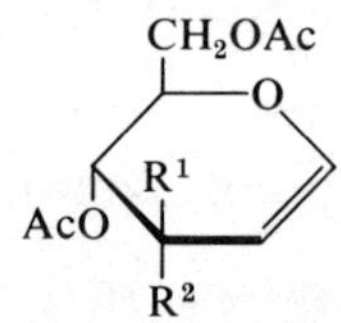

	R^1	R^2
(11)	H	N_3
(12)	N_3	H

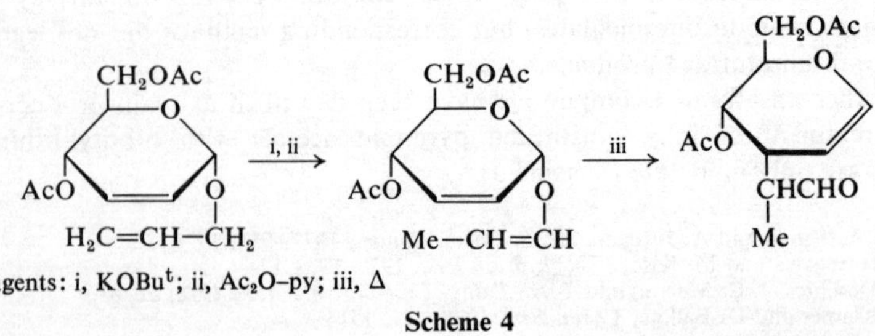

Reagents: i, $KOBu^t$; ii, Ac_2O–py; iii, Δ

Scheme 4

mainly products with the glycal structure. 2,3-Unsaturated glycosyl thiocyanates and xanthates were then prepared and allylically rearranged, and the approach was also applied to the preparation of a branched-chain compound (Scheme 4) and an unsaturated octose derivative (Scheme 5).[7]

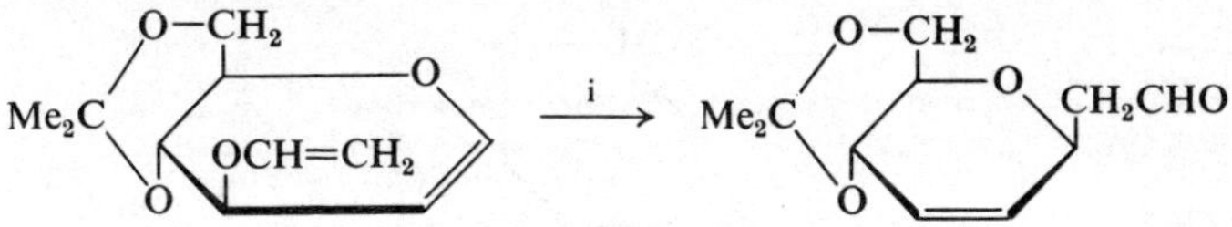

Reagent: i, Δ

Scheme 5

Irradiation of 3,4-di-*O*-acetyl-D-xylal in acetone-1,3-dioxolane afforded compounds (13)—(15) isolated in 29, 10, and 9% yield, respectively (together with 33% unchanged starting material).[8]

Enones, based on 2-hydroxyglycals, were prepared as illustrated in Scheme 6; they undergo addition of nucleophilic reagents to give glycosid-3-uloses.[9]

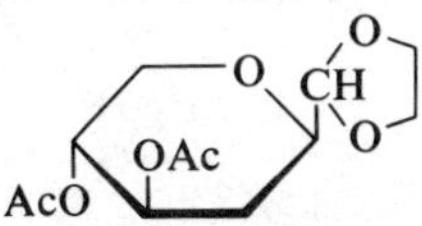

(13)

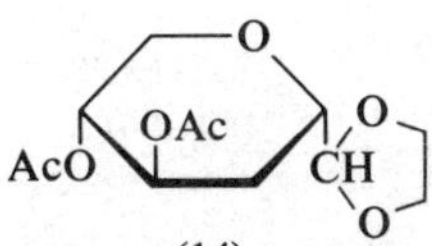

(14)

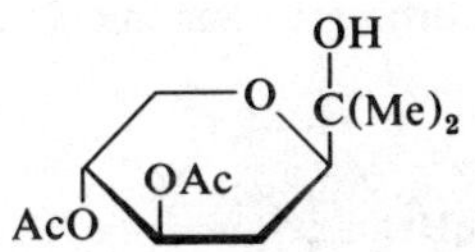

(15)

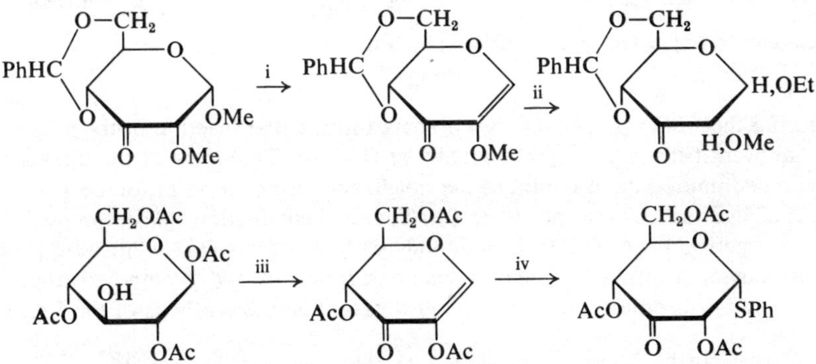

Reagents: Et_3N; ii, EtOH; iii, DMSO–Ac_2O; iv, PhSH

Scheme 6

References are made to the use of glycals also in Chapters 3 and 7.

2 Other Unsaturated Derivatives

Treatment of the product of periodate oxidation of methyl 4,6-*O*-benzylidene-α-D-glucopyranoside with *p*-toluenesulphonylhydrazine then with sodium borohydride in ethanol gave an *N*-toluene-*p*-sulphonyl-2,3-epimine which on

[7] K. Heyns and R. Hohlweg, *Chem. Ber.*, 1978, **111**, 1632.
[8] Y. Araki, K. Nishiyama, K. Matsuura, and Y. Ishido, *Carbohydrate Res.*, 1978, **63**, 288.
[9] J. Weissmüller and H. Kunz, *Tetrahedron Letters*, 1978, 3807.

further reduction with this reagent in 1,4-dioxan afforded the corresponding 2,3-dideoxy-hex-2-enose derivative.[10] 2,3-Unsaturated glycosyl units of antibiotics are referred to in Chapter 18.

Preparation of the 3,4-unsaturated anhydro-sugar (16) was effected from

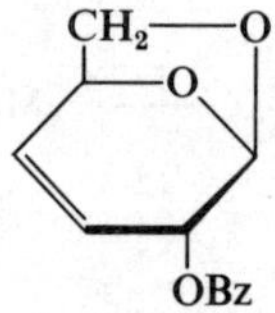

(16)

1,6-anhydro-β-D-galactopyranose by use of the 3,4-thionocarbonates.[11] For a 3,4-unsaturated compound obtained by pyrolysis of polysaccharides see Chapter 4. 4,5-Unsaturated glycosyl units of antibiotics are referred to in Chapter 18.

The first 4,5-unsaturated pyranoid 3-ulose derivative has been prepared according to Scheme 7; the following unexpected findings were encountered

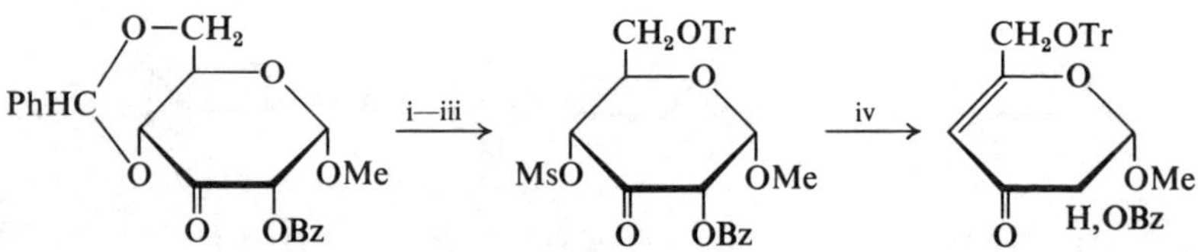

Reagents: i, H^+; ii, TrCl–py; iii, MsCl-py; iv, DBN

Scheme 7

during the work: (i) The 4,6-diol prepared in the first reaction in the scheme underwent benzoylation preferentially at O-4. (ii) Treatment of the mesylate with sodium iodide in butanone led not to the introduction of iodine at C-4, but of hydrogen at that position. (iii) The final elimination was accompanied by epimerization at C-2.[12] Compounds with structures (17)—(20) were prepared by appropriate elimination reactions applied to 5-*C*-bromo-derivatives, and the 5-*C*-bromo-D-glucose esters also permitted the synthesis of the 5-enes

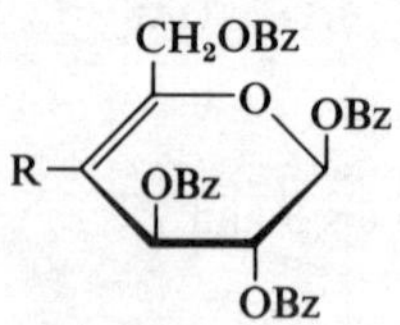

(17) R = H
(18) R = BzO

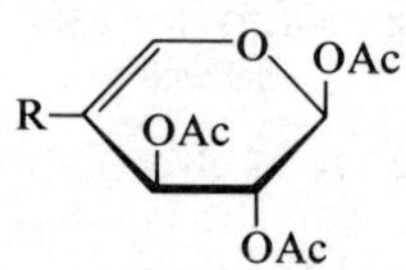

(19) R = H
(20) R = OAc

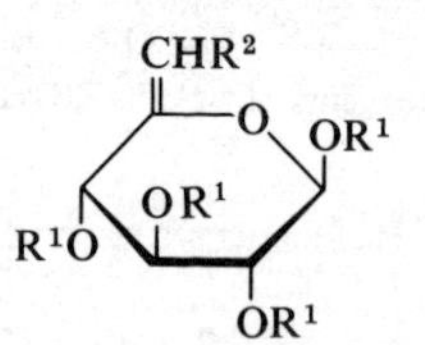

	R^1	R^2
(21)	Ac	H
(22)	Bz	OBz
(23)	Bz	H

[10] S. Honda, K. Kakehi, and S. Oguri, *Carbohydrate Res.*, 1978, **64**, 101.
[11] N. W. H. Cheetham, *Carbohydrate Res.*, 1978, **63**, 293.
[12] M. B. Yunker and B. Fraser-Reid, *Canad. J. Chem.*, 1978, **56**, 2221.

(21)—(23).[13] 2,6-Anhydroheptonic acid derivatives gave access to alkenes with similarly disposed double bonds (Scheme 8).[14]

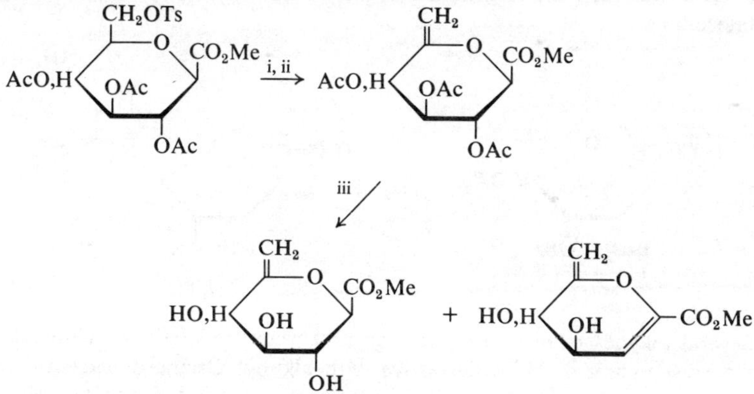

Reagents: i, NaI–Ac_2O; ii, NaF–py; iii, NaOMe

Scheme 8

Reaction of 1,2:3,5-di-*O*-methylene-6-*O*-toluene-*p*-sulphonyl-α-D-glucofuranose with potassium fluoride in ethylene glycol gave the corresponding fluoride and the alkene (24) which was derived from this first product. Various alternative methods of obtaining the alkene were developed based on dehydrohalogenation of 6-halogeno-compounds. Methanolysis of compound (24) gave methyl 6-deoxyglycosiduloses. Potassium t-butoxide in DMSO applied to other tosylates afforded corresponding alkenes.[15] A series of geminal difluoro-derivatives, *e.g.* (25) and (26), were prepared by application of Wittig reactions, and were examined by ^{13}C and ^{1}H n.m.r. spectroscopy.[16]

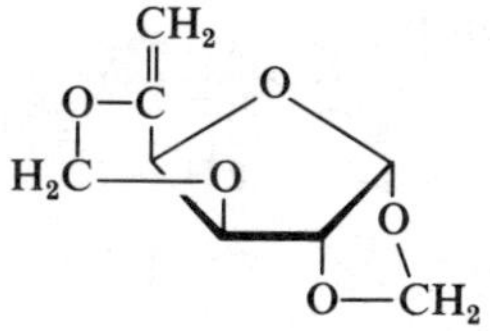

(24)

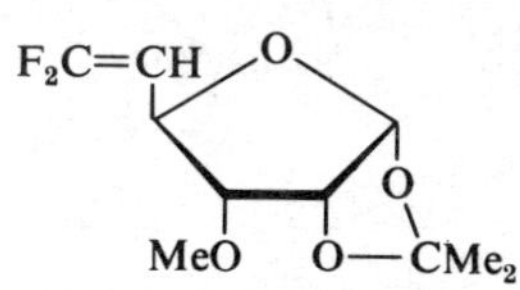

(25)

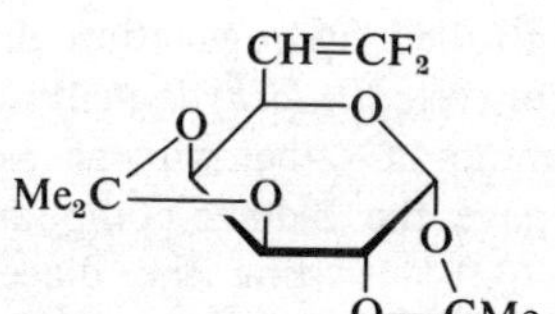

(26)

[13] R. J. Ferrier and P. C. Tyler, *J.C.S., Chem. Comm.*, 1978, 1019.
[14] C. Knapp and J. Lehmann, *Carbohydrate Res.*, 1978, **63**, 257.
[15] H. C. Srivastava and V. K. Srivastava, *Carbohydrate Res.*, 1978, **60**, 210.
[16] J. M. J. Tronchet, A. Bonenfant, and F. Barbalat-Rey, *Carbohydrate Res.*, 1978, **67**, 564.

Condensation between the alkenes (27), which were prepared by standard methods from D-allose, and 6-benzamidochloromercuripurine gave access to the nucleoside (28), the enantiomer of which was synthesized similarly from D-talose.[17]

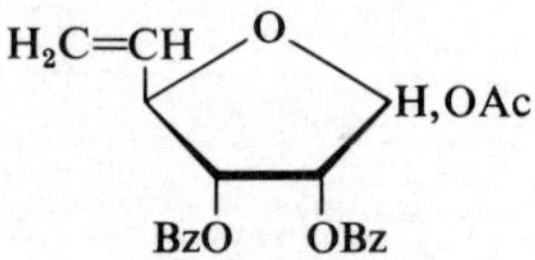

(27)

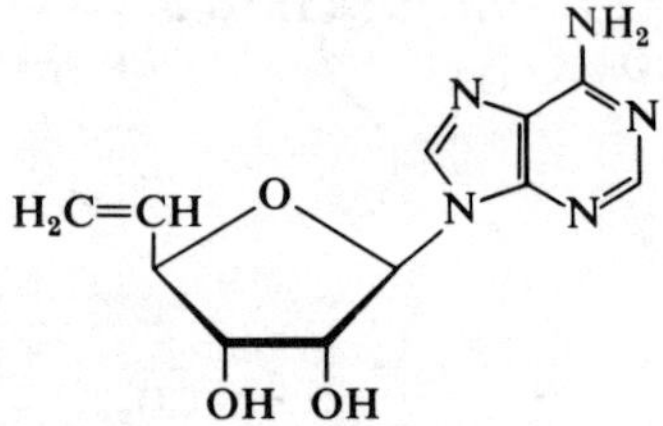

(28)

Several alkynes related to compound (29) were prepared by treatment of the corresponding dialdose derivative with alkynyl Grignard reagents, and by reduction and oxidation processes they were converted into a series of octose derivatives, *e.g.* reduction gave 7,8-dideoxy compounds and oxidation alkynyl glycos-6-uloses.[18]

Significant developments have been reported in the area of acyclic alkyne derivatives. 2,3-*O*-Isopropylidene-D-glyceraldehyde condensed with the Grignard reagent (30) gave the epimers (31) which, on hydrogenation, gave the 2,3-dideoxy analogues and these on mild ethanolysis afforded the dideoxyglycosides. Methanolysis of (31) gave the unprotected acyclic dimethylacetals,

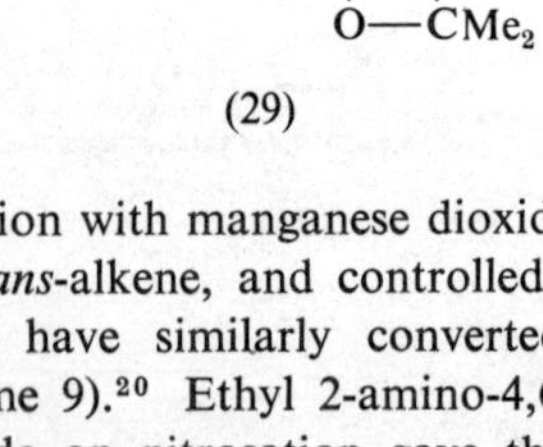

(29)

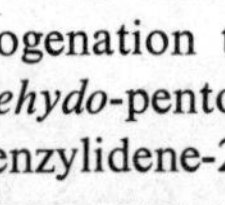

(30)

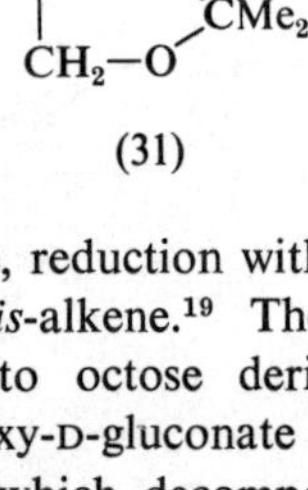

(31)

oxidation with manganese dioxide the 4-oxo-derivative, reduction with LAH the *trans*-alkene, and controlled hydrogenation the *cis*-alkene.[19] The same group have similarly converted *aldehydo*-pentoses to octose derivatives (Scheme 9).[20] Ethyl 2-amino-4,6-*O*-benzylidene-2-deoxy-D-gluconate hydrochloride on nitrosation gave the 2-diazo-compound which decomposed in alkali to the acetylene (32). Heating the diacetate of the intermediate afforded a mixture of the geometric isomers of the alkenes (33) and (34).[21]

[17] L. M. Lerner, *J. Org. Chem.*, 1978, **43**, 2469.
[18] A. González, A. Llamas, and R. Mestres, *Carbohydrate Res.*, 1977, **59**, 598.
[19] K. Heyns and K.-M. Grühn, *Tetrahedron Letters*, 1978, 2861.
[20] K. Heyns, K.-M. Grühn, and W. Rust, *Tetrahedron Letters*, 1978, 4015.
[21] Y. Gelas-Mialhe and D. Horton, *J. Org. Chem.*, 1978, **43**, 2307.

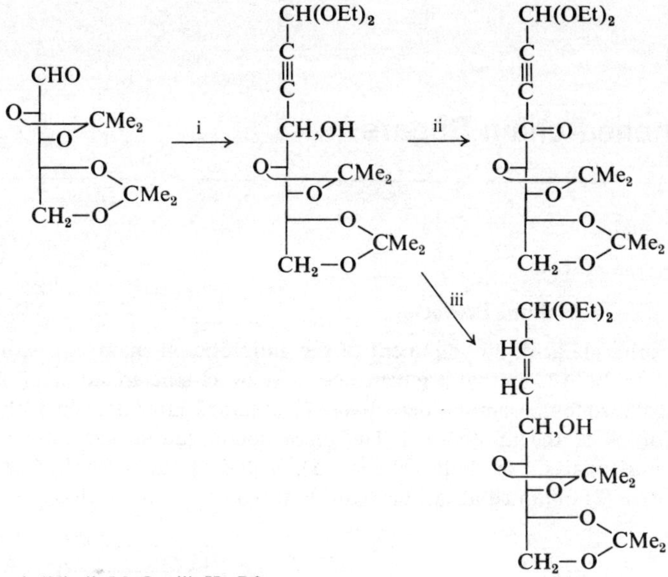

Reagents: i, (30); ii, MnO_2; iii, H_2–Pd

Scheme 9

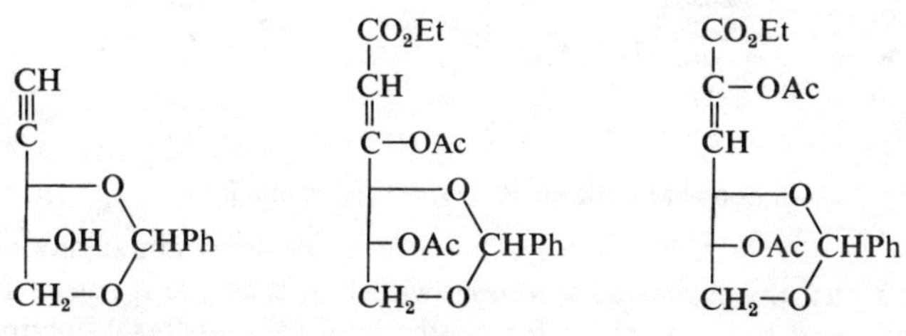

(32) (33) (34)

An acyclic unsaturated compound used in the synthesis of 2-deoxypentoses is noted in Chapter 11.

Unsaturated branched-chain sugars are mentioned in Chapters 13 and 16, and reactions of nitro-alkene carbohydrates with sulphur ylides are covered in Chapter 9.

13
Branched-chain Sugars

1 Natural Products

D-Moenuronic acid, a component of the antibiotic moenamycin A, has been shown to be 4-*C*-methyl-D-glucuronic acid by characterization of its ethyl furanosidurono-6,3-lactone derivative (1) obtained after acid hydrolysis and ethanolysis of the antibiotic.[1] The *gluco* configuration was established by ^{1}H n.m.r. analysis of derivatives of (1), including the 1,6-anhydrofuranose derivative (2) obtained after borohydride treatment of the lactone.

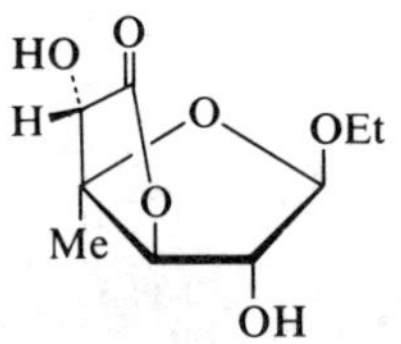

(1)

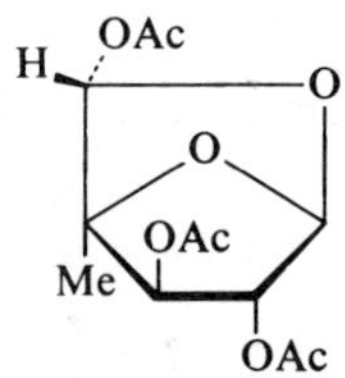

(2)

2 Compounds with an R^{1}—C—OR^{2} Branch

A review lecture including the synthesis of branched-chain sugars using the dithiane route and glycoside synthesis with branched-chain sugars has been published [2] and full reports on the synthesis of the 'γ-octose' obtained from isoquinocycline A [3] and of related 4-*C*-hydroxyethylhexose derivatives [4] have now appeared (see Vol. 10, p. 98).

An improved synthesis of methyl α-L-mycaroside from L-rhamnose includes the part-sequence shown in Scheme 1.[5] A new synthesis of methyl 4,6-*O*-

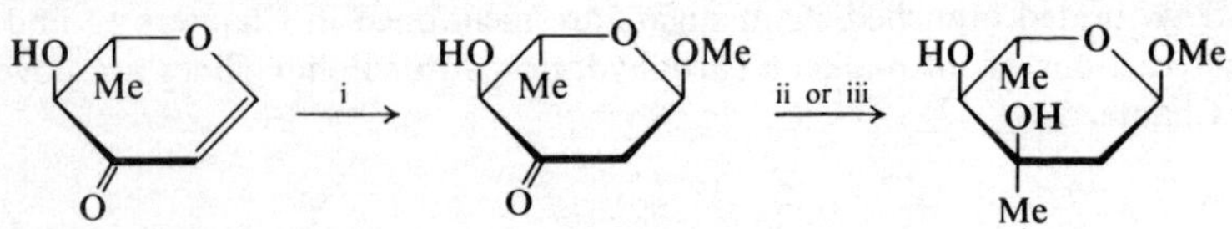

Reagents: i, NaOMe; ii, MeLi; iii, MeMgI

Scheme 1

[1] N. Langenfeld and P. Welzel, *Tetrahedron Letters*, 1978, 1833.
[2] H. Paulsen, *Pure Appl. Chem.*, 1977, **49**, 1169.
[3] H. Paulsen and V. Sinnwell, *Chem. Ber.*, 1978, **111**, 869.
[4] H. Paulsen and V. Sinnwell, *Chem. Ber.*, 1978, **111**, 879.
[5] J. Thiem and J. Elvers, *Chem. Ber.*, 1978, **111**, 3514.

benzylidene-2-deoxy-3-*C*-methyl-α-D-*arabino*-hexopyranoside (3), with the epimeric *ribo*-isomer as a minor product, utilized the Wittig-epoxidation sequence on the corresponding 3-ulose derivative (Scheme 2) which is stereochemically

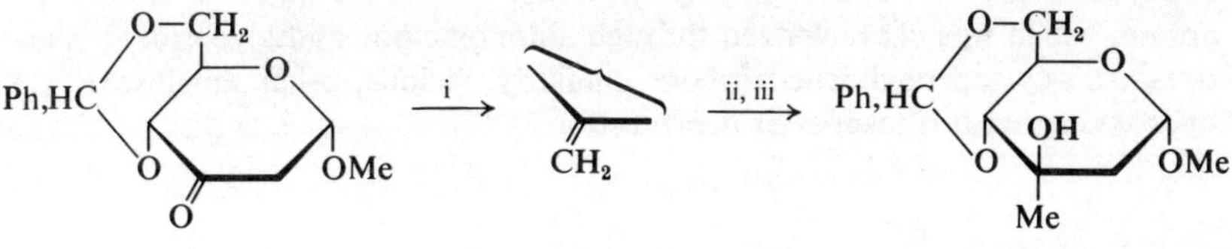

(3)

Reagents: i, $Ph_3P{=}CH_2$; ii, *m*-Cl·C_6H_4·CO_3H; iii, LAH

Scheme 2

complementary to the route using Grignard reagents; an alternative route to (3) by lithium aluminium hydride reduction of the oxiran (4) gave a 1 : 1 mixture of the isomeric 2- and 3-deoxy-3-*C*-methyl-D-*arabino*-hexopyranosides; D-evermicose (2,6-dideoxy-3-*C*-methyl-D-*arabino*-hexose) was obtained from (3) by standard procedures.[6]

5-Deoxy-3-*C*-hydroxymethyl-DL-lyxose and ribose derivatives have been synthesized from 2-methyl-3-furoic acid by a procedure analogous to that used for preparing DL-apiose from 3-furoic acid (see Vol. 7, p. 115).[7] L-Mycarose (5) has also been prepared from non-carbohydrate precursors by the

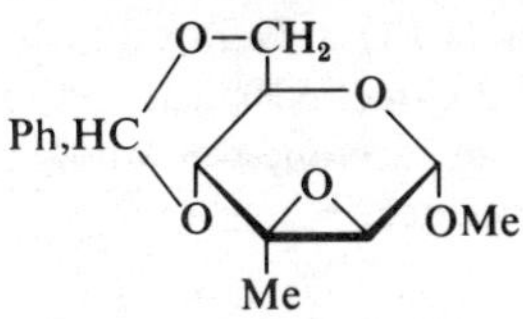

(4)

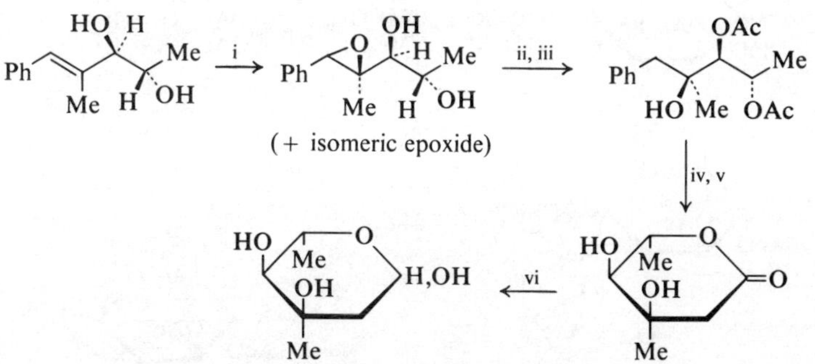

(5)

Reagents: i, *m*-Cl·C_6H_4·CO_3H; ii, Ac_2O–py; iii, H_2–Pd; iv, O_3; v, NaOH; vi, $AlH(Bu^i)_2$

Scheme 3

[6] M. Funabashi, N.-G. Hong, H. Kodama, and J. Yoshimura, *Carbohydrate Res.*, 1978, **67**, 139.

[7] T. Kinoshita and T. Miwa, *Bull. Chem. Soc. Japan*, 1978, **51**, 225.

sequence shown in Scheme 3 and L-olivomycose was similarly prepared from the isomeric epoxide.[8]

Aldol condensation of 2,3-*O*-ethylidene-D-erythrofuranose with formaldehyde gave the 2-*C*-hydroxymethyl-erythrose derivative (6), an isomer of apiose, which was characterized through aldonolactone and glycoside derivatives; 2,3-*O*-isopropylidene-D-ribose similarly yielded D-hamamelose (2-*C*-hydroxymethyl-D-ribose) after deprotection.[9]

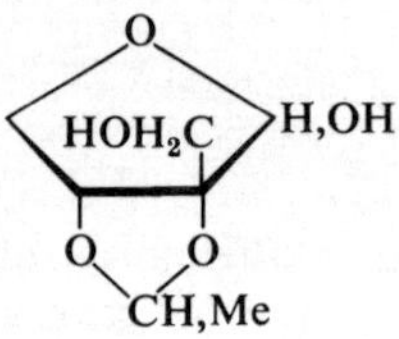

(6)

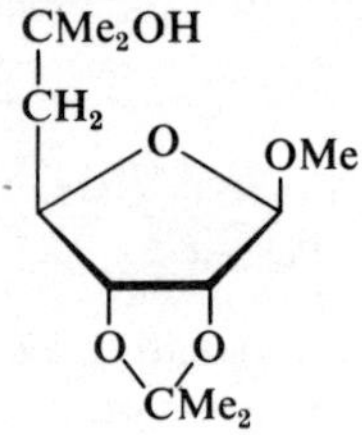

(7)

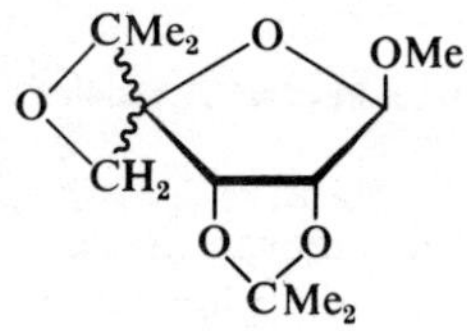

(8)

Photochemically induced addition of propan-2-ol and acetone to methyl 5-deoxy-2,3-*O*-isopropylidene-β-D-*erythro*-pent-4-enofuranoside yielded the branched-chain sugars (7) and (8) respectively as the major product in each case.[10] Similarly, photolysis of 2-acetoxy-3,4,6-tri-*O*-acetyl-D-glucal in acetone–propan-2-ol(1 : 200) yielded a mixture of the 2-*C*-methyl-3,7-anhydro-D-*glycero*-D-*gulo*-, D-*glycero*-D-*ido*-, and D-*glycero*-D-*galacto*-octitol derivatives (9)—(11) respectively. In (1 : 1) solvent, the doubly-branched chain 4-*C*-(1-hydroxy-1-methylethyl)-2-*C*-methyl sugar orthoacetate (12) was an additional product, whereas in acetone-t-butanol the 2-*C*-methyl-oct-4-ulose derivative (13) was also obtained.[11]

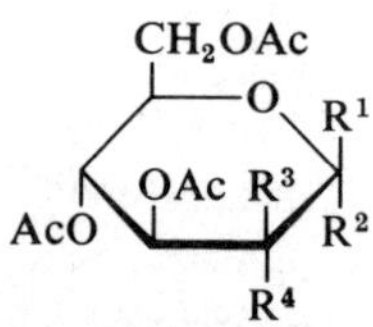

(9) $R^1 = CMe_2OH$, $R^4 = OAc$, $R^2 = R^3 = H$
(10) $R^2 = CMe_2OH$, $R^4 = OAc$, $R^1 = R^3 = H$
(11) $R^1 = CMe_2OH$, $R^3 = OAc$, $R^2 = R^4 = H$

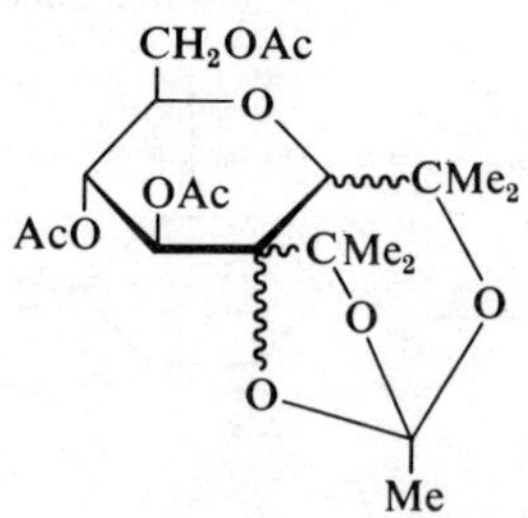

(12)

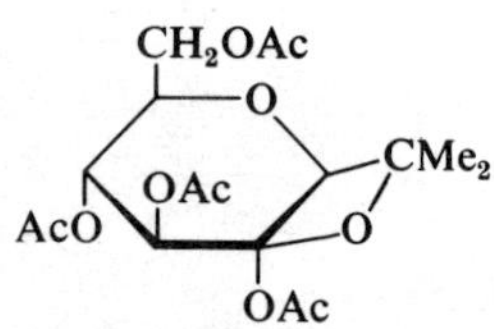

(13)

[8] C. Fuganti and P. Grasselli, *J.C.S. Chem. Comm.*, 1978, 299.
[9] P.-T. Ho, *Tetrahedron Letters*, 1978, 1623.
[10] Y. Araki, K. Senna, K. Matsuura, and Y. Ishido, *Carbohydrate Res.*, 1978, **65**, 159.
[11] Y. Araki, K. Senna, K. Matsuura, and Y. Ishido, *Carbohydrate Res.*, 1978, **64**, 109.

Reaction of methoxyvinyl lithium with 1,2;5,6-di-*O*-isopropylidene-α-D-*ribo*-hexofuranos-3-ulose gave the methoxyvinyl branched-chain sugar (14) and hence the derivatives shown in Scheme 4.[12] Similar reactions were carried

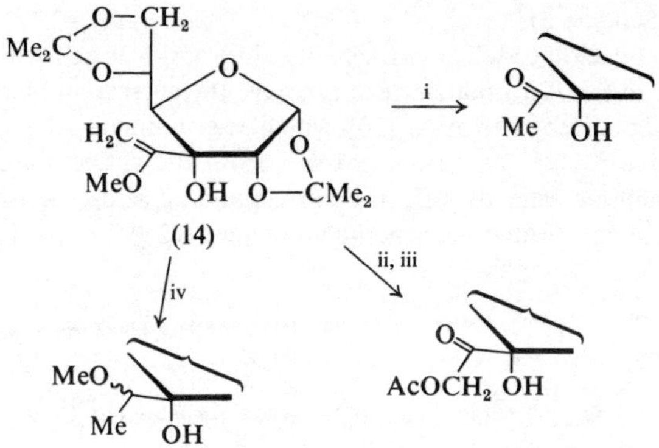

Reagents: i, H_3O^+; ii, *m*-Cl·C_6H_4·CO_3H or OsO_4–py; iii, Ac_2O; iv, H_2–Pt

Scheme 4

out on methyl 2,3,6-trideoxy-α-L-*glycero*-hexopyranosid-4-ulose leading to 4-*C*-acetyl and 4-*C*-hydroxyacetyl branched-chain sugar analogues.[13]

Condensation of 2,3-*O*-isopropylidene-D-glyceraldehyde with the anion formed from 1,1-diethoxyprop-2-ene gave a 7 : 3 mixture of the D-*erythro* and D-*threo*-isomers of the 2-deoxy-2-*C*-methylene pentose derivatives (15) which could be converted to the corresponding pentopyranosides (16) and (17) (Scheme 5); the glycoside (16) was further converted to methyl hamamelo-

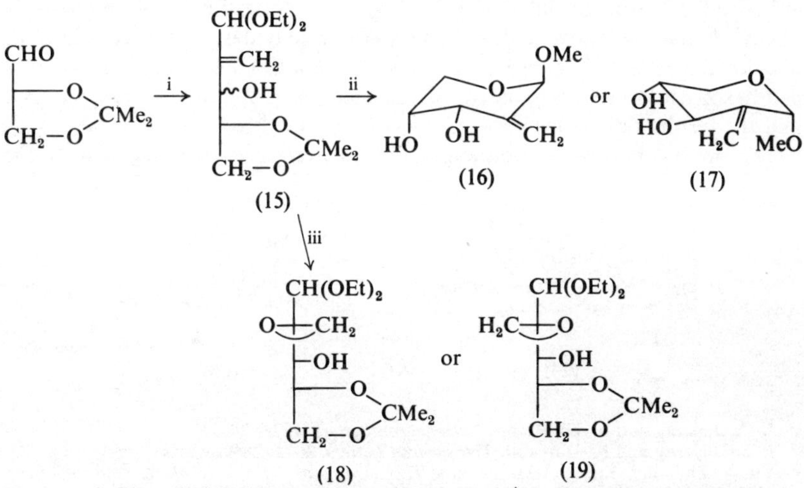

Reagents: i, CH_2=C(Li)CH(OEt)$_2$; ii, HCl–MeOH; iii, Bu^tO_2H–VO(acac)$_2$

Scheme 5

[12] J. S. Brimacombe and A. M. Mather, *Tetrahedron Letters*, 1978, 1167.

[13] J. S. Brimacombe, A. M. Mather, and R. Hanna, *Tetrahedron Letters*, 1978, 1171.

side.[14] In a further study, it was found that the acyclic unsaturated sugars (15) could be epoxidized stereospecifically by t-butyl hydroperoxide in the presence of a vanadium complex to give the oxiran derivatives (18) and (19) which could be hydrolysed to give the D-*arabino*- and D-*xylo*- isomers of hamamelose respectively (Scheme 5).[15]

1,2,5,6-Di-*O*-cyclohexylidene-α-D-*ribo*hexofuranos-3-ulose with γ-bromocrotonic ester in the Reformatski reaction gave the corresponding unsaturated branched-chain allose derivative (20) which was reduced to the 4-hydroxybutyl derivative (21).[16] The same hexos-3-ulose derivative similarly reacted with bromomalonic ester to yield the corresponding allose derivative (22) or the geminally disubstituted branched-chain sugar (23).[17]

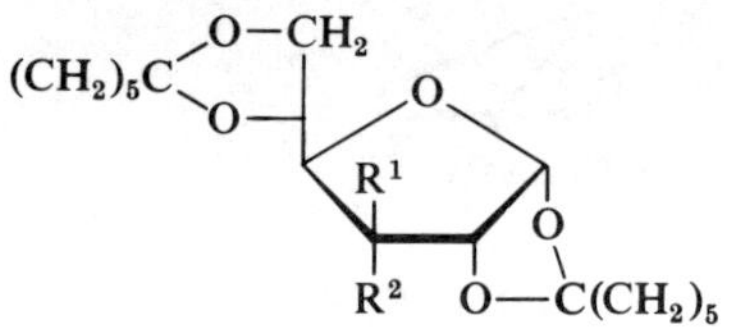

(20) $R^1 = \cdot CH_2CH{=}CHCO_2Et$, $R^2 = OH$
(21) $R^1 = \cdot CH_2CH_2CH_2CH_2OH$, $R^2 = OH$
(22) $R^1 = \cdot CH(CO_2Et)_2$, $R^2 = OH$
(23) $R^1 = R^2 = \cdot CH(CO_2Et)_2$

The influence of the nature of the reagent and reaction conditions on Grignard additions to 1,2,5,6-di-*O*-isopropylidene-α-*ribo*-hexofuranos-3-ulose has been studied using ethyl or phenyl magnesium halides; the 3-*C*-alkyl or aryl-allose derivative always predominates, but the nature and amounts of byproducts vary considerably; the 3-*O*-benzyl ether derivatives of the branched-chain sugars were also prepared without difficulty, despite the tertiary hydroxy-group, and hydrogenation of the *C*-phenyl derivative over a rhodium catalyst yielded the *C*-cyclohexyl analogue.[18] Phenyl magnesium bromide with 1,6-anhydro-2,3-*O*-isopropylidene-β-D-*lyxo*-hexopyranosid-4-ulose gave the 4-*C*-phenyl talose derivative (24).[18] The stereoselectivities of the reactions of diazomethane and Grignard reagents with methyl 4,6-*O*-benzylidene-α- and β-D-hexopyranosid-3-ulose derivatives isomeric at C-2 have also been studied and the results rationalized.[19]

The doubly branched-chain sugar (25), a derivative of a segment of the

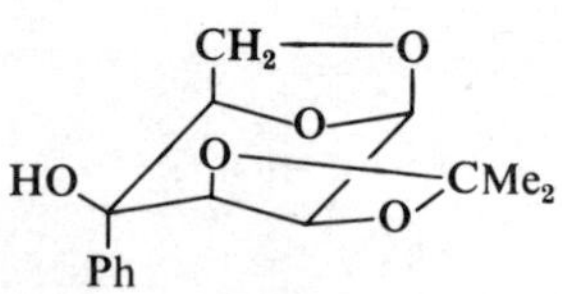

(24)

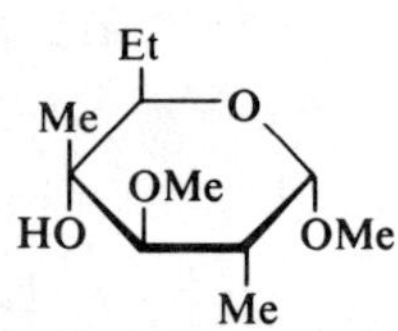

(25)

[14] J.-C. Depezay and Y. Le Merrer, *Tetrahedron Letters*, 1978, 2865.
[15] J.-C. Depezay and A. Dureault. *Tetrahedron Letters*, 1978, 2869.
[16] Yu. A. Zhdanov, Yu. E. Alekseev, and V. A. Tyumenev, *Doklady Akad. Nauk S.S.S.R.*, 1977, **237**, 109 (*Chem. Abs.*, 1978, **88**, 51 102j).
[17] Yu. A. Zhdanov, Yu. E. Alekseev, and S. S. Doroshenko, *Doklady Acad. Nauk S.S.S.R.*, 1978, **238**, 868 (*Chem. Abs.*, 1978, **88**, 170 407x)
[18] J.-C. Fischer and D. Horton, *Carbohydrate Res.*, 1977, **59**, 477.
[19] K. Sato and J. Yoshimura, *Bull. Chem. Soc. Japan*, 1978, **51**, 2116.

macrolide ring in erythronolide A, has been synthesized from a 2-deoxy-2-*C*-methyl-D-glucose derivative by standard transformations.[20]

^{13}C N.m.r. measurements have been used to assign the configuration of the branch carbon in pairs of 3-*C*-hydroxymethyl and 3-*C*-methyl branched-chain 1,2;4,6-di-*O*-acetalated hexopyranose derivatives by comparison with the *Z* and *E* isomers of 4-t-butyl-1-hydroxymethylcyclohexanol and with corresponding diacetals of D-glucose and D-allose.[21]

The 1-deoxy-D-*erythro*-pentulose derivative (26) with phenyl magnesium bromide gave the 2-*C*-phenyl-D-arabinitol (27), whereas the silyl ether of the pentulose (26*a*) gave exclusively the corresponding 2-*C*-phenyl-D-ribitol derivative (27*a*); hydrogenolysis of the 2-*O*-phenylcarbamate of the arabinitol (27) gave the 1,2-dideoxy-2-*C*-phenyl-D-*erythro*-pentitol (28) with inversion at C-2 (Scheme 6);[22] these results were supported by *X*-ray crystal structure data on compounds (27) and (28).[22a] The L-dendroketose derivatives (29 and 30)

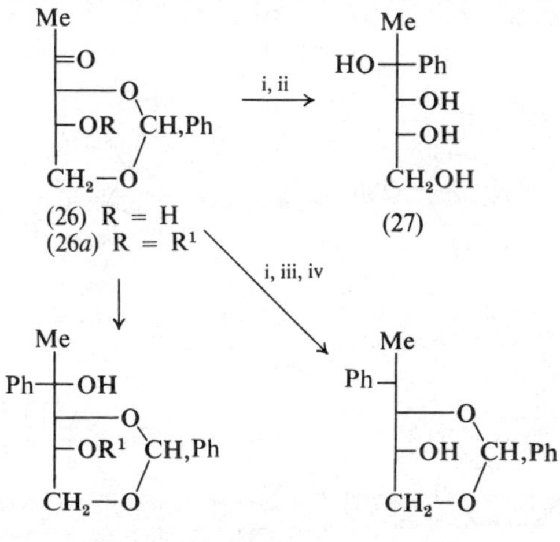

Reagents: i, PhMgBr; ii, H^+–H_2O; iii, PhNCO; iv, H_2–Pd

Scheme 6

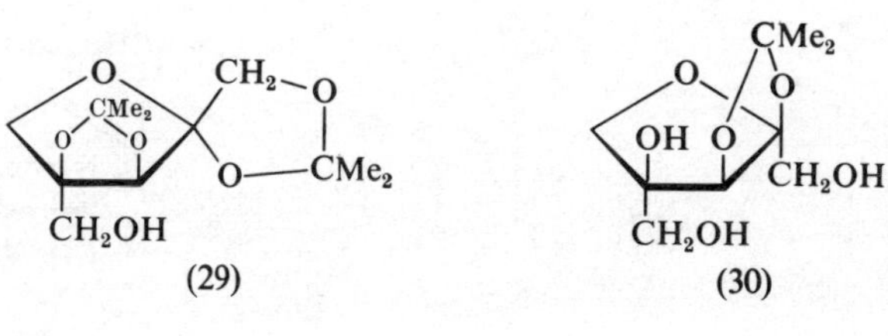

(29) (30)

[20] M. Miljkovic and D. Glisin, *Gals. Hem. Drus. Beogradu*, 1977, **42**, 659 (*Chem. Abs.*, 1978, **88**, 62 542d).

[21] P. M. Collins and V. R. N. Munasinghe, *Carbohydrate Res.*, 1978, **62**, 19.

[22] J.-C. Fischer, D. Horton, and W. Weckerle, *Carbohydrate Res.*, 1977, **59**, 459.

[22a] D. C. Rohner, J.-C. Fischer, D. Horton, and W. Weckerle, *Canad. J. Chem.*, 1978, **56**, 2915.

have been prepared stereospecifically from 2-*C*-hydroxymethyl-D-*erythro*-pentitol or D-*threo*-pentulose by conventional procedures.[23]

An adenine derivative of 3-*C*-methyl-L-*threo*-furanose is mentioned in Chapter 19.

3 Compounds with an R—C—N Branch

D-Evernitrose (31) has been synthesized as shown in Scheme 7.[24]

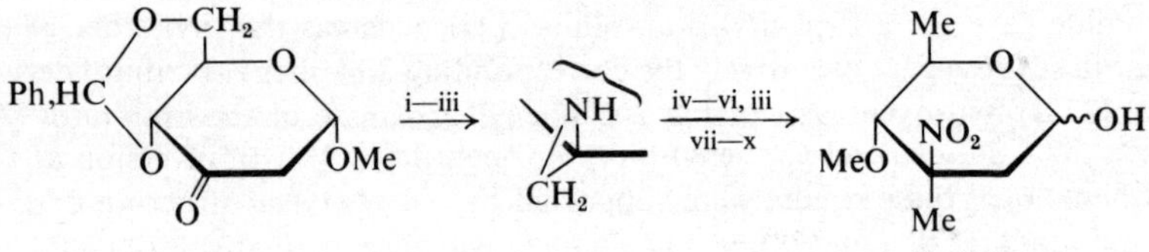

(31)

Reagents: i, HCN; ii, MSCl–py; iii, LAH; iv, H_2–Ni; v, AC_2O–py; vi, NBs; vii, NaH–MeI; viii, KOH; ix, *m*-Cl·$C_6H_4CO_3H$; x, H^+–H_2O

Scheme 7

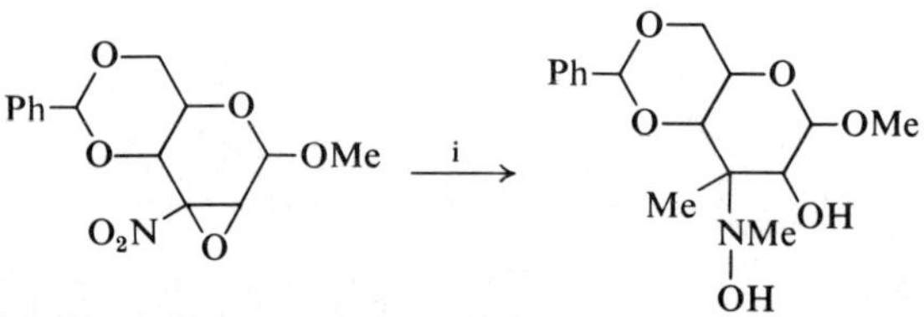

Reagents: i, NaN_3–HN_3–DMF; ii, H_2–Pd; iii, HOAc; iv, $NaIO_4$; v, $NaBH_4$; vi, BzCl–py; vii, CF_3CO_2H; viii, Ac_2O–py

Scheme 8

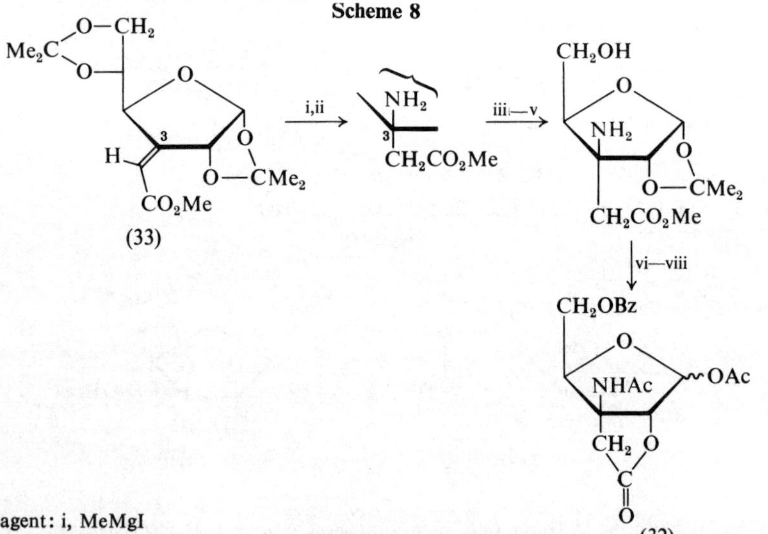

(33)

(32)

Reagent: i, MeMgI

Scheme 9

[23] H. C. Jarrell, W. A. Szarek, and J. K. N. Jones, *Carbohydrate Res.*, 1978, **64**, 283.
[24] J. Yoshimura, M. Matsuzawa, and M. Funabashi, *Bull. Chem. Soc. Japan*, 1978, **51**, 2064.

Carbohydrate α-nitroepoxides react with methylmagnesium iodide to yield branched-chain amino-sugars as outlined in Scheme 8, but the stereochemistry of the process has not been elucidated.[25]

The branched-chain amino-sugar derivative (32) has been prepared from the unsaturated branched-chain sugar (33) by the sequence shown in Scheme 9;[25a] these were then used to prepare analogues of the nucleoside moiety of the polyoxins which are referred to in Chapter 19.

4 Compounds with an R^1—C—R^2 Branch

1,2,4,6-Tetra-*O*-acetyl-3-deoxy-α- and -*β*-D-*erythro*-hex-2-enopyranose undergo photochemical addition of 1,3-dioxolane to yield mixtures of the corresponding 3-deoxy-3-(2-dioxalanyl)hexose stereoisomers (34); irradiation of the α-anomer in acetone–propan-2-ol (1 : 100) yielded the branched-chain sugar (35) whilst using a 1 : 1 ratio of solvents, (35) was obtained along with the novel ortho-acetate (36).[26]

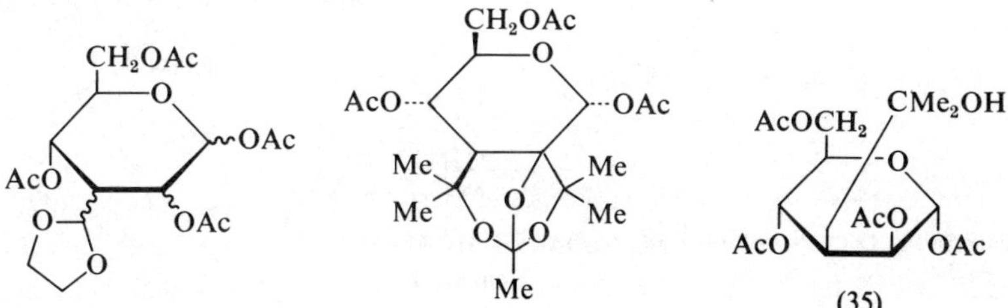

(34) (36) (35)

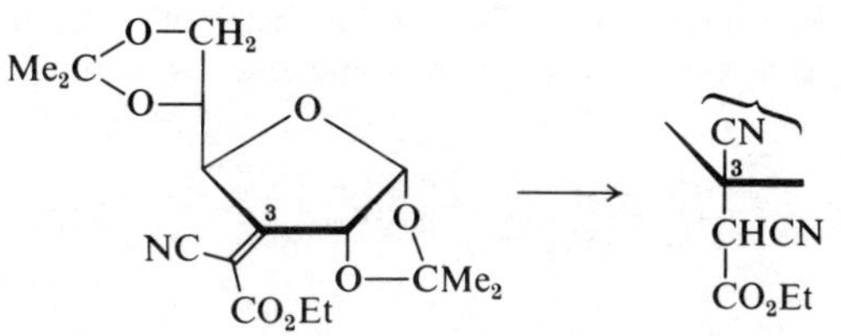

(37) (38)

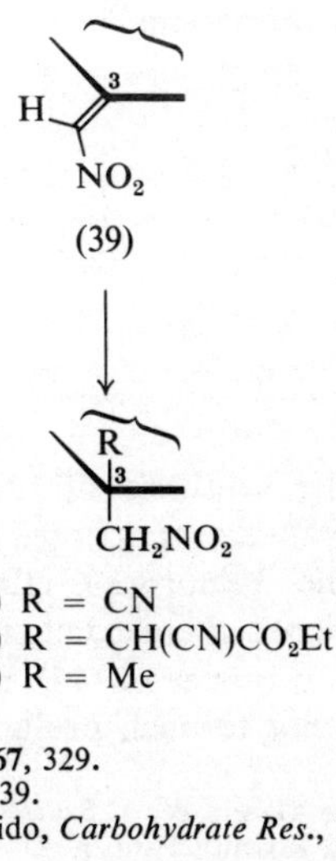

(39)

(40) R = CN
(41) R = $CH(CN)CO_2Et$
(42) R = Me

[25] H. H. Baer and C. B. Madukulu, *Carbohydrate Res.*, 1978, **67**, 329.
[25a] A. Rosenthal and M. Ratcliffe, *Carbohydrate Res.*, 1978 **60**, 39.
[26] Y. Araki, K. Nishiyama, K. Senna, K. Matsuura, and Y. Ishido, *Carbohydrate Res.*, 1978, **64**, 119.

Treatment of the 3-*C*-substituted methylene branched-chain sugar (37) with cyanide using phase-transfer catalysis has led to the *gem*-di-substituted derivative (38); the 3-*C*-nitro-methylene analogue (39) furnished the Michael adducts (40—42).[27]

Rosenthal's group have reported further syntheses of sugars incorporating amino-acid branched chains, using the lactone procedure to prepare the glycine derivatives shown in Scheme 10, the configurations being assigned by n.m.r. and c.d. measurements.[28]

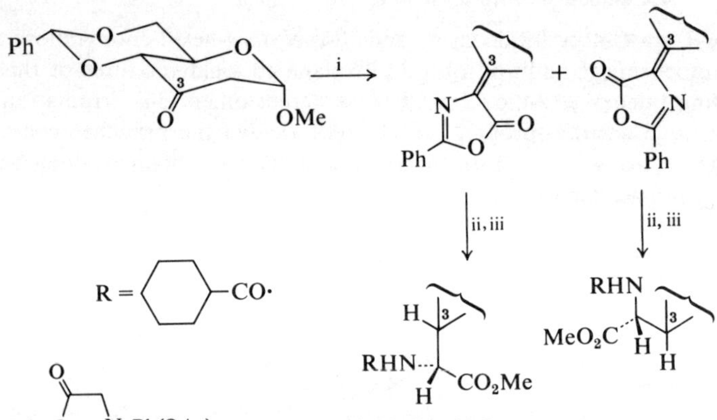

Reagents: i, 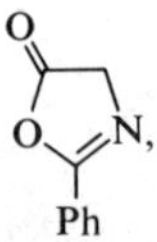 N, $Pb(OAc)_2$; ii, NaOAc; iii, H_2–Rh–Al_2O_3

Scheme 10

Methyl 2,4-dideoxy-4-*C*-carboxymethyl-α-D-*ribo*-hexopyranoside 2′ → 3 lactone (43), required as a key intermediate for the synthesis of optically active thromboxanes, has been synthesized from D-glucose by standard transformations.[29]

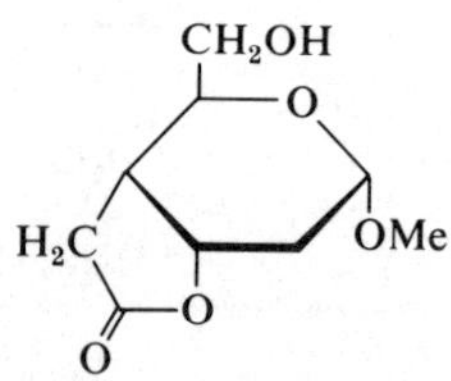

(43)

The synthesis of branched-chain sugars by application of the Claisen rearrangement is noted in Chapter 12.

The 3-fluoro-3-*C*-ethoxydicarbonyl-D-glucose derivative (44) yielded the branched-chain compounds (45) and (46) in a Wittig reaction (Scheme 11); the *cis*-isomer of (46), required for the synthesis of showdomycin analogues, was not formed, presumably for steric reasons; some product (47) with the

[27] Y. Ali and W. A. Szarek, *Carbohydrate Res.*, 1978, **67**, C17.
[28] A. Rosenthal and K. Dooley, *Carbohydrate Res.*, 1978, **60**, 193.
[29] H. Ohrui and S. Emoto, *Agric. Biol. Chem.* (*Japan*), 1977, **41**, 1773.

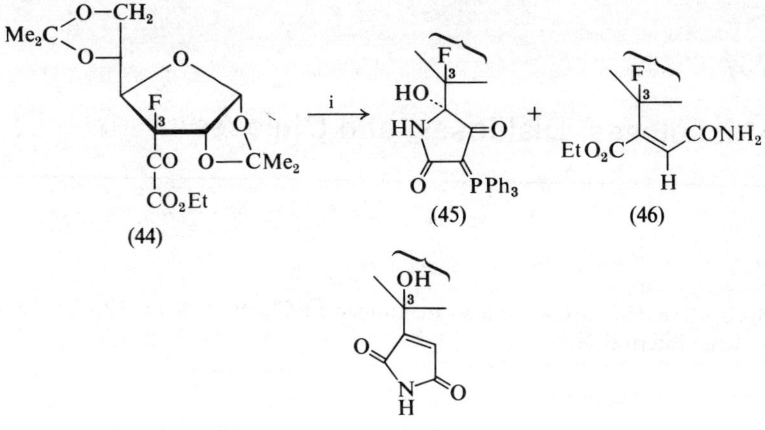

(44) (45) (46) (47)

Reagent: i, $Ph_3P{=}CHCONH_2$

Scheme 11

desired maleimide unit was obtained from the corresponding non-fluorinated glucose derivative.[30]

The synthesis of 3-*C*-acetamidomethyl-3-deoxy-D-ribofuranose and related branched-chain sugars described by Abdel-Rahman (see Vol. 11, p. 124) has been claimed to be a copy of an earlier publication by Lourens [31] (see Vol. 3, p. 122).

The rearrangement of a 2-deoxy-2-*C*-methyl-glycosulose derivative is mentioned in Chapter 14.

[30] J. C. A. Boeyens, A. J. Brink, and A. Jordaan, *South African J. Chem.*, 1978, **31**, 7 (*Chem. Abs.*, 1978, **89**, 197 865c).
[31] G. J. Lourens, *Bull. Chem. Soc.* (*Japan*), 1978, **51**, 955.

14
Aldosuloses, Dialdoses, and Diuloses

Glycosulose derivatives are also mentioned in Chapters 8, 11, 12, and 13 as synthetic intermediates.

1 Synthesis

Pyridinium chlorochromate in benzene has been recommended for the oxidation of carbohydrates; good to excellent yields were obtained in a few hours when isolated hydroxyl-groups in some standard compounds were oxidized to carbonyl functions.[1] Excellent yields have also been reported for such oxidations by brief treatment with chromium trioxide–pyridine complex in the presence of acetic anhydride.[2] Another variant on the use of dimethyl sulphoxide as an oxidant utilizes acetyl chloride as the co-reagent in dichloromethane at low temperature for several hours; yields of 60—75% of some standard glycosuloses [3] were claimed. Binkley's group have now successfully applied their photolytic oxidation procedure using pyruvate esters (see Vol.10, p. 171; Vol. 11, p. 129) to some nucleoside derivatives; 5′-*O*-trityl-thymidine yielded the corresponding ketone in good yield; other oxidation procedures readily overoxidize the unstable carbonyl products.[4]

The oxidation of glycosides to glycosid-4-uloses using *Acetobacter suboxydans* reported last year (see Vol. 11, p. 129) has been further investigated; the α-anomers of methyl D- and L-arabinopyranoside and D-lyxopyranoside were slowly oxidized at C-4 with moderate efficiency, whilst the β-anomers of methyl L-arabino- and D-ribopyranoside were oxidized at C-2; the specificity of the enzyme system towards methyl glycopyranosides was discussed, and ^{13}C n.m.r. data on the glycosidulose *O*-methyl oximes recorded.[5]

The isomeric octodialdose derivatives (1) have been prepared by the sequence of reactions outlined in Scheme 1; these compounds are novel *trans*-decalin type carbohydrate derivatives of interest as analogues of the octodiose units in apramycin and oxyapramycin.[6]

Sequential hydrolysis and methanolysis of the enol-acetal (2) yielded methyl 6-deoxy-D-*xylo*-hexo-furanosid-5-ulose (3); treatment of the 2,3-di-*O*-toluene-*p*-sulphonyl derivative of (3) with DBU gave a mixture of 2-methoxy-5-

1 D. H. Hollenberg, R. S. Klein, and J. J. Fox, *Carbohydrate Res.*, 1978, **67**, 491.
2 P. J. Garegg and B. Samuelson, *Carbohydrate Res.*, 1978, **67**, 267.
3 A. Skibinski, *Polish J. Chem.*, 1978, **52**, 1841.
4 R. W. Binkley, D. G. Hehemann, and W. W. Binkley, *J. Org. Chem.*, 1978, **43**, 2573.
5 G. W. Schnarr and W. A. Szarek, *Canad. J. Chem.*, 1978, **56**, 1752.
6 H. C. Jarrell and W. A. Szarek, *Canad. J. Chem.*, 1978, **56**, 144.

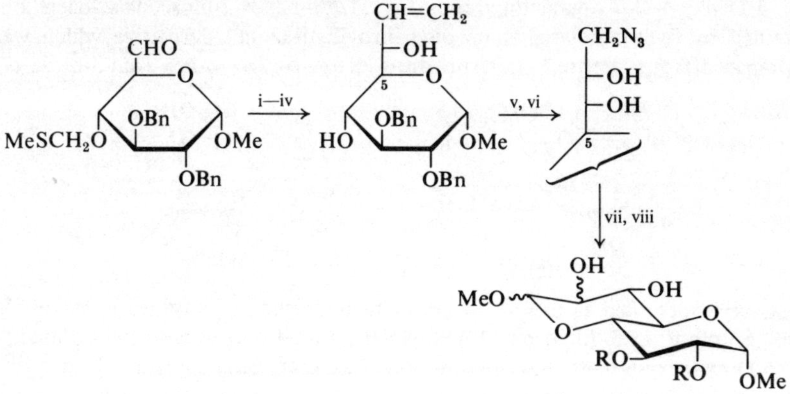

(1) R = Bn or H

Reagents: i, CH_2=CHMgBr; ii, Ac_2O–py; iii, $HgCl_2$; iv, MeO^-; v, *m*-$ClC_6H_4CO_3H$; vi, N_3^-; vii, $h\nu$; viii, H^+–MeOH

Scheme 1

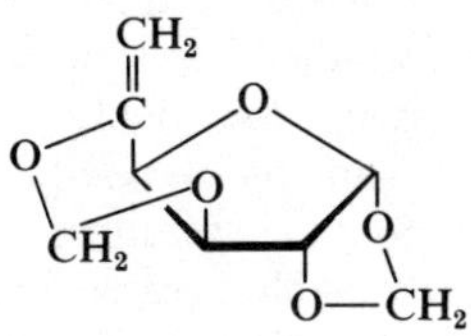

(2)

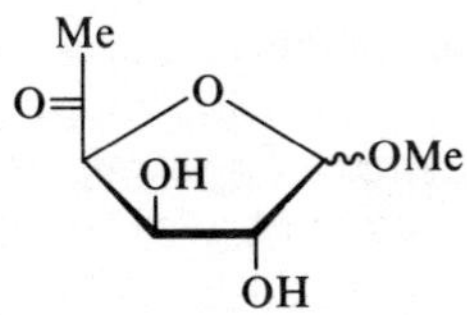

(3)

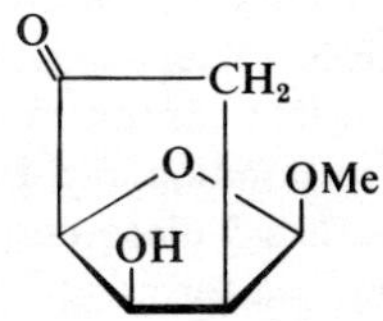

(4)

acetoxyfuran and a carbonyl derivative assigned the structure (4) from mass spectral evidence, although its n.m.r. spectrum was inconsistent with this structure.[7]

Tetrahydropyranyl ethers provide useful blocking groups for isolated hydroxy-groups during lithium-alkyl mediated reactions, which is illustrated for the synthesis of methyl 2,6-dideoxy-4-*O*-(tetrahydropyran-2-yl)-α-L-*erythro*-hexopyranosid-3-ulose (5) in scheme 2; similarly the 1-*O*-tetrahydropyranyl ether of 2,3:4,5-di-*O*-isopropylidene-*β*-D-fructopyranose yielded the hex-5-enulopyranose derivative (6) with lithium di-isopropylamide.[8]

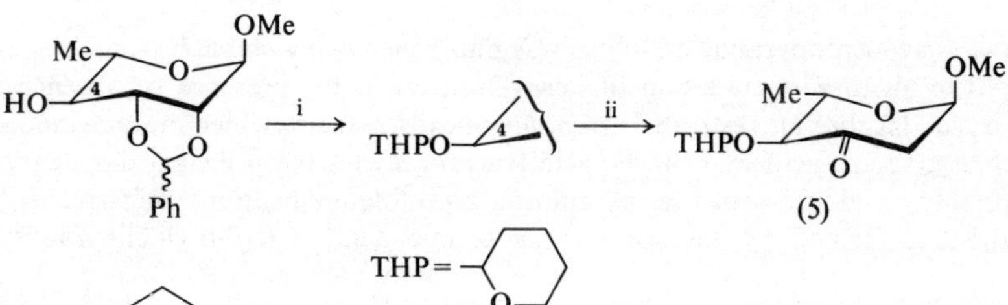

(5)

Reagents: i, 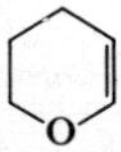 TsOH; ii, Bu^sLi

Scheme 2

[7] R. J. Ferrier and V. K. Srivastava, *Carbohydrate Res.*, 1977, **59**, 333.
[8] A. Klemer and D. Balkau, *J. Chem. Research* (S), 1978, 303; (M), 1978, 3823.

3-Deoxy-4-*O*-β-D-galactopyranosyl-D-*erythro*-hexos-2-ulose has been obtained by hydrazinolysis of its bis(benzoylhydrazone) derivative which was prepared from lactose.[9] 1,5-Anhydro-4-deoxy-D-*glycero*-hex-1-en-3-ulose (7)

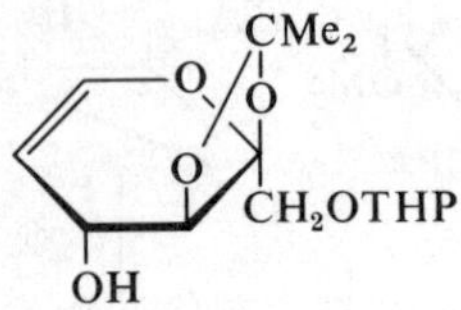

(6)

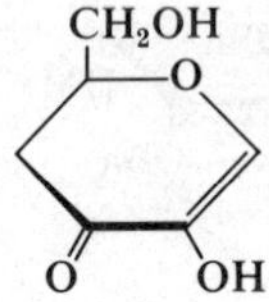

(7)

has been identified as one of the components in the tar obtained by pyrolysis of cellulose; 3,5-dihydroxy-2-methyl-4*H*-pyran-4-one was also isolated.[10] 3,5-Dihydroxy-6-methyl-2*H*-pyran-2-one has been isolated from the tar obtained by heating glucose with diethylammonium tartrate; cyclic aminoreductones were also characterized as components of the tar.[11]

2 Reactions

The photodecarbonylation of di-*O*-isopropylidene-D-*ribo*-hexos-3-ulose derivatives possessing sugar rings of different sizes has been studied; the septanose form lost carbon monoxide with ring contraction to give a low yield of the corresponding α-D-ribopyranose derivative, and an isomeric D-*xylo*-hexoseptanos-5-ulose derivative behaved analogously; by contrast, furanose and pyranose forms gave no ring-contracted products, presumably because these would be too strained.[12] In another study t-butyl-3,4-*O*-isopropylideneglycopyranosid-2-uloses were shown to undergo Norrish Type 1 cleavage with loss of carbon monoxide leading to ring-contracted or acyclic-unsaturated products (Scheme 3); the corresponding phenyl 3,4-*O*-isopropylidene-α-

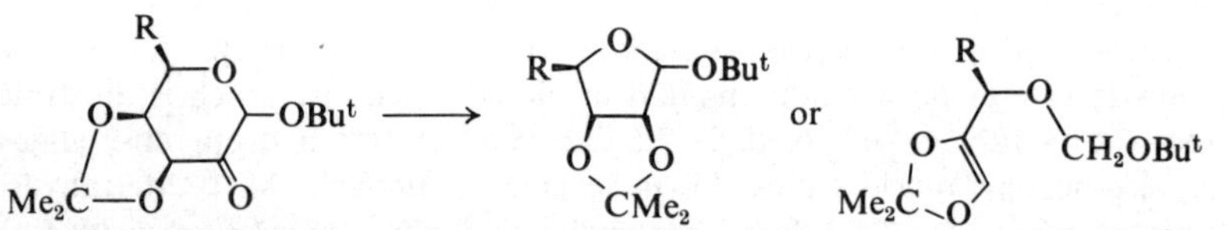

R = H or Me

Scheme 3

L-*erythro*-pentopyranosid-2-ulose was photochemically stable.[13]

The alkaline degradation of hexos-2-uloses in the presence or absence of oxygen has been investigated; D-*arabino*-hexos-2-ulose yielded more mannonic than gluconic acid by benzylic acid rearrangement, the principal degradation pathway yielding pentoses, pentuloses, and 3-deoxy-pentonic acids from the ene-diol formed by loss of C-1 as formic acid; 4-*O*-β-D-glucopyranosyl-

[9] M. M. A. Rahman, E. S. H. El Ashry, and N. Rashed, *Carbohydrate Res.*, 1978, **64**, 89.
[10] F. Shafizadeh, R. H. Furneaux, T. T. Stevenson, and T. G. Cochran, *Carbohydrate Res.*, 1978, **67**, 433.
[11] H. Pehr, G. Manninger, and T. Severin, *Carbohydrate Res.*, 1978, **64**, 1.
[12] K. Heyns, R. Nesk, and M. Paal, *Tetrahedron Letters*, 1978, 4011.
[13] P. M. Collins, R. Iyer, and A. S. Travis, *J. Chem. Research* (S), 1978, 446; (M), 1978, 5344.

D-*arabino*-hexos-2-ulose additionally underwent elimination of glucose to give an unstable 1,2,3-tricarbonyl intermediate which led to either 3,4-dihydroxybutanoic acid or 3-deoxypentonic acids; it was concluded that the formation of aldonic acid end groups *via* terminal aldosuloses is unimportant in the oxygen–hydrogen carbonate treatment of (1 → 4) linked polysaccharides.[14] (See also Vol. 11, p. 131.)

Reduction of the pentosid-2-ulose (8) by sodium borohydride in ethanol gave the dimer (9) as the major product, the expected lyxoside (10) being only a minor product (Scheme 4), whereas the β-anomer of (8) gave only the β-

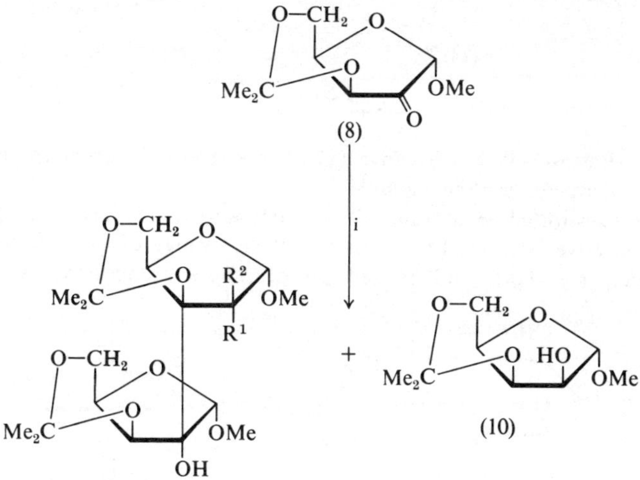

(9) R^1 = OH, R^2 = H
(11) R^1,R^2 = O=

Reagents: i, $NaBH_4$–EtOH

Scheme 4

anomer of (10); treatment of (8) with acetic anhydride and triethylamine gave the 2-ulose dimer (11), the presumed precursor of (9).[15] The lithium aluminium hydride reduction of levo-glucenone (1,6-anhydro-3,4-dideoxy-β-D-*glycero*-hex-3-enopyranos-2-ulose) (12) has been shown [16, 17] to give the D-*threo*-hex-2-enopyranose derivative (13), and not the D-*erythro* isomer previously claimed,[18] and its 3,4-dihydro analogue (14) similarly gives the 3,4-dideoxy-D-*threo*-hexose derivative (15) (Scheme 5).

3-*O*-Acetyl-1,6-anhydro-2-deoxy-2-*C*-methyl-β-D-xylopyranos-4-ulose (16) rearranged in presence of methylenetriphenylphosphorane to give the isomeric

[14] L. A. Lindstrom and O. Samuelson, *Carbohydrate Res.*, 1978, **64**, 57.
[15] J. C. A. Boevens, A. J. Brink, and A. Jordaan, *J. Chem. Research* (S), 1978, 187; (M), 1978, 2437.
[16] J. S. Brimacombe, F. Hunedy, and L. C. N. Tucker, *Carbohydrate Res.*, 1978, **60**, C11.
[17] J. Pecka and M. Cěrny, *Coll. Czech. Chem. Comm.*, 1978, **43**, 1720.
[18] F. Shafizadeh and P. P. S. Chin, *Carbohydrate Res.*, 1977, **58**, 79.

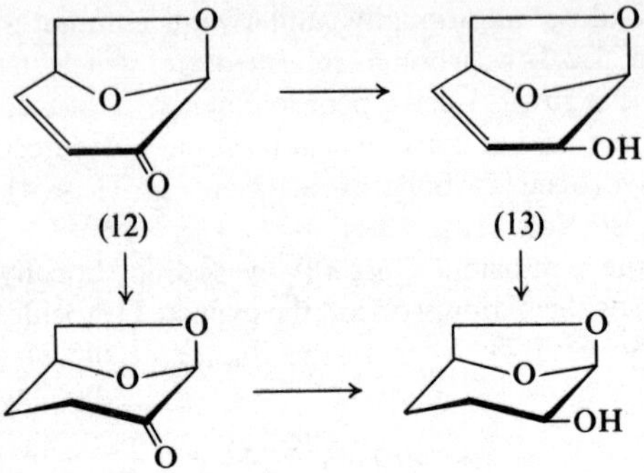

(12) (13) (14) (15)

Scheme 5

4-*O*-acetylhexos-3-ulose derivative (17), the suggested mechanism involving enediol intermediates (Scheme 6).[19]

The 6-substituted-hexofuranos-5-ulose derivatives (18) have been prepared and these have been used to synthesize a wide range of other 6-substituted analogues, *e.g.* (19), and related glycofuranose compounds, *e.g.* (20).[20]

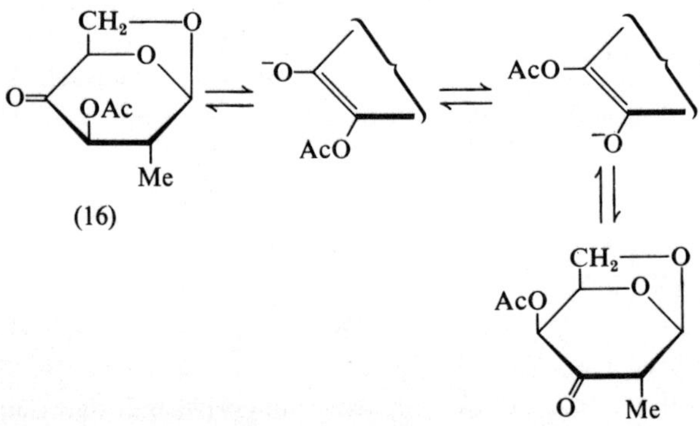

(16) (17)

Scheme 6

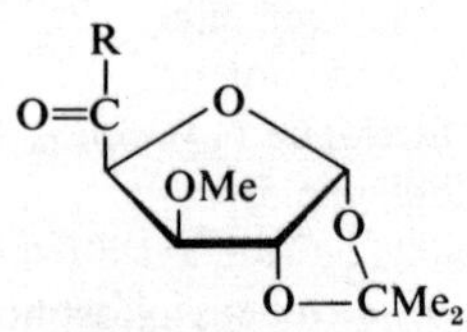

(18) R = CH_2SBn or CH_2Cl
(19) R = $CH_2S(O)Bn$, CCl_2SO_2Bn, CBr_2SO_2Bn, CBr=CHPh, or CH(SMe)COMe

[19] A. Ya. Shmyrina, A. S. Shashkov, A. F. Sviridor, O. S. Chizhov, and N. K. Kochetkov, *Bioorg. Khim.*, 1977, **3**, 1349 (*Chem. Abs.*, 1978, **88**, 38 086n).
[20] J. M. J. Tronchet and H. Eder, *Helv. Chim. Acta*, 1978, **61**, 2254,

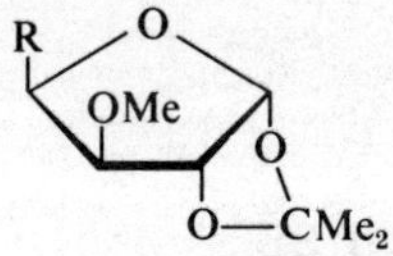

(20) R = $C(SBn){=}CH_2$, $CH{=}SBnMe$, or 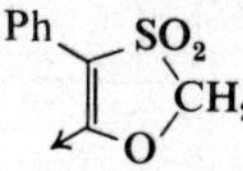

Additions of carbon nucleophiles to enones and the consequent synthesis of *C*-glycosyl compounds are noted in Chapter 3.

15
Sugar Acids and Lactones

1 Aldonic Acids

In alkaline media, and at room temperature, in the presence of platinum or rhodium, free sugars give excellent yields of aldonic acids. Reaction kinetics were studied for a range of hexoses, and a mechanism of reaction was proposed based on adsorption of the sugar on the metal.[1] A range of reports have appeared on the oxidation of free sugars; often the kinetics of the processes were given more attention than characterization of the products, and often C—C cleavage occurred to give multiple products:

Sugar studied	*Oxidizing agent*	*Ref.*
Arabinose	Cu^{2+}, H^+	2
Arabinose, xylose	Cu^{2+}, NH_4OH	3
Xylose	Manganese pyrophosphate, H^+	4
Glucose	Hg^{2+}; Ag^+, H^+	5
Glucose	Cu^{2+}; Fe^{3+}; Ce^{4+}, H^+	6
Glucose	Nessler's reagent ($Hg\,I_3^-$)	7
Glucose, galactose	Cu^{2+}, NH_4OH	8
Galactose	Fe^{3+}, H^+	9
Fructose, sorbose	Cu^{2+}, NH_4OH	10

Silver carbonate on Celite in methanol causes the oxidation of 3,4,6-tri-, 3,4-di-, and 3-*O*-methyl-D-fructose to the corresponding ethers of D-arabinonic acid. The 1,4,6-, 1,4,5-, and 1-ethers cleave at C-2—C-3, whereas the 1,3,4,6- and 1,3,4-substituted compounds are resistant to reaction under these conditions.[11] A suitable method for oxidizing 2,3:5,6-di-*O*-isopropylidene-α-D-mannofuranose and β-D-allofuranose to the corresponding lactones involves

[1] G. de Wit, J. J. de Vlieger, A. C. Kock-van Dalen, A. P. G. Kieboom, and H. van Bekkum, *Tetrahedron Letters*, 1978, 1327.
[2] V. I. Krupenskii, *Zhur. priklad. Khim.*, 1978, **51**, 1420 (*Chem. Abs.*, 1978, **89**, 110 171b).
[3] M. P. Singh, V. Tripathi, A. K. Singh, and R. K. Singh, *Indian J. Chem., Sect. A*, 1978, **16A**, 78 (*Chem. Abs.*, 1978, **88**, 170 406w).
[4] R. P. Bhatnagar and A. G. Fadnis, *Monatsh.*, 1978, **109**, 319 (*Chem. Abs.*, 1978, **89**, 12 726s).
[5] V. I. Krupenskii and T. V. Dolgaya, *Khim. Drev.*, 1978, 84 (*Chem. Abs.*, 1978, **89**, 6483u).
[6] V. I. Krupenskii and T. V. Dolgaya, *Khim. Drev.*, 1978, 81 (*Chem. Abs.*, 1978, **89**, 6482t).
[7] M. P. Singh and M. Kumar, *Nat. Acad. Sci. Letters* (*India*), 1978, **1**, 177 (*Chem. Abs.*, 1978, **89**, 129 824p).
[8] M. P. Singh, A. K. Singh, V. Tripathi, and R. K. Singh, *Indian J. Chem., Sect. A*, 1978, **16A**, 205 (*Chem. Abs.*, 1978, **89**, 44 007z).
[9] V. I. Krupenskii, *Zhur. priklad. Khim.*, 1978, **51**, 1420 (*Chem. Abs.*, 1978, **89**, 110 172c).
[10] M. P. Singh, A. K. Singh, and V. Tripathi, *J. Phys. Chem.*, 1978, **82**, 1222.
[11] H. Hammer and S. Morgenlie, *Acta Chem. Scand.*, 1978, **B32**, 343.

the use of ruthenium(VIII) oxide. Over-oxidation, however, occurs to some extent at C-4 and products, *e.g.* (1), with hydroxy-groups at that position were also obtained.[12] An unusual isomerization of 3,4,6-tri-*O*-methyl-D-mannose

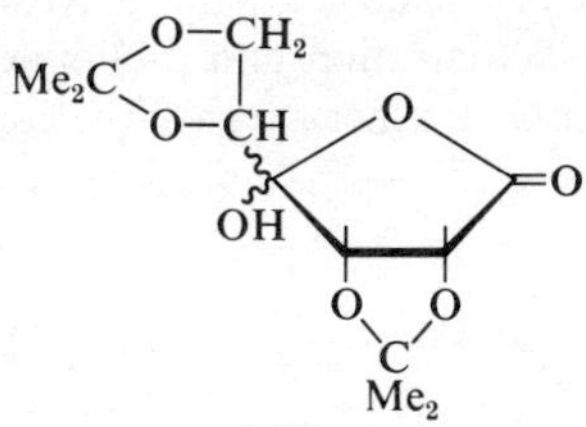

(1)

to 2-deoxy-3,4,6-tri-*O*-methyl-D-*arabino*-hexonic acid is reported to occur on heating with lead(II) hydroxide in aqueous solution.[13]

By use of hydroxy- and keto-acids as model compounds, the Ruff degradation of aldonic acids with hydrogen peroxide and iron(III) has been examined, and the pathway illustrated in Scheme 1 proposed. A remaining difficulty,

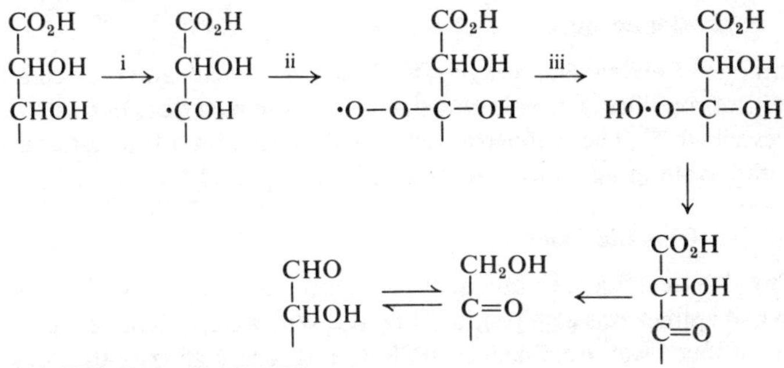

Reagents: i, $HO^{\bullet}$; ii, O_2; iii, $Fe_3{}^{+}$–H^+

Scheme 1

however, relates to the observation that single epimers of the aldose products are found, whereas this scheme would lead to the expectation that mixtures would result.[14]

A review (in German) of dichloromethyl methyl ether as a reagent in carbohydrate chemistry outlines its use in the conversion of sugar acids into the corresponding acid chlorides.[15]

L-Xylono-1,4-lactone can be obtained from the acid by storage over phosphorus pentoxide.[16] The hydrolysis of the lactones of 4,6-dideoxy-L-*xylo*-, L-*lyxo*-, L-*ribo*-, and L-*arabino*-hexonic acids, and methyl ethers thereof, has

12 R. H. Hall and K. Bischofburger, *Carbohydrate Res.*, 1978, **65**, 139.

13 R. A. Gakhokidze, *Soobshch. Akad. Nauk Gruz. S.S.R.*, 1978, **89**, 601 (*Chem. Abs.*, 1978, **89**, 101 147y).

14 K. J. Buchanan, A. Goosen, and J. D. Lovelock, *S. African J. Chem.*, 1977, **30**, 191 (*Chem. Abs.*, 1978, **89**, 110 165c).

15 H. Gross, I. Farkas, and R. Bognar, *Z. Chem.*, 1978, **18**, 201.

16 A. Skibinski, *Pol. J. Chem.*, 1978, **52**, 1307 (*Chem. Abs.*, 1978, **89**, 180 266k).

been studied kinetically, and the rates have been interpreted in terms of such factors as steric effects, intramolecular hydrogen bonding, and relative strengths of the acids.[17]

Acetylation of 4,6-*O*-benzylidene-2-benzyloxycarbonylamino-2-deoxy-D-gluconic acid resulted in lactonization and β-elimination to give compound (2),[18] and similarly, D-glucono-1,5-lactone gave the corresponding unsaturated

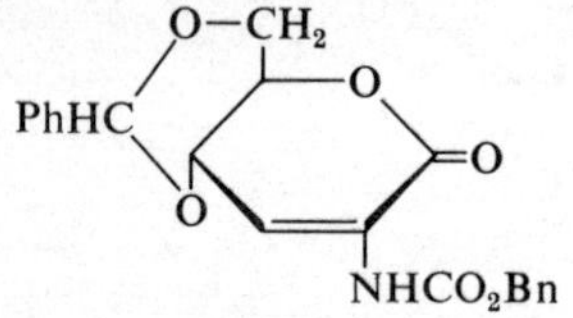

(2)

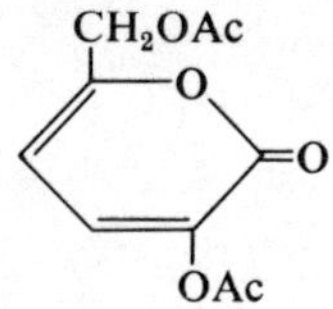

(3)

product as well as the tetra-acetate at room temperature on treatment with acetic anhydride in pyridine. On prolonged reaction, or at elevated temperature, further elimination occurred to give a product described as 3-acetoxy-6-acetoxymethylpyran-2-one (3).[19]

2 Aldaric Acids

The syntheses of several dialkyl xylarates from xylaric acid, and their conversion to 2,4-*O*-isopropylidene acetals and thence 3-*O*-methacryloyl derivatives have been described.[20] The copper(II) catalysed oxidation of tartaric acid by hexacyanoferrate in alkaline medium has been reported.[21]

3 Ulosonic Acids

4-*O*-Methyl-*N*-acetylneuraminic acid was synthesized as shown in Scheme 2, and the C-4 epimer was also produced by the Wittig step of the sequence. The primary product was obtained crystalline, and characterized by mass spectrometry and by proton and ^{13}C n.m.r. spectroscopy.[22] 5-*O*-(α-L-Rhamnopyranosyl)-, (β-D-glucopyranosyl)-, (2-acetamido-2-deoxy-β-D-glucopyranosyl)-, and (β-D-glucopyranosyluronic acid)-3-deoxyoctulosonic acids were prepared in low yields by condensation between the four 2-*O*-glycopyranosyl-D-arabinoses and oxalacetic acid.[23]

A specially prepared palladium on charcoal catalyst permits the oxidation of L-sorbose to L-*xylo*-hexulosonic acid in 50% yield and is of potential value in the industrial production of vitamin C.[24] The inhibitory effect of various

[17] L. Helešic, K. Kekurt, and J. Jarý, *Coll. Czech. Chem. Comm.*, 1978, **43**, 1241.
[18] G. Deak, K. Gall-Istok, J. Hasko-Breuer, and P. Sohar, *Acta Chim. Acad. Sci. Hung.*, 1977, **92**, 421 (*Chem. Abs.*, 1978, **88**, 23 286d).
[19] C. R. Nelson and J. S. Gratzl, *Carbohydrate Res.*, 1978, **60**, 267.
[20] V. L. Lapenko, B. I. Mikhant'ev, and A. I. Slivkin, *Izvest. Vyssh. Uchebn. Zaved., Khim. Khim. Tekhnol.*, 1978, **21**, 659 (*Chem. Abs.*, 1978, **89**, 129 851v).
[21] S. S. Srivastava and P. C. Mathur, *Indian J. Chem., Sect. A*, 1977, **15A**, 788 (*Chem. Abs.*, 1978, **88**, 89 958f).
[22] J.-M. Beau, P. Sinay, J. P. Kamerling, and J. F. G. Vliegenthart, *Carbohydrate Res.*, 1978, **67**, 65.
[23] R. S. Sarfati and L. Szabo, *Carbohydrate Res.*, 1978, **65**, 11.
[24] A. Lengyel-Mészáros, B. Losonczi, J. Petró, and I. Rusznák, *Acta Chim. Acad. Sci. Hung.*, 1978, **97**, 213.

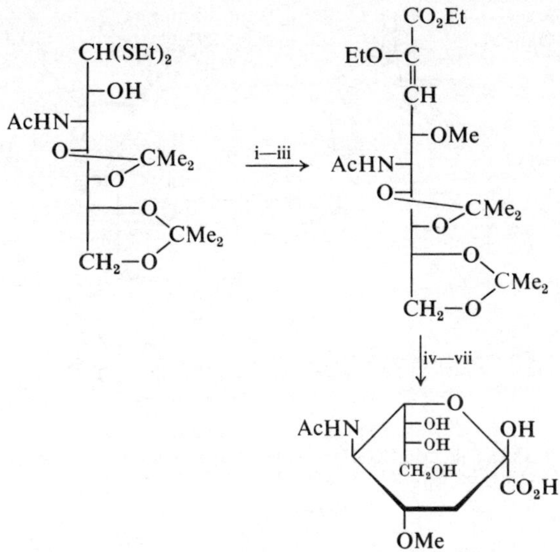

Reagents: i, MeI–$Ba(OH)_2$–DMF; ii, HgO–H_2O; iii, $Ph_3P{=}C\langle^{OEt}_{CO_2Et}$; iv, EtOH–$Hg(OAc)_2$; v, $NaBH_4$; vi, CF_3CO_2H; vii, H^+

Scheme 2

anions on the electrochemical oxidation of di-*O*-isopropylidene-L-sorbose in alkaline solution has been reported.[25]

D-*Arabino*-hexulosonic acid has been obtained crystalline from a *Pseudomonas fluorescens* growth medium.[26]

4 Uronic Acids

A synthesis of riburonic acid derivatives from simple precursors is outlined in Scheme 3; resolution of the intermediate acids was effected and both D- and L-products were obtained.[27] Other studies in the D-riburonic acid series reported the glycosyl bromide (4) and thence the corresponding azide; the amide (5) led to the unusual *N*-heterocyclic derivative (6).[28] Beckmann rearrangement of the oxime (7) in the presence of copper(II) acetate gave the amide (8), dehydration of which yielded the nitrile (9).[29]

An improved method of synthesis of benzyl 2,3,4-tri-*O*-benzyl-D-glucopyranuronate of value in the synthesis of glucuronosides has been reported,[30] and compounds (10) and (11) were prepared as inhibitors of β-D-glucuronidase. The former was made from D-glucaro-1,4-lactone, and the latter by application

[25] T. E. Pechatnova, I. A. Avrutskaya, and M. Ya. Fioshin, *Elektrokhimiya*, 1978, **14**, 946 (*Chem. Abs.*, 1978, **89**, 147 163b).
[26] H. Faubl and C. A. Kolb, *Carbohydrate Res.*, 1978, **63**, 315.
[27] R. R. Schmidt and A. Lieberknecht, *Angew. Chem. Internat. Edn.*, 1978, **17**, 769.
[28] R. R. Schmidt, K.-H. Jung, and P. Hermentin, *Chem. Ber.*, 1978, **111**, 3311.
[29] Yu. A. Zhdanov, Yu. E. Alekseev, and T. P. Sudareva, *Doklady Akad. Nauk S.S.S.R.*, 1978, **238**, 580 (*Chem. Abs.*, 1978, **88**, 152 880s).
[30] D. Keglević and D. Ljevákovič, *Carbohydrate Res.*, 1978, **64**, 319.

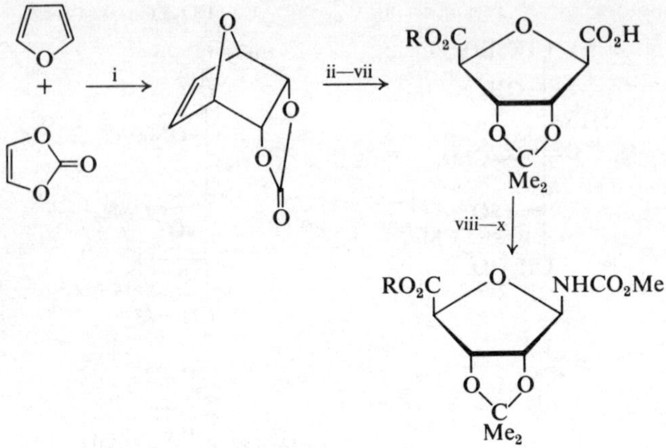

Reagents: i, Δ; ii, OsO_4; iii, Me_2CO–H^+; iv, $Ba(OH)_2$; v, $KMnO_4$; vi, Ac_2O; vii, ROH; viii, $ClCO_2Et$; ix, Me_3SiN_3; x, MeOH

Scheme 3

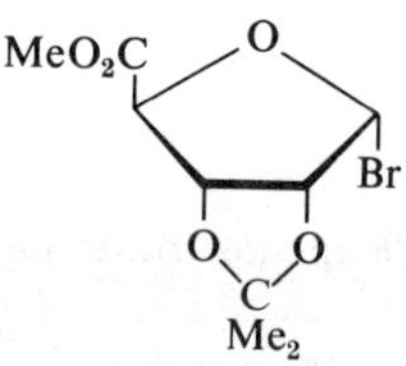

(4)

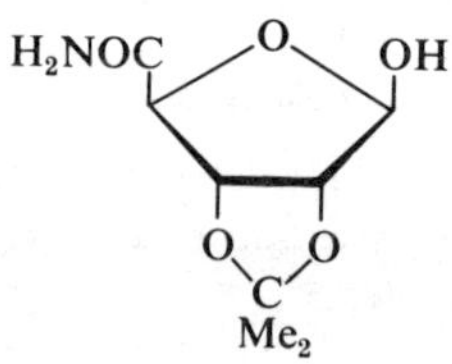

(5)

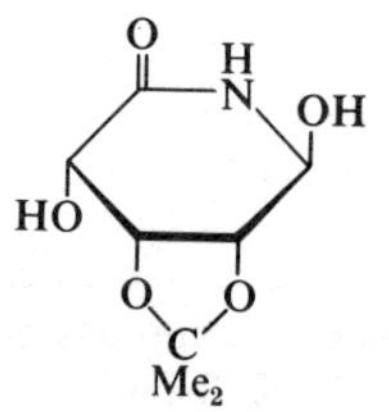

(6)

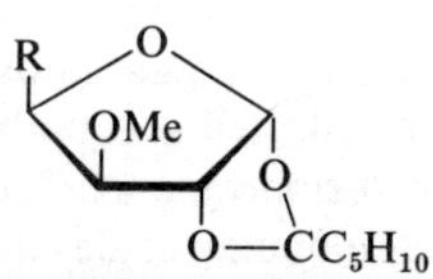

(7) R = CH=NOH
(8) R = $CONH_2$
(9) R = CN

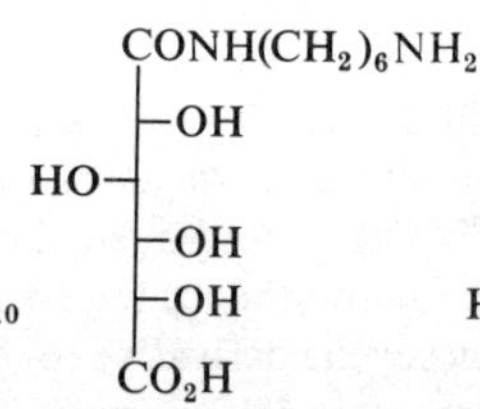

(10)

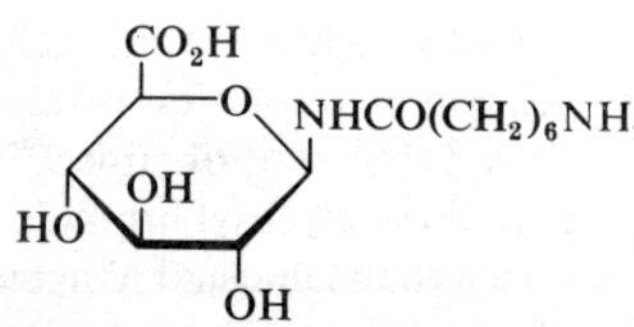

(11)

of the well-known acetylated glycosyl halide of the methyl uronate to give the azide which was reduced and acylated.[31]

Phenyl α-L-idopyranosiduronic acid and the 4-methylumbelliferyl analogue, required for the assay of α-L-iduronidase, were obtained from α-L-idopyranose penta-acetate by standard condensations followed by saponification of the ester groups and platinum-catalysed oxidation at C-6.[32] The synthesis of 5-azido-5-deoxy-L-idofuranurono-6,3-lactone derivatives is mentioned in

[31] R. T. Lee and Y. C. Lee, *Carbohydrate Res.*, 1978, **64**, 302.
[32] R. M. Srivastava, N. Hudson, F. R. Seymour, and B. Weissman, *Carbohydrate Res.*, 1978, **60**, 315.

Chapter 9. A substrate used in the assay of L-iduronic acid 2-sulphate sulphatase was obtained from sodium heparinate by methanolysis followed by nitrosation. Disaccharide monosulphates presumed to contain *O*-(methyl α-L-idopyranosiduronate-2-sulphate)-(1 → 4)-2,5-anhydro-D-mannose were isolated by gel filtration.[33] Brucine L-guluronate and D-mannuronate, on the other hand, were obtained by treatment with brucine of the barium salts of the acids formed on hydrolysis of alginic acid.[34]

Photochemical cleavage of the glycosidic linkage of uronic acid-containing compounds was studied using methyl (methyl D-glucopyranosid)uronate, and compounds (12)—(16) were identified amongst the products. The first three are believed to be derived from a radical produced at C-5 by C-5—C-6 fission,

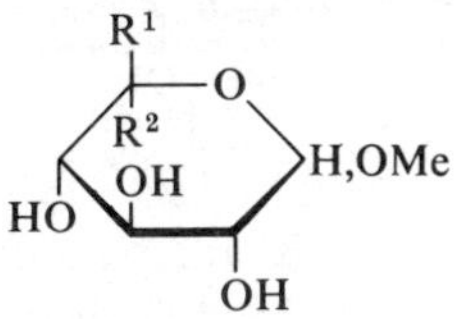

	R^1	R^2
(12)	H	H
(13)	CH_2OH	H
(14)	H	CH_2OH

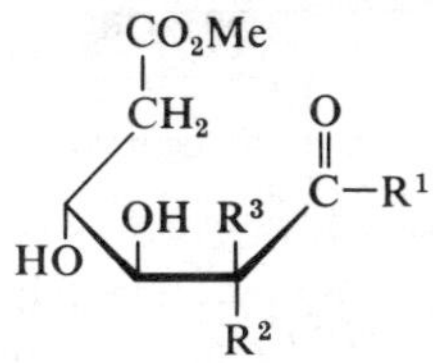

	R^1	R^2	R^3
(15)	H	OH	H
(16)	OMe	H	H

and the last two by photoreduction following hydrogen abstraction by the carbonyl group.[35] A related investigation examined the degradation of glucuronides and galacturonides with lead tetra-acetate followed by alkali (a method which causes substitution of acetoxy-groups at C-5 and consequent decarboxylation) then de-esterification to give 1,5-dialdehydes which can be trapped with nitromethane in base and converted into nitrocyclitols. While α- and β-glucuronides and β-galacturonides underwent cleavage in this way, the α-galacturonides were resistant to oxidation.[36]

Dissociation constants of D-glucuronic acid, D-galacturonic acid, their methyl ethers, and the corresponding 4,5-unsaturated derivatives have been measured and rationalized in terms of various operating electronic factors.[37]

Observations on details of the reduction of uronic acid esters to hexopyranosides, on nitrogen-containing derivatives, and on branched-chain uronic acids are made in Chapters 3, 9, and 13, respectively.

5 Ascorbic Acids

Isotopically labelled L-ascorbic acids with ^{14}C at C-4 or C-6, as well as uniformally labelled material, have been made using the 1,2-*O*-isopropylidene furanose derivatives of [3-^{14}C], [1-^{14}C], and D-[*U*-^{14}C]glucose (Vol. 5, p. 118).[38]

[33] L. C. Ginsberg, D. T. Di Ferrante, and N. Di Ferrante, *Carbohydrate Res.*, 1978, **64**, 225.
[34] I. R. Siddiqui, *Carbohydrate Res.*, 1978, **63**, 312.
[35] A. G. W. Bradbury and C. von Sonntag, *Carbohydrate Res.*, 1978, **60**, 183.
[36] I. Kitagawa, M. Yoshikawa, and A. Kadota, *Chem. and Pharm. Bull (Japan)*, 1978, **26**, 484.
[37] R. Kohn and P. Kováč, *Chem. Zvesti*, 1978, **32**, 478.
[38] M. Williams and F. A. Loewus, *Carbohydrate Res.*, 1978, **63**, 149.

Treatment of the anion of L-ascorbic acid with tetra-*O*-acetyl-α-D-glucopyranosyl bromide led to β-glucosylation at C-3, and subsequent methylation gave access to compound (17). When these two substitution processes were reversed the 2-β-D-glucopyranosyl 3-methyl ether (18) was obtained.[39] Conversion of the acid into the 5,6-*O*-isopropylidene derivative followed by sul-

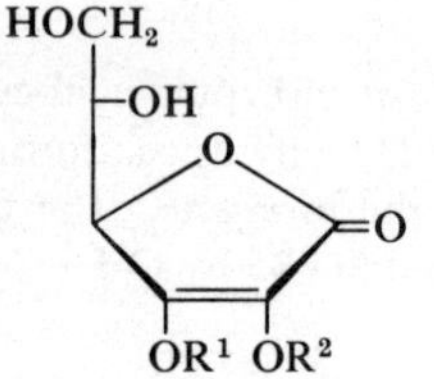

	R¹	R²
(17)	β-D-glucopyranosyl	Me
(18)	Me	β-D-glucopyranosyl

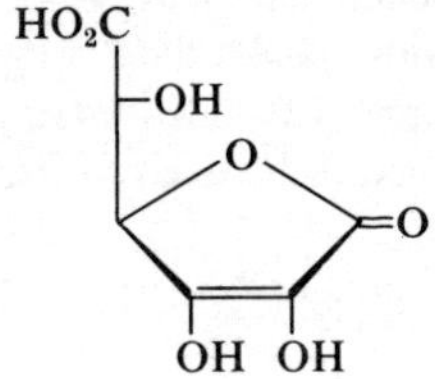

(19)

phation using pyridine–sulphur trioxide, acid hydrolysis, and treatment with potassium hydroxide gave the 2-sulphate 2,3-dipotassium salt,[40] whereas direct sulphation with strong sulphuric acid gave the 6-ester and with pyridine–sulphur trioxide the 5-sulphate. The biological properties of the 2-sulphate were also reviewed,[41] and this ester has been utilized as a protecting group for the enediol while oxidation to give the carboxylic acid 'saccharoascorbic acid' (19) was effected.[42] Using the 5,6-*O*-isopropylidene derivative, other workers have obtained the 2-phosphate in high yield as well as the 2,2′-di-substituted phosphate.[43]

A simple preparation of pure dehydro-L-ascorbic acid was developed using oxygen in methanol with active charcoal as catalyst.[44] Other studies of this oxidation have examined the role of various metal ions and have found Cu^{II} to be most effective in promoting the process,[45] and have used Cu^{II} and hydrogen peroxide,[46] Cu^{II} and oxygen,[47] and Mn^{III} in aqueous perchloric acid.[48] A polarographic study of this oxidation in basic media showed two oxidation waves each associated with a 2-electron transfer process, the first giving the dehydroacid and the second a product formed by further reaction at C-4.[49, 50]

[39] W. A. Szarek and K. S. Kim, *Carbohydrate Res.*, 1978, **67**, C13.
[40] M. Tsujimura, H. Yoshikawa, T. Hasegawa, and T. Suzuki, *Joshi Eiyo Daigaku Kijo*, 1975, **6**, 35 (*Chem. Abs.*, 1978, **88**, 23 294e).
[41] D. W. Lillard jun. and P. A. Seib, A.C.S. Symposia, No. 77 (Carbohydrate Sulphates), ed. R. G. Schweiger, American Chemical Society, 1978, p. 1.
[42] H. A. Stuber and B. M. Tolbert, *Carbohydrate Res.*, 1978, **60**, 251.
[43] C. H. Lee, P. A. Seib, Y. T. Liang, R. C. Hoseney, and C. W. Degoe, *Carbohydrate Res.*, 1978, **67**, 127.
[44] M. Ohmori and M. Takagi, *Agric. and Biol. Chem.* (*Japan*), 1978, **42**, 173.
[45] K. Takamura and M. Ito, *Chem. and Pharm. Bull.* (*Japan*), 1977, **25**, 3219.
[46] E. V. Shtamm and Yu. I. Skurlatov, *Zhur. fiz. Khim.*, 1977, **51**, 3132 (*Chem. Abs.*, 1978, **88**, 62 548k).
[47] E. V. Shtamm, A. P. Purmal, and Yu. I. Skurlatov, *Zhur. fiz. Khim.*, 1977, **51**, 3128 (*Chem. Abs.*, 1978, **88**, 62 547j).
[48] E. Pelizzetti, E. Mentasti, and E. Pramauro, *J.C.S. Dalton*, 1978, 61.
[49] J. J. Ruiz, A. Aldza, and M. Dominguez, *Canad. J. Chem.*, 1978, **56**, 1533.
[50] M. Dominguez, A. Aldza, and F. Sanchez-Burgos, *Anales de Quim.*, 1978, **74**, 199 (*Chem. Abs.*, 1978, **89**, 163 880p).

Diarylthallium hydroxides in pyridine give stable radical anions following complexing with L-ascorbic acid. E.s.r. studies showed that these species were formed immediately, and their concentrations built up over 3 h.[51]

The reaction of ascorbic acid with substituted amines is mentioned in Chapter 9.

[51] H. B. Stegmann, K. Scheffler, and P. Schuler, *Angew. Chem. Internat. Edn.*, 1978, **17**, 365.

16

Inorganic Derivatives

1 Carbon-bonded Compounds

The phosphonic acid (1) having a methylene group replacing O-1 of D-fructose-1-phosphate was prepared as indicated in Scheme 1.[1] The phosphorus-containing branched-chain compounds (2)—(4) were made by treatment of the corresponding exocyclic alkene with phosphorus nucleophiles,[2] and the acyclic compounds (5) and (6) (*cis*- and *trans*-isomers) by reaction between penta-*O*-acetyl-D-gluconyl chloride with trimethyl and tri-isopropyl

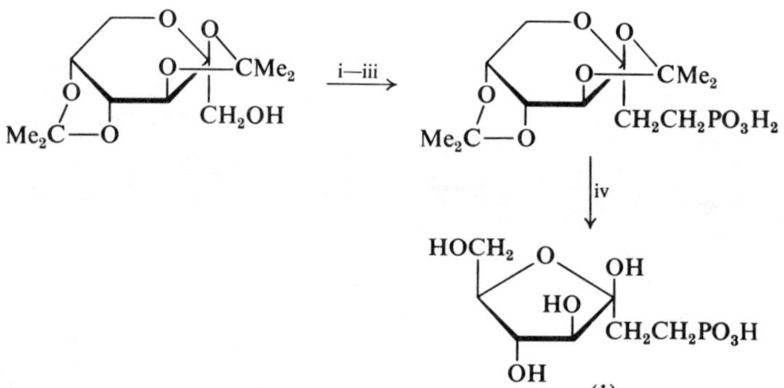

(1)

Reagents: i, DMSO–DCC; ii, $Ph_3P{=}CH\overset{O}{\overset{\|}{P}}(OPh)_2$; iii, H_2–Pt; iv, H^+

Scheme 1

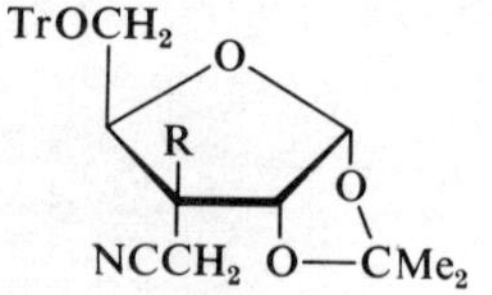

R
(2) PPh_2
(3) $PO(OEt)_2$
(4) $POPh_2$

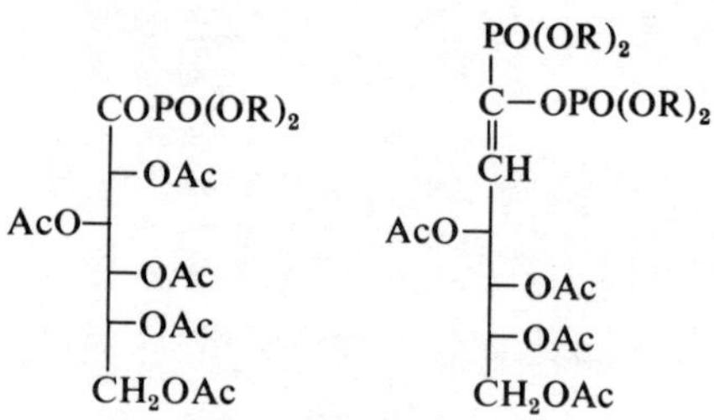

(5) R = Me or Pr^i (6)

[1] J.-C. Tang, B. E. Tropp, and R. Engel, *Tetrahderon Letters*, 1978, 723.
[2] J. M. J. Tronchet, J.-R. Neeser, and E. J. Charollais, *Helv. Chim. Acta*, 1978, **61**, 1942.

phosphites.[3] Hydrogenation of unsaturated phosphonates, *e.g.* (7), gave the 2,3-dideoxy adducts, but also the 2,3,4-analogues formed by hydrogenolysis at C-4. ^{13}C-1–^{31}P Coupling constants are considerably larger when the phosphorus atoms are equatorial, so that anomeric configurations can be determined. The unsaturated compounds do not conform with Hudson's isorotation rule.[4] Hydrolysis of the furanose derivative (8) followed by acetylation afforded isomeric compounds (9) one of which was characterized specifically by *X*-ray diffraction analysis.[5]

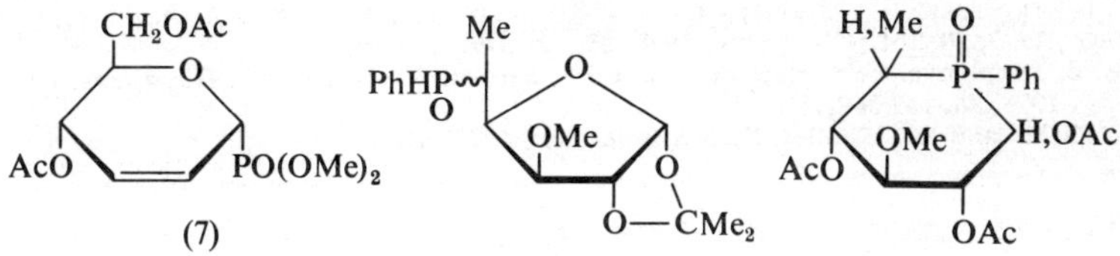

(7) (8) (9)

Arsenic derivatives of thiosugars and selenosugars, *e.g.* compounds (10) and (11), have been prepared from the corresponding thiol and selenol by treatment with dimethylarsenic chloride.[6]

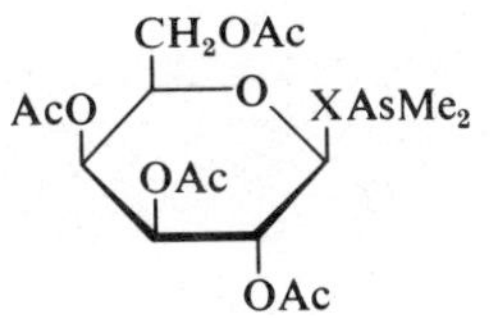

	X
(10)	S
(11)	Se

2 Oxygen-bonded Compounds

Several instances of the use of silicon and tin derivatives have been noted earlier (Chapters 4 and 6).

A review has appeared on the synthetic, analytical, and biological uses of borate complexes of carbohydrates.[7] The interaction of molybdate and tungstate with mannitol and glucitol have been further studied in dilute solution by potentiometric methods.[8]

Work with cations includes examination of the viscosity effects of sodium and potassium halides on solutions of xylose (structure-making and structure-breaking trends were studied),[9] calorimetric analysis of the interaction of calcium ions with D-galactose, *myo*-inositol, and lactose,[10] and conductance

[3] Yu. A. Zhdanov, L. A. Uzlova, and Z. I. Glebova, *Zhur. obshchei Khim.*, 1978, **48**, 227 (*Chem. Abs.*, 1978, **89**, 24 674m).

[4] J. Thiem, B. Meyer, and H. Paulsen, *Chem. Ber.*, 1978, **111**, 3325.

[5] M. Yamashita, Y. Nakatsukasa, S. Inokawa, K. Hirotsu, and J. Clardy, *Chem. Letters*, 1978, 871.

[6] J. R. Daniel and R. A. Zingaro, *Phosphorus Sulphur*, 1978, **4**, 179 (*Chem. Abs.*, 1978, **89**, 44 001t).

[7] W. Voelter, *Themen Chem. Bors*, 1976, 141 (*Chem. Abs.*, 1978, **88**, 121 552g).

[8] M. Mikešová and M. Bartušek, *Coll. Czech. Chem. Comm.*, 1978, **43**, 1867.

[9] Vishnu and V. P. Misra, *Carbohydrate Res.*, 1978, **60**, 29.

[10] W. J. Evans and V. L. Frampton, *Carbohydrate Res.*, 1977, **59**, 571.

studies on the interaction of sucrose with a range of singly and doubly charged cations.[11] Stability constant measurements have been recorded for the interactions between ascorbic acid and Fe^{II}, Mn^{II}, Co^{II}, Ni^{III}, Be^{II}, and Pb^{II},[12] glucose and xylose and Cr^{III}[13] and several monosaccharides and Zr^{IV} and Hf^{IV}.[14]

Organotin, organolead, and organogermanium complexes have been prepared from phthalate or succinate acid esters derived from sucrose.[15]

11 Vishnu and A. K. Singh, *Carbohydrate Res.*, 1978, **60**, 19.
12 K. P. Dubey and S. Parveen, *Current Sci.*, 1978, **47**, 415 (*Chem. Abs.*, 1978, **89**, 90 252y).
13 V. I. Krupenskii, T. V. Dolgaya, and I. I. Korol'kov, *Izvest. Vyssh. Uchebn. Zaved., Lesn. Zhur.*, 1977, **20**, 161 (*Chem. Abs.*, 1978, **88**, 121 579w).
14 Ts. B. Konunova, L. S. Kachkar, and G. F. Serova, *Koord. Khim.*, 1978, **4**, 391 (*Chem. Abs.*, 1978, **88**, 159 369r).
15 A. Parkin and R. C. Poller, *Carbohydrate Res.*, 1978, **62**, 83.

17

Alditols and Cyclitols

1 Alditols

A review on the preparation, properties, reactions, and uses of polyhydric sugar alcohols has appeared.[1]

A new antibiotic which is a 1,6-diester derivative of D-mannitol is mentioned in Chapter 18.

A method for the electrochemical reduction of D-xylose to 2-deoxy-D-*threo*-pentitol has been described.[2] The homogeneous hydrogenation of sugars using tris-triphenylphosphine ruthenium chloride is improved in the presence of hydrogen chloride, which inhibits competitive decarbonylation of the sugar.[3]

L-(2,3)-Threitol is easily prepared from (+)tartaric acid by lithium aluminium hydride reduction of the 2,3-*O*-isopropylidene derivative of diethyl tartrate, followed by acid hydrolysis of the resultant ketal.[4]

Conventional procedures have been used to convert methyl 4-azido-4-deoxy-α-D-glucopyranoside to the 1,4-diazido-1,4-deoxy-D-glucitol derivative (1) which was condensed with 4-deoxy-4-proprionamido-glucopyranosyl chloride (2) and the product reduced to yield the α- and β-linked alditol amino-glycosides (3) which are regioisomers of antibiotic P-2563(P)[5] (see Vol. 11, p. 155). Other references to antibiotics incorporating 1,4-diamino-1,4-dideoxy-D-glucitol occur in Chapter 18.

Methyl 2,2′,3,3′4′,6′-hexamethyl-β-maltoside (4) has been converted to the mixture of alditol glycosides (5) as outlined in Scheme 1, which can form the basis of a method for the specific degradation of uronic acid-containing polysaccharides, mass spectrometry showing the site of linking, and n.m.r. spectroscopy the glycosidic configuration.[6]

A set of modified conditions has allowed some control over the complex formose reaction so that the branched-chain alditols (6)—(8) represented the major products formed; initial catalysis was effected with calcium hydroxide, which was removed at the end of the induction period with oxalate, and the reaction was then adjusted to pH 12 with potassium hydroxide.[7]

[1] F. R. Benson, 'Kirk–Othmer Encylcopaedia of Chemical Technology', 3rd edn., Wiley, 1978, Vol. 1., p. 754.
[2] H. Berbalk, K. Eichinger, and P. Fricko, *Synthesis*, 1978, 42.
[3] W. M. Kruse and L. W. Wright, *Carbohydrate Res.*, 1978, **64**, 293.
[4] J. Bourson and L. Oliveros, *Bull. Soc. chim. France*, 1977, 1241.
[5] M. M. Pompipom, R. Bugianesi, E. Walton, and T. Y. Shen, *Carbohydrate Res.*, 1978, **65**, 121.
[6] G. O. Aspinall and K.-G. Rosell, *Canad. J. Chem.*, 1978, **56**, 680.
[7] Y. Shigemasa, O. Nagal, C. Sakazawa, R. Nakashima, and T. Matsuura, *J. Amer. Chem. Soc.*, 1978, **100**, 1309.

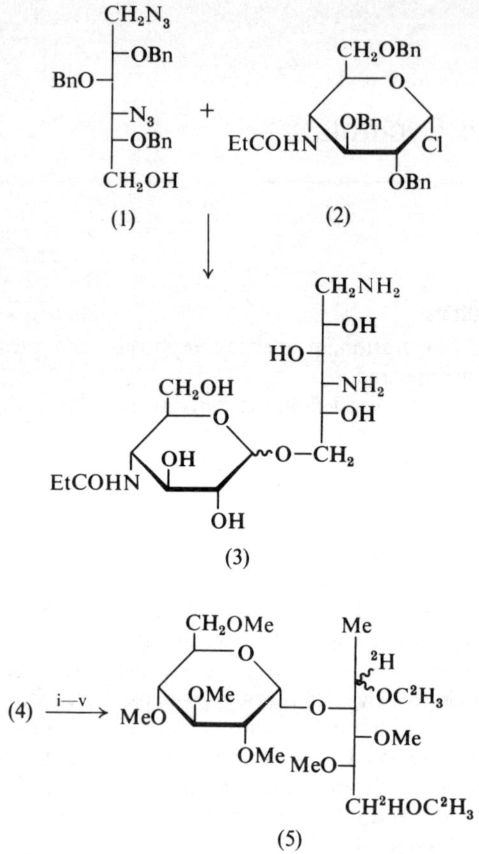

Reagents: i, Ph_3P–MeI; ii, DBU; iii, H^+; iv, $NaBD_4$; v, C^2H_3I–$MeSOCH_2Na$

Scheme 1

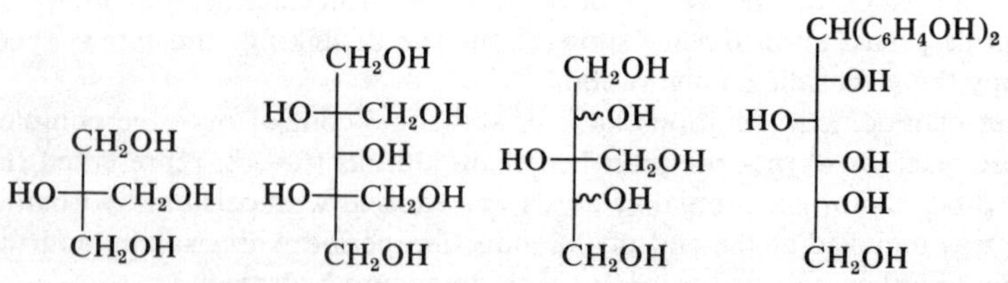

(6) (7) (8) (9)

The di-aryl D-glucitol derivative (9) was one of the products isolated from reaction of D-glucose with phenol in acid solution[8] (see also Chapter 3).

Acid-catalysed dehydration of D-mannitol gave a mixture of anhydrides which, after sequential isopropylidenation and tritylation, yielded crystalline 2,5-anhydro-1,3-*O*-isopropylidene-6-*O*-trityl-D-glucitol (10), providing an

[8] P.-A. Pernemalm, *Acta Chem. Scand.*, 1978, **32B**, 72.

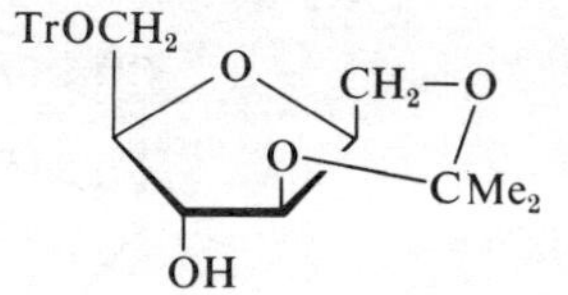

(10)

improved route to 2,5-anhydro-D-glucitol; (10) and a number of its 4-*O*-substituted derivatives were shown to adopt an E_1(D)-conformation.[9] 1,5-Anhydro-3,4-di-*O*-toluene-*p*-sulphonyl-D-mannitol was obtained from 3,4-di-*O*-toluene-*p*-sulphonyl-D-mannitol with one equivalent of toluene-*p*-sulphonyl chloride in pyridine at 0 °C.[10]

The 2,5-di-*O*-sulphonyl esters of 1,4:3,6-dianhydro-D-mannitol (11) underwent exchange with sodium iodide to yield a 1 : 1 mixture of the 2,5-dideoxy-2,5-di-iodo-D-glucitol and D-iditol derivatives (12) and (13) respectively, the 2-deoxy-2-iodo-5-*O*-methanesulphonyl D-glucitol (14) and D-mannitol (15)

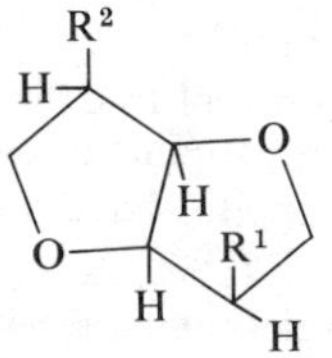

(11) $R^1 = R^2$ = OMs or OTs
(15) R^1 = I, R^2 = OMs

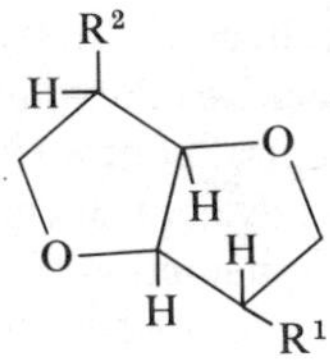

(12) $R^1 = R^2$ = I
(14) R^1 = I, R^2 = OMs
(16) $R^1 = R^2$ = OMs

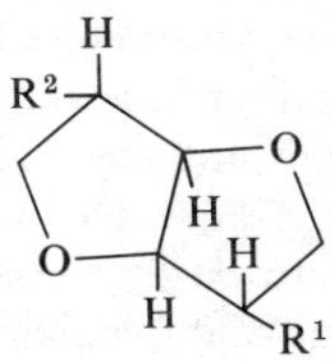

(13) $R^1 = R^2$ = I
(17) R^1 = OMs, R^2 = I

derivatives being detected as intermediates; each set of separated isomers could be rapidly interconverted by fast iodine–iodine substitution; by contrast, the D-glucitol diester derivative (16) only yielded the 5-deoxy-5-iodo-D-iditol ester (17), without isomerization. The results were explained by conformational analysis of the bicyclic ring system.[11]

2,5-Anhydro-1-deoxy-1-nitro-D-altritol and D-allitol derivatives, prepared by cyclization of corresponding 1-deoxy-1-nitro-D-altritol or D-allitol derivatives in dimethyl sulphoxide, undergo equilibration on heating in aqueous solvents *via* acyclic nitroalkene intermediates; whereas the *altro* isomer (18) predominated with 3,4-*O*-isopropylidene derivatives, the *allo* isomer (19) was preferred for the monocyclic derivative.[12]

Aldoses and alditols can be rapidly oxidized by hydrogen peroxide under certain conditions leading to the formation of formic acid with evolution of oxygen, and a mechanism was proposed involving catalysis by superoxide radicals generated from traces of transition metal salts; in Fenton's reaction ferrous ion is considered to play an analogous catalytic role.[13]

[9] T. A. W. Koerner, R. J. Voll, and E. S. Younathan, *Carbohydrate Res.*, 1977, **59**, 403.
[10] B. Matsuhiro and A. B. Zanlungo, *Carbohydrate Res.*, 1978, **63**, 297.
[11] J. Kuszmann and G. Medgyes, *Carbohydrate Res.*, 1978, **64**, 135.
[12] T. Takamoto, H. Omi, T. Matsuzaki, and R. Sudoh, *Carbohydrate Res.*, 1978, **60**, 97.
[13] H. S. Isbell and H. L. Frush, *Carbohydrate Res.*, 1977, **59**, C25.

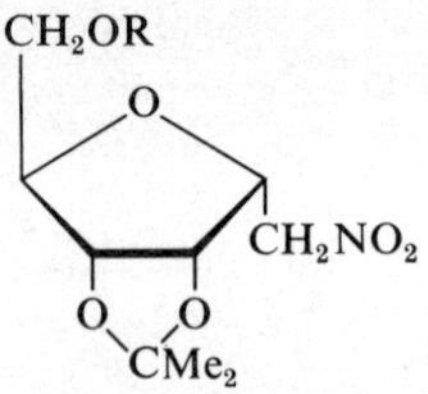

(18) R = H or Bz

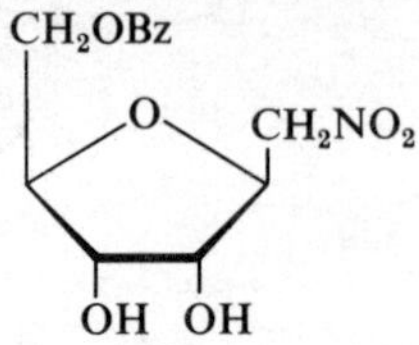

(19)

The acid-catalysed mono-acetalation of 2-deoxy-alditols with butanol is referred to in Chapter 5.

Thermal parameters affecting the crystallization of sorbitol have been studied; depending on conditions, two crystalline modifications can arise, and a process for the rapid crystallization of molten sorbitol was proposed.[14] Compression parameters of crystalline xylitol have also been reported.[15] The viscosity of concentrated aqueous solutions of sodium and potassium halides in presence of D-mannitol has been studied; the viscosity was related to the structure-making or structure-breaking effect of the salt in solution.[16]

Studies of specific heat capacities and volumes suggest that polyols bind to proteins and stabilize their natural conformations,[17] and studies of α-amino-acids in aqueous solutions of glycerol, mannitol, or glucitol have also been made to clarify this structural interaction.[18]

Ring and open-chain derivatives of L-erythritol 1,4-dicinnamate have been irradiated yielding products of photochemical asymmetric synthesis, in which intramolecular 2 + 2 cycloaddition occurs to give L-erythritol esters of cyclobutane-1,2-dicarboxylic acid; the 2,3-di-*O*-methyl derivative of L-erythritol gave a high degree of asymmetric induction, whereas the 2,3-*O*-isopropylidene derivative was much less stereospecific, and gave the opposite cyclobutane enantiomer.[19]

The analysis of hexopyranosyl-2-acetamido-2-deoxy-hexitols by g.c.–m.s. is mentioned in Chapter 22, and acetal derivatives of hexitols are referred to in Chapter 5.

2 Inositols

The irradiation of alkenyl acetals derived from monosaccharides has provided a new route to inositol derivatives, as illustrated in Scheme 2; the diethyl acetal (20) yielded the epimeric ethyl ether derivatives (21), whilst the corresponding dithioacetal (22) gave the inosose dithioacetal (23).[20] The 2,5-di-deoxystreptamine derivative (24) suitably protected for use as an intermediate

[14] J. Sztatisz, S. Gal, L. Fodor, and E. Pungar, *J. Therm. Analysis*, 1977, **12**, 351 (*Chem. Abs.*, 1978, **88**, 136 869q).
[15] A. V. Reznikovskaya, E. R. Leikin, and G. D. Soboleva, *Gidroliz. Lesokhim. Prom.-st.*, 1977, 7 (*Chem. Abs.*, 1978, **88**, 7253k).
[16] V. Misra and V. P. Misra, *Carbohydrate Res.*, 1978, **63**, 29.
[17] G. Dipaola and B. Belleau, *Canad. J. Chem.*, 1978, **56**, 848.
[18] G. Dipaola and B. Belleau, *Canad. J. Chem.*, 1978, **56**, 1827.
[19] B. S. Green, A. T. Hagler, Y. Rabinsohn, and M. Rejto, *Israel J. Chem.*, 1977, **15**, 124 (*Chem. Abs.*, 1978, **88**, 23 288f).
[20] A. A. Othman, N. A. Al-Masudi, and U.S. Al-timari, *J. Antibiotics*, 1978, **31**, 1007.

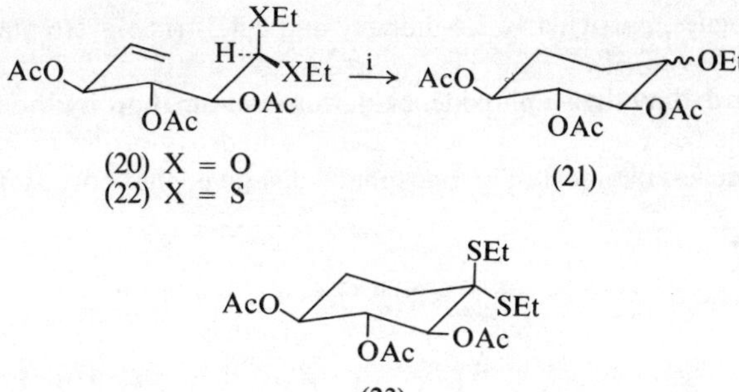

(20) X = O
(22) X = S

(21)

(23)

Reagents: i, *hν*

Scheme 2

for the synthesis of 5-deoxy-amino glycoside antibiotics has been prepared from *cis*-cyclohexanedi-epoxide as outlined in Scheme 3; the amino-ester was resolved and both enantiomers of (24) were obtained.[21] Other reactions of *cis*- or *trans*-cyclohexane di-epoxide led to 1,3-diazido-4,6-dihydroxycyclohexane, which was used to prepare a 5-deoxyneamine derivative *via* 2,5-dideoxystreptamine, and to mono- and bis-episulphides which were used in turn to prepare

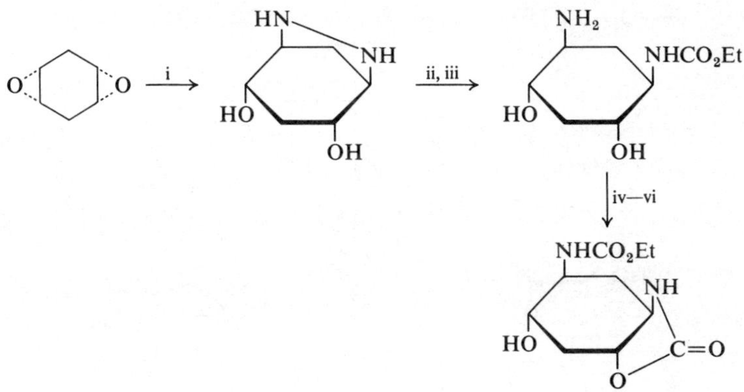

(24)

Reagents: i, N_2H_4; ii, H_2–Pd; iii, $ClCO_2Et$; iv, $ClCO_2Ph$; v, HO^-; vi, NEt_3

Scheme 3

a range of cyclohexanethiol derivatives, *e.g.* (25).[22] Thio analogues of 2,5-dideoxystreptamine have been prepared from the parent compound by standard procedures.[23]

Sulphurylchloride in pyridine has been used to convert inositol derivatives to corresponding chloro-deoxy compounds, which could then be reduced to

[21] G. Kavadias, S. Velkof, and B. Belleau, *Canad. J. Chem.* 1978, **56**, 404.
[22] F. Haviv and B. Belleau, *Canad. J. Chem.*, 1978, **56**, 2677.
[23] G. Kavadias and R. Droghini, *Canad. J. Chem.*, 1978, **56**, 2743.

deoxy analogues, particularly 2,4-dideoxy and 2,4,5-trideoxystreptamines, *e.g.* (26) → (27) → (28).[24]

A standard dimethylsulphoxide oxidation–sodium borohydride reduction sequence has been used to epimerize the C-4 hydroxy group in 2,3:5,6-di-*O*-cyclohexylidene-*myo*-inositol 1-phosphate to give the corresponding *epi*-isomer.[25]

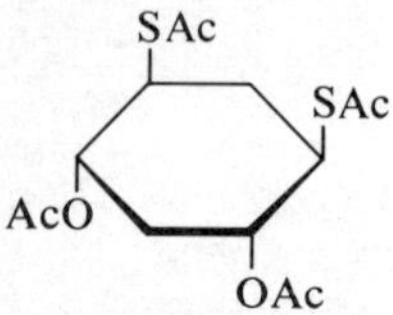

(25)

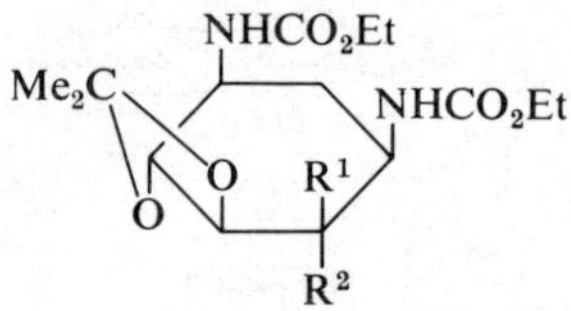

(26) R^1 = H, R^2 = OH
(27) R^1 = Cl, R^2 = H
(28) R^1 = R^2 = H

A series of pseudo-disaccharides has been prepared by condensing appropriate deoxy-inositol and glycal derivatives to give α-linked 3′-deoxy-cyclitol glycosides, *e.g.* (29) which were further transformed to amino-deoxy-cyclitol glycosides, *e.g.* (30).[26]

A number of tetrakis-methane-sulphonyl esters of *myo*-inositol have been prepared; the diacetate (31) underwent azide displacement with acetate participation leading to 3-amino-3-deoxy-*muco*-inositol derivatives, *e.g.* (32).[27]

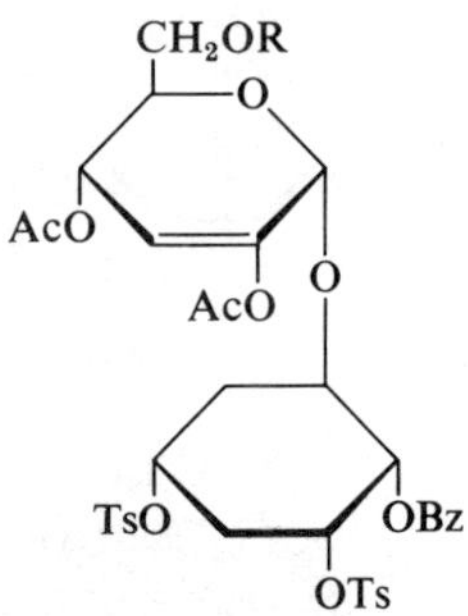

(29)

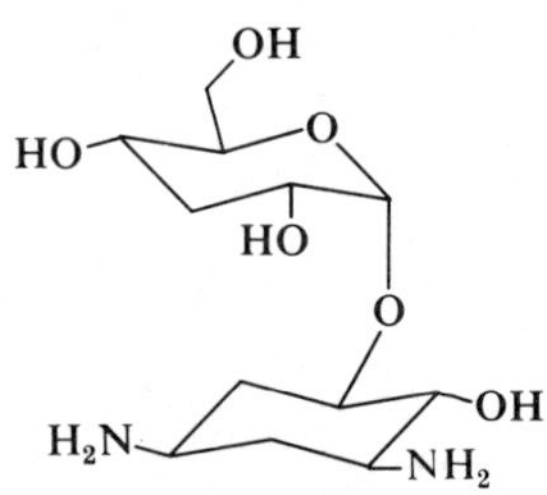

(30)

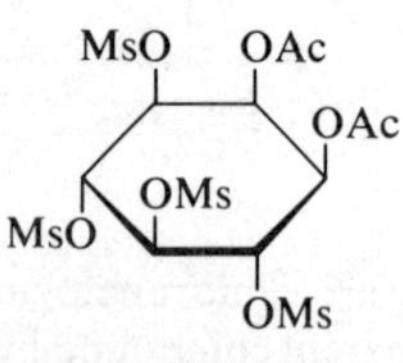

(31)

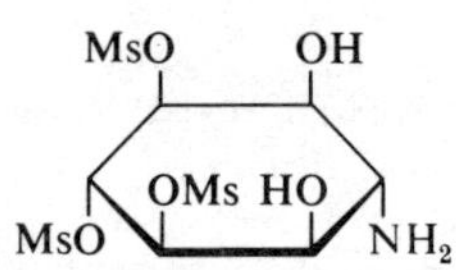

(32)

[24] S. Ogawa, Y. Hongo, H. Fujimon, K. Iwata, A. Kasuga, and T. Suami, *Bull. Chem. Soc. Japan*, 1978, **51**, 2957.

[25] V. P. Shevchenko, B. V. Rozynov, and Yu. G. Molotkovskii, *Bioorg. Khim.*, 1978, **4**, 383 (*Chem. Abs.*, 1978, **89**, 44 024c).

[26] J. Cléophax, D. K. Duc, J.-M. Delauményy, S. D. Gero, and A. Rolland, *J.C S. Chem. Comm.*, 1978, 771.

[27] M. del. C. P. de Friere and R. A. Cadenas, *Carbohydrate Res.*, 1978, **63**, 157.

Analogues of prumycin incorporating amino-deoxy cyclopentitol units and a new antibiotic, funiculosin, containing an arylcyclopentanetetraol unit, are mentioned in Chapter 18. Syntheses of (−)kasuganobiosamine, an amino-glycoside of *chiro*-inositol, and 3-amino-3-deoxy-β-D-glucopyranosyl derivatives of D-pinitol are mentioned in Chapter 8.

18
Antibiotics

A new survey of antibiotics includes sections on aminoglycosides, lincosaminides, macrolides, nucleosides, oligosaccharides, polyenes, and tetracyclines.[1]

1 Aminoglycoside Antibiotics

Reviews have appeared on the synthesis of aminoglycoside antibiotics,[2] and on the formation of new aminocyclitol antibiotics by mutants of aminocyclitol-producing organisms fed with aminocyclohexanol or related subunits.[3]

The structures of eight minor components of the nebramycin complex elaborated by *Streptomyces tenebrarius* have been elucidated; besides neamine and related pseudo-disaccharides, the new components are pseudo-trisaccharides related to kanamycin B, and the 3′-hydroxy analogue of the principal component of the complex, apramycin (see Vol. 10, p. 130).[4]

A synthesis of the octodiose present in apramycin is referred to in Chapter 2. See also Chapter 14 for the synthesis of octadialdose derivatives.

There have been many syntheses reported this year in quest of compounds with improved biological activity. Syntheses which involve glycosylation of streptamine derivatives have included the preparation of 4-*O*-(2-amino-2-deoxy-α-D-glucopyranosyl)- and 4-*O*-(6-amino-6-deoxy-α-D-glucopyranosyl)-2,5-dideoxystreptamine, the latter showing remarkable antimicrobial activity,[5] 5-deoxykanamycin A, which had only half the activity of kanamycin A,[6] 3′,3″-dideoxy-butirosin A[7] and 5″-amino-3′,5″-dideoxybutirosin A,[8] both showing comparable activity to 3′-deoxybutirosin A, 6-*O*-(3-amino-3-deoxy-α- and β-D-glucofuranosyl)-neamine, which are furanosyl isomers of kanamycin B,[9] 6-*O*-(β-D-ribofuranosyl) and 6-*O*-(α-D-arabinofuranosyl)-paromamine, which are less effective antibacterial agents than the corresponding amino-sugar

1 'Kirk–Othmer Encyclopaedia of Chemical Technology', ed. M. Grayson and D. Eckroth, 3rd edn., Wiley, New York 1978, Vol. 2.
2 S. Umezawa, *Kagaku* (*Tokyo*), 1978, **33**, 283 (*Chem. Abs.*, 1978, **89**, 135 710q).
3 K. L. Rinehart, jun., *Pure Appl. Chem.*, 1977, **49**, 1361.
4 K. F. Koch, K. E. Merkel, S. C. O'Connor, J. L. Occolowitz, J. W. Paschal, and D. E. Dorman, *J. Org. Chem.*, 1978, **43**, 1430.
5 S. Ogawa, Y. Funaki, K. Iwata, and T. Suami, *Asahi Garasu Kogyo Gijutsu Shoreikai Kenkyu Hokoku*, 1976, **29**, 171 (*Chem. Abs.*, 1978, **89**, 129 844v).
6 G. Kavadias, P. Dextraze, R. Massé, and B. Belleau, *Canad. J. Chem.*, 1978, **56**, 2086.
7 T. Tsuchiya, I. Watanabe, F. Nakamura, M. Hamada, and S. Umezawa, *J. Antibiotics*, 1978, **31**, 933.
8 I. Watanabe, T. Tsuchiya, F. Nakamura, M. Hamada, and S. Umezawa, *J. Antibiotics*, 1978, **31**, 863.
9 R. D. Sitrin, D. J. Cooper, and J. A. Weisbach, *J. Org. Chem.*, 1978, **43**, 3048.

derivatives,[10] and a new synthesis of ribostamycin and of butirosin B from neamine.[11] 4-*O*-(3-Deoxy-α-D-*ribo*-hexopyranosyl)-6-deoxy-streptamine and its 5-*O*-β-D-ribofuranosyl derivative have also been prepared by appropriate glycosylation of 1,3-di-*O*-toluene-*p*-sulphonyl-2,6-dideoxy-inositol derivatives followed by standard transformations to the corresponding 6-deoxy-streptamine derivatives.[12] Syntheses involving modification of naturally occurring antibiotics have included 1-deamidino, 3-deamidino, and 1,3-di(deamidino)-dihydrostreptomycin from dihydrostreptomycin, all of which lacked anti-bacterial activity;[13] 3′-deoxyseldomycin Factor 5 from seldomycin Factor 5 using the Barton procedure, the deoxy-compound showing greatly enhanced antibiotic activity against 3′-*O*-phosphorylating, amino-glycoside resistant strains compared with the parent antibiotic [14] (3′-episeldomycin Factor 5 was also prepared incidentally) (see also Vol. 11, p. 153); 2′-*N*-methyl-gentamicin C_{1a} from gentamicin C_{1a},[15] 2′-*N*-ethyl-paromomycin from paromomycin,[16] and also analogues of paromomycin in which the 2-amino-2-deoxy-D-glucose unit is replaced by 2,6-diamino-2,6-deoxy-D-glucose, -D-galactose, and D-mannose units;[17] 3″-*N*-alkyl analogues of gentamicin factors and sisomicin by chemical or photochemical de-*N*-methylation of the 3″-amino group in the parent compound followed by its re-*N*-alkylation, all of which showed similar antibiotic activity to the parent compounds;[18] and the conversion of the dimeric antibiotic 66-40C (see Vol. 11, p. 154) into the monomer sisomicin, 6′-*N*-methyl-sisomicin (antibiotic G-52), and a series of 6′-*N*-substituted analogues of sisomicin.[19] Direct selective benzoylation of *N*-acylated kanamycin B yielded a 3′,4′, 2″,4″,6″-penta-*O*-benzoyl derivative, which allowed the remaining 5-hydroxy-group to be removed by sequential treatment with sulphuryl chloride (yielding the 5-chlorodeoxy epimer) followed by tributylstannane reduction, leading to 5-deoxykanamycin B.[20]

3′-Deoxybutirosins A and B have been prepared from the parent antibiotics by a route involving catalytic reduction of a 2′,3′-epimino intermediate which is summarized in Scheme 1; the isomeric 3′-amino-2′,3′-dideoxy-derivative was formed simultaneously.[21] An alternative sequence for the deoxygenation of these and other aminoglycoside antibiotics at various positions of the sugar

[10] S. Hanessian, T. Ogawa, and T. Takamoto, *Canad. J. Chem.*, 1978, **56**, 1500.
[11] V. Kumar and W. A. Remers, *J. Org. Chem.*, 1978, **43**, 3327.
[12] J. Cleophax, J.-M. Delauméný, S. D. Gero, A. Rolland, and N. Rolland, *J.C.S. Chem. Comm.*, 1978, 773.
[13] T. Usui, T. Tsuchiya, and S. Umezawa, *J. Antibiotics*, 1978, **31**, 991.
[14] R. E. Carney, J. B. McAlpine, M. Jackson, R. S. Stanaszek, W. H. Washburn, M. Cirovic, and S. L. Meuller, *J. Antibiotics*, 1978, **31**, 441.
[15] H. Matsushima and Y. Mori, *J. Antibiotics*, 1978, **31**, 621.
[16] G. Cassinelli, G. Franceschi, G. Di Colo, and F. Arcamone, *J. Antibiotics*, 1978, **31**, 380.
[17] G. Cassinelli, P. Julita, and F. Arcamone, *J. Antibiotics*, 1978, **31**, 382.
[18] T. L. Nagabhushan, J. J. Wright, A. B. Cooper, W. N. Turner, and G. H. Miller, *J. Antibiotics*, 1978, **31**, 43.
[19] D. H. Davies, A. K. Mallams, M. Counelis, D. Loebenberg, E. L. Moss, jun., and J. A. Waitz, *J. Medicin. Chem.*, 1978, **21**, 189.
[20] T. Suami, S. Nishiyama, Y. Ishikawa, and E. Umezawa, *Bull. Chem. Soc. Japan*, 1978, **51**, 2354.
[21] H. Fukase, N. Mizokami, and S. Horii, *Carbohydrate Res.*, 1978, **60**, 289.

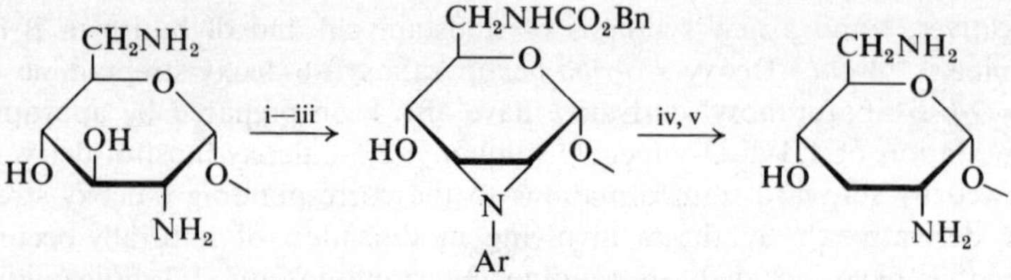

Reagents: i, p-$O_2N{\cdot}C_6H_4O{\cdot}CO_2Bn$; ii, p-$MeOC_6H_4CHO$–NEt_3, iii, Ph_3P–CCl_4; iv, H_2–Pd; v, H_2–Ni

Scheme 1

rings involves the reduction of appropriate dithiocarbonate precursors using tributylstannane.[22]

Periodate oxidation of derivatives of neomycin B, parmomomycin, and lividomycin B followed by base-catalysed β-elimination of the oxidized units has been used to obtain constituent pseudo-di- and -tri-saccharides of these antibiotics.[23] 5-Epi, 6-epi-, and 5,6-diepi-neamine have been prepared from appropriate neamine derivatives by standard sequences involving glycosulose or oxiran intermediates.[24]

Analogues of neamine were also prepared which contain a third amino-group in the cyclitol ring, *e.g.* (1) and (2), and these also showed antibiotic

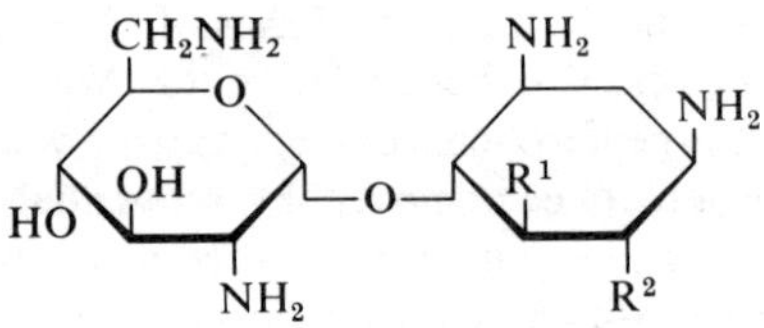

(1) $R^1 = NH_2$; $R^2 = OH$
(2) $R^1 = OH$, $R^2 = NH_2$

activity.[25] Similarly, paromamine has been converted to its 3′-epimer *via* a 3′-ketone intermediate, and this in turn was used to synthesize its 6′-amino-6′-deoxy analogue, *i.e.* 3′-epineamine, utilizing the azide displacement procedure on a 6′-bromo-6′-deoxy-3′-epi-paromamine intermediate; unfortunately both epimers were much less effective antibiotics than the parent compounds, even though they were not substrates for enzymes which deactivate this class of antibiotic.[26]

Specific complexing between metal ions and pairs of amino- and hydroxy-groups in aminoglycoside antibiotics has allowed the preparation of a range of specifically substituted derivatives; thus, a cobalt complex of sisomicin had the sites paired in (3) blocked and on acetylation afforded 3,2′,6′-tri-*N*-acetyl

[22] T. Hayashi, T. Iwaoka, N. Takeda, and E. Ohki, *Chem. and Pharm. Bull. Japan*, 1978, **26**, 1786.
[23] S. Hanessian, T. Takamoto, R. Marse, and G. Patil, *Canad. J. Chem.*, 1978, **56**, 1482.
[24] T. Suami, S. Nishiyama, Y. Ishikawa, and S. Katsura, *Carbohydrate Res.*, 1978, **65**, 57.
[25] S. Nishiyama, Y. Ishikawa, M. Yamazaki, and T. Suami, *Bull. Chem. Soc. Japan*, 1978, **51**, 555.
[26] S. Hanessian, R. Masse, and T. Nakagawa, *Canad. J. Chem.* 1978, **56**, 1509.

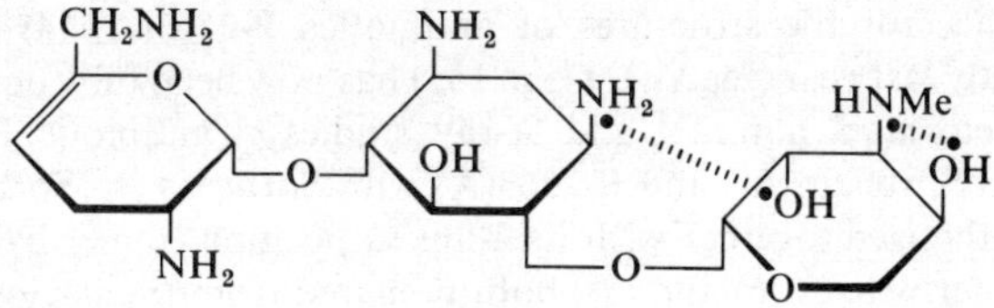

(3)

sisomicin.[27] Similarly copper(II) chelates of kanamycin A selectively block the 4″-hydroxy and the 3″-amino-groups, allowing selective acylation of the 1′-, 3-, and 6′-amino groups; the 1,3,6′-tri-*N*-acetyl derivative could be prepared in high yield using *N*-(benzyloxycarbonyloxy)-succinimide in the presence of cupric acetate.[28] The structure of the copper complex could be usefully studied from the line-broadening effect of the metal on the ^{13}C resonances of the antibiotic, the signals being affected in the order C-3″, C-4″ > C-1″ ≫ C-6″ ≫ C-6′.[29]

Model studies have established that carbohydrates containing vicinal *cis*-related and 4,6-*N*-benzyloxycarbonyl-aminol units can form *O,N*-benzylidene acetals analogous to the familiar diol derivatives, and accordingly penta-*N*-benzyloxycarbonyl paromomycin afforded the corresponding 4′;6′-*O*:4‴-*O*,6‴-*N*-dibenzylidene derivative on treatment with benzaldehyde and zinc chloride.[30]

^{13}C N.m.r. spectral data for amikacin and related compounds have been reported,[31] and the mass spectra of a number of partially *N*-acetylated derivatives of kanamycin A have been studied.[32]

Chemical ionization mass spectral (C.i.m.s.) data of aminoglycoside antibiotics are also referred to in Chapter 21.

3′,4′-Dideoxy, 6′-*N*-methyl, and 6′-*C*-methyl derivatives of butirosin have been prepared biosynthetically from 6′-*N*-methylneamine or gentamines using neamine-negative mutants of *Bacillus circulans*,[33–35] and the biosynthesis of neomycin subunits has been studied using pseudo-disaccharides prepared from 2,6-dideoxystreptamine instead of the natural 2-deoxystreptamine, a deoxystreptamine-idiotroph of *Streptomyces rimosis* forma *paromomycinus* incorporated 6-deoxyneamine but not 5-*O*-β-D-ribofuranosyl-2,6-dideoxystreptamine, suggesting an obligatory order of sub-unit assembly [36] (see also Vol. 10, p. 133).

27 T. L. Nagabhushan, A. B. Cooper, W. N. Turner, H. Tsai, S. McCombie, A. K. Mallams, D. Rane, J. J. Wright, P. Reichert, D. L. Boxler, and J. Weinstein, *J. Amer. Chem. Soc.*, 1978, **100**, 5253.
28 S. Hanessian and G. Patil, *Tetrahedron Letters*, 1978, 1035.
29 S. Hanessian and G. Patil, *Tetrahedron Letters*, 1978, 1031.
30 S. Hanessian, R. Massé, and G. Ekborg, *Canad. J. Chem.*, 1978, **56**, 1492.
31 S. Toda, S. Nakagawa, T. Naito, and H. Kawaguchi, *Tetrahedron Letters*, 1978, 3913.
32 A. M. Sepulchre, P. Varenne, B. C. Das, S. D. Gero, and H. E. Audier, *Nouv. J. Chim.*, 1978, **2**, 405 (*Chem. Abs.*, 1978, **89**, 185 987f).
33 K. Takeda, S. Okono, Y. Ohashi, and T. Furamai, *J. Antibiotics*, 1978, **31**, 1023.
34 K. Takeda, A. Kinumaki, H. Hayasaka, T. Yamaguchi, and Y. Ito, *J. Antibiotics*, 1978, **31**, 1031.
35 K. Takeda, A. Kinumaki, S. Okuno, T. Matsushita, and Y. Ito, *J. Antibiotics*, 1978, **31**, 1039.
36 C. J. Pearce, M. Akhtar, J. E. G. Barnett, D. Mercier, A.-M. Sepulchre, and S. D. Gero, *J. Antibiotics*, 1978, **31**, 74.

The evidence for the structures of antibiotics P-2563(A) (4) and (P) (5) reported briefly last year (see Vol. 11, p. 155) has now been fully documented,[37] including chemical,[38] n.m.r.,[39] and *X*-ray studies;[40] antibiotic P-2563(P) is identical with Sorbistin A_1, and P-2563(A) with Sorbistin B. Sorbistin A_1 (5) has been synthesized together with its 2-linked position isomer by condensing together the derivatives (6) and (7), both prepared from D-galactose, followed

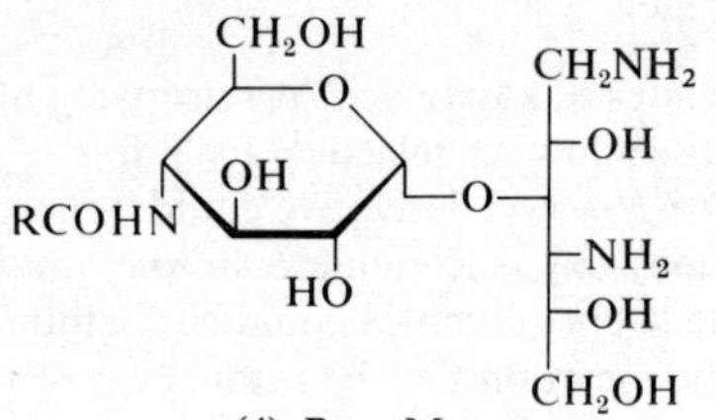

(4) R = Me
(5) R = Et

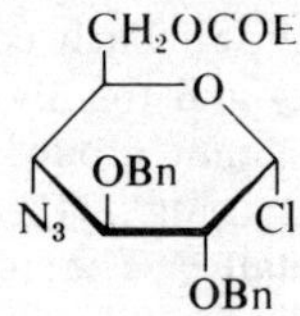

(6)

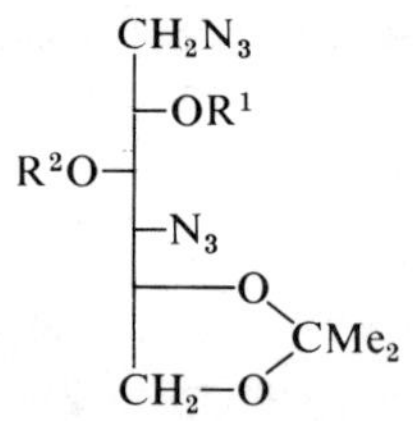

(7) R^1, R^2 = H, H or H, Bu_3Sn

by catalytic hydrogenation which occurs with concomitant $O \rightarrow N$ propanoyl migration.[41] This antibiotic has also been modified by conventional reactions to give a variety of 4′-*N*-acyl, 4′-*N*-propylamino-, and 6′-substituted derivatives, notably the 6′-chloro-6′-deoxy analogue.[42]

2 Macrolide Antibiotics

Revised structures for the constituents of the megalomycin complex have been proposed on the basis of *X*-ray and n.m.r. data; L-megosamine (2,3,6-trideoxy-3-dimethylamino-L-*ribo*-hexopyranose) (8) α-linked to the tertiary hydroxy-

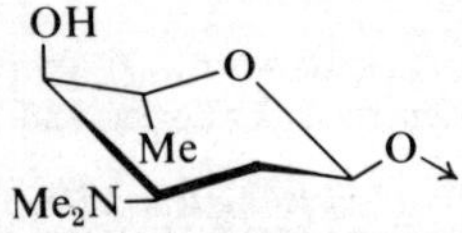

(8)

[37] K. Nara, Y. Sumino, K. Katamoto, S.-I. Akiyama, and M. Asai, *Chem. and Pharm. Bull. Japan*, 1978, **26**, 1075.
[38] K. Nara, K. Katamoto, S. Suzuki, S.-I. Akiyama, and E. Mizuta, *Chem. and Pharm. Bull. Japan*, 1978, **26**, 1083.
[39] K. Nara, K. Katamoto, S. Suzuki, and E. Mizuta, *Chem. and Pharm. Bull. Japan*, 1978, **26**, 1091.
[40] K. Kamiya, Y. Wada, and K. Nara, *Chem. and Pharm. Bull. Japan*, 1978, **26**, 2040.
[41] T. Ogawa, K. Katano, and M. Matsui, *Carbohydrate Res.*, 1978, **60**, C13.
[42] M. M. Pompipom, R. L. Bugionesi, and T.-Y. Shen, *J. Medicin. Chem.*, 1978, **21**, 221.

group at C-6 of the macrolide ring replaces D-rhodosamine (the D-*lyxo* amino-sugar epimeric at C-5) which was originally considered to be linked to the secondary hydroxy-group at C-11 [43] (see Vol. 7, p. 147). ^{13}C N.m.r. has also been used to show that in lucensomycin mycosamine is α-linked whereas in nystatin it is β-linked.[44] Mycoheptin has been shown to be a heptaene macrolide antibiotic containing mycosamine which is closely related to nystatin A and amphotericin B.[45] C.i.m.s. data have been reported for the macrolide antibiotics M-4365A_2 and G_2, which contain desosamine; the ions formed by cleavage of the sugar from the macrolide rings were particularly useful for structural assignment.[46]

3 Nucleoside Antibiotics

A brief review of nucleoside antibiotics has appeared which includes data on platenocidin, a new antibiotic from *S. platensis*,[47] and another review covers nucleoside analogues exhibiting selective antiviral activity.[48]

A new antibiotic nucleoside, mildiomycin, elaborated by *Streptoverticillium rimofaciens* B-98891, has been shown to have the structure (9).[49, 50]

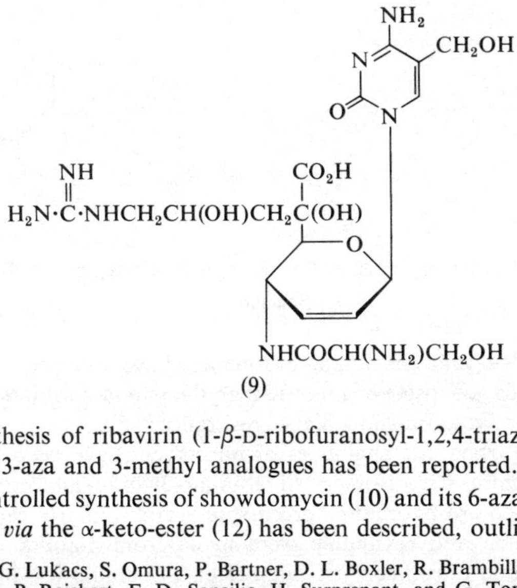

(9)

A new synthesis of ribavirin (1-β-D-ribofuranosyl-1,2,4-triazole-3-carboxamide) and its 3-aza and 3-methyl analogues has been reported.[51]

A stereocontrolled synthesis of showdomycin (10) and its 6-azapseudouridine analogue (11) *via* the α-keto-ester (12) has been described, outlined in Scheme

[43] T. T. Thang, G. Lukacs, S. Omura, P. Bartner, D. L. Boxler, R. Brambilla, A. K. Mallams, J. B. Morton, P. Reichert, F. D. Sancilio, H. Surprenant, and G. Tomalesky, *J. Amer. Chem. Soc.*, 1978, **100**, 663.
[44] A. Boicelli, *Tetrahedron Letters*, 1978, 1213.
[45] E. Borowski, J. Golik, J. Zielinski, L. Falkowshi, P. Kolodziejczyk, J. Pawlak, and Yu. Shenin, *J. Antibiotics*, 1978, **31**, 117.
[46] M. Suzuki, K.-I. Harada, N. Takeda, and A. Tatematsu, *Heterocycles*, 1977, **8**, 199 (*Chem. Abs.*, 1978, **88**, 105 712d).
[47] S. Nakamura and H. Kondo, *Heterocycles*, 1977, **8**, 583 (*Chem. Abs.*, 1978, **88**, 74 494c).
[48] E. De Clercq and P. F. Torrence, *J. Carbohydrates, Nucleosides, Nucleotides*, 1978, **5**, 187.
[49] S. Harada, E. Mizuta, and T. Kishi, *J. Amer. Chem. Soc.*, 1978, **100**, 4895.
[50] S. Harada and T. Kishi, *J. Antibiotics*, 1978, **31**, 519.
[51] J. T. Witkowski, L. F. Christensen, and R. K. Robins, *J. Carbohydrates, Nucleosides, Nucleotides*, 1978, **5**, 363.

2.[52] Standard transformations have been used to convert showdomycin to its 5′-deoxy-5′-halo derivatives.[53] A simple preparative route to the *C*-nucleoside antibiotics formycin and formycin B has been described.[54] Quantum mechanical calculations on the conformations of formycin, formycin B, oxoformycin, and showdomycin have been made, the predicted preferred conformations agreeing with those found experimentally.[55]

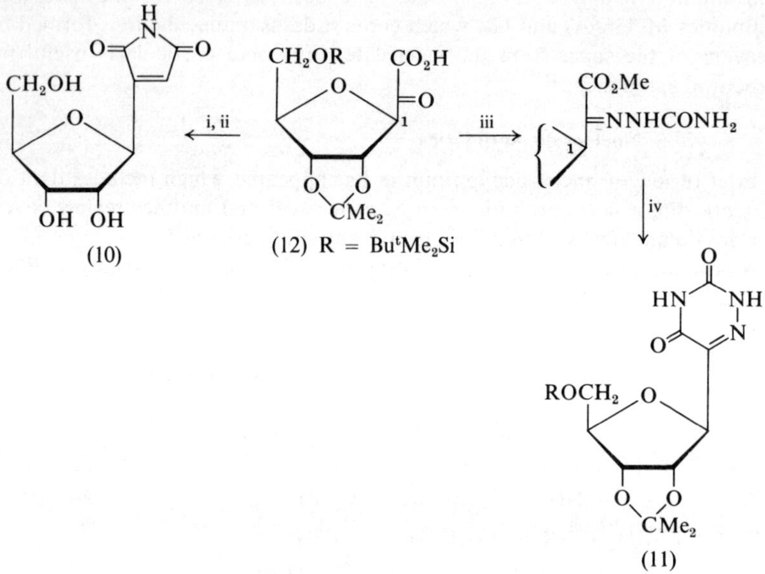

Reagents: i, $Ph_3PCHCONH_2$; ii, 50%TFA; iii, $H_2NCONHNH_2$; iv, NaOEt

Scheme 2

2′,3′- and 3′,5′-Cyclic phosphates of formycin have been prepared, which are more resistant to adenosine deaminase than the parent antibiotic; 1- and 2-*N*-isopropyl derivatives also show higher resistance.[56]

A series of carbocyclic analogues of puromycin have been prepared, *e.g.* (13), derived from stereoisomers of 1-amino-3-azido-2,4,5-trihydroxycyclopentane and 4-chloro-pyrimidine precursors,[57] and similarly carbocyclic analogues of uridine and thymidine were made from 1-amino tetrahydroxycyclopentanes and nitrourea.[58] A number of 9-(3-amino-2,4-dihydroxycyclopentyl)-adenine derivatives have been obtained from a bicyclic cyclopentene

[52] T. Sato, R. Ito, Y. Hayakawa, and R. Noyuri, *Tetrahedron Letters*, 1978, 1829.
[53] R. A. Earl and L. B. Townsend, *J. Carbohydrates, Nucleosides, Nucleotides*, 1978, **5**, 305.
[54] L. Kalvoda, *Coll. Czech. Chem. Comm.*, 1978, **43**, 1431.
[55] A. Saran, C. K. Mitra, and B. Pullman, *Int. J. Quantum Chem., Quantum Biol. Symp.*, 1977, 4 (Proc. Int. Symp. Quantum Biol. Quantum Pharmacol., 4th), p. 43 (*Chem. Abs.*, 1978, **88**, 74 512g).
[56] O. Makabe, A. Miyadera, M. Kinoshita, S. Umezawa, and T. Takeuchi, *J. Antibiotics*, 1978, **31**, 456.
[57] T. Suami, K. Tadano, M. Ayabe, and Y. Emori, *Bull. Chem. Soc. Japan*, 1978, **51**, 855.
[58] K. Tadano, S. Horiuchi, and T. Suami, *Bull. Chem. Soc. Japan*, 1978, **51**, 897.

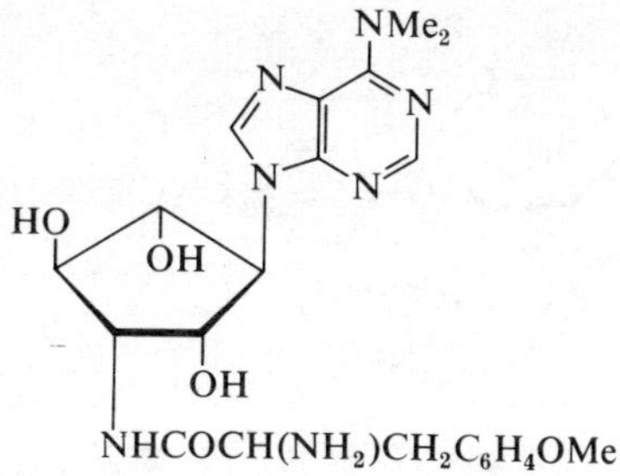

(13)

lactam.[59] The synthesis of puromycin analogues is also referred to in Chapter 19.

A range of 5′-*O*-acyl derivatives of 9-β-D-arabinofuranosyl adenine have been prepared as prodrugs of the parent nucleosides; the 5′-*O*-valeryl derivative was most effective as an antiviral agent.[60]

The biosynthesis of the nucleoside skeleton of polyoxins has been studied using ^{13}C- and ^{14}C-labelled glucose, glycerate, and uridine substrates; it was suggested that condensation of uridine as its 5′-aldehyde with phosphoenol pyruvate gives an octofuranuloseuronic acid nucleoside, which loses the two terminal carbon atoms by oxidative elimination followed by transamination to give the required uridyl-5-amino-5-deoxy-D-allofuranosyluronic acid structure.[61]

4 Miscellaneous Antibiotics

2,6-Dideoxy-L-*lyxo*-hexopyranosyl-ε-pyrromicinone and 2-deoxy-D-*erythro*-pentopyranosyl-carminomycinone,-daunomycinone, and -ε-pyrromicinone have been prepared from the appropriate mycinone and glycosyl bromide by Koenigs–Knorr procedures.[62]

Three closely related broad spectrum antibiotics from a species of *Nocardia* have been assigned the structures (14), (15), and (16) respectively.[63]

A new antifungal antibiotic, trichostatin C, has been shown to be the glucoside (17) of trichostatin A, representing the first example of a glucopyranosyl hydroxamate in nature.[65] Another new antibiotic substance, A 32390A, isolated from a culture of a species of *Pyrenochaeta*, NRRL-5786, has been shown to be 1,6-di-*O*-(2-isocyano-3-methyl-crotonyl)-D-mannitol.[66] A new antiviral and antifungal antibiotic, funiculosin, produced by *Penicillium*

[59] S. Daluge and R. Vince, *J. Org. Chem.*, 1978, **43**, 2311.
[60] D. C. Baker, T. H. Haskell, and S. R. Putt, *J. Medicin. Chem.*, 1978, **21**, 1218.
[61] K. Isono, T. Sato, K. Hirasawa, and S. Funayama, *Nucleic Acids Res., Spec. Publ.*, 1977, **3**, 91 (*Chem. Abs.*, 1978, **88**, 152 906c).
[62] H. S. El Khadem and D. L. Swartz, *Carbohydrate Res.*, 1978, **65**, C1.
[63] G. A. Ellestad, D. B. Cosulich, R. W. Broschard, J. H. Martin, M. P. Kunstmann, G. O. Morton, J. E. Lancaster, W. Fulmor, and F. M. Lovell, *J. Amer. Chem. Soc.*, 1978, **100**, 2515.
[64] K. Kakinuma and Y. Sakagami, *Agric. and Biol. Chem.* (*Japan*), 1978, **42**, 279.
[65] N. Tsuji and M. Kobayashi, *J. Antibiotics*, 1978, **31**, 939.
[66] G. G. Marconi, B. B. Molloy, R. Nagarajan, J. W. Martin, J. B. Deeter, and J. L. Occolowitz, *J. Antibiotics*, 1978, **31**, 27.

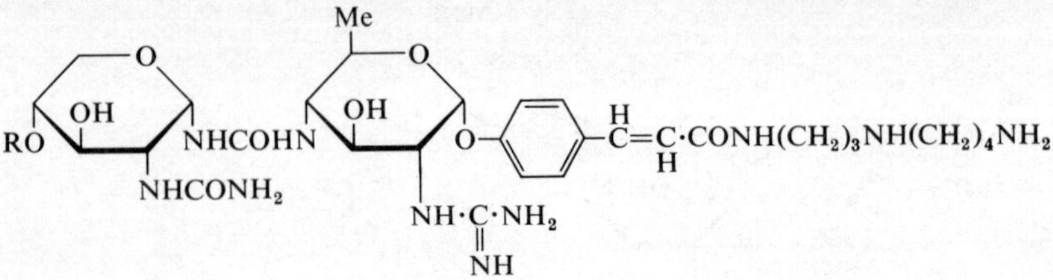

(14) R =

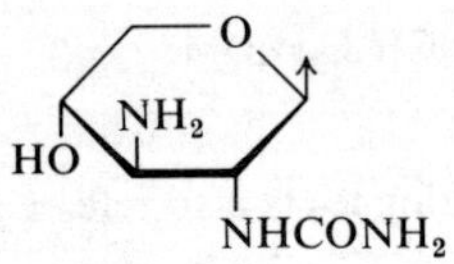

(15) R =

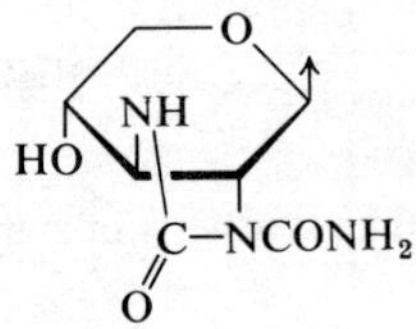

(16) R =

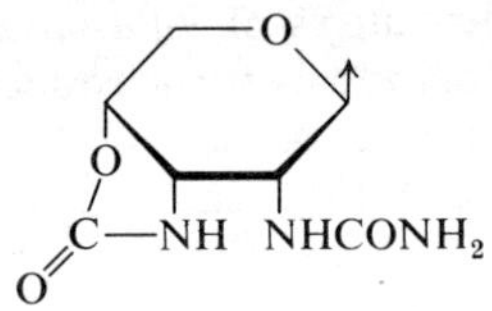

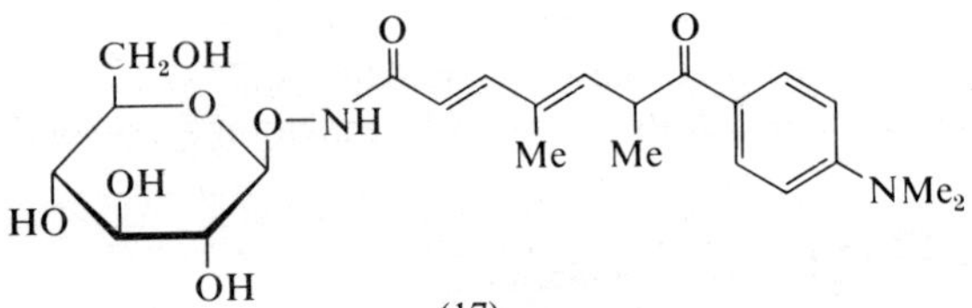

(17)

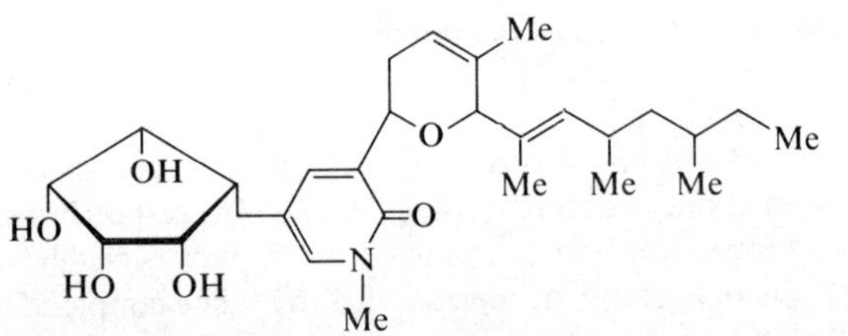

(18)

funiculosium THOM, is the all-*cis* cyclopentanetetraol derivative (18), a stereoform not previously found in nature.[67]

A minor revision of the structure of the aglycone portion of bleomycin and phleomycin has been proposed, a β-lactam ring being replaced by a corresponding open-chain amide unit.[68]

7-Deoxy- [69] and 9-deoxy-(4*R*)-dihydrospectinomycin as well as 9-deoxy-

[67] K. Ando, I. Matsuura, Y. Nawata, H. Endo, H. Sasaki, T. Okytomi, T. Saehi, and G. Tamura, *J. Antibiotics*, 1978, **31**, 533.

[68] T. Takita, Y. Muroka, T. Nakatani, A. Fujii, Y. Umezawa, H. Naganawa, and H. Umezawa, *J. Antibiotics*, 1978, **31**, 801.

[69] W. Rosenbrook, jun., R. E. Carney, R. S. Egan, R. S. Stanaszek, M. Cirovic, T. Nishinaga, K. Mochida, and Y. Mori, *J. Antibiotics*, 1978, **31**, 451.

spectinomycin [70] have been prepared from spectinomycin derivatives, but none showed any antibiotic activity. 9-Epi-(4*R*)-dihydrospectinomycin and 9-epispectinomycin have also been prepared from dihydrospectinomycin by a sulphonate–benzoate displacement sequence.[71] Acetonation of the (4*R*)-dihydrospectinomycin derivative (19) gave, besides the 4,4*a*-*O*-isopropylidene derivative, the rearranged acetonide (20) formed by reversible hemi-acetal

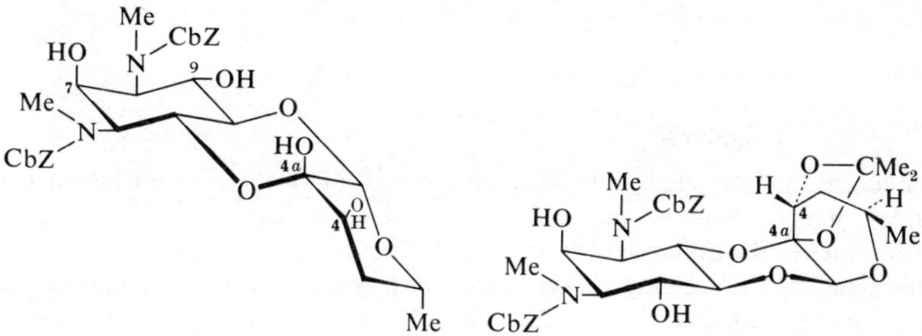

(19) Cbz = BnOCO· (20)

migration of the 4*a*-ketone group between the enantiotopic hydroxy-groups of the cyclitol ring.[72]

The interglycosidic links in the hetero-tetrasaccharide side-chain of ristomycin A have been investigated, and the structure determined to be *O*-*β*-D-arabinopyranosyl(1 → 2)-*O*-α-D-mannopyranosyl(1 → 2)-[*O*-α-L-rhamnopyranosyl-(1 → 6)]-D-glucose.[73] (See also Vol. 10, p. 131.) αα-Trehalose has been identified as the nonreducing disaccharide in parvulomycin, which also contains L-glutamic acid.[74]

The syntheses of prumycin and some isomeric compounds are covered in Chapter 8.

[70] L. Foley, J. T. S. Lin, and M. Weigele, *J. Antibiotics*, 1978, **31**, 979.
[71] L. Foley, J. T. S. Lin, and M. Weigele, *J. Antibiotics*, 1978, **31**, 985.
[72] L. Foley and M. Weigele, *J. Org. Chem.*, 1978, **43**, 4355.
[73] A. Neszmélyi, F. Sztaricskai, A. Lipták, and R. Bognar, *J. Antibiotics*, 1978, **31**, 974.
[74] M. Puskas, F. Sztaricskai, R. Bognar, V. N. Omel'chenko, and Yu. D. Shenin, *Antibiotiki* (*Moscow*), 1978, **23**, 26 (*Chem. Abs.*, 1978, **88**, 141 589c).

19

Nucleosides

1 Synthesis

A comparison between three standard methods of synthesis of purine nucleosides, namely fusion, condensation of acetylated glycosyl chlorides, and from 1-acetates in the presence of titanium(IV) chloride has been made. Thus when the glucosaminyl acetate (1) was fused with theophylline in the presence of *p*-toluenesulphonic acid and *p*-nitrophenol, a 45% yield of the nucleoside (2) was obtained. In contrast to the findings of Ishido *et al.* (*Carbohydrate Res.*, 1975, **44**, 215), lower yields were obtained when the activating agents were omitted. The nucleoside (3) synthesized in 32% yield from the glucosyl acetate (4) by fusion in the presence of *p*-toluenesulphonic acid gave physical constants different from those previously published and the yield was not significantly improved by use of activating agents. The fusion of the glycosyl chlorides (5), (6), and 2,3,4,6-tetra-*O*-benzoyl-αβ-D-glucopyranosyl chloride with 6-

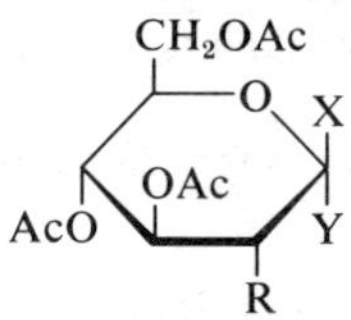

(1) R = NHAc, X = H, Y = OAc
(2) R = NHAc, X = theophyllin-7-yl, Y = H
(3) R = OAc, X = 6-benzamidopurin-9-yl, Y = H
(4) R = X = OAc, Y = H
(5) R = NHAc, X = H, Y = Cl
(6) R = OAc, X = H, Y = Cl

benzamido(chloromercuri)purine gave better yields of the β-nucleoside and was the method of choice. The use of benzyl groups in the sugar moiety produced, as expected, anomeric mixtures and it was found preferable to remove *N*-benzoyl groups prior to hydrogenolysis of benzyl groups.[1]

A new simplified synthesis which combines, in a one-pot reaction, silylation of the heterocyclic base, formation of the silylated condensation catalyst, and reaction of the base with peracylated sugar derivatives, giving nucleosides in high yield, is summarized in Scheme 1. Using this method, 5-methoxyuridine cytidine, and N^6-benzoyladenosine were synthesized in 60—70% yield, and 2′,3′,5′,-tri-*O*-benzoyluridine in 84% yield. When 1,2-dichloroethane was

[1] N. Pravdić and I. Franjić-Mihalić, *Carbohydrate Res.*, 1978, **62**, 301.

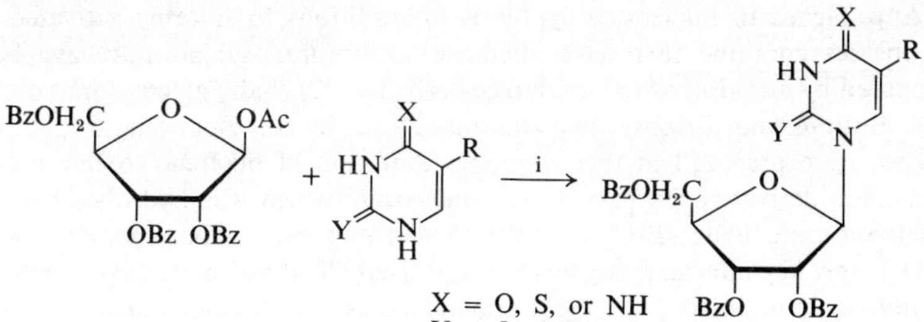

X = O, S, or NH
Y = O or S
R = H or OMe

Reagents: i, Me_3SiCl–$(Me_3Si)_2NH$–$NaClO_4$, NH_4ClO_4, or $C_4F_9SO_3K$–MeCN

Scheme 1

used as solvent in place of acetonitrile, two other products, N^3-nucleoside (30%) and N^1,N^3-bis-riboside (5—10%) were formed.[2] Transglycosylation has been used to synthesize 3′-deoxyuridine from cordycepin in 61% yield.[3] Yields in transglycosylation of pyrimidine to purine nucleosides have been improved by using (i) trimethylsilyl derivatives of *exo-N*-acylpurines as glycosyl acceptor and (ii) trimethylsilyl trifluoromethanesulphonate in acetonitrile–dichloroethane as the catalyst. The reaction was also successfully applied to the synthesis of purine analogues of polyoxin and octosyl acid nucleosides having labile amino- or anhydro-sugar uronic acid components.[4]

A simple two-stage conversion of nucleosides to their corresponding 5-deoxy nucleosides involves sequential treatment with thionyl chloride in HMPT, giving the 5′-chloro-5′-deoxy derivatives which were then reduced with tri-butyltin hydride and α,α′-azobis(isobutyronitrile) in dry THF. In this way 5′-deoxyadenosine, 5′-deoxyuridine, 5′-deoxythymidine, 5′-deoxytoyomycin, 5′-deoxysangivamycin, 2′,3′,5′-trideoxyadenosine, 3′5′-dideoxyadenosine, and 9-(5′-deoxy-β-D-arabinofuranosyl)adenine were prepared.[5]

The approach to nucleoside synthesis developed by Ogura *et al.* (*Heterocycles*, 1977, **8**, 125, and earlier references), which involved the reaction of glycosyl isothiocyanates with substituted hydrazines to give thiosemicarbazides which were cyclized with NBS, and finally desulphurized to give the required heterocycle, has now been applied to the synthesis of pyrazolopyrimidine nucleosides derived from D-glucopyranosyl, D-arabinopyranosyl, and D-ribofuranosyl isothiocyanates.[6] Syntheses of nucleoside analogues from reactions of glycosyl isothiocyanates with amino-acids, enamines, and diamines have been summarized.[7]

[2] H. Vorbruggen and B. Bennua, *Tetrahedron Letters*, 1978, 1339.
[3] R. J. Suhadolnik and T. Uematsu, *Carbohydrate Res.*, 1978, **61**, 545.
[4] T. Azuma and K. Isono, *Chem. and Pharm. Bull.* (*Japan*), 1977, **25**, 3347.
[5] Y. Wang, H. P. C. Hogenkamp, R. A. Long, G. R. Revankar, and R. K. Robins, *Carbohydrate Res.*, 1977, **59**, 449.
[6] H. Ogura, H. Takahashi, and E. Kudo, *J. Carbohydrates, Nucleosides, Nucleotides*, 1978, **5**, 329.
[7] H. Ogura and H. Takahashi, *Symposium Heterocycl.* [*Pap.*], 1977, 135 (*Chem. Abs.*, 1978, **89**, 180 279s).

Approaches to nucleoside synthesis by additions to suitably activated *N*-glycosides continue to receive attention. Ribofuranosyl purines have been prepared by addition of the imidazole derivative (7) to the glycosyl iminoether (8). Both α- and β-ribosyl hypoxanthines and the corresponding 6-thio analogues were prepared in this manner.[8] Addition of diketene to the amino-oxazoline derivative (9) gave a cyclonucleoside which was hydrolysed to the nucleoside analogue (10).[9] Cyclization of ribopyranosyldiaminomaleonitrile (11), a new key intermediate, by means of triethylorthoformate gave the dicyanoimidazole anomers (12), the β-anomer being the thermodynamic product.

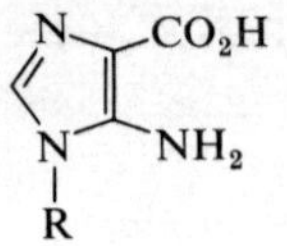

(7) R = H or Bn

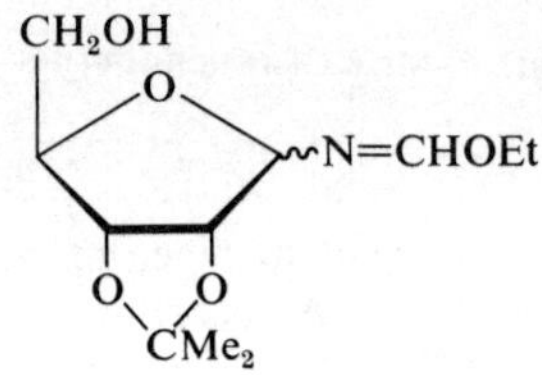

(8)

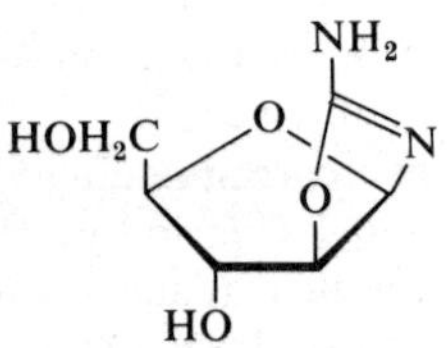

(9)

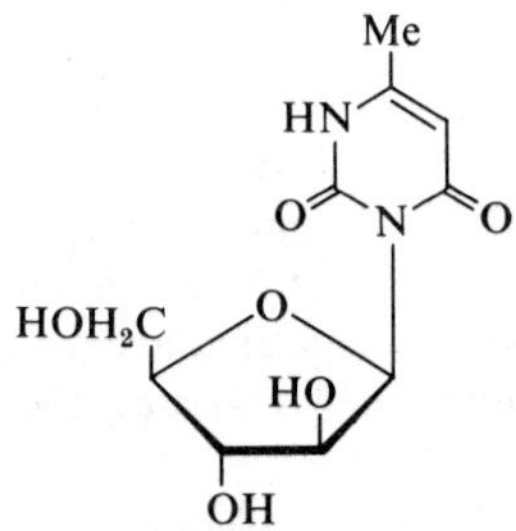

(10)

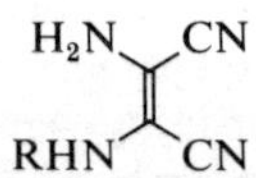

(11)

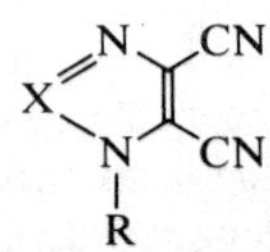

(12) X = CH
(13) X = N

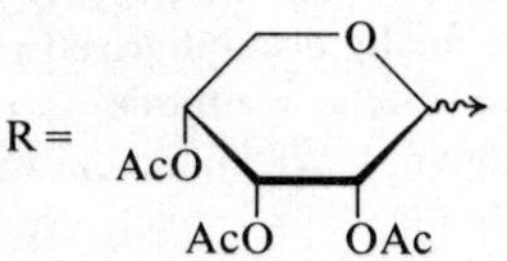

Similarly cyclization of (11) with isopentyl nitrile in methanol gave the dicyanotriazole nucleoside anomers (13). In fusion reactions of the 1,2,3-triazole (13) (R = H) with peracetylated ribopyranose and -furanose only the 2-substituted nucleosides were obtained.[10] Reaction of benzotriazole with tetra-*O*-acetyl-β-D-ribofuranose gave the 1- and 2-substituted isomers in 44 and 7%

[8] J. L. Montéro, A. Dhainaut, and J. L. Imbach, *J. Heterocyclic Chem.*, 1978, **15**, 929.
[9] I. Drizina, *Tezisy Doklady–Resp. Konf. Molodykh. Uch.–Khim.*, *2nd*, 1977, **1**, 11 (*Chem. Abs.*, 1978, **89**, 180 273k).
[10] J. P. Ferris and H. C. Huang, *J.C.S. Chem. Comm.*, 1978, 1094.

yield respectively. Similar reaction using tetra-*O*-acetyl-α-D-glucopyranosyl bromide gave 84% of the 1-glucosyl compound.[11]

Full details of the synthesis of adenine and imidazole nucleosides of D-mannofuranose from 2,3:5,6-di-*O*-isopropylidene-D-mannofuranosylamine have now appeared [12] (see Vol. 11, p. 171). The same group has also synthesized the arabinofuranosyl 5-aminoimidazole-4-carboxylate (14) by condensation of 2,3,5-tri-*O*-benzyl α-D-*arabino*-furanosyl chloride with the parent base followed by hydrogenation.[13] 1,3-Dipolar cycloaddition of glycosyl azides to 1,4-naphthoquinone or 2-methyl-1,4-benzoquinone gave the nucleosides (15), (16), and (17), which were shown to exhibit some cytostatic activity.[14]

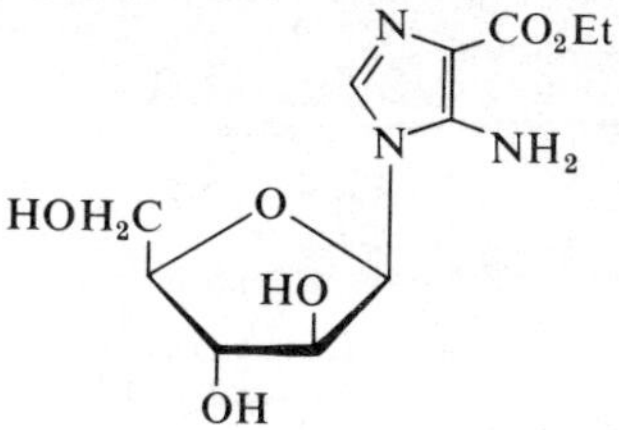

(14)

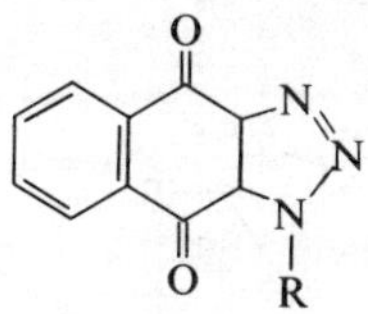

(15) R = 2,3,4,6-Tetra-*O*-acetyl-β-D-Glc*p*
(16) R = 2,3,5-Tri-*O*-acetyl-β-D-Rib*f*

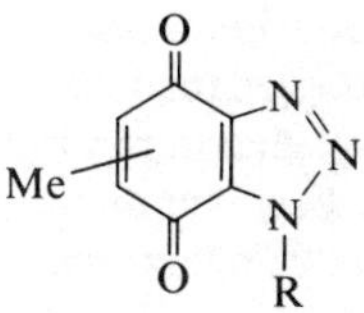

(17) R = 2,3,4,6-Tetra-*O*-acetyl-β-D-Glc*p*

The fusion reaction of 2-deoxy-3,5-di-*O*-*p*-nitrobenzoyl-D-*erythro*-pentofuranosyl chloride with the TMS derivative of 6-azathymine gave 86% yield of the β-6-azathymidine nucleoside (18) obtained without resort to chromato-

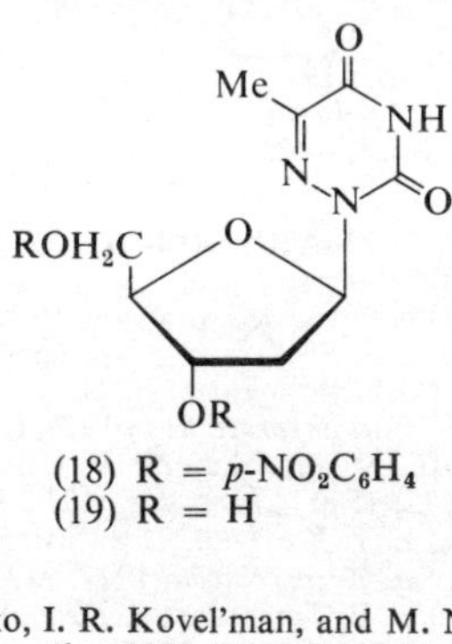

(18) R = p-$NO_2C_6H_4$
(19) R = H

[11] I. A. Korbukh, F. F. Blanko, I. R. Kovel'man, and M. N. Preobrazhenskaya, *Zhur. org. Khim.*, 1978, **14**, 1101 (*Chem. Abs.*, 1978, **89**, 110 255g).
[12] G. Mackenzie and G. Shaw, *J.C.S. Perkin I*, 1978, 1381.
[13] G. Mackenzie and G. Shaw, *J.C.S. Chem. Comm.*, 1978, 882.
[14] G. Alonso, M. Fuertes, M. T. Garcia-Lopez, F. G. De Las Heras, J. M. Infante, and M. Stud, *European J. Medicin. Chem.-Chim. Ther.*, 1978, **13**, 155 (*Chem. Abs.*, 1978, **89**, 60 025j).

graphy. Treatment with methanolic ammonia gave the parent nucleoside (19).[15] Condensation of 1,2,3,5-tetra-*O*-acetyl-β-D-ribose with 3-cyano-4-nitropyrazole or its 3-ethoxycarbonyl analogue gave the corresponding isomeric pyrazole nucleosides formed by reaction at either ring nitrogen. Small amounts of the α-anomers were also obtained with the cyano compound.[16] Analogous derivatives have similarly been obtained from 4-acetoxymethyl-3-methoxycarbonylpyrazole.[17]

Standard procedures have been used to synthesize 1-deazaguanosine and 1-deaza-6-thioguanosine.[18] The latter compound and its 6-methylthio-analogue have also been synthesized by another group.[19] The ribonucleosides of 7-substituted pyrazolo [4,3-*d*]pyrimidines, (20) and (21), have been prepared by

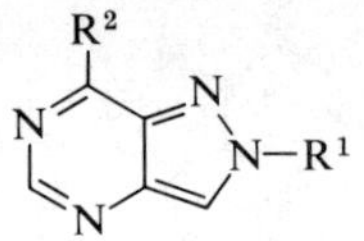

(20)

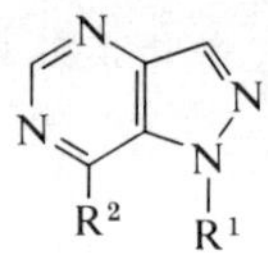

(21)

R^1 = α- or β-D-Rib*f*, R^2 = MeS, OH, or NH_2

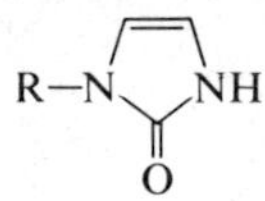

(22) R = β-D-Glc*p*
(23) R = β-D-Rib*f*

standard methods.[20] Vorbrüggen conditions (protected base and glycosyl bromide) have been used to synthesize the 1-(β-D-glycosyl)-4-imidazolin-2-ones (22) and (23).[21] The 7- and 9-β-D-glucofuranosides and -pyranosides of zeatin and 6-benzylaminopurine have been synthesized by standard methods, and compared with plant metabolites of cytokinins, which were thus shown to be the 7- and 9-β-D-glucopyranosides.[22]

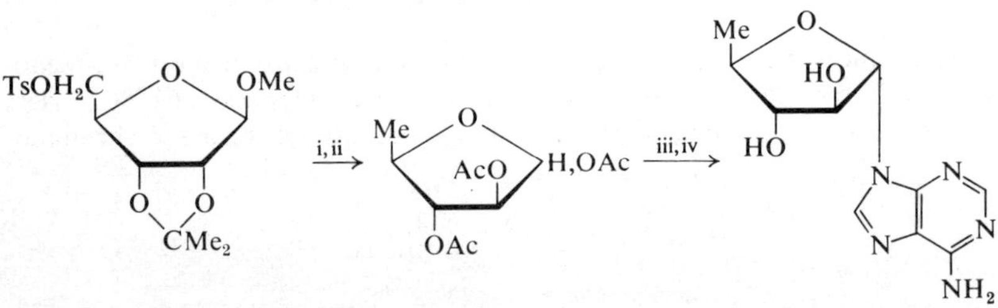

Reagents: i, NaH–DMSO; ii, Ac_2O–AcOH–H_2SO_4; iii, 6-Benzamido-9-chloromercuripurine–$TiCl_4$; iv, MeONa

Scheme 2

[15] G. T. Shiau and W. H. Prusoff, *Carbohydrate Res.*, 1978, **62**, 175.
[16] I. A. Korbukh, O. V. Budanova, N. G. Yakunina, V. I. Seraya, and M. N. Preobrazhenskaya, *Zhur. org. Khim.*, 1976, **12**, 1560 (*Chem. Abs.*, 1978, **88**, 38 104s).
[17] A. A. Akhrem, N. I. Garbuz, E. I. Kvasyuk, I. A. Mikhailopulo, and N. E. Pupeiko, *Vestsi Akad. Navuk B.S.S.R., Ser. Khim. Navuk*, 1977, 97 (*Chem. Abs.*, 1978, **88**, 7277w).
[18] B. L. Cline, R. P. Panzica, and L. B. Townsend, *J. Heterocyclic Chem.*, 1978, **15**, 839.
[19] R. D. Elliott and J. A. Montgomery, *J. Medicin. Chem.*, 1978, **21**, 112.
[20] O. V. Goryunova, I. A. Korbukh, and M. N. Preobrazhenskaya, *Zhur. org. Khim.*, 1978, **14**, 651 (*Chem. Abs.*, 1978, **89**, 110 216v).
[21] J. J. K. Novak, *Coll. Czech. Chem. Comm.*, 1978, **43**, 1511.
[22] D. E. Cowley, C. C. Duke, A. J. Liepa, J. K. MacLeod, and D. S. Letham, *Austral. J. Chem.*, 1978, **31**, 1095.

Ribonucleosides of 5- and 6-aminoimidazo[4,5-*b*]pyridine have been prepared by conventional methods.[23] The reaction sequence shown in Scheme 2 has been used to prepare 9-(5-deoxy-α-D-arabinofuranosyl)adenine from D-ribose.[24]

The triazine 2-deoxynucleoside (24) and its ribosyl analogue (25) have been synthesized by application of the silyl ether modification of the Hilbert–Johnson reaction. Compound (25) was shown to possess antiviral and antibacterial activity.[25] Ribosylation of 6,7-diphenyl-2-thiolumazine (26) has been

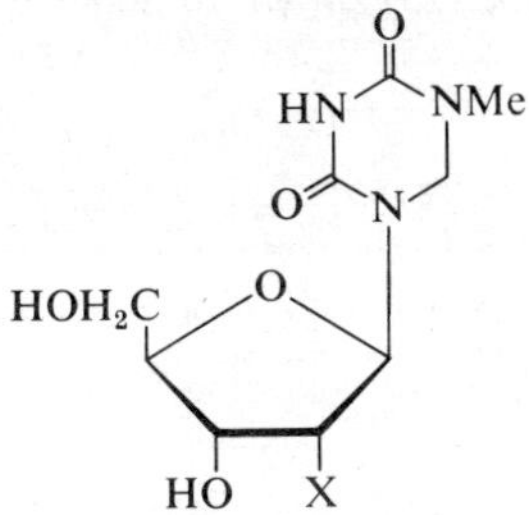

(24) X = H
(25) X = OH

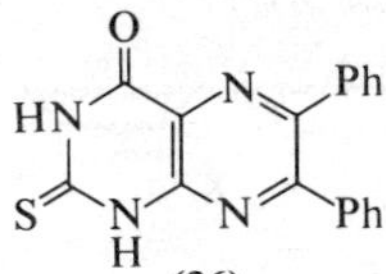

(26)

effected under various conditions. The silyl procedure with esters of the glycosyl halide gave predominantly *S*-bonded products with some *O*-linked isomers, whereas the mercuric salt method gave mainly *N*-glycosides. Mercury(II) bromide catalysed the isomerization of *S*-linked products to *N*-bonded compounds.[26] The useful dicyano-intermediates (27) and (28) have been

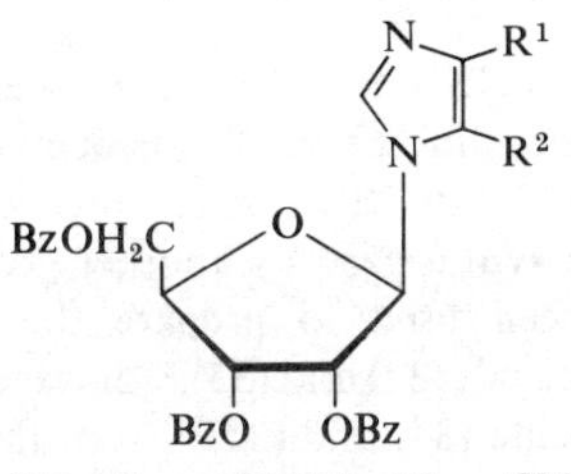

(27) R^1 = CH_2CN; R^2 = CN
(28) R^1 = CN; R^2 = CH_2CN

prepared by tin(IV) chloride catalysed reaction of 1-*O*-acetyl-2,3,5-tri-*O*-benzoyl-β-D-ribofuranose with the trimethylsilyl derivative of the base. Various ring closures of the side-chains led to a series of 3-deazapurine nucleoside derivatives.[27] Condensation of 6-nitro-indoline with 5-*O*-trityl-L-arabinose, D-fucose, D-arabinose, or L-rhamnose gave the corresponding 1-glycosyl-6-nitroindolines, from which, after acetylation, dehydrogenation, and

[23] T. Itoh, J. Inaba, and Y. Mizuno, *Heterocycles*, 1977, **8**, 433 (*Chem. Abs.*, 1978, **88**, 105 703b).
[24] L. M. Lerner, *J. Org. Chem.*, 1978, **43**, 161.
[25] H. Skulnick, *J. Org. Chem.*, 1978, **43**, 3188.
[26] I. W. Southon and W. Pfleiderer, *Chem. Ber.*, 1978, **111**, 2571.
[27] P. D. Cook and R. K. Robins, *J. Org. Chem.*, 1978, **43**, 289.

removal of protecting groups, 1-α- and 1-β-L-arabinofuranosyl, 1-β-D-fucopyranosyl, 1-α-D-arabinopyranosyl, and 1-α- and 1-β-L-rhamnopyranosyl derivatives of 6-nitroindole were obtained. The configurations of the arabinose and fucose derivatives were established by ^1H n.m.r. Comparison of the ^1H n.m.r. and c.d. of the products of sequential periodate oxidation, borohydride reduction, and acetylation allowed the configurational assignments of the rhamnose derivatives.[28] The syntheses of several imidazo[4,5-*d*]pyridazine nucleosides related to inosine have been reported. These include the preparation of the interesting di-ribosyl nucleoside (29) shown in Scheme 3.[29]

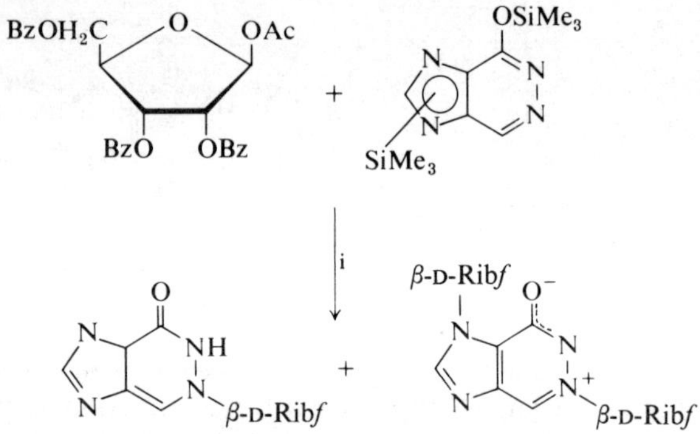

(29)

Reagent: i, $SnCl_4$

Scheme 3

The thionobenzazole nucleosides (30) have been prepared by standard methods from the corresponding base. The benzimidazole base also gave the 2,2′-anhydronucleoside (31) by the fusion procedure.[30] Compound (30) (X = NH) has also been synthesized by another group of workers.[31] Tin(IV) chloride catalysis has been used to prepare the (2,3,5-tri-*O*-benzoyl-β-D-ribofuranosyl)thioxoazauracils (32) and (33),[32] and the *N*-bridgehead analogues of guanosine and xanthosine (34) and (35)[33] from the TMS derivatives of the appropriate bases. Tri-n-butylstannylated pyrimidines have been treated with peracylated β-D-ribofuranose in the presence of tin(IV) chloride to afford protected pyrimidine nucleosides in 63—97% yields.[34] The pyrimidine nucleosides of tri-*O*-benzoyl-β-D-xylose (36) have been prepared in 90% yields by the Hilbert–Johnson procedure. The products were converted to the 6-thio-

[28] T. D. Miniker, V. I. Mukhanor, N. D. Chkanikov, I. V. Yartzeva, and M. N. Preobrazhenskaya, *Carbohydrate Res.*, 1978, **64**, 17.
[29] P. D. Cook, P. Dea, and R. K. Robins, *J. Heterocyclic Chem.*, 1978, **15**, 1.
[30] G. Gosselin, H. F. Loukil, A. Mathieu, and A. Mesli, *J. Heterocyclic Chem.*, 1978, **15**, 657.
[31] I. W. Southon and W. Pfleiderer, *Chem. Ber.*, 1978, **111**, 996.
[32] C. Cristescu, *Monatsh. Chem.*, 1977, **108**, 1455 (*Chem. Abs.*, 1978, **88**, 89 969k).
[33] S.-H. Kim, D. G. Bartholomew, L. B. Allen, R. K. Robins, G. R. Revankar, and P. Dea, *J. Medicin. Chem.*, 1978, **21**, 883.
[34] T. Ogawa and M. Matsui, *J. Organometallic Chem.*, 1978, **145**, C37.

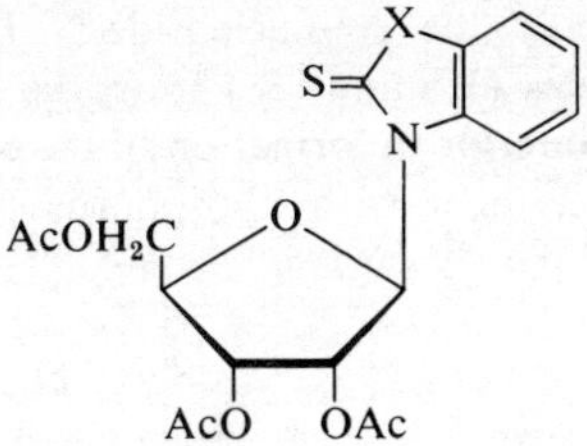

(30) X = O, S, or NH

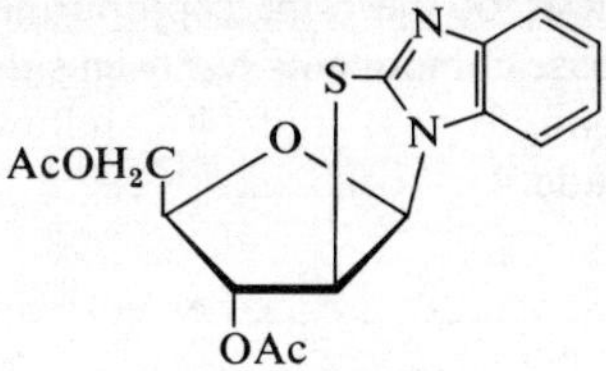

(31)

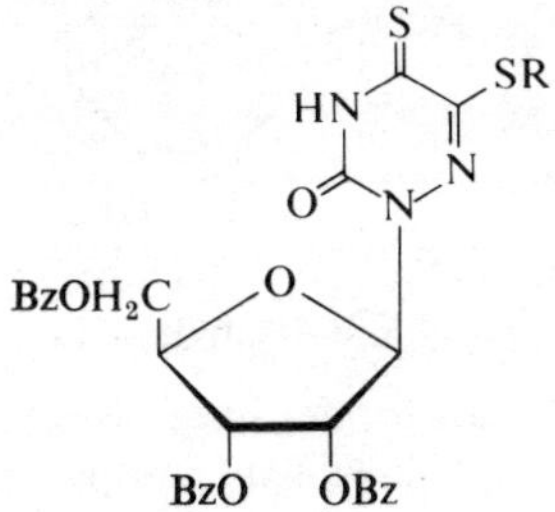

(32) R = Me
(33) R = Bn

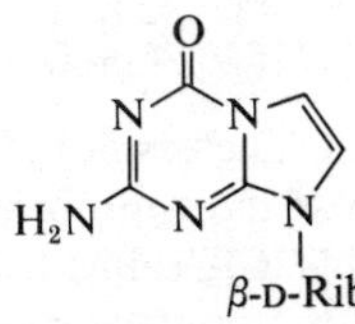

(34)

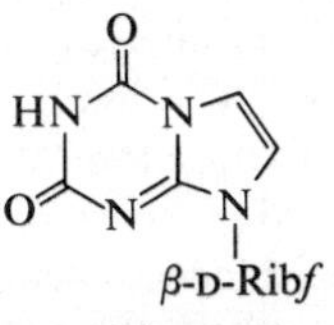

(35)

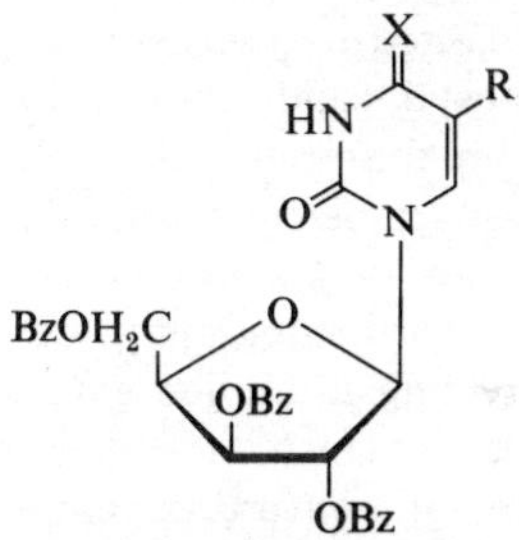

(36) R = H, Me, or F
X = O

derivatives (36; X = S) by treatment with phosphorus pentasulphide. All products were also debenzoylated.[35]

A report on the use of 2-chloro-3-ethylbenzoxazolinium tetrafluoroborate as a condensing agent for the synthesis of *trans*-nucleosides from derivatives of free sugars and heterocylces has appeared.[36] Nucleoside analogues derived from 3-deazaguanine have been synthesized *via* glycosylimidazole carboxamide precursors. Both 7- and 9-(5-deoxy-β-D-ribofuranosyl)-3-deazaguanine products were prepared.[37] Treatment of 5-methylthiouracil with 1-*O*-acetyl-2,3,5-tri-*O*-benzoyl-β-D-ribofuranose gave the N^1- and N^3-nucleosides, whereas the

[35] E. M. Kaz'mina, R. M. Kenbaeva, and B. N. Stepanenko, *Zhur. obshchei Khim.*, 1978, **48**, 1892 (*Chem. Abs.*, 1978, **89**, 180 287t).

[36] T. Mukaiyama, S. Shoda, T. Nakatsuka, and K. Narasaka, *Chem. Letters*, 1978, 605.

[37] P. D. Cook, L. B. Allen, D. G. Streeter, J. H. Huffman, R. W. Sidwell, and R. K. Robins, *J. Medicin. Chem.*, 1978, **21**, 1212.

corresponding sulphone derivative gave only the N^1-nucleoside.[38] Direct synthesis of the 6-mercaptopurine nucleosides (37) and (38) from the sugar and base components has been shown to be inferior to formation of the 6-oxo-compounds (39) and (40), followed by treatment with phosphorus pentasulphide.[39]

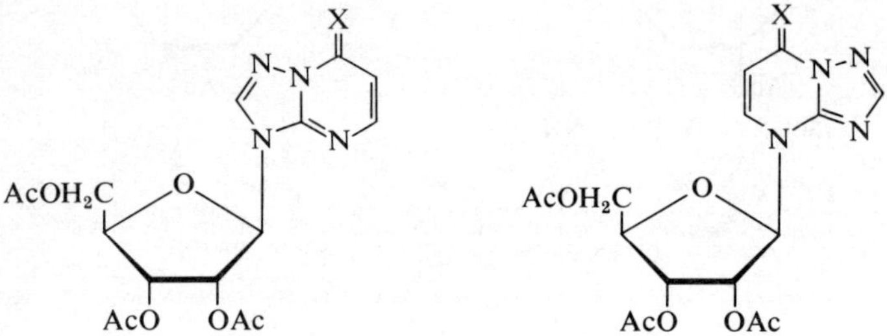

(37) X = S
(39) X = O

(38) X = S
(40) X = O

The [2-^{14}C]nucleosides of ribose and 2-deoxyribose have been prepared by incubation of the appropriate ribofuranosyl 1-phosphate and pyrimidine with *E. coli*.[40] [4-^{13}C]Uridine has been synthesized from the labelled bis(trimethylsilyl)uracil by standard means.[41]

Exploitation of the epimerization at C-2 when 2,3-*O*-isopropylidenefuranosides are subjected to acetolysis has enabled the syntheses of 9-(6-deoxy-β-L-galactofuranosyl) and 9-(6-deoxy-α-D-altrofuranosyl)adenine: the syrup resulting from the acetolysis of the L-taloside (41) was condensed with 6-benzamido(chloromercuri)purine in the presence of titanium(IV) chloride to give a mixture of the α-L-*talo*- and β-L-*galacto*-nucleosides which were separated by chromatography after hydrolysis. A similar procedure with the corresponding D-alloside gave a mixture of the α-D-*altro*- and β-D-*allo*-nucleoside from which the latter was eliminated by periodate oxidation.[42] A procedure for converting ribonucleosides to blocked compounds of the type (42) has been reported. These are useful for preparation of 2-*O*-methoxy-

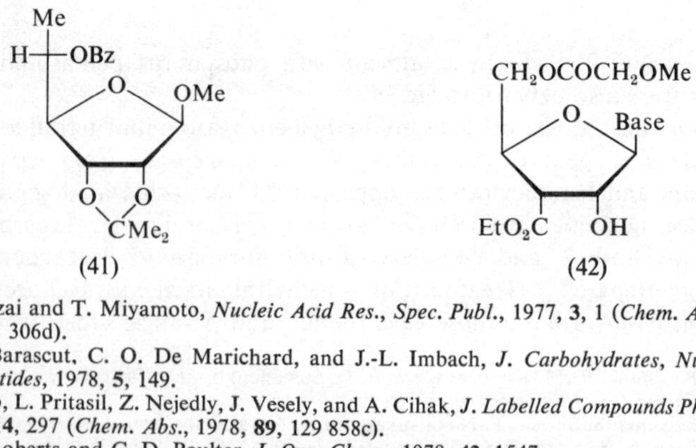

(41) (42)

[38] K. Anzai and T. Miyamoto, *Nucleic Acid Res., Spec. Publ.*, 1977, **3**, 1 (*Chem. Abs.*, 1978, **88**, 191 306d).
[39] J.-L. Barascut, C. O. De Marichard, and J.-L. Imbach, *J. Carbohydrates, Nucleosides, Nucleotides*, 1978, **5**, 149.
[40] J. Filip, L. Pritasil, Z. Nejedly, J. Vesely, and A. Cihak, *J. Labelled Compounds Pharmacol.*, 1978, **14**, 297 (*Chem. Abs.*, 1978, **89**, 129 858c).
[41] J. L. Roberts and C. D. Poulter, *J. Org. Chem.*, 1978, **43**, 1547.
[42] L. M. Lerner, *J. Org. Chem.*, 1978, **43**, 962.

tetrahydropyranyl derivatives, which are key intermediates for synthesis of oligoribonucleotides.[43] Another procedure for making these key intermediates involves selective hydrazinolysis of 2′,3′,5′-tri-*O*-acetyl-β-D-ribofuranosyl nucleosides to give 3′,5′-di-*O*-acetyl products.[44]

A synthesis of 5′(*R*)-[^{2}H]adenosine is outlined in Scheme 4.[45] In another report, an 80 : 20 mixture of 5′(*R*) and 5′(*S*)-monodeuteroadenosine was

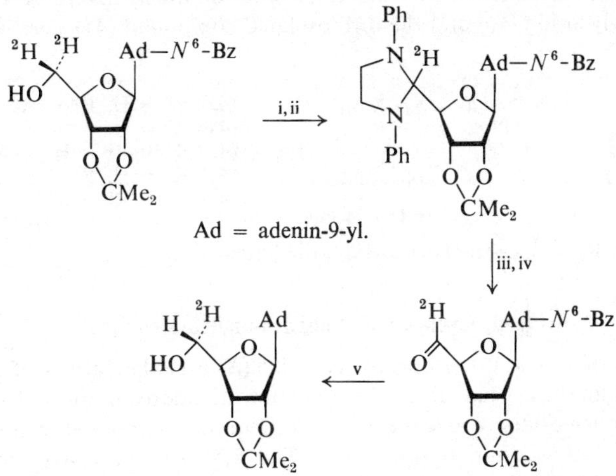

Reagents: i, Me_2SO–DCC–Cl_2HCCO_2H; ii, *N,N′*-diphenylethylene diamine; iii, Resin(H^+); iv, C_6H_6-azeotrope; v, (−)-α-pinene–9-borabicyclo[3,3,1]nonane

Scheme 4

obtained by reduction of 5′-aldehydo-*N*-benzoyladenosine with lithium aluminium deuteride followed by deblocking.[46]

The syntheses of 9-(5-deoxy-β-D-xylo- and -α-L-arabino-hexofuranosyl)-adenine have been reported.[47] Treatment of cytidine with 2,4-dinitrobenzoyl chloride gave the 2′,3′,5′-N^4-tetrabenzoate which with aqueous pyridine followed by methanolic ammonia furnished spongocytidine in 38% yield.[48] Various means of alkylating guanines and guanosines have been described.[49]

Attempted synthesis of 3-methylinosine by treatment of *N,N*-dimethyl-adenosine with excess methyl iodide failed because glycosyl cleavage occurred. Although the synthesis was successful with the 2′,3′,5′-tri-*O*-benzyl derivative, hydrogenolysis to deblock the product resulted in reduction of the purine ring. The problem was overcome by use of the 2′,3′,5′-tribenzoate. The ribosyl bond in 3-methylinosine was found to be unusually susceptible to acid

[43] J. H. Van Boom, P. M. J. Burgers, C. A. G. Haasnoot, and C. B. Reese, *Rec. Trav. chim.*, 1977, **96**, 91 (*Chem. Abs.*, 1978, **88**, 7266s).
[44] R. J. Gregoire and T. Neilson, *Canad. J. Chem.*, 1978, **56**, 487.
[45] R. J. Parry, *J.C.S. Chem. Comm.*, 1978, 294.
[46] M. Dupre and A. Gaudemer, *Tetrahedron*, 1978, **34**, 2783.
[47] W. A. Szarek, R. G. S. Ritchie, and D. M. Vyas, *Carbohydrate Res.*, 1978, **62**, 87.
[48] G. R. Niaz, F. Khan, and S. M. Ifzal, *Pakistan J. Scil. Ind. Res.*, 1976, **19**, 214 (*Chem. Abs.*, 1978, **89**, 110 224w).
[49] T. Itaya and K. Ogawa, *Tetrahedron Letters*, 1978, 2907.

hydrolysis; at 40 °C and pH 3 in water, hydrolysis was complete in 24 h, whereas at pH 1 at the same temperature the reaction was complete in 10 min.[50]

When the 5-*N*-2-acetoxyethyl derivative (43) was ribosylated using the silyl method the *N*-ribosyl nucleoside (44) was obtained. Removal of the acetyl group and photocleavage of the 2-hydroxyethyl side-chain gave 6,7-diphenyllumazine, which was also synthesized by direct ribosylation of the silylated base. When the same reaction was carried out on the hydroxyethyl compound (45), the glycoside (46) and the diribosylated compound (47) were obtained.[51]

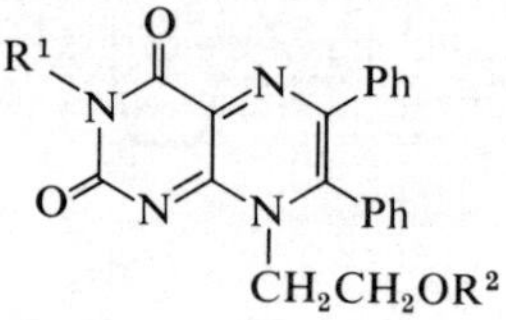

(43) $R^1 = H$, $R^2 = Ac$
(44) $R^1 = R^3$
(45) $R^1 = R^2 = H$
(46) $R^1 = H$, $R^2 = R^3$
(47) $R^1 = R^2 = R^3$

R^3 = 2,3,5-tri-*O*-benzoyl-β-D-ribofuranosyl

2 Bridged, Cyclo-, and Anhydro-nucleosides

Syntheses of the 2′,3′-*O*-isopropylidene-8,5′(*S*)-cycloderivatives of adenosine and guanosine have been achieved by selenium dioxide oxidation of the 5′-deoxy-8,5′-cyclonucleosides to give the 5′-oxo compounds, followed by stereoselective reduction with sodium borohydride. The 8,5′-(*R*) series was obtained by inversion of the (*S*) isomer *via* the 5′-mesyloxy derivative. Hydrolytic deamination of the adenosine derivatives gave the inosine counterparts. Their c.d. spectra reflected the chirality at the 5′-position.[52]

8,5′-(*R*)- and -(*S*)-Cycloadenosines have been synthesized using photolysis of the 8-phenylthio-nucleoside.[53] The cyclouridines (48) and (49) in alkaline solution are readily converted by aerial oxidation into the ketone (50), which in turn isomerizes at pH 8—9 to furnish the *xylo*-nucleoside (51). The *ribo*- and *xylo*-isomers (50) and (51) differ considerably in their degrees of hydration

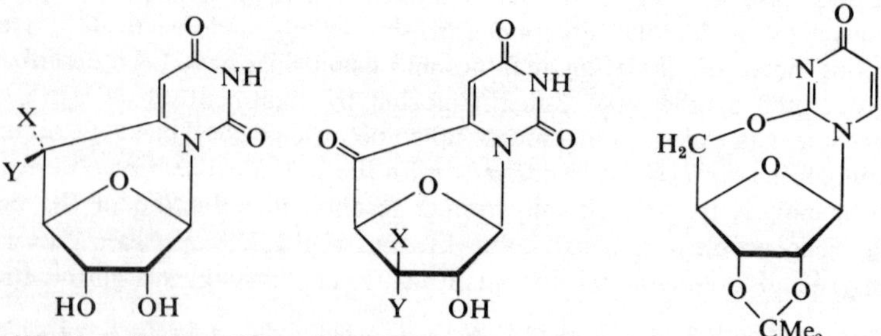

(48) X = H, Y = OH (50) X = H, Y = OH
(49) X = OH, Y = H (51) X = OH, Y = H (52)

[50] T. Itaya and H. Matsumoto, *Tetrahedron Letters*, 1978, 4047.
[51] K. Ienaga and W. Pfleiderer, *Chem. Ber.*, 1978, **111**, 2586.
[52] A. Matsuda, M. Tezuka, K. Niizuma, E. Sugiyama, and T. Ueda, *Tetrahedron*, 1978, **34**, 2633.
[53] A. Matsuda, M. Tezuka, and T. Ueda, *Tetrahedron*, 1978, **34**, 2449.

in aqueous solution, u.v. spectrum, and in the stereoselectivity of their reduction by sodium cyanoborohydride; (50) gives only the 5′(*S*) product whereas both 5′(*S*) and (*R*) are produced from (51).[54] Radiation of 8,5′-cycloadenosine 5′-phosphate gave the C-5′ epimer (*i.e.* L-*talo* compound) characterized by n.m.r.[55] Butan-1-ol solutions of trifluoroacetic acid have been found to detritylate 2,3′-anhydro-1-(2,5-di-*O*-trityl-β-D-xylofuranosyl)uracil at room temperature without rupture of the anhydro-ring. The method was shown to be generally applicable.[56] The 2,5′-cyclonucleoside (52) on methanolysis gave the blocked 5′-*O*-methyl uridine in good yield. Treatment of the latter with diphenylphosphate–boron trifluoride gave the 5′-phosphate.[57] The 2,2′-cyclotriazine nucleoside (53), an analogue of a naturally-occurring compound having antibiotic and antiviral activity, has been prepared as shown in Scheme 5. Several aspects of the synthesis of anhydronucleosides and related com-

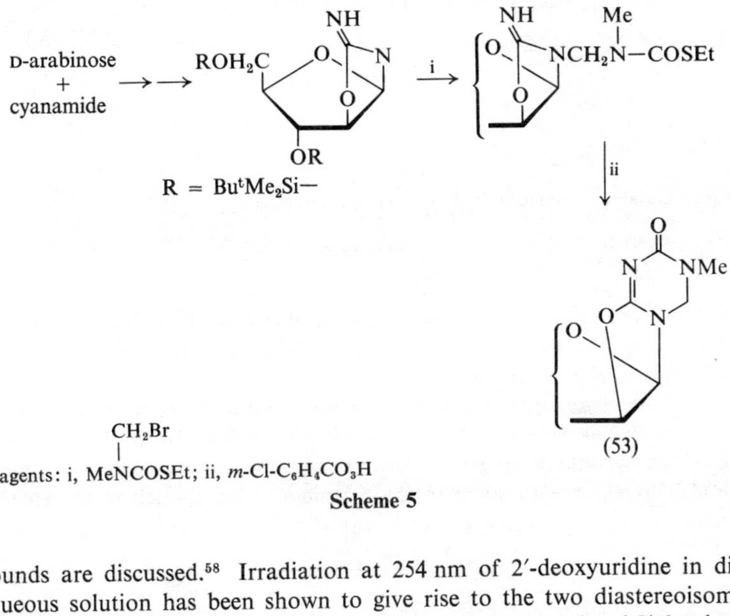

(53)

Reagents: i, $MeN(CH_2Br)COSEt$; ii, *m*-Cl-$C_6H_4CO_3H$

Scheme 5

pounds are discussed.[58] Irradiation at 254 nm of 2′-deoxyuridine in dilute aqueous solution has been shown to give rise to the two diastereoisomeric hydration adducts (54) as the major products and the two related 5′,6-anhydro-derivatives (55).[59] The synthesis of the 2,2′-anhydro-β-D-arabinofuranosyl triazine nucleoside (56), which possesses immunosuppressive activity, has been carried out as depicted in Scheme 6. The product (56) was also converted to the trione (57).[60]

54 B. A. Otter, E. A. Falco, and J. J. Fox, *J. Org. Chem.*, 1978, **43**, 481.
55 J. A. Raleigh and B. J. Blackburn, *Biochem. Biophys., Res. Comm.*, 1978, **83**, 1061.
56 M. MacCoss and D. J. Cameron, *Carbohydrate Res.*, 1978, **60**, 206.
57 J. Kimura and O. Mitsunobu, *Bull. Chem. Soc. Japan*, 1978, **51**, 1903.
58 W. Wierenga and J. A. Woltersom, *J. Org. Chem.*, 1978, **43**, 529.
59 J. Cadet, L.-S. Kan, and S. Y. Wang, *J. Amer. Chem. Soc.*, 1978, **100**, 6715.
60 W. Wierenga, B. E. Loughman, A. J. Gibbons, and H. E. Renis, *J. Medicin. Chem.*, 1978, **21**, 558.

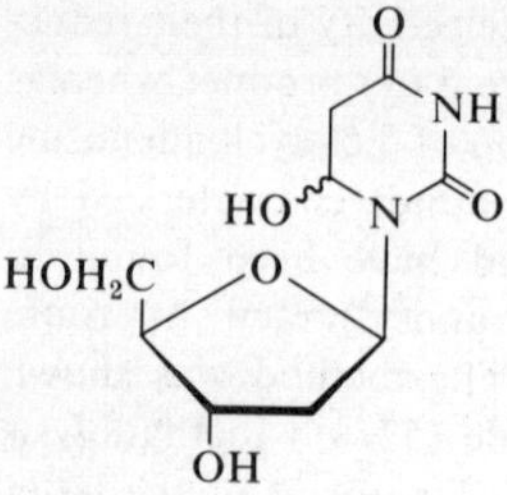

(54)

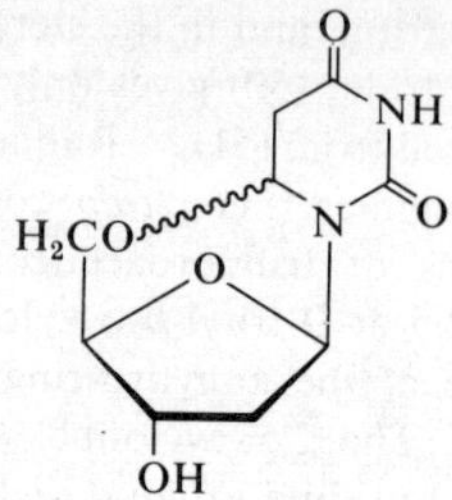

(55)

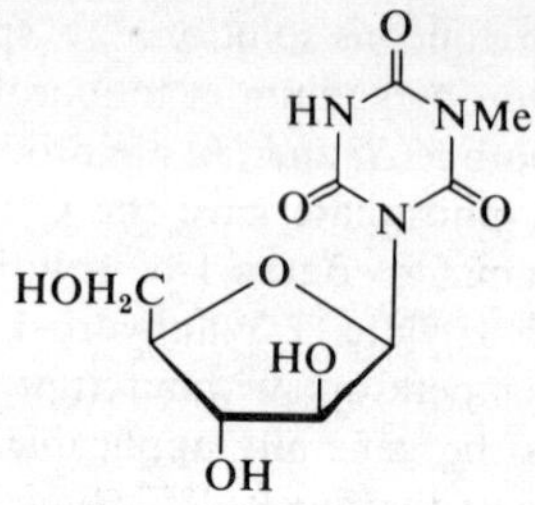

(57)

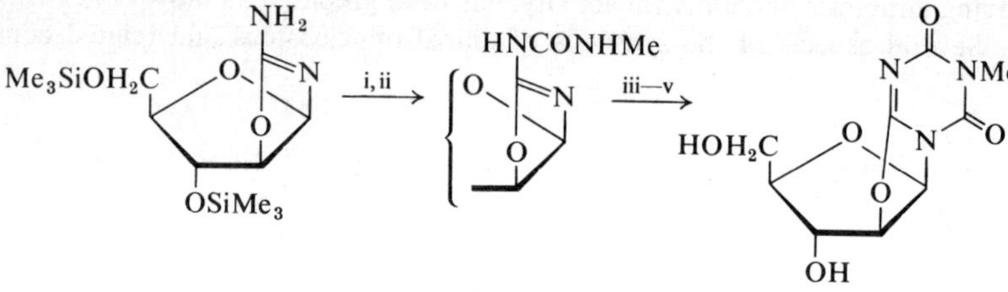

(56)

Reagents: i, NaH; ii, MeNCO; iv, 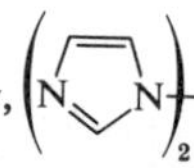 CO; v, H_3O^+

Scheme 6

Cyclocondensation of cytidine with acetylsalicyloyl chloride gave the corresponding 2,2′-anhydro-D-arabinofuranosyl-nucleoside, which was hydrolysed to give *arabino*-cytidine.[61]

Novel nucleoside analogues have been prepared by thermal intramolecular addition of the 5′-azido group of the uridine derivative (58) to the 5,6-double bond of the pyrimidine ring. The main product was shown to be the macrocycle (59) together with some of the 6-amino-6,5′-cyclonucleoside. Heating

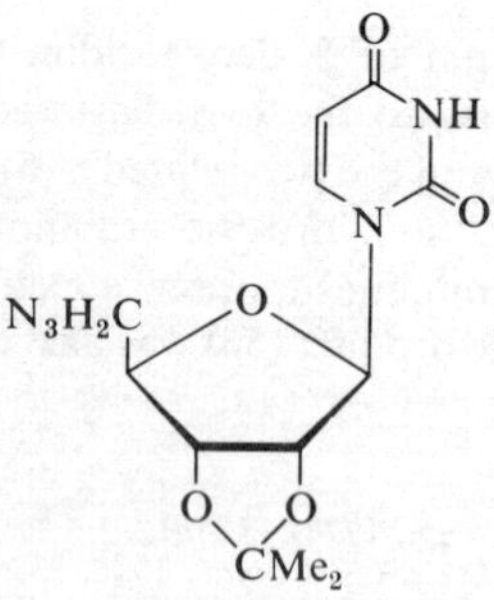

(58)

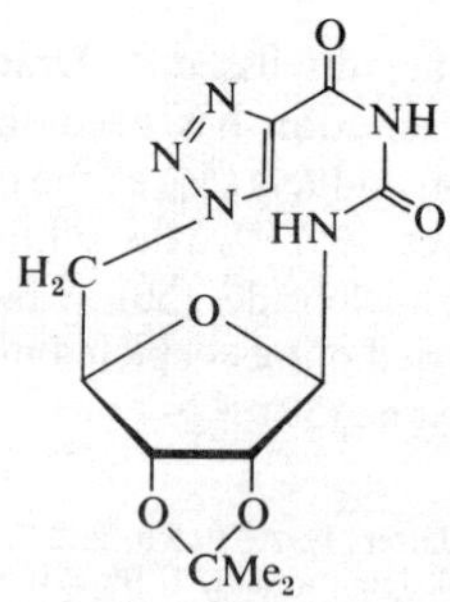

(59)

[61] E. N. Kalinchenko and E. I. Kvasyuk, *Tezisy Doklady-Resp. Konf. Molodykh. Uch.-Khim., 2nd*, 1977, **1**, 23 (*Chem. Abs.*, 1978, **89**, 110 239c).

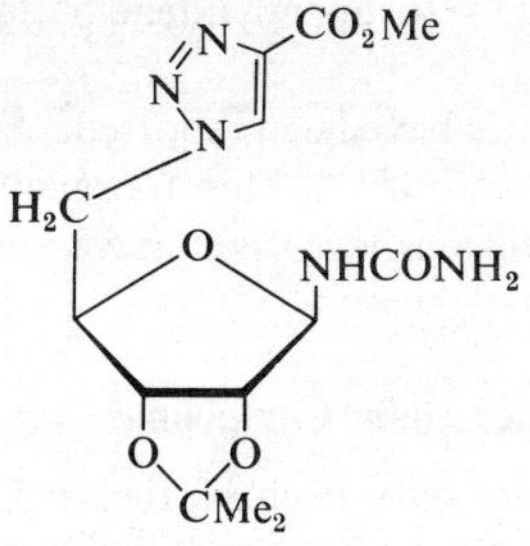

(60)

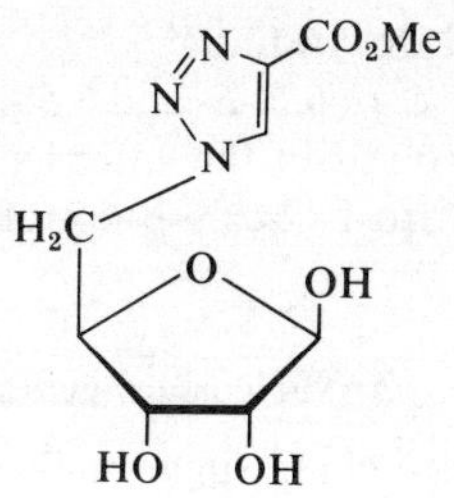

(61)

(59) in methanol gave (60) which was converted into the reversed triazole nucleoside (61) by diazotization followed by acid-catalysed hydrolysis.[62]

A series of 8,5′-cyclo-8-aminopurine nucleosides has been synthesized from the 8-bromo-5′-*O*-tosyl nucleoside (62) illustrated in Scheme 7 for the adenosine analogue.[63]

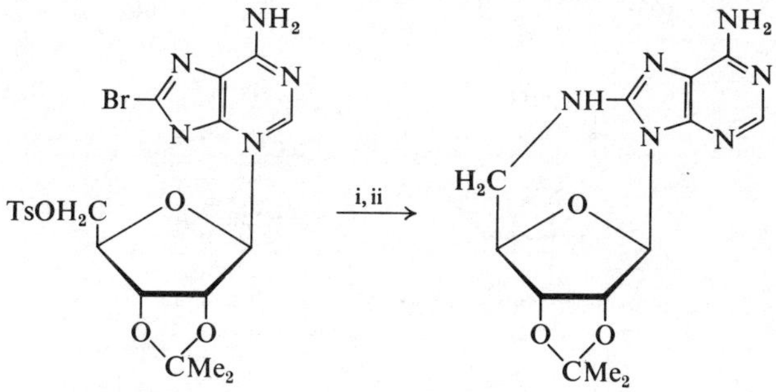

(62)

Reagents: i, NH_2NH_2; ii, I_2O_5

Scheme 7

Reaction of the 2-*N*-benzoylisocytidine derivative (63) with potassium phthalimide in dioxane gave *N*^{2},5′-anhydro-2′,3′-*O*-isopropylidene-isocytidine, while the corresponding *N*^{2},4′-anhydroisocytidine was prepared from the 4-enose-derivative (64), itself made by elimination from (63) with potassium

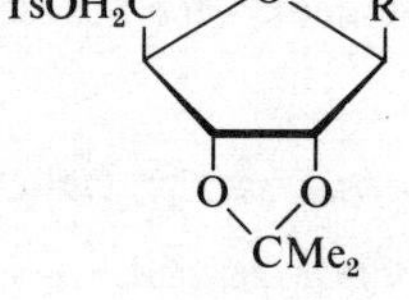

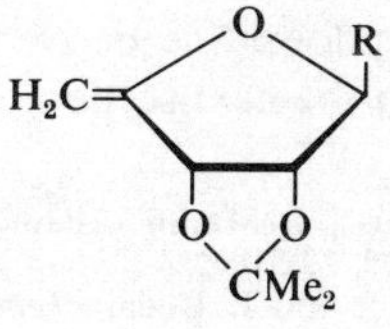

(63) R = N^2-benzoylisocytosine (64)

[62] T. Sasaki, K. Minamoto, T. Suzuki, and T. Sugiura, *J. Amer. Chem. Soc.*, 1978, **100**, 2248.
[63] T. Sasaki, K. Minamoto, and H. Itoh, *J. Org. Chem.*, 1978, **43**, 2320.

t-butoxide. Reduction of (64) gave 2′,3′-*O*-isopropylidene-5′-deoxylyxofuranosyl isocytidine.[64]

Syntheses of 6,5′-*S*- and 6,5′-*N*-cyclouridines have been reported. Attempted acid hydrolysis of the latter led to decomposition.[65] 2-Thio-6-azacytidine and 2,2′-anhydro-1-(2-thio-β-D-arabinofuranosyl)-6-azacytosine have been prepared.[66]

3 Nucleosides containing Branched-chain Compounds

Branched-chain sugar nucleosides of apiose have been synthesized from the derivatives (65) and (66) and the TMS derivatives of uracil, thymine, 5-bromouracil, or 5-iodo-uracil using tin(IV) chloride catalysis.[67] The synthesis of

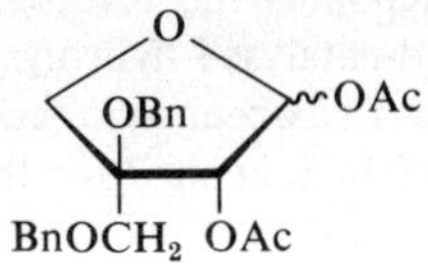

(65)

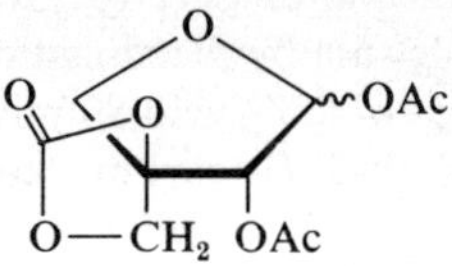

(66)

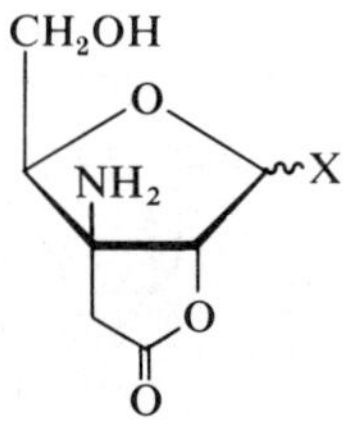

X = adenin-9-yl

(67)

9-(3-*C*-methyl-α-L-threofuranosyl)adenine has been accomplished by a standard route from the branched-chain sugar. The deamination of this nucleoside with adenosine deaminase was found to be about two hundred times slower than is the case with adenosine itself.[68]

The anomeric analogues (67) of the nucleoside moiety of the polyoxins have been prepared. C.d. was used to show the stereochemistry of the products.[69] The synthesis shown in Scheme 8 has been used to prepare the 4′-*C*-allyl nucleoside (68), and related analogues of nucleocidin were also synthesized.[70]

A pseudouridine derived from a 5-*C*-methyl-6-deoxy-*ribo*-hexose is mentioned in the next section.

[64] V. Skaric, J. Matulic-Adamic, and D. Skaric, *Heterocycles*, 1977, 7, 179 (*Chem. Abs.*, 1978, **88**, 23 310g).
[65] H. Inoue and T. Ueda, *Chem. and Pharm. Bull. (Japan)*, 1978, **26**, 2664.
[66] Y. L. Fu, R. Parthasarathy, and M. Bobek, *J. Carbohydrates, Nucleosides, Nucleotides*, 1978, **5**, 79.
[67] D. K. Parikh and R. R. Watson, *J. Medicin. Chem.*, 1978, **21**, 706.
[68] J. M. J. Tronchet and J. Tronchet, *Carbohydrate Res.*, 1977, **59**, 594.
[69] A. Rosenthal and M. Ratcliffe, *Carbohydrate Res.*, 1978, **60**, 39.
[70] J. A. Secrist and W. J. Winter, *J. Amer. Chem. Soc.*, 1978, **100**, 2554.

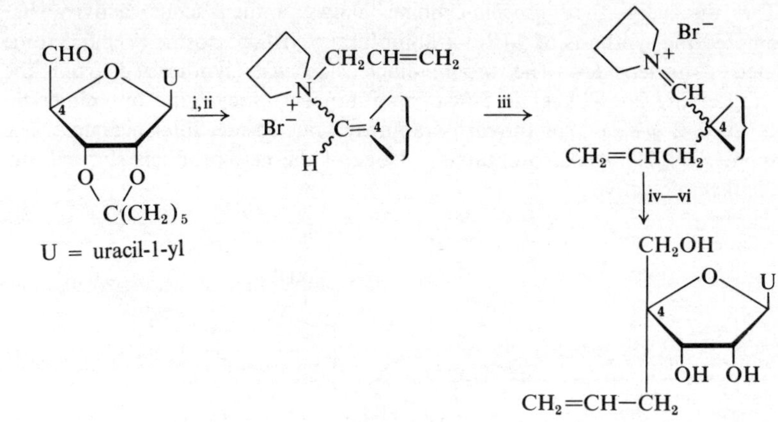

(68)

Reagents: i, Pyrrolidine; ii, $CH_2{=}CHCH_2MgBr$; iii, heat; iv, H_2O; v, $NaBH_4$; vi, H_3O^+

Scheme 8

4 *C*-Nucleosides

A stereocontrolled general synthesis of *C*-nucleosides from non-carbohydrate materials has been described, illustrated in Scheme 9 for pseudo-uridine derivatives (69) (R^1 = H, Me, C_5H_{11}, or Ph; R^2 = H); a branched-chain sugar derivative (69) (R = R^2 = Me) was analogously prepared.[71] 2-Thiopseudouridine and pseudoisocytidine were also obtained from the intermediate (70) (R^1 = R^2 = H) using thiourea and guanidine respectively in place of urea, and this intermediate was also converted to the maleimido *C*-nucleoside, showdomycin.[72]

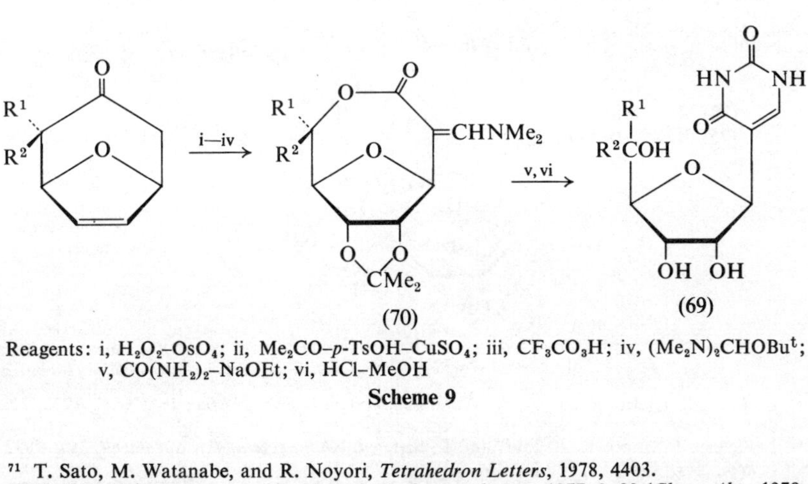

(70) (69)

Reagents: i, H_2O_2–OsO_4; ii, Me_2CO-*p*-TsOH–$CuSO_4$; iii, CF_3CO_3H; iv, $(Me_2N)_2CHOBu^t$; v, $CO(NH_2)_2$–NaOEt; vi, HCl–MeOH

Scheme 9

[71] T. Sato, M. Watanabe, and R. Noyori, *Tetrahedron Letters*, 1978, 4403.

[72] T. Sato, Y. Hayakawa, and R. Noyori, *Nucleic Acids Res.*, 1977, **3**, 29 (*Chem. Abs.*, 1978, **88**, 152 905d).

The discovery that *arabino*-cytidine shows antileukaemic activity has prompted the synthesis of 5-(β-D-arabinofuranosyl)-isocytosine (71) and some related *C*-nucleosides. The *arabino*-nucleoside was synthesized from the corresponding antileukaemic 5-(β-D-ribofuranosyl)-isocytosine by epimerization at C-2 *via* a 2′,4-anhydro-arabinosyl nucleoside intermediate. The α-anomer of (71) was also obtained. None of the new products showed any antileukaemic activity.[73]

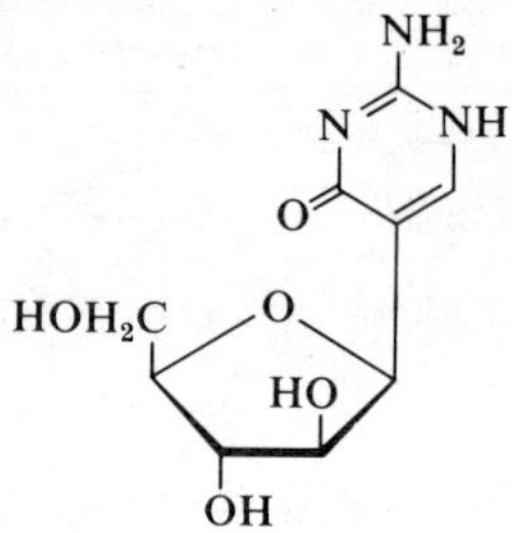

(71)

Addition reactions to unsaturated *C*-glycosides have continued to be utilized to synthesize *C*-nucleosides. Thus 2,3,5-tri-*O*-benzoyl-β-D-ribofuranosyl cyanide with sodium azide–ammonium chloride in DMF gave the tetrazole (72) in 72% yield. Treatment of (72) with acetic anhydride gave the 1,3,4-oxadiazole nucleoside (73). The benzothiazole (74) was obtained from the ribosyl cyanide by reaction with 2-aminothiophenol. Analogous compounds were derived from β-D-ribo-, β-D-xylo-, and β-D-galactopyranosyl cyanides.[74] A similar reaction has been used by a different group to synthesize α-D-arabinosyl tetrazole (75).[75] Reaction of 3,4-di-*O*-benzoyl-2,5-anhydro-D-

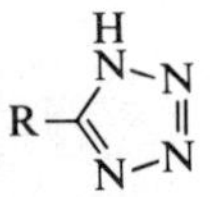

(72) R = β-D-Rib*f*
(75) R = α-D-Ara*f*

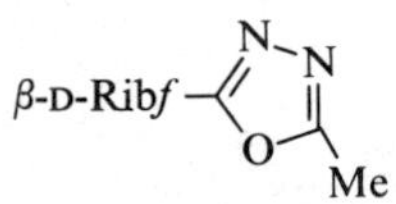

(73)

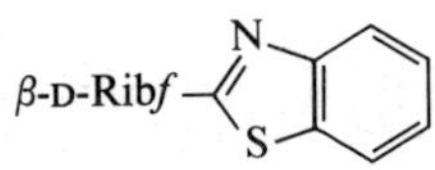

(74)

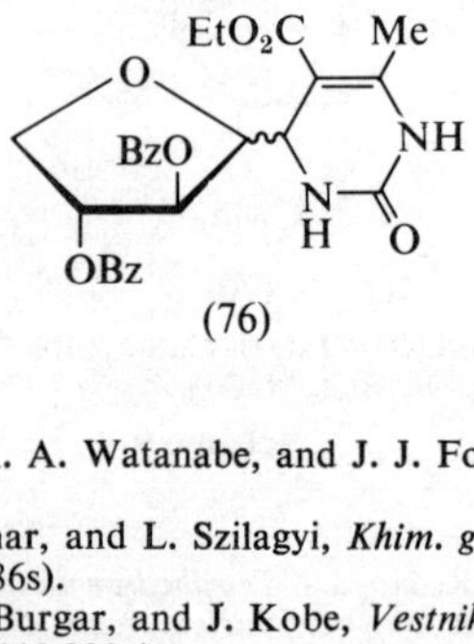

(76)

[73] C. K. Chu, U. Reichman, K. A. Watanabe, and J. J. Fox, *J. Medicin. Chem.*, 1978, **21**, 96.

[74] I. Farkas, I. Szabo, R. Bognar, and L. Szilagyi, *Khim. geterotsikl. Soedinenii*, 1978, 893 (*Chem. Abs.*, 1978, **89**, 180 286s).

[75] M. Hohnjec, M. Japelj, M. Burgar, and J. Kobe, *Vestnik Slovensk. kem. Drustva*, 1978, **25**, 1 (*Chem. Abs.*, 1978, **89**, 110 222u).

xylose with ethyl acetoacetate and urea (Biginelli reaction) gave the *C*-glycosyl dihydropyrimidine (76).[76] The syntheses of 3-phenyl-4-(α- and β-D-ribofuranosyl)-pyrazoles (77) have been reported (Scheme 10).[77]

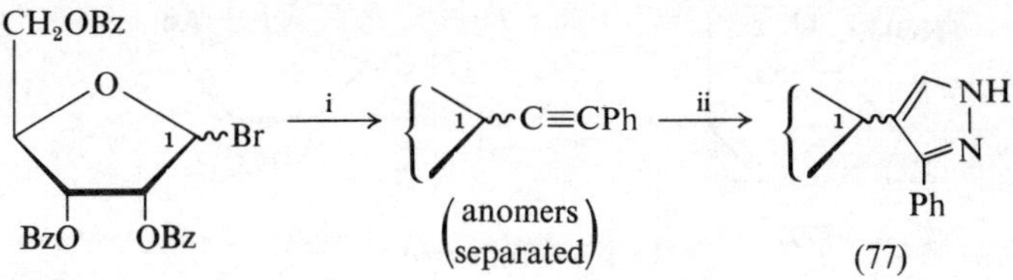

(77)

Reagents: i, $Hg(C{\equiv}CPh)_2$; ii, CH_2N_2

Scheme 10

Pseudouridine has been converted into pseudoisocytidine in a one-step reaction with guanine.[78]

Dehydration of the hydroxyalkyl chain of 1-phenyl-3-(D-*arabino*-tetritol-1-yl)pyrazolo[3,4-*b*]quinoxaline gave the *C*-nucleoside (78) in 82% yield.[79]

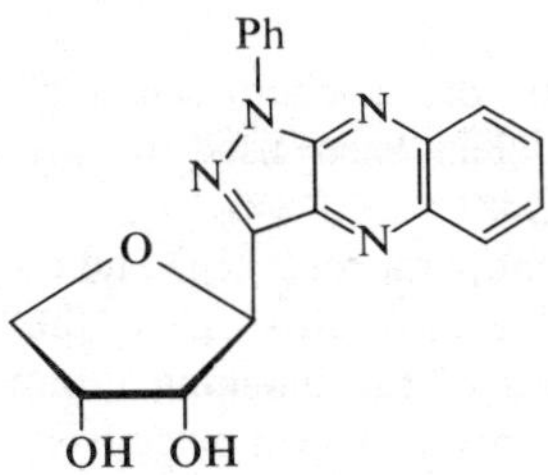

(78)

The relationship between the rate of tautomeric equilibrium of formycin and its enzymic reactions has shown that the former is faster so that the rate of any process which is specific to the minor tautomer is not limited by the rate of tautomerism.[80] Several references to *C*-bonded nucleosides appear in Chapter 3.

5 Unsaturated Nucleosides

The unsaturated nucleoside analogue (79) has been prepared from β-D-glucopyranosyluracil by a sequence involving elimination from the corresponding 2,2′-anhydro-3-*O*-mesyl nucleoside catalysed by sodium iodide.[81]

[76] J. A. Lopez Sastre and J. Molina Molina, *Anales de Quím.*, 1978, **74**, 353 (*Chem. Abs.*, 1978, **89**, 163 889y).
[77] N. Hamamichi, T. Miyasaka, and K. Arakawa, *Chem. and Pharm. Bull* (*Japan*), 1978, **26**, 898.
[78] K. Hirota, K. A. Watanabe, and J. J. Fox, *J. Org. Chem.*, 1978, **43**, 1192.
[79] M. A. E. Sallam, *Carbohydrate Res.*, 1978, **67**, 79.
[80] F. X. Cole and P. R. Schimmel, *J. Amer. Chem. Soc.*, 1978, **100**, 3957.
[81] T. Yamazaki, H. Sugiyama, K. Matsuda, and S. Seto, *Bull. Chem. Soc. Japan*, 1977, **50**, 3423.

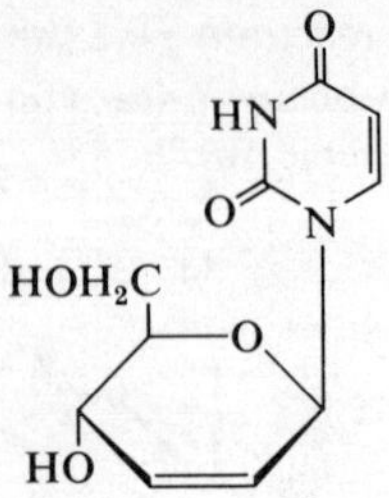

(79)

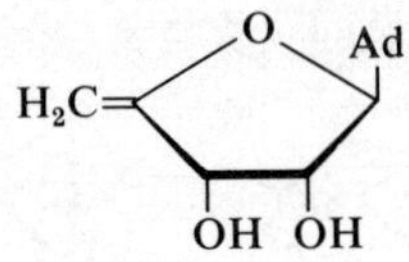

(80)

Ad = adenin-9-yl

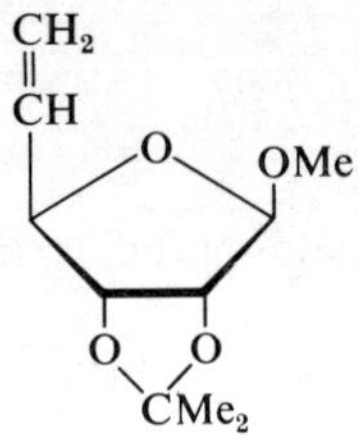

(81)

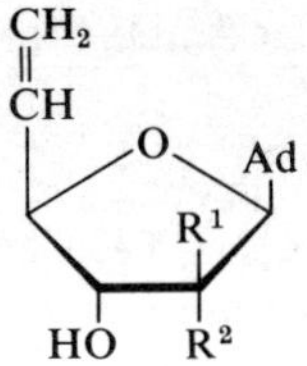

(82) R^1 = H, R^2 = OH
(83) R^1 = OH, R^2 = H

Thermal fragmentation of 5′-phenylseleno-5′-deoxyadenosine in the presence of triethylamine has been used to prepare the 4,5-unsaturated compound (80) in high yield.[82]

Acetolysis of the hex-5-enopyranoside (81) and condensation of the product with 6-benzamido(chloromercuri)purine gave the unsaturated nucleosides (82) and (83) after removal of the blocking groups. The D-*ribo*-nucleoside (82), which was the major product, was selectively destroyed with periodate to facilitate isolation of (83). The α-L-isomer of (83) was prepared by the same route and also by an improved method starting from D-galactose

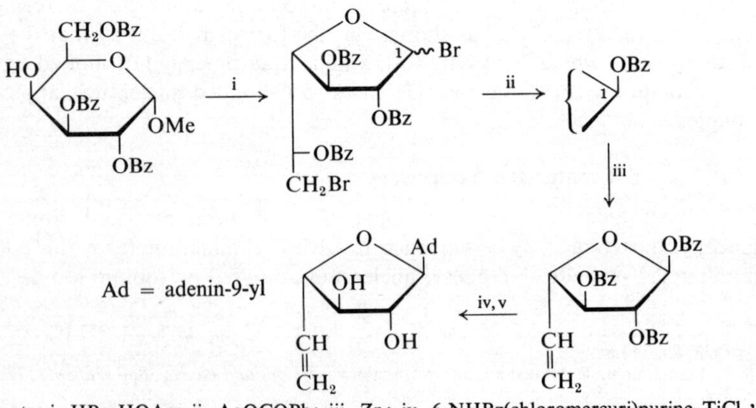

Reagents: i, HBr–HOAc; ii, AgOCOPh; iii, Zn; iv, 6-NHBz(chloromercuri)purine–$TiCl_4$,
v, MeONa

Scheme 11

[82] N. Zylber, J. Zylber, and A. Gaudemer, *J.C.S. Chem. Comm.*, 1978, 1084.

(Scheme 11). 9-(6-Deoxy-β-D-galactofuranosyl)adenine was also prepared.[83] A study of the sodium hydroxide-induced elimination of hydrogen chloride from 3′,5′-dichloro-2′,3′,5′-trideoxy-D-*threo*-pentosyl nucleosides (84) has been carried out. Whereas the adeninyl compound gave 2-adeninyl-5-methyl furan, the thyminyl and uracilyl analogues gave the dienes from direct elimination (85).[84]

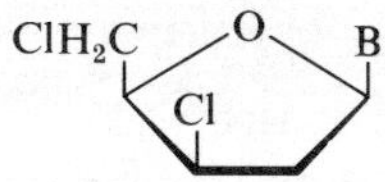

(84) B = adenin-9-yl, thymin-1-yl, or uracil-1-yl

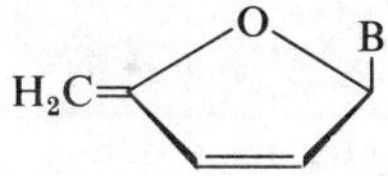

(85) B = thymin-1-yl or uracil-1-yl

The acetylenic nucleosides 5′,6′-dideoxy-*ribo*-hex-5′-ynosyl uracil and 2′,5′,6′-trideoxy-*erythro*-hex-5-ynosyl thymine have been prepared by treatment of the appropriate 5′-aldehydonucleoside sequentially with $Ph_3P{=}CBr_2$, acid, and butyl-lithium.[85]

6 Halogeno-sugar Nucleosides

2′-Deoxy-2′-halogeno-adenosines have been prepared from appropriate *arabino*-adenosine derivatives using the sulphonate ester displacement procedure with lithium or tetra-alkyl ammonium halides.[86, 87] A simple synthesis of 2′-deoxy-2′-fluorocytidine in 40% yield by nucleophilic substitution of 2,2′-anhydrocytidine with potassium fluoride–crown ether in anhydrous conditions has appeared; the use of hydrogen fluoride in dioxane was unsuccessful.[88] Treatment of 2′-deoxy-adenosine with thionyl chloride in HMPT gave 3′,5′-dichloro-2′,3′,5′-trideoxy-nucleoside (84) (B = adenin-9-yl) [89] [for reactions of (84) see previous section]. A synthesis of 2′,5′-dideoxy-5′-fluorouridine derivatives has been reported.[90] The reaction of uridine with thionyl chloride–HMPT gave an isomeric mixture of 5′-chloro-5′-deoxy-2′,3′-*O*-sulphinyluridines in 46 and 15% yields. An analogous reaction was performed on 5-fluoro-uridine. Sodium methoxide treatment of the cyclic sulphites gave the corresponding 5′-chloro-desulphited products. The corresponding 5′-deoxynucleosides were obtained by reduction of the chlorodeoxy compounds with tributyltin hydride.[91] The series of 2′-halogeno and 2′-deoxy analogues

83 V. K. Srivastava and L. M. Lerner, *Tetrahedron*, 1978, **34**, 2627.
84 Y. Wang and H. P. C. Hogenkamp, *J. Org. Chem.*, 1978, **43**, 3324.
85 R. A. Sharma and M. Bobek, *J. Org. Chem.*, 1978, **43**, 367.
86 M. Ikehara and H. Miki, *Chem. and Pharm. Bull* (*Japan*), 1978, **26**, 2449.
87 M. Ikehara, T. Maruyama, and H. Miki, *Tetrahedron*, 1978, **34**, 1133.
88 R. Mengel and W. Guschlbauer, *Angew. Chem. Internat. Edn.*, 1978, **17**, 525.
89 Y. Wang and H. P. C. Hogenkamp, *J. Org. Chem.*, 1978, **43**, 998.
90 G. Herrmann, R. Staske, and D. Cech, *Z. Chem.*, 1978, **18**, 258 (*Chem. Abs.*, 1978, **89**, 197 866d).
91 H. Hrebabecky and J. Beranek, *Nucleic Acids Res.*, 1978, **5**, 1029 (*Chem. Abs.*, 1978, **89**, 44 059t).

(86) and the *manno*-isomer (87) of 6′-deoxy-6′-fluoro-β-D-glucopyranosyl-thymine have been prepared from the 2′-*O*-tosyl derivative (88) *via* 2,2′-anhydro-nucleoside intermediates.[92] Treatment of 5′-*O*-tritylthymidine with triphenylphosphine–diethyl azocarboxylate and methyl iodide gave 2,3-dideoxy-3-iodo-5-*O*-trityl-β-D-*threo*-pentofuranosyl thymine as well as 5′-*O*-trityl-O^2,3′-cyclothymidine.[93]

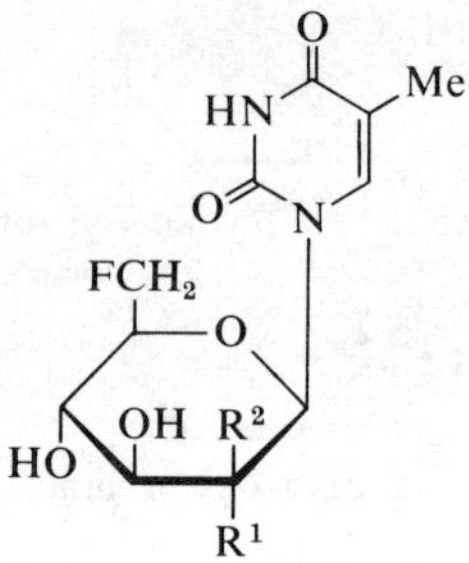

(86) R^1 = F, Cl, Br, I, or H; R^2 = H
(87) R^1 = H, R^2 = OH
(88) R^1 = OTs, R^2 = H

7 Ketonucleosides and Nucleosides containing Uronic Acid Components

The oxidation of isolated hydroxy-groups to carbonyl groups in carbohydrates by the photolysis of pyruvate esters (see Vol. 10, p. 171; Vol. 11, p. 129) has been extended to nucleosides. 5′-*O*-Tritylthymidine gave the crystalline 3′-oxo-nucleoside in 61% yield.[94]

Direct condensation using the silyl or silver salt methods has been used to prepare *N*- and *O*-glucuronic acid derivatives of 5-fluoro-uracil. The *O*-glucosidic derivatives were also prepared by the silver salt method. The antitumour properties of the products against L 1210 lymphoid leukaemia were studied.[95] Reaction of methyl acetobromoglucuronate with 6-mercapto-purine gave the *S*-substituted compound (89). The silver salt of (89) with more acetobromoglucuronate mercury(II) bromide gave the diglucuronide (90). Alkaline treatment of (89) to deprotect the nucleoside also gave the

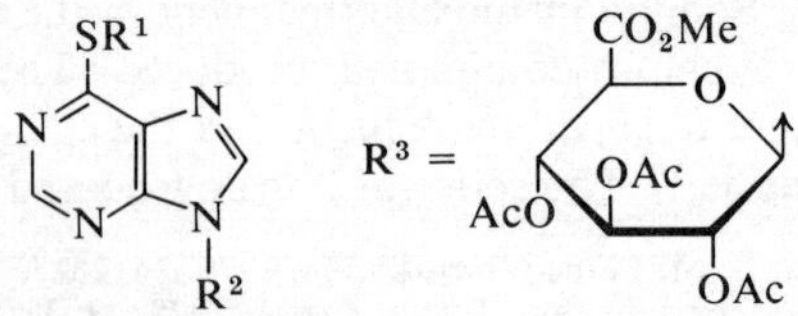

(89) R^1 = R^3, R^2 = H
(90) R^1 = R^2 = R^3

[92] M. Von Janta-Lipinski, G. Etzold, and P. Langen, *J. prakt. Chem.*, 1978, **320**, 157 (*Chem. Abs.*, 1978, **88**, 191 312c).
[93] H. Loibner and E. Zbiral, *Annalen*, 1978, 78.
[94] R. W. Binkley, D. G. Hehemann, and W. W. Binkley, *J. Org. Chem.*, 1978, **43**, 2573.
[95] M. Kaneko, M. Kimura, H. Tanaka, F. Shimizu, M. Arakawa, and B. Shimizu, *Nucleic Acids Res., Spec. Publ.*, 1977, **3**, 35 (*Chem. Abs.*, 1978, **88**, 191 307e).

corresponding 4,5-unsaturated compound by elimination of acetic acid.[96]

Platinum-catalysed oxidation of the terminal hydroxymethyl group of the parent nucleoside has led to the syntheses of various uronic acid analogues of uracil and cytosine nucleosides. By the transformation sequence acid–ester–amide–nitrile, the 5′-cyano-5′-deoxynucleosides were also prepared.[97]

Treatment of 2-chloro-tetrahydrofuran-5-carboxylic acid esters with 2,4-bis(trimethylsiloxy)-5-substituted-pyrimidines gave the corresponding uracil-substituted nucleosides (91) after saponification.[98]

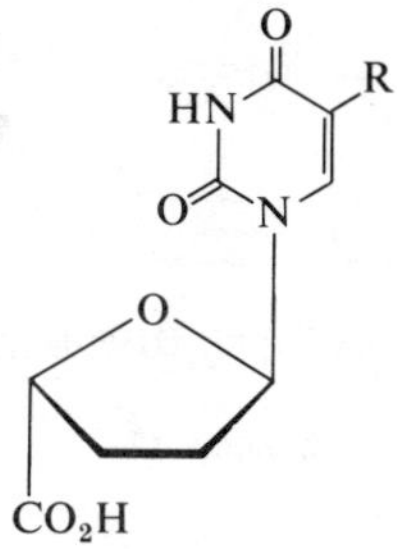

(91) R = H, F, or Me

The biosynthesis of the uronic acid component of polyoxins has been studied by means of ^{13}C n.m.r using D-[1-^{13}C]glucose as precursor.[99]

8 Nucleosides containing Amino-sugar Components

The antibiotic 2′-amino-2′-deoxyguanosine,[100] 2′-amino-2′-deoxyadenosine,[87] and 3′-amino-3-deoxy-*xylo*-adenosine [87] have been synthesized.

Using a reaction sequence involving dibutylstannylation and displacement of triflate, new 2′-substituted-2′-deoxyarabinofuranosyl adenine derivatives have been prepared (Scheme 12); the products (92) and (93) were converted into the corresponding 2′-amino- and 2′-thio-*arabino*-nucleosides.[101] A communication, with no synthetic details, on the preparation and use of 2′-azido- and 2′-amino-2-deoxy-β-D-arabinofuranosyl cytidine as cytotoxic agents has been published.[102]

2′-Amino-2′-deoxyarabinofuranosyl adenine has been synthesized *via* the aziridine (94) by reaction with ethyl trifluorothioloacetate; it was found to be less active against murine leukaemia cells than the corresponding 2′-hydroxy compound.[103] Opening of the anhydro-ring of 2,2′-anhydro-1-D-arabinofuranosyl-5-fluorocytosine with amines gave the 2′-substituted-amino-D-

[96] J. Oehlke, *Pharmazie*, 1978, **33**, 250 (*Chem. Abs.*, 1978, **89**, 129 865c).
[97] R. F. Schinazi, M. S. Chen, and W. H. Prusoff, *J. Medicin. Chem.*, 1978, **21**, 1141.
[98] V. Silina, *Tezisy Doklady-Resp. Konf. Molodykh. Uch.-Khim., 2nd*, 1977, **1**, 10 (*Chem. Abs.*, 1978, **89**, 110 238d).
[99] K. Isono, T. Sato, K. Hirasawa, S. Funayama, and S. Suzuki, *J. Amer. Chem. Soc.*, 1978, **100**, 3937.
[100] M. Ikehara and T. Maruyama, *Chem. and Pharm. Bull. (Japan)*, 1978, **26**, 240.
[101] R. Ranganathan and D. Larwood, *Tetrahedron Letters*, 1978, 4341.
[102] M. Bobek, Y. C. Cheng, and A. Bloch, *J. Medicin. Chem.*, 1978, **21**, 597.
[103] M. J. Robins and S. D. Hawrelek, *Tetrahedron Letters*, 1978, 3653.

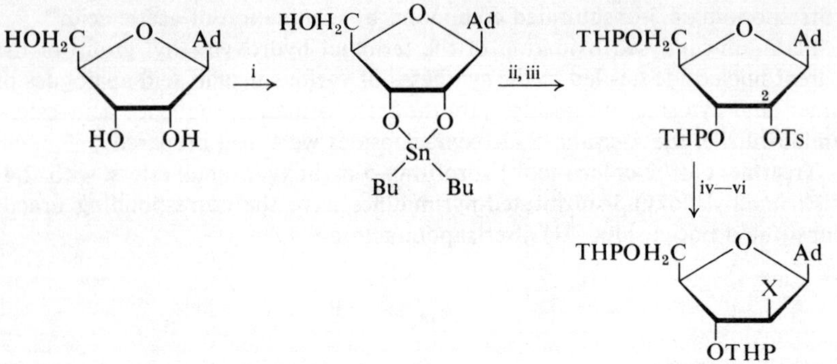

(92) X = N_3
(93) X = SAc

Reagents: i, Bu_2SnO–DMF; ii, $TsCl$–NEt_3; iii, DHP–H^+; iv, Na–Hg; v, NaH–CF_3SO_2Cl; vi, MX–HMPA

Scheme 12

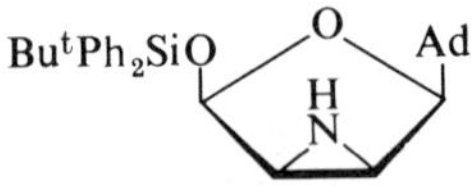

(94)
Ad = adenin-9-yl

ribosyl derivatives. A similar reaction using either sodium hydrosulphide or hydrogen sulphide led mainly to specific defluorination of the pyrimidine ring.[104]

An alternative synthesis of 1-(2-amino-2-deoxy-β-D-glucopyranosyl)uracil has been achieved by the silyl method using 3,4,6-tri-*O*-acetyl-2-deoxy-2-trifluoroacetamido-α-D-glucopyranosyl bromide followed by deprotection.[105]

Nucleosides have been modified at C-3′ by means of the reaction with triphenylphosphine–diethylazo carboxylate. Thus 5′-*O*-tritylthymidine and 5′-*O*-trityldeoxyadenosine with the reagent and hydrazoic acid gave the 3′-azido-*threo*-analogues (95) and (96) respectively, inversion occurring at

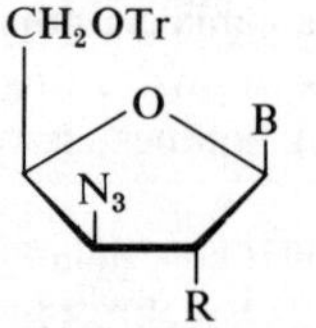

(95) R = H, B = thymin-1-yl
(96) R = H, B = adenin-9-yl
(97) R = OTr, B = uracil-1-yl

[104] A. F. Cook and M. J. Holman, *J. Org. Chem.*, 1978, **100**, 4200.
[105] H. G. Garg and R. W. Jeanloz, *Carbohydrate Res.*, 1978, **62**, 185.

C-3′. Similar reactions were performed on 2′,5′-*O*-ditrityluridine to give the 3′-azido-*xylo*-nucleoside (97) whereas 3′,5′-*O*-ditrityluridine afforded only O^2,2′-cyclouridine.[93]

A preparation of 3′-azido-3′-deoxyadenosine and hence 3′-amino-3′-deoxyadenosine has been reported.[106]

The readily available anhydronucleoside (98) undergoes ring-opening with lithium azide to give 3′-azido-2′,3′-dideoxythymidine, whose 5′-*O*-acetyl derivative when condensed with silylated N^6-octanoyladenine in the presence of trimethylsilyl trifluoromethane sulphonate underwent transglycosylation to give the α- and β-anomers of 3′-azido-2′,3′-dideoxyadenosine. The corresponding 3′-azido-9-substituted-guanine nucleosides were prepared in a like

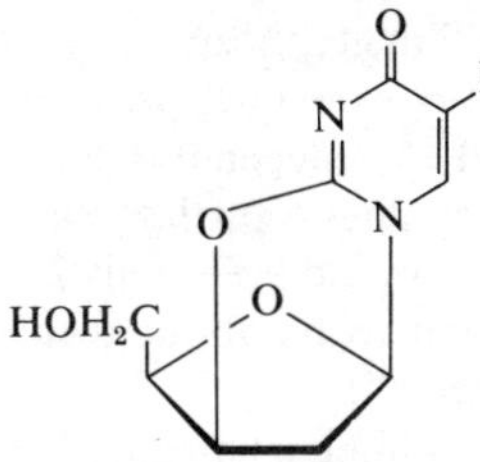

(98)

manner and all azido-products were reduced to the 3′-amino-3′-deoxynucleosides.[107]

An analogue of puromycin, in which the carbonyl group of the *N*-phenylalanyl residue is replaced by methylene, has been prepared by *N*-alkylation of 3′-amino-3′-deoxy-*N*,*N*-dimethyladenosine to the corresponding 3′-*N*-(L-2-amino-3-phenylpropyl) derivative.[108]

Condensation of 2-azido-2-deoxy-α,β-D-arabinofuranosyl chlorides (see Chapter 7) with silylated uracil gave a mixture of the α- and β-uracil derivatives, which were separated. The β-anomer was reduced to give the 2′-amino-2′-deoxy-*arabino*-uridine.[109]

Standard methods have been used to convert 5-halogeno-uridines into 5′-amino-2′,5′-dideoxy-5-halogeno-uridines.[110] 5′-Amino-5′-deoxynucleosides have furnished 5′-sulphamino-5′-deoxynucleosides on reaction with trimethylaminosulphonic acid.[111]

3′,5′-Diamino-3′,5′-dideoxythymidine has been prepared by standard procedures in nine steps from thymidine and shown to have no use as an antiviral or antineoplastic agent, in contrast to the monoamino derivatives.[112]

A series of derivatives of 7-(4-amino-4-deoxy-α-D-talopyranosyl)theophylline and 7-(6-amino-6-deoxy-α-D-mannopyranosyl)theophylline have been

[106] A. V. Azhayev and J. Smrt, *Coll. Czech. Chem. Comm.*, 1978, **43**, 1520.
[107] M. Imazawa and F. Eckstein, *J. Org. Chem.*, 1978, **43**, 3044.
[108] T. Kato and J. Zemlicka, *Tetrahedron Letters*, 1978, 4741.
[109] M. Bobek and V. Martin, *Tetrahedron Letters*, 1978, 1919.
[110] T.-S. Lin and W. H. Prusoff, *J. Medicin. Chem.*, 1978, **21**, 106.
[111] W. S. Mungall, L. J. Lemmen, K. L. Lemmen, J. K. Dethmers, and L. L. Norling, *J. Medicin. Chem.*, 1978, **21**, 704.
[112] T.-S. Lin and W. H. Prusoff, *J. Medicin. Chem.*, 1978, **21**, 109.

synthesized using standard methods. The products were converted into the *N*′-(*N*-benzyloxycarbonyl-L-tyrosine) derivatives.[113]

9 Phosphates and Related Esters of Nucleosides

A review of the chemical synthesis of oligo- and poly-nucleotides by the phosphotriester approach has appeared.[114]

The selective phosphorylation of the 5′-hydroxy-group of nucleosides has been achieved using tris(quinolin-8-yl)phosphate. Removal of the blocking group was accomplished by treatment with copper(II) chloride. The 5′-phosphates of adenosine, guanosine, uridine, and cytidine were prepared by this route.[115] The nucleotides of N^1- and N^8-pteridine (lumazine and noxamthopterine, respectively) have been prepared using the method of Yoshikawa [116] (see Vol. 4, p. 42). Trimetaphosphate in water has been reported to phosphorylate nucleosides to give nucleoside triphosphates. The yields obtained varied between 10 and 15% depending on the acidity of the hydroxy-groups. With 2′- and 3′-triphosphates a further reaction to furnish 2′,3′-cyclic phosphates occurred and hence to yield eventually 2′- and 3′-monophosphates. The 3′-hydroxy-group was found to be about twice as reactive towards this reagent as the 5′-hydroxy-group.[117]

The 5′-monophosphates and -diphosphates of 2-methyladenosine, 2-ethyladenosine, and 2-isopropylidene adenosine have been synthesized and converted enzymically to the corresponding polyadenylic acids, which have been shown by c.d. to possess singly stranded, strongly stacked structures which reorientate to stable double helix in acidic media.[118] Pteridine 5′-monophosphates have been converted into the 5′-diphosphates using the Moffatt–Khorana procedure (*J. Amer. Chem. Soc.*, 1961, **83**, 649).[119] To introduce phosphate groups at the 3′- and 5′- ends of oligonucleotide chains, *O*-aryl-*S*-methylphosphorochloridothioates have been used. Thus the blocked uridine (99) was converted into the thiophosphate (100) which was selectively hydrolysed to yield the terminal phosphate (101). Applications of this method towards the preparation of oligonucleotides were also given.[120] Alkylation of glycosyl phosphates to yield triesters has been achieved by means of tetrabutylammonium fluoride-catalysed reaction with dimethyl or diethyl sulphate. Cyclic phosphates gave the cyclic triester; the base was also alkylated. The compounds produced are important for the study of mutagenic or lethal reactions in DNA and RNA.[121] 5′-*O*-Monomethoxytrityl-3′-*O*-t-butyldimethylsilyluridine has been shown to be a useful intermediate for subsequent phosphorylation by *p*-chlorophenylphosphate in pyridine.[122] Elemental

113 F. Leclercq, J. Jumulet-Bach, and K. Antonakis, *Carbohydrate Res.*, 1978, **62**, 73.
114 C. B. Reese, *Tetrahedron*, 1978, **34**, 3143.
115 H. Takaku in 'Nucleic Acid Chemistry', ed. L. B. Townsend and R. S. Tipson, Wiley, New York, 1978, Vol. 2, p. 865 (*Chem. Abs.*, 1978, **89**, 197 870a).
116 M. Hattori and W. Pfleiderer, *Annalen*, 1978, 1780.
117 E. Etaix and L. E. Orgel, *J. Carbohydrates, Nucleosides, Nucleotides*, 1978, **5**, 91.
118 M. Hattori, K. Ienaga, and W. Pfleiderer, *Annalen*, 1978, 1796.
119 M. Hattori and W. Pfleiderer, *Annalen*, 1978, 1788.
120 C. B. Reese and L. Yan, *J.C.S. Chem. Comm.*, 1978, 1050.
121 K. K. Ogilvie, S. L. Beaucage, and M. F. Gillen, *Tetrahedron Letters*, 1978, 3203.
122 K. L. Sadana and P. C. Loewen, *Tetrahedron Letters*, 1978, 5095.

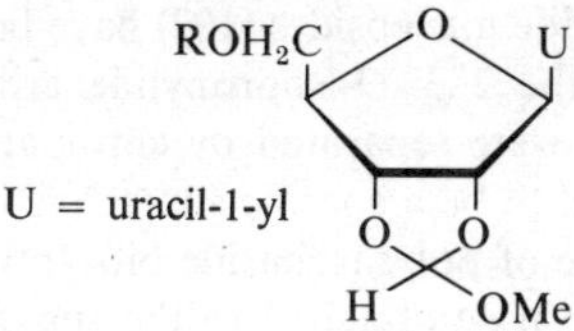

(99) R = H

(100) R = $-P(=O)(SMe)(OAr)$

(101) R = $-P(=O)(OH)(O^-)$

sulphur adds to phosphite triesters in pyridine solution to give phosphorothiate triesters in quantitative yields. The reaction was applied to the synthesis of thiono-analogues of dinucleoside monophosphates.[123]

Butyldiphenylchlorosilanes have been shown to react with deoxy- and *N*-protected-deoxymononucleotides to give, in each case, a high yield of 3′-*O*-butyldiphenylsilyl ether, useful as a protecting group which allows rapid extraction of synthetic intermediates and reverse phase h.p.l.c. isolation. The group is rapidly removed on treatment with fluoride ion in pyridine at room temperature.[124]

Appropriate dichlorophosphonates have been used to synthesize the phosphorus stereoisomers of uridine 2′,3′-trityl-[125] and t-butyl-phosphonates.[126] N.m.r. was used to determine the stereochemistry at phosphorus. The nitro-benzyl group has been found to be a useful blocking group for cyclic phosphate since it can be easily removed *in situ* by u.v. irradiation.[127] A range of 3′,5′-cyclic-phosphate and -phosphite derivatives of thymidine, including amide and thiono analogues, have been prepared by conventional methods, and their configuration at phosphorus determined by ^{31}P n.m.r.[128]

Thiophosphorylation of 3′-amino-3′-deoxyadenosine led to a series of 2′,3′- and 3′,5′-cyclic thiophosphate derivatives which were obtained as diastereoisomers at phosphorus.[129]

10 Other Derivatives

An analogue of adenosylhomocysteine, 5′-deoxy-5′-isobutylthio-3-deaza-adenosine, has been prepared and its biological and biochemical properties

[123] P. M. J. Burgers and F. Eckstein, *Tetrahedron Letters*, 1978, 3835.
[124] R. A. Jones, H.-J. Fritz, and H. G. Khorana, *Biochemistry*, 1978, **17**, 1268.
[125] I. D. Shingarova, S. Ya. Mel'nik, I. V. Yartseva, A. A. Borisenko, and M. N. Preobrazhenskaya, *Bioorg. Khim.*, 1977, **3**, 1641 (*Chem. Abs.*, 1978, **88**, 105 705d).
[126] I. D. Shingarova, S. Y. Mel'nik, and M. N. Preobrazhenskaya, *Carbohydrate Res.*, 1978, **65**, 288.
[127] J. Engels and R. Reidys, *Experientia*, 1978, **34**, 14.
[128] G. S. Bajwa and W. G. Bentrude, *Tetrahedron Letters*, 1978, 421.
[129] M. Morr and L. Ernst, *Chem. Ber.*, 1978, **111**, 2152.

studied.[130] The disaccharide nucleosides (102) have been obtained *via* Königs–Knorr glycosylation of the 2′,3′-*O*-isopropylideneribofuranosyl nucleosides. The anomeric glycosides were separated by chromatography. The 5′-*O*-β-D-glucosylinosine was found to be a specific site-active inhibitor of maltodextrin phophorylase, the enzyme of polysaccharide biosynthesis.[131]

Isonucleosides with the base attached to the sugar at positions other than C-1 have received attention. The 2′-isonucleoside (103) has been obtained by the route shown in Scheme 13. An analogous route was used to synthesize

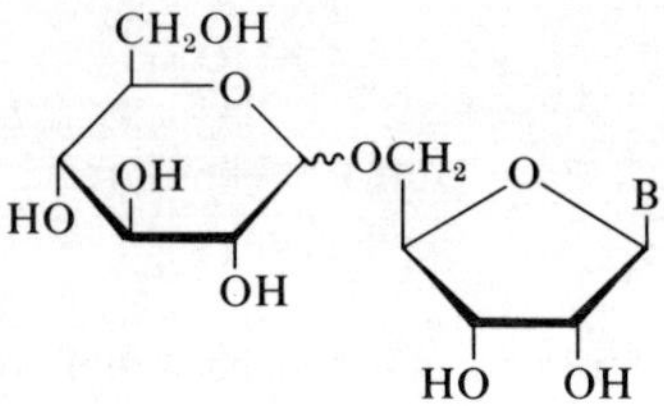

(102) B = uracil-1-yl, adenin-9-yl, or inosin-9-yl

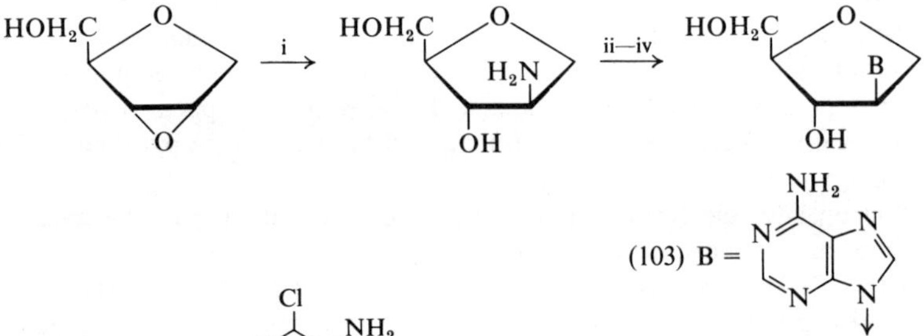

(103) B =

Reagents: i, NH_4OH; ii, 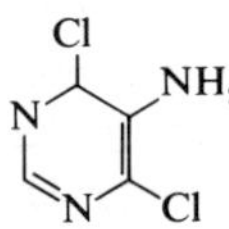 -Et_3N; iii, $HC(OEt)_3$–H^+; iv, NH_3

Scheme 13

the corresponding uracil isonucleoside.[132] The terminal triazole isonucleosides (104) have been prepared by 1,3-dipolar addition of phenyl azide to acetylenic sugar derivatives, and mesityl nitrile oxide similarly gave 3-mesityl-isoxazole derivatives.[133]

Reaction of adenosine with triphenylphosphine and diethyl azodicarboxylate gave the 2′,3′-phosphorane (105) and not the 3′,5′-phosphorane as previously reported (Mitsunobu *et al.*, *Chem. letters*, 1974, 1473). The structure was confirmed from ^{1}H and ^{31}P n.m.r. measurements. With *arabino*-adenosine the 2′,3′-*lyxo*-epoxide (106) was obtained in near quantitative yield; *xylo*-adenosine also gave (106) but much more slowly and in only 60% yield.[134]

[130] P. K. Chiang, G. L. Cantoni, J. P. Bader, W. M. Shannon, H. J. Thomas, and J. A. Montgomery, *Biochem. Biophys. Res. Comm.*, 1978, **82**, 417.
[131] F. W. Lichtenthaler, Y. Sanemitsu, and T. Nohara, *Angew. Chem. Internat. Edn.*, 1978, **17**, 772.
[132] J. A. Montgomery and H. J. Thomas, *J. Org. Chem.*, 1978, **43**, 541.
[133] D. Horton and J.-H. Tsai, *Carbohydrate Res.*, 1978, **67**, 357.
[134] R. Mengel and M. Bartke, *Angew. Chem. Internat. Edn.*, 1978, **17**, 679.

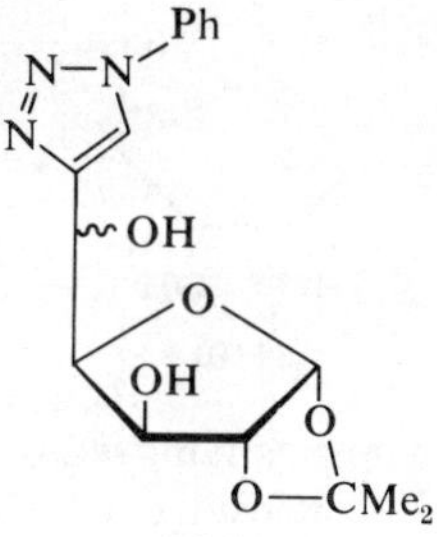

(104)

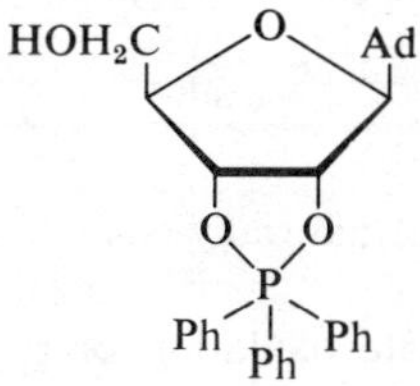

(105)

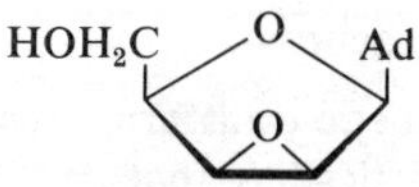

(106)

Ad = adenin-9-yl

A new class of bridged pyrimidine nucleosides has been obtained by the action of diazomethane on the cyclic keto-nucleoside (107) in methanol; a mixture of the cyclonucleosides (108) and (109) was obtained. Some conversions of (108) were discussed.[135]

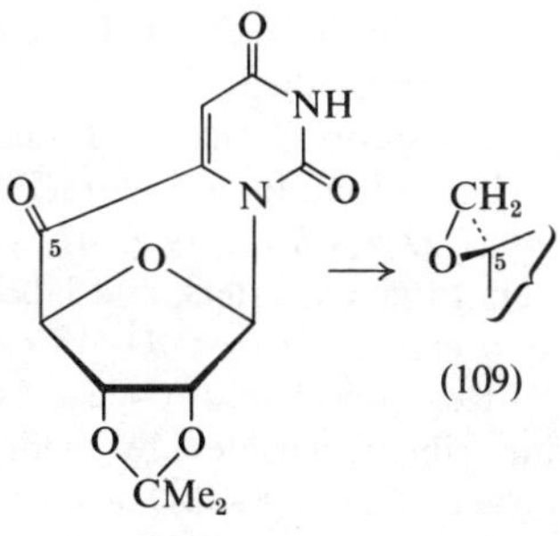

(107)

(109)

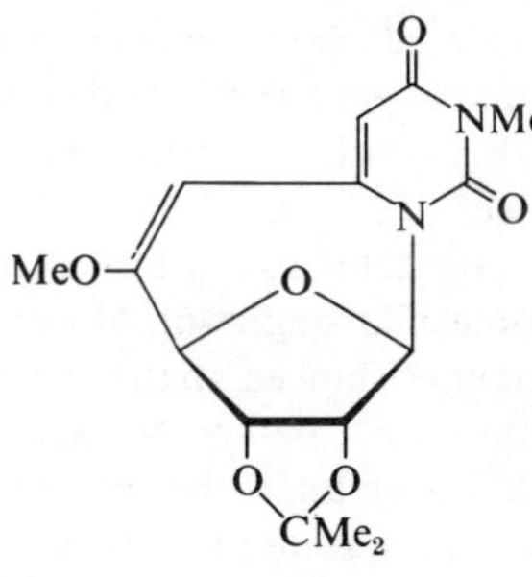

(108)

Purine nucleoside analogues with thiazolidine ring systems replacing the sugar have been made.[136]

4′-Thiocordycepin has been synthesized.[137] 4′-Thio-D-*erythro*-furanosyl nucleosides (110) have been prepared by reaction of the corresponding 4-thioerythrose-2,3-phenylboronate with purine derivatives in the presence of *p*-toluenesulphonic acid or with silylated pyrimidines in the presence of tin(IV)

[135] B. A. Otter and E. A. Falco, *Tetrahedron Letters*, 1978, 4383.
[136] M. Iwakawa, B. M. Pinto, and W. A. Szarek, *Canad. J. Chem.*, 1978, **56**, 326.
[137] R. G. S. Ritchie, D. M. Vyas, and W. A. Szarek, *Canad. J. Chem.*, 1978, **56**, 794.

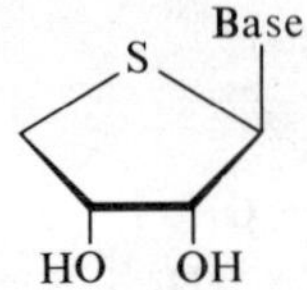

(110)

chloride. The blocking boronate was removed by treatment with propan-1,3-diol. 4′-Thio-*threo*-furanosyl nucleosides were also made from 2,3-di-*O*-benzoyl-L-threose. An interesting feature with both the *threo*-dibenzoate and the *erythro*-phenylboronate was that the pyrimidines gave only the pyrimidin-3-yl nucleoside.[138]

11 Reactions

A review of synthetic transformations of naturally occurring nucleosides and nucleoside antibiotics has appeared.[139]

The optimum conditions for periodate oxidation of ribonucleosides and ribonucleotides have been investigated.[140]

Trimethylanilinium hydroxide or methoxide at 200 °C have been found to be useful reagents for permethylation of nucleosides. At 100 °C partial methylation occurs.[141] The same group has investigated the use of trimethyl phosphate as a methylating agent for ribonucleosides. Below pH 11 methylation of the base occurs, while increasing the pH results in methylation of the sugar, with the more acidic O-2′ being preferentially etherified.[142] Heating of pyrimidines or their nucleosides in DMF with trimethyloxosulphonium hydroxide has been demonstrated to lead mainly to *N*-methylated products but cytidine gave the 2′-*O*-methyl-ether as the main product.[143]

Further work has been carried out on the disproportionation found on treatment of 3′,5′-di-*O*-aroyl derivatives of 1-β-D-arabinofuranosyluracil with sodium benzoate in hot DMF (T. Sasaki *et al.*, *J. Org. Chem.*, 1976, **41**, 3138). Intermolecular migration of ester group occurs to give triesters, and labelling experiments showed that the migrating group can move from O-5′ or O-3′ depending on whether N^3 was protected.[144] Regioselective 2′-*O*-deacylation of fully acetylated purine and pyrimidine ribonucleosides by hydroxyammonium acetate in pyridine has been reported. The 3′,5′-di-*O*-acyl derivatives were obtained in 53—85% yields, while an excess of reagent led to 5′-*O*-acetylribonucleosides.[145] The reaction of 8-bromoadenosine with methylhydrazine gives the expected 8-methylhydrazinoadenosine, whereas when

138 J. E. McCormick and R. S. McElhinney, *J.C.S. Perkin I*, 1978, 500.
139 M. J. Robins, *Bioorg. Chem.*, 1977, **3**, 221 (*Chem. Abs.*, 1978, **89**, 6477v).
140 A. I. Busev and V. Zaharans, *Khim.-Farm. Zhur.*, 1978, **12**, 137 (*Chem. Abs.*, 1978, **89**, 152 782g).
141 G. R. Pettit, P. Brown, J. J. Einck, K. Yamauchi, and R. M. Blazer, *Synthesis Comm.*, 1977, **7**, 449.
142 K. Yamauchi and M. Kinoshita, *J.C.S. Perkin I*, 1978, 762.
143 K. Yamauchi, N. Nakamura, and M. Kinoshita, *J. Org. Chem.*, 1978, **43**, 1593.
144 T. Sasaki, K. Minamoto, and T. Okugawa, *J. Org. Chem.*, 1978, **43**, 350.
145 Y. Ishido and N. Sakairi, *Nucleic Acids Res., Spec. Publ.*, 1977, **3**, 13 (*Chem. Abs.*, 1978, **88**, 136 890q).

8-bromo-2′-*O*-toluene-*p*-sulphonyladenosine is used, the bridged compound (111) is formed. Oxidation of (111) with mercury(II) acetate furnished the hydrazono-cyclonucleoside (112). Hydrazine reacts in an analogous manner.[146]

Rearrangement by a 1,3-intramolecular shift from N^3 to N^9 occurs with retention of anomeric configuration in the nucleoside (113) during heating and g.c.–m.s.[147]

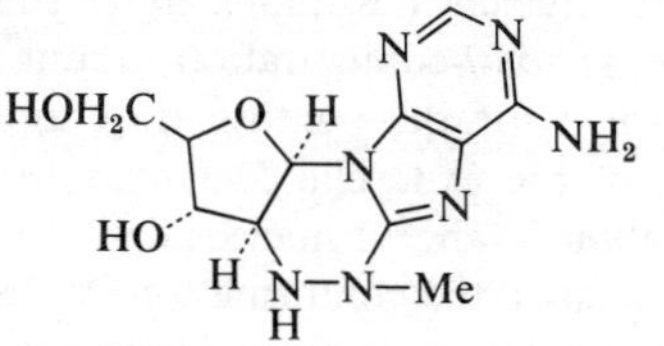

(111)

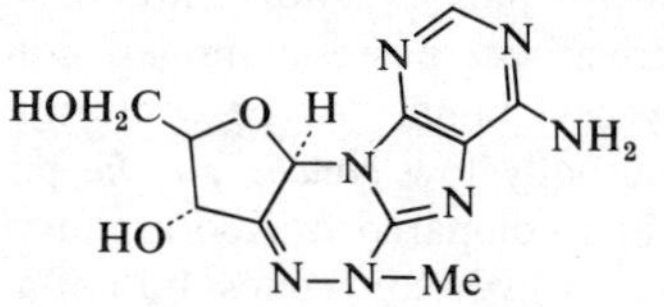

(112)

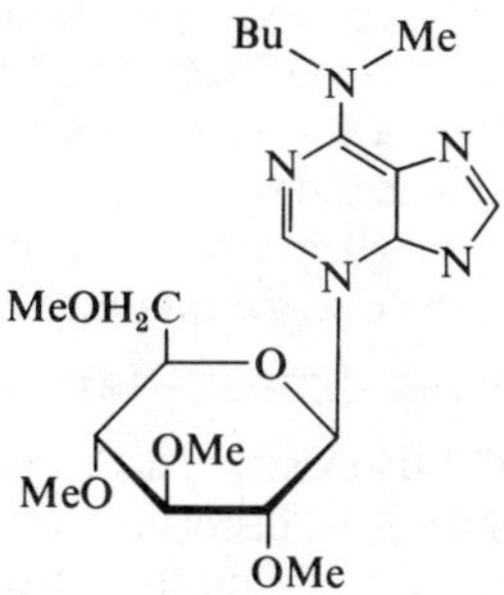

(113)

A study of the effect of alkylation of the bases on the stability of the glycosyl linkage in nucleosides has been carried out.[148]

12 Physical Measurements and Kinetics

An *ab initio* SCF-MO computation has been conducted on uridine.[149]

The barrier to rotation about the *N*-glycosyl bond of β-D-glucopyranosylamines and nucleosides of bases possessing 2-fold symmetry has been studied as a function of solvent, acylation of the sugar moieties, and substitution on the aglycone at positions removed from the glycosidic bond. The results indicated that the main barriers to rotation are interactions between O-2′ and O-5′ and the carbonyl groups of the bases adjacent to the *N*-glycosyl bond.[150] The co-ordination geometry and hydration of cobalt complexes and nucleotides have been studied.[151] Crystal structural analysis has shown

146 J. B. Chattopadhyaya and C. B. Reese, *J.C.S. Chem. Comm.*, 1978, 86.
147 J. V. Eichholzer, J. K. Macleod, and R. E. Summons, *Austral. J. Chem.*, 1977, **31**, 893.
148 B. Singer, M. Kröger, and M. Carrano, *Biochemistry*, 1978, **17**, 1246.
149 A. Pullman and H. Berthod, *Internat. J. Quantum Chem., Quantum Biol. Symp.*, 1977, **4**, 327 (*Chem. Abs.*, 1978, **88**, 23 306k).
150 J. C. Jochims, H. von Voithenberg, and G. Wegner, *Chem. Ber.*, 1978, **111**, 1693.
151 D. M. L. Goodgame, G. A. Leach, and Z. Warnke, *Biochem. Biophys. Res. Comm.*, 1978, **82**, 92.

that the glycosyl torsion angle in 3-deaza-6-azauridine is in the 'high-anti' region, in marked contrast to those found in uridine, 3-deazauridine, and other pyrimidine nucleosides.[152] The crystal structure of N^6-(*N*-threonylcarbonyl)adenosine, a constituent of tRNA, has been determined as part of a study of conformation in the anticodon loops.[153]

^{1}H N.m.r. studies of various *O*′-methyl 9-β-D-xylofuranosyl adenines using the Karplus equation showed that the free nucleosides and their *O*′-methyl derivatives possess similar conformations: *anti*-configuration about the glycosyl bond, *N*-type conformer, *C*(3′)-*endo* for the pentose ring, and an unusually low *gauche–gauche* population of the exocyclic 5′-hydroxymethyl group compared to ribofuranosyl and arabinofuranosyl nucleosides. Ionization of hydroxy-groups did not appear to change the conformation.[154] In the same publication relative acidities by elution sequence from Dowex (OH^-) column were shown to be 5′-OH < 3′-OH < 2′-OH. The multiplicity of the H-4′ resonance of 2′,3′-*O*-isopropylidine ribonucleosides has been demonstrated to be diagnostic of α- or β-configuration. The former gives an apparent triplet whereas the latter gives a higher multiplet. Exceptions to the general rule are N^6-linked adenine nucleosides.[155]

Using purine nucleosides labelled with ^{14}C and ^{3}H in the base and ^{2}H at C-1′, it has been shown for the enzyme-catalysed reaction:

$$\text{nucleoside} + \text{phosphate} \rightleftharpoons \text{base} + \text{ribose 1-phosphate}$$

that a significant k_H/k^{2}_H of 1.05 exists. This suggests that in binding to the enzyme the nucleoside undergoes distortion of the sugar ring with C-1′ undergoing a partial change in hybridization from sp^3 to sp^2 state.[156] The acid-catalysed hydrolyses of inosine and adenosine have been examined using the [1′-^{2}H]compounds. The isotope ratio ($k_H/k^{\cdot}_H$ = 1.2) led to the conclusion that the protonated species have marked oxocarbonium character. Details of possible mechanism were discussed.[157] The kinetics of alkaline hydrolysis of the acyl groups of *N*,3′-*O*-diacyl derivatives of 2′-deoxycytidine 5′-phosphates have been determined.[158]

Differential scanning calorimetry has been used to obtain information on the bond free-energies which maintain the fibrous structure in nucleic acids, on susceptibility to thermal degradation by homolytic cleavage of the glycosidic bond, and on deamination of the bases. A mechanism for thermolysis of the glycosidic bond in the melt was considered.[159]

152 B. J. Graves and D. J. Hodgson, *Biochim. Biophys. Acta*, 1978, **520**, 229.
153 R. Parthasarathy, J. M. Ohrt, and G. B. Chheda, *Biochemistry*, 1977, **16**, 4999.
154 I. Ekiel, E. Darzynkiewicz, L. Dudycz, and D. Shugar, *Biochemistry*, 1978, **17**, 1530.
155 M. MacCoss, M. J. Robins, B. Rayner, and J.-L. Imbach, *Carbohydrate Res.*, 1977, **59**, 575.
156 R. L. Stein, R. Romero, H. G. Bull, and E. H. Cordes, *J. Amer. Chem. Soc.*, 1978, **100**, 6249.
157 R. Romero, R. Stein, H. G. Bull, and E. H. Cordes, *J. Amer. Chem. Soc.*, 1978, **100**, 7620.
158 H. Seliger and H.-H. Görtz, *Chem. Ber.*, 1978, **100**, 3732.
159 P. G. Olafsson and A. M. Bryan, *J. Therm. Analysis*, 1977, **11**, 359 (*Chem. Abs.*, 1978, **88**, 74 511f).

20

N.M.R. Spectroscopy and Conformational Features

The Fourier transform ^{15}N n.m.r. of amino-sugars has been reviewed.[1] A new method for determining the absolute configuration at secondary hydroxylic sites in organic compounds uses glucosylation and examination of the consequent ^{13}C chemical shifts exhibited by the alcohol and the carbohydrate. Extensive tables of ^{13}C chemical shifts for the different atoms of a wide range of glucosides and their peracetates are provided for comparison.[2] From a compilation of n.m.r. data for twenty-four 2,3-anhydro-pento- and -hexosides, configurational isomers of the epoxides have been differentiated by means of their ^{13}C spectra, whereas a combination of ^{1}H and ^{13}C n.m.r. can provide a qualitative idea of their conformations. A general conclusion from this work was that 2,3-anhydrohexosides prefer the $^{0}H_5$ conformation.[3] The ^{1}H n.m.r. spectra of molybdate complexes of aldoses have shown that lyxose and its homomorphs exist in a skew conformation in the complex with the arrangement of hydroxy-groups shown in (1). Ribose and its homo-

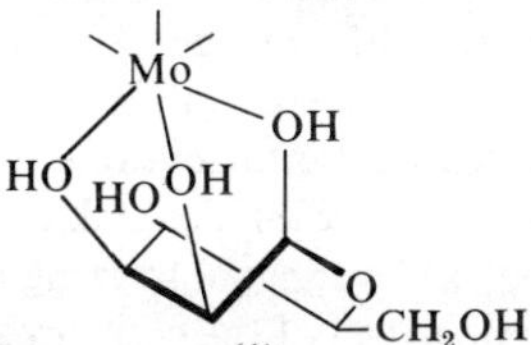

(1)

morphs, although strongly complexing, provided considerably broadened ^{1}H signals incapable of analysis. Little change in ^{1}H n.m.r. spectra occurred with arabinose and xylose and their homomorphs.[4]

The influence of orbital effects on equilibrium conformation of terminally unsaturated glycenose derivatives has been calculated and compared with results obtained by ^{1}H coupling constants. Glycenoses and alkenes having the general formula RCH(OR′)CH=CHX (*E*) preferentially adopt, in solution, a conformation in which the alkoxy-group eclipses the double bond. This conformation is particularly favoured by resonance when X is an electron-

[1] B. Coxon, *Pure Appl. Chem.*, 1977, **49**, 1151.
[2] S. Seo, Y. Tomita, K. Tori, and Y. Yoshimura, *J. Amer. Chem. Soc.*, 1978, **100**, 3331.
[3] M. Chmielewski, J. Mieczkowski, W. Priebe, A. Zamojski, and H. Adamowicz, *Tetrahedron*, 1978, **34**, 3325.
[4] J. Alfoldi, L. Petrus, and V. Bilik, *Coll. Czech. Chem. Comm.*, 1978, **43**, 1476.

withdrawing group; it is less favoured or totally unfavoured when X is electron-donating, which may be explained by interorbital interactions.[5]

The unsaturated phosphonates typified in (2) do not conform with Hudson's isorotation rule;[6] the conformational free energy for the dimethylphosphono group was found to be 8.34 kJ mol^{-1} and its anomeric effect was calculated to be 2.33 kJ mol^{-1}.

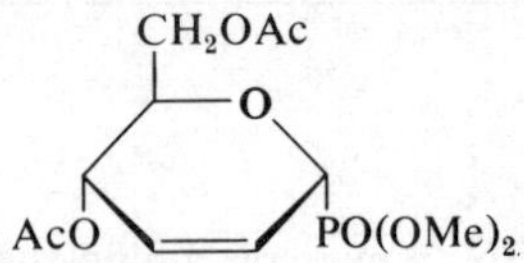

(2)

Glycosides have been shown to fall into two groups by analysis of *X*-ray crystal structures; these are the *α-gauche–gauche* and *β-trans–gauche* in relation to dimethoxymethane. Mean geometries for the two groups were generated from the analyses and the generalized anomeric effect was clearly demonstrated. The coupling of bond angles to torsion angles as noted by Sundaralingam was also observed.[7]

1 Acyclic Systems

The conformations of the phenylhydrazones, 4-nitrophenylhydrazones, and 2,4-dinitrophenylhydrazones of D-erythrose, D-arabinose, D-galactose, D-fructose peracetates, and of D-erythrose 4-nitrophenylhydrazone were studied by ^{1}H n.m.r. The evidence supported the predominance of the zig-zag or sickle forms in solution.[8] A study of the ^{1}H n.m.r. spectrum of tetra-*O*-acetyl-D-ribose di-isobutyl dithioacetal at 250 MHz over the temperature range +20 to −120 °C suggests a conformationally flexible molecule at room temperature and progressively more of the sickle conformation favoured as the temperature is lowered, although no freezing-out of the preferred conformation occurred.[9] A comparison of the conformations of D-ribose diphenyl and diethyl dithioacetals in the crystal and in solution (using 250 or 300 MHz ^{1}H n.m.r.) has shown differences between the two states. In solution the compounds adopt a sickle conformation with torsion about C-3–C-4, *i.e.* $_3G^4$, whereas in the crystal intermolecular hydrogen-bonds not present in solution cause the compound to adopt a $_2G^4$ conformation which includes an O-4–C-1 *β*-interaction.[10]

^{1}H N.m.r. spectra of potassium and calcium salts of gluconic acid in the presence of europium salts have shown that the former are selectively bound to the europium, a function of conformation.[11]

5 J. M. J. Tronchet and T. N. Xuan, *Carbohydrate Res.*, 1978, **67**, 469.
6 J. Thiem, B. Meyer, and H. Paulsen, *Chem. Ber.*, 1978, **111**, 3325.
7 S. Perez and R. H. Marchessault, *Carbohydrate Res.*, 1978, **65**, 114.
8 J. Alfoldi, R. Palovcik, C. Peciar, and K. Linek, *Chem. Zvesti*, 1978, **32**, 238.
9 M. Blanc-Muesser, J. Defaye, and D. Horton, *J. Org. Chem.*, 1978, **43**, 3053.
10 A. Ducruix, D. Horton, C. Pascard, J. D. Wander, and T. Prange, *J. Chem. Research* (S), 1978, 470; *J. Chem. Research* (M), 1978, 5438.
11 T. Taga, Y. Kuroda, and K. Osaki, *Bull. Chem. Soc. Japan*, 1977, **50**, 3079.

2 Pyranoid Systems

Equations for the calculation of the number of allowable conformational states for *O*-methylhexopyranosides have been developed.[12] Coupling constants for ^{1}H–^{1}H in glycals have been calculated, using an FP–SCF–INDO procedure, for the two half-chair conformations of arabinal and xylal. Comparison between the calculated and observed values show that the former are higher for *geminal* and all-*trans* couplings although they are of the correct order of magnitude and correct relative values; the other couplings were reasonably well reproduced. The dependence of coupling on substituent orientation and the point to which this can be calculated are discussed.[13]

The ^{1}H n.m.r. spectra of *N*-acetyl and *N*-benzoyl derivatives of β-D-xylo-, β-D-lyxo-, and α-L-arabinopyranosylamines in d_6-dimethyl sulphoxide and d_5-pyridine have shown that the amide possesses a *trans* conformation.[14] In a separate publication the same author concludes that the ^{1}H n.m.r. absorptions of the methyl 2-acetamido-2-deoxyglycoses and *N*-acetylglycopyranosylamines are independent of axial and equatorial orientation of the acetamido group and cannot be used to determine the conformation of the sugar.[15] Acetylation of *N*-*p*-nitrophenyl-D-xylopyranosylamine with acetic anhydride in the presence of zinc chloride gave the known β-*N*-acetyl derivative in 90% yield and a new α-*N*-acetyl derivative in 2% yield. Spectroscopic data indicated that the former adopts a $^4C_1$ conformation and the latter a $^1C_4$ conformation.[16] Application of ^{13}C n.m.r. to the pyridinium salts of peracetylated D-glucopyranose 2-, 3-, 4-, and 6-sulphates has shown that the technique can be used to determine the position of the acid group.[17] N.m.r. data and bond and rotational energies have been used to calculate the conformations of thirty mono-, di-, and tri-*O*-methyl-α,β-D-glucopyranosides. An equation for determining the number of possible allowed conformations was developed.[18]

Calculated rotamer states for the acetoxymethyl group at C-5 of gluco- and galacto-pyranosyl rings have been compared with those obtained from *X*-ray crystallography and found to be in good agreement.[19] Molecular orbital calculations on dimethoxymethane, a model for the anomeric centre of methyl pyranosides, have produced results consistent with the anomeric and the *exo*-anomeric effects. The results were compared with evidence obtained by neutron diffraction.[20] Dynamic ^{1}H n.m.r. methods applied to the

[12] V. P. Panov, V. N. Gritsan, and R. G. Zhbankov, *Doklady Akad. Nauk B.S.S.R.*, 1978, **22**, 238 (*Chem. Abs.*, 1978, **89**, 44 093z).
[13] J. Santoro, *Anales de Quim.*, 1978, **74**, 538 (*Chem. Abs.*, 1978, **89**, 180 243a).
[14] A. S. Cerezo, *Anales Asoc. quim. argentina*, 1976, **64**, 351 (*Chem. Abs.*, 1978, **89**, 129 845w).
[15] A. S. Cerezo, *Anales Asoc. quim. argentina*, 1976, **64**, 356 (*Chem. Abs.*, 1978, **89**, 129 846x).
[16] D. Piskorska and J. Sokolowski, *Pol. J. Chem.*, 1978, **52**, 329 (*Chem. Abs.*, 1978, **89**, 44 098e).
[17] A. F. Pavlenko, N. I. Belogortseva, V. V. Isakov, and Yu. S. Ovodov, *Bioorg. Khim.*, 1978, **4**, 664 (*Chem. Abs.*, 1978, **89**, 110 148z).
[18] V. P. Panov, V. N. Gritsan, and R. G. Zhbankov, *Doklady Akad. Nauk B.S.S.R.*, 1977, **21**, 1106 (*Chem. Abs.*, 1978, **88**, 121 647s).
[19] S. Perez, J. St-Pierre, and R. H. Marchessault, *Canad. J. Chem.*, 1978, **56**, 2866.
[20] G. A. Jeffrey, J. A. Pople, J. S. Binkley, and S. Vishveshwara, *J. Amer. Chem. Soc.*, 1978, **100**, 373.

calculation of the barriers to rotation about the glycosidic bonds of β-D-glycopyranosyl isocyanuric acids, *e.g.* (3), have shown that the barrier depends on the steric repulsion between carbonyls at C-2 and C-6 of the aglycone and the O-2′ and O-5′ of the sugar. The studies were applied to 2-deoxy-β-D-glucopyranosides, 6-substituted β-D-glucopyranosides, β-D-galactopyranosides, β-D-xylopyranosides, and β-D-ribopyranosides.[21]

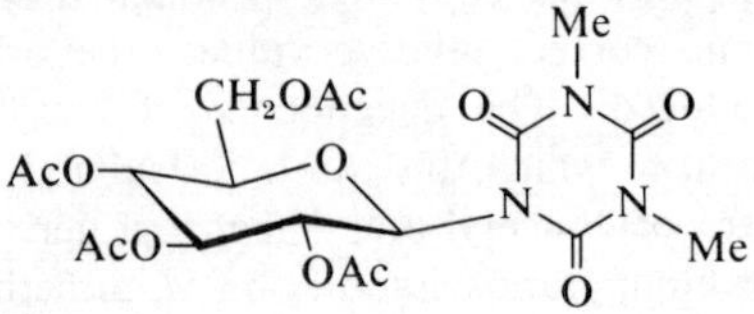

(3)

A range of glycenose derivatives incorporating a terminal 2,2-difluorovinyl group, whose preparation is mentioned in Chapter 12, have been studied by ^{13}C and ^{1}H n.m.r. and by m.s.[22]

Chiroptical parameters for esters of L-altruronic acid in various solvents have been shown to reflect the conformational mobility and rotation of the chromophoric group about the C-5–C-6 bond. The conformational equilibria are sensitive to the anomeric configuration, and to the pattern of *O*-methylation in methylated derivatives. In the permethylated methyl altruronates the α-anomer exists predominantly in the $^{4}C_{1}$ conformation (4) while the β-anomer prefers the $^{1}C_{4}$ conformation (5), thus reducing the *syn*-axial interactions involving the glycosidic methoxyl in each case.[23]

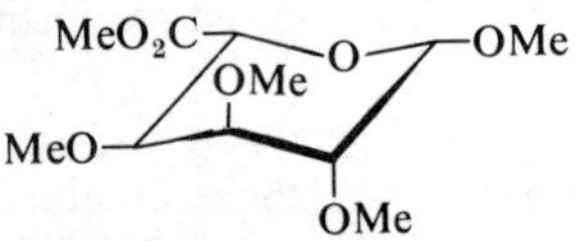

(4)

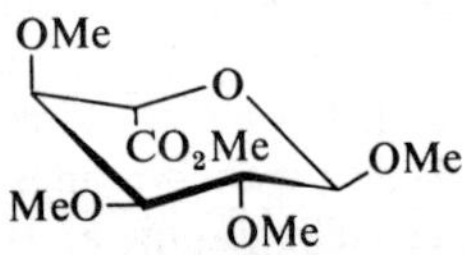

(5)

3 Furanoid Systems

The conformational analysis of a range of 1,2-*O*-isopropylidene-α-D-gluco- and -allo-furanose derivatives (6) has been reported, using ^{1}H n.m.r., and the 3,5-phenylboronate of the *gluco* isomer was also examined. Rotamers about the C-5–C-6 bond were distinguished.[24] The ^{1}H n.m.r. spectra of α- and β-anomers of phenyl, thiophenyl, and 1-naphthyl 2,3,5-tri-*O*-acetyl-D-ribofuranosides have shown that the preferred conformation is C_3-*endo*. Changing

[21] J. C. Jochims, H. von Voithenberg, and G. Wegner, *Chem. Ber.*, 1978, **111**, 2745.
[22] J. M. J. Tronchet, A. Bonenfaut, and F. Barbalat-Rey, *Carbohydrate Res.*, 1978, **67**, 564.
[23] S. Bystricky, T. Sticzay, and P. Kovac, *Coll. Czech. Chem. Comm.*, 1978, **43**, 2062.
[24] A. A. Akhrem, I. A. Mikhailopulo, and A. S. Fridman, *Vestsi Akad. Navuk B.S.S.R., Ser. khim. Navuk*, 1978, 92 (*Chem. Abs.*, 1978, **89**, 44 002u).

the aglycone did not appear to affect the conformation of the sugar ring in any of the β-glycosides.[25]

The n.m.r. analysis of the anhydropentose *p*-nitrophenylhydrazones (7) and their deacetylated derivatives indicated a rapid interconversion between the favoured $^3T_4$ and less favoured $^4T_3$ forms. The ring structure was intermediate between the $^3T_4$ and E_4 envelope conformations.[26]

Nucleoside analogues (8) derived from nicotinamide and their reduced forms have been studied by 220 MHz ^{1}H n.m.r.[27]

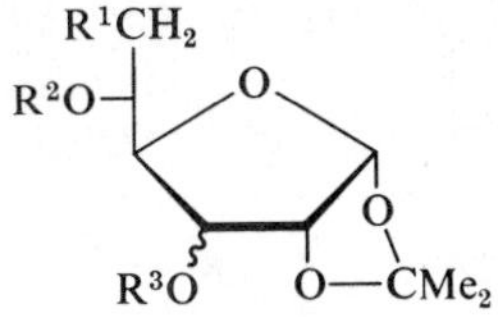

(6) R^1 = OAc or Cl; R^2 = H or Ac;
R^3 = Ts, Ac, CONHPh, Bz, or Bn

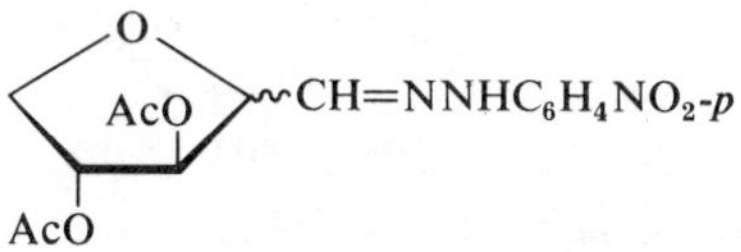

(7)

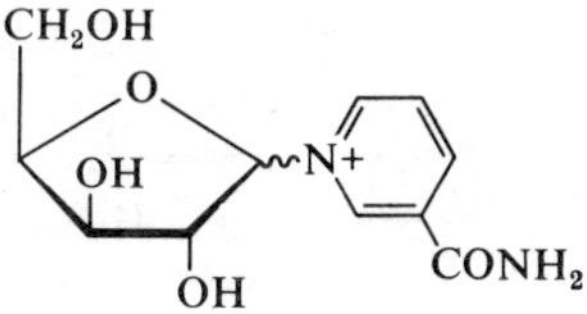

(8)

A theoretical interpretation of the *cis*-vicinal ^{1}H–^{1}H coupling constants in the pentofuranosyl rings of nucleosides has been made.[28]

Pulsed n.m.r. methods have been used to probe the conformations adopted by 2′-amino-2′-deoxyadenosine, 3′-amino-3′-deoxyadenosine, and puromycin. Longitudinal proton relaxation time and nuclear Overhauser enhancement measurements were carried out to characterize orientation of the base relative to ribose.[29] The latter two compounds have a preferred *N-anti-g*$^+$ conformation, while 2′-amino-2′-deoxyadenosine adopts the *S-syn-g*$^+$*/t* family of conformations. These findings corroborate the previous proposal relating the *N* state of ribose with *anti*-orientation of the base and the *S* state with *syn* orientation. Nuclear Overhauser enhancements have also been used to determine the site of glycosylation in nucleosides of *S*-triazole; the n.O.e. of H-1′ due to 3-Me or 5-Me showed that the structure of the major product of fusion of 1-*O*-acetyl-2,3,5-tri-*O*-benzoyl-β-D-ribofuranose with 3,5-dimethyl-1,2,4-triazole was (9) and that of the minor was (10).[30] High resolution

[25] W. Z. Antkowiak and W. J. Krzyzosiak, *Bull. Acad. polon. Sci., Sér. Sci. chim.*, 1977, **25**, 421 (*Chem. Abs.*, 1978, **88**, 7299c).
[26] A. Rockenbauer, J. Hemela, G. Toth, and A. Gerecs, *Magyar, Kem. Folyóirat*, 1977, **83**, 442 (*Chem. Abs.*, 1978, **88**, 170 402s).
[27] N. J. Oppenheimer, *J. Carbohydrates, Nucleosides, Nucleotides*, 1978, **5**, 251.
[28] A. Jaworski, I. Ekiel, and D. Shugar, *J. Amer. Chem. Soc.*, 1978, **100**, 4357.
[29] H. Plach, E. Westhof, H. D. Luedemann, and R. Mengel, *European J. Biochem.*, 1977, **80**, 295 (*Chem. Abs.*, 1978, **88**, 89 962c).
[30] J. Kobe and J. C. Valdes, *Carbohydrate Res.*, 1978, **65**, 278.

^1H n.m.r. spectra at 220 MHz and 100 MHz of 6-methyl-2′-deoxyuridine and its 3′- and 5′-phosphates were examined and compared with the data for the thymine analogues. The conformations and rotamer states about the anomeric centre and about the C-4′–C-5′ bonds were established.[31] ^1H n.m.r. studies at 220 MHz of nucleoside cyclic monophosphates (11) and (12) in ^2H_2O have shown that the adopted conformations are ^3E and ^2E respectively,

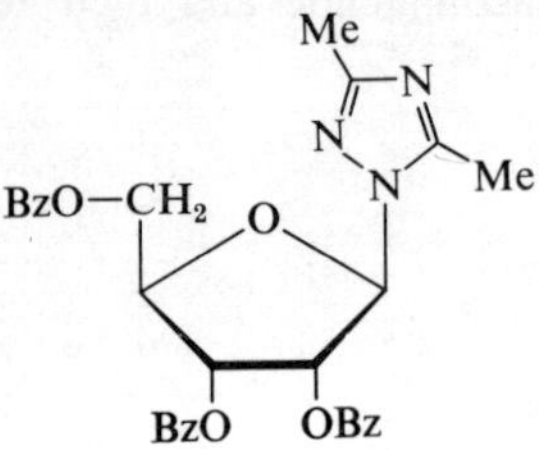

(9)

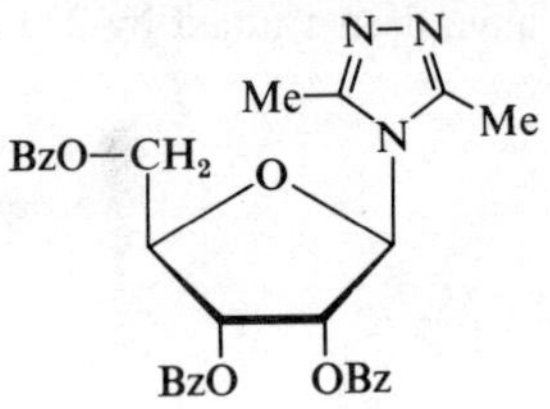

(10) or α-anomer

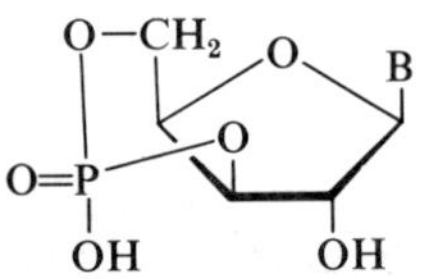

(11)

B = adenin-9-yl

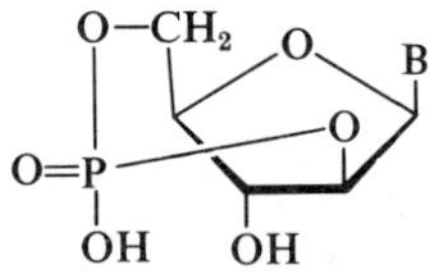

(12)

with the phosphate rings in a chair form. An unusual $^4J_{P,H}$ of 2.4 Hz is observed between H-4′ and P in the spectrum of the *xylo*-compound. The spectrum of the *arabino*-nucleoside (12) contains a *vic*-$J_{P,H}$ of 30.8 Hz between H-5′ and P. An assessment of the use of the Karplus equation showed that the suggested constants of Lee and Sarma (*J. Amer. Chem. Soc.*, 1976, **98,** 3541) were not applicable.[32]

Force-field calculations have led to the conclusion that the furanose ring of nucleic acids is extensively flexible. The activation energy for distortion of C-3′-*endo* to C-2′-*endo* was found to be only 2.09 kJ mol^{-1}. The wider implications of the flexibility were discussed.[33]

The ^{14}N nuclear spin-lattice relaxation times and proton coupling constants for adenosine, guanosine, and related compounds have been reported.[34]

4 Oligosaccharides

In a structural study of the products from reaction of methyl β-D-cellobioside with benzylidene chloride in pyridine, simultaneous relaxation and double-resonance experiments were carried out. This novel technique allowed couplings to be assigned to protons normally superimposed by multiplets

[31] A. L. George, F. E. Hruska, K. K. Ogilvie, and A. Holy, *Canad. J. Chem.*, 1978, **56,** 1170.
[32] M. MacCoss, F. S. Ezra, M. J. Robins, and S. S. Danyluk, *Carbohydrate Res.*, 1978, **62,** 203.
[33] M. Levitt and A. Warshel, *J. Amer. Chem. Soc.*, 1978, **100,** 2607.
[34] M. E. Moseley and P. Stilbs, *Canad. J. Chem.*, 1978, **56,** 1302.

and complete assignments for the spectra of products (13) and (14) were achieved. In a typical experiment a delay of 0.33 s between the 90° and 180° pulses changed the multiplet due to 4′-H, 5′-H, and 3-H at δ 3.60 to a triplet for 3-H which further reduced to a doublet on simultaneous irradiation of H-2. N.O.e. measurements were used to determine the configuration of C-7 of the eight-membered ring which adopts a 'chair–tub' conformation rather than 'crown' (chair–chair) or 'butterfly' (tub–tub) forms. Other products from this reaction were the expected methyl penta-*O*-acetyl-4′,6′-*O*-benzylidene-β-D-cellobioside and methyl 2,3,3′-tri-*O*-acetyl-2′,6:4′,6′-di-*O*-benzylidene-β-D-cellobioside.[35]

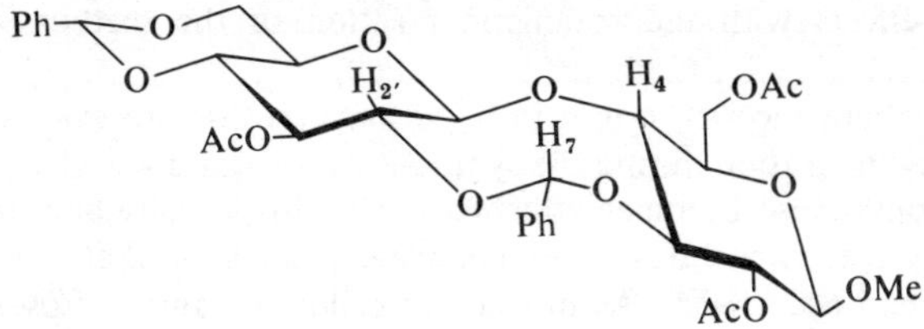

(13)

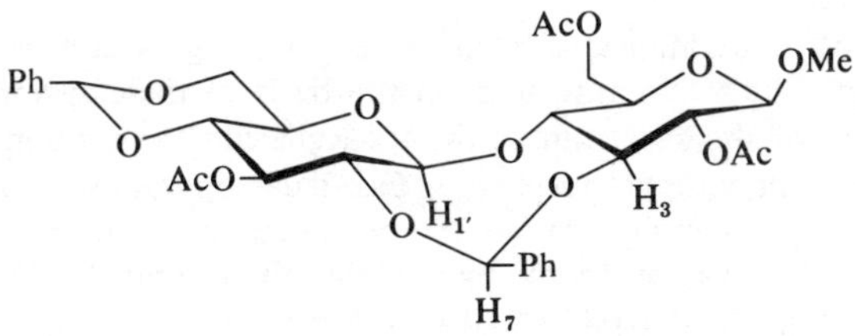

(14)

Detailed analysis of the ^{1}H n.m.r. spectrum of 3,6-anhydro-α,α-trehalose hexa-acetate using methyl 3,6-anhydro-α-D-glucoside acetates and methyl 2,3,4,6-tetra-*O*-acetyl-α-D-glucoside as models has shown that the anhydro-ring is in a distorted $^{1}C_{1}$(D) conformation and the other ring is in a $^{4}C_{1}$(D) form.[36]

5 ^{13}C N.M.R. Spectroscopy

The full paper on the ^{13}C n.m.r. of the glycosides derived from *Stevia paniculata* LAG has appeared [37] (see Vol. 10, p. 182), and the structures of two sweet glucosides from *S. rebaudiana* Bertoni, rebaudioside-D and -E, have been studied by means of ^{13}C n.m.r. and enzymic and chemical hydrolysis.[38]

A report of a study of ^{13}C n.m.r. spectra of a wide range of pyranoside epoxides has confirmed that the presence of the epoxide ring leads to an upfield

35 K. Bock, B. Meyer, and J. Thiem, *Angew. Chem. Internat. Edn.*, 1978, **17**, 447.
36 J. Defaye and D. Horton, *Carbohydrate Res.*, 1978, **65**, 283.
37 K. Yamasaki, H. Kohda, T. Kobayashi, N. Kaneda, R. Kasai, O. Tanaka, and K. Nishi, *Chem. and Pharm. Bull. (Japan)*, 1977, **25**, 2895.
38 I. Sakamoto, K. Yamasaki, and O. Tanaka, *Chem. and Pharm. Bull. (Japan)*, 1977, **25**, 3437.

shift of the oxiran carbon atom signals of 10—23 p.p.m. relative to the parent diol compound, due to the α-effect. Distinction between 2,3- and 3,4-anhydro-sugars may be made by means of the chemical shift of C-1 which is shifted upfield by 4—8 p.p.m. (β-effect) or 0.5—3 p.p.m. (γ-effect) respectively.[39] The ^{13}C n.m.r. spectrum of 3-*O*-methyl-D-fructose has been studied in detail and resonances for all four cyclic forms were assigned in order to provide parameters for the assignment of D-fructose itself. The confusion in the literature on this subject was surveyed.[40]

In order to be able to apply ^{13}C n.m.r. for the identification of glucuronate residues in polysaccharides, the spectra of the complete range of methyl-(methyl *O*-methyl-α-D-glucopyranosid)uronates have been studied. A correlation of β-effects with the rotameric position of the methoxy-groups was made.[41]

Natural abundance ^{13}C n.m.r. in ^2H_2O has been used to examine the possibility of acyl migration during the synthesis of di- and tri-saccharides in which α-D-galactopyranose becomes attached to O-4 in partially acylated galactose units. Expected structures were confirmed, showing that no 6 → 4 acyl migration had occurred.[42] As models for cellobiose and maltose, twenty-one simple derivatives of cyclohexyloxytetrahydropyran have been studied by ^{13}C and ^{1}H n.m.r.[43]

L-Rhamnose-containing disaccharides and their peracetyl derivatives have been examined by ^{13}C n.m.r., and on the basis of these spectra a complete interpretation of the spectrum of the trisaccharide *O*-α-L-rhaminopyranosyl-(1 → 3)-*O*-α-L-rhamnopyranosyl-(1 → 6)-D-galactopyranose was possible. The difficulty of assignment of the anomeric configuration using the chemical shift value of C-1 was overcome by studying the chemical shifts of the other carbon atoms, particularly C-3 and C-5.[44] Tetradeca-*O*-acetyl-stachyose and an isomer with the terminal galactose unit β- instead of α- linked have been examined by ^{13}C n.m.r.[45] Published tables of ^{13}C chemical shifts of many glucosides and their peracetates have already been mentioned in another connection.[2] Complete ^{13}C n.m.r. data on all possible tetra-*O*-acetyl-β-D-glucopyranoses, and on 1,3,4,6-tetra-*O*-acetyl-α-D-glucopyranose and many D-mannofuranosyl- and D-mannopyranosyl-D-glucose derivatives have been obtained and tabulated.[46]

Inter-residue ^{13}C–^{1}H coupling in disaccharides has been used to determine the orientation of the glycosidic linkage. Coupling over three bonds between

[39] A. S. Shashkov, A. Ya. Shmyrina, A. F. Sviridov, Kh. A. Arifkhodzhaev, and O. S. Chizhov, *Bioorg. Khim.*, 1977, **3**, 1503.
[40] T. A. W. Koerner, R. J. Voll, L. W. Cary, and E. S. Younathan, *Biochem. Biophys. Res. Comm.*, 1978, **82**, 1273.
[41] A. S. Shashkov, A. F. Sviridov, O. S. Chizov, and P. Kovac, *Carbohydrate Res.*, 1978, **62**, 11.
[42] D. D. Cox, E. Kurt Metzner, L. W. Cary, and E. J. Reist, *Carbohydrate Res.*, 1978, **67**, 23.
[43] F. Lafuma and C. Quivoron, *Canad. J. Chem.*, 1978, **56**, 2065.
[44] C. Lafitte, A. M. N. P. Du, F. Winternitz, R. Wylde, and F. Pratviel-Sosa, *Carbohydrate Res.*, 1978, **67**, 105.
[45] T. Suami, T. Otake, S. Nishiyama, R. Adachi, and K. Nakamura, *Bull. Chem. Soc. Japan*, 1978, **51**, 1826.
[46] H. Komura, A. Matsuno, Y. Ishido, K. Kushida, and K. Aoki, *Carbohydrate Res.*, 1978, **65**, 271.

natural abundance ^{13}C-1 and ^{1}H-4′, and ^{13}C-4′ and ^{1}H-1 was measured in maltose, cyclohexa-amylose, and related compounds and found to be close to 3 Hz, suggesting that in aqueous solution more staggered conformations are present than in the solid state. Data for sucrose, maltosan, α,α-trehalose, and other three-bond couplings are also given.[47] The ^{13}C chemical shifts in $^{2}H_2O$ for central D-glucosyl residues in unmodified (1 → 4)-linked β-D-gluco-oligosaccharides obtained by acetolysis of cellulose have been shown to be independent of chain length. The signal due to C-4 which occurs at 80 p.p.m. provides the most useful indicator for detection of β-(1 → 4)-linked units in a glucan as it appears free from other signals in the spectrum.[48] Oligosaccharides derived from alginic acid have been investigated by ^{13}C n.m.r. and the resonance due to C-1 in mannuronic acid has been distinguished from that due to C-1 in guluronic acid. In order to identify these signals several model compounds were examined including sodium D-mannuronate, sodium L-guluronate, and homo and mixed di- and tri-saccharides.[49]

^{13}C N.m.r. has been used to distinguish between 7- and 9-purine substitution in ribofuranosyl nucleosides; useful signals are those of C-1′ which differ for the two types and those of the carbon atoms of the purine moiety which are characteristic for the two types of substitution.[50] The anomeric configuration of mycosamine in the polyene antifungal antibiotics lucensomycin and nystatin has been determined by ^{13}C n.m.r.; the former contains an α-mycosaminyl residue while the latter is a β-mycosaminide.[51]

The effect of *O*-deuteriation upon the proton-decoupled ^{13}C n.m.r. spectra of carbohydrates in [$^{2}H_6$]DMSO has shown that there is an upfield shift of 0.1—0.2 p.p.m. in the ^{13}C resonance relative to the undeuteriated compounds.[52] Analysis of the stereochemical dependence of ^{13}C–^{1}H and ^{13}C–^{13}C coupling constants of mono- and di-saccharides enriched with ^{13}C at C-1, C-1′, or C-6 (prepared from D-[1-^{13}C]glucose or D-[6-^{13}C]glucose) has shown that a Karplus-type equation operates; $^{1}J_{C\text{-}1,\,H\text{-}1}$ and $^{2}J_{C\text{-}1,\,H\text{-}2}$ values are related to the established anomeric configuration; $^{1}J_{C\text{-}6,\,H\text{-}6}$ for the pro-*R* and the pro-*S* methylene protons differ and were assigned by means of specific monodeuteriation. Also given are the values of $^{2}J_{C-O-C}$ through the glycosidic bond. The use of the derived modified Karplus equation for $^{1}J_{^{13}C-H}$ and $^{1}J_{^{13}C-^{13}C}$ allowed an estimate of the dihedral angles ϕ and ψ which were compared with available solid-state *X*-ray data.[53]

^{13}C N.m.r. spectroscopy has been used in correlation with rates of a second-order S_N2 inversion occurring at an α-pyranosyl bromide using tetraethylammonium chloride: the introduced chlorine causes a shift in C-1 resonance which was found to be proportional to ln *k* (the rate constant) in the case of glucosyl halides but no useful correlation was noted for the galactosyl series.

[47] A. Parfondry, N. Cyr, and A. S. Perkin, *Carbohydrate Res.*, 1977, **59**, 299.
[48] Y. Inoue and R. Chuzo, *Carbohydrate Res.*, 1978, **60**, 367.
[49] J. Boyd and J. R. Turvey, *Carbohydrate Res.*, 1978, **61**, 223.
[50] P. Fischer, G. Losch, and R. R. Schmidt, *Tetrahedron Letters*, 1978, 1505.
[51] A. Boicelli, *Tetrahedron Letters*, 1978, 1213.
[52] S.-C. Ho, H. J. Koch, and R. S. Stuart, *Carbohydrate Res.*, 1978, **64**, 251.
[53] D. Y. Gagnaire, R. Nardin, F. R. Taravel, and M. R. Vignon, *Nouv. J. Chim.*, 1977, **1**, 423 (*Chem. Abs.*, 1978, **88**, 89 948c).

The syntheses of the compounds used are dealt with in Chapter 7 and the paper contains a large amount of ^1H and ^{13}C n.m.r. data.[54]

Long-range ^{13}C–^1H spin–spin coupling has been used to determine the conformation if L-ascorbic acid in deuterium oxide solution: the favoured conformation proposed is shown in (15).[55]

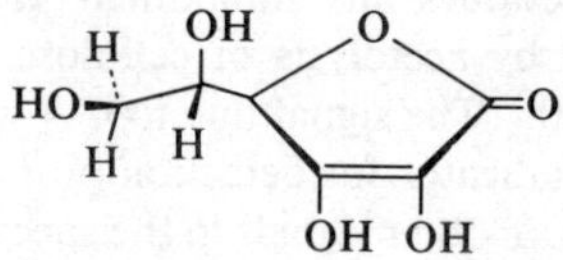

(15)

Line-broadening of ^{13}C signals on addition of copper sulphate solution to kanamycin A and to ribostamycin provides a structural and functional probe; increasing Cu^{2+}/antibiotic concentration ratios selectively changes the signals due to the carbon atoms attached to the complexing groups. Thus in kanamycin A the signals were shown to be affected in the order C-3″, C-4″ > C-1″ ⩾ C-6″ ⋙ C-6′. The remaining sites appear not to be primary sites for Cu^{2+} bonding. In the case of ribostamycin the ^{13}C nuclei of the pseudo-disaccharide were selectively affected.[56]

The methods of Perkin (*Canad. J. Chem.*, 1975, **53**, 1872) using $^2J_{C-C-H}$ coupling have been applied to some amino-sugars: 2-amino-2-deoxy-D-[1-^{13}C] glucose and -mannose hydrochlorides have shown qualitative similarities to the non-aminated [1-^{13}C]hexoses. The C-1–C-3 coupling was resolvable only for β-anomers whereas C-1–C-5 coupling was only observed for α-anomers. Similar dihedral angle dependence for C-1–H-2 couplings to those for the hexoses was noted, these being negative for α- and β-anomers of both sugars.[57]

6 Lanthanide-shift Reagents

Lanthanide-induced ^1H n.m.r. shifts for glyceric, gluconic, and lactobionic acid, measured in ^2H_2O at various pH values, have demonstrated that positions 1, 2, and 3 in all three compounds are involved in the complex formed.[58]

A comparison of the bound chemical shifts in the presence of $Eu(fod)_3$ of the anhydro sugar (16) with those of sugars not containing the anhydro-ring

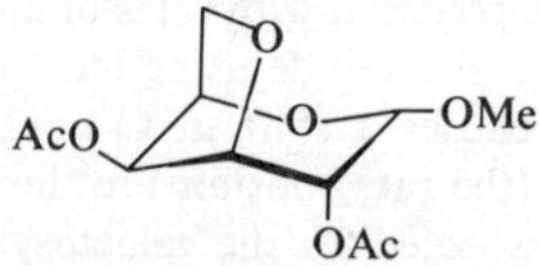

(16)

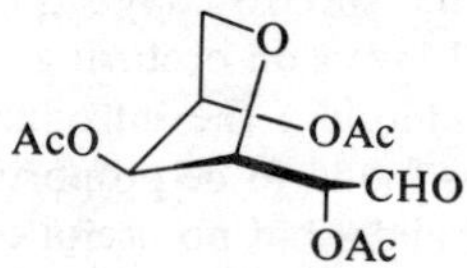

(17)

[54] H. Paulsen, A. Richter, V. Sinnwell, and W. Stenzel, *Carbohydrate Res.*, 1978, **64**, 339.
[55] T. Ogawa, J. Uzawa, and M. Matsui, *Carbohydrate Res.*, 1977, **59**, c32.
[56] S. Hanessian and G. Patil, *Tetrahedron Letters*, 1978, 1031.
[57] T. E. Walker, R. E. London, R. Barker, and N. A. Matwiyoff, *Carbohydrate Res.*, 1978, **60**, 9.
[58] T. Taga, Y. Kuroda, and M. Ohashi, *Bull. Chem. Soc. Japan*, 1978, **51**, 2278.

and with the 3,6-anhydro-D-galactose derivative (17) has shown that a 1 : 1 adduct is formed between the 2- and 4-acetoxy-groups and the europium. With (16) complexing is stronger to the 2-acetoxy than to the 4-acetoxy-group. The spectra of mixtures of (16), (17), and peracetylated galactoses showed that the former complexed most strongly with the reagent and the authors suggest that this may provide a means of investigating mixtures of the two compounds, *e.g.* in hydrolysates of carageenans.[59]

7 Other Nuclei

Thioanalogues of adenosine phosphates have been studied by ^{31}P n.m.r. and the chemical shifts were compared to those of the oxygenated compounds. The effects of changing pH and the concentration of added Mg^{2+} led to the conclusion that chemical shift data cannot yield unequivocal information concerning the absolute structure of the metal complexes of nucleosides but can be used to monitor changes in chelation, for example, in binding to enzymes.[60]

Configurational assignments for 2,3-di- and 2,3,4-tri-deoxy phosphonates of the type (18), formed by hydrogenation of compounds of type (2), have been

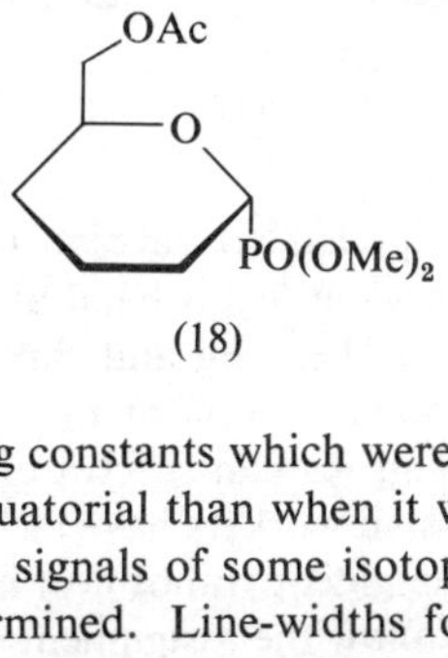

(18)

made using $^{1}J_{C-1,P}$ coupling constants which were found to be larger when the phosphonate group was equatorial than when it was axial.[6]

Parameters of ^{17}O n.m.r. signals of some isotope-enriched monosaccharide derivatives have been determined. Line-widths for ^{17}O were large (230—600 Hz) compared to chemical shifts, even at 100 °C, for primary alcohols, linear secondary alcohols, five-membered ring alcohols, and equatorial and axial alcohol groups on six-membered rings. The signal due to O-1 of 2,3,4,6-tetra-*O*-methyl-α-D-mannopyranose showed significant deshielding, that of O-2 of D-mannose-2-^{17}O diethylthioacetal was deshielded by the two thioacetal groups. Replacement of OH by OMe resulted in a shielding of 21 p.p.m. for the primary O-1 of D-glucitol-1-^{17}O, and 9 p.p.m. for the axial O-3 of methyl β-D-allopyranoside-3-^{17}O. The technique is unlikely to be of general use, since compounds containing 2-atom-percent ^{17}O are needed; these are not available and are difficult to synthesize.[61]

[59] K. Izumi, *Carbohydrate Res.*, 1978, **62**, 368.
[60] E. K. Jaffe and M. Cohn, *Biochemistry*, 1978, **17**, 652.
[61] P. A. J. Gorin and M. Mazurek, *Carbohydrate Res.*, 1978, **67**, 479.

21
Other Physical Methods

1 I.R. Spectroscopy

Intramolecular hydrogen bonding in 30 model compounds, *e.g.* (1), has been examined in carbon tetrachloride as part of an investigation into associations in the cellulose and amylose molecules.[1]

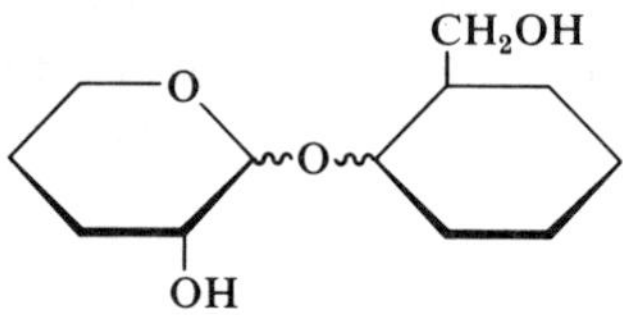

(1)

Laser-Raman techniques applied to several 1-thio-β-D-glucopyranosides and -galactopyranosides have identified a band at 891 cm^{-1} which is characteristic of axial anomeric C—H bending and can be used for distinguishing between anomers. Various other aspects of the spectra were discussed.[2] A theoretical paper on the i.r. and Raman spectra of α- and β-D-glucopyranose has appeared,[3] and other Russian workers have reconsidered the i.r. spectra of their penta-acetates (and some glucopyranosides) in the 50—800 cm^{-1} region. Differences were noted which allow the assignment of anomeric configuration.[4] An examination of fifteen sulphonate esters of α-D-glycopyranosides shown by n.m.r. spectroscopy to be in the $^4C_1$ conformation indicated that absorptions in the 800—900 cm^{-1} region cannot be used reliably for characterization of the axial or equatorial nature of the ester groups.[5] Ascorbic acid was examined in aqueous solution over the concentration range 0.02—1.0 M, and pH range 1.8—13.5 and hydrogen bonding within the ionized and unionized species was discussed.[6]

[1] F. Lafuma and C. Quivoron, *Canad. J. Chem.*, 1978, **56**, 2076.
[2] A. T. Tu, J. Lee, and Y. C. Lee, *Carbohydrate Res.*, 1978, **67**, 295.
[3] V. V. Sivchik and R. G. Zhdankov, *Zhur. priklad. Spektroskopii*, 1978, **28**, 1038 (*Chem. Abs.*, 1978, **89**, 110 170a).
[4] K. A. Asankozhoev, K. K. Koshoev, M. A. Salimov, and V. M. Tul'chinskii, *Izvest. Akad. Nauk Kirg, S.S.R.*, 1977, 53 (*Chem. Abs.*, 1978, **88**, 89 939a).
[5] R. C. Chalk, M. E. Evans, F. W. Parrish, and J. A. Sousa, *Carbohydrate Res.*, 1978, **61**, 549.
[6] N. P. Evtushenko, *Biol. Aspekty Koord. Khim.* (*Lektsii Doklady Shk. Bioneorg. Khim.*), 1st, 1975 (pub. 1977), 148 (*Chem. Abs.*, 1978, **88**, 79 032k).

2 U.V. Spectroscopy

Several studies on the c.d. of carbohydrate derivatives, reported later in this Chapter, have included u.v. absorption data.

3 Mass Spectroscopy

In the area of new techniques, a procedure for examining thermally sensitive, involatile compounds has been exemplified by application to sucrose and lactose. The sample is heated very rapidly (1200 °C in less than 0.2 s) in the electron beam, and all ions produced are detected simultaneously on a photographic plate. Ions corresponding to $(M + H)^+$ as well as several formed by loss of water were observed.[7] Addition of a small quantity of sodium iodide or potassium iodide to oligosaccharides or glycolipids produced very large enhancements in the corresponding $[M + C]^+$ ions (C = cation) in field desorption spectra.[8] Negative ion spectra of methylated methyl pentosides and hexosides have been reported and the observed fragmentation patterns discussed.[9]

Examination of the trimethylsilyl ethers of D-glucose, D-galactose, and D-mannose by gas chromatography followed by chemical ionization mass spectrometry using ammonia and 2-methylpropane as reagent gases showed that anomers, epimers, and pyranose–furanose isomers could all be distinguished.[10] A similar study has been made on the mixed aldononitrile peracetates derived from rhamnose, fucose, arabinose, xylose, mannose, galactose, and glucose using isobutane as reagent gas.[11] Several reports describe work with glycosides. Permethylated derivatives of α-D-glucose and β-D-galactose having single deuterium labelling at all specific positions in the sugar rings were examined to clarify the structures and mechanisms of formation of fragment ions with m/e 187, 145, 111, and 101.[12,13] The same Russian group then reported on chemical ionization studies of methyl tetra-*O*-methyl-β-D-galactopyranoside, its α-anomer, and deuteriated and deoxy derivatives.[14] Whereas simple methyl hexopyranoside derivatives give prominent $[M-31]^+$ ions, long-chain fatty acid ester derivatives give $[M-32]^+$ species. The detailed mechanism of the loss of methanol from methyl 2,3,4-tri-*O*-acetyl-6-*O*-palmitoyl-α-D-glucopyranoside was reported.[15] A thorough investigation into the fragmentation of methyl 2,3,5,6-tetra-*O*-methyl-D-glucofuranoside led to a means of identifying pentoses

[7] W. R. Anderson, W. Frick, and G. D. Daves, *J. Amer. Chem. Soc.*, 1978, **100**, 1974.
[8] G. Puzo and J. C. Prome, *Adv. Mass Spectrometry*, 1978, **7B**, 1596 (*Chem. Abs.*, 1978, **89**, 147 156b).
[9] L. A. Baltina, V. S. Fal'ko, V. I. Khvostenko, and G. A. Tolstikov, *Zhur. obshchei Khim.*, 1977, **47**, 2379 (*Chem. Abs.*, 1978, **88**, 89 973g).
[10] T. Murata and S. Takahashi, *Carbohydrate Res.*, 1978, **62**, 1.
[11] B. W. Li, T. W. Cochran, and J. R. Vercellotti, *Carbohydrate Res.*, 1977, **59**, 567.
[12] A. Ya. Ott, B. M. Zolotarev, and O. S. Chizhov, *Izvest. Akad. Nauk S.S.S.R., Ser. khim.*, 1978, 377 (*Chem. Abs.*, 1978, **88**, 191 339s).
[13] B. M. Zolotarev, A. Ya. Ott, and O. S. Chizhov, *Adv. Mass Spectrometry*, 1978, **7B**, 1371 (*Chem. Abs.*, 1978, **89**, 180 295n).
[14] V. I. Kadentsev, A. A. Solov'yov, and O. S. Chizhov, *Adv. Mass Spectrometry*, 1978, **7B**, 1465.
[15] G. Puzo, J. L. Tichadon, and J. C. Prome, *Org. Mass Spectrometry*, 1978, **13**, 51.

and hexoses as pyranosides or furanosides. An extensive range of specific trideuteriomethyl analogues were used in the study.[16]

Detailed schemes have been outlined for the fragmentation of compounds (2) and (3) and the 1,6-anhydro-β-pyranose and β-furanose analogues as well as 2,3-epithio-β-D-allofuranose.[17] A range of disaccharides specifically labelled with ^{13}C was synthesized for mass spectrometric analysis; they were mainly

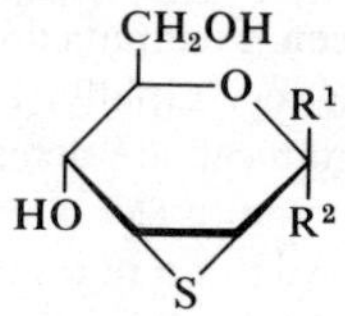

	R^1	R^2
(2)	H	OMe
(3)	OMe	H

1,6-β-, 1,3-α-, and 1,3-β-linked compounds.[18] The position of linkage of permethylated hexopyranosyl-2-acetamido-2-deoxyhexitols and hence of the initial disaccharides can be assigned on the basis of the primary fragmentation ions of the alditol units. For trimethylsilyl derivatives, other fragmentation pathways complicate the interpretations.[19]

The spectra of 14 trimethylsilylated flavanoid mono- and di-glycosides all showed well-defined molecular ion peaks, and the spectra of the *O*-glycosides allow the assignment of the position of sugar attachment, the sugar sequence, and the nature of the interunit linkage.[20] Similar work has been done on several *O*- and *C*-glycosylflavones using permethyl derivatives,[21] and 20 naturally occurring flavonoid mono-, di-, and triglycosides have been examined as isopropylidene, permethyl derivatives. The natures of interglycoside bonding and the ring sizes of the glycosidic units could be determined, and isomeric sugar units could be distinguished.[22]

Comparison of the mass spectra of TMS derivatives of hexuronic acids and lactones with those of free sugars showed specific effects due to the carboxy function. Detailed analyses of the spectra of β-D-glucopyranuronic acid, α-D-galactofuranuronic acid, and β-D-glucofuranurono-6,3-lactone ethers were given.[23] A simple procedure for identifying methylated uronic acids which can distinguish between stereoisomers has been proposed,[24] and the same group

[16] E. G. de Jong, W. Heerma, B. Dujardin, J. Haverkamp, and J. F. G. Vliegenthart, *Carbohydrate Res.*, 1978, **60**, 229.
[17] T. Yamaguchi and M. Kojima, *Org. Mass Spectrometry*, 1978, **13**, 113.
[18] C. Bosso, F. Tavarel, J. Ulrich, and M. Vignon, *Org. Mass Spectrometry*, 1978, **13**, 477.
[19] I. Mononen, J. Finne, and J. Kärkkäinen, *Carbohydrate Res.*, 1978, **60**, 371.
[20] H. Schels, H. D. Zinsmeister, and K. Pfleger, *Phytochemistry*, 1978, **17**, 523 (*Chem. Abs.*, 1978, **89**, 110 246e).
[21] M. L. Bouillant, A. Besset, J. Favre-Bonvin, and J. Chopin, *Phytochemistry*, 1978, **17**, 527.
[22] O. Seligmann and H. Wagner, *Tetrahedron*, 1978, **34**, 3299.
[23] J. F. Kennedy and S. M. Robertson, *Carbohydrate Res.*, 1978, **67**, 1.
[24] V. Kováčik, V. Mihálov, and P. Kováč, *Chem. Zvesti*, 1978, **32**, 258.

has reported on the spectra derived from permethylated aldobiuronic and pseudoaldobiuronic acids and their products of β-elimination.[25]

In the area of biologically active natural products, m.s. has been used to identify the position of acylation of partially acetylated kanamycins[26] (see Chapter 18 for further reference to relevant antibiotic work) and in an examination of pyrimidine 5′,6-anhydronucleosides.[27] The Field Desorption spectra of 13 nucleoside antibiotics all exhibited intense molecular ions [and $(M + 1)^+$ ions] and fragment ions characteristic of their chemical structures.[28]

Several of the papers noted in the g.l.c. section of Chapter 22 contain mass spectral data.

4 X-Ray and Neutron Diffraction Crystallography

A review of the types of hydrogen bonding found in crystals of carbohydrates describes the finding, by neutron diffraction methods, of seven types of interaction ranging from the strong with hydrogen–oxygen distances of 1.73 Å to the weak distances of 2.6 Å. Hydrogen bonding patterns are strongly influenced by the 'co-operative' and 'anomeric' effects.[29] A further contribution from Jeffrey's laboratory established, as expected from predictions, that situations with donor–acceptor interactions between hydroxy-groups leading to 'chains' of hydrogen-bonded systems are more common than those found with a hydroxy-group acting only as a donor and, in the former cases, hydrogen bond distances are shorter. Secondly, anomeric hydroxy-groups are invariably involved in hydrogen bond donation and give strong bonds of length less than 1.95 Å. The oxygen atoms act rarely as acceptors and when they do the bonds formed are weak. The issue of chains of hydrogen bonds was considered further.[30] A further analysis of *X*-ray structures of glycosides has considered the rotamer states of the aglycones and related them to bond lengths.[31]

Specific crystal structures have been reported as follows (neutron diffraction studies are identified by the letters 'n.d.').

Free Sugars and Simple Derivatives Thereof.—α-L-Rhamnopyranose (n.d.),[32] 2,6-dideoxy-β-D-*ribo*-hexopyranose (β-D-digitoxose),[33] β-D-*glycero*-D-*gulo*-heptopyranose,[34] α-maltose,[35] α,β-melibiose,[36] and 3-*O*-α-D-glucopyranosyl-β-D-fructopyranose (turanose).[37, 38]

[25] V. Kováčik, V. Mihálov, J. Hirsch, and P. Kováč, *Biomed. Mass Spectrometry*, 1978, **5**, 136 (*Chem. Abs.*, 1978, **89**, 129 849a).
[26] D. C. DeJongh, M. S. Balachandran Nagar, G. Patil, and S. Hanessian, *Tetrahedron*, 1978, **34**, 1869.
[27] G. Puzo, K. H. Schram, J. G. Liehr, and J. A. McCloskey, *J. Org. Chem.*, 1978, **43**, 767.
[28] K. Fukushima and T. Arai, *J. Antibiotics*, 1978, **31**, 377.
[29] G. A. Jeffrey and S. Takagi, *Accounts Chem. Res.*, 1978, **11**, 264.
[30] G. A. Jeffrey and L. Lewis, *Carbohydrate Res.*, 1978, **60**, 179.
[31] S. Pérez and R. H. Marchessault, *Carbohydrate Res.*, 1978, **65**, 114.
[32] S. Takagi and G. A. Jeffrey, *Acta Cryst.*, 1978, **B34**, 2551.
[33] J. A. Kanters, L. M. J. Batenburg, W. P. J. Gaykema, and G. Roelofsen, *Acta Cryst.*, 1978, **B34**, 3049.
[34] J. A. Kanters, A. J. H. M. Kock, and G. Roelofsen, *Acta Cryst.*, 1978, **B34**, 3285.
[35] F. Takusagawa and R. A. Jacobson, *Acta Cryst.*, 1978, **34B**, 213.
[36] M. E. Gress, G. A. Jeffrey, and D. C. Rohrer, *Acta Cryst.*, 1978, **34B**, 508.
[37] A. Neuman, D. Avenel, and H. Gillier-Pandraud, *Acta Cryst.*, 1978, **34B**, 242.
[38] J. A. Kanters, W. P. J. Gaykema, and G. Roelofsen, *Acta Cryst.*, 1978, **B34**, 1813.

Glycosides and Simple Derivatives Thereof.—Methyl α-D-xylopyranoside,[39] methyl α-L- and β-L-arabinopyranoside,[40] methyl β-D-ribopyranoside (*X*-ray and n.d.),[41] methyl β-D-galactopyranoside,[42] methyl tetra-*O*-acetyl-β-D-glucopyranoside,[43] 1-naphthyl 2,3,4,6-tetra-*O*-acetyl-β-D-glucopyranoside,[44] 1-acetyl-3-benzamido-4-(2,3,4,6-tetra-*O*-acetyl-β-D-glucopyranosyloxy)-Δ^3-2-pyrrolinone,[45] 8[2-(β-D-glucopyranosyloxy)isopropyl]-9-hydroxy-8,9-dihydro-angelium,[46] methyl β-D-laminaribioside hepta-acetate,[47] and α-D-allopyranosyl α-D-allopyranoside, $CaCl_2$.[48]

Amino-sugar Derivatives.—Methyl 4,6-*O*-benzylidene-2,3-dideoxy-2,3-phenyl-azo-α-D-mannoside,[49] 2,4-bis(*N*-acetyl-*N*-benzoylamino)-3,6-di-*O*-benzoyl-2,4-dideoxy-α-D-idopyranoside,[50] 1-(*N*-benzyl-*N*-methyl)amino-1-deoxy-α-D-*lyxo*-hexulopyranose,[51] 3,4,6-tri-*O*-acetyl-2-(*N*-acetyl)acetamido-1,2-dideoxy-D-*arabino*-hex-1-enopyranose,[52] 7-acetamido-6,7,8-trideoxy-1,2:3,4-di-*O*-iso-propylidene-α-D-*glycero*-D-*galacto*-octopyranose,[53] and α-*N*,*N*′-diacetylchito-biose.[54] Further amino-derivatives are noted in the 'acid' section below.

Anhydro-sugars and Derivatives.—1,4:3,6-Dianhydro-α-D-glucopyranose,[55] 1,6-anhydro-1-thio-β-D-glucopyranose,[56] and 2,7-anhydro-α-L-*glycero*-D-*manno*-octulopyranose (4).[57]

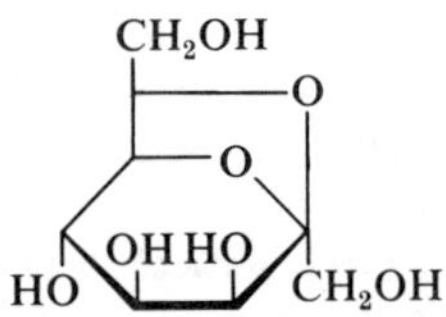

(4)

Acid Derivatives.—Calcium DL-glycerate,[58] potassium D-glucarate,[59] calcium L-ascorbate (comparison of two analyses),[60] 2,3:4,6-di-*O*-isopropylidene-L-

[39] S. Takagi and G. A. Jeffrey, *Acta Cryst.*, 1978, **B34**, 3104.
[40] S. Takagi and G. A. Jeffrey, *Acta Cryst.*, 1978, **B34**, 1591.
[41] V. J. James, J. D. Stevens, and F. H. Moore, *Acta Cryst.*, 1978, **34B**, 188.
[42] S. Takagi and G. A. Jeffrey, *Acta Cryst.*, 1978, **B34**, 2006.
[43] P. Zugenmaier and G. Rappenecker, *Acta Cryst.*, 1978, **34B**, 164.
[44] M. W. Makinen and N. W. Isaacs, *Acta Cryst.*, 1978, **B34**, 1584.
[45] A. Mercer and J. Trotter, *Acta Cryst.*, 1978, **B34**, 1596.
[46] B. F. Pedersen and J. Karlsen, *Acta Cryst.*, 1978, **B34**, 1893.
[47] H. Takeda, T. Kaiya, N. Yasuoka, and N. Kasai, *Carbohydrate Res.*, 1978, **62**, 27.
[48] J. Ollis, V. J. James, S. J. Angyal, and P. M. Pojer, *Carbohydrate Res.*, 1978, **60**, 219.
[49] G. M. Sheldrick, B. E. Davison, and J. Trotter, *Acta Cryst.*, 1978, **B34**, 1387.
[50] P. Luger and H. Paulsen, *Acta Cryst.*, 1978, **B34**, 1254.
[51] S. Pérez, A. López-Castro, and R. Márquez, *Acta Cryst.*, 1978, **B34**, 2341.
[52] B. Kojić-Prodić and V. Rogić, *Acta Cryst.*, 1978, **B34**, 858.
[53] J. C. A. Boeyens, E. B. Rathbone, and G. R. Woolard, *Carbohydrate Res.*, 1978, **62**, 39.
[54] F. Mo and L. H. Jensen, *Acta Cryst.*, 1978, **B34**, 1562.
[55] J. Kopf and P. Köll, *Acta Cryst.*, 1978, **B34**, 2502.
[56] S. Takagi and G. A. Jeffrey, *Acta Cryst.*, 1978, **B34**, 816.
[57] H. J. Hecht, P. Luger, and R. Reinhardt, *J. Cryst. Mol. Struct.*, 1977, **7**, 161 (*Chem. Abs.*, 1978, **89**, 155 923q).
[58] T. Taga, M. Ohashi, and K. Osaki, *Bull. Chem. Soc. Japan*, 1978, **51**, 1697.
[59] T. Taga, Y. Kuroda, and K. Osaki, *Bull. Chem. Soc. Japan*, 1977, **50**, 3079.
[60] S. C. Abrahams, J. L. Bernstein, C. E. Bugg, and J. Hvoslef, *Acta Cryst.*, 1978, **B34**, 2981.

xylo-hexulosonic acid,[61] 2-acetamido-2,3-dideoxy-D-*erythro*-hex-2-enono-1,4-lactone,[62] 2-acetamido-2,3-dideoxy-5,6-*O*-isopropylidene-D-*threo*-hex-2-enono-1,4-lactone,[63] sodium D-glucuronate,[64] and 5-azido-5-deoxy-1,2-*O*-isopropylidene-3-*O*-methanesulphonyl-β-L-iduronic acid diethylamide.[65]

Acyclic Derivative.—2(*S*)-4-*O*-Acetyl-3,5-*O*-benzylidene-1,2-dideoxy-2-*C*-phenyl-D-*erythro*-pentitol.[66]

Nucleosides, Nucleotides, Derivatives, and Related Compounds.—(Data derived from *X*-ray analysis of 99 compounds containing the *N*-β-D-pentofuranosyl unit have been correlated. The questions of conformational interchange *via* ring flattening or pseudorotation and the orientation of groups bonded to C-5′ were considered in detail.[67])

5-Aminouridine,[68] 5-vinyl-2′-deoxyuridine,[69] 5-ethynyl-2′-deoxyuridine,[70] 5-nitro-1-(β-D-riburonosyl)uracil,[71] uridine 5′-methylphosphate.[72]

6-Azathymidine,[73] 5(*S*)-5-hydroxy-5,6-dihydrothymidine.[74]

Cytidine 2′,3′-cyclicphosphate,[75] polymeric complexes of Cd^{II} and Cu^{II} and cytidine 5′-phosphate,[76] 1-(β-D-arabinofuranosyl)cytosine Pt^{II} complex.[77]

Adenosine complexed with 5-bromouracil,[78] *S*-methyl-5′-thioadenosine,[79] 2′,3′-*O*-isopropylideneadenosine,[80] 2′,3′-*O*-(2-carboxyethyl)ethylideneadenosine,[81] the amino-acid adenosine derivatives (5) and (6) which are constitutents of tRNA,[82] 8-[(2-aminoethyl)amino]adenosine 3′,5′-cyclic phosphate.[83]

8-Iodoguanosine,[84] 2-*N*-methylguanosine,[85] a guanosine Hg^{II} complex,[86] 2′,3′,5′-tri-*O*-acetyl-6-*O*-(mesitylenesulphonyl)-guanosine,[87] guanosine 5′-phosphate, Cu^{II} complex.[88]

[61] S. Takagi and G. A. Jeffrey, *Acta Cryst.*, 1978, **B34**, 2932.
[62] Z. Ruzić-Toroš and F. Lazarini, *Acta Cryst.*, 1978, **B34**, 854.
[63] Z. Ruzić-Toroš and I. Leban, *Acta Cryst.*, 1978, **B34**, 1226.
[64] L. J. DeLucas, G. L. Gartland, and C. E. Bugg, *Carbohydrate Res.*, 1978, **62**, 213.
[65] R. Klaska, O. Jarchow, C. Günther, and H. Paulsen, *Acta Cryst.*, 1978, **34B**, 226.
[66] D. C. Rohrer, J.-C. Fischer, D. Horton, and W. Weckerle, *Canad. J. Chem.*, 1978, **56**, 2915.
[67] P. Murray-Rust and S. Motherwell, *Acta Cryst.*, 1978, **B34**, 2534.
[68] E. Egert, H. J. Lindner, W. Hillen, and H. G. Gassen, *Acta Cryst.*, 1978, **B34**, 2204.
[69] T. A. Hamor, M. K. O'Leary, and R. T. Walker, *Acta Cryst.*, 1978, **B34**, 1627.
[70] P. J. Barr, T. A. Hamor, and R. T. Walker, *Acta Cryst.*, 1978, **B34**, 2799.
[71] T. Srikrishnan and R. Parthasarathy, *Acta Cryst.*, 1978, **B34**, 1363.
[72] J. D. Hoogendorp and C. Romers, *Acta Cryst.*, 1978, **B34**, 2724.
[73] A. Banerjee and W. Saenger, *Acta Cryst.*, 1978, **B34**, 1294.
[74] A. Grand and J. Cadet, *Acta Cryst.*, 1978, **B34**, 1524.
[75] B. S. Reddy and W. Saenger, *Acta Cryst.*, 1978, **B34**, 1520.
[76] G. R. Clark and J. D. Orbell, *Acta Cryst.*, 1978, **B34**, 1815.
[77] S. Neidle, G. L. Taylor, and A. B. Robins, *Acta Cryst.*, 1978, **B34**, 1838.
[78] K. Aiba, T. Hata, S. Sato, and C. Tamura, *Acta Cryst.*, 1978, **B34**, 1259.
[79] N. Borkakoti and R. A. Palmer, *Acta Cryst.*, 1978, **B34**, 867.
[80] S. Sprang, D. C. Rohrer, and M. Sundaralingham, *Acta Cryst.*, 1978, **B34**, 2803.
[81] D. Adamiak, M. Noltemeyer, W. Saenger, and F. Seela, *Z. Naturforsch., C: Biosci.*, 1978, **33c**, 169 (*Chem. Abs.*, 1978, **89**, 24 718d).
[82] R. Parthasarathy, J. M. Ohrt, and G. B. Chheda, *Biochemistry*, 1977, **16**, 4999.
[83] W. S. Sheldrick and E. Rieke, *Acta Cryst.*, 1978, **B34**, 2324.
[84] J. H. Al-Mukhtar and H. R. Wilson, *Acta Cryst.*, 1978, **34B**, 337.
[85] R. A. G. de Graaff, F. B. Martens, and C. Romers, *Acta Cryst.*, 1978, **B34**, 3012.
[86] M. Authier-Martin, J. Hubert, R. Rivest, and A. L. Beauchamp, *Acta Cryst.*, 1978, **34B**, 273.
[87] S. Neidle, A. Achari, G. M. Sheldrick, C. B. Reese, and P. K. Bridson, *Acta Cryst.*, 1978, **B34**, 2794.
[88] G. R. Clark, J. D. Orbell, and K. Aoki, *Acta Cryst.*, 1978, **B34**, 2119.

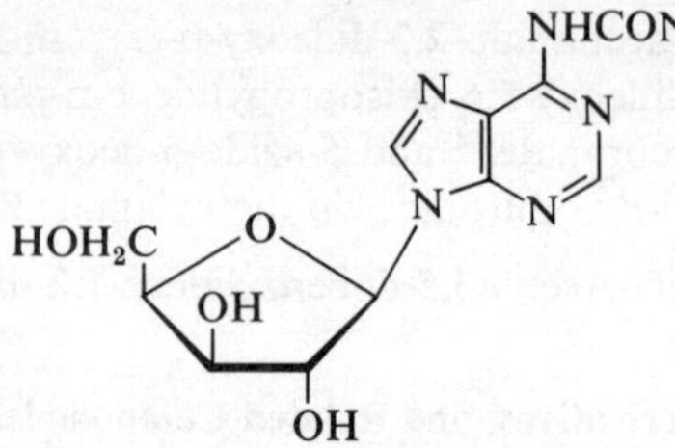

R
(5) H
(6) CHOH.Me

1-Phenyl-4,5-(D-*glycero*-L-*gluco*-heptofurano)imidazolidine-2-thione,[89] 1-*p*-tolyl-4-(*β*-D-erythrofuranosyl)imidazoline-2-thione,[90] 2,3-diphenyl-5-(α-D-lyxofuranosyl)tetrazolium bromide,[91] 1-(2,3,4,6-tetra-*O*-acetyl-α-D-mannopyranosyl)-*N*,*N′*-dimethylisocyanuric acid.[92]

Antibiotics.—8-Azatubercidin,[93] fortimicin B (7),[94] toyocamycin (8),[95] and 1,4-diamino-1,4-dideoxy-3-*O*-4-deoxy-4-propanoylamino-α-D-glucopyranosyl) -D-glucitol.[95a]

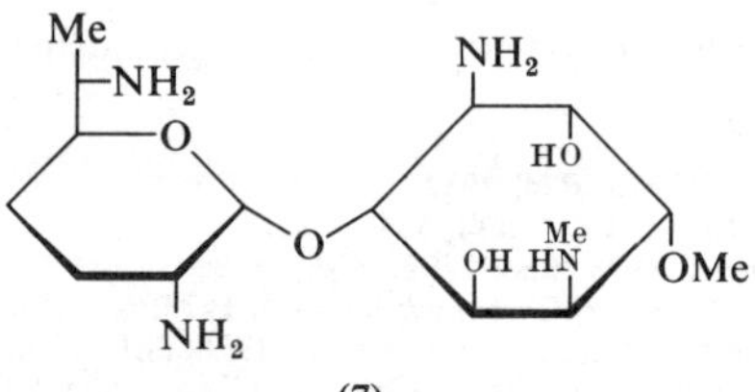

(7)

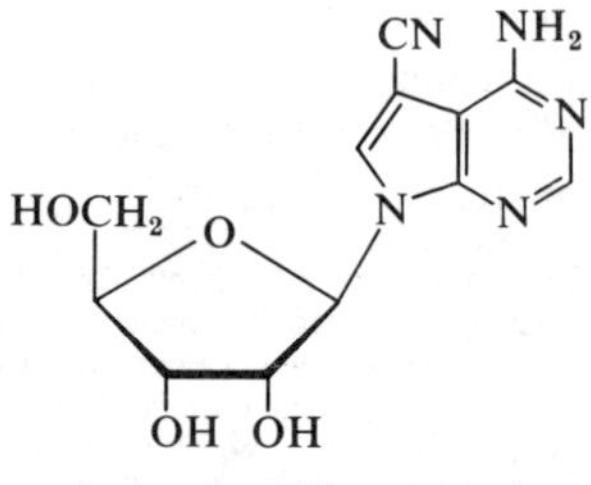

(8)

Other Derivatives.—1,2,4-Tri-*O*-acetyl-5,6-dideoxy-3-*O*-methyl-5-*C*-[(*S*)phenylphosphinyl]-*β*-D-glucopyranose (9)[96] and 1-*C*-(1,3-dithian-2-yl)-2,3:5,6-di-*O*-isopropylidene-*β*-L-gulofuranose (10).[97]

[89] R. Jiménez-Garay, P. Villares, A. López-Castro, and R. Márquez, *Acta Cryst.*, 1978, **34B**, 184.
[90] I. Barragán, A. López-Castro, and R. Márquez, *Acta Cryst.*, 1978, **34B**, 295.
[91] K. B. Lindberg, M. Czugler, V. Zsoldos-Mády, I. Pintér, and A. Messmer, *Acta Cryst.*, 1978, **B34**, 3484.
[92] W. Depmeier, H. von Voithenberg, and J. C. Jochims, *Chem. Ber.*, 1978, **111**, 2010.
[93] S. Sprang, R. Scheller, D. Rohrer, and M. Sundaralingam, *J. Amer. Chem. Soc.*, 1978, **100**, 2867.
[94] N. Hirayama, K. Shirahata, Y. Ohashi, Y. Sasada, and J. R. Martin, *Acta Cryst.*, 1978, **B34**, 2648.
[95] P. Prusiner and M. Sundaralingham, *Acta Cryst.*, 1978, **34B**, 517.
[95a] K. Kamiya, Y. Wada, and K. Nara, *Chem. and Pharm. Bull.* (*Japan*), 1978, **26**, 2040.
[96] M. Yamashita, Y. Nakatsukasa, S. Inokawa, K. Hirotsu, and J. Clardy, *Chemistry Letters*, 1978, 871.
[97] H. Ogura, K. Furuhata, H. Takahashi, and Y. Iitaka, *Chem. and Pharm. Bull* (*Japan*), 1978, **26**, 2782.

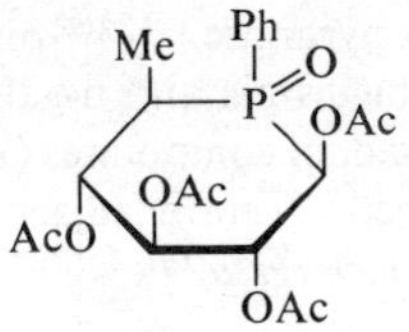

(9)

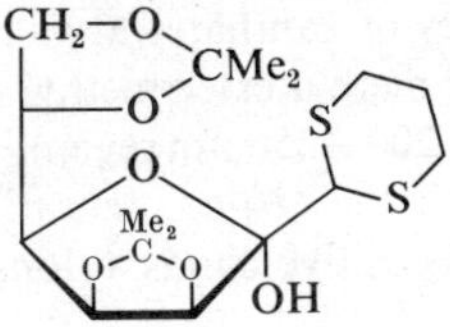

(10)

5 E.S.R. Spectroscopy

Radicals formed during the oxidation of D-glucose, D-glucitol, and methyl α-D-glucopyranoside by Co^{III} were examined by e.s.r. methods which showed that they were formed by C–H bond cleavage. The formation of an intermediate complex was also established.[98] E.s.r. studies performed on species produced during oxidation of ascorbic acid by diarylthallium hydroxide showed that the radical ions involved could be enhanced by addition of lead oxide. Observed splittings were interpreted in terms of both hydrogen and metal coupling.[99]

6 Polarimetry and Circular Dichroism Studies

Studies of the rotational strengths of 5-thio- and 1,5-dithio-D-glucopyranosides have shown how they can lead to deductions of anomeric configurations and molecular conformations.[100]

The first vibrational c.d. spectra of carbohydrates have appeared, the technique offering possible means of focussing on individual chiral centres within molecules which have several such centres. The C—H stretching regions of the spectra of methyl α- and β-D-glucopyranoside showed significant sign changes for several bands which were therefore tentatively assigned to methyl and anomeric C—H stretchings.[101] On the other hand, far-u.v. c.d. data for various sugars, oligosaccharides, sialic acids and various derivatives have been reported.[102]

A theoretical paper on the c.d. spectra of *vicinal* diacylaminodideoxy sugars described the prediction of bands of opposite sign at 180—200 nm ($\pi \rightarrow \pi^*$) and a small band at 210 nm ($n \rightarrow \pi^*$). The signs and magnitudes depend upon the orientations of the amide groups with respect to the sugar rings, and this leads to a method for determining the preferred rotamer states of such compounds as 2,3-diacetamido-2,3-dideoxy-D-glucose and the compound representing the linkage point of glycopeptides: 2-acetamido-1-*N*-aspartyl-2-deoxy-β-D-glucopyranosylamine.[103] Two reports have appeared on c.d., o.r.d., and

[98] V. I. Kurlyankina, V. N. Shadrin, E. N. Kazbekov, and V. A. Molotkov, *Zhur. obshchei Khim.*, 1978, **48**, 433 (*Chem. Abs.*, 1978, **89**, 24 641y).
[99] H. B. Stegmann, K. Scheffler, and P. Schuler, *Angew. Chem. Internat. Edn.*, 1978, **17**, 365.
[100] H. Meguro, E. Ohtaki, K. Tuzimura, N. A. Hughes, and L. Hough, *Tetrahedron Letters*, 1978, 3371.
[101] C. Marcott, H. A. Havel, J. Overend, and A. Moscowitz, *J. Amer. Chem. Soc.*, 1978, **100**, 7088.
[102] C. A. Bush, *Jerusalem Symposium Quantum Chem. Biochem.*, 1977, **10**, 209 (*Chem. Abs.*, 1978, **89**, 197 834s).
[103] C. A. Bush and A. Duben, *J. Amer. Chem. Soc.*, 1978, **100**, 4987.

u.v. studies of xanthates at C-1 and C-6 of glucopyranose,[104, 105] one on c.d. spectra of methyl ester methyl glycosides of D-glucuronic and D-galacturonic acid in the 200—250 nm region,[106] and one on the c.d. of compounds (11)—(15). Those with the (*R*)-configuration at the asymmetric centre adjacent to the rings gave positive bands at long wavelength and *vice versa*.[107]

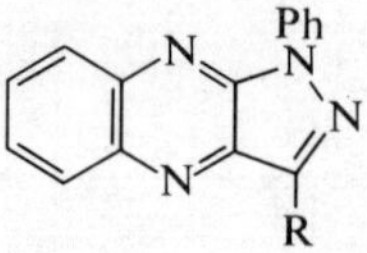

(11) dihydroxy-D-*glycero*-ethyl
(12) dihydroxy-L-*glycero*-ethyl
(13) trihydroxy-D-*erythro*-propyl
(14) tetrahydroxy-D-*arabino*-butyl
(15) tetrahydroxy-D-*lyxo*-butyl

[104] K. Satsumabayashi, Y. Nomoto, and H. Nakahara, *Nippon Shika Daigaku Kiyo*, 1977, **6**, 79 (*Chem. Abs.*, 1978, **88**, 7242f).
[105] K. Satsumabayashi, Y. Nomoto, and H. Nakahara, *Nippon Shika Daigaku Kiyo*, 1978, **7**, 133 (*Chem. Abs.*, 1978, **89**, 197 833r).
[106] T. Sticzay, S. Bystrický, C. Peciar, and J. Hirsch, *Coll. Czech. Chem. Comm.*, 1978, **43**, 1496.
[107] M. A. E. Sallam, *Carbohydrate Res.*, 1978, **66**, C4.

22

Separatory and Analytical Methods

1 Chromatographic Methods

Gas–Liquid Chromatography.—From a study of reactions of D- and L-sugars with chiral alcohols it was concluded that the diastereoisomeric products, after acetylation, could be separated and that the best method of separating the original enantiomers and thus determining their absolute configuration involved the use of butan-2-ol.[1] Analogous work used trimethylsilyl ethers of (−)-butan-2-ol glycosides, and showed that absolute configurational analysis of sugar mixtures could be effected in this way.[2]

Other reports have appeared on the g.l.c. of partially methylated alditol acetates,[3] partially methylated mannononitrile peracetates,[4] TMS ethers of methyl glycosides derived from mucopolysaccharides,[5] and TMS ethers and butylboronate TMS ethers of polyhydroxyalkylpyrazines[6] (note that several of these papers include reports of mass spectral studies).

Ion-exchange and Cellulose Column Chromatography.—After separation on anion exchange columns, several sugars, sugar alcohols, and sugar acids were detected directly by their absorptions at *ca.* 200 nm.[7] Cellulose columns (and papers) impregnated with tungstate allow good separations of polyhydroxy-compounds.[8]

Thin Layer Chromatography.—Silica gel compacted to a density comparable with that of h.p.l.c. columns allowed good separations of a range of malto-oligosaccharides up to DP12.[9]

High Pressure Liquid Chromatography.—Reverse phase h.p.l.c. using octadecyl bonded groups has been developed for the rapid and efficient separation of nucleosides, nucleotides, and protected oligonucleotides on both analytical and preparative scales.[10]

[1] K. Leontein, B. Lindberg, and J. Lönngren, *Carbohydrate Res.*, 1978, **62**, 359.
[2] G. J. Gerwig, J. P. Kamerling, and J. F. G. Vliegenthart, *Carbohydrate Res.*, 1978, **62**, 349.
[3] H. Parolis and D. McGarvie, *Carbohydrate Res.*, 1978, **62**, 363.
[4] G. R. Woolard, E. B. Rathbone, and A. W. Dercksen, *S. African J. Chem.*, 1977, **30**, 169 (*Chem. Abs.*, 1978, **88**, 7237h).
[5] M. Yokota and T. Mori, *Carbohydrate Res.*, 1977, **59**, 289.
[6] H. Tsuchida, K. Kitamura, M. Komoto, and N. Akimori, *Carbohydrate Res.*, 1978, **67**, 549.
[7] L. A. T. Verhaar and J. M. H. Dirkx, *Carbohydrate Res.*, 1978, **62**, 197.
[8] H. J. F. Angus, J. Briggs, N. A. Sufi, and H. Weigel, *Carbohydrate Res.*, 1978, **66**, 25.
[9] D. Nurok and A. Zlatkis, *Carbohydrate Res.*, 1978, **65**, 265.
[10] H.-J. Fritz, R. Belagaje, E. L. Brown, R. H. Fritz, R. A. Jones, R. G. Lees, and H. G. Khorana, *Biochemistry*, 1978, **17**, 1257.

2 Other Analytical Methods

The direct detection of a variety of simple carbohydrate compounds by u.v. methods is mentioned in reference 7 above. Spectrophotometry has also been used to determine ketoses (as furan derivatives[11] and glucose by its reduction of 3,5-dinitrosalicylic acid to the 3-amino-5-nitro analogue.[12]

[11] C. Laluce, A. de Carvalho, and R. Molinari, *Ecletica Quim.*, 1976, **1**, 7 (*Chem. Abs.*, 1978, **88**, 38 088q).

[12] W. T. Smith jun. and C. Cheng, *Anal. Letters*, 1978, **B11**, 191 (*Chem. Abs.*, 1978, **89**, 110 149a).

23
The Synthesis of Optically Active Non-carbohydrate Compounds

The streptolidine lactam (1), a component of the streptothricin group of antibiotics, has been synthesized as outlined in Scheme 1.[1] D-Glucose has been used as an optically active source material for the synthesis of α-multistriatin (2),[2] a beetle hormone, which is summarized in Scheme 2, and of (−)-*cis*-rose oxide (3), shown in Scheme 3.[3]

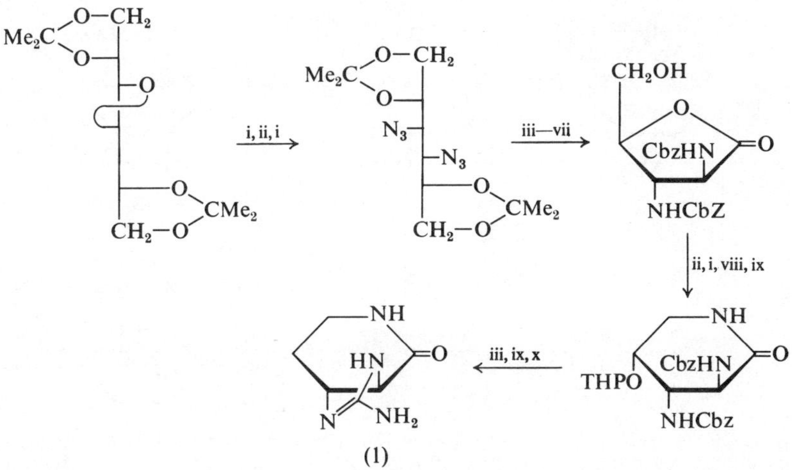

(1)

Reagents: i, N_3^-; ii, MsCl–py; iii, H_2–Pd; iv, BnO_2CCl; v, H_2O–HOAc; vi, $NaIO_4$(1 mol); vii, Br_2; viii, RaneyNi–H_2; ix, DHP–H^+; x, CNBr; xi, H_3O^+

Scheme 1

The configuration of the mould metabolite canadensolide (4) has been established as 2*S*,3*R*,4*R* by the sequence shown in Scheme 4, the intermediate (5) being radily prepared from D-glucose.[4]

[1] M. Kinoshita and Y. Suzuki, *Bull. Chem. Soc. Japan*, 1977, **50**, 2375.
[2] P.-E. Sum and L. Weiler, *Canad. J. Chem.*, 1978, **56**, 2700.
[3] T. Ogawa, N. Takasaka, and M. Matsui, *Carbohydrate Res.*, 1978, **60**, C4.
[4] R. C. Anderson and B. Fraser-Reid, *Tetrahedron Letters*, 1978, 3233.

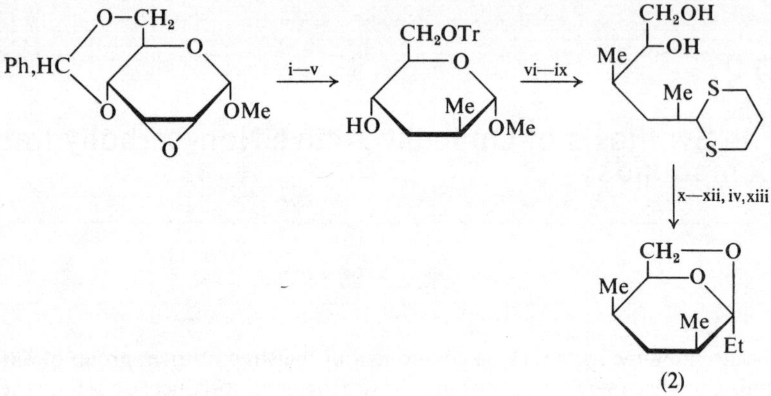

(2)

Reagents: i, Me_2CuLi; ii, NaH–CS_2–MeI; iii, Bu_3SnH; iv, H^+; v, TrCl; vi, CrO_3–py; vii, $Ph_3P{=}CH_2$; viii, H_2; ix, $HS(CH_2)_3SH$–BF_3; x, $Me_2C(OMe)_2$–H^+; xi, BuLi; xii, EtI; xiii, Hg^{2+}

Scheme 2

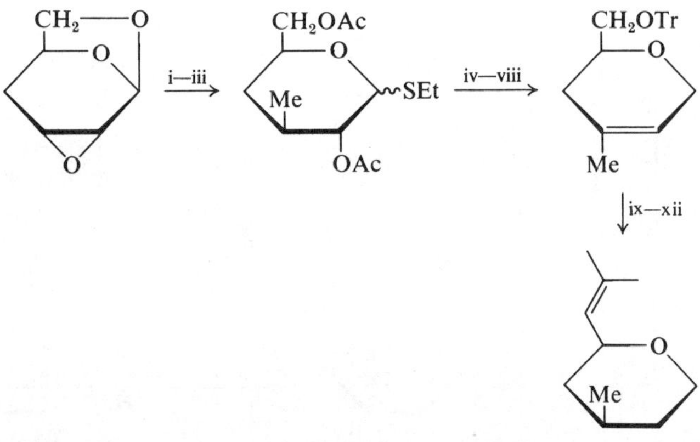

(3)

Reagents: i, Me_2CuLi; ii, Ac_2O–BF_3; iii, Bu_3SnSEt–$SnCl_4$; iv, MeO^-; v, Ni; vi, TrCl–py; vii, MsCl–py; viii, NaI–HMPT; ix, H_2–Pd; x, *aci*-Resin; xi, $C_5H_5NHCrO_3Cl$; xii, $Ph_3P{=}CMe_2$

Scheme 3

D-Xylose has been used as the reference molecule for a stereospecific synthesis of tetrahydrocerulenin (6), outlined in Scheme 5; the enantiomer of (6) was also synthesized from the same D-xylose derivative (7) by a related sequence involving the isomeric 2,3-anhydro-D-lyxofuranose intermediate (8), and established that natural (+)-cerulenin has the 2*R*,3*S* configuration shown in (6) and not the enantiomeric configuration previously claimed.[5]

[5] J.-R. Pougny and P. Sinay, *Tetrahedron Letters*, 1978, 3301.

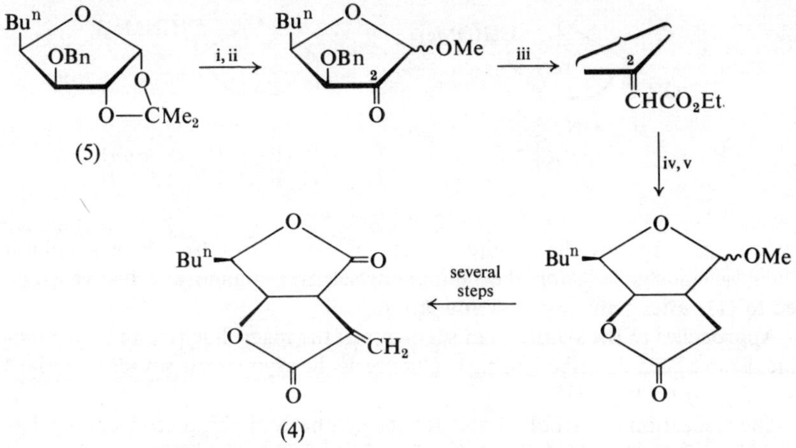

Reagents: i, MeOH–HCl; ii, $C_5H_5NHCrO_3Cl$; iii, $Ph_3P{=}CHCO_2Et$; iv, H_2–Ni; v, H_2–Pd

Scheme 4

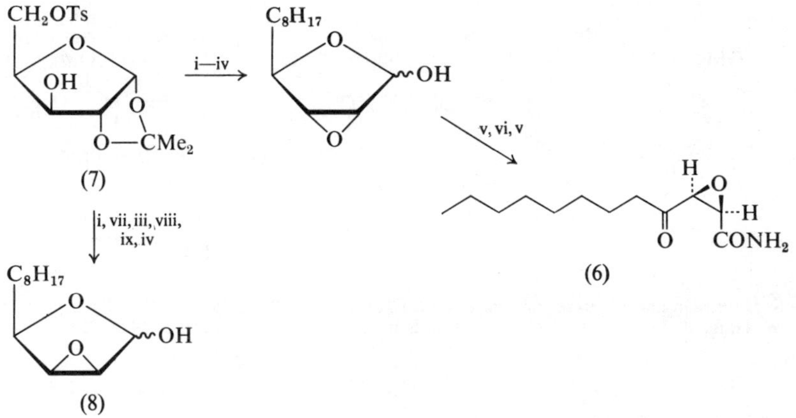

Reagents: i, $C_7H_{15}MgBr$–Li_2CuCl_4; ii, MsCl–Et_3N; iii, CF_3CO_2H–H_2O; iv, NaOMe; v, CrO_3–py; vi, NH_4OH; vii, BzCl–py; viii, MsCl–Pr^i_2NH; ix, Ag_2O–$AgClO_4$

Scheme 5

The 2,5-anhydro-L-idose 3,6-disulphonate ester derivative (9), prepared from D-glucose in good yield, had been used to synthesize both D-epiallomuscarine (10)[6] and D-isoepiallomuscarine (11),[7] both sequences involved the 3,4-anhydro-talose derivative formed from (9) which with lithium aluminium

[6] P.-C. Wang, Z. Lysenko, and M. M. Jouillié, *Tetrahedron Letters*, 1978, 1657.
[7] P.-C. Wang, Z. Lysenko, and M. M. Jouillié, *Heterocycles*, 1978, **9**, 753.

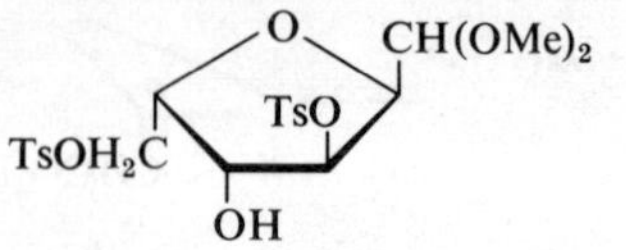

(9)

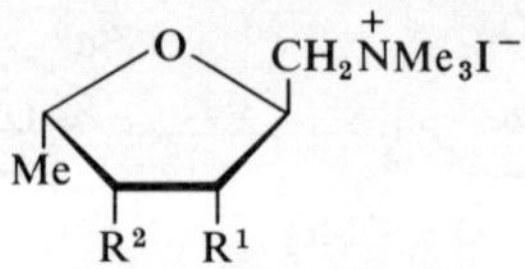

(10) R^1 = H, R^2 = OH
(11) R^1 = OH, R^2 = H

hydride gave mainly the 3,6-dideoxy sugar leading to (10), whereas sodium phenylselenide gave mainly the 4,6-diphenylselenyl L-mannose derivative which led to (11) after reduction and amination.

Approaches to the syntheses of segments of the macrolide ring in erythronolide A have been described, using D-glucose as the asymmetric source material [8] (see also Vol. 11, p. 231).

The utilization of carbohydrates for the synthesis of chiral crown ethers has been further explored, giving derivatives with the general structures (12) and (13) using 4,-6-*O*-benzylidene derivatives of methyl α-D-galacto-, gluco-, and manno-pyranoside in a base-catalysed reaction with diethylene glycol bis-toluene-*p*-sulphonate.[9]

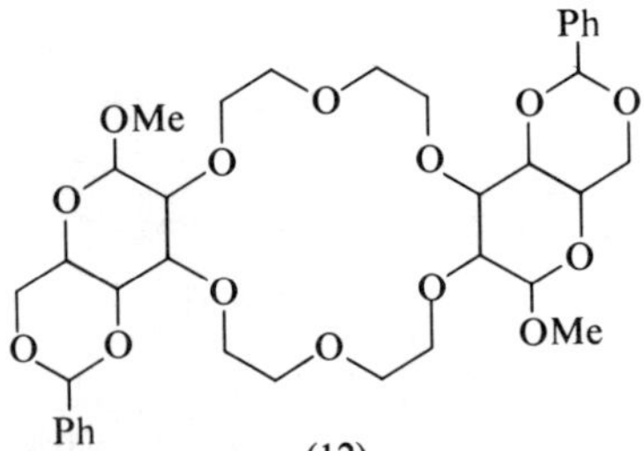

(12)

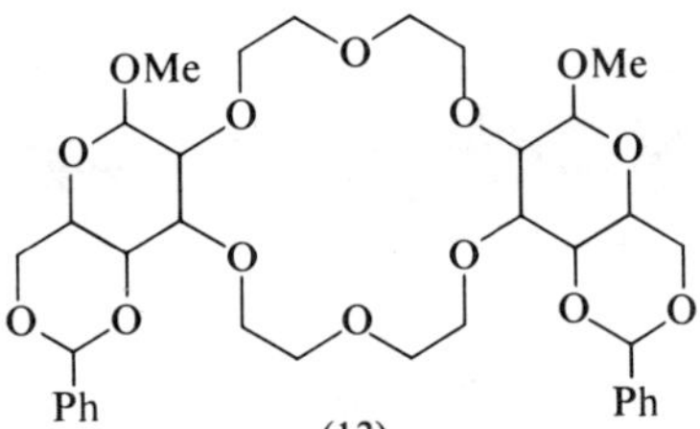

(13)

[8] S. Hanessian and G. Rancourt, *Pure Appl. Chem.*, 1977, **49**, 1201.
[9] W. Hain, R. Lehnert, H. Röllele, and G. Schröder, *Tetrahedron Letters*, 1978, 625,

Part II

MACROMOLECULES

By

J. F. Kennedy
I. M. Morrison
C. M. Sturgeon
R. J. Sturgeon

1
Introduction

BY J. F. KENNEDY

This Part II Report has now been produced annually for twelve years, and whereas the original objectives stated in Volume 1, Part II have been retained, further objectives have from time to time been added. With the recent changes in Senior Reportership, it is timely to state for the benefit of both readers and prospective readers the principles which underlie the current selection of literature for inclusion in the Report.

1 Scope and Coverage of the Report

The Report is concerned with the chemistry, structure–biological relationships, and biochemical aspects of macromolecular carbohydrates including polysaccharides, the carbohydrate moieties of glycoproteins and proteoglycans, and enzymes which cleave carbohydrate linkages. References to carbohydrate synthases and transferases are included in the relevant subsections dealing with biosynthesis. Although they contain carbohydrate, nucleic acids are not the subject of this Report; they are covered in the parallel Specialist Periodical Reports Series entitled 'Macromolecular Chemistry' and 'Organophosphorus Chemistry'.

Some two hundred journals, publishers' book lists, etc. are regularly drawn upon for the compilation of the Report. The continued increase in production costs has made it necessary to restrict the extent of information given for each citation, but it is intended at all times that the coverage of the literature shall be as comprehensive as possible. Suggestions from readers will always be welcomed.

The chapter on General Methods in Carbohydrate Chemistry (Chapter 2) now summarizes new methods of component analysis and of structural determination. The following chapters on Plant and Algal Polysaccharides (Chapter 3) and Microbial Polysaccharides (Chapter 4) deal largely with the application of classical and new techniques to their isolation and structural elucidation. The chapter on Glycoproteins, Glycopeptides, Proteoglycans, and Animal Polysaccharides (Chapter 5) (including lectins) not only emphasizes the carbohydrate chemistry but also draws on references which describe the involvement of such molecules in the processes of life, maintenance of health, and disease. Enzymes, the subject of the following chapter (Chapter 6), which are relevant to carbohydrate cleavage are being discovered at an almost alarming rate, and thus the Report seeks to bring references to them under a classification based on the Enzyme Commission (E.C.) system, but suitably extended to incorporate newly encountered substrate specificities. Mutant forms of the enzymes are also given attention. Both enzymes, and Glycolipids and Gangliosides (Chapter 7), are also given attention from the viewpoint of the ways in which they can be assayed conveniently for the purpose of diagnosis of a wide range of clinical conditions. The final chapter, on Chemical Synthesis and Modification of Oligosaccharides, Polysaccharides, Enzymes, and Glycolipids (Chapter 8), covers the (attempted) synthesis of polysaccharides, glycoproteins, and enzymes, but the greater bulk of literature cited falls under the heading of derivatization of these macromolecules. Such chemical derivatizations are discussed specifically from the viewpoint of their final application – a wide scope which enhances such aspects as gel formation, graft co-polymerization, enzyme immobilization, immunoadsorption, affinity chromatography, *etc.*

2 Organization, Nomenclature, and Use of the Report

To facilitate identification by the reader of references in particular areas, it is worthwhile at this juncture to collate various principles of organization of, and nomenclature used in, this Report.

Each chapter has its own set of references, each starting from unity. Readers may check for any further information on a reference which may be cited in more than one chapter by use of the author index. Would readers please note that the listing of references given in Part I is completely separate from the listing in Part II (and *vice versa*).

As in earlier volumes, citations involving ^{13}C n.m.r. spectroscopy are located in the chapters dealing with the principal subjects of the citations, and therefore not necessarily in Chapter 2 (General Methods).

The naming of microbial organisms has continued to present problems, particularly in view of the re-classification and renaming which are being advocated on taxonomic grounds. In this Report, the author's designation of the organism has generally been used, normally no account being taken of strains, but readers are asked to look for references under all possible names for an organism. In the case of plants, where possible, the common trivial name has been given in addition to the Latin. However, it should be borne in mind that often more than one common name is used for the same species.

It is now apparent that many enzyme molecules contain carbohydrate and are therefore also glycoproteins. For the sake of convenience, reports dealing with such are included under enzymes (Chapter 6) rather than under glycoproteins (Chapter 5). Concerning the nomenclature of the nine-carbon sugar units which are common features of glycoproteins and gangliosides, in this work, the use of the systematic term 5-amino-3,5-dideoxy-α-D-*glycero*-D-*galacto*-2-nonulopyranosonic acid for neuraminic acid is equivalent to the use of 2-amino-2-deoxy-D-glucose for glucosamine. Thus where possible any additional substituent is specified. The corresponding term 'sialic acid' is not used in this work since it is essentially synonymous, but the adjective 'desialylized' is employed as an acceptable, and brief, way of describing a glycoprotein, *etc.* from which a 5-amino-3,5-dideoxy-α-D-*glycero*-D-*galacto*-2-nonulopyranosylonic acid units(s) or derivative thereof has been removed.

The assignment of an enzyme to a particular section in the Chapter is largely based on the information given in the original paper, but it must be recognized that, *e.g.* various galactosidases are also active against rhamnosides. References to glycoside hydrolases active against a unique glycoside-type rather than a polysaccharide-type substrate are included in the relevant general section: *e.g.* galactosidases active only against glycolipids are classified under galactosidases. In spite of the Enzyme Commission's distinction between β-D-2-acetamido-2-deoxyhexosidases and β-D-2-acetamido-2-deoxy-glucosidases and -galactosidases, all references to such enzymes are included in the one section since it is not always possible to determine from the information available whether the author's assignment is correct.

The handbook of the Enzyme Commission, 'Enzyme Nomenclature', updated in 1972 [1, 2] and supplemented by information based on the literature up to 1975 [3] has continued to be used for enzyme naming. However, with continued discoveries, it is inevitable that not all enzymes reported have an E.C. number. The problem of enzyme classification has continued since more is now known about the substrate specificities and selectivities of carbohydrate cleaving enzymes. In this work, further unifications of the subsection headings of Chapter 6 have therefore been necessary. The sections on glycoside hydrolases continue to deal with enzymes with broad specificities and which react with the (substituted-)phenyl glycosides in the familiar fashion. Enzymes which are specific for polysaccharides are dealt with in the section on polysaccharide hydrolases and lyases, with classification where possible according to their *exo*- and *endo*-action pattern. A particular problem in this respect lies in there being several enzymes which are active against a homoglycan, *e.g.* with glucanases. Where it has not been possible to make an E.C. assignment the reference has been placed in the relevant miscellaneous section. The term 'amylase' has been used when the type of amylase was not specified, or when the combined α- and β-amylase activity was measured. However, in many cases, the problem of identifying an enzyme could be eliminated by the author's quotation of the relevant E.C. number.

The nomenclature of β-D-fructofuranosidases and sucrose-hydrolysing enzymes now demands special attention. The attention of readers is drawn to the classification of β-D-fructofuranosidase, sucrose α-D-glucohydrolase, and *exo*-β-D-fructosidase as separate enzymes.

Whereas a sub-section has been set aside to deal with relevant aspects of glycoproteinaceous glycosyl transferases *per se*, where the transferase activity is related to a known glycoside hydrolase activity both activities are discussed under the heading of the latter. Further aspects of glycosyl transferases are dealt with in Chapter 5 under the headings of biosyntheses.

When making use of the chapter on enzymes, readers must take into account the fact that, whereas the Enzyme Commission numbering system suggests that there is only one enzyme per number, it is becoming increasingly realized that there are many variants of each enzyme. Such a situation will only be fully clarified when full structural and substrate specificity data are available.

1 J. F. Kennedy, in 'Carbohydrate Chemistry', A Specialist Periodical Report, ed. J. S. Brimacombe, The Chemical Society, London, 1975, Vol. 7, p. 201.
2 J. F. Kennedy, in ref. 1, p. 356.
3 J. F. Kennedy, in 'Carbohydrate Chemistry', A Specialist Periodical Report, ed. J. S. Brimacombe, The Chemical Society, London, 1978, Vol. 10, p. 355.

A full index of all the enzymes reported in Chapter 6 is appended to the Chapter itself.

In view of the progress in the field of immobilized or insolublized enzymes, definition of the terms employed is required. The term 'immobilized' applies to enzymes *etc.* which have been reacted with another molecule to give a product of greater molecular size, *i.e.* the mobility of the enzyme, *etc.* in solution has been reduced. However, such derivatives may be water soluble or insoluble. By international agreement the use of the term 'immobilized' is recommended rather than use of the original term 'insolubilized' which does not encompass the water-soluble macromolecular derivatives.

Tables have again been made use of in Chapter 8 (Chemical Synthesis and Modification of Oligosaccharides, Polysaccharides, Glycoproteins, Enzymes, Gangliosides, and Glycolipids) to facilitate easy reference to both enzymes and carbohydrate-containing macromolecules derivatized by insolubilization and the use of a cyclic imidocarbonate derivative of macroproous agarose as a matrix for insolubilization. These Tables follow the format adopted previously. In this Chapter the use of the prefix 'tri' for the substitution of glycans has not been used since it is very frequently applied by authors to situations where the degree of substitution is less than three, *e.g.* 'cellulose triacetate (d.s. 2.3.)'. The prefix 'per' is only applied in the proper sense of the term. Since agarose in macroporous form is marketed by a number of companies, it is referred to in Chapter 8 as 'macroporous agarose'; the name 'Sepharose®' is not used in reference to modified agarose gels. The name 'Sephadex®', however, has been retained since it refers to a particular type of dextran derivative.

The popularity of use of cyanogen bromide-activated agarose for the immobilization of enzymes, affinants, *etc.* has continued but further work on the mechanism of activity of the matrix has suggested that the derivative is an isourea rather than the heretofore assumed cyclic imidocarbonate. Since other references have claimed to have proved the existence of the cyclic imidocarbonate ring, the use of the term 'agarose cyclic imidocarbonate' has been retained in this volume but readers should familiarize themselves with the underlying chemistry.[4, 5]

3 Significant Advances in Macromolecular Carbohydrate Chemistry

Recent noteworthy advances and trends in macromolecular carbohydrate chemistry include the following:

Numerous publications have been devoted to studies on the synthesis of α-D-glucans by strains of *Streptococcus* species and the relationship of these organisms to dental caries.

The fact that 'hydroxy-groups at positions' C-2, C-3, C-4, or C-6 in the D-glucose moiety of α-D-glucopyranosyl phosphate are not essential for its substrate specificity in the transglycosylation reaction with α-(1 → 4)-D-glucan phosphorylase in plant polysaccharides was probably an important observation. Recent discoveries on the distribution of side chains in plant polysaccharides

[4] J. F. Kennedy, in ref. 1, p. 511.
[5] J. F. Kennedy, in ref. 3, p. 433.

include the presence of D-glucuronic acid residues linked to contiguous D-xylose residues in larchwood hemicellulose and the occurrence of D-galactose residues distributed in pairs or triplicate in guar gum. Studies on *Ascophyllum nodosum* and *Fucus vesiculosus* indicate that the structural components of the brown algae may be very complex with much intermolecular bonding.

Studies on the cell surface glycoproteins of numerous viruses have been reported and the methods of glycosylation and transport of some viral proteins have been established. Investigations into cell surface glycoproteins continue to be popular. Two new techniques have been developed for the specific labelling of carbohydrate moieties of membrane-bound glycoproteins. In one method D-galactosyl residues are introduced into the glycoproteins by use of a D-galactosyl transferase, and in the other method, mild periodate oxidation followed by sodium borotritide reduction is used to label neuraminic acid residues.

The recent work on the cellulase enzymes of white-rot fungi also deserves mention.

Advances in glycolipid chemistry include the isolation of new types of gangliosides from human erythrocyte membranes and the elucidation of information on their blood-group activities. The importance of polyisoprenol-sugar derivatives continues to attract much effort and the enzyme work on rat liver and yeast membranes are worth consideration. Another advance is the discovery that neuraminidases which are also able to hydrolyse the appropriate bond in the ganglioside G_{M1} do exist. This bond was thought to be resistant to enzymic attack due to steric hindrance.

Considerable effort continues to be devoted to the development of new methods of enzyme immobilization and of novel affinity chromatography media. Immunoadsorption has been applied to the isolation of numerous biologically important molecules. Oligosaccharides co-polymerized with acrylamide have been exploited with much success for the isolation of many lectins from different sources. Active immobilized cells have been prepared and their potential applications for industrial processes and in medicine have been demonstrated.

Recent volumes of *Carbohydrate Research* have been issued in honour of various persons with international reputations in carbohydrate chemistry and their achievements, *viz.* Professor Dexter French and his work on starch and glycogen,[6] Dr. Allene R. Jeanes[7] and her work on microbial polysaccharides particularly dextrin and xanthan, and Dr. Elizabeth E. Percival[8] and her work on various algal polysaccharides.

4 Conclusions and Readership

Undoubtedly Part II continues to serve as a rapid route to the primary literature in the field, and the Series continues in conjunction with relevant reviews to provide complete surveys of the literature on carbohydrate-type macromolecules. Whereas it is never the intention of the Report to overlap

6 J. J. Marshall and W. L. Whelan, *Carbohydrate Res.*, 1978, **61**, 1.
7 P. A. Sandford, *Carbohydrate Res.*, 1978, **66**, 3.
8 H. Weigel, *Carbohydrate Res.*, 1978, **66**, 7.

unnecessarily with the more biologically orientated publications it is intended to be of regular use not only to the chemist, but also to the biochemist, medical biochemist, and clinical chemist. Such persons are therefore encouraged to refer to the Report annually for a comprehensive listing of references on the Chemical foundations to their work.

2
General Methods

BY R. J. STURGEON

1 Analysis

A technique for the controlled thermal degradation of polysaccharides by Curie-point pyrolysis with analysis of the pyrolysis products by field ionization m.s. has been developed.[1] Integrated ion recording by photographic detection gives reproducible fingerprints in a method which appears to be a useful tool for the characterization and identification of microgram quantities of polysaccharides.

The concentrations of dextrans eluted from gel chromatography columns have been estimated using an interferometric method.[2] Negligible errors due to increments of molecular weight and optical density were reported, although corrections must be made for absorbance and non-linear response factors. Neutral polysaccharides, alone or in the presence of DNA, have been determined by means of alternating current polarography, using dextran as a reference compound and measuring the capacitance peak at -1650 mV.[3]

Pre-natal diagnosis of mucopolysaccharidoses has been attempted by measurement of the levels of glycosaminoglycans in blood, urine, and amniotic fluid.[4] The polysaccharides form insoluble complexes with quinacrine.

Conjugates of alcohol dehydrogenase and lactate dehydrogenase, when bound covalently to concanavalin A, have been reported to retain their abilities to bind with glycoproteins and to catalyse oxidation–reduction reactions with NAD in the presence of appropriate substrates.[5] This dual capacity renders these protein conjugates useful reagents for the location of glycoprotein bands on polyacrylamide gels: precipitation of formazan dye occurs as a result of re-oxidation of the NADH produced by the dehydrogenase activity.

A method for screening urine for abnormalities in the excretion of glycoconjugates has been described.[6] Glycoproteins, glycolipids, glycosaminoglycans, and oligosaccharides isolated from the urine are hydrolysed to the monosaccharide components which are then determined by g.l.c. as their TMS-methyl glycoside derivatives. The results have been applied to a mentally handicapped population and to patients with some other diseases.[7] A general procedure, involving a combination of cation- and anion-exchange chromatography and charcoal–Celite column chromatography, has been

[1] H. R. Schulten and W. Gortz, *Analyt. Chem.*, 1978, **50**, 428.
[2] L. Hagel, *Analyt. Chem.*, 1978, **50**, 569.
[3] B. Malfoy and J. A. Reynaud, *Analyt. Biochem.*, 1978, **84**, 1.
[4] S. K. Mitra and K. Blau, *Clin. Chim. Acta*, 1978, **89**, 127.
[5] G. Avigad, *Analyt. Biochem.*, 1978, **86**, 443.
[6] P. Purkiss, R. C. Hughes, and R. W. E. Watts, *Clinical Chem.*, 1978, **24**, 669.
[7] P. Purkiss, R. C. Hughes, and R. W. E. Watts, *Clinical Chem.*, 1978, **24**, 714.

used in the isolation of glycopeptides and acidic oligosaccharides from normal and pathological urine.[8]

Neutral oligosaccharides, excreted in the urine of patients with mannosidosis, have been separated by an improved t.l.c. method.[9] Mono- and disaccharides, isolated by t.l.c. from urine after treatment with mixed-bed resins, have been identified after separation by t.l.c.[10] A convenient, on-line removal of inorganic salts from mixtures of oligosaccharides in an automated h.p.l.c. system has been described.[11] Analytical grade cation-exchange resins have been used in the separation of mixtures of cellodextrins and other oligosaccharides by h.p.l.c.[12] Gel-filtration chromatography has been used to separate series of homologous malto- and isomalto-oligosaccharides and their corresponding alditols.[13] Quantitative chromatography of the reduced oligosaccharides permits the analysis of commercial hydrogenated 'glucose' syrups.

A diphenylamine–aniline–phosphoric acid reagent has been used in the differentiation of (1 → 2)-linked reducing disaccharides from other disaccharides after their separation by paper chromatography or t.l.c.[14] The addition of an h.p.l.c. amine modifier to impregnate silica columns *in situ* has been shown to give acceptable separations without the use of expensive chemically bonded amine column packing materials.[15] Good separations of fifteen monosaccharide peracetates and twelve disaccharide peracetates have been achieved using h.p.l.c.[16] For detection either a refractometer or a u.v. photometer was used, depending on the eluant. A reagent containing vanadium(V) ions in sulphuric acid has been used in the quantitative analysis of sucrose, D-glucose, and D-fructose.[17] The yellow vanadium(V) ion is reduced to the blue vanadium(IV) ion and either the consumption of vanadium(V) ion or the production of vanadium(IV) ion can be measured.

A non-corrosive aqueous reagent containing 2,2′-bis-cinchoninate has been adapted for the automated analysis of monosaccharides following chromatography of their borate complexes on ion-exchange systems.[18] Owing to the high baseline stability of the detection system achievable with this reagent, the analyser could be operated at a high sensitivity. The colour formation is linear between 0.1 and 30 μg of sugar. D-Glucose, D-fructose, D-mannose, and their respective alcohols have been separated by ion-exchange chromatography by eluting with borate buffers.[19] The ratio of amounts of sodium tetraborate to boric acid and the presence of acetic acid in the reagent system had an important influence on the order of elution of the sugars. Direct

[8] G. Strecker and A. Lemaire-Poitau, *J. Chromatog.*, 1977, **143**, 553.
[9] R. B. Friedman, M. A. Williams, H. W. Moser, and E. H. Kolodvy, *Clinical Chem.*, 1978, **24**, 1576.
[10] W. Prinz, W. Meldrum, and L. Wilkinson, *Clin. Chim. Acta.*, 1978, **82**, 229.
[11] L. E. Fitt, *J. Chromatog.*, 1978, **152**, 243.
[12] M. R. Ladisch, A. L. Huebner, and G. T. Tsao, *J. Chromatog.*, 1978, **147**, 185.
[13] F. Schmidt and B. S. Enevoldsen, *Carbohydrate Res.*, 1978, **61**, 197.
[14] T. Toba and S. Adachi, *J. Chromatog.*, 1978, **1534**, 110.
[15] K. Aitzetmüller, *J. Chromatog.*, 1978, **156**, 354.
[16] J. Thiem, J. Schwenther, H. Karl, A. Sievers, and J. Reimer, *J. Chromatog.*, 1978, **155**, 107.
[17] K. M. Haldorsen, *J. Chromatog.*, 1978, **150**, 485.
[18] M. Sinner and J. Puls, *J. Chromatog.*, 1978, **156**, 197.
[19] L. A. T. Verhaar and J. M. H. Dirkx, *Carbohydrate Res.*, 1977, **59**, 1.

detection of sugars, sugar alcohols, and sugar acids, following ion-exchange chromatography, has been achieved using u.v. spectrometry.[20]

Improved chromatographic separations of sugars on anion-exchange resins have been achieved using a resin-conditioning procedure involving treatment with ethanol.[21] Monosaccharides in plasma and whole blood have been separated by t.l.c. on silica gel and estimated by direct densitometry of the sugars after location with 4-aminobenzoic acid.[22] The separation of D-fructose and D-glucose on X-type zeolites with aqueous methanol as eluant has been reported.[23] The adsorption capacity and the selectivity of adsorption of the resins for these and other monosaccharides, disaccharides, and related alditols are dependent on the cationic form of the zeolite, the extent of cation exchange, and the nature of the solvent.

The determination of monosaccharides commonly occurring in cerebrospinal fluid has been reported using a liquid chromatography sugar analyser.[24] High sensitivity (0.05–0.23 μg per compound) and good resolution permitted the routine detection of carbohydrates in biological fluids without prior extraction and concentration. Reducing sugars and polyols have been estimated simultaneously in an autoanalyser by anion-exchange chromatography using borate buffers.[25] Improved separations of acetylated hexoses on silica gel using h.p.l.c. have been reported.[26]

A comparison of the use of Aminex-A15 anion-exchange resin and diethylaminoethyl derivatives of glycol methacrylate macroporous gel for the separation of sugars by h.p.l.c. has been made.[27] The latter support was used as a packing for automated sugar analysis, when a decrease in distribution constants without a decrease in efficiency was achieved. H.p.l.c. data for unsubstituted monosaccharides, unsubstituted glycosides, partially methylated and completely acetylated monosaccharides have been recorded.[28]

A rapid one-step method for the formation of TMS ethers of sugars and sugar phosphates in the presence of DMSO and cyclohexanone has been reported.[29] The concentration of TMS derivatives of sugars has been achieved by partition into hexamethyldisiloxane, prior to g.l.c.[30] Relative g.l.c. retention times and relative intensities of fragments from mass spectra have been recorded for the TMS derivatives of a number of mono- and di-saccharides.[31] Using standardized conditions, the procedure was applied successfully to the determination of D-glucose in blood and the ratio of monosaccharides in pectic acid and gum arabic. Resolution of mono- and oligo-saccharides by g.l.c. has been achieved in short analysis times using high initial temperatures,

[20] L. A. T. Verhaar and J. M. H. Dirkx, *Carbohydrate Res.*, 1978, **62**, 197.
[21] K. Mopper, *Analyt. Biochem.*, 1978, **87**, 162.
[22] I. S. Menzies, J. N. Mount, and M. J. Wheeler, *Ann. Clinical Biochem.*, 1978, **15**, 65.
[23] Th. M. Wortel and H. van Bekkum, *Rec. Trav. chim.*, 1978, **97**, 156.
[24] R. Seuffer, W. Voelter, and H. Bauer, *J. Clinical Chem. Clinical Biochem.*, 1977, **15**, 663.
[25] M. H. Simatupang, M. Sinner, and H. H. Dietrichs, *J. Chromatog.*, 1978, **155**, 446.
[26] J. Thiem, H. Karl, J. Schwentner, and J. Reimer, *J. Chromatog.*, 1978, **147**, 491.
[27] Z. Chytilová, O. Mikeš, J. Farkaš, P. Štrop, and P. Vrátný, *J. Chromatog.*, 1978, **153**, 37.
[28] G. D. McGinnis and P. Fang, *J. Chromatog.*, 1978, **153**, 107.
[29] D. J. Leblanc and A. J. S. Ball, *Analyt. Biochem.*, 1978, **84**, 574.
[30] M. Martinez, D. Nurok, and A. Zlatkis, *Analyt. Chem.*, 1978, **50**, 1226.
[31] K. Jankowski and D. Gaudin, *Biomed. Mass Spectrometry*, 1978, **5**, 371.

thermally stable stationary phases, and a multi-step time programme.[32] TMS-Oxime derivatives of a variety of monosaccharides have been separated by g.l.c. and detected with a selective and sensitive alkaline earth metal salt flame ionization detector.[33] Monosaccharides and monosaccharide derivatives from seminal plasma have been identified by g.l.c.–m.s. of their TMS-oxime derivatives.[34]

The reproducibility of recoveries of monosaccharides, amino-sugars, and amino-acids obtained from acid hydrolyses of glycoproteins has been improved by addition of glycerol prior to concentration by freeze-drying or rotary evaporation.[35]

Improvements in methods for the analysis of the carbohydrate components of glycoproteins have been the subject of numerous reports. Neutral- and amino-sugars have been separated on an autoanalyser using anion- and cation-exchange columns respectively.[36] Problems encountered in the use of D-mannitol and other alditols as internal standards in the estimation of the TMS-derivatives of methyl glycosides by g.l.c. have been evaluated.[37] After methanolysis better recoveries of the methyl glycosides were obtained by using an anion-exchange resin instead of silver carbonate for neutralization of the acid. Methanolysis followed by trifluoroacetylation to generate methyl *N,O*-trifluoroacetyl glycosides has been advocated for the analysis of the carbohydrate portion of glycoproteins.[38–40] By use of capillary columns and an electron-capture detector monosaccharides in the picomole range can be detected. The monosaccharide compositions of some glycoproteins have been determined by g.l.c. of the TMS-ether methyl glycoside derivatives.[41] No tailing on the chromatograms was observed when trimethylsilylimidazole was used in the silylation step. The stability of reducing sugars towards trifluoroacetolysis has been studied.[42] Under appropriate conditions 2-acetamido-2-deoxyhexose residues from the reducing ends of oligosaccharides may be eliminated. The monosaccharide composition of acidic glycosaminoglycans derived from animal tissues has been determined by g.l.c. of the TMS ethers of the methyl glycosides.[43] *N*-Acetylneuraminic acid, released from glycoproteins with acid or neuraminidase, has been estimated by monitoring the protonated molecular ions of the fully silylated neuraminic acid in g.l.c.–m.s. using isobutane ionization.[44]

A rapid analytical system based on anion-exchange chromatography has been developed for the determination of amino-sugars; the method is without interference from neutral sugars or amino-acids and has a sensitivity in

32 G. A. Janauer and P. Englmaier, *J. Chromatog.*, 1978, **153**, 539.
33 H. Morita and W. G. Montgomery, *J. Chromatog.*, 1978, **155**, 195.
34 P. Størset, O. Stokke, and E. Jellum, *J. Chromatog.*, 1978, **145**, 351.
35 R. Dawson and K. Mopper, *Analyt. Biochem.*, 1978, **84**, 186.
36 D. Bladier, G. Perret, J. Pre, and P. Cornillot, *J. Chromatog.*, 1978, **154**, 330.
37 S. J. Rickert and C. C. Sweeley, *J. Chromatog.*, 1978, **147**, 317.
38 M. M. Wrann and C. W. Todd, *J. Chromatog.*, 1978, **147**, 309.
39 D. G. Pritchard and W. Niedermeier, *J. Chromatog.*, 1978, **152**, 487.
40 M. Tomana, W. Niedermeier, and C. Spivey, *Analyt. Biochem.*, 1978, **89**, 110.
41 A. Cahour and L. Hartmann, *J. Chromatog.*, 1978, **152**, 475.
42 B. Nilsson and S. Svensson, *Carbohydrate Res.*, 1978, **65**, 169.
43 M. Yokota and T. Mori, *Carbohydrate Res.*, 1977, **59**, 289.
44 J. Roboz, R. Suzuki, and J. G. Bekesi, *Analyt. Biochem.*, 1978, **87**, 195.

the nanomole range.[45] Improvements in the separation, and an increase in the sensitivity of detection of 2-amino-2-deoxy-D-glucose, 2-amino-2-deoxy-D-galactose, and their corresponding aminohexitols, have been reported.[46] Amino-acids and amino-sugars have been separated on columns of silylated controlled-pore glass by ligand exchange chromatography employing Cu(II) ions.[47] The disadvantages of this system are the relatively short lifetimes of the columns and the leaching of the copper from the stationary phase. Hexitols and aminohexitols, derived from the monosaccharide components of glycoproteins by reduction with sodium borotritide, have been assayed *via* separation using high-voltage electrophoresis in borate buffer.[48]

A hexokinase–sodium borotritide procedure for the determination of neutral sugars arising from the acid hydrolysis of glycoproteins has been developed.[49] The hexitols D-glucitol, D-mannitol, and D-galactitol are notoriously difficult to separate by paper chromatographic systems. However, treatment of the respective hexoses with ATP in the presence of hexokinase converts D-glucose and D-mannose (but not D-galactose) to the corresponding hexose 6-phosphates. After reduction of the resultant sugar mixture with sodium borotritide the two phosphorylated hexitols are readily separated from D-galactitol.

Uronic acids have been separated on anion-exchange resins in a simple system involving no elevated column temperatures or changes in buffer composition.[50] A highly sensitive non-corrosive detection reagent, 2,2′-bis-cinchoninate, which did not react with amino-acids or neutral monosaccharides, was employed.

Two of the standard enzymic methods for the determination of D-glucose (the D-glucose oxidase and the D-glucose dehydrogenase procedures) have been compared for routine estimation of that sugar in blood.[51, 52] A continuous-flow method for the determination of D-glucose in blood using the hexokinase–D-glucose 6-phosphate dehydrogenase method has been reported.[53] D-Glucose oxidase and peroxidase have been separately immobilized by reaction with diazotized polyaminostyrene beads, for the measurement of D-glucose in an autoanalyser.[54] A computerized, feedback-control system for the dynamic control of D-glucose levels in the blood of diabetics has incorporated an on-line D-glucose analyser which measures electrochemically the hydrogen peroxide generated from D-glucose and D-glucose oxidase.[55] D-Glucose, in serum, has been reacted using columns containing D-glucose

[45] R. Dawson and K. Mopper, *Analyt. Biochem.*, 1978, **84**, 191.
[46] P. W. Chen and T. F. Boat, *Analyt. Biochem.*, 1978, **85**, 276.
[47] R. G. Masters and D. E. Leyden, *Analyt. Chim. Acta*, 1978, **98**, 9.
[48] M. Tomana, W. Niedermeier, C. Spivey, and J. Gerard, *Microchem. J.*, 1978, **23**, 93.
[49] Y. Kawai and R. G. Spiro, *J. Biol. Chem.*, 1977, **252**, 6236.
[50] K. Mopper, *Analyt. Biochem.*, 1978, **86**, 597.
[51] W. Stein, I. Mildner, and R. Maulbetsch, *J. Clinical Chem. Clinical Biochem.*, 1978, **16**, 225.
[52] G. Heinemann, *J. Clinical Chem. Clinical Biochem.*, 1978, **16**, 235.
[53] B. Lloyd, J. Burrin, P. Smythe, and K. G. M. M. Alberti, *Clinical Chem.*, 1978, **24**, 1724.
[54] J. N. Miller, B. F. Rocks, and D. T. Burns, *Analyt. Clin. Acta*, 1977, **93**, 353.
[55] E. J. Fogt, L. M. Dodd, E. M. Jenning, and A. H. Clemens, *Clinical Chem.*, 1978, **24**, 1366.

oxidase immobilized on controlled pore glass.[56] The heat produced in this primary reaction was measured directly on the column without any need for coupling reactions and allowed the determination of concentrations of 10–450 μM of D-glucose. Coupling of a simple differentiating circuit with an oxygen-electrode assembly gives a rapid and inexpensive system for the estimation of D-glucose in body fluids.[57] D-Glucose has been determined in a closed flow-through system employing D-glucose oxidase, in which oxygen depletion is monitored by a three-electrode amperometric system.[58] A method for the estimation of D-glucose, using a hydrogen peroxide analyser-coupled electrode, has been evaluated using three commercially available analysers.[59] The pre-deproteinization of blood has been shown to be unnecessary in the determination of D-galactose by D-galactose dehydrogenase in a centrifugal analyser.[60]

Xylitol and D-glucitol have been measured using D-glucitol dehydrogenase.[61] Mixtures of polyhydroxy compounds have been separated as their tungstate complexes.[62] Polyol profiles in cerebrospinal fluid have been measured by g.l.c. of the TMS ethers.[63]

Acetic acid, liberated from the *O*- and *N*-acetyl groups of bacterial lipopolysaccharides, has been determined by g.l.c. analysis.[64]

The absorbance of the red quinonamine dye formed in the reaction of phenol and 4-amino-1,5-dimethyl-2-phenyl-3-pyrazolone (4-aminoantipyrine) with periodate ions has been used in the estimation of periodate consumption in the oxidation of starch and other carbohydrates.[65] The evaluation of the use of pyridine as a solvent in the oxidation of water-insoluble carbohydrates by periodate has been reported.[66] The three types of dialdehydes produced by this oxidation were converted to the TMS ethers of the dithioacetals and measured by g.l.c.[67] Oxidation proceeded in a normal Malapradian fashion but more slowly than it does in water.

The multiple peak patterns obtained in the g.l.c. of the trimethylsilylated (−)-2-butyl glycosides[68] or acetylated (+)-2-octyl glycosides[69] of D- and L-monosaccharides have been used for the assignment of the absolute configuration of the sugars.

Glycosyltransferase activities have been measured after termination of the reaction with phenol which results in the partition of protein into the phenol

[56] B. Danielsson, K. Gadd, B. Mattiasson, and K. Mosbach, *Clin. Chim. Acta,* 1977, **81**, 163.
[57] H. L. J. Makin, P. J. Warren, and J. D. Edridge, *Clin. Chim. Acta*, 1978, **84**, 137.
[58] C. M. Wolff, and H. A. Mottola, *Analyt. Chem.*, 1978, **50**, 94.
[59] W. W. Spencer, D. Sylvester, and G. H. Nelson, *Clinical Chem.*, 1978, **24**, 386.
[60] M. Gabrielli, *Clinical Chem.*, 1978, **24**, 1990.
[61] K. H. Bassler, K. Wagner, and B. Schonerstedt, *J. Clinical Chem. Clinical Biochem.*, 1978, **16**, 547.
[62] H. J. F. Angus, J. Briggs, N. A. Sufi, and H. Weigel, *Carbohydrate Res.*, 1978, **66**, 25.
[63] S. L. Smith, M. Novotny, and E. L. Weber, *Clinical Chem.*, 1978, **24**, 545.
[64] I. Fromme and H. Beilharz, *Analyt. Biochem.*, 1978, **84**, 347.
[65] H. Gallati, *Analyt. Chim. Acta*, 1978, **96**, 211.
[66] S. Honda, Y. Ohkaru, and K. Kakehi, *Analyt. Chim. Acta*, 1978, **98**, 85.
[67] S. Honda, Y. Fukuhara, and K. Kakehi, *Analyt. Chem.*, 1978, **50**, 55.
[68] G. J. Gerwig, J. P. Kamerling, and J. F. G. Vliegenthart, *Carbohydrate Res.*, 1978, **62**, 349.
[69] K. Leontein, B. Lindberg, and J. Lönngren, *Carbohydrate Res.*, 1978, **62**, 359.

phase and of carbohydrate into the aqueous phase.[70] Substrates and products were then assayed by g.l.c. of a suitable derivative. The method is limited since only oligosaccharides up to Dp = 4 can be resolved by the procedure.

2 Structural Methods

Aspects of chemical methods used in the structural elucidation of polysaccharides and complex carbohydrates have been reviewed.[81] In a critical examination of the use of g.l.c.–m.s. in the identification of TMS ethers of monosaccharides, a standardized method, which uses a medium resolution mass spectrometer and short chromatographic columns, has been proposed.[72] TMS Ethers of monosaccharides have been characterized by g.l.c.–chemical ionization m.s. with ammonia as reagent gas.[73] Molecular weights were determined, and fragment ions were produced in a quantity high enough to differentiate between stereoisomers (epimers and anomers). Disaccharides have been determined by permethylation followed by g.l.c.[74] The method has been used in the detection of carbohydrate intolerance secondary to intestinal disaccharidase deficiency.

Per-*O*-acetyl aldononitriles and methylated aldononitrile acetates have been separated by g.l.c. on a mixed liquid phase.[75] Chemical ionization m.s. showed the presence of peaks at M + 1 in each instance. The retention times of a number of partially methylated alditol acetates have been recorded on a variety of stationary phases.[76] Using a combination of silica gel chromatography and g.l.c., fifteen methyl ethers of methyl α-D-galactopyranoside have been isolated from the partial methylation products of D-galactose.[77]

Comparison of the mass spectra of the TMS derivatives of hexuronic acids and their lactones with those of other monosaccharides has revealed specific effects of the carboxy-group on the fragmentation patterns.[78] A detailed analysis was presented of the fragmentations of the TMS derivatives of β-D-glucopyranuronic acid, α-D-galactofuranuronic acid, and β-D-glucofuranurono-6,3-lactone.

Neuraminic acids have been reduced with sodium borohydride and the m.s. fragmentation patterns of the corresponding peracetylated methyl esters of 5-acylamino-3,5-dideoxy-nononic and -heptonic acids have been recorded.[79] A set of partially *O*-methylated derivatives of methyl *N*-acetyl-*N*-methyl-β-D-neuraminate methyl glycoside have been prepared as reference compounds

70 B. Darbyshire and R. J. Henry, *Analyt. Biochem.*, 1978, **90**, 234.
71 H. Yamaguchi, *Seikagaku*, 1977, **49**, 1316.
72 K. Jankowski and D. Gaudin, *Biomed. Mass Spectrometry*, 1978, **5**, 371.
73 T. Murata and S. Takahashi, *Carbohydrate Res.*, 1978, **62**, 1.
74 R. M. Thompson, R. D. Coffin, M. R. Glick, and J. F. Fitzgerald, *Clin. Chim. Acta*, 1978, **84**, 185.
75 B. W. Li, T. W. Cochran, and J. R. Vercellotti, *Carbohydrate Res.*, 1977, **59**, 567.
76 H. Parolis and D. McGarvie, *Carbohydrate Res.*, 1978, **62**, 363.
77 B. Fournet, J. M. Dhalluin, Y. Leroy, J. Montreuil, and H. Mayer, *J. Chromatog.*, 1978, **153**, 91.
78 J. F. Kennedy and S. M. Robertson, *Carbohydrate Res.*, 1978, **67**, 1.
79 G. P. Smirnova, N. V. Chekareva, O. S. Chizhov, B. M. Zolotarev, and N. K. Kochetkov, *Carbohydrate Res.*, 1977, **59**, 235.

for the methylation analysis of acyl neuraminic acids.[80] The g.l.c.–m.s. patterns for the *O*-TMS and *O*-acetyl derivatives of these compounds were described in detail.

Using the same technique as for glycolipids, a mass spectrometric sequence study of the oligosaccharides of human transferrin has been reported.[81] The method should be useful for microscale fingerprinting of mixtures of glycopeptides released from glycoproteins after proteolysis. Amino-acid sequence studies on the methylated glycopeptide derived from the antifreeze glycoproteins of an antarctic fish have shown that the attachment of the oligosaccharides to the peptide backbone is by *O*-glycosyl linkages to threonine residues.[82] No mass spectral peaks corresponding to L-threonine-bound carbohydrate were observed for the methylated glycopeptides, indicating that the elimination of oligosaccharide chains took place by base-catalysed elimination at the methylation stage. Tri-*O*-methyl derivatives of D-*glycero*-L-*manno*-heptose have been synthesized.[83]

The ^{13}C n.m.r. spectra of all the methyl ethers of methyl (methyl α-D-glucopyranosid)uronates have been interpreted.[84] ^{1}H N.m.r. spectroscopy has shown that molybdate complexes of aldoses of the series of sugars homomorphous to lyxose (D-lyxose, D-mannose, D-*glycero*-L-*manno*-heptose, D-*glycero*-D-*gulo*-heptose) exist in a skew conformation, having a 1a,2e,2q arrangement of hydroxy groups.[85] Aldoses of the ribose series (L-ribose, L-talose, D-*glycero*-D-*talo*-heptose) complexed with molybdate display substantially different spectra, shown in considerable shifts and broadening of proton signals, compared with the spectra of the corresponding aqueous solutions.

Aqueous solutions of maltose and cyclohexa-amylose have been shown to exhibit inter-residue ^{12}C–O–C–^{1}H couplings of close to 3 Hz.[86] In terms of torsional angles, ϕ and ψ, these findings suggest that in aqueous solution the molecules favour conformations that are appreciably more staggered than those known to exist in the solid state. Analogous measurements on *O*-acetyl derivatives suggest that ϕ is smaller and ψ larger than in maltose. High resolution ^{1}H- and ^{13}C-n.m.r. spectra of D-glucopyranose and 2-acetamido-2-deoxy-D-glucopyranose and their methyl glycosides have been interpreted and the changes in the chemical shift have been correlated with substitution patterns.[87]

An assessment has been made of ^{17}O-n.m.r. spectroscopy as an analytical tool in carbohydrate chemistry.[88] In terms of resolution of signals of ^{17}O-

[80] H. van Halbeek, J. Haverkamp, J. P. Kamerling, J. F. G. Vliegenthart, C. Versluis, and R. Schauer, *Carbohydrate Res.*, 1978, **60**, 51.

[81] K. A. Karlsson, I. Pascher, B. E. Samuelson, J. Finne, T. Krusius, and H. Rauvala, *F.E.B.S. Letters*, 1978, **94**, 413.

[82] H. R. Morris, M. R. Thompson, D. T. Osuga, A. I. Ahmed, S. M. Chaix, J. R. Vandenheede, and R. E. Feeney, *J. Biol. Chem.*, 1978, **253**, 5155.

[83] P. Szabó, *Carbohydrate Res.*, 1977, **59**, 580.

[84] A. S. Shashkov, A. F. Sviridov, O. S. Chizhov, and P. Kovač, *Carbohydrate Res.*, 1978, **62**, 11.

[85] J. Alfoldi, L. Petrus, and V. Bílik, *Coll. Czech. Chem. Comm.*, 1978, **43**, 1476.

[86] A. Parfondry, N. Cyr, and A. S. Perlin, *Carbohydrate Res.*, 1977, **59**, 299.

[87] S. J. Perkins, L. N. Johnson, D. C. Phillips, and R. A. Dwek, *Carbohydrate Res.*, 1977, **59**, 19.

[88] P. A. J. Gorin and M. Mazurek, *Carbohydrate Res.*, 1978, **67**, 479.

n.m.r. spectra of monosaccharides, the method suffers in comparison with those using ^{1}H, ^{13}C, and ^{2}H nuclei in that compounds containing at least 2 atom percent of ^{17}O are required.

The type of branching in complex asparagine-linked glycan chains of glycoproteins has been studied by 360 MHz ^{1}H n.m.r. spectroscopy.[89] The bi-, tri-, and tetra-antennary structures of asialo- and afuco-glycan chains were distinguished on the basis of chemical shifts of protons of the branched D-manno-oligosaccharide core and of the 2-acetamido protons of the 2-acetamido-2-deoxy-D-glucose residues.

A new method for the *N*-deacetylation of 2-acetamido-2-deoxy sugars is based on transamidation with a mixture of trifluoroacetic acid and trifluoroacetic anhydride, followed by removal of the resulting *O*- and *N*-trifluoroacetyl groups.[90] *N*-Deacetylation followed by nitrous acid deamination of per-*O*-methylated amino-oligosaccharides has afforded products which may be converted into derivatives suitable for characterization by g.l.c.–m.s.[91] The 2,5-anhydrohexoses produced have been converted into alditol trideuteriomethyl ethers, alditol acetates, and aldononitriles. The importance of side reactions leading to the formation of 2-deoxy-2-*C*-formylpentofuranosides was discussed. A method has been reported for the determination of the position of the linkage of the 2-acetamido-2-deoxy-D-glucosyl and 2-acetamido-2-deoxy-D-galactosyl residues in oligosaccharides and glycoproteins.[92] Sequential periodate oxidation and Smith degradation of the borohydride-reduced terminal residues were followed by identification of the fragments (*e.g.* 2-acetamido-2-deoxy-D-glycerol, 2-acetamido-2-deoxy-L-threitol, 2-acetamido-2-deoxy-L-arabinitol and 2-acetamido-2-deoxy-D-xylitol) by g.l.c. of their TMS- or TFA- derivatives.

Polysaccharides containing side chains of unsubstituted or carbohydrate-substituted uronic acid residues have been modified by elimination of the uronic acid residues using a sequence of acetylation, treatment with triethylamine (base-catalysed elimination) and deacetylation.[93] Selective cleavage of glycopyranosiduronamide linkages in methylated polysaccharides has produced pentodialdose derivatives to which any sugar residue external to a uronic acid residue remains attached, together with degraded methylated polysaccharide carrying aglyconic hydroxy-groups.[94]

Alkaline degradation of branched polysaccharides under controlled conditions has been used to eliminate the backbone of the molecule and convert branches linked through positions 6 of the backbone into oligosaccharides terminated with saccharinic acid residues.[95]

Information on the initial second-order rate coefficients of periodate-ion

[89] L. Dorland, J. Haverkamp, J. F. G. Vliegenthart, B. Fournet, G. Strecker, G. Spik, J. Montreuil, K. Schmid, and J. F. Binette, *F.E.B.S. Letters*, 1978, **89**, 149.
[90] B. Nilsson and S. Svensson, *Carbohydrate Res.*, 1978, **62**, 377.
[91] G. O. Aspinall, E. Przybylski, R. G. S. Ritchie, and C. O. Wong, *Carbohydrate Res.*, 1978, **66**, 225.
[92] M. A. E. Shaban, V. N. Reinhold, and R. W. Jeanloz, *Carbohydrate Res.*, 1977, **59**, 213.
[93] B. Lindberg, F. Lindh, and J. Lönngren, *Carbohydrate Res.*, 1978, **61**, 81.
[94] G. O. Aspinall and K. G. Rosell, *Canad. J. Chem.*, 1978, **56**, 685.
[95] R. A. Young and K. V. Sarkanen, *Carbohydrate Res.*, 1977, **59**, 193.

oxidation of a number of model compounds and polysaccharides has resulted in the postulation of selective Smith degradation of such molecules.[96] A rapid Smith degradation procedure has been developed for the production of glycerol from non-reducing terminal residues of (1 → 4)-α-D-glucans.[97] Since over oxidation does not interfere with the production of glycerol, higher temperatures may be used with advantage in the oxidation step. Products of the oxidation of glycosides with periodate have been converted into 2,4-dinitrophenylhydrazones and separated by h.p.l.c.[98] The possible use of this method for the determination of interglycosidic linkages in oligo- and polysaccharides is discussed.

Four commercial agarose type gels have been used at enhanced flow rates for the molecular weight distribution analysis of dextran.[99] The performances of the gels were discussed in terms of resolution, molecular weight correction factors, and variations in plate height as a function of flow rates.

Agarose columns which had been calibrated with commercially available dextran fractions have been used for the measurement of distributions of molecular radii of chondroitin sulphate and keratan sulphate.[100] The values obtained agreed with those obtained for these glycosaminoglycans by other, indirect methods. Problems encountered in estimating the molecular weights of glycoproteins by gel filtration have been highlighted.[101] For sialylated glycoproteins of relatively low molecular weight, such as fetuin and α_1-acid glycoprotein, gel filtration gives apparent molecular weights much larger than the known values. The corresponding asialoglycoproteins have apparent molecular weights lower than expected from the sole removal of sialic acid.

A bibliography of crystal structures of polysaccharides has been presented.[102] The conformations of some (1 → 6)-α-D-glucans have been plotted in terms of isoenergy contours covering the range of 0–12 kcal mol^{-1} of relative energies.[103] Theoretical isoenergy maps were in agreement with experimental data from crystal structures of oligosaccharides having the same linkage.

96 K. M. Aalmo, M. F. Ishak, and T. J. Painter, *Carbohydrate Res.*, 1978, **63**, C3.
97 S. Hizukuri and S. Osaki, *Carbohydrate Res.*, 1978, **63**, 261.
98 S. Honda and K. Kakehi, *J. Chromatog.*, 1978, **152**, 405.
99 L. Hagel, *J. Chromatog.*, 1978, **160**, 59.
100 R. H. Pearce and B. J. Grimmer, *J. Chromatog.*, 1978, **150**, 548.
101 J. A. Alhadeff, *Biochem. J.*, 1978, **173**, 315.
102 R. H. Marchessault and P. R. Sundararajan, in 'Advances in Carbohydrate Chemistry and Biochemistry', Vol. 33, ed. R. S. Tipson and D. Horton, Academic Press, New York, 1976, p. 387.
103 I. Tvaroska, S. Pérez, and R. Marchessault, *Carbohydrate Res.*, 1978, **61**, 97.

3
Plant and Algal Polysaccharides

BY I. M. MORRISON

1 Starch

The polysaccharide chains of crystalline B-amylose exist as double stranded helices which have a right-handed six-fold helical conformation of 20.8 Å.[1] The helices are anti-parallel and have an open channel in the hexagonal array which is filled with water molecules. Two helices (*i.e.* twelve D-gluose residues) exist in each unit cell which has the dimensions $a = b = 18.50$ Å and c, the fibre repeat distance, 10.40 Å with the angle γ being 120°. Similar studies on the crystalline polymorph A-amylose have shown that it also exists as a right-handed double helix in which the chains are parallel.[2] The helices pack in an antiparallel fashion into an orthorhombic unit cell ($a = 11.90$ Å, $b = 17.70$ Å, and $c = 10.52$ Å). The water content of A-amylose was only about 25% of the value for B-amylose.

Linear homodisperse amylose, branched polymers (formed by linking maltopentaose to amylose with hydrazone) and cross-linked products (formed by crosslinking amylose chains with epichlorhydrin) have been characterized by absorption and circular dichroism measurements of their iodine complexes.[3] The long-wave Cotton effect was split into a broad negative band and a positive band of smaller width for the amylose–iodine complex. The dichroic absorption had a maximum value for amylose molecules with chain lengths of 50 residues and the results are consistent with the formation of a stiff helical rod at these chain lengths. When amylose branches were linked densely to glycogen, stronger Cotton effects were observed indicating that an ordered arrangement existed. When amylose molecules were cross-linked with epichlorhydrin to give soluble gels, the segments between the cross-linkages formed imperfect helices although they did interact with iodine. The structural changes which occurred in the pre-gel phase when amylose chains were cross-linked using epichlorhydrin were measured by chemical analysis, viscosity, and light scattering measurements and the optical properties of the iodine complexes.[4] In the early stages of reaction, the glycerol diether groups were mainly present as intermolecular cross-links. With further reaction a decrease in particle density occurred possibly due to the association of amylose segments by hydrogen bonding in the weakly crosslinked products and the solubilizing effect of monoether substituents introduced at a later stage. After gelation occurred, when the density of cross-links can be estimated by equilibrium swelling in

[1] H.-C. H. Wu and A. Sarko, *Carbohydrate Res.*, 1978, **61**, 7.
[2] H.-C. H. Wu and A. Sarko, *Carbohydrate Res.*, 1978, **61**, 27.
[3] B. Pfannemüller, *Carbohydrate Res.*, 1978, **61**, 41.
[4] B. Pfannemüller and M. Sperlich, *Makromol. Chem.*, 1978, **179**, 1465.

water, the size of the segments between the cross-links decreased exponentially with the concentration of epichlorhydrin.[5] No unsubstituted pendant branches were present as shown by the low susceptibility to β-amylase. The resistance to α-amylase increased with epichlorhydrin concentration. The weakly cross-linked gel could still be stained with iodine suggesting that longer amylose segments were present in the gel than in the sol forms.

To study the light scattering and specific refractive increment behaviour of amyloses in methyl sulphoxide–water mixtures, it was found necessary to use dialysed solutions from which the 'micro-gel' had been removed.[6] Discrepancies in the published molecular weight–intrinsic viscosity relationships for amylose may be due to the preparation and treatment of the sample and possibly to the botanical source of the amylose. The maltodextrins maltotriose to maltohexaose and possibly even higher oligosaccharides were taken up in the gel phase of starch granules in excess of the amount expected on the basis of their molecular size.[7] The maximum adsorption of oligosaccharide occurred with maltotetraose. It was suggested that the interbranch region of branched molecules and segments of linear molecules are of sufficient length to form double helices with the linear oligosaccharides. That the phenomenon was unique to the maltodextrin series was confirmed by showing that other oligosaccharides were excluded from the gel phase.

No noteworthy differences were detected between the percentages of amylose and between the distribution of its unit chain lengths before and after isoamylase treatment of maize starches of the single mutants, *ae*, *wx*, and o_2, and the double mutant combinations *ae* o_2, and *wx* o_2.[8] Starch granules from the *ae* mutants have intermediate fractions which differed in their unit chain length from those of typical amylose and amylopectin.

The separation of malto-oligosaccharides by high performance liquid chromatography on silica gel has been reported.[9] Complete resolution could only be obtained with three oligomers but eleven oligomers could be identified. Similar oligosaccharides, up to DP 20, could be separated by multiple descending paper chromatography.[10] The branched saccharide of DP n moved between the linear oligosaccharides of DP n and $n + 1$.

D-Glucans from conventional corn syrups could not be measured by reaction with concanavalin A.[11] In a rapid technique developed for the measurement of total non-structural carbohydrates in plant tissue the starch is enzymically hydrolysed and the total reducing sugar is determined colorimetrically.[12] Repetitive determinations of amylase activity in serum (*i.e.* combined α-, β-, and glucoamylase activities) can be made by measuring the D-glucose released from starch with immobilized D-glucose oxidase in a flow-through system.[13]

[5] B. Pfannemüller and M. Sperlich, *Makromol. Chem.*, 1978, **179**, 1485.
[6] F. R. Dintzis and R. Tobin, *Carbohydrate Res.*, 1978, **66**, 71.
[7] S. A. Brown and D. French, *Carbohydrate Res.*, 1977, **59**, 203.
[8] Y. Ikawa, D. V. Glover, Y. Sugimoto, and H. Fuwa, *Carbohydrate Res.*, 1978, **61**, 211.
[9] D. Nurok and A. Zlatkis, *Carbohydrate Res.*, 1978, **65**, 265.
[10] K. Umeki and K. Kainuma, *J. Chromatog.*, 1978, **150**, 242.
[11] L. W. Doner, *J. Agric. Food Chem.*, 1978, **26**, 707.
[12] A. J. Da Silveira, F. F. Feitosa Teles, and J. W. Stull, *J. Agric. Food Chem.*, 1978, **26**, 770.
[13] D. P. Nikolelis and H. A. Mottola, *Analyt. Chem.*, 1978, **50**, 1665,

The textural (stickiness and consistency on cooking) and physiochemical properties of rice could not be explained on the basis of the total amylose content but these properties correlated well with the insoluble amylose content.[14] The protein content had no effect on these properties. A cross-linked amylose has been used as a substrate to determine the α- and β-amylase activities of amylolytic preparations.[15]

The molecular structures of unmodified starch and a 2-hydroxypropyl distarch phosphate derivative from tapioca have been determined by enzymic methods.[16] The modified starch was prepared by cross-linking the starch with phosphate groups and substituting each chain with a single 2-hydroxypropyl group. The enzymes used in this study were pullulanase, β-amylase, and glucoamylase either alone or in various combinations. The amylopectin in the unmodified starch had one chain of DP 45 to every 7.5 chains of DP 15. Debranching of the amylopectin, in the modified starch with pullulanase, released only about half of the chains of DP 15. β-Amylolysis of the debranched modified amylopectin to the β-limit dextran was incomplete although all of the debranched chains of DP 15 were converted to maltose and maltotriose. The overall results suggested that about half of the chains of DP 15 in the modified amylopectin contain no modifying groups. A model depicting the possible sites of 2-hydroxypropyl and/or phosphate groups in the modified amylopectin molecule was presented. The combined action of barley α- and β-amylases on barley starch granules was more extensive than that of the individual enzymes.[17] Even greater degradation was achieved using the total enzymic extract, indicating participation by other enzymes. The action of malted barley α-amylase on amylopectin yielded 6^3-α-D-glucosylmaltotriose (1) as the smallest oligosaccharide product but no panose.[18] The mode of action of this α-amylase was therefore similar to that of malted rye, human saliva, and porcine pancreas α-amylases.

α-D-Glc*p*
1
↓
6
α-D-Glc*p*-(1 → 4)-α-D-Glc*p*-(1 → 4)-D-Glc*p*

(1)

If mercuric chloride is used in the steeping procedure for the preparation of barley starch granules, a sulphydryl reagent must be used in the later stages to prevent inactivation of the α-amylase in barley malt.[19] The proteinaceous α-amylase inhibitor of beans (*Phaseolus vulgaris*) has been purified and is a glycoprotein (molecular weight 4.3×10^4) containing 15% carbohydrate.[20] It inhibits salivary and pancreatic α-amylases but has no effect on plant, fungal, or bacterial α-amylases nor on sweet potato β-amylase. In the breakdown of

[14] K. R. Bhattacharya, C. M. Sowbhagya, and Y. M. I. Swamy, *J. Sci. Food Agric.*, 1978, **29**, 359.
[15] M. A. Mateescu, *Biochimie*, 1978, **60**, 535.
[16] L. F. Hood and C. Mercier, *Carbohydrate Res.*, 1978, **61**, 53.
[17] I. Maeda, S. Kiribuchi, and M. Nakamura, *Agric. and Biol. Chem.* (*Japan*), 1978, **42**, 259.
[18] R. S. Hall and D. J. Manners, *Carbohydrate Res.*, 1978, **66**, 295.
[19] H. Taniguchi, I. Maeda, and M. Nakamura, *Agric. and Biol. Chem.* (*Japan*), 1978, **42**, 661.
[20] K.-H. Pick and G. Wöber, *Z. physiol. Chem.*, 1978, **359**, 1371.

starch in pea cotyledons, the enzyme activity levels are controlled by the amount of soluble sugar present in the cotyledon.[21] Amylolysis is delayed by gibberelic acid probably due to the effect of the latter in controlling the cotyledon metabolism. Starch breakdown in the spadix of *Arum maculatum* was not caused by α-amylase but the hydrolysis pattern produced on breakdown was consistent with the action of a thermo-labile *endo*-amylase.[22]

Hydroxy-groups at positions C-2, C-3, C-4, or C-6 in the D-glucose portion of α-D-glucopyranosyl phosphate are not essential for its substrate specificity in the transglycosylation reaction with α-(1 → 4)-D-glucan phosphorylase.[23] The affinity of plant α-(1 → 4)-D-glucan phosphorylase and the rate of hexose incorporation into α-(1 → 4)-D-glucan decreases in the following sequence: α-D-glucopyranosyl phosphate > 2-deoxy- > 6-deoxy > 4-deoxy- > 3-deoxy-D-glucopyranosyl phosphates. The deoxyglucosyl analogues of α-D-glucopyranosyl phosphate act as competitive inhibitors in the elongation reaction of the α-(1 → 4) chains of starch. It was found that more than one residue of 2-deoxy-D-glucose or 6-deoxy-D-glucose could be incorporated into the non-reducing terminus of α-(1 → 4)-D-glucan chains of starch.

Two major forms of the branching enzyme α-(1 → 4)-D-glucan : α-(1 → 4)-D-glucan 6-glucosyl transferase were detected in the developing kernels of *Zea mays*.[24] The first (A) accounted for 24% of the total branching activity based on phosphoryl stimulation but accounted for 74% of the total activity based on the branching of amylose as assayed by the amylose-I_2 method. This enzyme was not associated with starch synthase activity. The second enzyme (B) was not easy to purify being co-eluted from chromatographic colums with enzymes possessing both primed and unprimed starch synthase activities. Both enzymes could induce the introduction of further branches into amylopectin. Although both enzymes were monomeric in nature (molecular weight 7.0—9.0 × 10^4), enzyme B could be separated into two forms, I and II, one of which was more active than the other in putting additional branches into amylopectins. When sugary and amylose-extender (two non-linked mutant) varieties of maize were compared for the above starch-branching enzymes, enzyme BII was not detected in extracts from amylose-extender kernels while the branching enzymes A and BI appeared unaltered in that mutant.[25] It was also observed that unprimed starch synthase activity was low in fractions isolated by DEAE-cellulose chromatography from extracts of amylose-extender maize but higher activity levels could be regenerated by addition of any of the branching enzymes to the systems. The branching enzyme A from sugary kernels had a different pattern of activity from the non-mutant branching enzyme A when amylose was used as substrate.

The synthesis of α-(1 → 4) bonds in polysaccharides from ADP-D-glucose by α-(1 → 4)-D-glucan synthases prepared from sweet corn was dependent on added carbohydrate primer in solutions of low ionic strength.[26] However in

[21] A. Garcia-Luis and J. L. Guardiola, *Ann. Botany*, 1978, **178**, 337.
[22] P. V. Bulpin and T. Ap Rees, *Phytochemistry*, 1978, **17**, 391.
[23] J. Kemek, S. Kucar, and J. Zamocky, *European J. Biochem.*, 1978, **89**, 291.
[24] C. D. Boyer and J. Preiss, *Carbohydrate Res.*, 1978, **61**, 321.
[25] C. D. Boyer and J. Preiss, *Biochem. Biophys. Res. Comm.*, 1978, **80**, 169.
[26] S. Schiefer, E. C. Y. Lee, and W. J. Whelan, *Carbohydrate Res.*, 1978, **61**, 239.

citrate buffers of high ionic strength, synthesis was apparently independent of added primer. Treatment of the enzyme preparation with α-amylase and glucoamylase led to the abolition of the unprimed synthesis. This amylase-treated preparation then showed a normal dependence on α-(1 → 4)-D-glucan primer, branched primers being the most effective. The polysaccharide product of the unprimed reaction was glycogen-like having an average chain length of 14. These studies suggest that the phenomenon of unprimed synthesis in solutions of high ionic strength can be explained as an enhanced affinity of the enzyme for traces of primer in the enzyme preparation and not caused by a *de novo* synthesis of polysaccharide that occurs in the absence of a primer.

2 Fructans

The D-fructan (average DP = 51) which was isolated, along with pectic acid and a D-galactan, from garlic bulbs had an inulin-type structure, according to the results of methylation and Smith degradation analysis and the action of β-D-fructofuranosidase.[27] Perennial ryegrass contained a mixture of water-soluble polysaccharides from which two fructans were isolated.[28] The predominant one, as confirmed by ^{13}C n.m.r. spectroscopy, contained only levan-type, [β-(2 → 6)], linkages with a terminal D-glucosyl residue linked (1 → 2) to the reducing end of the D-fructan chain.A preliminary report on fructan metabolism in onions has been published.[29]

3 Cellulose

A historical review on some resolved and unresolved questions in wood chemistry, particularly related to cellulose and lignin, has been published.[30] A review on the basic trends in the present-day study of cellulose (and other polysaccharides) has been published.[31] The chemical modification of cellulose has been reviewed from a historical perspective.[32]

The crystal structure of native ramie cellulose was shown to be similar to that recently reported for *Valonia* cellulose.[33] The earlier conclusion that the ramie diffraction data could not be satisfied by a conventional cellulose model is refuted. R″ (an indication of diffraction error) values of 0.158, 0.185, and 0.175 were obtained for anti-parallel, parallel up, and parallel down alignments respectively. The author considers that the anti-parallel model provides the best accounting for the ramie data and is therefore probably the correct model for both cotton and ramie cellulose. The changes in crystallinity and physical characteristics of microcrystalline cellulose which occur on grinding have been studied.[34] Under one set of grinding conditions, the specific surface area rose

[27] N. N. Das and A. Das, *Carbohydrate Res.*, 1978, **64**, 155.
[28] J. Tomasic, H. J. Jennings, and C. P. J. Glaudemans, *Carbohydrate Res.*, 1978, **62**, 127.
[29] R. J. Henry and B. Derbyshire, *Proc. Austral. Biochem. Soc.*, 1978, **11**, 38.
[30] K. Freudenberg, *Cellulose Chem. Technol.*, 1978, **12**, 3.
[31] Z. A. Rogovin, *Cellulose Chem. Technol.*, 1978, **12**, 123.
[32] D. F. Durso, *Modif. Cellul.* (Symp. Cellul. Pap. Text. Div. Amer. Chem. Soc.), 1978, 23.
[33] A. D. French, *Carbohydrate Res.*, 1978, **61**, 67.
[34] Y. Nakai, E. Fukuoka, S. Nakajima, and K. Yamamoto, *Chem. and Pharm. Bull. (Japan)*, 1977, **25**, 2490.

threefold but no significant change in the pore size was observed. The recrystallization of amorphous cellulose was observed to be caused by solvent displacements to the same form as cellulose II. A lack of agreement between the results of the use of X-ray diffraction data and I_2-adsorption values for the measurement of crystallinity has been observed for acid-modified rayon and cotton cellulose fibres.[35]

High resolution electron microscopy showed that α-cellulose could be split into smaller fibrillar elements by heating and that an elementary fibril with a uniform diameter did not exist within the cell wall, although certain fibril diameters were dominant.[36] In addition to the cellulose fibrils, supramolecular structures of polyoses and lignin were detected in the holocellulose as coils either attached to cellulose chains or deposited within fibre bundles.

The reactions of dichlorotriazinyl dyes with cellulose have been studied and the diffusion coefficient was shown to decrease with an increase in pH.[37] The rate constants of the reaction were pseudo first order. The pH dependence was similar to that of hydrolysis of the dye.

A water-soluble cellulose acetate has been prepared by partial hydrolysis of cellulose acetate in a homogeneous medium.[38] The hydrolysis, in dilute acetic acid, is a first-order pseudomonomolecular reaction giving a product of DP 200—250. The acetyl groups are randomly substituted on O-2 and O-3 of the soluble product and this can be further reacted with epichlorhydrin to give cross-linked products. New groups of cellulose derivatives have been produced by the interaction of cellulose with inorganic acids.[39] Phosphorylation of cellulose with either meta- or pyro-phosphoric acids in a urea melt at 140—150 °C gave cellulose phosphate (D.S. 30—33 %) with an ion-exchange capacity of 4.5—5.5 mg equiv. g^{-1}. The reaction of cellulose with metaboric acid at 220 °C gave a product which contained 1.9 % boron. Both products showed good stability towards hydrolysis and could be used as sorbents for the stabilization of blood.

Cellulose can be converted into a hemiacetal derivative, 'methylol cellulose', by reacting with formaldehyde in methyl sulphoxide. Other aprotic solvents, such as pyridine, *NN*-dimethylformamide, *NN*-dimethylacetamide, *N*-methyl-2-pyrrolidinone, and thiolane-1-oxide can be substituted for methylsulphoxide for this conversion.[40] More concentrated solutions of 'methylol cellulose' could be obtained when cellulose and paraformaldehyde were heated together in the solvent than when formaldehyde gas was bubbled through a suspension of cellulose in the solvent. Solutions of 10 % cellulose in methylsulphoxide could be obtained by this method. During the swelling of cellulose in ethanolamine, ethylenediamine, and methyl sulphoxide, the solvent retention capacity of cellulose at 20—120 °C decreased by 10—20 %, depending on the solvent.[41]

[35] H. G. Shinouda, A. Kinawi and M. M. Abdelmoteleb, *Makromol. Chem.*, 1978, **179**, 455.
[36] D. Fengel, *Holzforschung*, 1978, **32**, 37.
[37] H. Motomura and Z. Morita, *Bull. Chem. Soc. Japan*, 1978, **51**, 1332.
[38] G. Rakhmanberdiev, G. A. Petropavlovskii, and Kh. U. Usmanov, *Cellulose Chem. Technol.*, 1978, **12**, 153.
[39] I. N. Ermolenko and N. K. Luneva, *Cellulose Chem. Technol.*, 1977, **11**, 647.
[40] T. J. Baker, L. R. Schroeder, and D. C. Johnson, *Carbohydrate Res.*, 1978, **67**, C4.
[41] W. Wagenknecht, H. Schleicher, and B. Phillip, *Cellulose Chem. Technol.*, 1978, **12**, 11.

The solvent retention capacity was lowest in all temperature ranges when the cellulose was swollen with methyl sulphoxide. There was no further decrease in swelling in any of the solvents above 120 °C.

During an alkaline pulping process, one third of the wood polysaccharides were converted to acidic products by β-alkoxy elimination reactions.[42] The effect of delignification, wood and pulp composition, and possiblities for stabilization of the carbohydrates during the pulping process were considered. Amongst the acidic products formed were D-*gluco*- and D-*xylo*-isosaccharinic acids as well as 2-hydroxybutanoic, lactic, acetic, and formic acids. The commercial feasibility of recovering these acids was discussed. The results of investigations on the oxidative degradation of native cellulose supported an earlier reaction scheme, put forward for the degradation of alkali cellulose, except that the fast pre-stage of degradation, involving labile bonds in the alkali cellulose, was absent.[43] Since all the bonds in native cellulose are of the same type, they are all of equal strength. The first step is the uptake of an oxygen atom and the second step, the scission proper, involves alkoxy elimination. The third, post oxidation, step is the removal of about three D-glucose residues.

The role of the redox cycle of quinone–hydroquinone in stabilizing carbohydrates during the alkaline treatment of wood chips was studied using cellobiose as a model compound.[44] In the absence of oxygen, the formation of glyconic acid was negligible. The operation of the redox cycle which started on addition of oxygen to the system gave an increased yield of glyconic acid at the expense of formation of saccharinic and lactic acids. Oxygen alone was a less efficient oxidant than oxygen–quinone.

During the bleaching of pulps with sulphur dioxide in alkaline solutions in the presence of oxygen, depolymerization of cellulose was retarded by the addition of Mn^{II} salts,[45] though the rate of delignification was hardly affected by the Mn^{II} salts. In the presence of iron, Mn^{II} salts caused more rapid delignification and retarded the depolymerization of cellulose. When cellobiose was used as a model compound for the alkaline depolymerization of cellulose in the presence of oxygen, the addition of small amounts of $MnCl_2$ promoted the oxidation of cellobiose and of the D-glucose liberated from the cellobiose by a β-elimination reaction.[46] Large amounts of $MnCl_2$ retarded both reactions due to precipitation of MnO. Addition of $MnSO_4$ selectively increased the rate of delignification of sulphite pulp during the bleaching reaction.[47] The presence of this salt also reduced the activation energy for delignification by nearly 6 kcal mol^{-1}, reduced the extent to which cellulose underwent oxidative degradation, and increased the pulp yield by 2.1%. An analysis of the acidic groups produced and the terminal sugar residues indicated that similar endwise degradation reactions occurred during the reaction of hydrocellulose with sodium bicarbonate as occurred with sodium hydroxide.[48] Some differences in

[42] E. Sjöström, *TAPPI*, 1977, **60**, 151.
[43] A. Palma, S. Jovanovic, and G. V. Schulz, *Makromol. Chem.*, 1978, **179**, 395.
[44] K. Ruoho and E. Sjöström, *TAPPI*, 1978, **61**, 87.
[45] M. Manoucheri and O. Samuelson, *Svensk Papperstidn.*, 1977, **80**, 381.
[46] M. Manoucheri and O. Samuelson, *Cellulose Chem. Technol.*, 1978, **12**, 105.
[47] G. L. Akim and M. A. Akezhev, *Cellulose Chem. Technol.*, 1977, **11**, 703.
[48] M. H. Johansson and O. Samuelson, *J. Appl. Polymer Sci.*, 1978, **22**, 615.

the products were noted, for example fragmentation reactions were favoured in bicarbonate solutions while benzilic acid type rearrangements were less favoured in bicarbonate. The formation of 2-deoxy-*erythro*-pentonic acid as end groups was favoured in bicarbonate.

During the alkaline degradation of cellulose and cellobiose, a yellow chromophore is formed whose spectral properties are dependent on pH;[49] *e.g.* λ_{max} 290 nm at pH 12 was shifted to 256 nm at pH 6 for both hydrocellulose and cellobiose. It is suggested that the chromophore is a β-hydroxy-α-enone which, at acid pH values, tautomerizes to an aliphatic β-diketone. It was concluded that the chromophore is a minor product since the molar absorption coefficients of such compounds are very large ($>1 \times 10^4$) and the apparent molar absorption coefficient for hydrocellulose was only about 900.

The effect of the heating rate and the degree of transformation on such kineti c parameters as the activation energy and reaction order in the thermal degradation of phenyl- or pyridyl-carbamoylmethyl cellulose, indicated that the degradation reaction was very complex.[50] Amidation of carboxymethylcellulose decreased the heat resistance and favoured the breaking of macromolecular chains so that the supramolecular structure affected the heat stability of the derivative and the mechanism of thermal degradation.

Products from the pyrolysis of cellulose in the absence of a catalyst were found to be mainly derivatives of 1,6-anhydro-D-glucose and other, unsaturated, products.[51] One product, formed in 1.4% yield, was found to be 1,5-anhydro-4-deoxy-D-*glycero*-hex-1-en-3-ulose. This product was also formed during the pyrolysis of amylose, amylopectin, and laminarin. A further product from cellulose, probably formed as an oxidation product, was 3,5-dihydroxy-2-methyl-4*H*-pyran-4-one. Sixty-three compounds were detected in the condensate of smoke produced by pyrolysis of cellulose.[52] These compounds included furans, cyclic ketones, lactones, benzene derivatives, aliphatic ketones, and aldehydes. One of the major products was 2-hydroxy-3-methyl-cyclopenten-1-one.

In further studies on the crystal structures of cyclodextrin complexes, the α-cyclodextrin : 4-nitrophenol (1 : 2) complex was found to have an orthorhombic unit cell.[53, 54] The α-cyclodextrin : methylsulphoxide : methanol (1 : 1 : 2) dihydrate complex had a monoclinic unit cell. In this structure, the methylsulphoxide molecule is on the side of the secondary hydroxy-groups in the α-cyclodextrin cavity such that it forms two hydrogen bonds with adjacent α-cyclodextrin molecules. One methanol molecule is found on the side of the primary hydroxy-group in the cavity while the other is located outside the cavity. The two water molecules are distributed in the space between the α-cyclodextrin molecules.

49 M. Lewin, I. Ziderman, N. Weiss, A. Barch, and A. Ettinger, *Carbohydrate Res.*, 1978, **62**, 393.
50 C. Simionescu, I. A. Schneider, N. Hurduc and S. Dumitriu, *Cellulose Chem. Technol.*, 1977, **11**, 577.
51 F. Shafizadeh, R. H. Furneaux, T. T. Stevenson, and T. G. Cochran, *Carbohydrate Res.*, 1978, **67**, 433.
52 H. Sakuma, N. Shimojima, and S. Sugawara, *Agric. and Biol. Chem. (Japan)*, 1978, **42**, 359.
53 K. Harata, H. Uedaira, and J. Tawaka, *Bull. Chem. Soc. Japan*, 1978, **51**, 1627.
54 K. Harata, *Bull. Chem. Soc. Japan*, 1978, **51**, 1644.

The simultaneous and selective modification of all six primary hydroxy-groups of the α-cyclodextrin molecule has been studied.[55] Using well-established and conventional procedures, such derivatives as the dodecabenzoate were prepared, in which substitution was on O-2 and O-3 of each D-glucose residue. Hexa-*O*-methyl- and hexakis-6-amino-6-deoxy-α-cyclodextrin were prepared where substitution was on the primary hydroxy-groups. Direct substitution was even easier when β-cyclodextrin was used. Derivatives such as the heptakis 6-azido-6-deoxy-β-cyclodextrin tetradeca-acetate could be prepared.

Dimeric β-cyclodextrin complexes have been used as model compounds to study membrane diffusion transport through lipid bilayer membranes.[56] α-Cyclodextrin is able to interact with some optically-active benzene derivatives.[57] The ability of the compound with the benzene ring to be inserted into the central cavity of the α-cyclodextrin molecule provided the driving force for complex formation. The effect of cyclodextrins on the dissociation of some azo dyes showed that the apparent pK_a values of the dyes increased with increasing cyclodextrin concentration.[58] The effect was explained as a hydrophobic interaction between the dye and the cyclodextrin molecules.

A kinetic study of the hydrolysis of woods and purified plant polysaccharides in 75% sulphuric acid was monitored by electrical conductivity measurements.[59] The coefficients of resistance of all polysaccharide substrates increased with increasing hydrolysis time. From these measurements, the degrees of polymerization of cellulose, xylan, and inulin were estimated to be 1900, 66, and 16 respectively. The polysaccharide components of various woods and pulps can be hydrolysed with trifluoracetic acid which, in contrast to sulphuric acid hydrolysis, does not require a neutralization step since trifluoracetic acid is volatile.[60] The presence of lignin in the wood samples impeded the hydrolysis of the polysaccharides, requiring longer reaction times and correction factors to compensate for loss of monosaccharide by degradation reactions.

A method for the separation of cellodextrins, using high performance liquid chromatography on a cation exchange resin in the Ca^{2+} form, and using water as the eluant, has been published[61] and a method for the separation of macro quantities of glucose oligomers has also appeared.[62]

Cellulose, which has been pre-swollen in cadoxene solutions, can be quantitatively hydrolysed to D-glucose by the cellulase complex from *Trichoderma reesei*.[63] The cadoxene solution is necessarily removed before enzymic treatment and can be re-used. Good recoveries of D-glucose could even be obtained from lignified agricultural wastes. The cellulase complex from *Penicillium funiculosum* has been used to determine the effect of cell-wall degrading enzymes on the morphology and fine structure of cellulosic substrates. When

[55] J. Boger, R. J. Corcoran, and J. M. Lehn, *Helv. Chim. Acta*, 1978, **61**, 2190.
[56] J. J. Stezowski, K. H. Jogun, E. Eckle, and K. Bartels, *Nature*, 1978, **274**, 617.
[57] A. Cooper and D. D. MacNicol, *J. C. S. Perkin II*, 1978, 760.
[58] Y. Matsui and K. Mochida, *Bull. Chem. Soc. Japan*, 1978, **51**, 673.
[59] D. Mueller, *Holzforschung*, 1978, **32**, 96.
[60] D. Fengel, G. Wegener, A. Heizmann and M. Przyklenk, *Cellulose Chem. Technol.*, 1978, **12**, 31.
[61] M. R. Ladisch, A. L. Huebner, and G. T. Tsao, *J. Chromatog.*, 1978, **147**, 185.
[62] S. Dziedzic and M. W. Kearsley, *J. Chromatog.*, 1978, **154**, 295.
[63] M. R. Ladisch, C. M. Ladisch, and G. T. Tsao, *Science*, 1978, **201**, 743.

wheat straw pulp was used as the cellulosic substrate, the crystallinity index was increased from 59 to 75% but there was a reduction in the crystallite dimensions from 57 to 39 Å in the direction perpendicular to the (002) plane without any change in the crystallite length or orientation.[64] Light and electron microscopy revealed that the microfibrillar width was reduced from 98 to 66 Å. When bagasse pulp and sodium chlorite-bleached sawdust from *Tectona grandis* were used as the cellulosic substrates similar results were obtained.[65] Electron microscopy revealed that cellulolysis proceeded preferentially along the (101) and (002) planes of the cellulose but the sawdust cellulose was more difficult to degrade.

An extracellular enzyme which utilizes molecular oxygen to oxidise cellodextrins to the corresponding aldonic acids has been isolated from culture filtrates of the white-rot fungus *Sporotrichum pulverulentum*.[66] The purified enzyme, named cellobiose oxidase, is a glycoprotein (molecular weight 9.3×10^4). Ultraviolet spectral data indicated that the enzyme is a hemoprotein, containing one flavin component per enzyme molecule. The possible role of this complex enzyme in cellulose degradation is discussed. The regulation of *endo*-β-1,4-D-glucanases in *S. pulverulentum* has been investigated using a technique based on the viscosity-lowering effect of the *endo*-β-1,4-D-glucanases on solutions of carboxymethylcellulose.[67] The effect of inducers and repressors can determine whether the enzymes are localized on the cell wall surfaces or are released into the surrounding medium. Cellobiose causes induction even at 1 mg l^{-1}, while D-glucose at 50 mg l^{-1} causes repression of the formation of enzyme. Similar studies with the mould *Trichoderma viride* show that cellobiose is not an efficient inducer in this organism. Other differences between the two organisms were observed. The *endo*-β-1,4-D-glucanases of *T. viride* were bound to the cell wall while those from *S. pulverulentum* were released into solution having first appeared bound to the wall. It was concluded that these two fungi in their natural environment assure their nutrition in different ways.

During the growth of *Phialophora malorum* on cellulose and carboxymethylcellulose, some of the cellulases were found bound to the cell wall while others were found in the cell-free solution.[68] The presence of both D-glucose and cellobiose repressed formation of cellulases but low carboxymethylcellulase activity was detected in the D-glucose-containing growth medium. An extracellular cellulase–hemicellulase complex has been isolated from cell-free culture solutions of *Phoma hibernica*.[69] The complex was able to degrade cellulose, carboxymethylcellulose, xylan, and galacto-, gluco-, and galactoglucomannans. The pH optimum for hydrolysis of cellulose and galactomannan was 4.5 but that for hydrolysis of the xylan was 4.5—5.5. The cellulose- and galactomannan-degrading enzymes were *endo*-enzymes whereas those degrading xylans were *exo*-enzymes.

[64] S. M. Betrabet and K. M. Paralikar, *Cellulose Chem. Technol.*, 1977, **11**, 615.
[65] S. M. Betrabet and K. M. Paralikar, *Cellulose Chem. Technol.*, 1978, **12**, 241.
[66] A. R. Ayer, S. B. Ayer, and K.-E. Eriksson, *European J. Biochem.*, 1978, **90**, 171.
[67] K.-E. Eriksson and S. G. Hamp, *European J. Biochem.*, 1978, **90**, 183.
[68] B. Berg, *Arch. Mikrobiol.*, 1978, **118**, 61.
[69] H. Urbanek, J. Zalewska-Sobczek and A. Borowinska, *Arch. Mikrobiol.*, 1978, **118**, 265.

The decomposition of lignocellulose by selected strains of *Streptomyces* species was mainly confined to the cellulose fraction.[70] Between 25 and 40% of the cellulose was degraded but the maximum degradation of lignin was only 3.5%. A number of thermophilic and thermotolerant fungi have been compared for their ability to degrade cellulose and ligno-cellulosic substrates. *Chrosporum pruinosum* and *S. pulverulentum* were both able to degrade lignin and cellulose while *Thielavia thermophilia* was only able to degrade cellulose.

Mammalian cellulases are very rare but cellulose was found to be digested in the mid-gut of the termite, *Macrotermes natalensis*.[72] However, the termite is a fungus feeder and the cellulase enzymes are acquired from the fungi and not secreted by the animal's own digestive tract.

The biosynthesis of cellulose is inhibited by the presence of coumarin in the algae *Prototheca zopfii*.[73] The only step in the biosynthetic pathway which is inhibited is the transfer of oligosaccharide chains from the lipid intermediate to the protein acceptor. The transfer of mannose residues by a similar reaction is not inhibited.

The cellulose content of the bark from the woods *Picea abies*, *Pinus sylvestris*, *Fagus sylvatica*, and *Quercus robus* was only about 50% of that from the corresponding xylem tissue.[74] Of the other polysaccharide components, bark tissue contained a higher content of xylan but a lower content of glucomannan than the xylem tissue. The major difference between the two types of tissue was the high pectic content of bark, arabinose being a major component of this fraction. Cellulosic walls, similar in composition to those obtained by homogenizing plant tissue in buffer solutions, could be obtained from excised root tissue of a variety of plants by extraction with detergent solutions.[75] The residue obtained after detergent extraction of the stalks of *Dictyostelium discoideum* contained cellulose (57%), protein (15%), and a non-cellulosic heteropolymer in a covalently bound matrix which also contained fatty acids.[76] This composition was similar to the residue obtained from the sheath of the same plant.

4 Hemicelluloses

An extensive investigation of the polysaccharides found in cambial tissue of *Populus tremuloides* and *Tilia americana* has been carried out.[77] The tissue from poplar contained pectic material (40%), arabinogalactan (12%), xylan (10%), xyloglucan (6%), glucomannan (1%), cellulose (21%), and protein (10%). The tissue from basswood had a very similar composition except that it contained more xyloglucan and less xylan. Except for the xylan and glucomannan content, the cambial tissue from these two woods had compositions very similar

[70] D. L. Crawford, *Appl. Environ. Microbiol.*, 1978, **35**, 1041.
[71] S. L. Rosenberg, *Mycologia*, 1978, **70**, 1.
[72] M. M. Martin and J. S. Martin, *Science*, 1978, **199**, 1453.
[73] H. E. Hopp, P. A. Romero, and R. P. Lezica, *F.E.B.S. Letters*, 1978, **86**, 259.
[74] H. H. Dietrichs, K. Garves, D. Behrendorf, and M. Sinner, *Holzforschung*, 1978, **32**, 60.
[75] N. Cathala, M. H. Ghorbal, A. Lamant, and L. Salsac, *Compt. rend. Acad. Sci. Ser. D.*, 1978, **286**, 1025.
[76] H. Freeze and W. F. Loomis, *Biochim. Biophys. Acta*, 1978, **539**, 529.
[77] B. W. Simson and T. E. Timell, *Cellulose Chem. Technol.*, 1978, **12**, 39.

to that of sycamore suspension-culture cell walls. The xyloglucans from cambial tissue of both species of wood contained L-fucose, D-galactose, D-xylose, and D-glucose residues in the ratio 4 : 23 : 28 : 45.[78] The structure of the polysaccharides was based on a main chain of β-1,4-linked D-glucose residues, about 67% of which carried side chains of D-xylose residues linked at O-6 of the D-glucose residues. The xyloglucan from poplarwood had a number-average molecular weight of 6.2×10^4 whilst that from basswood showed a wide molecular weight distribution ($1—10 \times 10^4$) with more molecules being nearer the lower end of the range. The arabinogalactan from cambial tissue of poplarwood contains D-galactose, L-arabinose, L-rhamnose, and D-glucuronic acid (or its 4-*O*-methyl ether) residues in the ratio 10 : 4 : 1 : 1.[79] The main chain consisted of β-1,3-linked D-galactopyranose residues, most of which were further substituted at O-6. This polysaccharide was covalently linked to protein in which L-hydroxyproline was the major amino-acid constituent. The protein was probably linked to the main chain of the galactan. The glycoprotein was polydisperse (molecular weight *ca.* 2×10^5). The xylan from cambial tissue of poplarwood was a 4-*O*-methyl-glucuronoxylan which had the same structure and physical properties as a similar xylan from the wood except that it contained only half of the acidic side chains.[80] Between 40 and 50% of the cambial tissue of poplarwood and 50—60% of the tissue of basswood was pectin. The pectin consisted of a main chain of α-1,4-linked D-galacturonic acid residues which were interdispersed with α-1,2-linked L-rhamnose residues. The L-rhamnose residues often contained branch points at C-4. Long chains of β-1,4-linked D-galactose residues and α-1,5-linked L-arabinose residues were also found as well as some terminal non-reducing L-fucose residues. These pectins were similar in structure to those isolated from alfalfa, soybeans, and lemon peel, as well as phloem tissue from *Abies amabilis*.

A number of sub-fractions have been isolated from the hemicellulose B fraction of the husks of Sorghum grain.[81] Three of these sub-fractions were each found to contain residues of L-arabinose, D-xylose, D-galactose, D-glucose, and D-glucuronic acid and its 4-*O*-methyl ether. Structural studies confirmed that the three polysaccharides were xylans with the usual β-1,4-linked main chain. There were considerable differences in the D.P. of these polysaccharides, the values obtained being 993, 2380, and 851.

Two series of acidic oligosaccharides were isolated from the partial acid hydrolysate of a xylan from larchwood.[82] The first series contained only one uronic acid residue per molecule while the second series contained two uronic acid residues and between two and four D-xylose residues per molecule. The structures of the four oligosaccharides are shown, (2)—(5). It was noted that in each oligosaccharide the two uronic acid residues were attached to contiguous D-xylose residues. Since more of the second series than the first was isolated, it was suggested that a far higher proportion of contiguous side chains are

[78] B. W. Simson and T. E. Timell, *Cellulose Chem. Technol.*, 1978, **12**, 51.
[79] B. W. Simson and T. E. Timell, *Cellulose Chem. Technol.*, 1978, **12**, 63.
[80] B. W. Simson and T. E. Timell, *Cellulose Chem. Technol.*, 1978, **12**, 79.
[81] G. R. Woolard, E. B. Rathbone, and L. Novellie, *Carbohydrate Res.*, 1977, **59**, 547.
[82] K. Shimizu, M. Hashi, and K. Sakurai, *Carbohydrate Res.*, 1978, **62**, 117.

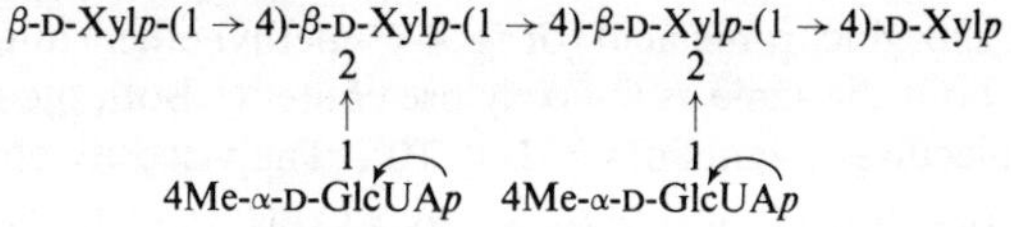

(2)

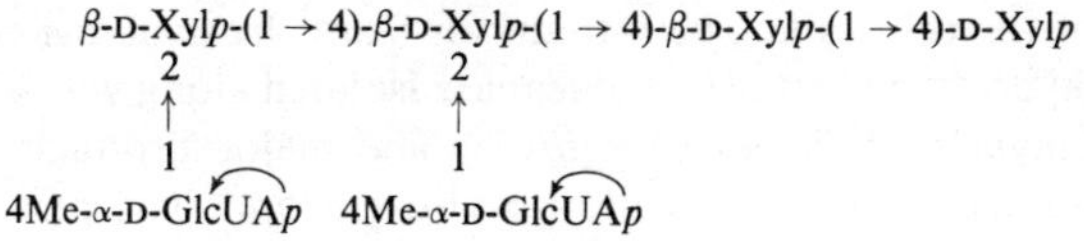

(3)

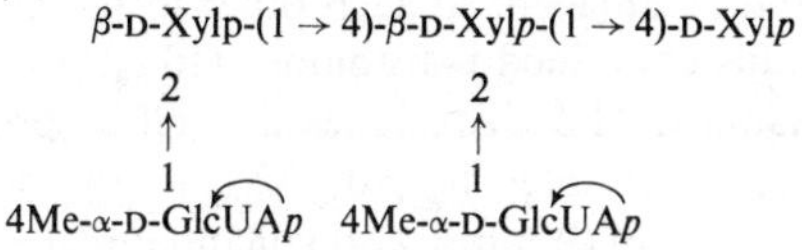

(4)

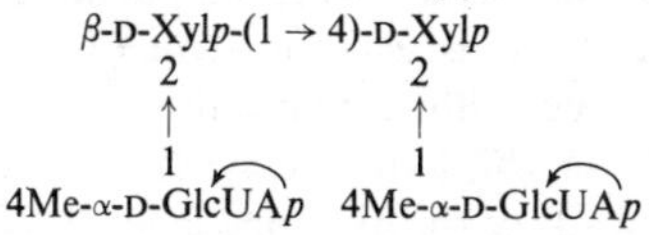

(5)

present than was previously reported for larchwood xylan. The hemicellulosic fractions from Quebec black spruce and balsam fir have been isolated from pulps obtained from these two woods by various pulping methods.[83] The xylans and glucomannans which had the least contamination were obtained from the holocelluloses. Similar polysaccharides obtained from kraft pulps and from the original woods had similar molecular sizes but the polysaccharides from the bisulphite pulps had a much lower molecular size.

Enzymic hydrolysis of spruce wood with a commercial cellulase preparation gave a lignin fraction which, according to spectroscopic evidence, was similar to the milled wood lignin obtained from the same woods.[84] It did differ from milled wood lignin in that the lignin was still covalently bound to carbohydrate. During digestion of ryegrass in the bovine rumen, a lignin–carbohydrate complex is released in a soluble form from the cell wall.[85] Once dissolved, the complex does not appear to undergo any further hydrolysis either in the rumen or along the gastro-intestinal tract. Lignin, synthesized from coniferyl alcohol and hydrogen peroxide by cell walls of *Pinus elliottii* tissue cultures, was found to be covalently bound to both carbohydrate and hydroxyproline-containing protein.[86]

The principal classes of polysaccharides found in the non-cellulosic matrix of *Avena sativa* coleoptile cell walls are glucuronoarabinoxylans and hemicellulosic β-D-glucans.[87] Pectin is a minor component. One of the xylans con-

[83] K. D. Sears, W. J. Alexander, O. Goldschmid, and J. K. Hamilton, *TAPPI*, 1978, **61**, 105.
[84] B. Bezuch and J. Polcin, *Cellulose Chem. Technol.*, 1978, **12**, 473.
[85] M. J. Nielson and G. N. Richards, *J. Sci. Food Agric.*, 1978, **29**, 513.
[86] F. W. Whitmore, *Phytochemistry*, 1978, **17**, 421.
[87] S. Wada and P. M. Ray, *Phytochemistry*, 1978, **17**, 923.

tained 5—10% D-glucuronic acid (or its 4-*O*-methyl ether) and the L-arabinose content was about the same as the D-xylose content. Both the D-glucan and the xylan had molecular weights of *ca.* 2 × 10^5. The viscosity characteristics and the gel-forming ability of the β-D-glucan were evidence of appreciable molecular interactions which suggest that this polysaccharide is an important structural component of the cell wall. The β-D-glucan from *Avena sativa* coleoptile cell walls can be hydrolysed by the β-D-glucanase isolated along with and separated from the α-amylase of *Bacillus subtilis*.[88] The major products were 3-*O*-β-cellobiosyl-D-glucose and 3-*O*-β-cellotriosyl-D-glucose. The presence of these oligosaccharides confirmed that the β-(1 → 4) linkage was present in this glucan and that these linkages could be adjacent to each other in the polymer chain.

Barley endosperm cell walls contained L-arabinose (10%), D-xylose (13%), D-glucose (74%), and D-mannose (2.5%) residues as well as protein (5%).[89] The mixed linkage β-D-glucan accounted for 70—72% of the D-glucose residues, the remainder being present in cellulose and glucomannans. Some of the β-D-glucan was water-soluble and the remainder was alkali-soluble but the ratio of 1 → 3 to 1 → 4 linkages in both fractions was the same (3 : 7). During the germination of barley grains, it has been shown that although starch provided 81.5% of the carbohydrate supply to the living tissue the remaining 18.5% was provided by the gums and cell walls of the aleurone layer.[90] A comparison has been made of non-cellulosic β-D-glucans isolated from the cell walls of five members of the *Gramineae* family, *Zea mays*, *Hordeum vulgare*, *Triticum vulgare*, *Secale cereale*, and *Sorghum bicolor*.[91] Treatment with specific β-D-glucanases from *Bacillus subtilis* or *Rhizopus sp.* yielded virtually identical oligosaccharide fragments. The two predominant products were 3-*O*-β-cellobiosyl-D-glucose and 3-*O*-β-cellotriosyl-D-glucose suggesting that these polysaccharides had a rather regular repeating sequence. The ratio of the tri- to the tetra-saccharide suggested that 30—31% of the linkages were β-(1 → 3). The yields of cell wall β-D-glucan ranged from 4.1% of the wall of *Hordeum* to 9.7% of *Sorghum*.

The β-D-glucan synthase activity of *Pisum sativum* epicotyls was less in particulate fractions than in tissue slices, in the synthesis of alkali-insoluble β-D-glucans.[92] The D-glucans contained both β-(1 → 3) and β-(1 → 4) linkages but the ratio was not constant, the β-(1 → 4) linkage being preferred at low substrate concentrations. A suppressing and an activating factor have both been found in *Pisum sativum* extracts which moderate β-D-glucan synthase activity.[93] The suppressing factor affects the biosynthesis of β-(1 → 4) linkages more than that of β-(1 → 3) linkages while the activator affects both linkages.

The distribution of xyloglucan in the cotyledons and hypocotyls of *Phaseolus aureus* seeds has been studied.[94] The plant tissue was extracted with increasing concentrations of potassium hydroxide and the xyloglucan in the extracts was

[88] R. Yamamoto and D. J. Nevins, *Carbohydrate Res.*, 1978, **67**, 275.
[89] G. M. Ballance and D. J. Manners, *Carbohydrate Res.*, 1978, **61**, 107.
[90] P. Morrall and D. E. Briggs, *Phytochemistry*, 1978, **17**, 1495.
[91] D. J. Nevins, R. Yamamoto, and D. J. Huber, *Phytochemistry*, 1978, **17**, 1503.
[92] Y. Raymond, G. B. Fincher, and G. A. Maclachlan, *Plant Physiol.*, 1978, **61**, 938.
[93] H.-Y. Chao and G. A. Maclachlan, *Plant Physiol.*, 1978, **61**, 943.
[94] Y. Kato and K. Matsuda, *Plant Cell. Physiol.*, 1977, **18**, 1089.

determined by the amyloid reaction after incubation with cellulase. Most of the xyloglucan was found in the 24% potassium hydroxide extract and the properties of the polysaccharide were the same for both types of tissue. This work confirmed similar studies on etiolated bean hypocotyls. An arabinoxyloglucan has been isolated from the leaves of *Nicotinia tabacum*.[95] From the results of methylation and partial acetolysis analysis and the action of cellulase, it was concluded that the structure of the polysaccharide is based on a β-(1 → 4)-D-glucan backbone with side chains of α-D-xylose and α-L-arabinofuranosyl-(1 → 2)-α-D-xylose linked to C-6 of some of the D-glucose residues. About 60% of the D-glucose residues are substituted and about 40% of these are further substituted with L-arabinose residues.

A cellulase, extracted from *Pisum sativum* tissue, was able to reduce the viscosity of solutions of the xyloglucan from *Capucine*.[96] The mode of action of this cellulase complex was different to that observed when 'Onozuka' cellulase complex was used. Seeds of *Tropaeolum sp.* and *Tamarindus sp.* both contain high concentrations of xyloglucans which are associated with sucrose and D-galactosyl-sucroses of the raffinose series.[97] In contrast, *Balsamine sp.* seeds, which do not contain much xyloglucan, do not contain oligosaccharides of the raffinose series. The *Balsamine* seeds do contain planteose (6-*O*-D-fructosyl-sucrose).

The rates of degradation in sodium hydroxide of a series of glucomannans, galactomannans, and plant gums having a D-mannan backbone were determined.[98] From these data, rate constants were derived for the depolymerization reaction involving β-elimination from the reducing end of the polysaccharide chains and for the termination of this reaction. The rates of degradation for polysaccharides which had a bicomponent backbone (*e.g.* glucomannans) were between those of the pure D-mannan and D-glucan polymers. Although the rates of degradation of the two pure polysaccharides were of the same magnitude, the presence of side chains, attached to the D-mannan backbone, as in carob and guar gum, considerably lowered the rates of reaction.

Crystals of ivory nut D-mannan have been grown on microfibrils from *Valonia sp.* cellulose.[99] These crystals were in the form of perpendicularly orientated lamellae with widths of *ca.* 10^3 Å as revealed by electron microscopy. The structure has been designated a 'shish-kebab' type structure. Precipitation of the D-mannan on beaten softwood pulp fragments also gave 'shish-kebab' type structures, indicating that the cellulosic fibre cell wall is a good substrate for orientated crystallization of D-mannan.

A water-soluble acidic polysaccharide is present in the outer layer of the cell walls of *Tremella fuciformis* which contains residues of D-glucuronic acid, D-mannose, and D-xylose (0.5 : 3.8 : 1.0).[100] Structural studies have shown that this polysaccharide has a (1 → 3)-linked D-mannan backbone to which are attached single residues of D-glucuronic acid, D-mannose, and D-xylose at O-2.

[95] S. Eda and K. Kato, *Agric. and Biol. Chem. (Japan)*, 1978, **42**, 351.
[96] R. Guignard and P.-E. Pilet, *Compt. rend. Acad. Sci. Ser. D.*, 1978, **286**, 855.
[97] A. D. Youcef, J. E. Courtois, and P. Le Dizet, *Carbohydrate Res.*, 1978, **61**, 169.
[98] R. A. Young and L. Liss, *Cellulose Chem. Technol.*, 1978, **12**, 399.
[99] H. Chanzy, M. Dube, and R. H. Marchessault, *TAPPI*, 1978, **61**, 81.
[100] Y. Sone and A. Misaki, *Agric. and Biol. Chem. (Japan)*, 1978, **42**, 825.

The alkali-soluble fraction from the cell walls contains a mixture of the same acidic polysaccharide and a β-D-glucan. This glucan contains β-(1 → 3) and β-(1 → 4)-linkages in the ratio 2 : 3.

The modes of action of β-D-mannanases of diverse origin on legume seed galactomannans have been reported.[101] The enzymes have different abilities to hydrolyse galactomannans especially when the backbone is highly substituted with D-galactose residues.

Periodate oxidation studies have been used to determine any regularities within the molecular structure of the arabinogalactan from *Larix occidentalis*.[102] This polysaccharide contains two fractions, one with molecular weight 1.8×10^4 and the other with molecular weight 7.8×10^4. Blocks of about twelve β-1,3-linked D-galactopyranosyl residues were found in these polysaccharides, three such blocks occurring in the lower molecular weight fraction and more in the one with a higher molecular weight. These arabinogalactans were similar to those found in *Acacia sp.* gums. An arabinogalactan has been isolated from black gram (*Phaseolus mungo*).[103] This polysaccharide has a molecular weight of 1.44×10^5 and has a high viscosity (*ca.* ten times that of larch arabinogalactan) especially between pH values of 5 and 7, which decreases with increasing temperature. Residues of L-arabinose, D-galactose, D-galacturonic acid, and L-rhamnose (30 : 20 : 3 : 5) were found in the polysaccharide but the function of the latter two sugars is not known. Arabinogalactan–protein complexes have been isolated from the female reproductive tissues of *Gladiolus gandavensis*.[104]

Cross-linking of larch arabinogalactan chains using epichlorhydrin has produced a new affinity matrix for the purification of *Ricinus communis* lectin.[105] This matrix has an affinity for the lectin fifty times greater than any other matrix investigated. The reactions of various agglutinins with arabinogalactans isolated from wheat and ryegrass have been investigated.[106] Only a mouse anti-galactan protein had a strong agglutinating activity while the plant agglutinins from clam, sponge, and anemone possessed only weak activity. Neither abrin or ricin lectin had any observed activity but vertebrate agglutinins had strong activity. The removal of L-arabinose side chains from the wheat polysaccharide increased the ability of this polysaccharide to react. The authors suggest that the differences in activity are due to differences in the linkages and in the number ratio of L-arabinose to D-galactose residues.

Polysaccharides have been isolated from the crude enzyme complex, papain, and two of them contain the same sugar residues, D-glucuronic acid, D-glucose, D-galactose, L-arabinose, and L-rhamnose but in the ratios 4 : 1 : 12 : 10 : 4 and 13 : 1 : 40 : 26 : 12.[107] The polysaccharides seriously affected the aggrega-

[101] B. V. McCleary, *Proc. Austral. Biochem. Soc.*, 1978, **11**, 33.
[102] S. C. Churms, E. H. Merrifield, and A. M. Stephen, *Carbohydrate Res.*, 1978, **64**, C1.
[103] N. S. Susheelamma and M. V. L. Rao, *J. Agric. Food Chem.*, 1978, **26**, 1434.
[104] P. A. Gleeson, A. E. Clarke, M. A. Jermyn, and R. B. Knox, *Proc. Austral. Biochem. Soc.*, 1978, **11**, 33.
[105] T. Majumdar and A. Surolia, *Experientia*, 1978, **34**, 979.
[106] B. A. Baldo, H. Neukom, B. A. Stone, and G. Uhlenbruck, *Austral. J. Biol. Sci.*, 1978, **31**, 149.
[107] E. V. Chandrasekaran, J. N. BeMiller, and S. C. D. Lee, *Carbohydrate Res.*, 1978, **60**, 105.

tion and alignment of normal human fibroblasts but had no effect on mouse embryo fibroblasts.

A series of polysaccharides have been isolated from the sea grass, *Phyllospadix torreyi*, which are unusual in containing residues of D-apiose (2—28%).[108] One of these polysaccharides had the composition, L-rhamnose, L-arabinose, D-xylose, D-apiose, D-galactose, and D-mannuronic acid in the molar ratio 4 : 1 : 5 : 2 : 4 : 9. Of the D-apiose residues 80% were oxidized by periodate and the major neutral sugar residues found in the hydrolysate of the methylated polysaccharide were 2,3,5-tri-*O*-methyl-L-arabinose, 2,5-di-*O*-methyl-L-arabinose, 3,4-di-*O*-methyl-L-rhamnose, 2,3-di-*O*-methyl-D-xylose, 2,3,6-tri-*O*-methyl-D-galactose, and 2,4-di-*O*-methyl-D-galactose in the ratio 1 : 1 : 6 : 6 : 3 : 3.

A novel polysaccharide gel, which has been isolated from the fruit of *Thaumatococcus danielli*, contains residues of L-arabinose, D-xylose, D-glucuronic acid and its 4-*O*-methyl ether in the ratio 1.0 : 7.2 : 1.91 : 0.66.[109] All of the uronic acid residues were oxidized by periodate and, from the results of methylation analysis, the structure was found to be highly branched. Many of the uronic acid residues were removed by the action of β-D-glucuronidase but the gel was not completely depolymerized by this treatment.

Cell wall material from *Solanum tuberosum* has been purified and its structure has been investigated by the use of methylation analysis.[110] The general features of the cell wall material were discussed in the light of these results.

The synthesis and assembly of plant cell walls, including possible mechanisms for the control of this synthesis, have been reviewed.[111] During the production of slime polysaccharide by the membranes of maize root cap cells, two glycoproteins and a slime polysaccharide were identified, and isolated.[112] The glycoproteins were considered to be precursors of the slime polysaccharide in that the carbohydrate portion of each was similar. The linkage region in one of the glycoproteins involved a D-xylose-L-threonine bond. The other sugar residues present in the carbohydrate portion were D-galactose, D-glucose, D-mannose, L-arabinose, D-xylose, and L-fucose. These glycoproteins are probably involved in the transport of the carbohydrate portion across the cell wall. Three soluble polysaccharides as well as a L-hydroxyproline-containing glycoprotein are secreted by suspension-cultured sycamore cells.[113] Labelled L-fucose was only incorporated into the fucoxyloglucan and L-arabinose only into the arabinogalactan while labelled D-glucose was a precursor of both polysaccharide and protein. When the rate of secretion of polysaccharide had reached a steady state, the rate could be increased by adding polyelectrolytes to the growth medium. These cations probably induce secretion by causing a

[108] G. R. Woolard and J. K. N. Jones, *Carbohydrate Res.*, 1978, **63**, 327.
[109] S. K. Adesina and J. D. Higginbotham, *Carbohydrate Res.*, 1977, **59**, 517.
[110] S. G. Ring and R. R. Selvendran, *Phytochemistry*, 1978, **17**, 745.
[111] D. H. Northcote, in 'Synthesis, Assembly, and Turnover of Cell Surface Components', Cell Surface Reviews Vol. 4, ed. G. Poste and G. L. Nicolson, Elsevier North-Holland Biomed. Press, New York, 1977, 717.
[112] J. R. Green and D. H. Northcote, *Biochem. J.*, 1978, **170**, 599.
[113] M. R. Morris and D. H. Northcote, *Biochem. J.*, 1977, **166**, 603.

depolymerization of the negative electrical potential at the cell surface and encourage fusion of the secretory vesicles with the plasma membrane. Fusion of the vesicles with the membrane is probably the rate-limiting step and may be a potential control point. The decomposition of cell wall material, when stimulated by auxin, could be caused by an influx of Ca^{2+} ions which cause fusion of the vesicles with the plasma membrane.

Konjac mannan contains acetate residues linked to the polysaccharide by ester bonds.[114] One acetate group is present for every 19 residues of D-mannose and acetic acid is the only acid present. A series of water-soluble methylated glucomannans from Konjac have been prepared.[115] The polysaccharides with the maximum viscosity had a degree of substitution of 0.45.

Recent developments in the industrial uses of hemicelluloses have been reviewed.[116] The isolation and purification of these polysaccharides as well as the results of structural analyses and the industrial applications have all been considered.

The cell wall components of the pollens of some woody and herbaceous plants have been investigated.[117] The cellulose content of the *Cruciferae* was high and that of the *Ericaceae* was low with a mean value of 0.5%. The hemicellulose content of the *Ranunculus sp.* was high while that of *Acer*, *Cruciferae*, and *Fagus sp.* was low with a mean value of 7.0%.

The monosaccharide composition of peat hydrolysates was successfully determined by g.l.c. of the trimethylsilylated oxime derivatives.[118] Amongst the sugars recognized were D-xylose, L-arabinose, D-ribose, L-rhamnose, D-galactose, and D-glucose.

5 Pectins

A hitherto unobserved component of the primary cell walls of dicotyledonous plants, rhamnogalacturonan II, has been isolated and partially characterized.[119] It is a very complex polysaccharide containing residues of ten different monosaccharides including D-apiose, 2-*O*-methyl-D-xylose, and 2-*O*-methyl-L-fucose. The polysaccharide, which accounts for 3—4% of the primary cell walls of suspension-cultured sycamore cells, is also characterized by the presence of 2-linked D-glucosyluronic acid, 3,4-linked L-fucosyl, and 3-linked L-rhamnosyl residues. These linkages have not previously been detected in polysaccharides of sycamore primary cell walls. Evidence was also presented that similar polysaccharides are present in the primary cell walls of pea, pinto bean, and tomato.

The neutral oligosaccharides (6)—(11) have been isolated from a partial acid hydrolysate of the water-soluble pectic polysaccharide of the opium poppy.[120] A partial structure of this polysaccharide has already been proposed.

[114] K. Maekaji, *Agric. and Biol. Chem.* (*Japan*), 1978, **42**, 177.
[115] N. Kishida and S. Okimasu, *Agric. and Biol. Chem.* (*Japan*), 1978, **42**, 669.
[116] R. L. Whistler and R. N. Shah, *Modif. Cellul.* (Symp. Cellul., Pap., Text. Div. Amer. Chem. Soc.), 1978, 341.
[117] A. R. McLellan, *Ann. Botany*, 1977, **41**, 1225.
[118] H. Morita and W. G. Montgomery, *J. Chromatog.*, 1978, **155**, 195.
[119] A. G. Darvill, M. McNeil, and P. Albersheim, *Plant Physiol.*, 1978, **62**, 418.
[120] B. S. Paulsen, B. Johansen, and J. K. Wold, *Carbohydrate Res.*, 1978, **65**, 320.

L-Ara*f*-(1 → 3)-L-Ara (6)   L-Ara*p*-(1 → 3)-L-Ara (7)   α-L-Rha*p*-(1 → 4)-D-Xyl*p* (8)

β-D-Xyl*p*-(1 → 4)-β-D-Xyl*p*-(1 → 4)-D-Xyl*p*

(9)

α-L-Rha*p*-(1 → 4)-β-D-Xyl*p*-(1 → 4)-D-Xyl*p*

(10)

β-D-Xyl*p*-(1 → 4)-β-D-Xyl*p*-(1 → 4)-β-D-Xyl*p*-(1 → 4)-D-Xyl*p*

(11)

From an examination of the polygalacturonate fractions obtained from unripe and ripe apple fruit cortical tissue, it was found that one fraction was wall bound and the other was not.[121] The results suggested that the wall-bound fraction was degraded during the ripening process but the soluble fraction was not. Esterification of the pectic material using diazomethane helped to solubilize it as did esterification using acidic methanol although the use of the latter reagent caused hydrolysis of L-arabinofuranosyl linkages. Both reagents reduced the cohesion of ethanol-extracted apple tissue. The presence of free carboxy-groups in the polygalacturonate probably maintains cell cohesion through co-operative binding of Ca^{2+} ions. The integrity of apple primary cell wall structure was thought to depend on non-covalent bonding between cellulose, protein, and polygalacturonate.

The metabolism of polymethylgalacturonate has been followed in ripening apples.[122] After the onset of ethylene production, when the fruit started to soften, the total polygalacturonate content decreased but the water-soluble fraction increased. There was no change in the overall degree of esterification but the ester content of the water-soluble fraction increased. Methyl groups continued to be incorporated into the polygalacturonate from L-methionine during ripening but the rate of incorporation of *myo*-inositol decreased. Separation of cells during the ripening process is probably caused by the removal of polygalacturonate with a low methoxy content from the middle lamella by the action of exopolygalacturonase enzymes. The author suggests that the anomalous incorporation of L-methionine may be due to a continuing synthesis of esterified polysaccharide.

A series of *O*-acetyl derivatives of pectic acid has been prepared to study the intramolecular binding of Ca^{2+} ions to pectic acids.[123] The electrostatic free enthalpy of dissociation of these polyacids and the rate of exchange of the counter-ion from Ca^{2+} to K^+ were determined by potentiometric titration measurements. The binding of Ca^{2+} ions was characterized by the production of single ion affinity coefficients determined in solutions of the calcium salts of *O*-acetyl pectic acids and by the circular dichroism measurements of calcium and potassium salts of the pectate derivatives. The intramolecular binding of Ca^{2+} ions to isolated pectate macromolecules is purely of electrostatic character in contrast to the intermolecular chelate binding of Ca^{2+} ions which takes place when the pectate macromolecules aggregate.

[121] M. Knee, *Phytochemistry*, 1978, **17**, 1257.
[122] M. Knee, *Phytochemistry*, 1978, **17**, 1261.
[123] R. Kohn and A. Malovikova, *Coll. Czech. Chem. Comm.*, 1978, **43**, 1709.

The influence of temperature, ionic strength, pH and various additives on the circular dichroism spectra of pectic solutions has been reported.[124] On a pH change from 7.5 to 2.0, hyper- and hypso-chromic effects were observed. The molecular ellipticity (θ) contribution from the free carboxy-group was the same as that from the methyl esterified carboxy-group. The gelling of pectins caused by a change in the solvent composition led to an increase in θ. If the ionization level was kept constant, θ for the carboxy chromophore could be used as a probe for the interpretation of intermolecular interactions.

In contrast to cellulose and xylans, no anhydrosugars or lactones were produced during the thermal decomposition of pectic substances from tobacco leaves.[125] By far the major volatile product from de-esterified pectin was 2-furaldehyde (12) while both the original pectin and its re-esterified form gave (12), methyl-5-formyl-2-furoate (13), and methyl-2-furoate (14) as the major products.

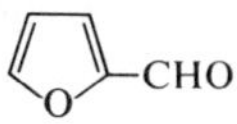

(12)

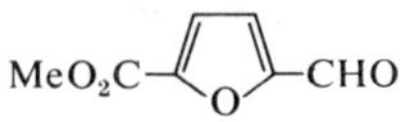

(13)

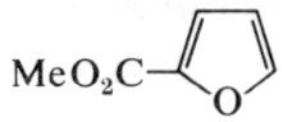

(14)

6 Gums and Mucilages

The gum from *Acacia mabellae* contains two polysaccharide components.[126] These polysaccharides, designated A and B, have molecular weights of 8×10^4 and 1.6×10^5 respectively. The composition and physical constants of the two polysaccharides were very similar (D-galactose 77 and 78%, L-arabinose 15 and 13%, L-rhamnose 3 and 3%, uronic acid 5 and 6%, optical rotation $+1.5°$ and $+6.5°$ respectively) and each consumed 6.0 mmol g^{-1} of sodium periodate. The products from the Smith-degradation of the polysaccharides contained D-galactose 96 and 97% and L-arabinose 4 and 3% and both had molecular weights of 3.5×10^4. Polysaccharide B is considered to be composed of two chains of polysaccharide A. The bond between the chains is not considered to be ionic since complete deionization had no effect on the molecular weight of B.

The polysaccharide component of the stigmatic exudate of *Lilium longiflorum* contained residues of L-arabinose (26%), L-rhamnose (6%), D-galactose (57%), and D-glucuronic acid (11%).[127] This highly branched polysaccharide is very similar to the acidic polysaccharide exudate of the gum from *Araucaria bidwillii* in that the main chain of 1,3-linked D-galactose residues is substituted at O-6. The outer chains are terminated at the non-reducing end by residues of L-arabinofuranose, L-rhamnopyranose, D-galactopyranose, and D-glucuronic acid. All of the L-rhamnose non-reducing end groups are linked in the partial sequence (15) which also occurs in gum arabic.

The distribution of the D-galactose side chains in guar gum has been determined by a specific degradation scheme outlined in (16).[128] The methylated

[124] I. G. Plaschina, E. E. Braudo, and V. B. Tolstoguzov, *Carbohydrate Res.*, 1978, **60**, 1.
[125] K. Ohnishi, E. Takagi, and K. Kato, *Carbohydrate Res.*, 1978, **67**, 281.
[126] S. C. Churms, E. H. Merrifield, and A. M. Stephen, *Carbohydrate Res.*, 1978, **63**, 337.
[127] G. O. Aspinall and K.-G. Rosell, *Phytochemistry*, 1978, **17**, 919.
[128] J. Hoffman and S. Svensson, *Carbohydrate Res.*, 1978, **65**, 65.

L-Rha*p*-(1 → 4)-D-GlcUA*p*-(1 → 6)-D-Gal*p*-1 →
3
↑

(15)

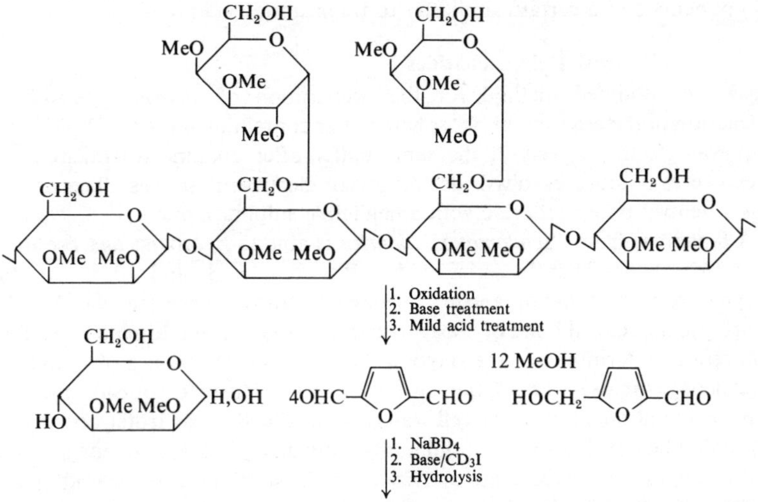

2,3-Di-*O*-methyl-1,4,5,6-tetra-*O*-deuteriomethyl-D-mannitol

(16)

guar gum, in which the hydroxy-groups on C-6 of both D-galactose and D-mannose residues are unsubstituted was subjected to the reaction sequence of periodate oxidation, base-catalysed β-elimination, and mild acid hydrolysis. The D-galactose residues in the side chains and any unsubstituted D-mannose residues in the main chain were degraded, leaving intact any branched D-mannose residues which were adjacent to another branched D-mannose residue. The results of methylation analysis of the degradation mixture confirmed that the D-galactose side chains were distributed on the main chain in pairs or triplets. A procedure to separate peracetylated mono- and di-saccharides by h.p.l.c. on silica gel has been used to quantify the constituents of guar gum.[129]

The gum exudate from *Khaya grandifoliola* has a limited solubility in water.[130] About 50% of this acidic gum was in the form of the free acid while the remainder was present as the calcium salt. Ca^{2+} ions could be removed from the gum by dissolving in sodium carbonate solution but this procedure significantly reduced the viscosity of the gum.

An unusual form of holocellulose was isolated from *Sphagnum sp.* moss which retained most of the cation-exchange properties of that species.[131] The exchange property survived the delignification reaction and only a small portion could be removed from the holocellulose by alkaline extraction. The

[129] J. Thiem, J. Schwentner, H. Karl, A. Sievers, and J. Reimer, *J. Chromatog.*, 1978, **155**, 107.
[130] M. Aslam, G. Pass, and G. O. Phillips, *J. Sci. Food Agric.*, 1978, **29**, 563.
[131] T. J. Painter and N. A. Sørensen, *Carbohydrate Res.*, 1978, **66**, C1.

holocellulose accounted for *ca.* 50% by weight of the moss and was resistant to the action of both cellulase and pectinase enzymes. The authors suggest that the holocellulose is a cellulose-bound gum or mucilage similar to those found in quinces and the seeds of mustard, flax, and cress. The non-cellulosic components had a certain similarity to tragacanthic acid.

7 Algal Polysaccharides

Agar.—A modified method for the preparation of agarose involved the extraction of the granules in phosphate buffer containing 4M urea.[132] The next step was electrophoresis in the same buffer after alkaline treatment in the presence of sodium borohydride to prevent electroendosmosis. The product was a neutral polysaccharide with a negligible sulphur content.

A β-agarase from *Pseudomonas atlantica* is able to hydrolyse agarose to the neoagaro-oligosaccharide series with DP's of 4, 6, 8, and 10.[133] Another enzyme from the same organism, β-neoagarotetraose hydrolase, was found to hydrolyse the central bond of neoagarotetraose and the penultimate bond at the non-reducing terminus of the oligosaccharides with DP values of 6 and 8 to give neoagarobiose. This latter enzyme was located at or outside the cytoplasmic membrane but in the cell wall region. The β-neoagarotetraose hydrolase could be purified so that it did not contain any β-agarase, α-neoagarobiose hydrolase, or transglycosidase activity. Similar studies on the degradation of agarose gels by a bacterial agarase were also carried out with a *Pseudomonas*-like bacterium.[134] The agarose was degraded to 72% by this enzyme in 48 h but mechanical agitation could increase this value to 95%. The hydrolysis was considered to be a two-stage procedure involving an initial rapid hydrolysis of the gel to polysaccharide chains of molecular weight 10^4—10^5. The second stage, the rate-determining step, was a slower hydrolysis to oligosaccharides, the predominant products being neoagarotetraose and neoagarohexaose. An un-named Gram-negative marine bacterium was shown to liquify agar by producing an α-agarase which cleaved the α-(1 → 3) linkage between 3,6-anhydro-L-galactose and D-galactose to give the agaro-oligosaccharide series. From the reaction products, agarohexaitol and agarotetraitol were prepared and characterized.[135]

The proton n.m.r. spectra of some 3,6-anhydro-D-galactose derivatives in the presence of a europium shift reagent showed that the shift reagent had a strong affinity for these derivatives.[136] This phenomenon could be used to analyse for residues of 3,6-anhydro-D-galactose or its enantiomorph in agar- and carrageenan-like polysaccharides.

Alginates.—In a historical series of articles on the development of the seaweed industry, the first article has considered the use of alginates.[137]

The stoicheiometry of the Ca^{2+} ion-induced chelation of alginate chains has

132 M. S. Karve and N. R. Kale, *Prep. Biochem.*, 1977, **7**, 403.
133 D. Groleau and W. Yaphe, *Canad. J. Microbiol.*, 1977, **23**, 672.
134 M. Malmquist, *Carbohydrate Res.*, 1978, **62**, 337.
135 K. S. Young, S. S. Bhattacharjee, and W. Yaphe, *Carbohydrate Res.*, 1978, **66**, 207.
136 K. Izumi, *Carbohydrate Res.*, 1978, **62**, 368.
137 E. Booth, *Chem. Ind.* (*Eng.*), 1977, **13**, 528.

been investigated by the use of circular dichroism spectroscopy and equilibrium dialysis measurements.[138] The circular dichroism intensity in the region of the spectra where the carboxy-group absorbs increased to 50% of that required when based on the poly-L-guluronate content. The amount of bound Ca^{2+} ion, on swamping with Na^+ ions, was not altered. In terms of the 'egg-box' model, it was suggested by the authors that dimerization of the poly-L-guluronate segments of the alginate chains into a regular two-fold conformation was related to that already characterized for the free acid with tight interchain chelation occurring on the interior faces of each dimer. This hypothesis was consistent with the results of electron microscopy studies.

The oxidation of alginate chains with periodate has a marked effect on the intrinsic viscosity of the alginate but the authors consider that previous values may be incorrect due to errors in measuring the molecular weight of the original alginate.[139]

A poly-α-L-guluronate lyase is produced by *Klebsiella aerogenes* which is able to degrade blocks containing L-guluronic acid and D-mannuronic acid residues in alginic acid to oligosaccharides.[140] An analysis of these oligosaccharides has indicated that the poly(mannuronate–guluronate) blocks of alginates do not have a strictly alternating sequence of the two uronic acid residues. The degradation of algal alginates by the enzyme to leave polymannuronate blocks intact suggested that these blocks have a DP of at least 24. The oligosaccharides derived by hydrolysis of alginic acid could be determined by ^{13}C n.m.r. spectroscopy.[141] The analysis relies on the chemical shifts of the anomeric C-atom and can be used to distinguish both L-guluronic and D-mannuronic acid residues as well as the 4,5-unsaturated acid produced by lyase action on the uronic acid residues.

Polarimetry of the brucine salts of L-guluronic acid and D-mannuronic acid released from alginates by acid hydrolysis has been used to measure the ratio of these acids in the native alginate.[142] Nitroxide spin-labels have been investigated as molecular probes to determine the microviscosity of aqueous solutions of carbohydrates such as alginate.[143]

Carrageenans.—The carrageenans from *Euchema* and related algal species have been isolated and analysed after fractionation into KCl-soluble and -insoluble components.[144] Their immunochemistry was investigated using anti-(κ-carrageenan) antibody. This antibody only showed a small degree of cross-reactivity with ι-type carrageenans so the κ- and ι-types could be distinguished. The data yielded information on the carrageenan–anti-carrageenan interaction as well as details of the fine structure of these polysaccharides. Immunochemical and chemical differences between the λ-type carrageenans from certain red algae have been investigated and in some samples, such as those from *Gigartina* and *Petrocelis*, were attributed to the low content of 6-sulphate groups on the

[138] E. R. Morris, D. A. Rees, D. Thom, and J. Boyd, *Carbohydrate Res.*, 1978, **66**, 145.
[139] I. L. Andersen, T. J. Painter, and O. Smidsrød, *Carbohydrate Res.*, 1977, **59**, 563.
[140] J. Boyd and J. R. Turvey, *Carbohydrate Res.*, 1978, **66**, 187.
[141] J. Boyd and J. R. Turvey, *Carbohydrate Res.*, 1978, **61**, 223.
[142] I. R. Siddiqui, *Carbohydrate Res.*, 1978, **67**, 289.
[143] J. D. Aplin and L. D. Hall, *Carbohydrate Res.*, 1977, **59**, C20.
[144] V. Di Ninno and E. L. McCandless, *Carbohydrate Res.*, 1978, **66**, 85.

4-linked D-galactose residues in these polysaccharides.[145] These carrageenans were, to a large extent, ξ-like but did have some λ-like features. Differences between the two carrageenans were observed in their i.r. spectra. The λ-carrageenans from *Irideae cordata* and *Rhodoglossum californicum* differed from that of *Chondrus crispus* to a small extent and these differences were not due to the 6-sulphate content.

The addition of D-glucose, sucrose, or D-glucose syrups to dilute solutions of κ-carrageenan increased the apparent viscosity of the solution but reduced the value of the intrinsic viscosity.[146] In the presence of D-glucose syrups, acidic polysaccharides behaved like neutral macromolecules. The authors suggest that the inorganic cation contaminants of the syrup, by interacting with these polyanions, prevent dissociation of the ionic groups in the polysaccharides.

Carrageenans were partially hydrolysed when sterilized in a synthetic milk salt solution.[147] The reduction in the molecular size of these polysaccharides did not occur when heated in aqueous solution at neutral pH values and was caused by a lowering of the pH which occurred with heating in the milk salt solution.

The ^{13}C n.m.r. spectra of κ- and ι-carrageenans as well as agar exhibit peaks with chemical shifts between 103.2 and 103.6 p.p.m. which can be attributed to C-1 of the D-galactose and D-galactose 4-sulphate residues of these polymers.[148] Chemical shifts below 100 p.p.m. were assigned to other groups in the molecules.

Other Algal Polysaccharides.—Laminaran, fucans, and alginic acid have been isolated from *Desmarestia ligulata* and *D. firma*.[149] The amount of fucan was low in both species but each contained residues of D-glucuronic acid, D-galactose, D-xylose, and L-fucose while that from *D. ligulata* also contained residues of D-mannose. The laminaran from *D. ligulata* contained chains terminated at the reducing end by D-mannitol residues and D-glucose residues while the laminaran from *D. firma* only contained chains terminated at the reducing end by D-glucose residues. The residue after sequential extraction of *D. ligulata* with water, acid, and alkali contained cellulose, a uronan, and protein.

The L-fucose-containing polysaccharides which were isolated from the brown alga, *Dictyota dichotoma*, were fractionated into water-soluble and water-insoluble components.[150] The latter contained a high proportion of protein. Although D-glucose was the major sugar component in this fraction, all the sugar residues found in the water-soluble fraction were also found in this fraction. The D-glucan could be completely removed from the water-soluble fraction by fractional precipitation to give a glucan-free sulphated polysaccharide which contained residues of D-glucuronic acid, D-galactose, and L-fucose. This polymer had a higher anti-coagulating activity than heparin against either sheep or human plasma.

[145] V. Di Ninno and E. L. McCandless, *Carbohydrate Res.*, 1978, **67**, 235.
[146] A. M. Elfak, G. Pass, and G. O. Phillips, *J. Sci. Food Agric.*, 1978, **29**, 557.
[147] S. Badui, N. Desai, and P. M. T. Hansen, *J. Agric. Food Chem.*, 1978, **26**, 675.
[148] S. S. Bhattacharjee, W. Yaphe, and G. K. Hamer, *Carbohydrate Res.*, 1978, **60**, C1.
[149] G. E. Carlberg, E. Percival, and M. A. Rhaman, *Phytochemistry*, 1978, **17**, 1289.
[150] A. F. Abdel-Fattah, M. M.-D. Hussein, and S. T. Fouad, *Phytochemistry*, 1978, **17**, 741.

The structure of the polysaccharide from *Pachymenia carnosa* has been determined from the results of methylation analysis of the native and desulphated material.[151] In the native polysaccharide, the ratio of (1 → 4) : (1 → 3) linkages was 1 : 2.26. During the desulphation reaction, depolymerization of the polymer chain took place specifically at the 1,3-linked D-galactose residues. The sulphate groups are located at positions 2,4 and 2,6 of the 1,3-linked D-galactose residues. Desulphation occurred when the native polymer was methylated by the Purdie procedure in methylsulphoxide as solvent with the O-2-sulphate being more labile than the O-4- or O-6-sulphate. No desulphation was observed when the polysaccharide was methylated by the Hakomori procedure.

L-Fucose-containing polysaccharides were also isolated from the brown algae *Ascophyllum nodosum* and *Fucus vesiculosus*.[152] *A. nodosum* contained 'Ascophyllum', two 'complexes', and a galactofucan while *F. vesiculosus* contained an 'Ascophyllum'-like polymer and a single 'complex'. On mild acidic hydrolysis, each complex could be separated into a faster and a slower moving component by electrophoresis. Although the complex from *F. vesiculosus* contained more L-fucose residues than either complex from *A. nodosum*, neither of the algae contained a pure L-fucan sulphate in the native state. The galactofucan from *A. nodosum* contained residues of L-galactose and L-fucose in ratio 1.0 : 1.1 but it also contained D-glucuronic acid residues (8 %), sulphate (15 %), and protein (7 %).[153] The results of methylation analysis and oxidation with periodate showed that D-glucuronic acid was present as non-reducing terminal residues and sulphate was also present on non-reducing terminal residues. L-Galactose was present in (1 → 4)-linked chains while L-fucose was attached to the main chain by (1 → 2)-linkages and was mainly present as non-reducing terminal residues. The L-fucose-containing complexes from *A. nodosum* were hydrolysed by mild acid to give an L-fucan and an 'Ascophyllan'-like fraction.[154] From a time-course investigation of the acidic hydrolysis, pronase digestion, and base-catalysed hydrolysis, it was concluded that the complexes were composed of an L-fucan type backbone to which 'Ascophyllan'-like molecules were attached by acid-labile linkages. The primary uronic acid in the 'Ascophyllan'-like portion of the complex was D-mannuronic acid whereas the 'free Ascophyllan' isolated in this investigation contained residues of D-mannuronic, L-guluronic, and D-glucuronic acids. L-Hydroxyproline was present in the complexes but did not appear to be part of the major carbohydrate–protein linkages: the involvement of L-threonine and L-serine was more likely from the results of base-catalysed hydrolysis.

From intrinsic viscosity measurements, the β-(1 → 4)-linked D-mannan found in the cell walls of the alga, *Codium*, had a DP of *ca.* 375 and the chain length distribution was nearly monodisperse.[155] The DP was independent of

151 H. Parolis, *Carbohydrate Res.*, 1978, **62**, 313.
152 D. G. Metcalf and B. Larsen, *Carbohydrate Res.*, 1977, **59**, 531.
153 D. G. Metcalf, T. L. Schneider, and R. W. Barnett, *Carbohydrate Res.*, 1978, **66**, 167.
154 D. G. Metcalf and B. Larsen, *Carbohydrate Res.*, 1977, **59**, 539.
155 M. Marxfigini and M. Coundematus, *Makromol. Chem.*, 1978, **179**, 1541.

the stage of growth which suggested that the biopolymerization was genetically, not age, controlled.

The reserve D-glucan stored by the marine diatom *Skeletonema costatum* is based on a β-(1 → 3)-D-glucan backbone with occasional branch points at C-2 and C-6.[156] Although this structure is called a 'Chrysolaminaran-type', none of the polysaccharide chains were terminated at their reducing end by residues of D-mannitol. The molecular weight of these polymers was 1.3×10^4 with an average chain length of *ca.* 11 residues.

A carbohydrate-containing polymer present in the mucilage tubes of the diatom *Berkeleya rutilans* may be a glycoprotein.[157] The carbohydrate component contains residues of D-mannose, D-mannuronic acid, D-xylose, and 3,6-anhydro-D-glucose along with sulphate ester. The D-mannose and D-mannuronic acid residues are present in 1,3-linked form and each is also probably present as branch points with branches at C-2. D-Xylose is present as non-reducing terminal groups and linked either by (1 → 2) or (1 → 4) bonds. The 3,6-anhydro-D-glucose residues are joined by (1 → 4) or (1 → 5) bonds while sulphate groups are linked to O-2 or O-4 of sugar residues in the main chain.

Chitin.—Block and random copolymers of 2-acetamido-2-deoxy-D-glucose and 2-amino-2-deoxy-D-glucose residues in chitin have been obtained by heterogeneous and homogeneous hydrolysis.[158] Alkaline hydrolysis of chitin under heterogeneous conditions caused *N*-deacetylation which reached an upper limit at low reaction temperatures. The products of hydrolysis were all water-soluble and those from which up to 80% of the acetate groups were removed still remained crystalline. Hydrolysis of the *N*-acetyl groups probably occurred in the amorphous region of the polysaccharide to give a block copolymer. Homogeneous alkaline hydrolysis produced water-soluble products even when only 50% of the *N*-acetyl groups had been removed. The crystallinity of these products decreased with increasing reaction time in such a way that when only 40% of the *N*-acetyl groups had been removed the products were amorphous and random co-polymers.

Polyelectrolytic complexes have been prepared by the reaction of chitosan with sodium carboxymethyldextran.[159] Water-soluble precipitates were formed whose composition was affected by the reaction pH, order of mixing of the polyelectrolytes, and the molar ratio of amino-sugar to Na^+ ion. There was no measureable interaction between carboxy- and hydroxy-groups in the polysaccharides. The number of binding sites in the complex prepared at pH 3.0 was lower than that in the complex prepared at pH 6.5 but both were able to suppress the coagulation of blood.

The centric diatom *Cyclotella* contains a special apparatus for the synthesis of chitin fibrils.[160]

[156] B. S. Paulsen and S. Myklestad, *Carbohydrate Res.*, 1978, **62**, 386.
[157] B. S. Paulsen, A. Haug, and B. Larsen, *Carbohydrate Res.*, 1978, **66**, 103.
[158] K. Kurita, T. Sannan, and Y. Iwakura, *Makromol. Chem.*, 1977, **178**, 3197.
[159] H. Fukuda and Y. Kikuchi, *Bull. Chem. Soc. Japan*, 1978, **51**, 1142.
[160] H. Herth, *Naturwiss.*, 1978, **65**, 260.

4
Microbial Polysaccharides

BY R. J. STURGEON

1 Teichoic Acids

When *Bacillus licheniformis* is grown under conditions of phosphate limitation, teichoic acid is partially replaced by teichuronic acid irrespective of the growth conditions.[1] Repression of the synthesis of the enzymes involved in the synthesis of teichoic acid was observed when the intracellular phosphate level decreased, and CDP-glycerol pyrophosphorylase became inactivated under the phosphate limiting conditions. An unusually high molecular weight teichuronic acid (5×10^5) having a narrow size distribution has been isolated from *B. megaterium*.[2] Large differences between the extended chain length of the polysaccharide and its observed molecular shape were noted, suggesting a well-ordered structure. Further evidence for the structures of the teichoic acids from *B. stearothermophilus* B65 and *B. subtilis* var. *niger* WM has been reported.[3] Data obtained from the ^{13}C and ^{31}P n.m.r. spectra were consistent with a poly(D-glucosylglycerol phosphate) type structure. The results of periodate oxidation and ^{13}C n.m.r. spectroscopic studies on the teichoic acid of *B. subtilis* var. *niger* WM showed that for this organism D-glucose is not an integral part of the polymer chain.

No evidence for the presence of glycerol teichoic acids, as judged by immunochemical or chemical means, has been found for certain oral streptococci.[4] Immunoelectrophoresis studies have revealed that cell walls of *Streptococcus faecalis* and *S. faecum* contain teichoic acid and lipoteichoic acid, both of which are active against group D antisera, whilst those of *S. suis* contain only lipoteichoic acid.[5] The structure of the C-polysaccharide isolated from the cell walls of *S. pneumoniae*, and composed of equimolar amounts of 2-acetamido-2-deoxy-D-galactose, 2-acetamido-4-amino-2,4,6-trideoxyhexose, ribitol phosphate, and choline phosphate, has been studied.[6] After studies involving a combination of hydrolyses with acid and alkali, periodate oxidation, and methylation analysis, the repeating unit (1) was proposed. Antibodies to phosphoryl choline present in the serum of a horse immunized with *S. pneumoniae* VII have shown a much higher affinity for species-specific pneumococcal C teichoic acid than for phosphoryl choline or an unrelated capsular polysaccharide

[1] H. Hussey, S. Sueda, S. C. Cheam, and J. Baddiley, *European J. Biochem.*, 1978, **82**, 169.
[2] R. J. Ivatt and C. Gilvarg, *Biochemistry*, 1978, **17**, 3997.
[3] W. R. De Boer, J. T. M. Wouters, A. J. Anderson, and A. R. Archibald, *European J. Biochem.*, 1978, **85**, 433.
[4] B. Rosan, *Science*, 1978, **201**, 918.
[5] S. D. Elliott, M. McCarty, and R. C. Lancefield, *J. Exp. Med.*, 1977, **145**, 490.
[6] I. R. Poxton, E. Tarelli, and J. Baddiley, *Biochem. J.*, 1978, **175**, 1033.

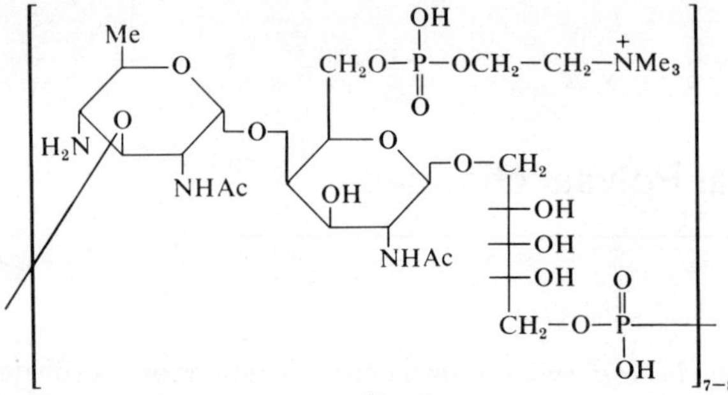

(1)

which also contains phosphoryl choline. This implies that the polysaccharide part of these molecules is also involved in the binding of the antibody.[7]

Lipid analysis of several glycerol teichoic acid preparations has indicated that covalently bound lipid is not required for spontaneous adsorption of glycerol teichoic acid to erythrocyte membranes.[8] A glycolipid–lipoteichoic acid complex has been isolated from *Lactobacillus casei*.[9] Two components in the products of mild acid treatment of the complex were identified as 1,2-di-*O*-acyl-3-*O*-[β-D-Glc*p*-(1 → 6)-α-D-Gal*p*-(1 → 2)-α-D-Glc*p*]-glycerol and 1,2-di-*O*-acyl-3-*O*-[β-D-Glc*p*-(1 → 6)-α-D-Gal*p*-(1 → 2)-(6-*O*-acyl-α-D-Glc*p*)]-glycerol. Some properties of a lipoteichoic acid antigen from *S. pyogenes* have been described.[10] The presence of fatty acids was reflected in the behaviour of the teichoic acid as an antigen. Excretion of lipoteichoic acid when Group A streptococci are treated with penicillin has been detected in the resting phase of growth as well as in the previously reported logarithmic phase, and is accompanied by a loss of ability of the organisms to adhere to human oral mucosal cells.[11] The binding of epithelial cells to Group A streptococci has been reported to involve the lipoteichoic acid which is located on the fimbriae of these organisms.[12]

The chemistry and biochemistry of the linkage units between teichoic acid and peptidoglycan in bacterial cell walls have been reviewed.[13] Changes in the cell walls of mutants of *Bacillus subtilis* induced by temperature have been reflected in the production of different proportions of teichoic acid to peptidoglycan.[14]

Evidence has been presented to show that at least three lipid intermediates participate in the synthesis of the three glycerol phosphate residues which represent the linkage unit between the teichoic acids and peptidoglycan of *Staphylococcus aureus* H and a *Micrococcus* species.[15] The structures of the

[7] B. N. Manjula and C. P. J. Glaudemans, *Immunochemistry*, 1978, **15**, 269.
[8] H. R. Cooper, F. W. Chorpenning, and S. Rosen, *Infection and Immunity*, 1978, **19**, 462.
[9] M. Nakamo and W. Fischer, *Z. physiol. Chem.*, 1978, **359**, 1.
[10] A. B. Rudczynski and R. W. Jackson, *Immunochemistry*, 1978, **15**, 83.
[11] M. L. Alkan and E. H. Beachey, *J. Clinical Invest.*, 1978, **61**, 671.
[12] E. H. Beachey and I. Opek, *J. Exp. Med.*, 1976, **143**, 759.
[13] J. Coley, E. Tarelli, A. R. Archibald, and J. Baddiley, *F.E.B.S. Letters*, 1978, **88**, 1.
[14] M. A. Shiflett, D. Brooks, and F. E. Young, *J. Bacteriol.*, 1977, **132**, 681.
[15] H. A. I. McArthur, F. M. Roberts, I. C. Hancock, and J. Baddiley, *F.E.B.S. Letters*, 1978, **86**, 193.

three lipids were P^1-2-acetamido-2-deoxy-D-glucopyranosyl-P^2-polyisoprenyl pyrophosphate, P^1-2-acetamido-2-deoxy-D-glucopyranosyl-phosphorylglycerol-P^2-polyisoprenyl pyrophosphate, and P^1-2-acetamido-2-deoxy-D-glucosyl-*bis*-(phosphoryl glycerol)-P^2-polyisoprenyl pyrophosphate. No evidence has been obtained for the existence of a linkage unit between teichoic acid and peptidoglycan in a bacteriophage-resistant mutant of *Staph. aureus.*[16] It was suggested that the mutational defect in this organism is in the synthesis of the P^1-2-acetamido-2-deoxy-D-glucopyranosyl-P^2-polyisoprenyl pyrophosphate intermediate. Some of the changes in normal cell wall function and structure that accompany rod to sphere transition in temperature-sensitive mutants of *B. subtilis* 168 have been related to quantitative differences in the ratio of teichoic acid to peptidoglycan.[17]

2 Peptidoglycans

The stability of mucopeptide to hydrogen fluoride has made it possible to isolate the peptidoglycan portions of lipid intermediates from bacterial cells without degradation and free from contaminating breakdown products from the wall itself.[18] The peptidoglycans from a number of Gram-negative thermophilic bacteria which utilize hydrocarbons have been isolated.[19] The bacteria can be subdivided into four distinct groups on the basis of the amino-acid and amino-sugar composition of these peptidoglycans. Intact cells of *Agrobacterium tumefaciens* retain their characteristic shape and contain 40% peptidoglycan after extraction with sodium dodecyl sulphate and digestion with proteolytic enzymes.[20] Membrane preparations from *Bacillus cereus* have been incubated with UDP-2-acetamido-2-deoxy-D-glucose and three lipid-linked intermediates of peptidoglycan synthesis isolated from the resulting digests.[21] These intermediates were identified as a 2-acetamido-2-deoxy-D-glucosyl-phosphoryl undecaprenol, a similarly constituted intermediate differing only in the prenol grouping, and a 2-acetamido-2-deoxy-D-glucosyl-pyrophosphoryl undecaprenol.

An assay has been described for measurement of the trans-glycosylation reaction which promotes the formation of uncrosslinked peptidoglycan; the lipid intermediate Glc*N*Ac-Mur*N*Ac-P-P-lipid is used as a substrate and a particulate fraction from *Escherichia coli* K12 is used as donor.[22] The site of inhibition of peptidoglycan synthesis during the stringent response in *E. coli* has been determined in strains which were auxotrophic for both L-lysine and diaminopimelic acid.[23] The reaction which is most likely to be inhibited during the stringent response is the terminal one involving the peptidoglycan. A successful reconstitution of an ordered membranous structure for *E. coli* from purified outer membrane components including protein and lipopolysaccharide,

[16] R. Bracha, R. Davidson, and D. Mirelman, *J. Bacteriol.*, 1978, **134**, 412.
[17] M. A. Shiflett, D. Brooks, and F. E. Young, *J. Bacteriol.*, 1977, **132**, 681.
[18] E. Fuchs and C. Gilvarg, *Analyt. Biochem.*, 1978, **90**, 465.
[19] G. J. Merkel, S. S. Stapleton, and J. J. Perry, *J. Gen. Microbiol.*, 1978, **109**, 141.
[20] D. Levanon, *Experientia*, 1978, **34**, 354.
[21] S. Yamamori, N. Murazumi, Y. Araki, and E. Ito, *J. Biol. Chem.*, 1978, **253**, 6516.
[22] Y. van Heijendort, M. Derrien, and J. van Heijenoort, *F.E.B.S. Letters*, 1978, **89**, 141.
[23] W. D. Ramey and E. E. Ishiguro, *J. Bacteriol.*, 1978, **135**, 71.

and peptidoglycan, has been reported.[24] Two new proteins associated with peptidoglycan have been located in the outer membrane of *E. coli* K12.[25]

The activities of DD-carboxypeptidase and LD-carboxypeptidase of *Gaffkya homari* have been measured during biosynthesis of peptidoglycan.[26] A method for the large-scale preparation of 2-acetamido-2-deoxy-D-glucose 6-phosphate from the cell walls of *Micrococcus lysodeikticus* has been reported.[27] The hypothesis that the biosynthesis of the *N*-glycolyl group of muramic acid occurring in peptidoglycans is catalysed by an *N*-acetylhydroxylase has been confirmed. ^{18}O became incorporated into the *N*-acetylmuramic acid component of the peptidoglycan of *Mycobacterium phlei* when grown in the presence of pure oxygen enriched with ^{18}O, possibly by hydroxylation of UDP-*N*-acetylmuramic acid.[28]

Inhibition of autolysis of the cell envelope peptidoglycan of *Neisseria gonorrhoeae* has been accomplished by the addition of Mg^{2+} or antibiotics.[29] Hydrolysis of the peptidoglycan layer occurred in the presence of *N*-2-hydroxyethyl piperazine *N'*-2-ethane sulphonic acid buffers at alkaline pH values with solubilization of both peptide and glycan groupings. The peptidoglycan of *Staphylococcus aureus* has been found capable of activating both the classical and alternative pathways of the complement system.[30] The fluorescent properties of the lipid intermediate, undecaprenyl pyrophosphoryl-Mur*N*Ac-L-Ala-D-Glu-L-Lys-(*N*ε-dimethylaminonaphthalene sulphonyl)-D-Ala-D-Ala in membrane fragments from *S. aureus* have been used in a study on the biosynthesis of peptidoglycan.[31] The undecaprenyl pyrophosphoryl pentapeptide is immobilized within a hydrophobic environment close to the membrane surface defined by membrane-associated enzymes involved in glycan synthesis and within 4—6 Å of the lipid matrix.

Patterns of the synthesis of cell wall peptidoglycan and cellular autolytic capacity have been examined in synchronized populations of *Streptococcus faecium*.[32] The data presented were consistent with the role of autolytic enzyme activity in a previously proposed model for cell division of *S. faecium*. A revision of previous work has been reported for information on the secretion of lipids which is induced by the inhibition of peptidoglycan synthesis in Streptococci.[33] The peptidoglycan of some strains belonging to the genus *Thermus* has been reported to contain significant amounts of glycine and L-ornithine.[34] The cell walls of seven strains of *Methano*-bacterium have been shown to be devoid of the normal type of peptidoglycan.[35] The sacculi of all

[24] H. Yamada and S. Mizushima, *J. Bacteriol.*, 1978, **135**, 1024.
[25] N. Hazumi, H. Yamada, and S. Mizushima, *F.E.M.S. Microbiol. Letters*, 1978, **4**, 275.
[26] W. P. Hammes and H. Seidel, *European J. Biochem.*, 1978, **84**, 141.
[27] K. Kato, S. Iwata, T. Matsuda, and S. Kotani, *Biken J.*, 1978, **21**, 63.
[28] L. Essers and H. J. Schoop, *Biochim. Biophys. Acta*, 1978, **544**, 180.
[29] W. S. Wegener, B. H. Hebeler, and S. A. Morse, *Infection and Immunity*, 1977, **18**, 210.
[30] P. K. Peterson, B. J. Wilkinson, Y. Kim, D. Schmeling, S. D. Douglas, P. G. Quie, and J. Verhoef, *J. Clinical Invest.*, 1978, **61**, 597.
[31] W. A. Weppner and F. C. Neuthaus, *J. Biol. Chem.*, 1978, **253**, 472.
[32] R. P. Hinks, L. Daneo-Moore, and G. D. Shockman, *J. Bacteriol.*, 1978, **134**, 1074.
[33] D. Horne, R. Hakenbeck, and A. Tomasz, *J. Bacteriol.*, 1978, **133**, 1050.
[34] R. A. Pask-Hughes and R. A. D. Williams, *J. Gen. Microbiol.*, 1978, **107**, 65.
[35] O. Kandler and H. König, *Arch. Microbiol.*, 1978, **118**, 141.

strains contained L-lysine, L-alanine, L-glutamic acid, and either 2-acetamido-2-deoxy-D-glucose or 2-acetamido-2-deoxy-D-galactose; muramic acid and D-amino-acids were absent. This evidence supports the view that, within the methanogens, different cell wall polymers characteristic of particular groups of organisms may have changed during evolution and supports the hypothesis that the evolution of the methanogens was separated from that of the procaryotic organisms at a very early stage.

3 Lipopolysaccharides

By capillary g.l.c. of the TMS(−)-2-butyl glycosides of the sugars, the D- and L-configurations of the monosaccharide components of the lipopolysaccharide of *Salmonella typhimurium* have been assigned.[36] Some 3-deoxy-2-octulosonic acids substituted at position 5 with either α-L-rhamnopyranose, β-D-glucopyranose, or 2-acetamido-2-deoxy-β-D-glucopyranose have been synthesized and their behaviour in the thiobarbiturate reaction, for following acid hydrolysis, recorded.[37] The different responses measured may be useful in giving preliminary information on the nature of the sugar attached to the 3-deoxy-2-octulosonic acid. Human complement has been shown to give superior results in the measurement of bacterial lipopolysaccharide when compared with guinea pig complement.[38] The acetyl groups present in lipopolysaccharides have been measured by direct g.l.c. of the liberated acetic acid.[39]

Electron microscopic studies have indicated that in wild-type cells, lipopolysaccharide–protein complexes are stabilized by either divalent cations or polyamines.[40] The binding of the cationic polypeptide polymyxin B to the lipid A portion of bacterial lipopolysaccharides has resulted in the generation of higher molecular weight aggregates with decreased isopycnic density.[41]

The lipopolysaccharide of the marine bacterium *Alteromonas haloplanktos* is composed of residues of D-glucose, D-galactose, 2-amino-2-deoxy-D-glucose, 2-amino-2-deoxy-D-galactose, heptose, and two unidentified sugar components.[42] The lipopolysaccharide of *Agrobacterium tumifaciens* has been shown to contain hexose, amino-sugar, and 3-deoxy-2-keto-octulosonic acid, but no heptose or 2-amino-2-deoxy-D-galactose.[43] Comparable amounts of two different polysaccharides have been isolated from the endotoxin of *Bordetella pertussis*.[44] Both represent entire specific polysaccharide core-regions and possess, as the reducing terminal sugar, a single molecule of 3-deoxy-2-octulosonic acid; in one case the sugar is phosphorylated in position 5 and presumably substituted in position 4, and in the other case the octulosonic acid is not phosphorylated but is substituted in position 5. By a combination of methylation analysis and periodate oxidation studies the structure (2) of

[36] G. J. Gerwig, J. P. Kamerling, and J. F. G. Vliegenthart, *Carbohydrate Res.*, 1978, **62**, 349.
[37] R. S. Sarfati and L. Szabó, *Carbohydrate Res.*, 1978, **65**, 11.
[38] J. N. Saddler and A. C. Wardlaw, *Experientia*, 1978, **34**, 1227.
[39] I. Fromme and H. Beilharz, *Analyt. Biochem.*, 1978, **84**, 347.
[40] L. Van Alphen, A. Verkleij, J. Leunissen-Bijvelt, and B. Lugtenberg, *J. Bacteriol.*, 1978, **134**, 1089.
[41] D. C. Morrison and D. M. Jacobs, *Immunochemistry*, 1976, **13**, 813.
[42] J. M. D. Rienzo and R. A. MacLeod, *J. Bacteriol.*, 1978, **136**, 158.
[43] M. Salkinoja-Salonen and R. Boeck, *J. Gen. Microbiol.*, 1978, **105**, 119.
[44] A. Le Dur, M. Caroff, R. Chaby, and L. Szabó, *European J. Biochem.*, 1978, **84**, 579.

```
2)-β-D-Manp-(1→2)-β-D-Manp-(1→2)-β-D-Manp-(1→2)-β-D-Manp-(1→3)-α-D-GlcpNAc-(1→
           3                    3
           ↑                    ↑
           1                    1
   α-2-OAcAbe              α-D-Glcp
```

(2)

the O-antigenic polysaccharide from a *Citrobacter* species has been assigned.[45]

An ^{125}I-labelled lipopolysaccharide having unaltered biophysical, immunological, and biological properties has been prepared from *Escherichia coli.*[46] No differences have been detected in the compositions of the lipopolysaccharides of a wild-type and a mutant-type *E. coli* K12.[47] On the basis of the chemical analysis of the lipopolysaccharides of bacteriophage-resistant mutants of *E. coli* K12, in conjunction with phage typing data, it has been possible to deduce the residue(s) involved at the receptor site for each of the phages used, and to determine which residues hinder the action of these receptors.[48] Lipopolysaccharide from a thermosensitive mutant of *E. coli* K12 has been subjected to dephosphorylation, enzymic hydrolysis, Smith degradation, and methylation analysis.[49] The core fragment of the lipopolysaccharide was assigned the structure (3).

```
D-Glcp-(1→2)-D-Glcp-(1→3)-D-Glcp-(1→3)-Hep-(1→3)-Hep→KDO
                              6         [ 7 ]     [ 7 ]
                              ↑         [ ↑ ]     [ ↑ ]
                              1         [ 1 ]     [ 1 ]
                          α-D-Galp      [Hep]x    [Hep]y
```

Hep = heptose
KDO = 3-deoxy-2-octulosonic acid
when $x = 1, y = 0; x = 0, y = 1$

(3)

From studies of the biosynthesis of the O-8 antigen of *E. coli*, it has been established that, as for the corresponding O-9 D-mannan, the first stage in the synthesis is the substitution of a hydrophobic carrier molecule with D-glucose followed by transfer of D-mannose residues.[50] This results in the production of D-mannan chains with terminal reducing D-glucose residues. Based on the carbohydrate composition of their lipopolysaccharides, a number of strains of *Fusobacterium nucleatum* have been classified into six chemotypes,[51] and some strains of *Veillonella* have been classified into four chemotypes.[52] An immunohistochemical method, used to localize lipopolysaccharide antigens at an ultrastructural level, has indicated that the lipopolysaccharide of *Bacteroides oralis* is bound to a complex which includes D-glucans in the outer membrane of the cell wall.[53] Electron microscopic examination of the cell wall

[45] B. Jann, P. Prehm, and K. Jann, *J. Bacteriol.*, 1978, **134**, 462.
[46] R. J. Ulevitch, *Immunochemistry*, 1978, **15**, 157.
[47] D. Blache, M. Bruneteau, and G. Michel, *F.E.B.S. Letters*, 1977, **84**, 327.
[48] R. N. Picken and I. R. Beacham, *J. Gen. Microbiol.*, 1977, **102**, 305.
[49] M. Quabonzi, M. Bruneteau, and G. Michel, *Biochimie*, 1978, **60**, 639.
[50] H. C. Flemming and K. Jann, *F.E.M.S. Microbiol. Letters*, 1978, **4**, 203.
[51] G. Fredriksen and T. Hofstad, *Acta Path. Microbiol. Scand.*, 1978, **86B**, 41.
[52] T. Hofstad, *Acta Path. Microbiol. Scand*, 1978, **86B**, 47.
[53] G. Dahlen, H. Nygren, and H. A. Hansson, *Infection and Immunity*, 1978, **19**, 265.

complex of *Methylomonas albus* has revealed goblet-shaped bodies attached to the outermost surface of the lipopolysaccharide.[54]

The immunological activities of lipopolysaccharides, capsular polysaccharides, and outer membrane proteins of *Neisseria meningitidis* have been located by polyacrylamide gel electrophoresis and an immunoperoxidase assay technique.[55] Examination of the chemical composition of the lipopolysaccharides of *N. meningitidis* groups B and C, and of *N. perflava*, *N. sicca*, *N. gonorrhoeae*, and *N. catarrhalis*, has indicated that no common structural pattern exists among the lipopolysaccharides of the Neisseriaceae.[56] The cell envelopes of *Paracoccus denitrificans* have been shown to contain 3-deoxy-2-octulosonic acid and 3-hydroxy fatty acids, the expected components of lipopolysaccharides.[57] However the low level of other neutral or amino-sugars suggest that the lipopolysaccharide may lack long saccharide chains.

Isopycnic density gradient centrifugation has been used in the characterization of the lipopolysaccharide from *Pasturella multocida*.[58] Cerulenin, an inhibitor of fatty acid biosynthesis, has been used to induce changes in the fatty acid composition of the lipopolysaccharide of *Proteus mirabilis*.[59, 60]

The chemical composition of the lipopolysaccharide of a rough strain of *Pseudomonas aeruginosa* has been shown to differ from that of a smooth strain of the organism in that it lacks residues of D-glucose, L-rhamnose, and 2-amino-2,6-dideoxy-L-galactose.[61] These results, together with analysis of the polysaccharide fraction, indicated that the rough strain lacks side chain material and is defective in its inner core region. The reaction of lysozyme with the lipopolysaccharide of *P. aeruginosa* has been shown to involve two stages: an initial interaction, followed by a slow sedimentation of the complex.[62] The effect of lipopolysaccharide composition on the ultrastructure of *P. aeruginosa* has been reported; polymerized O-antigenic side chains in the lipopolysaccharide appear to be required to produce the wild-type appearance of the outer membrane of the organism.[63] In the lipid A component of the lipopolysaccharide of *P. diminuta*, 2-amino-2-deoxy-D-glucose is replaced by 2,3-diamino-2,3-dideoxy-D-glucose, as has been shown previously in two *Rhodopseudomonas* species.[64] A comparative study on the location of a number of O-methyl sugars in various O-antigens of photosynthetic bacteria, as well as in two low molecular weight polysaccharides of *R. gelatinosa*, has been reported.[65] The results of methylation analysis using deuteriomethyl iodide showed

[54] P. Jeffries and J. F. Wilkinson, *Arch. Mikrobiol.*, 1978, **119**, 227.
[55] J. T. Poolman, C. T. P. Hopman, and H. C. Zanen, *F.E.M.S. Microbiol. Letters*, 1978, **4**, 245.
[56] G. A. Limjuco, Y. D. Karkhanis, J. Y. Zeltner, R. Z. Maigetter, J. J. King, and D. J. Carlo, *J. Gen. Microbiol.*, 1978, **104**, 187.
[57] B. J. Wilkinson, *F.E.M.S. Microbiol. Letters*, 1977, **2**, 285.
[58] M. Phillips and P. A. Rebers, *Analyt. Biochem.*, 1978, **85**, 265.
[59] S. Rottem, O. Markowitz, and S. Razin, *European J. Biochem.*, 1978, **85**, 445.
[60] S. Rottem, O. Markowitz, and S. Razin, *European J. Biochem.*, 1978, **85**, 451.
[61] K. Jarrell and A. M. Kropinski, *Microbios.*, 1977, **19**, 103.
[62] D. F. Day, M. L. Marceau-Day, and J. M. Ingram, *Canad. J. Microbiol.*, 1978, **24**, 196.
[63] P. M. Meadow, P. L. Wells, M. Salkinoja-Salonen, and E. L. Nurmiaho, *J. Gen. Microbiol.*, 1978, **105**, 23.
[64] S. G. Wilkinson and D. P. Taylor, *J. Gen. Microbiol.*, 1978, **109**, 367.
[65] R. N. Tharanathan, H. Mayer, and J. Weckesser, *Biochem. J.*, 1978, **171**, 403.

that the *O*-methyl sugars may occur at non-reducing termini or at internal positions in chains or both.

The compositions of the lipopolysaccharides of eight wild-type strains of *Rhodospirillum tenue* have been reported.[66] Structural studies on the lipid A portion of the lipopolysaccharide of *R. tenue* 2761 have revealed an unusual structure (4) in which a (1 → 6)-linked 2-amino-2-deoxy-D-glucosyl disaccharide is substituted at the non-reducing end by 4-amino-4-deoxy-L-arabinopyranose 1-phosphate and at the reducing end glycosidically by D-arabinofuranose 1-phosphate.[67] The amino-sugar at the reducing end of the central disaccharide is branched, carrying a third 2-amino-2-deoxy-D-glucose residue having a (1 → 4) linkage.

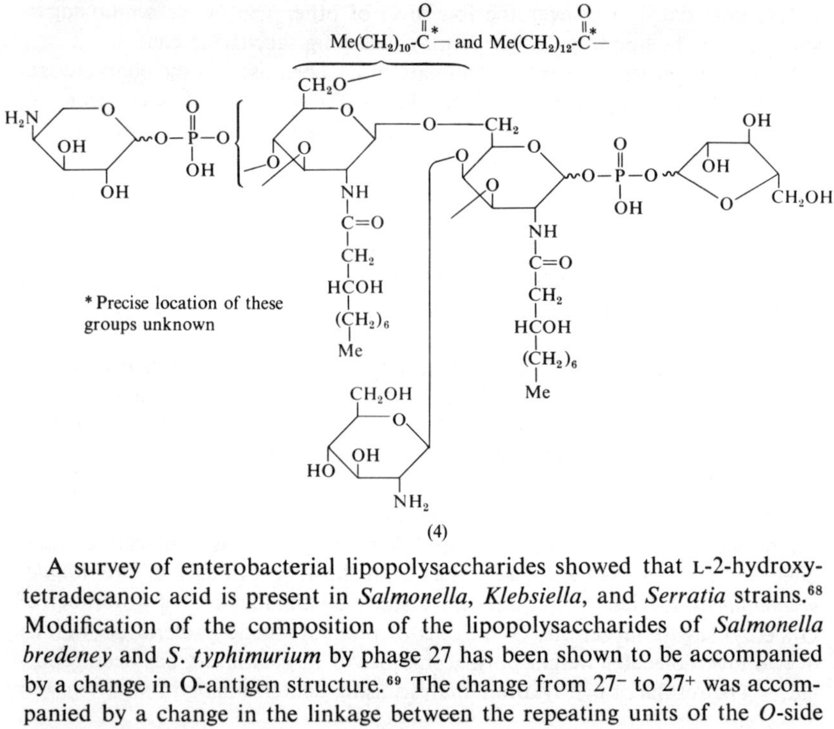

(4)

A survey of enterobacterial lipopolysaccharides showed that L-2-hydroxytetradecanoic acid is present in *Salmonella*, *Klebsiella*, and *Serratia* strains.[68] Modification of the composition of the lipopolysaccharides of *Salmonella bredeney* and *S. typhimurium* by phage 27 has been shown to be accompanied by a change in O-antigen structure.[69] The change from 27^- to 27^+ was accompanied by a change in the linkage between the repeating units of the *O*-side chain from D-galactose-(1 → 2)-D-mannose to D-galactose-(1 → 6)-D-mannose. A mutant of *S. typhimurium*, defective in the synthesis of 3-deoxy-D-*manno*-octulosonic acid 8-phosphate, has been used in a study of the biosynthesis of the lipid A component of the lipopolysaccharide.[70] A lipid A intermediate, consisting of a poly(diphosphorylated 2-amino-2-deoxy-D-glucose) backbone substituted by two 3-hydroxymyristate groups, accumulates and can serve as a

[66] J. Weckesser, G. Drews, R. Indira, and H. Mayer, *J. Bacteriol.*, 1977, **130**, 629.
[67] R. N. Tharanathan, J. Weckesser, and H. Mayer, *European J. Biochem.*, 1978, **84**, 385.
[68] K. Bryn and E. T. Rietschel, *European J. Biochem.*, 1978, **86**, 311.
[69] A. A. Lindberg, C. G. Hellerqvist, G. Bagdian-Motta, and P. H. Mäkelä, *J. Gen. Microbiol.*, 1978, **107**, 279.
[70] V. Lehmann, J. Redmond, A. Egan, and I. Minner, *European J. Biochem.*, 1978, **86**, 487.

direct acceptor for the incorporation of 4-amino-4-deoxy-L-arabinose, phosphorylethanolamine, and 3-deoxy-D-*manno*-octulosonic acid. The lipopolysaccharide of *S. typhimurium* has been cleaved by a phage-derived *endo*-rhamnanase to produce an octasaccharide.[71] Antibodies raised to the oligosaccharide–bovine serum albumin conjugate showed that the octasaccharide retained full O-antigenic activity. Defects in the lipopolysaccharide core of a number of mutants of *S. typhimurium* which are resistant to the Felix O phage but retain the smooth character have been reported.[72]

A re-investigation of the structures of the O-antigens of all known serotypes and subserotypes of *Shigella flexneri* has been reported.[73] The results support the hypothesis that these antigens are composed of the basic tetrasaccharide repeating unit → 3)-β-D-Glc*p*NAc-(1 → 2)-α-L-Rha*p*-(1 → 2)-α-L-Rha*p*-(1 → 3)-α-L-Rha*p*. The immunological determinant responsible for a number of factors was identified as an α-D-glucopyranosyl group linked to different positions on the repeating unit, whilst other factors were due to the presence of an *O*-acetyl group linked to position 2 of the 3-substituted α-L-rhamnopyranosyl residue. By comparing the results of methylation analysis and periodate oxidation studies on the lipopolysaccharides of the phase I, phase II, and D-galactose-deficient R mutants of *S. sonnei* a common structure (5) has been proposed for their hexose regions.[74] The results of serological studies indicated that 3-deoxy-D-*manno*-octulosonic acid is the immunodominant sugar.

D-Gal*p*-(1→2)-D-Gal*p*-(1→2)-D-Glc*p*-(1→3)-D-Glc*p*-(1→)-Hep-dOclA region
6 ↑ 2 (dOclA)$_m$ [on first Gal*p*]; 6 ↑ 2 (dOclA)$_n$ [on second Gal*p*]; 3 ↑ 1 D-Glc*p*-(3←)-(Phase I O-side chain) [on Glc*p*]

Hep = heptose
dOclA = 3-deoxy-D-*manno*-octulosonic acid
Phase I: $m = n = 0$; Phase II $m = 0$ or 1, $n = 0$ or 1, $m+n = 1$ or 2

(5)

A lipopolysaccharide of composition similar, but not identical, to that of the intact cell wall of *Spirillum serpens* has been located in a superficial layer of the wall.[75] The major constituent sugars of the polysaccharide portion of the lipopolysaccharide of *Thiocapsa roseopersicina* have been identified as D-mannose, L-rhamnose, 3-amino-3,6-dideoxy-D-galactose, and D-glucose.[76] A conspicuous feature of the lipid A fraction is the D-mannose content. The polysaccharide portion of the lipopolysaccharides of two antigenically different strains of *Vibrio cholerae* have been fractionated into a high molecular weight polysaccharide corresponding to the side chain and a lower molecular weight

[71] S. B. Svenson and A. A. Lindberg, *J. Immunol.*, 1978, **120**, 1750.
[72] H. P. Hudson, A. A. Lindberg, and B. A. D. Stocker, *J. Gen. Microbiol.*, 1978, **109**, 97.
[73] L. Kenne, B. Lindberg, K. Petersson, E. Katzenellenbogen, and E. Romanowska, *European J. Biochem.*, 1978, **91**, 279.
[74] T. Kontrohr and B. Kocsis, *European J. Biochem.*, 1978, **88**, 267.
[75] I. R. Chester and R. G. E. Murray, *J. Bacteriol.*, 1978, **133**, 932.
[76] R. E. Hurlbert, J. Weckesser, R. N. Tharanathan, and H. Mayer, *European J. Biochem.*, 1978, **90**, 241.

polysaccharide corresponding to the core region.[77] Residues of D-glucose, D-mannose, D-galactose, D-xylose, L-arabinose, L-fucose, D-mannuronic acid, 2-amino-2-deoxy-D-glucose, and 2-amino-2-deoxy-D-galactose have been identified in the polysaccharide isolated from the lipopolysaccharide from a non-agglutinable *Vibrio* species.[78] Mild acid hydrolysis has been used to release 4-amino-4,6-dideoxy-D-mannose (D-perosamine) and 4-amino-4-deoxy-L-arabinose from the lipopolysaccharides of *V. cholerae* and related *Vibrio* species.[79] Both sugars were reported to be highly unstable and to undergo complex changes in neutral and basic media. Lipid A has been isolated from the lipopolysaccharide of *Yersinia pseudotuberculosis* and its composition compared with that from *Y. pestis*.[80]

The isolation and purification of the enterobacterial common antigen from *Salmonella montevideo* has been reported.[81] Chemical analysis identified the antigen as a linear polymer of alternating sequences of (1 → 4)-linked 2-acetamido-2-deoxy-D-glucose and 2-acetamido-2-deoxy-D-mannopyranosyluronic acid units, esterified to a small extent by palmitic and acetic acids. Esterification of the carboxylic acid groups of the 2-acetamido-2-deoxy-D-mannopyranosyluronic acid residues was followed by a reversible loss in the antigen's precipitating capacity with antibody, suggesting that this sugar is an essential part of the serological determinant of the antigen.[82, 83] The antigen has been reported to be covalently linked to the hexose region of the lipopolysaccharide only in an immunogenic strain of *Escherichia coli* and not in a non-immunogenic strain.[84]

The components of enterobacterial common antigen isolated from *Shigella sonnei* have been identified as 2-acetamido-2-deoxy-D-mannopyranosyluronic acid and 2-acetamido-2-deoxy-D-glucose (2 : 1).[85] Fatty acids (3.2%) and *O*-acetyl groups were also detected. Biosynthesis of the aminouronic acid component of the *Salmonella typhimurium* antigen was shown to require UDP-2-acetamido-2-deoxy-D-mannopyranosyluronic acid (UDP-D-Man*p*NAcUA) which is synthesized from UDP-2-acetamido-2-deoxy-D-glucose (Scheme 1).[86] The loss of UDP-D-Man*p*NAc dehydrogenase activity in *rff* mutants in *S. typhimurium* results in the inability of the organism to synthesize the enterobacterial common antigen.

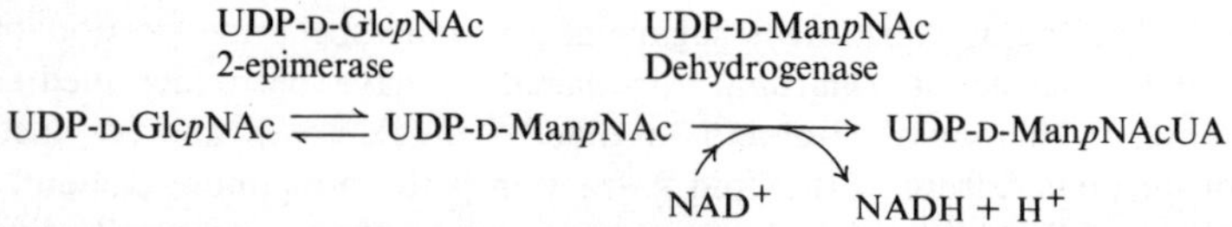

Scheme 1

[77] S. Raziuddin, *Indian J. Biochem. Biophys.*, 1977, **14**, 262.
[78] A. K. Sen and A. K. Mukherjee, *Carbohydrate Res.*, 1978, **64**, 215,
[79] J. W. Redmond, *Biochim. Biophys. Acta*, 1978, **542**, 378.
[80] I. N. Krasikova, V. I. Gorbach, T. F. Solov'eva, and Y. S. Ovodov, *European J. Biochem.*, 1978, **89**, 287.
[81] D. Männel and H. Mayer, *European J. Biochem.*, 1978, **86**, 361.
[82] D. Männel and H. Mayer, *European J. Biochem.*, 1978, **86**, 371.
[83] D. Männel, P. H. Mäkelä, and H. Mayer, *J. Bacteriol.*, 1978, **135**, 348.
[84] P. Kiss, J. Rinno, G. Schmidt, and H. Mayer, *European J. Biochem.*, 1978, **88**, 211.
[85] C. Lugowski, and E. Romanowska, *European J. Biochem.*, 1978, **91**, 89.
[86] H. C. Lew, H. Nikaido, and P. H. Mäkelä, *J. Bacteriol.*, 1978, **136**, 227.

4 Capsular Polysaccharides

A new method established for the *N*-deacetylation of 2-acetamido-2-deoxyhexoses involves transamidation with trifluoracetic anhydride.[87] Very little depolymerization of the capsular polysaccharide of *Diplococcus pneumoniae* type 14 occurred, and on subsequent treatment with nitrous acid it was confirmed that the polysaccharide contains D-glucose and D-galactose (1 : 2). Side chains of some capsular polysaccharides consisting of single or substituted uronic acid residues have been eliminated in a sequence involving acetylation, treatment with acetic anhydride–triethylamine, and deacetylation.[88] A method for the selective cleavage of glycopyranosiduronamide linkages in methylated polysaccharides has been applied to a study of the structure of the capsular polysaccharide of *Klebsiella* K47.[89]

Using established methods, the structures (6—15) of the following capsular polysaccharides have been established: *Cryptococcus neoformans* (6),[90]

```
3)→D-Manp-(1→3)-D-Manp-(1→3)-D-Manp-(1→
    2            2            2
    ↑            ↑            ↑
    1            1            1
    R            R            R
```

R = D-Xyl*p* (predominant) or D-Glc*p*UA

(6)

→3)-α-D-Gal*p*-(1→3)-α-D-Gal*p*UA-(1→3)-α-L-Fuc*p*-(1→

(7)

```
→4)-α-D-Galp-(1→2)-α-D-Ribf-(1→
                        3
                        ↑
                        1
                      α-Tyv
```

Tyv = tyvelose

(8)

```
                                   Me   CO2H
                                     \ /
                                      C
                                     / \
                                    6   4
→3)-α-L-Fucp-(1→3)-β-D-Glcp-(1→3)-β-D-Manp-(1→4)-α-D-GlcpUA-(1→
```

(9)

```
→3)-α-L-Rhap-(1→3)-β-D-Galp-(1→4)-α-D-Glcp-(1→
                                        3
                                        ↑
                                        1
α-D-Glcp-(1→4)-β-D-GlcpUA-(1→2)-α-L-Rhap
```

(10)

```
                     →3)-α-L-Rhap-(1→3)-β-D-Glcp-(1→
                               2
                               ↑
                               1
β-D-GlcpUA-(1→6)-α-D-Glcp
```

(11)

[87] B. Nilsson and S. Svensson, *Carbohydrate Res.*, 1978, **62**, 377.
[88] B. Lindberg, F. Lindh, and J. Lönngren, *Carbohydrate Res.*, 1978, **61**, 81.
[89] G. O. Aspinall and K. G. Rosell, *Canad. J. Chem.*, 1978, **56**, 685.
[90] E. H. Merrifield, A. M. Stephen, and J. Vandenende, *S. African J. Sci.*, 1978, **74**, 146.

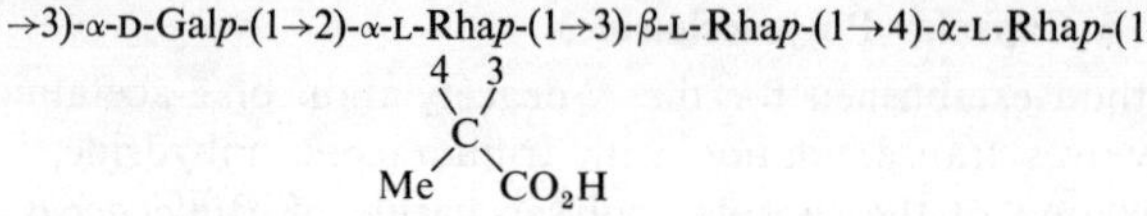

(12)

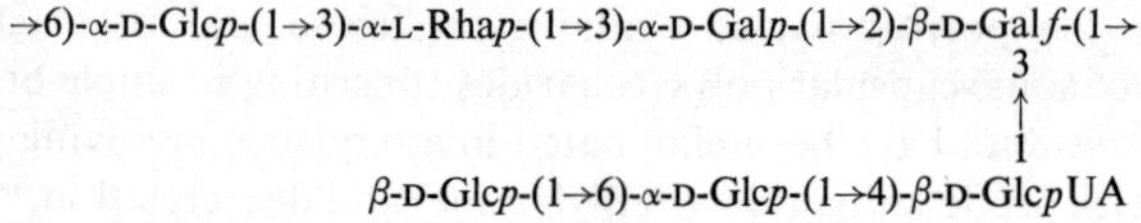

(13)

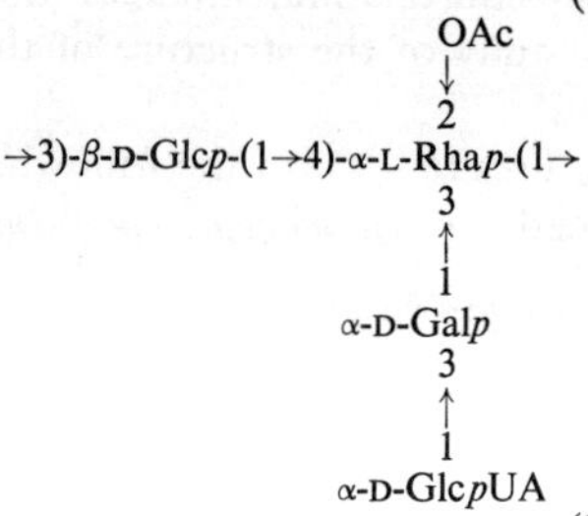

(14)

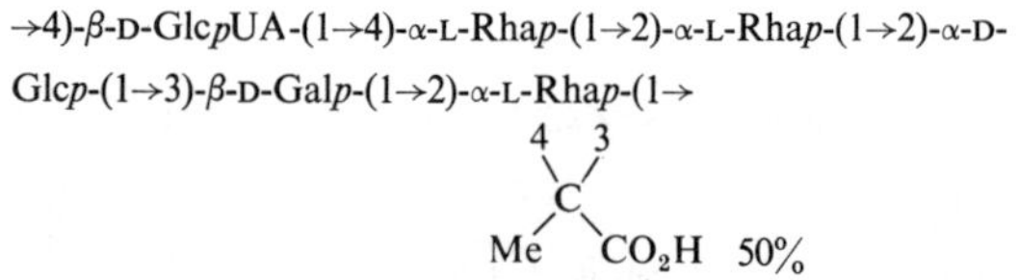

(15)

Escherichia coli K42 (7),[91] *Eubacterium suburreum* L32 (8),[92] and *Klebsiella*: K6 (9),[93] K18 (10),[94] K23 (11),[95] K32 (12),[96] K41 (13),[97] K55 (14),[98] and K70 (15).[99]

The capsular polysaccharide of *Diplococcus pneumoniae* has been shown to possess a linear pentasaccharide repeating unit containing residues of D-glucopyranosyluronic acid, D-galactofuranose, and L-rhamnopyranose (1 : 2 : 2).[100] The presence of 2,3-linked D-galactofuranosyl residues in the polysaccharide of *Klebsiella* K41 has been confirmed by the characterization of 5,6-di-*O*-methyl-D-galactofuranose as a product in the methylation of the polysaccharide, and of the presence of L-arabinose in the Smith degradation

[91] H. Niemann, A. K. Chakraborty, H. Friebolin, and S. Stirm, *J. Bacteriol.*, 1978, **133**, 390.
[92] J. Hoffman, B. Lindberg, T. Hofstad, and N. Skaug, *Carbohydrate Res.*, 1978, **66**, 67.
[93] U. Elsässer-Beile, H. Freibolin, and S. Stirm, *Carbohydrate Res.*, 1978, **65**, 245.
[94] G. G. S. Dutton, K. L. Mackie, and M. T. Yang, *Carbohydrate Res.*, 1978, **65**, 251.
[95] G. G. S. Dutton, K. L. Mackie, A. V. Savage, and M. D. Stephenson, *Carbohydrate Res.*, 1978, **66**, 125.
[96] G. M. Bebault, G. G. S. Dutton, N. A. Funnell, and K. L. Mackie, *Carbohydrate Res.*, 1978, **63**, 183.
[97] J. P. Joseleau, M. Lapeyre, M. Vignon, and G. G. S. Dutton, *Carbohydrate Res.*, 1978, **67**, 197.
[98] G. M. Bebault and G. G. S. Dutton, *Carbohydrate Res.*, 1978, **64**, 199.
[99] G. G. S. Dutton and K. L. Mackie, *Carbohydrate Res.*, 1978, **62**, 321.
[100] N. Roy, *Carbohydrate Res.*, 1978, **63**, 333.

products of the polysaccharide.[101] A pentasaccharide repeating unit has been proposed for the capsular polysaccharide of *Klebsiella* K61, which is composed of residues of D-galactose, D-glucose, D-mannose, and D-glucuronic acid (1 : 2 : 1 : 1).[102] Although some of the basic structural features were identified from the results of methylation analysis, an unequivocal structure could not be established.

The polysaccharide antigen of *Haemophilus influenzae* type B has been labelled with tritium by use of a microculture technique using 6-[^{3}H]-D-glucose.[103] Fibre *X*-ray diffraction patterns have been interpreted for the polysaccharide from *Klebsiella* K57, and results indicate that the molecule has a three-fold helix with an axially projected repeat unit of 1.143 nm which correlates directly with the chemical repeat unit.[104]

The capsular polysaccharide antigen of *Neisseria meningitidis* serogroup 29-e, composed of equimolar amounts of 2-acetamido-2-deoxy-D-galactose and 3-deoxy-D-*manno*-octulosonic acid (KDO), has been identified, using a combination of chemical and ^{13}C n.m.r. spectroscopic techniques, as a linear polysaccharide having alternative sequences of the two residues (16).[105]

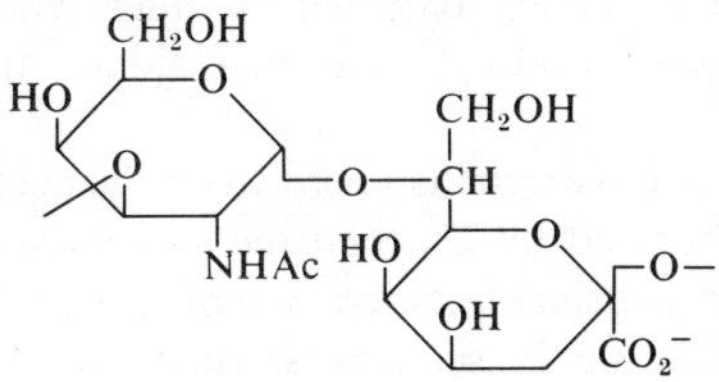

(16)

A purified capsular polysaccharide antigen from *Streptococcus* type 87 and polysaccharide antigens from two capsular pneumococcal strains isolated after transformation with the DNA of the same streptococcal strain have been shown to be immunochemically identical.[106] Each polysaccharide was composed of residues of L-rhamnose (the immunodominant sugar), D-glucose, 2-amino-2-deoxy-D-galactose, and phosphate. Further evidence has been presented to substantiate the view that the type-specific capsular substance of group B type Ia streptococci possesses terminal neuraminic acid residues and a D-galactosyl oligosaccharide as the two major immunodeterminants.[107]

5 Extracellular and Intracellular Polysaccharides

Light-scattering measurements and the specific refractive increment behaviour of dextran in DMSO–water mixtures have been reported.[108] The advantages

101 G. Chambat, J. P. Joseleau, M. Lapeyre, and A. Lefebvre, *Carbohydrate Res.*, 1978, **63**, 323.
102 A. S. Rao, N. Roy, and W. Nimmich, *Carbohydrate Res.*, 1978, **67**, 449.
103 P. Anderson, *J. Immunol.*, 1978, **120**, 866.
104 D. H. Isaac, K. H. Gardner, E. D. T. Atkins, U. Esasser-Beile, and S. Stirm, *Carbohydrate Res.*, 1978, **66**, 43.
105 A. K. Bhattacharjee, H. J. Jennings, and C. P. Kenny, *Biochemistry*, 1978, **17**, 645.
106 P. S. Ventkateswaran and N. Stanton, *Infection and Immunity*, 1978, **20**, 194.
107 J. A. Kane and W. W. Karakawa, *Infection and Immunity*, 1978, **19**, 983.
108 F. R. Dintzis and R. Tobin, *Carbohydrate Res.*, 1978, **66**, 71.

of using a dialysed solution and pair of solvents to obtain values free from errors caused by impurities and solute–solvent association were stressed. No concentration dependence of the specific refractive index increment for dextran dissolved in water or DMSO has been detected.[109]

In an ^{1}H n.m.r. spectroscopic study of the extracellular glucans of *Streptococcus mutans*, it has been possible to assign the doublet resonances to the (1 → 6)- and (1 → 3)-α-D-glucopyranosyl linkages occurring in homogeneous, heterogeneous, and branched configurations, although estimates of the different proportions of these linkages could not be made.[110]

Extracellular water-insoluble α-D-glucans of several strains of *Streptococcus* species have been examined using methylation analysis, and enzymic synthesis and degradation.[111] All the glucans were highly branched and contained chains having sequences of (1 → 3)- and (1 → 6)-linked D-glucose residues. A hypothesis was advanced to explain the role of glucosyltransferases in the synthesis of highly branched glucans containing both types of chain.

Kinetic evidence for hemiacetal formation during the oxidation of dextran by aqueous sodium periodate has been interpreted in terms of intra-residual hemiacetal formation by singly oxidized residues, and inter-residual hemiacetal formation between doubly oxidized residues and intact adjacent D-glucose residues.[112]

Five tetrasaccharides have been isolated from the acetolysis products of the insoluble dextran elaborated by *Leuconostoc mesenteroides* NRRL B-1299.[113] Structural identification of the oligomers demonstrated that the (1 → 3)-linked α-D-glucosyl residues occur in sequences, thus supporting the theory that there is a relationship between lower solubility and sequences of (1 → 3)-linked residues in the dextrans. A number of dextrans have been hydrolysed by an *exo*-isomaltohydrolase (EC 3.2.1.94) from *Arthrobacter globiformis* to produce, in addition to isomaltose, one or two pentasaccharides in which an α-D-glucosyl residue is substituted at O-2, O-3, or O-4 by an isomaltosyl unit.[114]

The relationship between the structures of six native dextrans and their effects on non-specific resistance to infection in mice has been studied.[115] The activity of the dextrans generally increased with increase in the extent of branching and was possibly related to the lengths of the outer chains of the dextrans. Reactions of dextransucrase of *Leuconostoc mesenteroides* and sucrose in the presence of sugars of low molecular weight have been shown to give a low molecular weight dextran and a series of oligosaccharides.[116] It is proposed that the acceptor reaction proceeds by nucleophilic displacement of D-glucosyl and dextranosyl groups from a covalent enzyme complex by a specific acceptor hydroxy-group, and that this reaction effects a glycosidic linkage between the

[109] A. M. Basedon, K. H. Ebert, and U. Ruland, *Makromol. Chem.*, 1978, **179**, 1351.
[110] T. S. Meyer, B. L. Lamberts, and R. S. Egan, *Carbohydrate Res.*, 1978, **66**, 33.
[111] M. D. Hare, S. Svensson, and G. J. Walker, *Carbohydrate Res.*, 1978, **66**, 245.
[112] M. K. Ishak and T. J. Painter, *Carbohydrate Res.*, 1978, **64**, 189.
[113] T. Watanabe, K. Shisido, M. Kobayashi, and K. Matsuda, *Carbohydrate Res.*, 1978, **61**, 119.
[114] T. Sawai, T. Tohyama, and T. Natsume, *Carbohydrate Res.*, 1978, **66**, 195.
[115] M. E. Preobrazhenskaya, E. L. Rosenfeld, and L. Kandra, *Carbohydrate Res.*, 1978, **66**, 213.
[116] J. F. Robyt and T. F. Walseth, *Carbohydrate Res.*, 1978, **61**, 433.

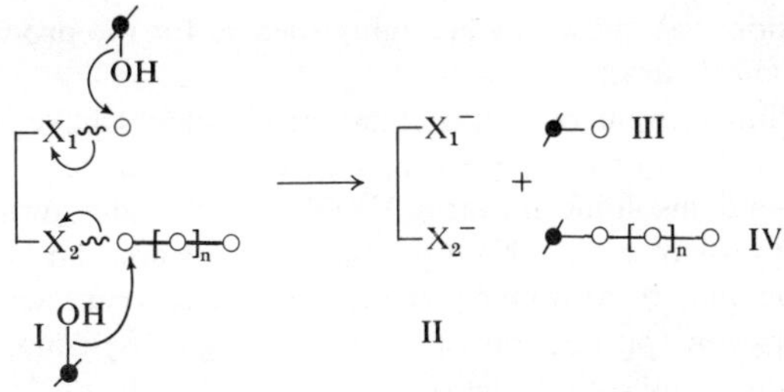

I—labelled acceptor displacing D-glucosyl and dextranosyl groups from the charged enzyme. —O—O— two (1 → 6)-linked α-D-glucosyl residues, ●—OH D-[^{14}C]glucose, D-[^{14}C]-fructose, or α-D-[^{14}C]glucosyl-(1 → 4)-D-glucose.

II—X_1 and X_2 represent nucleophilic groups at the active site of the enzyme.

III—oligosaccharide obtained by displacement of D-glucosyl group.

IV—dextran obtained by displacement of dextranosyl group.

D-glucosyl and dextranosyl groups and the acceptor (Scheme 2). The acceptor reaction appears to serve in the termination of polymerization of dextran by displacing the growing polysaccharide chain from the active site of the enzyme. Both glucoamylase and endodextranase have been found to inhibit the synthesis of dextran with the dextransucrases of *L. mesenteroides*, by a mechanism which, in one case, involves competition between the glucoamylase and dextransucrase for the D-glucosyl groups at the non-reducing end of the dextran, and in the other case may result in a rapid hydrolysis of acceptor dextrans.[117] These results support the view that the synthesis of dextran occurs by the addition of D-glucosyl groups (which become residues) at the non-reducing end of the growing dextran.

The synthesis of extracellular glucans by a number of strains of two types of *Streptococcus mutans* has been examined.[118] Differing amounts of polysaccharide and differing ratios of soluble to insoluble polysaccharide were obtained with the two types. Extracellular water-soluble and cell-associated glucans from a wild-type strain of *S. mutans*, and from mutants defective in the synthesis of glucans, have been subjected to methylation analysis.[119] Two distinct α-D-glucan classes were identified from the wild-type, one rich in (1 → 3)-linked- and the other rich in (1 → 6)-linked-D-glucopyranosyl residues. The mutants exhibited a diminution of virulence which is ascribable to a failure to form the polysaccharide with the high level of (1 → 3)-glucosidic linkages.

Lysophosphatidyl choline and other phosphoglycerides have been used to stimulate the production of both water soluble- and water insoluble-glucans by the dextransucrase of *S. mutans*,[120] and addition of non-ionic detergents has resulted in elevated levels of glucosyl transferase being produced by *S. salivarius* cells.[121] Dextransucrase activity has been isolated from the insoluble dextran-cell complex produced by *S. mutans* K1R.[122] The dextran

[117] M. Kobayashi and K. Matsuda, *Carbohydrate Res.*, 1978, **66**, 277.
[118] K. Trautner, F. Gehring, and D. Lohmann, *Arch. Oral Biol.*, 1978, **23**, 175.
[119] M. Freedman, D. Birkhed, and K. Granath, *Infection and Immunity*, 1978, **21**, 17.
[120] S. K. Harlander and C. F. Schachtele, *Infection and Immunity*, 1978, **19**, 450.
[121] C. L. Wittenberger, A. J. Beaman, and L. N. Lee, *J. Bacteriol.*, 1978, **133**, 231.
[122] M. M. McCabe and E. E. Smith, *Carbohydrate Res.*, 1978, **63**, 223.

acceptor reaction was shown to be highly specific for the production of high molecular weight dextran.

The extracellular D-glucosyltransferases of *S. mutans* have been separated into two enzyme fractions which act on sucrose to produce as the major products soluble and insoluble D-glucan.[123] Oligosaccharides terminated at the reducing end by sucrose have been prepared from starch and used in growth studies on *S. mutans*.[124] Although growth and production of acid were normal, cellular aggregation, production of D-glucan, D-glucosyltransferase activity, plaque formation, and adherence of cells to glass surfaces were diminished. Further results have been presented to support the theory that neither D-glucosyltransferase production nor D-glucan synthesis from sucrose is required for, or promotes, the initial attachment of *S. mutans* cells to teeth or to hydroxyapatite surfaces treated with saliva.[125]

The presence of bacterial species other than *S. mutans* in smooth surface dental plaque has been shown to be due partly to the contact of cells with D-glucan in the developing plaque and not to the binding of cell-free D-glucosyltransferase and the *in situ* synthesis of glucan.[126] The binding of D-glucosyltransferase from the culture supernatant of *S. mutans* to the cells of *S. mutans*, the synthesis of cell-bound D-glucan from sucrose by the bound enzyme, and the subsequent adherence of the cells to a glass surface have been described.[127]. The binding of the D-glucosyltransferase to the cell surface of *S. mutans* has been suggested to involve multiple binding sites with different types of protein–polysaccharide interactions.[128] Alternatively, the binding sites of the free enzyme have been considered to be the surface D-glucans of *S. mutans* and it has been suggested that the cell-associated and extracellular D-glucosyltransferases are most likely to be alternate states of the same enzyme protein.[129]

Preformed, cell-associated water-insoluble D-glucan of *S. mutans* has been shown to have no significant adherent property.[130] Adherence only occurred as a result of active *de novo* D-glucan biosynthesis in the presence of sucrose, cells and extracellular or cell-bound D-glucosyltransferases. The relative proportion of D-glucosyltransferase isoenzymes elaborated by mutants of *S. mutans*, insofar as they affect the physicochemical properties of the D-glucans produced, has been shown to determine the adherence abilities of the cells.[131] Again, the importance of *de novo* D-glucan biosynthesis in the adherence process was stressed. Further evidence has indicated that the D-glucosyltransferase enzymes of *S. mutans* are constitutive cell products synthesized regardless of the carbon source being utilized. [132] The dextran produced by *S. sanguis*

[123] J. E. Ciardi, A. J. Beaman, and C. L. Wittenberger, *Infection and Immunity*, 1977, **18**, 237.
[124] T. Ikeda, T. Shiota, J. R. McGhee, S. Otake, S. M. Michalek, K. Ochiai, M. Hirasawa, and K. Sugimoto, *Infection and Immunity*, 1978, **19**, 477.
[125] W. B. Clark, L. C. Bammann, and R. J. Gibbons, *Infection and Immunity*, 1978, **21**, 681.
[126] S. Hamada, S. Tai, and H. D. Slade, *Infection and Immunity*, 1978, **21**, 213.
[127] S. Hamada, Y. Kobayashi, and H. D. Slade, *Microbiol. and Immunol.*, 1978, **22**, 279.
[128] H. K. Kuramitsu, and L. Ingersoll, *Infection and Immunity*, 1978, **20**, 652.
[129] S. Hamada and M. Torii, *Infection and Immunity*, 1978, **20**, 592.
[130] S. Hamada, *Microbios. Letters*, 1977, **5**, 141.
[131] T. Koga and M. Inoue, *Infection and Immunity*, 1978, **19**, 402.
[132] W. M. Janda and H. K. Kuramitsu, *Infection and Immunity*, 1978, **19**, 116.

has been shown to be a factor that influences the organism's ability to adhere to surfaces and to stimulate non-bacterial endocarditis.[133] The action of antibodies, directed against both extracellular and cell-bound D-glucosyltransferase of *S. sanguis*, on the kinetics of the enzyme reaction has shown that anti-D-glucosyltransferase antibody acts as a non-competitive inhibitor with respect to sucrose.[134] This inhibition is diminished upon addition of primer dextran and the antibodies act as a mixed type inhibitor with this substrate.

Two dextran-specific myeloma proteins have been isolated from mice.[135] Both anti-dextrans were specific for (1 → 6)-linked-α-D-glucosyl residues but differed with respect to their relative reactivities with dextrans containing various proportions of (1 → 6)-, (1 → 4)-, and (1 → 3)-linked α-D-glucosyl residues.

The cyclodextrin D-glucanotransferase (EC 2.4.1.19) from *Bacillus macerans* has been purified to homogeneity and shown to be composed of two subunits.[136] A procedure for the preparation of cyclohexa- and cyclohepta-amyloses and branched cycloamyloses from different starches using the cyclodextrin D-glucanotransferase of *B. macerans* in the presence of sodium dodecyl sulphate has been reported.[137] Hydroxyethyl cycloamyloses were produced by the action of the enzyme on hydroxyethyl starch and hydroxyethyl amylose, when the main product was identified as monohydroxyethyl cyclohexa-amylose. At high concentrations of the substrate cyclohexa-amylose, and low concentrations of the acceptors maltose or D-glucose, the cyclodextrin D-glucanotransferase from *Klebsiella pneumoniae* M5 has been shown to synthesize amylose-like (1 → 4)-α-D-glucopyranosyl chains having an average degree of polymerization of 45.[138] Significantly larger proportions of longer chains and of cyclohepta-amylose were formed using maltose as the acceptor.

A mechanism for the *de novo* synthesis of glycogen in *Escherichia coli* has been proposed.[139] Two different D-glucosyltransferases are involved; one transfers D-glucose to a protein acceptor either from UDP-D-glucose or ADP-D-glucose to produce an oligosaccharide linked to protein of a certain minimal size, and the other is an ADP-D-glucose-dependent enzyme exclusively involved in the elongation of the oligosaccharide–protein intermediate. The participation of a branching enzyme then produces a polysaccharide with the characteristics of glycogen. The pyridoxal phosphate allosteric activator site and the ADP-D-glucose-protected pyridoxal phosphate binding site of the ADP-D-glucose-synthetase of *E. coli* B have been characterized.[140] An intracellular polysaccharide (mol. wt. 2—4.5 × 10^5) accumulating in a methane utilizing bacterium, *Methylococcus* sp., has been characterized as a (1 → 4)-, (1 → 6)-α-D-glucan, with average chain length and average interior and exterior chain lengths similar to those of glycogen.[141]

[133] W. Scheld, J. A. Valone, and M. A. Sande, *J. Clinical Invest*. 1978, **61**, 1394.
[134] J. P. Klein, M. Scholler, and R. M. Frank, *Infection and Immunity*, 1978, **20**, 619.
[135] A. M. Wu, E. A. Kabat, and M. G. Weigert, *Carbohydrate Res.*, 1978, **66**, 113.
[136] S. Kobayashi, K. Kainuma, and S. Suzuki, *Carbohydrate Res.*, 1978, **61**, 229.
[137] S. Kobayashi, K. Kainuma, and S. Suzuki, *J. Agric. Chem. Soc. Japan*, 1977, **51**, 691.
[138] H. Bender, *Carbohydrate Res.*, 1978, **65**, 85.
[139] R. Barengo and C. R. Krisman, *Biochim. Biophys. Acta*, 1978, **540**, 190.
[140] T. F. Parsons and J. Preiss, *J. Biol. Chem.*, 1978, **253**, 7638.
[141] J. D. Linton and R. E. Cripps, *Arch. Mikrobiol.*, 1978, **117**, 41.

An oligosaccharide, composed of residues of D-glucose and D-galactose, with succinate and pyruvate groups, has been isolated from the culture filtrate of *Alcaligenes faecalis* var. *myxogenes*.[142] Its similarity in properties to an oligosaccharide isolated from the parent succinoglucan using a succino-D-glucan polymerase suggests that the oligosaccharide is the repeating unit of the polysaccharide. A complex between Congo Red and curdlan, the gel-forming (1 → 3)-α-D-glucan of *A. faecalis*, has been studied by c.d.[143] The ordered structure of the polysaccharide is observed in the gel state.

Extracellular polysaccharides composed of D-glucose, pyruvic acid, and succinic acid have been isolated from a number of strains of *Agrobacterium*.[144] The conformational behaviour of the extracellular gel-forming polysaccharides from *Arthrobacter* species has been investigated by nuclear magnetic resonance relaxation and optical rotation techniques.[145] A number of the polysaccharides studied showed evidence of an ordered conformation which could be melted out on heating. *O*-Deacetylation destroyed the ordered conformation of two *A. viscosus* polysaccharides, whilst *O*-deacylation of the *A. stabilis* polysaccharide stabilized the ordered structure.

The rate of production of extracellular polysaccharide synthesized by *Azotobacter vinelandii* was quite unaffected by most growth-limiting nutrients.[146] An extracellular polysaccharide, isolated by anion-exchange chromatography from the viscous culture media of *Bacillus amyloliquefaciens*, has been shown to have high molecular weight (1×10^6) and high intrinsic viscosity.[147] It contains residues of D-galactose, D-mannose, D-glucuronic acid, 2-acetamido-2-deoxy-D-glucose, 2-acetamido-2-deoxy-D-galactose, and 4-acetamido-2-amino-2,4,6-trideoxy-D-glucose.

The biosynthesis of the *O*-acetylated alginic acid-type of polysaccharide by *Pseudomonas aeruginosa* under a variety of culture conditions has been reported.[148] An acylated slime polysaccharide, composed of D-glucosyl-, D-galactosyl-, D-rhamnosyl-, and D-mannosyl residues, has been isolated from a *Pseudomonas* species.[149] Using a combination of paper chromatography and g.c.—m.s., 6-*O*-methyl-D-galactose, 3-*O*-methyl-D-glucose, 6-deoxy-L-talose and seven other aldoses have been identified in the acid hydrolysates of an extracellular polysaccharide of a *Rhizobium* species.[150] The intrinsic viscosity, molecular weight, and radii of gyration of the levan of *Streptomyces salivarius* have been measured in different solvents.[151] Data indicated that, in water, levan is not aggregated by hydrogen bonds and that values for both the refractive index and radius of gyration are lower in this solvent than in aqueous

[142] M. Hisamatsu, J. Abe, A. Amemura, and T. Harada, *Carbohydrate Res.*, 1978, **66**, 289.
[143] K. Ogawa and M. Hatano, *Carbohydrate Res.*, 1978, **67**, 527.
[144] M. Hisamatsu, K. Sano, A. Amemura, and T. Harada, *Carbohydrate Res.*, 1978, **61**, 89.
[145] A. Darke, E. R. Morris, D. A. Rees, and E. J. Welsh, *Carbohydrate Res.*, 1978, **66**, 133.
[146] T. R. Jarman, L. Deavin, S. Slocombe, and R. C. Righelato, *J. Gen. Microbiol.*, 1978, **107**, 59.
[147] B. Samor, G. Spik, D. Delebassée, M. Béthencourt, B. Fournet, P. Sicard, B. Colein, and J. Montreuil, *Compt. rend.*, 1977, **285D**, 1259.
[148] F. A. Mian, T. R. Jarman, and R. C. Righelato, *J. Bacteriol.*, 1978, **134**, 418.
[149] A. G. Williams and J. W. T. Wimpenny, *J. Gen. Microbiol.*, 1978, **104**, 47.
[150] L. D. Kennedy, *Carbohydrate Res.*, 1978, **61**, 217.
[151] S. S. Stivala and W. S. Bahary, *Carbohydrate Res.*, 1978, **67**, 17.

solutions of urea. Interactions of an anti-levan myeloma protein with the cell-associated levan of *Actinomyces viscosus* indicate that the levan contains (2 → 6)- and (2 → 1)-linked β-D-glucosyl units.[152]

The separation of xanthan gums of differing pyruvate content has been achieved by fractional precipitation with alcohol.[153] Under steady-stage conditions in a chemostat, *Xanthomonas campestris* has been shown to elaborate considerable amounts of xanthan gum.[154] Growth of the organism under conditions of phosphate and magnesium limitation gave rise to polysaccharides with reduced pyruvate content, suggesting a direct effect of these elements on the availability of phosphoenolpyruvate and the pyruvylation mechanism. A typical native xanthan sample has a molecular weight of 15×10^6 and is a rod-like molecule with a stiffness similar to that of native DNA.[155] The absolute values of the intrinsic viscosity indicate that the polysaccharide has a mass per unit length of about 1.9×10^3 daltons nm^{-1}, which is twice the figure obtained for the single-stranded structure proposed from *X*-ray work. Xanthan has been found to acquire cross-reactivity with horse anti-pneumococcal type VIII serum after mild acid treatment.[156] Inhibition studies showed that the combining sites of antibody cross-precipitated by the polysaccharide optimally accommodate a cellodextrin structure consisting of a terminal or internal sequence of two (1 → 4)-linked β-D-glucose residues.

A polysaccharide isolated from a *Xanthomonas* species, and composed of D-glucose, D-mannose, 3-deoxy-2-octulosonic acid (KDO), and an unknown sugar component, resembles cell-wall lipopolysaccharide in containing KDO, although lipid is absent.[157] Structural studies showed the polysaccharide to contain (1 → 3)- and (1 → 4)-linked hexopyranosyl residues with branching, but the position of the KDO and the unknown sugar was not established.

6 Miscellaneous Polysaccharides

The long chain polyenoic fatty acids α- and β-paranaric acid have been shown to form 1 : 1 complexes with the 6-*O*-methyl-D-glucose-containing polysaccharide of *Mycobacterium smegmatis*.[158] They can then be displaced from the complexes by palmitoyl-CoA which in turn forms a 1 : 1 complex with the polysaccharide. The interaction depends on the structural integrity of the polysaccharide, since complex formation is related to the chain length of the polysaccharide, and unmethylated D-glucans do not form complexes. The toxin of *M. ulcerans* has been characterized as a phospholipoprotein–polysaccharide complex.[159] Toxic activity is destroyed by treatment with proteolytic enzymes, phospholipase, amylase, α- and β-D-glucosidases, and cellulase. Type-specific anti-pneumococcal sera, which are specific for α-L-rhamnosyl

[152] T. N. Warner and C. H. Miller, *Infection and Immunity*, 1978, **19**, 711

[153] P. A. Sandford, P. R. Watson, and C. A. Knutson, *Carbohydrate Res.*, 1978, **63**, 253.

[154] I. W. Davidson, *F.E.M.S. Microbiol. Letters*, 1978, **3**, 347.

[155] G. Holzwarth, *Carbohydrate Res.*, 1978, **66**, 173.

[156] P. Z. Allen, *Arch. Biochem. Biophys.*, 1978, **188**, 376.

[157] T. Yadomae, H. Yamada, T. Miyazaki, T. Omori, and T. Hirota, *Carbohydrate Res.*, 1978, **60**, 129.

[158] K. K. Yagusaki and C. E. Ballou, *Proc. Nat. Acad. Sci., U.S.A.*, 1978, **75**, 691.

[159] W. T. Hockmeyer, R. E. Krieg, M. Reich, and R. D. Johnson, *Infection and Immunity*, 1978, **21**, 124.

linkages, have been used in cross-precipitation reactions with a polysaccharide fraction from *Mycoplasma pneumoniae*.[160]

The interaction of an alkali-stable polysaccharide from the cell surface of staphylococci with human fibrinogen has been modified by prior treatment of the polysaccharide with D-glucose oxidase, α-D-galactosidase, or β-D-2-acetamido-2-deoxyglucosidase.[161]

→4)-β-D-Glc*p*-(1→4)-β-D-Glc*p*-(1→4)-β-D-Gal*p*-(1→
6
↑
1
β-D-Glc*p*
4
↑
1
β-D-Gal*p*

(17)

→3)-6-deoxy-L-Tal*p*-(1→3)-L-Rha*p*-(1→3)-D-Gal*p*-(1→2)-L-Rha*p*-(1→
4
↑
1
D-Glc*p*UA

(18)

The structures of a diheteroglycan (17) from the cell walls of *Streptococcus faecalis* and a tetraheteroglycan (18) from the cell walls of *S. bovis* have been reported.[162] The two polysaccharides are immunogenic substances located on the cell surface of the organisms.

→2/3)-α-L-Rha*p*-(1→3)-α-L-Rha*p*-(1→2)-α-L-Rha*p*-(1→
2
↑
1
α-D-Glc*p*
6
↑
1
α-D-Glc*p*

(19)

An antigenic diheteroglycan (19) has been isolated from the cell walls of *S. bovis*.[163] In a typical molecule, there are 16 of such units alternately linked by (1 → 3)-α- and (1 → 2)-α-L-rhamnosyl residues, with isomaltosyl side-chains as the immunodominant groups of the glycan.

7 Fungal Polysaccharides

The agglutinins from wheat germ, soya bean, and peanut have been used in binding studies to Penicillia and Aspergilli.[165] The three lectins inhibited incorporation of various precursors into fungal hyphae, indicating interference with fungal growth. The findings support an earlier suggestion that one role

[160] P. Z. Allen and B. Prescott, *Infection and Immunity*, 1978, **20**, 421.
[161] K. Yoshida, T. Ohtomo, and Y. Usui, *Experientia*, 1978, **34**, 885.
[162] J. H. Pazur and L. S. Forsberg, *Carbohydrate Res.*, 1978, **60**, 167.
[163] J. H. Pazur, D. J. Dropkin, and L. S. Forsberg, *Carbohydrate Res.*, 1978, **66**, 155.
[164] J. H. Pazur, K. L. Dreher, and L. S. Forsberg, *J. Biol. Chem.*, 1978, **253**, 1832.
[165] R. Barkai-Golan, D. Mirelman, and N. Sharon, *Arch. Mikrobiol.*, 1978, **116**, 119.

of lectins in plants is to provide protection against fungal pathogens. Cross-reactivity of an antiserum to *Absidia cylindrospora* with other fungi has indicated that this antiserum has specificity only for Mucoraceae or Mucorales.[166]

Cytochemical determination of polysaccharides, after periodate oxidation procedures, has been used in the identification of eight constitutive layers in the cell wall of the blastospore of *Candida albicans*.[167] Each layer was assigned a probable chemical composition on the basis of its reactivity at the ultra-structural level. The taxonomy of the genus *Dipodascus* has been examined with respect to the carbohydrate composition of intact cells.[168] Data obtained support the classification of *D. uninucleatus* in a separate genus *Dipodascopsis*.

Glucans.—A mutation in *Aspergillus nidulans* has led to a loss of both melanin and the major cell-wall polysaccharide, (1 → 3)-α-D-glucan.[169] The mutant walls contained increased amounts of (1 → 3)-β-D-glucan and D-galactose polymers, and electron microscopy indicated that they had lost the outermost wall layer. Such walls were more readily digested by lytic enzymes, and this increased susceptibility to hydrolysis was shown to be due to the absence of (1 → 3)-α-D-glucan and not to the absence of melanin. Using a combination of *X*-ray diffraction analysis, enzymolysis, and electron microscopy it has been shown that the majority of the nigeran, a (1 → 3),(1 → 4)-α-D-glucan, is exposed to the cell's exterior in the hyphal walls of *A. niger* and *A. awamori*.[170] Two further fractions of the polysaccharide were only detectable after aqueous or alkaline extraction of the walls.

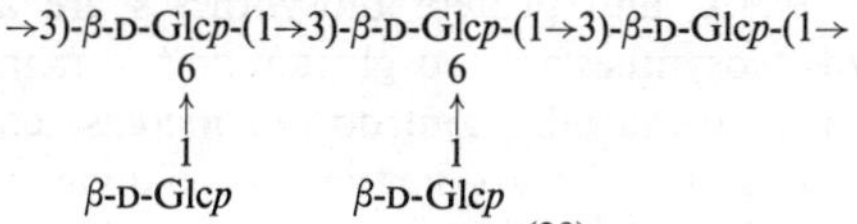

(20)

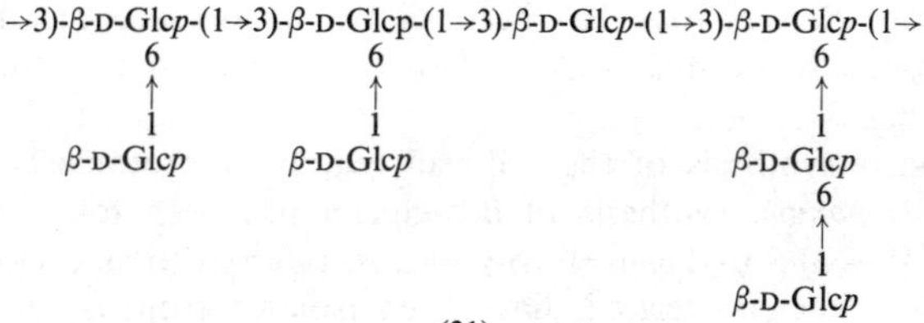

(21)

The structures of a water-soluble D-glucan (20) and an alkali-soluble D-glucan (21) from the fruit body of *Auricularia auricula-judae* have been established.[171] The structural resemblance of polysaccharides from *Tremella fuciformis* to those of other *Tremella* species and the group *Cryptococcus* is

[166] O. Hayashi, T. Yadomae, H. Yamada, and T. Miyazaki, *J. Gen. Microbiol.*, 1978, **108**, 345.
[167] D. Poulain, C. Tronchin, J. F. Dubremetz, and J. Biguet, *Ann. Microbiol.*, 1978, **129*A***, 141.
[168] A. C. M. Weijman, *J. Microbiol. Serol.*, 1977, **43**, 323.
[169] I. Polacheck and R. F. Rosenberger, *J. Bacteriol.*, 1977, **132**, 650.
[170] T. F. Bobbitt, J. H. Nordin, M. Roux, J. R. Revol, and R. H. Marchessault, *J. Bacteriol.*, 1977, **132**, 691.
[171] Y. Sone, M. Kakuta, and A. Misaki, *Agric. and Biol. Chem. (Japan)*, 1978, **42**, 417.

supported by immunochemical studies which indicate very close phylogenic relationships.

Four D-glucans have been isolated from the sclerotia of *Grifora umbellata* by extraction with water, then alkali.[172] Common structural features include (1 → 3)-β- and (1 → 6)-β-linkages both in the backbone and in the linear part of branches. In a similar study using the same fungal source, the isolated D-glucans, which demonstrate anti-tumour activities, were reported to be composed of a backbone containing (1 → 6)- and (1 → 3)-linked β-D-glucosyl residues with two types of branch points involving (1 → 6)- and (1 → 4)-D-glucosyl linkages.[173]

The water-soluble polysaccharide of *Lampteromyces japonicus* has been identified as a β-D-glucan having a backbone of (1 → 3)-linkages and branches at O-6 with gentiobiosyl groups.[174] An α-D-glucan, containing (1 → 3)- and (1 → 4)-linkages (5.3 : 1) and having a low degree of branching, has been isolated from the fruit bodies of *Lentinus edodes*.[175] Gel filtration studies on the D-glucan after degradation using amylolytic enzymes or Smith degradation suggested that the 1,4-linked D-glucopyranosyl residues are not interspersed randomly throughout the molecule but are present in regions close to the non-reducing ends. The hyphal cell wall of *Leptomitus lacteus* has been shown to contain D-glucan with β-(1 → 3)- and β-(1 → 6)-linkages (74%), cellulose (7%) and chitin (14.5%).[176]

An inositol-dependent strain of *Saccharomyces cerevisiae* synthesizes D-glucan although DNA, RNA, and protein biosyntheses are appreciably inhibited.[177] The rates of biosynthesis of D-glucan and D-mannan in *S. cerevisiae* have been shown to be variable and do not increase continuously through the entire cell cycle of growth.[178] At the time of cell division and in the pre-budding phase the rate of synthesis of both polysaccharides is reduced. D-Glucan from *S. cerevisiae* has been shown to enhance the non-specific protection against septicemia by *Staphylococcus aureus* in normal and leukemic mice.[179]

During the re-synthesis of the cell-wall region by protoplasts of *Schizosaccharomyces pombe*, synthesis of β-D-glucan has been found to precede synthesis of (1 → 3)-α-D-glucan.[180] D-Glucans, referred to as elicitors, but so far structurally uncharacterized, have been isolated from the cell walls of *Phytophthora infestans* and found to cause the agglutination and inactivation of protoplasts from potato leaf tissue.[181] There appear to be specific elicitor receptor sites on the outer surface of the potato leaf plasma membranes. The

172 K. Kato, K. Mutoh, T. Egashira, M. Hiura, and Y. Ueno, *Agric. and Biol. Chem. (Japan)*, 1978, **42**, 1073.
173 T. Miyazaki, N. Oikawa, H. Yamada, and T. Yamodae, *Carbohydrate Res.*, 1978, **65**, 235.
174 K. Fukuda and A. Hamada, *Biochim. Biophys. Acta*, 1978, **538**, 580.
175 M. Shida, T. Uchida, and K. Matsuda, *Carbohydrate Res.*, 1978, **60**, 117.
176 J. M. Aronson and C. C. Lin, *Mycologia*, 1978, **70**, 363.
177 A. Dominguez, J. R. Villanueva, and R. Sentandreu, *J. Microbiol. Serol.*, 1978, **44**, 25.
178 P. Biely, *Arch. Mikrobiol.*, 1978, **119**, 213.
179 N. R. DiLuzio and D. L. Williams, *Infection and Immunity*, 1978, **20**, 804.
180 D. R. Kreger and M. Kopecka, *J. Gen. Microbiol.*, 1978, **108**, 269.
181 B. M. Peters, D. H. Cribbs, and D. A. Stelzig, *Science*, 1978, **201**, 364.

mechanisms of the resistance of plants to micro-organisms have been reviewed.[182]

A method for the determination of glycogen in whole yeast cell walls has involved only a pre-treatment to remove alkali-labile constituents of the wall, without prior disruption of the cell walls.[183] The glycogen is then converted to D-glucose by glucoamylase, and the liberated D-glucose measured using conventional procedures. Changes in the structure of glycogen and the activities of phosphorylase and debranching enzyme from a strain of *S. cerevisiae* which is capable of accumulating large amounts of glycogen have been examined.[184] The values for the average chain length (18.0) and β-amylolysis limit (57.9%) indicate that the glycogen has larger outer chains than that from a normal strain of *Saccharomyces* and has a similar structure to that of amylopectin. The results of studies involving measurement of the two enzyme levels, first under D-glucose-limiting conditions and then during growth in D-glucose, suggested (i) that the phosphorylase activity is regulated by physiological conditions and plays an important role in the regulation of glycogen degradation, but (ii) that the debranching enzyme controls the rate-limiting step in the degradation.

Treatment of *Agaricus bisporus* with α-amylase has shown the presence and absence of glycogen in the hyphae and the spores respectively.[185] The α-D-glucan isolated from *Cyttaria harioti* has been subjected to partial acid hydrolysis, acetolysis, and enzymic degradation using glucoamylase and pullulanase.[186] A structure based on maltotriose units connected by $(1 \rightarrow 6)$-α-D-glucosyl linkages is proposed for the D-glucan. This appears to be the first intracellular pullulan-type of D-glucan to be isolated from a fungus, although the possibility that the polysaccharide is elaborated by an internal yeast cannot be excluded since whole stroma were used in the study.

The UDP-D-glucose: glycogen 4-α-D-glucosyltransferase of *Neurospora crassa* has been purified and shown to be a trimer (mol. wt. 2.7×10^5) and to be dependent on D-glucose 6-phosphate.[187] The enzyme catalyses the transfer of D-glucose from UDP-D-glucose to glycogen and its phosphorylase limit dextrin, but not to its β-amylase limit dextrin. Limited proteolysis converts the enzyme to a D-glucose 6-phosphate independent form. Although UDP-D-glucose has been found to be the most active D-glucosyl donor in the synthesis of glycogen in *N. crassa*, in the presence of a primer of the $(1 \rightarrow 4)$-linked α-D-glucan-type it can be replaced by ADP-D-glucose.[188]

Pullulan has been hydrolysed with an enzyme isolated from *Thermoactinomyces vulgaris* which acts differently from pullulanase, isopullulanase, or glucoamylase in cleaving the $(1 \rightarrow 4)$-α-D-glucosyl linkages in pullulan and

[182] P. Albersheim, A. R. Ayers, B. S. Valent, J. Ebel, M. Hahn, P. Wolpert, and J. Carlson, *J. Supramol. Struc.*, 1977, **6**, 599.
[183] J. U. Becker, *Analyt. Biochem.*, 1978, **86**, 56.
[184] S. Tabata, T. Kiyosaki, R. Tsunamoto, and S. Hizukuri, *J. Soc. Ferm. Technol.*, 1978, **56**, 110.
[185] J. Lendemann and D. Rast, *Trans. Brit. Mycol. Soc.*, 1978, **71**, 146.
[186] N. Waksman, R. M. de Lederkremer, and S. Cerezo, *Carbohydrate Res.*, 1977, **59**, 505.
[187] H. Takahara and K. Matsuda, *Biochim. Biophys. Acta*, 1978, **522**, 363.
[188] J. D. Fontana and C. R. Krisman, *Biochim. Biophys. Acta*, 1978, **540**, 183.

liberating panose as the main product.[189] Chemical studies on the extracellular β-D-glucan elaborated by *Botrytis cinerea* have indicated the presence of a backbone of β-1,3-linked D-glucosyl residues with considerable branching.[190] Hydrolysis of the polysaccharide with an *exo*-(1 $\rightarrow$ 3)-β-D-glucanase yields D-glucose and gentiobiose.[191] The polysaccharide functions as a reserve material when growth of *B. cinerea* occurs in a hexose-depleted medium.

An α-D-glucan composed of maltotriose units joined by a single α-(1 $\rightarrow$ 3)-linkage has been isolated from *Elsinoe leucospila*.[192] Periodate oxidation, Smith degradation, and enzymic hydrolysis studies have indicated that an extracellular polysaccharide from *Monilinia fructigena* has a molecular weight of 7×10^4 and a main chain of (1 $\rightarrow$ 3)-linked β-D-glucosyl residues with side chains of single D-glucosyl groups linked β-1,6 to every second D-glucose residue of the main chain.[193] In contrast, a polydisperse extracellular D-glucan (mol. wt. 5×10^3—8×10^4) has been isolated from the same organism.[194] (1 $\rightarrow$ 4)-Linkages predominate in the glucan, as evidenced by the results of periodate oxidation and the susceptibility of the polysaccharide to both α-amylase and a (1 $\rightarrow$ 4)-β-D-glucanase. Some *X*-ray diffraction results for this D-glucan have been reported.[195] That two completely different polysaccharides have been synthesized by the same fungus may result from the use of different growth media.

A temperature-sensitive strain of *Aspergillus nidulans* has been characterized as producing a more thermolabile phospho-D-mannose mutase than the wild-type strain.[196] Together with other results, this suggests a lack of D-mannose 1-phosphate which is normally involved in the synthesis of cell-wall D-mannan. By use of conventional electron microscopy and interactions with concanavalin A, D-mannan polymers have been shown to extend throughout the wall of the blastospore of *Candida albicans*.[197] The D-mannan of *C. guilliermondii* has been subjected to acetolysis to yield a homologous series of (1 $\rightarrow$ 2)-linked. α-D-mannosyl oligosaccharides (DP 1 $\rightarrow$ 5).[198] Together with methylation analysis and g.c.—m.s. results, the data demonstrated the presence of a polysaccharide with a (1 $\rightarrow$ 6)-linked D-mannopyranosyl main chain having (1 $\rightarrow$ 2)-linked D-mannopyranosyl side chains. The surface-localized polysaccharide of *C. tropicalis*, which was grown on a hydrocarbon substrate, has been identified as a D-mannan with fatty acid (4%) covalently attached to it.[199]

A D-mannosylphosphate-transferase has been isolated from *Saccharomyces cerevisiae* X2180 and shown to catalyse the transfer of D-mannose-1-phosphate

[189] Y. Sakano, M. Kogure, T. Kobayashi, M. Tamura, and M. Suekane, *Carbohydrate Res.*, 1978, **61**, 175.
[190] D. Dubourdieu, B. Fournet, A. Bertrand, and P. Ribereau-Gayon, *Compt. rend.*, 1978, **286D**, 229.
[191] D. Dubourdieu, B. Pucheut-Plante, M. Mercier, and P. Ribereau-Gayon, *Compt. rend.*, 1978, **287D**, 571.
[192] Y. Tsumuraya, A. Misaki, S. Takaya, and M. Torii, *Carbohydrate Res.*, 1978, **66**, 53.
[193] F. Santamaria, F. Reyes, and R. Lahoz, *J. Gen. Microbiol.*, 1978, **109**, 287.
[194] S. A. Archer, J. R. Clamp, and D. Migliore, *J. Gen. Microbiol.*, 1977, **102**, 157.
[195] S. A. Archer, E. D. T. Atkins, and K. C. H. Gardner, *F.E.B.S. Letters*, 1978, **92**, 343.
[196] B. P. Valentine and B. W. Bainbridge, *J. Gen. Microbiol.*, 1978, **109**, 155.
[197] A. Cassone, E. Mattia, and L. Boldrini, *J. Gen. Microbiol.*, 1978, **105**, 263.
[198] K. H. Rademacher and G. Reuter, *Z. Allgem. Mikrolbiol.*, 1978, **18**, 63.
[199] O. Kappeli, M. Müller, and A. Fiechter, *J. Bacteriol.*, 1978, **133**, 952.

units from GDP-D-mannose to a reduced (1 → 2)-linked [^{3}H]-α-D-mannotetraose.[200] The product is analogous in structure to the phosphorylated D-mannan side chains of the cell wall which suggests that the enzymic activity is involved in D-mannan-protein biosynthesis in the intact cell. Assays have been developed for some transfer reactions involved in the biosynthesis of *S. cerevisiae* D-mannan-proteins.[201] Phosphomannans from *S. cerevisiae* X2180, *Hansenula holstii*, and some of the products of their partial acid hydrolysis have been used to inhibit pinocytosis of human β-D-glucuronidase by human fibroblasts.[202] The inhibitory properties are considered to be due to the presence of the 6-phosphorylmonoester groups of the D-mannosyl groupings. D-Mannan from bakers' yeast has been separated into four fractions.[203] Three of the fractions exhibited similar precipitation patterns with concanavalin A and with homologous anti-*S. cerevisiae* serum. The differences in immunoreactivity shown by the fourth fraction are probably a result of a smaller number of branch points or a shorter average chain length. Differences in serological activities of some D-mannans from *S. cerevisiae* have been shown to be related to the phosphate content of these polysaccharides.[204] The nature of some of the participating glycolipids involved in the biosynthesis of D-mannan-proteins of *S. cerevisiae* has been reported.[205] Isolated particulate membrane preparations were capable of transferring D-mannose from GDP-D-mannose into dolichyl-phosphoryl-D-mannose, dolichyl-pyrophosphoryl oligosaccharides, and glycoproteins.

α-D-Man*p*
1
↓
2
α-D-Man*p*
1
↓
2
α-D-Man*p*
1
↓
2
→6)-α-D-Man*p*-(1→6)-α-D-Man*p*-(1→

(22)

The D-mannan (22) isolated from *Lampteromyces japonicus* resembles that from *S. cerevisiae* except for the absence of (1 → 3)-linkages.[174] A closer structural resemblance exists with the D-mannans of *Saccharomyces rouxii*, *Trichosporon aculeatum*, and *Candida albicans*.

A model for the distribution of the active and zymogen forms of chitin synthetase in the apical and sub-apical regions of the hyphae of *Aspergillus nidulans* has been reported.[206] Large differences in the cell wall composition,

[200] E. M. Karson and C. E. Ballou, *J. Biol. Chem.*, 1978, **253**, 6484.
[201] M. V. Elorza, G. Larriba, J. R. Villanueva, and R. Sentandreu, *J. Microbiol. Serol.*, 1977, **43**, 129.
[202] A. Kaplan, D. Fischer, and W. S. Sly, *J. Biol. Chem.*, 1978, **253**, 647.
[203] Y. Okubo and S. Suzuki, *Carbohydrate Res.*, 1978, **62**, 135.
[204] Y. Okubo, T. Ichikawa, and S. Suzuki, *J. Bacteriol.*, 1978, **136**, 63.
[205] A. Dominguel, J. R. Villanueva, and R. Sentandreu, *J. Microbiol. Serol.*, 1978, **44**, 183.
[206] S. Isaac, N. S. Ryder, and J. F. Peberdy, *J. Gen. Microbiol.*, 1978, **105**, 45.

especially in the levels of chitin, have been demonstrated at different developmental phases of the growth of *Coprinus cinereus*.[207]

Mutants of *S. cerevisiae*, unable to synthesize 2-amino-2-deoxy-D-glucose owing to a defect in the enzyme 2-amino-2-deoxy-D-glucose 6-phosphate ketol isomerase (amino-transferring), have shown aberrations, both in sporulation and in vegetative growth, which are coupled with an inability to synthesize chitin.[208] Three different polysaccharides, each containing 2-acetamido-2-deoxy-D-glucosyl residues, have been identified in the hyphal wall of *Mucor mucedo*: chitin, a heteropolysaccharide composed of residues of 2-acetamido-2-deoxy-D-glucose and 2-amino-2-deoxy-D-glucose, and a heteropolysaccharide with peptide attached to some of the 2-acetamido-2-deoxy-D-glucosyl residues.[209] Chitin synthetase of *Mucor rouxii* has been located inside the permeability barrier of the cell in the form of an inactive precursor.[210] The regulation of the enzyme in this dimorphic fungus involves a destroying factor, probably a protease, located in a separate compartment in the cell. The synthesis and proteolytic activation of chitin synthetase in *Phycomyces blakesleeanus* have been reported.[211]

The biosynthesis of amino-sugars and their incorporation into polysaccharides and glycoproteins have been studied in a number of species of fungi.[212] Levels of amino-sugars were quantitated after hydrolysis, using an automated amino-acid analyser, programmed for the separation of amino-sugars, or by g.l.c. analysis.

The glycosyltransferases of particulate membrane fractions from *S. cerevisiae* have incorporated 2-acetamido-2-deoxy-D-glucose from UDP-2-acetamido-2-deoxy-D-glucose into lipid intermediates before transfer into endogenous protein.[213]

8 Miscellaneous Fungal Polysaccharides

An extracellular polysaccharide produced by *Aspergillus nidulans* is composed of residues of 2-amino-2-deoxy-D-galactose (70%) and D-galactose (15%).[214] A polysaccharide isolated from the ascocarps of *Cordyceps sinensis*, and containing residues of D-galactose and D-mannose (1 : 1), has been shown to contain a backbone of (1 → 2)-linked α-D-mannopyranosyl residues with branches composed of (1 → 3), (1 → 5)-, and (1 → 6)-linked D-galactofuranosyl residues and (1 → 4)-linked D-galactopyranosyl residues.[215] The absence of D-xylose and pigment has been demonstrated in the cell wall of an albino mutant of *Drechslera sorokiniana*; they are present in the wall of the wild type.[216]

207 R. Marchant, *J. Gen. Microbiol.*, 1978, **106**, 195.
208 C. E. Ballou, S. K. Maitra, J. W. Walker, and W. L. Whelan, *Proc. Nat. Acad. Sci. U.S.A.*, 1977, **74**, 4351.
209 R. Datema, J. G. H. Wessels, and H. van den Ende, *European J. Biochem.*, 1977, **80**, 621.
210 R. Sentandreu and J. Ruiz-Herrera, *Current Microbiol.*, 1978, **1**, 77.
211 A. J. Van Laere and A. R. Carlier, *Arch. Mikrobiol.*, 1978, **116**, 181.
212 T. W. Cochran and J. R. Vercellotti, *Carbohydrate Res.*, 1978, **61**, 529.
213 F. Reuvers, C. Habets-Willems, A. Reinking, and P. Boer, *Biochim. Biophys. Acta*, 1977, **486**, 541.
214 J. A. Leal and R. Ruperez, *Trans. Brit. Mycol. Soc.*, 1978, **70**, 115.
215 T. Miyazaki, N. Oikawa, and H. Yamada, *Chem. and Pharm. Bull.* (*Japan*), 1977, **25**, 3324.
216 K. H. Al-Rikabi and R. Bonaly, *Canad. J. Microbiol.*, 1977, **23**, 1508.

Extracellular polysaccharide fractions from the cellular slime mould *Dictyostelium discoideum* have been shown to exhibit strong B-cell mitogenic activities, which were destroyed on oxidation of the polysaccharide with periodate.[217] The polysaccharides also enhanced the *in vitro* antibody response against sheep red blood cells. The relative proportions of L-rhamnose, L-fucose, D-glucose, D-mannose, D-galactose, 2-amino-2-deoxy-D-glucose, and 2-amino-2-deoxy-D-galactose in the cells and plasma membranes of *D. discoideum* have been found to vary as a function of development.[218] Thus D-galactose occurred in the vegetative cell (2%) and in the fruiting bodies (12%) whilst 2-amino-2-deoxy-D-galactose was only detected in the fruiting bodies.

A polysaccharide (mol. wt. 9.5 × 10^5) containing residues of D-galactose, D-mannose, and L-fucose (12 : 3 : 4) has been purified from the fruiting bodies of *Flammulina velutipes*.[219] The structure (23) is based on the results of methylation analysis, i.r. spectroscopy, and interaction studies with concanavalin A. Every third (1 → 6)-linked α-D-galactopyranosyl residue is substituted with either an L-fucopyranosyl group or a 3-*O*-α-D-mannopyranosyl-L-fucopyranosyl unit.

Inorganic pyrophosphate and a glycoglycuronan containing residues of L-fucose, D-mannose, D-galactose, and D-glucuronic acid (5 : 1 : 1 : 6) have been isolated from the hyphal wall of *Mucor mucedo*.[220]

A polysaccharide containing D-galactose (89%), sulphate (9.6%), and phosphate (2.5%) has been isolated from the nuclei of *Physarum polycephalum*.[221] Apart from the high sulphate content, the polysaccharide appears to have a similar structure to an extracellular polysaccharide produced by the myxomycete. Both contain mainly (1 → 4)-linked D-galactosyl residues, with small amounts of (1 → 3)- and (1 → 6)-linkages and with one branch point every 13 residues. The D-galactan has been shown to be associated chiefly with chromatin in the nuclei and to a smaller extent in the cytoplasm and some vacuoles.[222] Localization of the polysaccharide was followed by electron microscopy using a specific 'sandwich technique', for which sections of the microplasmodia of *P. polycephalum* were incubated with the D-galactose-specific lectin from *Ricinus communis*. The bound lectin is localized with gold granules labelled with asialoceruloplasmin which itself contains non-reducing terminal D-galactosyl groups. The specificity of the method was assessed by marking, under the same conditions, the galactomannan of the cell wall of *Schizosaccharomyces pombe*.

The chemical composition of slime as well as the secretion kinetics of unlabelled and radioactively labelled slimes during naturally occurring and artificially induced spherulation of *Physarum polycephalum* have been reported.[223]

[217] H. Yamada, I. Suzuki, Y. Kumasawa, Y. Kawamura, K. Mizunoe, Y. Aramaki, and T. Miyazaki, *Biochim. Biophys. Acta*, 1978, **538**, 627.
[218] D. C. Kilpatrick and J. L. Stirling, *Biochim. Biophys. Acta*, 1978, **543**, 357.
[219] T. Mukumoto and H. Yamaguchi, *Carbohydrate Res.*, 1977, **59**, 614.
[220] R. Datema, H. van den Ende, and J. G. H. Wessels, *European J. Biochem.*, 1977, **80**, 611.
[221] D. R. Farr and M. Horisberger, *Biochim. Biophys. Acta*, 1978, **539**, 37.
[222] M. Horisberger, D. R. Farr, and M. Vonlanthen, *Biochim. Biophys. Acta*, 1978, **542**, 308.
[223] K. Zaar, *Arch. Mikrobiol.*, 1978, **117**, 303.

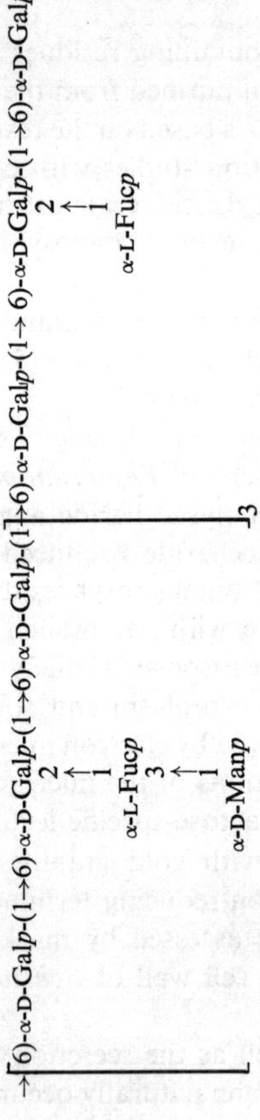

(23)

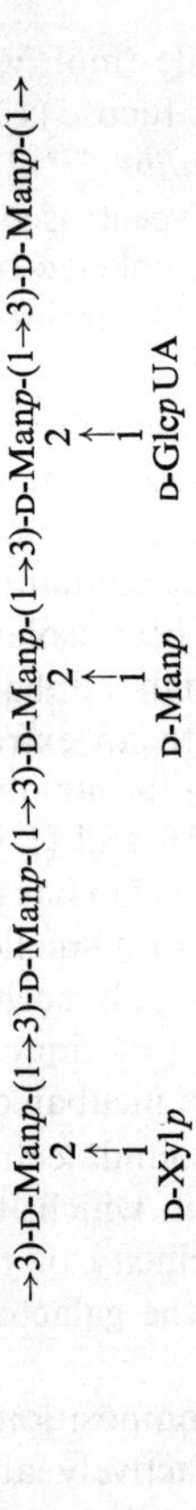

(24)

Methylation of a polysaccharide containing 2-amino-2-deoxy-D-galactose (88%), protein (6.8%), and phosphate (4.7%) has indicated a linear structure of (1 → 4)-linked amino-sugar residues.[224] An L-rhamno-D-mannan containing single unit α-L-rhamnopyranosyl side chains has been identified in isolated conidia of *Sporothrix schenckii* and differs from the di-L-rhamnosyl-D-mannan synthesized by the hyphae.[225]

On the basis of the ^{13}C n.m.r. spectra, various structural assignments have been made to polysaccharides of *S. schenkii*, including a neutral mono-L-rhamnosyl-D-mannan and an acidic L-rhamno-D-mannan.[226] Many strains from species of *Sporothrix*, *Ceratocystis*, *Europhium*, and *Graphium* have exhibited serological cross-reactions with *S. schenckii*, and contain polysaccharides composed of residues of L-rhamnose and D-mannose.[227] The D-galacto-D-mannan of *Schizosaccharomyces pombe* has been located at the periphery of the cell wall and near the plasmalemma.[228] No evidence was obtained for the presence of chitin.

Proteolysis followed by aqueous extraction of the cell walls of *Tremella fuciformis* has been used to solubilize a polysaccharide containing residues of D-mannose, D-xylose, and D-glucuronic acid.[229] Methylation analysis indicates a structure (24) composed of a backbone of (1 → 3)-linked D-mannosyl residues, some of which are substituted at O-2 with single, or short chains of, D-xylosyl, D-mannosyl, and D-glucopyranosyluronic acid residues. An alkali-insoluble polysaccharide closely resembling this structure was also identified, together with a β-D-glucan containing both (1 → 3)- and (1 → 6)-linkages. A neutral polysaccharide (mol. wt. 8×10^3), composed of residues of D-xylose, D-mannose, D-galactose, and D-glucose (2 : 4 : 5 : 35), has been isolated from alkali solubilization of *T. fuciformis*.[230] The results of methylation analysis, periodate oxidation, Smith degradation, and partial acid hydrolysis indicate that the polysaccharide contains a backbone composed mainly of (1 → 6)-linked D-glucopyranosyl residues with branch points at C-3 to D-xylopyranosyl- and D-glucopyranosyl end-groups.

A polysaccharide–protein complex isolated from the cell walls of *Torulopsis candida* contains residues of D-mannose and 2-acetamido-2-deoxy-D-glucose.[231] The D-mannose units are attached singly and as (1 → 2)-linked di- and trisaccharides to L-serine residues of the peptide chain by *O*-glycosidic linkages, while D-mannan chains are terminated at the reducing end by units of chitobiose which in turn are attached to the peptide backbone by *N*-glycosidic linkages.

[224] D. R. Farr, A. Schuler-Hoyanessian, and M. Horisberger, *Carbohydrate Res.*, 1977, **59**, 151.
[225] L. R. Travassos and L. Mendonca-Previato, *Infection and Immunity*, 1978, **19**, 1.
[226] L. R. Travassos, L. Mendonca-Previato, and P. A. J. Gorin, *Infection and Immunity*, 1978, **19**, 1107.
[227] H. Ishizaki, R. W. Wheat, D. P. Kiel, and N. F. Conant, *Infection and Immunity*, 1978, **21**, 585.
[228] M. Horisberger, M. Vonlanthen, and J. Rosset, *Arch. Mikrobiol.*, 1978, **119**, 107.
[229] Y. Sone and A. Misaki, *Agric. and Biol. Chem.* (*Japan*), 1978, **42**, 825.
[230] S. Ukai, T. Kiho, C. Hara, and K. Hirose, *Chem. and Pharm. Bull.* (*Japan*), 1978, **26**, 1707.
[231] J. B. Leleu, B. Fournet, J. P. Morilhat, R. Bonaly, and J. Montreuil, *Biochimie*, 1977, **59**, 687.

Glycopeptides having varying levels of carbohydrate (42—82%) have been isolated from *Trichophyton mentagrophytes*, *T. rubrum*, and *Microsporum canis*.[232] Each species contains both D-glucopeptides and D-mannopeptides but only the latter elicit a delayed hypersensitivity reaction in skin tests with animals.

A water-soluble α-D-glucan, containing (1 → 3)- and (1 → 4)-linkages, has been isolated from the lichen *Stereocaulon japonicum*.[233]

[232] S. A. Moser and J. D. Pollack, *Infection and Immunity*, 1978, **19**, 1031.
[233] I. Yokota and S. Shibata, *Chem. and Pharm. Bull. (Japan)*, 1978, **26**, 2668.

5

Glycoproteins, Glycopeptides, Proteoglycans, and Animal Polysaccharides

BY R. J. STURGEON

1 Glycoproteins of Micro-organisms

Further evidence has shown that glycosylation of surface glycoproteins of *Halobacterium salinarium* occurs *via* a cyclic pathway involving lipid intermediates.[1] Heterosaccharides are formed at the cytoplasmic side of the membrane by sequential transfer of sugars to isoprenylpyrophosphate, and then transferred at the exterior surface of the cell to the protein acceptor. The resulting lipid pyrophosphate is dephosphorylated to yield the isoprenyl monophosphate which again acts as an acceptor. Bacitracin was used to block the dephosphorylation, presumably by complexing with the lipid pyrophosphate at the outer cell surface, thus inhibiting further glycosylation.

The capsular antigen of *Streptococcus* type 87 contains (in addition to amino-acids) D-glucose, D-galactose, 2-amino-2-deoxy-D-glucose, and phosphate.[2] Reaction with homologous antiserum identified two closely associated types of molecules which may represent a glycoprotein and the corresponding apoprotein. A glycopeptide, containing residues of D-glucose, D-mannose, and 2-amino-2-deoxy-D-glucose, has been isolated after proteolytic digestion of the adenosine triphosphatase from the membranes of *Micrococcus lysodeikticus*.[3] Biosynthetic studies following the incorporation of D-[^{14}C]-glucose has confirmed this observation.[4] In addition, a low molecular weight glycopeptide fraction containing residues of D-glucose, D-galactose, and D-mannose (1 : 1 : 2) was identified.

Ribonuclease L, produced by an *Aspergillus* species and having different enzymic properties from the corresponding ribonucleases M, R, and T_3, is a glycoprotein, containing 6% carbohydrate.[5] L-Rhamnose and D-mannose have been identified as the principal sugars in a mixture of phytotoxic glycopeptides (83% carbohydrate, 7% peptide) produced by cultures of *Ceratocystis ulmi*.[6] The main structural components were identified as a backbone of 1,6-linked α-D-mannopyranosyl residues with singly-linked L-rhamnopyranosyl groups attached to O-3 of almost all the D-mannosyl residues. The carbohydrate–peptide linkages involve *O*-L-seryl and *O*-L-threonyl residues.

[1] M. E. Mescher and J. L. Strominger, *F.E.B.S. Letters*, 1978, **89**, 37.
[2] P. S. Venkateswaran, N. Stanton, C. Buettger and R. Austrian, *Infection and Immunity*, 1978, **20**, 180.
[3] J. M. Andreu, R. Warth, and E. Munoz, *F.E.B.S. Letters*, 1978, **86**, 1.
[4] A. Guerraro, E. Muhoz, and J. M. Andreu, *Current Microbiol.*, 1978, **1**, 129.
[5] Y. Eto, T. Kikuchi, H. Horitsu, and M. Tomoyeda, *Agric. and Biol. Chem. (Japan)*, 1978, **42**, 393.
[6] G. Strobel, N. van Alfen, K. D. Hapner, M. McNeil, and P. Albersheim, *Biochim. Biophys. Acta*, 1978, **538**, 60.

A membrane-bound D-mannosyltransferase from *Penicillium charlesii* has catalysed the incorporation of D-mannose from GDP-D-mannose into both the polysaccharide and oligosaccharide regions of a peptidogalactomannan.[7] Three enzymes secreted by whole cells and protoplasts of *Pichia polymorpha*, an *endo*-(1 → 3)-β-D-glucanase and two *exo*-(1 → 3)-β-D-glucanases, have been identified as glycoproteins, each containing different ratios of D-glucose and D-mannose.[8] The results support the hypothesis that glycosylation takes place before or during secretion of these enzymes into the periplasmic space, and that no further glycosylation occurs during their secretion into the supernatant.

Multiple forms of carboxypeptidase Y from *Saccharomyces cerevisiae* have been characterized as glycoproteins (mol. wt. 5.3—6.5 × 10^4) with variable carbohydrate content (9—20%).[9] Variations in the mechanism of catalysis by the enzyme were possibly due to differences in conformation of these glycoproteins arising from their different carbohydrate contents. Carboxypeptidase Y (mol. wt. 6.1 × 10^4) is synthesized *via* a precursor (mol. wt. 6.7 × 10^4) which is not an inhibitor–enzyme complex as previously described, but a glycoprotein that is converted by trypsin or by yeast proteinase B to a molecule of the same size as the final enzyme.[10] The incorporation of D-[1-^{14}C]-mannose into the precursor indicates that the glycosylation takes place before or immediately after completion of the synthesis of the polypeptide chain. A membrane-associated glycoprotein (mol. wt. 1.9 × 10^5) of *S. cerevisiae* accounting for less than 5% of the total cell β-D-fructofuranosidase (invertase) activity appears to be a precursor form (mol. wt. 2.4 × 10^5) of the glycoprotein bound to the external surface of the cell wall.[11]

Surface glycoproteins have been identified only in non-aggregating vesicles from the plasma membrane of *Saccharomyces cerevisiae*.[12] Structural studies have identified the glycoprotein from the outer cell wall of *Torulopsis candida* as having a polypeptide backbone to which D-mannose, 2-*O*-D-mannopyranosyl-D-mannose, and D-mannopyranosyl-(1 → 2)-D-mannopyranosyl-(1 → 2)-D-mannose are attached *via* L-serine or L-threonine residues.[13] In addition, a D-mannan is attached *via* a chitobiose bridge to L-asparagine. Glycopeptides composed of carbohydrate (42—82%), protein (4—11%) and aminosugars (0.3—0.6%) have been isolated from whole cultures of *Trichophyton mentagrophytes*, *T. rubrum*, and *Microsporum canis*.[14] Each strain contained D-glucopeptides and D-mannopeptides, but only the D-mannopeptides of the two *Trichophyton* species elicited a delayed hypersensitivity response in animals.

The plasma membrane glycoproteins of *Dictyostelium discoideum* seem to be restricted to classes which bind concanavalin A and/or wheat germ agglutinin.[15]

[7] J. E. Gander, L. R. Drewes, F. Fang, and A. Lui, *J. Biol. Chem.*, 1977, **252**, 2187.
[8] T. G. Villa, V. Notario, and J. R. Villanueva, *J. Gen. Microbiol.*, 1978, **109**, 371.
[9] H. C. Margolis, Y. Nakagawa, K. T. Douglas, and E. T. Kaiser, *J. Biol. Chem.*, 1978, **253**, 7891.
[10] A. Hasilik and W. Tanner, *European J. Biochem.*, 1978, **85**, 599.
[11] P. Babczinski and W. Tanner, *Biochim. Biophys. Acta*, 1978, **538**, 426.
[12] G. H. Rank, J. H. Gerlach, A. J. Robertson, and R. P. van Hoeven, *Nature*, 1978, **273**, 682.
[13] J. B. Leleu, B. Fournet, J. P. Morihat, R. Bonaly, and J. Montreuil, *Biochimie*, 1977, **59**, 687.
[14] S. A. Moser and J. D. Pollack, *Infection and Immunity*, 1978, **19**, 1031.
[15] C. M. West, D. McMahon, and R. S. Molday, *J. Biol. Chem.*, 1978, **253**, 1716.

Twenty-five macromolecules, apparent mol. wts. 8×10^3—9.5×10^4, were identified after electrophoresis and staining with fluorescent wheat germ agglutinin. Some of these receptors are distinct from those which bind concanavalin A. Six differences, five of which are in carbohydrate-containing molecules, have been detected between the plasma membranes of a mutant and a wild-type *D. discoideum* strain.[16] During acquisition of aggregation competence in *D. discoideum* cells, a new antigen appears on the surface.[17] This constituent has been purified and identified as a concanavalin A-binding glycoprotein.

An adenovirus type 2-induced polypeptide (mol. wt. 2×10^4) has been identified as a glycopeptide on the basis of labelling studies with 2-amino-2-deoxy-D-[^{3}H]glucose and binding studies with concanavalin A.[18]

The tetrameric nature of neuraminidase isolated from a strain of influenza virus has been confirmed.[19] The tetramer is composed of two disulphide linked dimers, themselves associated by non-covalent interactions. The high carbohydrate content (46%), which includes residues of 2-acetamido-2-deoxy-D-glucose, 2-acetamido-2-deoxy-D-galactose, D-galactose, D-mannose, and L-fucose, is thought to explain discrepancies in molecular weight values and also the exceptional resistance of the molecule to digestion by proteolytic enzymes. In contrast, the carbohydrate side-chains of the neuraminidase of B/Lee/40 influenza virus contain only 2-acetamido-2-deoxy-D-glucose, D-galactose, D-mannose, and L-fucose, and these are attached by amino-sugar–L-asparagine linkages in an intact glycoprotein (mol. wt. 2.8×10^5) and sub-unit (mol. wt. 7×10^4).[20]

A haemagglutinin from influenza virus has been identified as a glycoprotein containing 16% carbohydrate including residues of D-galactose, D-mannose, L-fucose, and 2-amino-2-deoxy-D-glucose (6 : 11 : 2 : 5).[21] Oligosaccharides (mol. wt. 2.7×10^3) isolated following proteolytic digestion contained D-galactose and L-fucose at non-reducing ends of the molecules. Two glycopeptides have been isolated by treatment of influenza virus haemagglutinin with bromelain.[22] Antibodies produced against either the intact virus or against purified undissociated haemagglutinin reacted against the two purified glycopeptides.[23] The haemagglutinins synthesized by the virus in the presence of 2-amino-2-deoxy-D-glucose have been shown to be heterogeneous with respect to the extent of glycosylation.[24] Virus particles when grown in the presence of D-*arabino*-hexose contained glycoproteins of different electrophoretic mobility and exhibited lower haemagglutinating activity, whilst particles without haemagglutinating activity were produced by tunicamycin-treated cells.

[16] S. Hoffman and D. McMahon, *J. Biol. Chem.*, 1978, **253**, 278.
[17] K. Müller and G. Gerish, *Nature*, 1978, **274**, 445.
[18] Y. H. Jeng, W. S. M. Wold, and M. Green, *J. Virol.*, 1978, **28**, 314.
[19] N. P. Groome, G. Belyavin, A. Lansdell, and D. Ashford, *Biochim. Biophys. Acta*, 1977, **495**, 58.
[20] A. K. Allen, J. J. Skehel, and V. Yuferov, *J. Gen. Virol.*, 1977, **37**, 625.
[21] J. K. Collins and C. A. Knight, *J. Virol.*, 1978, **26**, 457.
[22] K. Polakova, G. Russ, and B. Styk, *Acta Virol.*, 1978, **22**, 362.
[23] G. Russ, B. Styk, and K. Polakova, *Acta Virol.*, 1978, **22**, 371.
[24] K. Nakamura and R. W. Compans, *Virology*, 1978, **84**, 303.

The oligosaccharide moieties of the WSN strain of influenza viral glycoproteins have been analysed after extensive digestion of either virions or isolated viral glycoproteins with proteolytic enzymes followed by gel filtration.[25] Two distinct types of glycopeptides (types I and II) were identified when the virus was grown in any of three cell types. Type I contains 2-amino-2-deoxy-D-glucose, D-mannose, D-galactose, and L-fucose, whereas type II contains only 2-amino-2-deoxy-D-glucose and D-mannose. The presence of glycopeptides similar to these two types has been demonstrated in fowl plague virus grown in chick embryo fibroblasts.[26]

As a result of the identification of some amino-acid sequences of the two major component polypeptides of the hepatitis B surface antigen, it has been proposed that the antigen contains a single major immunogenic peptide component (mol. wt. 2.2×10^4), part of which is modified by the addition of carbohydrate to give rise to a glycopeptide (apparent mol. wt. 2.8×10^4).[27]

A glycoprotein containing 2-amino-2-deoxy-D-glucose and representing the intact envelope glycoprotein of avian leukosis-sarcoma virus has been isolated using lectin affinity chromatography.[28]

Four different glycoprotein precursors of the internal virion core proteins of murine leukaemia virus have been identified.[29] A minor glycoprotein (mol. wt. 5.2×10^4) purified from Rauscher murine leukaemia virus has been identified as a product of the proteolytic cleavage of a glycoprotein (mol. wt. 7×10^4).[30]

Two glycoproteins have been isolated from measles virus; the larger glycoprotein is composed of subunits (mol. wt. 7.6×10^4) that are bound by disulphide bonds, and the smaller glycoprotein contains a portion rich in 2-amino-2-deoxy-D-glucosyl residues linked to an otherwise unglycosylated protein by disulphide bonds.[31] Measles viral envelope glycoproteins have been isolated and purified by an immunoprecipitation technique.[32]

Viral glycoproteins (mol. wt. 7×10^4) from transformed cultures of Moloney murine type C virus have been shown to contain more neuraminic acid than the corresponding glycoproteins from non-transformed cells.[33] The envelope glycoproteins from a wide range of endogenous type C and type D viruses have been isolated and characterized using immunological methods.[34] A class of immunological determinants unique to, and broadly reactive among, the two types of virus isolates was recognized. Intracellular precursors of the major glycoprotein of avian oncoviruses have been isolated from virus-infected chicken embryo fibroblasts.[35] Information on nearest-neighbour interactions

[25] K. Nakamura and R. W. Compans, *Virology*, 1978, **86**, 432.
[26] R. T. Schwarz, M. F. G. Schmidt, U. Anwer, and H. D. Klenk, *J. Virol.*, 1977, **23**, 217.
[27] D. L. Peterson, I. M. Roberts, and G. M. Vyas, *Proc. Nat. Acad. Sci. U.S.A.*, 1977, **74**, 1530.
[28] E. J. Smith, L. B. Crittenden, and A. K. Whitson, *Virology*, 1978, **84**, 331.
[29] L. H. Evans, S. Dresler, and D. Kabat, *J. Virol.*, 1977, **24**, 865.
[30] M. C. Kemp, K. S. Wise, L. E. Edlund, R. T. Acton, and R. W. Compans, *J. Virol.*, 1978, **28**, 84.
[31] J. M. Hardwick and R. H. Bussell, *J. Virol.*, 1978, **25**, 687.
[32] T. W. Fenger, J. W. Smith, and C. Howe, *J. Virol.*, 1978, **28**, 292.
[33] J. Forchhammer and G. Turnock, *Virology*, 1978, **88**, 177.
[34] S. G. Devare, R. E. Hanson, and J. R. Stephenson, *J. Virol.*, 1978, **26**, 316.
[35] D. L. Buchhagen and H. Hanafusa, *J. Virol.*, 1978, **25**, 845.

of the major RNA-tumour-virus glycoprotein with the surfaces of non-producing murine cells has been reported.[36] Comparative studies with intact virions indicate that the glycoprotein molecules have similar nearest-neighbour interactions in both systems.

A radioimmunoassay procedure has been developed for the estimation of glycoprotein gp 47 from murine mammary tumour virus.[37]

Treatment of the envelope glycoprotein from avian myeloblastosis virus with glycosidases has shown that an intact carbohydrate chain is essential for immunological activity.[38] In contrast, digestion of B77 avian sarcoma virus with glycosidases resulted in no loss of antigenic activity. Changes in the antigenic activity of the former glycoprotein and an avian myeloblastosis-associated virus arise from the different patterns of glycosylation which occur when the viruses are grown in different host cells.[39] The B77 avian sarcoma virus glycoproteins (mol. wts. 8.5×10^4 and 3.7×10^4) are formed by proteolytic cleavage of a glycosylated precursor polypeptide (mol. wt. 9.5×10^4).[40] The site of cleavage appears to be the rough endoplasmic reticulum, from where the products are transported to the plasma membrane of the cell for incorporation into the budding virus.

A viral envelope glycoprotein (mol. wt. 8.5×10^4) has been isolated from Rous sarcoma virus by treatment with 2-mercaptoethanol, while a second surface glycoprotein (mol. wt. 3.5×10^4) remains associated with the membrane of intact virus particles and is probably embedded in a lipid bilayer.[41] The outer layer of calf retrovirus contains a glycoprotein with non-reducing terminal D-glucose or D-mannose groups as concluded from studies of interactions with concanavalin A.[42] Lectins with specificities for D-galactose, 2-acetamido-2-deoxy-D-glucose, and neuraminic acid gave no reaction with the virus.

Purification of some retroviruses has been achieved by chromatography on columns of Sepharose 4B®.[43] In a single step, the method allows greater recovery of infectious viral particles, conserves the viral glycoproteins, and allows removal of high concentrations of serum.

A polymeric form of the glycoprotein (mol. wt. 4×10^5) of rabies virus has been isolated and purified.[44] In vaccination experiments in animals this material afforded protection against a lethal challenge infection from rabies virus.

The attachment of two amphiphilic membrane glycoproteins of Semliki forest virus by their *C*-termini to a lipid bilayer has been described.[45] The

[36] L. J. Takemoto, C. F. Fox, F. C. Jensen, J. H. Elder, and R. A. Lerner, *Proc. Nat. Acad. Sci. U.S.A.*, 1978, **75**, 3644.
[37] P. F. Zangerie, C. M. Calberg-Bacq, C. Colin, P. Franchmont, C. Francois, L. Gosselin, S. Kozma, and P. M. Osterrieth, *Cancer Res.*, 1977, **37**, 4326.
[38] L. J. Van Eldik, J. C. Paulson, R. W. Green, and R. E. Smith, *Virology*, 1978, **86**, 193.
[39] L. J. Van Eldick and R. E. Smith, *Virology*, 1978, **90**, 80.
[40] M. Hayman, *Virology*, 1978, **85**, 475.
[41] G. Pauli, W. Rohde, and E. Harms, *Arch. Virol.*, 1978, **58**, 61.
[42] J. Cohen, R. Maget-Dana, A. C. Roche, and M. Monsigny, *F.E.B.S. Letters*, 1978, **87**, 26.
[43] M. McGrath, O. Witte, T. Pincus, and I. L. Weissman, *J. Virol.*, 1978, **25**, 923.
[44] B. Dietzschold, J. H. Cox, L. G. Schneider, T. J. Wiktor, and H. Koprowski, *J. Gen. Virol.*, 1978, **40**, 131.
[45] H. Garoff and H. Sönderlund, *J. Mol. Biol.*, 1978, **124**, 535.

similarity of the locations of Semliki forest virus glycoproteins in cells treated with tunicamycin and in control cells has led to the suggestion that the correct insertion of the glycoproteins is into the rough microsomes, despite the fact that these proteins are synthesized under conditions in which glycosylation is inhibited.[46] In order to study how the proteins from this virus are inserted into the membrane, cleaved, and glycosylated, a cell-free system has been constructed in which the viral proteins are translated from the messenger RNA in the presence of microsomes from dog pancreas.[47] Insertion into the membrane, proteolytic cleavage, and glycosylation failed to occur if membranes were added after a critical time, shortly after the commencement of synthesis of about 100 amino-acid residues of another protein (C protein).

D-Mannose, D-galactose, L-fucose, and 2-acetamido-2-deoxy-D-glucose have been identified as the constituent sugars of two envelope glycoproteins (mol. wts. 6.7×10^4 and 6.5×10^4) of Sendai virus.[48] One of the glycoproteins, which has haemagglutinating and neuraminidase activities, contains two types of carbohydrate chains and the other, which is involved in haemolytic and cell fusion activities, contains a single type of carbohydrate chain.

Low concentrations of sodium chloride in chicken embryo cell culture medium infected with Sindbis virus have made it possible for the virus envelope glycoprotein to penetrate the cell membrane.[49] The carbohydrate portions of Sindbis virus glycoproteins have been compared with the carbohydrate portions of cell surface glycoproteins from uninfected host cells.[50] From the size of glycopeptides, as determined by gel chromatography, and by a comparison of sugar linkages, the oligosaccharides added to the virus glycoproteins by both chick and baby hamster kidney cells do not appear to be unique, Sindbis-specific structures but possess structures similar to some of those found in uninfected host cells.

The synthesis and intracellular processing and membrane assembly of vesicular stomatitis viral glycoprotein has been reviewed.[51] An article has described the glycoprotein and protein precursors to plasma membranes in HeLa cells infected by vesicular stomatitis virus.[52]

Based on measurements using quasi-elastic light scattering and a resistive-pulse technique, the glycoprotein chains of vesicular stomatitis virus have been shown to extend almost twice as far out of the membrane as do those of Sindbis virus, suggesting that these viruses must have quite different protein conformations.[53] The soluble glycoprotein antigen of the vesicular stomatitis virus which is released into the culture medium has been demonstrated to originate from the envelope glycoprotein by proteolytic cleavage at the cell

[46] H. Garoff and R. T. Schwarz, *Nature*, 1978, **274**, 487.
[47] H. Garoff, K. Simons, and B. Dobberstein, *J. Mol. Biol.*, 1978, **124**, 587.
[48] T. Kohama, K. Shimizu, and N. Ishida, *Virology*, 1978, **90**, 223.
[49] J. W. Bell, R. F. Garry, and M. R. F. Waite, *J. Virol.*, 1978, **25**, 764.
[50] K. Keegstra and D. Burke, *J. Supramol. Struct.*, 1977, **7**, 371.
[51] F. N. Katz, J. E. Rothman, D. M. Knipe, and H. F. Lodish, *J. Supramol. Struct.*, 1977, **7**, 353.
[52] P. H. Atkinson, *J. Supramol. Struct.*, 1978, **8**, 89.
[53] B. I. Feuer, E. E. Uzgiris, R. W. Deblois, D. H. Cluxton, and J. Lenard, *Virology*, 1978, **90**, 156.

surface.[54] Structural analysis of the membrane glycoprotein showed it to contain two similar carbohydrate chains which are composed of three outer chain units, a D-mannose-rich 'core,' and a peptide linkage unit with the structure (1).[55]

When biosynthesized *in vitro*, glycoprotein G of vesicular stomatitis virus is protected from proteolytic digestion by being inserted into the microsomal membrane vesicles.[56] The glycoprotein does not appear to be completely transported across the membrane but a portion of the polypeptide chain remains on the cytoplasmic face of the membrane. In a study of the assembly of the viral membrane, tunicamycin has been used to prevent migration of the glycoprotein from the rough endoplasmic reticulum to the smooth intracellular membranes.[57]

The heterosaccharide moiety of the product of the initial glycosylation of the G glycoprotein occurring at the rough endoplasmic reticulum has been shown to contain a large oligomannosyl core [(D-Man)$\geqslant 7 \rightarrow$ (D-Glc*N*Ac)$_2$].[58] Either during or shortly before the addition of the distal branch sugars at smooth internal membranes some D-mannosyl residues are removed to provide the virus oligomannosyl core D-Man$_3$-(D-Glc*N*Ac)$_2$. This mannose-rich oligosaccharide has been shown to be transferred from a lipid carrier to the G protein.[59] The structure (2) of this lipid intermediate has been established by studies involving degradation with glycosidases, methylation analysis, acetolysis, and Smith degradation.[60]

After transfer of this oligosaccharide to the G protein, biosynthesis continues to give rise to the completed complex oligosaccharide unit (1).[61] This is achieved by the successive enzymic removal of the three D-glucosyl residues (groups) and six of the nine D-mannosyl residues (groups) from structure (2). Evidence for the existence of an α-D-mannosidase in crude membrane extracts and its dependance on 2-acetamido-2-deoxy-D-glucosyl transferase activity, which is involved in the attachment of the 2-acetamido-2-deoxy-D-glucosyl residues in a late stage of processing of the oligosaccharide, has been reported.[62]

The inhibition by an isolated vesicular stomatitis virus glycoprotein of the incorporation of nucleosides into cellular DNA and RNA has indicated that this glycoprotein is at least in part responsible for the inhibition which has been found previously to be associated with the virion.[63] In addition, the glycoprotein exhibits haemagglutinatory activity which, for both the virion and its isolated glycoprotein, is host-cell dependent.[64]

[54] S. P. Little and A. S. Huang, *J. Virol.*, 1978, **27**, 330.
[55] C. L. Reading, E. E. Penhoet, and C. E. Ballou, *J. Biol. Chem.*, 1978, **253**, 5600.
[56] F. Toneguzzo and H. P. Ghosh, *Proc. Nat. Acad. Sci. U.S.A.*, 1978, **75**, 715.
[57] T. G. Morrison, C. O. McQuain, and D. Simpson, *J. Virol.*, 1978, **28**, 368.
[58] L. A. Hunt, J. R. Etchison, and D. F. Summers, *Proc. Nat. Acad. Sci. U.S.A.*, 1978, **75**, 754.
[59] I. Tabas, S. Schlesinger, and S. Kornfeld, *J. Biol. Chem.*, 1978, **253**, 716.
[60] E. Li, I. Tabas, and S. Kornfeld, *J. Biol. Chem.*, 1978, **253**, 7762.
[61] S. Kornfeld, E. Li, and I. Tabas, *J. Biol. Chem.*, 1978, **253**, 7771.
[62] I. Tabas and S. Kornfeld, *J. Biol. Chem.*, 1978, **253**, 7779.
[63] J. J. McSharry and P. W. Choppin, *Virology*, 1978, **84**, 172.
[64] J. J. McSharry, C. A. Ledda, H. J. Freiman, and P. W. Choppin, *Virology*, 1978, **84**, 183.

α-NeuNAc-(2 → 3)-β-D-Gal*p*-(1 → 4)-β-D-Glc*p*NAc

1 → 4

α-NeufNAc-(2 → 3)-β-D-Gal*p*-(1 → 4)-β-D-Glc*p*NAc-(1 → 2)-α-D-Man*p*-(1 → 6)-β-D-Man*p*-(1 → 4)-β-D-Glc*p*NAc-(1 → 4)-D-Glc*p*NAc-L-Asn-

3 ← 1 (on β-D-Man*p*); 6 ← 1 α-L-Fuc*p* (on D-Glc*p*NAc)

α-NeufNAc-(2 → 3)-β-D-Gal*p*-(1 → 4)-β-D-Glc*p*NAc-(1 → 2)-D-Man*p*

(1)

α-D-Man*p*-(1 → 2)-α-D-Man*p*-(1 → 6)-α-D-Man*p*-(1 → 6)-β-D-Man*p*-(1 → 3/4)-β-D-Glc*p*NAc-(1 → 4)-D-Glc*p*NAc-P-P-dolichol

3 ← 1 α-D-Man*p* 2 ← 1 α-D-Man*p* (on α-D-Man*p*)

3 ← 1 α-D-Man*p* 2 ← 1 α-D-Man*p* 2 ← 1 α-D-Man*p* 3 ← 1 α-D-Glc*p* 3 ← 1 (D-Glc*p*)$_n$ 2 ← 1 (D-Glc*p*)$_m$ (on β-D-Man*p*)

(2)

$n = 0$ or 1; if $n = 0$, $m = 0$; if $n = 1$, $m = 0$ or 1

2 Plant Glycoproteins

Hydrazinolysis has been used to destroy the peptide moieties in glycoproteins, allowing the isolation of almost intact carbohydrate chains.[65] The method has been applied to pineapple stem bromelain, for which the structure of a glycopeptide (3) was assigned as a result of methylation analysis.

$$\begin{array}{llll}
\alpha\text{-D-Man}p\text{-}(1 \rightarrow 6)\text{-}\beta\text{-D-Man}p\text{-}(1 \rightarrow 4)\text{-}\beta\text{-D-Glc}p\textit{N}\text{Ac-}(1 \rightarrow 4)\text{-D-Glc}p\textit{N}\text{Ac-L-Asn} \\
\quad 2 & & \quad 3 & \\
\quad \uparrow & & \quad \uparrow & \\
\quad 1 & & \quad 1 & \\
\quad \text{R} & & \quad \text{R}' &
\end{array}$$

R = α-L-Fuc*p* or β-D-Xyl*p*; R′ = β-D-Xyl*p* or α-L-Fuc*p*

(3)

Isocitrate lyase has been purified from the glyoxysomes of cucumber (*Cucumis sativus*) cotyledons.[66] The glycoproteinaceous nature of the enzyme was demonstrated by periodate oxidation studies, and by the *in vivo* incorporation of 2-amino-2-deoxy-D-[^{3}H]glucose into a species which was precipitable by antibodies to isocitrate lyase.

An L-arabino D-gluco-D-galactan–protein complex, containing 97% carbohydrate and 3% protein, has been isolated from the female reproductive tissues of *Gladiolus gandavensis*.[67] The carbohydrate component, composed of residues of D-galactose, L-arabinose, and D-glucose (13 : 2 : 3), possessed a backbone of 1,3-linked β-D-galactosyl residues carrying branches of 1,6-linked D-galactopyranosyl residues and 1,6-linked D-glucosyl residues some of which are terminated by L-arabinofuranosyl groups. This L-arabino D-gluco-D-galactan–protein is structurally related to, but distinct from, the soluble L-arabino D-galactan which occurs in the leaves and petals of the same plant.

Each of the three glycoproteins which have been isolated from the seeds of *Lupinus augustifolius* contains D-mannosyl-D-galactosyl-, and 2-amino-2-deoxy-D-glucosyl- residues, in different ratios.[68] Two types of protein–polysaccharide linkages have been described for glycoproteins from *Phaseolus vulgaris*; all of the L-serine and most of the L-hydroxyproline residues are linked to L-arabinose, D-galactose, and D-glucose, but some hydroxyproline residues are attached to L-arabinose only.[69]

The glycoproteins present in growing leaves of *P. vulgaris* possibly play a structural or enzymic role rather than serving as reserve material to be metabolized as they do in seeds.[70] However, the ten-fold increase in the levels of glycoproteins associated with the chloroplast upon addition of 2-amino-2-deoxy-D-glucose indicates that the biosynthesis of this insoluble glycoprotein is limited by endogenous precursors. Mannolipid has been shown to function as an intermediate in the glycosylation of protein in cotyledons of *P. vulgaris*.[71] The lipid moiety has been characterized as dolichol type.

[65] M. Fukada, T. Kondo, and T. Osawa, *J. Biochem.* (*Japan*), 1976, **80**, 1223.
[66] J. Frevert and H. Kindl, *European J. Biochem.*, 1978, **92**, 35.
[67] P. A. Gleeson, A. E. Clarke, M. A. Jermyn, and R. B. Knox, *Proc. Austral. Biochem. Soc.*, 1978, **11**, 33.
[68] C. A. Eaton-Mordas and K. G. Moore, *Phytochemistry*, 1978, **17**, 619.
[69] R. G. Brown and W. C. Kimmins, *Phytochemistry*, 1978, **17**, 29.
[70] D. Racusen and M. Foote, *Canad. J. Bot.*, 1978, **56**, 209.
[71] D. P. Delmer, C. Kulow, and M. C. Ericson, *Plant Physiol.*, 1978, **61**, 25.

2-Amino-2-deoxy-D-[^{14}C]glucose has been incorporated into the glycoproteins of developing pea (*Pisum sativum*) cotyledons.[72] Part of the radiolabel becomes incorporated into the glycopeptide linkage region, as demonstrated by the isolation of N^4-(2-acetamido-2-deoxy-β-D-[^{14}C]glucopyranosyl) hydrogen L-asparaginate *via* partial acid hydrolysis. When UDP-2-acetamido-2-deoxy-D-[^{14}C]glucose is used as precursor, the majority of the label appears in a glycoprotein fraction of cotyledons from *P. sativum*, whereas very little D-mannose is transferred from GDP-D-[^{14}C]mannose into the same fraction.[73] Several glyoxysomal proteins from the castor bean (*Ricinus communis*) have been shown to contain covalently bound carbohydrate; this contained residues of L-fucose, D-mannose, D-galactose, and 2-amino-2-deoxy-D-mannose.[74]

Evidence has been obtained for a two-step mechanism for the synthesis of a (1 → 4)-α-D-glucan–protein in potato tubers.[75] Initially UDP-D-glucose: protein transglucosylase binds one D-glucosyl residue to an aminoacyl residue of a protein acceptor molecule to form the 'primer'. Then the transfer of D-glucosyl residues from UDP-D-glucose, ADP-D-glucose, or D-glucose 1-phosphate occurs in the presence of D-glucosyl transferases to enlarge the 'primer' molecule.

Two glycoprotein fractions have been isolated from membranes of the root-cap cells of *Zea mays* after their incubation with L-[1-^{3}H]fucose.[76] The glycoproteins appear to be precursors of the slime polysaccharides; their carbohydrate moieties have compositions similar to that of the free slime. For one of the glycoproteins the linkage between carbohydrate and protein was shown to be a D-xylose-threonine bond.

The biosynthesis of paramylon, a (1 → 3)-β-D-glucan of *Euglena gracilis*, has been shown to involve a protein primer[77] In the presence of solubilized membranes from *E. gracilis*, D-glucose has been incorporated from UDP-D-glucose into a fraction of high molecular weight which is insoluble in trichloroacetic acid. The product had the properties of a (1 → 3)-β-D-glucan covalently linked to protein by a bond labile to acid.

3 Lectins

The properties of plant lectins as carbohydrate-binding proteins have been reviewed.[78] The endogenous role of lectins isolated from vertebrate tissues has also been assessed.[79] In a review of the dimensions and specificities of carbohydrate recognition sites on lectins and antibodies, a comparison has been made of the specificity of the combining sites, with emphasis on those reacting with blood group determinants.[80]

Colloidal gold, ferritin, and peroxidase have been used as electron markers in the double labelling of a number of lectin-binding sites.[81]

72 S. K. Browder and L. Beevers, *F.E.B.S. Letters*, 1978, **89**, 145.
73 L. Beevers and R. M. Mense, *Plant Physiol.*, 1977, **60**, 703.
74 R. B. Mellor, L. Bowden, and J. M. Lord, *F.E.B.S. Letters*, 1978, **90**, 275.
75 J. S. Tandecarz and C. E. Cardini, *Biochim. Biophys. Acta*, 1978, **543**, 423.
76 J. R. Green and D. H. Northcote, *Biochem. J.*, 1978, **170**, 599.
77 A. D. Tomos and D. H. Northcote, *Biochem. J.*, 1978, **174**, 283.
78 H. Rudiger, *Naturwiss.*, 1978, **65**, 239.
79 D. L. Simpson, D. R. Thorne, and H. H. Loh, *Life Sciences*, 1978, **22**, 727.
80 E. A. Kabat, *J. Supramol. Struct.*, 1978, **8**, 79.
81 J. Roth and M. Binder, *J. Histochem. Cytochem.*, 1978, **26**, 163.

The ability of various fungi to bind fluorescein isothiocyanate derivatives of lectins has been examined.[82] Soya bean agglutinin and peanut agglutinin were shown to inhibit growth of certain Penicillia and Aspergilli, supporting the suggestion that lectins function as part of the defence mechanisms of plants against fungal pathogens.

The cytochemical distribution of D-mannose-containing glycoproteins in the rat alimentary tract has been observed using a concanavalin A–horse radish peroxidase fixation method.[83]

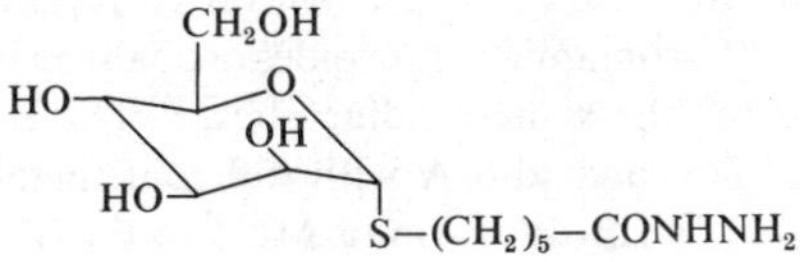

(4)

The synthetic hydrazide (4) has been attached covalently to aldehyde groups generated by periodate oxidation of cells such as erythrocytes thus enabling them to bind to concanavalin A.[84] This makes it possible to modify the antigenicity of a cell by introduction of simple sugars at the cell surface.

Amino-acid sequence studies have been reported for concanavalin A and lentil (*Lens culinaris*) lectin. Extensive homologies were observed.[85] Based on the alignment of sequences of the α- and β-chains of *L. culinaris* lectin with portions of the concanavalin A protein chain situated between residues 70—160, a close relationship between the two proteins was suggested. The results support a theory that the *L. culinaris* lectin is synthesized as a single polypeptide chain and cleaved subsequently into two or possibly three fragments, two of which would be the α- and β-subunits and the third a fragment homologous to residues 1—70 of concanavalin A. A common origin for the two lectins is suggested, as evidenced by the conservation of major features of the proteins, such as primary and secondary structure, hydrophobic regions, and the specificity of binding to carbohydrates.

A direct method involving fluorescent-light microscopy and sodium dodecyl sulphate gel electrophoresis has been developed for the study of the binding of lectins to glycoproteins and subunits thereof.[86] Fluorescein-labelled concanavalin A was shown to interact with both agglutinins (RCAI and RCAII) from *Ricinus communis*, rather than with RCAI only as suggested by other reports.

Ethylenedinitrilotetracetic acid has been shown to remove Ca^{2+} and Mn^{2+} ions from concanavalin A at pH 5.3 with a concomitant loss of saccharide-binding activity, which could be partly recovered by addition of either Ca^{2+} or Mn^{2+} or fully recovered by addition of both ions together.[87] Concanavalin A, when complexed with polysaccharide prior to treatment with ethylenedinitrilotetracetic acid, is largely resistant to the action of this agent. This

[82] R. Barkai-Golan, D. Mirelman, and N. Sharon, *Arch. Mikrobiol.*, 1978, **116**, 119.
[83] T. Katsuyama and S. S. Spicer, *J. Histochem. Cytochem.*, 1978, **26**, 233.
[84] G. A. Orr and R. R. Rando, *Nature*, 1978, **272**, 722.
[85] A. Forier, R. D. Neve, L. Kanarek, and A. D. Strosberg, *Proc. Nat. Acad. Sci. U.S.A.*, 1978, **75**, 1136.
[86] P. Maher and R. S. Molday, *F.E.B.S. Letters*, 1977, **84**, 391.
[87] R. J. Doyle and S. D. Stroupe, *Carbohydrate Res.*, 1978, **64**, 327.

suggests that the affinity of concanavalin A for Ca^{2+} and Mn^{2+} increases on binding of carbohydrate.

Conformational differences between the three-dimensional structures of concanavalin A and its demetallized form have been shown to affect many of the residues involved in the binding of saccharides.[88] Significant differences between the holo- and apo-proteins are confined to a reorganization and disordering of the loop containing residues 7—25 and which contains all the metal ligands of the protein.

The affinity of concanavalin A for Ca^{2+} has been reported to be lessened in the presence of methyl α-D-mannopyranoside, probably as a result of Ca^{2+} and glycoside competing for the same binding sites.[89] The existence of two conformational states of concanavalin A with different metal binding properties has been demonstrated.[90] Introduction of Mn^{2+} to the S1 site and Ca^{2+} to the S2 site of the apoprotein was shown to induce a conformational change in that protein. This was ascribed to a cis–trans isomerization of a peptide bond in the secondary structure, which results in tight binding of the metal ions. This induced conformation which is identical to that of the native protein is referred to as 'locked' and the initial conformation as 'unlocked'. Evidence has been provided indicating the existence of a relatively stable, locked, ternary Ca^{2+}–concanavalin A complex (having Ca^{2+} in both the S1 and S2 sites) that possesses properties similar to those of the native Ca^{2+}, Mn^{2+}–lectin.[91] The effects of added methyl α-D-mannopyranoside on the rate of replacement of Ca^{2+} by Mn^{2+} at the S1 site of the locked ternary Ca^{2+}–concanavalin A complex indicated that the latter complex binds saccharides as strongly as the locked Ca^{2+}, Mn^{2+}–concanavalin A complex, and suggests that the locked conformation of the lectin is primarily responsible for saccharide binding activity, the function of the bound metals being to maintain the protein in the locked conformation.

Identical rate constants have been reported for saccharide-binding by concanavalin A when Mn^{2+}, Co^{2+}, or Zn^{2+} occupy the S1 site and Ca^{2+} occupies the S2 site.[92] The affinity of the protein for saccharides is dependent upon the metal occupancies of both sites and is sensitive to the identity of the metal ion in site S2. Titrations of metal employing e.s.r. spectroscopy and equilibrium dialysis have revealed that the binding of Mn^{2+} to concanavalin A is co-operative in the presence of Ca^{2+} and non-co-operative in its absence.[93] The degree of co-operativity also increases with increasing pH. At pH 7.2 only Ca^{2+} and not a transition-metal ion is required for saccharide binding to concanavalin A, whereas at pH 5, at which most other studies have been carried out, both a transition metal and Ca^{2+} are necessary for binding.[94]

[88] G. N. Reeke, J. W. Becker, and G. M. Edelman, *Proc. Nat. Acad. Sci. U.S.A.*, 1978, **75**, 2286.
[89] R. J. Doyle and S. D. Stroupe, *Carbohydrate Res.*, 1978, **67**, 545.
[90] R. D. Brown, C. F. Brewer, and S. F. Koenig, *Biochemistry*, 1977, **16**, 3883.
[91] S. H. Koenig, C. F. Brewer, and R. D. Brown, *Biochemistry*, 1978, **17**, 4251.
[92] A. D. Sherry, A. E. Buck, and C. A. Peterson, *Biochemistry*, 1978, **17**, 2169.
[93] G. M. Alter, E. R. Pandolfino, D. J. Christie, and J. A. Magnuson, *Biochemistry*, 1977, **16**, 4034.
[94] D. J. Christie, G. M. Alter, and J. A. Magnuson, *Biochemistry*, 1978, **17**, 4425.

The fluorescence of 4-methylumbelliferyl α-D-mannobioside and 4-methylumbelliferyl α-D-mannotrioside (Vol. 9, p. 379, ref. 339) has been shown to change markedly on binding to concanavalin A.[95] From a study of constants for the association of the saccharide derivatives with the lectin, it appears that, in the complex formation, the position of the 4-methylumbelliferyl group depends on the length of the carbohydrate chain. The enthalpies of binding are independent of the number of D-mannopyranosyl residues in the ligands, suggesting that the 4-methylumbelliferyl glycosides bind to concanavalin A *via* a single D-mannopyranosyl residue.

E.s.r. spectroscopy studies have demonstrated the utility of europium(II) as a probe for Ca^{2+} binding sites in concanavalin A.[96]

The binding of lanthanide ions to concanavalin A has been studied using c.d., magnetic c.d., and fluorescence spectroscopy.[97] A Scatchard analysis of the fluorescence enhancement of Tb^{3+}-binding reveals the presence of two classes of sites. One strong binding site (S3), which is at least 39 Å from the S1 site and which is not affected by addition of transition-metal ions or Ca^{2+}, is present in each monomer. Lanthanide ions do not bind to site S2.

The self-association of solutions of concanavalin A at pH 7 has been shown to be temperature dependent; the lectin is essentially subject to a reversible dimer–tetramer transition below 23 °C.[98] From the measurement of various kinetic parameters of the transition and calculation of thermodynamic constants, it is concluded that the two dimers associate through a large hydrophobic binding region.

Optimal conditions have been reported for the preparation of concanavalin A covalently linked to ferritin by glutaraldehyde.[99]

A study of the specificity of *Bandeiraea simplicifolia* lectin I by competitive-binding assays with blood-group substances and with oligosaccharides derived from blood-groups A and B substances has been reported.[100] The results show that binding of human blood-group B substance by lectin I is inhibited best by some glycosides having terminal non-reducing 2-acetamido-2-deoxy-α-D-galactopyranosyl groups, as well as by those having non-reducing α-D-galactopyranosyl groups. The lectin is unusual in that the oligosaccharides studied could be classed into two groups having different slopes in the competitive-binding assay. The difference in slopes is explained in terms of a recent finding that the *B. simplicifolia* lectin I is a mixture of five isolectins, containing subunits A and B, in the forms (A_4, A_3B, A_2B_2, AB_3, and B_4).[101] It is proposed that subunit A is most specific for terminal non-reducing 2-acetamido-2-deoxy-α-D-galactopyranosyl groups but also reacts with terminal non-reducing α-D-galactopyranosyl groups, whereas subunit B tends to be more specific for terminal non-reducing α-D-galactopyranosyl groups.

[95] A. Van Landschoot, E. G. Loontiens, and C. K. De Bruyne, *European J. Biochem.*, 1978, **83**, 277.
[96] R. B. Homer and B. D. Mortimer, *F.E.B.S. Letters*, 1978, **87**, 69.
[97] C. E. Richardson and W. D. Behnke, *Biochim. Biophys. Acta*, 1978, **534**, 267.
[98] M. Huet and J. M. Claverie, *Biochemistry*, 1978, **17**, 236.
[99] U. Schirmbock and E. W. Rautenberg, *J. Immunol. Methods*, 1978, **20**, 61.
[100] E. C. Kisailus and E. A. Kabat, *Carbohydrate Res.*, 1978, **67**, 243.
[101] L. A. Murphy and I. J. Goldstein, *J. Biol. Chem.*, 1977, **252**, 4739.

An immunochemical study of the combining sites of *Bandeiraea simplicifolia* lectin II has indicated that the lectin is not blood-group specific.[102] Individual human B and H blood-group substances react to different extents according to their heterogeneity; those with terminal non-reducing 2-acetamido-2-deoxy-β-D-glucopyranosyl groups react well while those lacking such residues do not become precipitated. It is unusual that certain disaccharides having β-D-linkages react as well or better than others with α-D-linkages. Molecular models account for these findings; overall shape and contour of the molecule and hydrophobic bonds are the decisive factors in binding.

The 2-acetamido-2-deoxy-D-glucopyranosyl-binding lectin from *Bandeiraea simplicifolia* occurs as a tetramer and possesses one binding site per subunit.[103] Equilibrium dialysis and carbohydrate binding studies have shown that the combining sites are independent and non-interacting. This is in contrast to concanavalin A which interacts with α- and β-linked 2-acetamido-2-deoxy-D-glucopyranosyl residues, and differs from wheat germ agglutinin and potato lectin in that it does not bind to internal 1,4-linked 2-acetamido-2-deoxy-β-D-glucopyranosyl residues, nor does it appear to possess an extended binding site for 1,4-linked 2-acetamido-2-deoxy-β-D-glucopyranosyl residues in general.

The L-fucose-binding lectins from *Ulex europaeus* and *Lotus tetragonobolus* have been isolated following affinity chromatographic separations on columns of Sepharose to which L-fucose has been covalently attached.[104] Analysis of the lectins using haemagglutination, gel filtration, and polyacrylamide gel electrophoresis revealed the presence of isolectins or aggregated species. The lectin from *U. europaeus* has been isolated by affinity chromatography on columns of a synthetic co-polymer to which 4-aminophenyl α-L-fucopyranoside is attached.[105] The binding characteristics of the lectins from *U. europaeus* and *L. tetragonobolus* have been measured by haemagglutination inhibition analysis; results indicate that the lectin from the latter requires a lower concentration of saccharides to cause complete inhibition.[106] The binding site of the blood-group H-specific lectin I from *U. europaeus* has been studied by precipitation, inhibition of precipitation, and competitive binding assays using [^{3}H]-labelled hog mucin.[107] The lectin has the greatest affinity for an oligosaccharide from blood group H having the structure α-L-Fuc*p*-(1 $\rightarrow$ 2)-β-D-Gal*p*-(1 $\rightarrow$ 4)-β-D-Glc*p*NAc-(1 $\rightarrow$ 6)-R, where R = remainder of the glycoprotein molecule.

An affinity matrix produced by cross-linking arabinogalactan with epichlorohydrin has been used in the large scale preparation of peanut (*Arachis hypogaea*) lectin.[108] Many of the macromolecular properties of this lectin have been found to conform to general properties observed for numerous other plant lectins.[109] At pH 8, the lectin exists as a compactly folded tetramer (mol. wt. 9.8×10^4) but at low pH it exists as a globular dimer. In dissociating solvents,

102 C. Wood, E. A. Kabat, S. Ebisu, and J. Goldstein, *Ann. Immunol.*, 1978, **129C**, 143.
103 S. Ebisu, P. N. Iyer, and I. J. Goldstein, *Carbohydrate Res.*, 1978, **61**, 129.
104 H. J. Allen and E. A. Z. Johnson, *Carbohydrate Res.*, 1977, **58**, 253.
105 A. Racois and E. Boschetti, *Biochimie*, 1978, **60**, 193.
106 H. J. Allen, E. A. Z. Johnson, and K. L. Matta, *Immunol. Commun.*, 1977, **6**, 585.
107 M. E. A. Pereira, E. C. Kisailus, F. Gruezo, and E. A. Kabat, *Arch. Biochem. Biophys.*, 1978, **185**, 108.
108 T. Majumdar and A. Surolia, *Prep. Biochem.*, 1978, **8**, 119.
109 W. W. Fish, L. M. Hamlin, and R. L. Miller, *Arch. Biochem. Biophys.*, 1978, **190**, 693.

four equal-sized polypeptides are observable. Heterogeneity among the *A. hypogaea* lectins has been demonstrated by polyacrylamide gel electrophoresis.[110]

In a c.d. study of the conformation and conformational transitions of the lectins from *Arachis hypogaea*, *Lens culinaris*, *Glycine max*, and *Ricinus communis*, those of *A. hypogaea* and *L. culinaris* appear to possess a high level of pleated sheet structure in the β-conformation.[111] When the tertiary structure of each of the lectins became disorganized in the presence of sodium dodecyl sulphate, a new conformation was observed with a higher helical content than that found in the native protein. Dissociation into subunits may be one of the first steps in these transitions.

A fluorimetric method for the thermodynamic and kinetic interpretation of lectin–receptor reactions has been validated in a study of the binding of rabbit erythrocytes to *Lens culinaris* lectin conjugated with fluorescein.[112] It was demonstrated that under conditions when even a low number of receptors are saturated, most lectin molecules are monovalently bound to the receptor. The high affinity for the membrane receptor, contrasting with the low affinity for specific monosaccharides, is explained by the very slow dissociation reaction which requires twice the energy of the association reaction.

The lectin of lentil seeds (*Lens esculenta* Moench) has been dissociated into subunits upon incubation of its solution at temperatures above 50°C.[113] The dissociation is accompanied by precipitation of the heavy subunit, release of Mn^{2+} from the protein, and an almost complete loss of haemagglutinating activity. Addition of Mn^{2+} or D-glucose to solutions of the lectin has a partially protective effect on the dissociation and inactivation.

A glycopeptide (mol. wt. 1.3×10^3), containing L-fucose, D-mannose, D-glucose, and asparagine, has been isolated from the products of proteolysis of the lima bean lectin (*Phaseolus lunatus*).[114] Methylation analysis, Smith degradation, and subsequent degradation by specific glycosidases has resulted in the tentative proposal of a structure (5).

Two groups of isolectins have been isolated from *Phaseolus vulgaris* seeds after chromatography on immobilized fetuin.[115] Although the lectins in the two groups have similar amino-acid and sugar compositions, they share only a partial structural identity as demonstrated by immunochemical criteria.

Two lectins, having binding-specificities towards D-galactosyl residues in glycoproteins and glycolipids, have been isolated from the seeds of *Ricinus communis* using columns of cross-linked guar gum.[116] Temperature-jump relaxation kinetics have been used to study the binding of 4-nitrophenyl β-D-galactoside to one of the lectins, and turbidity data on the binding of the

[110] R. A. Newman, *Z. physiol. Chem.*, 1977, **358**, 1517.
[111] B. Jirgensons, *Biochim. Biophys. Acta*, 1978, **536**, 205.
[112] J. Holbeke, A. Foriers, A. B. Schreiber, and A. D. Strosberg, *Biochemistry*, 1978, **17**, 5000.
[113] I. Körbelová, M. Tichá, and J. Kocourek, *Acta biol. med. germ.*, 1978, **37**, 487.
[114] A. Misaki and I. J. Goldstein, *J. Biol. Chem.*, 1977, **252**, 6995.
[115] A. Pusztai and J. C. Stewart, *Biochim. Biophys. Acta*, 1978, **536**, 38.
[116] P. S. Appukuttan, A. Surolia, and B. K. Bachhawat, *Indian J. Biochem. Biophys.*, 1977, **14**, 382.

α-D-Man*p*-(1 → 2)-α-D-Man*p*-(1 → 6)-β-D-Man*p*-(1 → 3/4)-β-D-Glc*pN*Ac*p*-(1 → 4)-β-D-Glc*pN*Ac-L-Asn

β-D-Man*p*: 3 ↑ 1 α-D-Man*p*

β-D-Glc*pN*Ac*p*: ? ↑ 1 $(\alpha\text{-L-Fuc}f)_{1 \text{ or } 0}$

β-D-Glc*pN*Ac-L-Asn: ? ↑ 1 $(\alpha\text{-L-Fuc}f)_{0 \text{ or } 1}$

(5)

same lectin to a galactomannan have been reported.[117] The presence of small amounts of D-galactose or its derivatives causes a decrease in the initial rate of the precipitation reaction. The relevance of this finding to a lectin–receptor binding assay is discussed.

Short oligosaccharide chains containing three and four L-arabinofuranosyl residues attached to L-hydroxyproline have been isolated from the lectin of potato tubers *via* elimination by alkali.[118] No evidence was obtained for the existence of polysaccharide attached to the polypeptide backbone. The authors speculate that the lectin might carry sequences of L-hydroxyproline residues as found in some cell wall glycoproteins, and suggest that it is a sequence of consecutive L-proline residues which acts as a recognition site for the hydroxylation and glycosylation enzymes. A different sequence of amino-acids might then be the recognition site for hydroxylation and subsequent attachment of large oligosaccharides to the other glycoproteins containing L-hydroxyproline, typified by the extracellular polysaccharides of sycamore.

A glycopeptide (mol. wt. 3.2×10^4), which contained all the carbohydrate and L-hydroxyproline of the original glycoprotein, has been isolated by proteolytic digestion of a reduced and carboxymethylated derivative of potato lectin.[119] Two unusual features of this glycoprotein are the presence of β-L-arabinofuranosyl residues as mono- and oligo-saccharides linked to the L-hydroxyproline residues in the polypeptide chain, and the existence of a number of L-serine residues of the polypeptide chain that are substituted by α-D-galactopyranosyl residues. It is suggested that the potato lectin may be a representative of a class of soluble plant glycoproteins that would include precursors of the cell-wall glycoprotein extensin, which may also contain β-L-arabinofuranosyl residues linked to L-serine residues of the polypeptide chain.

Removal of 40% of the D-mannosyl residues of the lectin from *Dolichos biflorus* by digestion with α-D-mannosidase does not affect the agglutination of type A erythrocytes or the precipitation of blood group A and H substances, suggesting that complete integrity of the carbohydrate components of the lectin is not necessary for biological activity.[120]

Two mitogens from pokeweed, *Phytolacca americana*, have been shown to have the same carbohydrate binding specificities, binding primarily to the carbohydrate chains of band 3 glycoprotein of human erythrocyte membranes, possibly at the di-*N*-acetylchitobiosyl grouping.[121] Membrane receptors for the two mitogens have been prepared from erythrocytes by an affinity chromatography procedure utilizing the mitogens immobilized on agarose.

In the presence of trichloroethanol and ultraviolet irradiation, the two fluorescent L-tryptophan residues of each of the polypeptide chains of wheat germ agglutinin undergo photochemical modification, and the resultant protein loses its L-tryptophan fluorescence.[122] Concomitant with the photo-

[117] S. K. Podder, A. Surolia, and B. K. Bachhawat, *F.E.B.S. Letters*, 1978, **85**, 313.
[118] R. H. A. Muray and D. H. Northcote, *Phytochemistry*, 1978, **17**, 623.
[119] A. K. Allen, N. N. Desai, A. Neuberger, and J. M. Creeth, *Biochem. J.*, 1978, **171**, 665.
[120] S. L. Biroc and M. E. Etzler, *Biochim. Biophys. Acta*, 1978, **544**, 85.
[121] K. Yokoyama, T. Terao, and T. Osawa, *Biochim. Biophys. Acta*, 1978, **538**, 384.
[122] J. P. Privat and M. Charlier, *European J. Biochem.*, 1978, **84**, 79.

chemical reaction, a reduction in the haemagglutinating activity of the protein and its affinity towards oligomers of chitin is observed. A probable location for these L-tryptophan residues in the binding sites of the lectin was proposed. Results have been presented of an ^{1}H n.m.r. spectroscopic investigation of the binding of 1-*O*-methyl-di-*N*-acetylchitobioside and 1-*O*-methyl-tri-*N*-acetylchitotrioside to wheat germ agglutinin.[123] The use of these methyl glycoside derivatives greatly simplified the analysis by removing the contribution of the anomeric protons of the free oligosaccharides, and the results obtained support earlier theories about the binding of the oligosaccharides to three binding sub-sites on the lectin.

Phytohaemagglutinins from various species of Leguminosae have been shown to react with immunoglobulins IgG, IgA, and IgM from normal human serum.[124] Apart from differences observed between species, all the lectins react better with IgM and IgA than with IgG. The results obtained are discussed in relation to the heterogeneity of immunoglobulins.

Various lectins from the seeds and roots of plants have been purified; their chemical and biological properties are listed in Table 1.

A lectin, isolated from the seeds of *Amphicarpaea bracteata* by affinity chromatography on 2-acetamido-2-deoxy-D-galactose coupled to epoxy-activated agarose, has been shown to have characteristics common with *Dolichos biflorus* lectin, especially in relation to the agglutination of human A_1 erythrocytes.[125] It is assumed that the lectin interacts with non-reducing terminal 2-acetamido-2-deoxy-D-galactosyl residues on the red cell membrane of these erythrocytes.

A number of lectins have been purified by affinity chromatography on *O*-glycosyl polyacrylamide gels.[127] The physicochemical properties of well-characterized lectins such as *Glycine soja*, *Dolichos biflorus*, *Phaseolus lunatus*, *Helix pomatia*, and *Ricinus communis* lectins compared well with those of preparations obtained by alternative chromatographic systems.

The lectins from gorse (*Ulex europaeus*) and broom (*Sarothamnus scoparius*) seeds have been isolated by affinity chromatography on an L-fucose derivative of agarose.[131] Both lectins were characterized as glycoproteins (mol. wt. 6×10^4) having two sub-units and similar amino-acid compositions, and binding to human B lymphocytes. The lectin from *S. scoparius* agglutinated human O-type erythrocytes only weakly, and the agglutination was inhibited by low concentrations of L-fucose and 2-acetamido-2-deoxy-D-galactose. A 2-acetamido-2-deoxy-α-D-galactosyl derivative of polyacrylamide was also used in the isolation of the lectin from *S. scoparius*, yielding a product having twice

123 J. P. Grivet, F. Delmotte, and M. Monsigny, *F.E.B.S. Letters*, 1978, **88**, 176.
124 P. Rougé, C. Chatelain, and D. Pere, *Ann. pharm. franc.*, 1978, **36**, 143.
125 L. J. Blacik, M. Breen, H. G. Weinstein, R. A. Sittig, and M. Cole, *Biochim. Biophys. Acta*, 1978, **532**, 225.
126 V. Hořejši and J. Kocourek, *Biochim. Biophys. Acta*, 1978, **532**, 92.
127 V. Hořejši and J. Kocourek, *Biochim. Biophys. Acta*, 1978, **538**, 299.
128 E. I. Mbadiwe and S. I. O. Agogbua, *Phytochemistry*, 1978, **17**, 1057.
129 V. Hořejši, O. Chaloupecká, and J. Kocourek, *Biochim. Biophys. Acta*, 1978, **539**, 287.
130 V. Hořejši, C. Haškovec, and J. Kocoureck, *Biochim. Biophys. Acta*, 1978, **532**, 98.
131 L. G. Gürtler, *Biochim. Biophys. Acta*, 1978, **544**, 593.

Table 1 *Chemical and physical properties of purified lectins*

Source	Molecular weight	Sub-units	Carbohydrate (percentage) Neutral	Amino-sugar	Specificity Human blood type	Sugar	Ref.
Plant							
Amphicarpaea bracteata			2.0		A_1	D-Gal*N*Ac	125
Datura stramonium	1.2×10^5	2	28.0	4.5	ABO	chito-oligo-saccharide	126
Maclura pomifera	6×10^4		3.0	0	ABO	D-Gal, D-Gal*N*Ac, lactose	127
Macuna flagellipes	1.46×10^5			0.5	B		128
Ononis hircina Jacq.	1.1×10^5	4	7.2	tr.	O	D-Gal, D-Gal*N*Ac	129
Ononis spinosa	1.1×10^5	4	6.3	tr.	O	D-Gal, D-Gal*N*Ac, lactose, L-Ara	127
Robinia pseudocacia	1.1×10^5	4	7.2	0.47	ABO	D-Gal*N*Ac	130
Sarothamnus scoparius	6×10^4	2	0.8	2.6	O (weakly)	L-Fuc, D-Gal*N*Ac	131
	1.2×10^5	4	5.5	0.7	ABO[a]	D-Gal, D-Gal*N*Ac lactose, raffinose	127
Vicia cracca	1.2×10^5	4	1.6	tr.	ABO	D-Gal*N*Ac	127
Vigna radiata	1.6×10^5	4			—	D-Gal, D-Xyl	132
Viscum album	1.15×10^5	4?	10.5		ABO	D-Gal, lactose, raffinose	133
Microbial							
Clitocybe nebularis	7×10^5	4	2.6		ABO	D-Gal*N*Ac, lactose	127
Fomes fomentarius	6×10^5	3	25.0	tr.	B	D-Gal, D-Gal*N*Ac raffinose	127
Marasmius oreades	5.1×10^4	4	5.4	tr.	B, O		127
Animal							
Abramis brama	3.1×10^4	2?			B		134
Anguilla anguilla	5×10^4		0.2	0.	O	L-Fuc	127
Coregonus lavaretus/maraena	2.5×10^4		10.4		B	-L-Rha, D-Gal	135
Perca fluviatilis			2.0		O	D-Man, L-Fuc, D-Glc	134
Rutilis rutilis	2.5×10^4	2.6	2.6	0.4	ABO	L-Rha, D-Gal, L-Ara	134
Scardinus erythrophtalmus	2.2×10^4		3.6		ABO	D-Gal, L-Rha, L-Ara	134
Tachypleus tridentatus					AB	Neu*N*Ac, D-Glc*N*Ac	136
					ABO	D-Gal*N*Ac	
					ABO	D-Man*N*Ac	
					ABO	D-Gal*N*Ac	
Tridacna maxima	4.7×10^5	12[b]	5.7	1.3	O	D-Gal, D-Gal*N*Ac	137
Vimba vimba	2.5×10^4		8.4		ABO	L-Rha, D-Gal, L-Ara	134

the molecular weight, twice the number of sub-units and a different carbohydrate content.[127] Agglutination of human erythrocytes could only be demonstrated after treatment of the cells with trypsin. Varieties within the species of *Sarothamnus* may account for the observed differences.

A phytohaemagglutinin from mung beans (*Vigna radiata*, formerly known as *Phaseolus aureus*) has been co-purified with α-D-galactosidase using methods that would resolve components of significantly different size or charge.[132] The lectin and the enzyme possess indistinguishable specificities towards substrates and inhibitors and in their binding to red blood cells, and they are inhibited from binding by the same spectrum of sugars. The authors consider that legume lectins may, in general, be plant enzymes.

Hydrophilic water-insoluble gels suitable for affinity chromatography of lectins have been prepared by co-polymerization of acrylamide, *N,N'*-methylene bisacrylamide and alkenyl 1-thioglycosides.[138]

Free monosaccharides have been used for the direct glycosylation of Spheron®, a spherical macroporous hydroxyalkyl methacrylate–ethylene dimethacrylate co-polymer.[139] The derivatives were shown to be efficient affinity carriers in the isolation of a variety of lectins.

Water-soluble *O*-glycosyl derivatives of polyacrylamide have been prepared by co-polymerization of acrylamide and allyl glycosides of various sugars.[140] Quantitative precipitin curves obtained with these co-polymers and lectins were similar to those produced by lectins and naturally occurring polysaccharides; thus the co-polymers can serve as model substances for lectin interactions.

Extracts of seeds from seventy-nine plant species have been tested for agglutination and lytic activities with red cells of ten vertebrate species.[141] Some of the lectins were recognized as being potential diagnostic reagents in distinguishing blood of different animal species.

Two distinct developmentally regulated lectins in embryonic chick pectoral muscle have been described.[142] One is specifically inhibited by lactose and the other by 2-acetamido-2-deoxy-D-galactose.

The neuraminic acid-binding lectin, limulin, has been isolated from the haemolymph of the horseshoe crab, *Limulus polyphemus*.[143] The amino-acid sequence of 76 residues of this molecule, including 50 amino-acids from the amino-terminus, has been obtained by sequence analysis of the intact protein and its cyanogen bromide-derived fragments. No obvious sequence homology between the invertebrate lectin and vertebrate immunoglobulins was observed.

[132] C. N. Hankins and L. M. Shannon, *J. Biol. Chem.*, 1978, **253**, 7791.
[133] P. Ziska, H. Franz, and A. Kindt, *Experientia*, 1978, **34**, 123.
[134] A. Krajhanzl, V. Hořejši, and J. Kocourek, *Biochim. Biophys. Acta*, 1978, **532**, 215.
[135] A. Krajhanzl, V. Hořejši, and J. Kocourek, *Biochim. Biophys. Acta*, 1978, **532**, 209.
[136] S. Shimizu, M. Ito, and M. Niwa, *Biochim. Biophys. Acta*, 1977, **500**, 71.
[137] B. A. Baldo, W. H. Sawyer, R. V. Stick, and G. Uhlenbruck, *Biochem. J.*, 1978, **175**, 467.
[138] J. Pipková, V. Hořejši, and J. Kocourek, *Biochim. Biophys. Acta*, 1978, **541**, 515.
[139] K. Filka, J. Couper, and J. Kocourek, *Biochim. Biophys. Acta*, 1978, **539**, 518.
[140] V. Hořejši, P. Smolek, and J. Kocourek, *Biochim. Biophys. Acta*, 1978, **538**, 293.
[141] V. Bhalla, V. Gaur, and K. Bhatia, *Vox Sanguinis*, 1978, **35**, 241.
[142] F. J. Mir-Lechaire and S. H. Barondes, *Nature*, 1978, **272**, 256.
[143] R. Kaplan, S. S. Li, and J. M. Kehoe, *Biochemistry*, 1977, **16**, 4297.

Ordered conformational regions such as α-helix or β-structure were deemed absent on the basis of c.d. studies.[144]

At least three lectin species have been detected in lymphocytes isolated from the thymus and spleen of BALB/c mice.[145] Lectins specific towards β-D-galactosides and 2-acetamido-2-deoxy-D-galactosides are found mainly in thymus extracts, and those specific towards α-D-mannosides in spleen extracts.

An interferon-antagonistic substance has been extracted from mouse costal cartilage, and in addition to its anti-interferon effect, it has been shown to share many properties of plant lectins.[146] 2-Amino-2-deoxy-D-glucose and 2-amino-2-deoxy-D-galactose, but not their respective 2-acetamido-derivatives, D-glucopyranosyluronic acid and L-rhamnose, are efficient inhibitors of the anti-interferon-induced cell agglutination.

The toxic lectins abrin and ricin have been the subject of a review.[147] The sugar residues of ricin are supposedly of little importance for the biological properties of the lectin, as shown by periodate oxidation studies of the glycoprotein.[148] In the case of abrin, periodate oxidation reduces strongly its toxicity in mice, apparently due to the B chain which is glycoproteinaceous and contains residues of D-mannose and 2-acetamido-2-deoxy-D-glucose, although other evidence suggest that the effect may result from the oxidation of some amino-acid residues. Two constituent polypeptide chains with L-isoleucine and L-alanine residues at the respective *N*-termini have been isolated from ricin E and separated from each other.[149] Examination of the amino-acid composition and tryptic peptides revealed that the L-isoleucine-terminated chain is common to both ricin E and ricin D. The L-alanine-terminated chain of ricin E is a homologous protein having several amino-acid replacements compared to that of ricin D. Differences in toxicity of ricins E and D towards malignant cultured cells may be due to the differences in the formation of the L-alanine-terminated chain. The specific interaction of ricin from *R. sanguineus* with the ligands, D-galactose and lactose, has been studied by equilibrium dialysis, fluorescence polarization, and ultracentrifugation analysis.[150] Scatchard plots of equilibrium dialysis data show that two molecules of lactose but only one molecule of D-galactose bind per molecule of ricin. Together with other results, this indicated that the lectin contains two distinct saccharide binding sites.

Abrin- and ricin-resistant variants of established HeLa cell lines have been characterized.[151] The reduced number of binding sites on two of the cell variants is due to an increased level of neuraminic acid on oligosaccharide chains terminated at the non-reducing end by D-galactose. However, these over-sialylated variants are not cross-resistant to another plant toxin, modeccin,

144 A. C. Roche, J. C. Maurizot, and M. Monsigny, *F.E.B.S. Letters*, 1978, **91**, 233.
145 C. M. T. Kieda, D. J. Bowles, A. Ravid, and N. Sharon, *F.E.B.S. Letters*, 1978, **94**, 391.
146 F. Chany-Fournier, A. Pauloin, and C. Chany, *Proc. Nat. Acad. Sci., U.S.A.*, 1978, **75**, 2333.
147 S. Olsnes and A. Pihl, *Trends in Biol. Sciences*, 1978, **3**, 7.
148 K. Sandvig, S. Olsnes, and A. Pihl, *European J. Biochem.*, 1978, **84**, 323.
149 G. Funatsu, T. Mise, H. Matsuda, and M. Funatsu, *Agric. and Biol. Chem. (Japan)*, 1978, **42**, 851.
150 C. Zentz, J. P. Frénoy, and R. Bourrillon, *Biochim. Biophys. Acta*, 1978, **536**, 18.
151 S. Olsnes and K. Refsnes, *European J. Biochem.*, 1978, **88**, 7.

which also appears to bind to terminal D-galactosyl residues. Possibly the structure of the modeccin binding sites, which appear to involve the sequence β-D-Gal*p*-(1 → 3)-D-Gal*p*NAc, is not recognized by the sialyltransferase which in the two variants transfers neuraminic acid to the binding sites for abrin and ricin, forming the sequence β-D-Gal*p*-(1 → 4)-β-D-Glc*p*NAc-D-Man*p*. Treatment of HeLa cells with neuraminidase has increased the number of binding sites to abrin and ricin two-fold whereas in a toxin-resistant variant the number was increased ten-fold.[152] The desialylized toxin-resistant cells also became more sensitive to the toxins.

A simple model system has been constructed for study of the kinetics of lectin binding to particles containing immobilized protein receptors.[153] Rates of association of ^{125}I-labelled abrin, and the related abrus agglutinin lectins, with immobilized desialylized fetuin as the receptor molecule have been studied. Similar results were obtained for the binding of toxin to immobilized fetuin and to HeLa cells. The presence of unlabelled ligand accelerated the dissociation of bound labelled ligand in the artificial situation where the receptors are immobilized, and a mechanism involving a direct interaction of the unlabelled ligand with the ligand–receptor complex is suggested.

The binding of abrin to HeLa cells has been shown to decrease with increasing cell density although changes in the cell density do not seem to influence the number of available surface receptors.[154]

The roots of *Adenia digitata* contain the toxic lectin modeccin, which, like abrin and ricin, consists of two dissimilar peptide chains and acts by inactivating ribosomes.[155, 156] Terminal D-galactosyl residues on receptor molecules are involved in the binding sites of all three molecules.

The inhibition of protein synthesis in Chinese hamster cells by diphtheria toxin is antagonized by concanavalin A, succinylated concanavalin A, and wheat germ agglutinin.[157] The effects are reversed by methyl α-D-mannopyranoside in the first two cases and by 2-acetamido-2-deoxy-D-glucose in the last. Thus diphtheria toxin appears to have the saccharide binding properties of a lectin.

D-Mannosyl oligopeptides derived from ovalbumin and the cell wall polysaccharide from *Salmonella suis*, and containing terminal 2-acetamido-2-deoxy-D-glucosyl residues, are effective inhibitors of binding of the toxin to the cells. A disulphide-linked conjugate of concanavalin A and fragment A from trypsinized diphtheria toxin has been synthesized and shown to be a toxin for HeLa, Chinese hamster ovary and SV3T3(murine) cells.[158]

A non-toxic mutant protein from diphtheria toxin has been conjugated to

[152] K. Sandvig, S. Olsnes, and A. Pihl, *European J. Biochem.*, 1978, **82**, 13.
[153] K. Sandvig, S. Olsnes, and A. Pihl, *European J. Biochem.*, 1978, **88**, 307.
[154] K. Sandvig, *F.E.B.S. Letters*, 1978, **89**, 233.
[155] K. Refnes, T. Haylett, K. Sandvig, and S. Olsnes, *Biochem. Biophys. Res. Comm.*, 1977, **79**, 1176.
[156] S. Olsnes, T. Haylett, and K. Refsnes, *J. Biol. Chem.*, 1978, **253**, 5069.
[157] R. K. Draper, D. Chin, and M. I. Simon, *Proc. Nat. Acad. Sci., U.S.A.*, 1978, **75**, 261.
[158] D. G. Gilliland, R. J. Collier, J. M. Moehring, and T. J. Moehring, *Proc. Nat. Acad. Sci., U.S.A.*, 1978, **75**, 5319.

ricin to give a product having a greater toxicity in mice than either of the two toxins alone.[159]

4 Blood-group Substances

The biochemistry and genetics of some human blood groups have been the subject of a review.[160]

The inhibitor of influenza virus B haemagglutinin present in human saliva is associated with the mucous blood group glycoprotein fraction and is inactivated by neuraminidase.[161] Removal of the terminal D-galactosyl residue from blood group B substance by α-D-galactosidase, whilst decreasing blood group B activity and increasing blood group H activity, also increases the haemagglutinating activity of the glycoprotein. The terminal α-D-galactosyl group and probably the 2-acetamido-2-deoxy-D-galactosyl residues may interfere with access of the virus to the neuraminic acid binding sites on these or adjacent oligosaccharide chains of the glycoprotein.

Blood-group active ABH substances of both stomach secretion and saliva have been shown to originate entirely from the secreted glycoproteins and not from the glycolipids.[162, 163] Glycopeptides containing high proportions of D-galactose and 2-acetamido-2-deoxy-D-glucose have been isolated from human erythrocyte membranes on treatment with pronase.[164] The products are stable to alkali and exhibit blood-group A and B activities.

The glycoproteins of human erythrocyte membranes contain two classes of saccharide chains of differing molecular sizes and exhibiting blood-group ABH determinants.[165] Glycopeptides with blood-group M and N determinants have been degraded with cyanogen bromide and some of the fragments have been characterized immunochemically,[166] The location of M and N determinants in a fragment of red cell glycoprotein possessing genetically determined deficiencies in the polypeptide chain confirms earlier proposals that the M and N specificities are dependent on the amino-acid residues in the polypeptide chain and that a neuraminic acid-containing oligosaccharide(s) is an NM-specific part of these determinants.

Amino-acid sequencing studies have indicated that the MN blood-group antigen determinant of the major human erythrocyte membrane (MN) sialoglycoprotein is located on its *N*-terminal octasaccharideglycopeptide.[167] No clear evidence was obtained to indicate whether or not neuraminic acid residues linked to D-galactosyl- and/or 2-acetamido-2-deoxy-D-galactosyl-residues are involved in the binding areas. Isolated erythrocyte glycoproteins with blood group M and N activities have been subjected to periodate oxida-

[159] T. Uchida, M. Yamaizumi, and Y. Okada, *Biochem. Biophys. Res. Comm.*, 1978, **81**, 238.
[160] W. M. Watkins, *Proc. Roy. Soc.*, 1978, **202**, **B**, 31.
[161] T. F. Boat, J. Davis, R. C. Stern, and P. W. Cheng, *Biochim. Biophys. Acta*, 1978, **540**, 127.
[162] B. L. Slomiany and A. Slomiany, *European J. Biochem.*, 1978, **85**, 249.
[163] A. Slomiany, B. L. Slomiany, and G. B. J. Glass, *Biochim. Biophys. Acta*, **1978**, **540**, 278.
[164] J. Finne, T. Krusius, H. Rauvala, R. Kekomaki, and G. Myllyla, *F.E.B.S. Letters*, 1978, **89**, 111.
[165] S. Takasaki, K. Yamashita, and A. Kobata, *J. Biol. Chem.*, 1978, **253**, 6086.
[166] E. Lisowska and K. Wasniowska, *European J. Biochem.*, 1978, **88**, 247.
[167] W. Dahr and G. Uhlenbruck, *Z. physiol. Chem.*, 1978, **359**, 835.

tion.[168] Under mild oxidation conditions, extensive transformation of *N*-acetylneuraminic acid to its C_8 and C_7 analogues occurs, and under strong oxidation conditions this amino-sugar is completely destroyed. A greater proportion of D-galactosyl residues are oxidized in the N-specific glycoprotein than in the corresponding M-specific glycoprotein.

The amino-acid sequence of human erythrocyte glycophorin A^{MN} has been determined by Edman degradation, using both the direct Edman and dansyl-Edman procedures simultaneously for determination of the glycosylated amino-acid residues.[169] The polypeptide chain of 131 amino-acid residues contains sixteen oligosaccharide units attached to the *N*-terminal third of the molecule. Only one of the oligosaccharide units is attached to the polypeptide chain by an *N*-glycosidic linkage to L-asparagine; the remaining oligosaccharide units have *O*-glycosidic linkages to L-serine and L-threonine. Differences in the amino-acid sequences and *N*-terminal amino-acids are correlated with M or N blood-group activity in glycophorin A, the two forms of which (A^N and A^M) are coded for by genes at the M and N locus. Heterogeneity in the carbohydrate composition of glycophorin A seems to cause differences in sensitivity to different proteases. A second glycoprotein, glycophorin B, has an amino-acid sequence identical to that of glycophorin A^N in the first 23 positions and carries N-activity only.[170]

A human ductal mammary carcinoma cell line growing in a chemically defined medium spontaneously releases glycoproteins having M, N, T, and Tn blood-group activities.[171] These observations are in accord with the idea that early survival of the cells may depend on presentation of a cell surface similar to M and/or N erythrocyte membrane glycoproteins. A three-dimensional model has been constructed for a sialoglycopentapeptide located at the *N*-terminus of human erythrocyte glycoprotein having blood-group MM activity.[172]

Affinity chromatography of ^{125}I-labelled solubilized stroma from human erythrocytes and of ^{125}I-labelled band 3 (the major intrinsic membrane protein) on an anti-(blood group I) adsorbent column has indicated the association of blood-group I activity with band 3.[173]

A study of common ABH blood donors, and rare 'Bombay' and 'Para-bombay' individuals, has shown the presence of 2-α-L-fucosyltransferase activity in human sera and erythrocyte membranes.[174]

An oligosaccharide which inhibits wheat germ agglutinin haemagglutination more strongly than tri-*N*-acetyl-chitotriose has been isolated from blood-group H substance by alkaline sodium borohydride treatment of pig gastric mucosa.[175] Analysis indicated the presence of a hexasaccharide containing residues of D-galactose, 2-acetamido-2-deoxy-D-glucose, and 2-acetamido-2-

[168] G. F. Springer and H. J. Yang, *Vox Sanguinis*, 1978, **35**, 255.
[169] M. Tomita, H. Furthmayr, and V. T. Marchesi, *Biochemistry*, 1978, **17**, 4756.
[170] H. Furthmayr, *Nature*, 1978, **271**, 519.
[171] J. H. Anglin, M. P. Lerner, and R. E. Nordquist, *Nature*, 1977, **269**, 254.
[172] G. F. Springer, H. J. Yang, and P. R. Desai, *Naturwiss.*, 1978, **65**, 547.
[173] R. A. Childs, T. Feizi, M. Fukuda, and S. I. Hakomori, *Biochem. J.*, 1978, **173**, 333.
[174] C. Mulet, J. P. Carton, J. Badet, and C. Salmon, *F.E.B.S. Letters*, 1977, **84**, 74.
[175] M. Tuppy and M. Wrann, *Monatsh.*, 1978, **109**, 703.

deoxy-D-galactitol (2 : 3 : 1). The inhibitory activity is sensitive to the β-D-2-acetamido-2-deoxyglucosidase activity of *Trichomonas foetus*. Evidence for the binding of terminal non-reducing 2-acetamido-2-deoxy-D-glucosyl groups of the hexasaccharide (6) to wheat germ agglutinin was obtained *via* methylation analysis and Smith degradation.[176]

R_1-D-Glc*p*NAc
1
↓
3/6
D-Gal*p*NAcol
6/3
↑
1
R_2-D-Glc*p*NAc-(1 → 4)-D-Gal*p*

R_1 = D-Glc*p*NAc-(1 → 2)-D-Gal*p*-(1 → 4)-, R_2 = H, or *vice versa*.

(6)

Twenty individual reduced oligosaccharides, containing seven to eleven monosaccharide units, have been isolated from blood-group H substance of pig stomach linings by sodium borohydride degradation.[177] The oligosaccharide structures (7—26) were determined by means of periodate oxidation, methylation analysis, partial acid hydrolysis, and enzymic hydrolysis. All the oligosaccharides reported can be divided into four series (7—11), (12—16), (17—21), and (22—26). The oligosaccharides belonging to each series have a common oligosaccharide fragment to which terminal L-fucosyl and/or 2-acetamido-2-deoxy-D-glucosyl groups are attached. Comparison of all the proposed oligosaccharide structures, including those in earlier reports on tri-, penta-, and hexa-saccharides, shows that the lower oligosaccharides represent the structural element of the higher oligosaccharides.

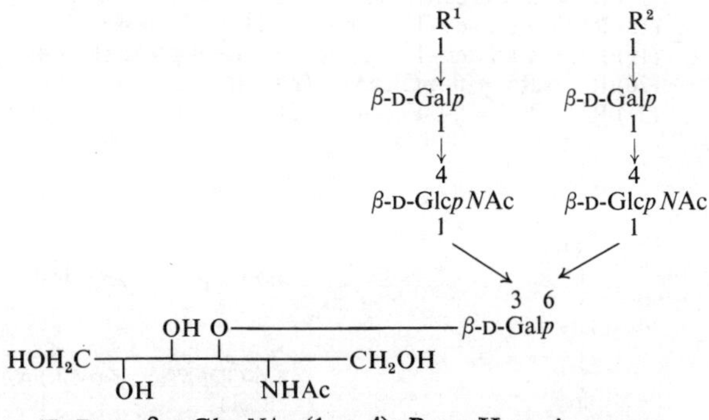

(7) R_1 = β-D-Glc*p*NAc-(1 → 4); R_2 = H or *vice versa*.
(8) R_1 = α-L-Fuc*p*-(1 → 2); R_2 = H or *vice versa*.
(9) R_1 = β-D-Glc*p*NAc-(1 → 4); R_2 = α-L-Fuc*p*-(1 → 2).
(10) R_1 = R_2 = β-D-Glc*p*NAc-(1 → 4).
(11) R_1 = R_2 = α-L-Fuc*p*-(1 → 2).

176 M. M. Wrann, and H. Tuppy, *European J. Biochem.*, 1978, **92**, 105.
177 V. A. Derevitskaya, N. P. Arbatsky, and N. K. Kochetkov, *European J. Biochem.*, 1978, **86**, 423.

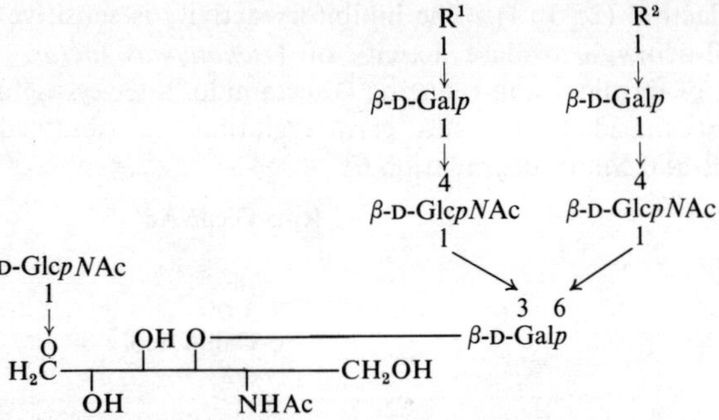

(12) R^1 = β-D-GlcpNAc-(1 → 4); R^2 = H or *vice versa*.
(13) R^1 = α-L-Fucp-(1 → 2); R^2 = H or *vice versa*.
(14) R^1 = β-D-GlcPNAc-(1 → 4); R^2 = H or *vice versa*.
(15) R^1 = R^2 = β-D-GlcpNAc-(1 → 4).
(16) R^1 = R^2 = α-L-Fucp-(1 → 2).

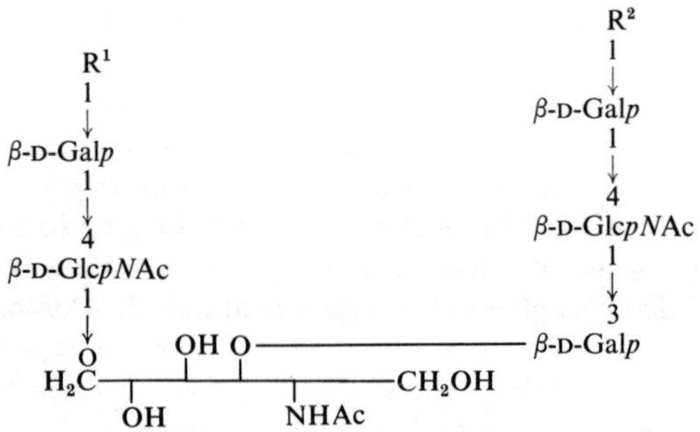

(17) R^1 = β-D-GlcpNAc-(1 → 4); R^2 = H or *vice versa*.
(18) R^1 = α-L-Fucp-(1 → 2); R^2 = H or *vice versa*.
(19) R^1 = α-L-Fucp-(1 → 2); R^2 = β-D-GlcpNAc-(1 → 4).
(20) R^1 = R^2 = β-D-GlcpNAc-(1 → 4).
(21) R^1 = R^2 = α-L-Fucp-(1 → 2).

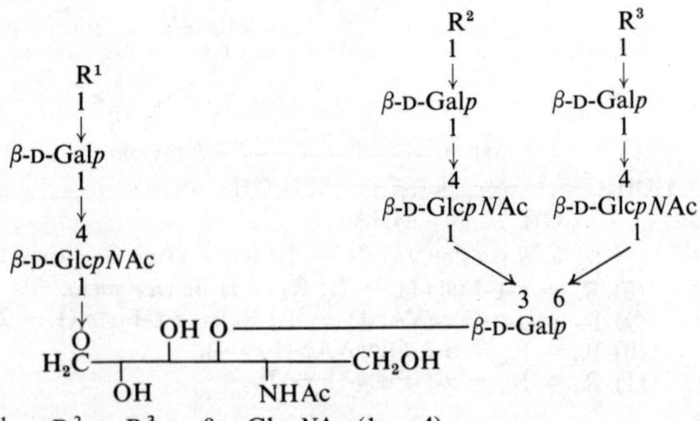

(22) R^1 = R^2 = R^3 = β-D-GlcpNAc-(1 → 4).
(23) R^1 = R^3 = α-L-Fucp-(1 → 2); R^2 = β-D-GlcpNAc-(1 → 4).
(24) R^1 = α-L-Fucp-(1 → 2), R^2 = H or *vice versa*; R^3 = β-D-GlcpNAc.

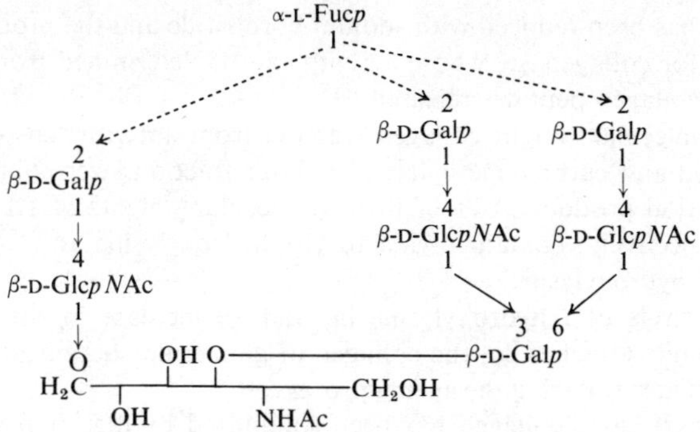

(25)

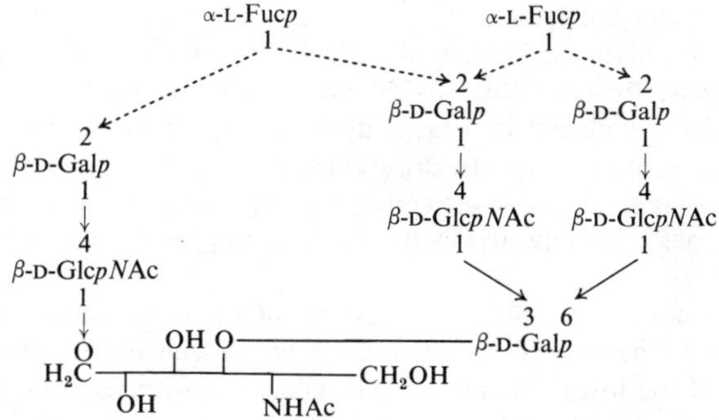

(26)

5 Collagen

A method for the analysis of L-hydroxylysine, D-glucosyl-D-galactosyl-L-hydroxylysine, and D-galactosyl-L-hydroxylysine in nmole amounts after alkaline hydrolysis of collagen samples requires preliminary separation by high voltage electrophoresis followed by detection using fluorescamine.[178] Alternatively these molecules can be separated by cation-exchange chromatography and detected using the orcinol–sulphuric acid assay.[179] The degree of occurrence of concomitant destruction of the glycopeptides during the alkaline hydrolysis was greater than previously reported; and conditions for reducing the degree of destruction were given. Mature bovine skin collagen was found to contain less D-glucosyl-D-galactosyl-L-hydroxylysine than embryonic skin collagen, whereas the differences in D-galactosyl-L-hydroxylysine were insignificant. Glycosylated L-hydroxylysines from urinary excretions have been separated on columns of Dowex-50 resins in an assessment of the activity of diseases affecting collagen metabolism.[180]

178 C. Cintron, B. D. Peczon, and C. L. Kublin, *Analyt. Biochem.*, 1978, **87**, 622.
179 C. H. Pearson, L. Ainsworth, and A. Chovelon, *Connective Tissue Res.*, 1978, **6**, 51.
180 W. C. Bisbee, and P. C. Kelleher, *J. Chromatog.*, 1978, **145**, 473.

Collagen has been reduced with sodium borotritide and the product used as a substrate for collagenase, whose activity can be determined from the levels of tritiated collagen peptides released.[181]

A high molecular weight collagen fraction from anterior lens capsule has been reduced and carboxymethylated.[182] Four fractions were obtained from the degradation products; two of them (molecular weights of 1.1×10^5 and 5×10^4 respectively) contain D-galactosyl-L-hydroxylysine and D-glucosyl-D-galactosyl-L-hydroxylysine.

Elevated levels of L-hydroxylysine in, and an increase in the number of saccharide units attached to, the collagen of glomerular basement membrane are observed in rats during the ageing process.[183]

Bovine heart valve collagen has been solubilized by digestion with pepsin followed by extraction with dithiothreitol under non-denaturing conditions.[184] It is implied that disulphide-bonded proteins may be partly responsible for the insolubility of this collagen.

The metachromatic properties of collagen on polyacrylamide gels stained with Coomassie Brilliant Blue R-250 have been investigated.[185] The metachromasy of the collagen is measured by changes in the degree of spectral shift, relative to orthochromatically stained proteins in the sample. Since the metachromasy is shared by most collagen polypeptides, the phenomenon may be due, in part, to some aspect of the repeating primary structure of this protein.

Some unusual *X*-ray diffraction features of rat tail tendon collagen have been shown to be related to the planar–crimp concept.[186] An attempted correlation of the low angle meridional reflections with features of the equatorial wide angle reflections has raised the possibility that the orientation of the banding in the fibrils and the orientation of tropocollagen molecules need not be uniquely correlated.

Small angle *X*-ray- and neutron-diffraction patterns obtained for the annulus fibrosus of porcine invertebral disc have highlighted differences between the collagen in this tissue and that in tendon.[187]

A collagen fraction from the body wall glycoproteins of the leech, *Hirudo medicinalis*, has been shown to contain a higher proportion of D-glucosyl-D-galactosyl-L-hydroxylysine units than other vertebrate and invertebrate collagens.[188]

The enzymic cleavage of the carboxy-terminal peptides from procollagen by carboxy-terminal procollagen protease which occurs on incubation of chick tendon fibroblasts is not directly inhibited by tunicamycin, although procollagen molecules which are underglycosylated act as a substrate for this

[181] C. L. Hu, G. Crombie, and C. Franzblau, *Analyt. Biochem.*, 1978, **88**, 638.
[182] S. N. Dixit, *F.E.B.S. Letters*, 1978, **85**, 153.
[183] J. R. Hoyer and R. G. Spiro, *Arch. Biochem. Biophys.*, 1978, **185**, 496.
[184] S. A. Jimenez and R. I. Bashey, *Biochem. J.*, 1978, **173**, 337.
[185] S. Micko and W. W. Schlaepfer, *Analyt. Biochem.*, 1978, **88**, 566.
[186] L. J. Gathercole and A. Keller, *Biochim. Biophys. Acta*, 1978, **535**, 253.
[187] C. Berthet, D. J. S. Humes, A. Miller, and P. A. Timmins, *Science*, 1978, **199**, 547.
[188] T. Biswas and A. K. Mukherjee, *Carbohydrate Res.*, 1978, **63**, 173.

enzyme.[189] Concanavalin A also inhibited the conversion of procollagen to collagen by fibroblasts in culture, and in an *in vitro* system, inhibits the procollagen protease; this inhibition can be reversed by methyl α-D-glucopyranoside. It is suggested that the protease is a glycoprotein, having oligosaccharide side chains which can bind to concanavalin A, and that tunicamycin affects the secretion, activity, or activation of the enzyme.

The incorporation of the L-proline analogue *cis*-4-hydroxy-L-proline into procollagen polypeptides which occurs on incubation with matrix-free chick embryo tendon cells prevents the polypeptides from folding into a triple-helical formation.[190] Secretion of the non-helical polypeptides is delayed, but contrary to previous observations, they are not degraded intracellularly. Instead they are secreted as non-helical pro-γ chains which contain increased amounts of glycosylated L-hydroxylysine.

Secretion by matrix-free chick embryo tendon cells of Type I procollagen containing 2-amino-2-deoxy-D-glucose and D-mannose has been demonstrated.[191] A peptide arising from the carboxy-terminus contains 2-amino-2-deoxy-D-glucose and D-mannose, whilst an *N*-terminal peptide contains only 2-amino-2-deoxy-D-glucose. Evidence has been provided to show that under conditions which inhibit the hydroxylation of L-lysine and glycosylation of L-hydroxylysine these two sugars are incorporated into unhydroxylated procollagen, at the level of the rough endoplasmic reticulum.

Under-hydroxylated basement membrane collagen has been shown to be synthesized by the rat embryonic parietal yolk sac system and secreted into the yolk sac medium at the same rate as the hydroxylated basement membrane collagen.[192] However, unlike the hydroxylated material, the under-hydroxylated basement membrane collagen is not deposited on the Reichert's membrane and is more susceptible to degradation by extracellular proteases. The synthesis of the 2-amino-2-deoxy-D-glucosyl group was shown to proceed even in the presence of inhibitors of the hydroxylation process. Incorporation of D-[2-^{3}H]mannose into the intracellular pro-α-chains of collagen has been shown to take place within the cisternae of the rough endoplasmic reticulum of matrix-free cells from leg tendons of chick embryos.[193]

Collagenase, secreted by cultured rheumatoid synovial cells in an inactive form, has been shown to bind to collagen fibrils.[194] The binding as well as the activation of the enzyme by proteases such as plasmin occurs in the presence of α_2 macroglobulin, the plasma protein responsible for anti-collagenase activity.

A peptide bond within the extra-helical carboxy-terminal sequence of the α1(I) chain of collagen is cleaved by cathepsin D.[195] This ability to depolymer-

[189] D. Duksin, J. M. Davidson, and P. Bornstein, *Arch. Biochem. Biophys.*, 1978, **185**, 326.
[190] V. J. Uitto, J. Uitto, W. W. Y. Kao, and D. J. Prockop, *Arch. Biochem. Biophys.*, 1978, **185**, 214.
[191] C. C. Clark and N. A. Kefalides, *J. Biol. Chem.*, 1978, **253**, 47.
[192] M. E. Maragoudakis, H. J. Kalinsky, and J. Wasvary, *Biochim. Biophys. Acta*, 1978, **538**, 139.
[193] H. Anttinen, A. Oikarinen, L. Ryhanen, and K. I. Kivirikko, *F.E.B.S. Letters*, 1978, **87**, 222.
[194] C. A. Vater, C. L. Mainardi, and E. D. Harris, *Biochim. Biophys. Acta*, 1978, **539**, 238.
[195] P. G. Scott and C. H. Pearson, *F.E.B.S. Letters*, 1978, **88**, 41.

ize cross-linked peptides of collagen, together with the extracellular occurrence of cathepsin D, indicate a possible involvement of this proteolytic enzyme in collagen degradation *in vivo*.

Chemical and physical analyses of the collagen isolated from cell lines derived from a mouse teratocarcinoma source or a mouse blastocyst source have identified a type V collagen, with distinct differences from types I, II, III, and IV collagen.[196]

When tropocollagen is treated with oxygen free-radicals changes are observed in the viscosity and sub-unit patterns.[197] The product behaves like collagen treated with pepsin in many of its properties, suggesting that oxidation of susceptible amino-acids and cleavage of non-helical regions of the tropocollagen molecule have taken place.

Studies on the mechanism of chick embryo collagen D-glucosyltransferase has revealed an ordered mechanism in which substrates are bound to the enzyme in the order Mn^{2+}, UDP-D-glucose, and collagen.[198] The order of release from the enzyme–substrate complex is D-glucosylated collagen, UDP, and Mn^{2+}, but Mn^{2+} need not leave the enzyme during each catalytic cycle. The acceptor specificity of the enzyme has been reinvestigated using preparations purified by affinity chromatography, and the enzyme has been shown to transfer D-glucose only to collagens.[199]

6 Glycogens

Glycogen levels in whole blood or plasma have been measured after prior removal of free D-glucose by the hexokinase–D-glucose 6-phosphate dehydrogenase reaction, followed by estimation of the D-glucose then liberated from the glycogen by glucoamylase.[200] Glycogen has also been converted by glucoamylase to D-glucose and the liberated D-glucose measured by D-glucose oxidase.[201]

Data obtained from low temperature ^1H n.m.r. spectra of glycogen have allowed the evaluation of the amount of water bound to the polysaccharide.[202] High molecular weight glycogen (1×10^9) exhibits water-binding properties analogous to those previously reported for other glycoproteins, but low molecular weight (1×10^7) glycogens show anomalous binding characteristics with large amounts of associated 'non-freezing' water. The levels reported are significantly higher than those found by hydrodynamic investigations.

Structural differences, mainly associated with the average chain lengths, have been reported for two piscine muscle glycogens from *Catla catla* and *Clarias batrachus*.[203]

Glycogen debranching enzyme, α- and β-sub-units of phosphorylase kinase, glycogen phosphorylase, and glycogen synthase have been shown to be

196 C. D. Little and R. L. Church, *Arch. Biochem. Biophys.*, 1978, **190**, 632.
197 K. Venkatasubramanian and K. T. Joseph, *Indian J. Biochem. Biophys.*, 1977, **14**, 217.
198 R. Myllylä, *European J. Biochem.*, 1976, **70**, 225.
199 H. Anttinen, R. Myllylä, and K. I. Kivirikko, *Biochem. J.*, 1978, **175**, 737.
200 B. Eisele and G. Heinzel, *Analyt. Biochem.*, 1978, **84**, 594.
201 M. L. Bruss and A. L. Black, *Analyt. Biochem.*, 1978, **84**, 309.
202 T. Brittain and R. Geddes, *Biochim. Biophys. Acta*, 1978, **543**, 258.
203 R. V. Krishnamoorthy and B. J. Begum, *Experientia*, 1978, **34**, 832.

associated with the protein–glycogen complex isolated from rabbit skeletal muscle.[204] Three other protein species, namely the heavy chains of myosin, actin, and calsequestrin, which were reported to be present in earlier preparations have been shown to be constituents of particulate material which has co-precipitated with the protein–glycogen complex.

[^{14}C]Glycogen has been synthesized *in vitro* from rabbit liver glycogen and D-[^{14}C]glucose in the presence of glycogen debranching enzyme, amylo-1,6-glucosidase-4-α-glucanotransferase.[205] The distribution of D-[^{14}C]glucose groups in the polysaccharide was determined and it is considered that the D-[^{14}C]glucose incorporation assay for amylo-1,6-glucosidase as normally carried out, is probably not specific just for the hydrolytic action of this enzyme, but that it depends indirectly on the transferase action as well.

Glycogen-depleted muscle fibres have been visualized microscopically after replenishment of the glycogen by incubation in the presence of UDP-D-glucose and α-D-glucan UDP-D-glucose glucosyltransferase.[206] The polysaccharide was then visualized by the periodate–Schiff reaction.

A study of the action of rabbit-muscle phosphorylase and glycogen synthase on structurally modified glycogens has confirmed that both these transglucosylases act less readily on debranched glycogen and on glycogen possessing elongated outer chains.[207] Glycogen phosphorylase *a* has been purified from rabbit liver and some of its kinetic properties have been reported.[208] A synergism in the joint inhibiting action of D-glucose and ATP was detected and the probability that their binding sites on the phosphorylase are sterically separated was considered.

A semicontinuous assay for glycogen phosphorylase *b* has been developed by measuring the release of phosphate from D-glucose 1-phosphate on incubation with glycogen and AMP.[209] The method can be used for the measurement of phosphorylase *a* if AMP is omitted from the incubation of glycogen and D-glucose 1-phosphate.

X-ray crystallographic data have been presented to describe the binding of maltoheptaose and other oligosaccharides at the glycogen storage site of rabbit muscle phosphorylase *a*.[210] Various kinetic mechanisms which require the presence of two oligosaccharide-binding sites, effector and catalytic, were tested for their fit to experimental data, obtained from a measurement of initial reaction rates carried out both in the absence of AMP, and in the presence of maltoheptaose and D-glucose 1-phosphate as substrates. Maltoheptaose (and presumably glycogen) appears to bind to a glycogen storage site distinct from the active site, and it is the saccharide–protein interactions at the storage site which are responsible for the activation of phosphorylase and

[204] B. Caudwell, J. F. Antoniw, and P. Cohen, *European J. Biochem.*, 1978, **86**, 511.
[205] D. H. Brown, D. J. Sprinkle, and B. I. Brown, *Carbohydrate Res.*, 1978, **61**, 265.
[206] C. W. Pool, Y. E. Donselaar, and P. A. M. Griep, *J. Histochem. Cytochem.*, 1978, **26**, 742.
[207] J. Carter and E. E. Smith, *Carbohydrate Res.*, 1978, **61**, 395.
[208] T. B. Eronina, G. V. Silonova, and N. B. Livanova, *Biochemistry* (*U.S.S.R.*), 1977, **42**, 975.
[209] I. T. Carney, R. J. Beynon, J. Kak, and N. Birket, *Analyt. Biochem.*, 1978, **85**, 321.
[210] P. J. Kasvinsky, N. B. Madsen, R. J. Fletterick, and J. Sygusch, *J. Biol. Chem.*, 1978, **253**, 1290.

heterotropic co-operativity between substrates. Glycogen, activated with cyanogen bromide, has been covalently bound to phosphorylase *b*.[211] The product showed a higher affinity than the native enzyme towards D-glucose 1-phosphate but a lower homotropic co-operativity with respect to activation of AMP.

7 Glycosaminoglycuronans, Glycosaminoglycans, and their Protein and Peptide Derivatives

Occurrence, Isolation, and Structures.—The structure of glycosaminoglycans, their binding properties for intra- and extra-cellular and cell surface components, and their biosynthesis have been reviewed.[212] An elementary short review on the isolation and separation of glycosaminoglycans has been published.[213]

The analysis of glycosaminoglycans from cases of genetic hyperglycosaminoglycanuria (mucopolysaccharidosis) has been achieved by fluorescence measurement of the products of the reaction of these polysaccharides with Acridine Orange.[214] The separation of 2-amino-2-deoxy-D-glucose and 2-amino-2-deoxy-D-galactose in an auto-analyser has been reported.[215] The method is suited to the analysis of proteoglycans and glycosaminoglycans, allowing one of the amino-sugars to be measured in the presence of large amounts of the other.

Glycosaminoglycans, extracted from tissue *via* proteolysis, have been purified by paper chromatography using a zinc acetate-based solvent at pH 5.0.[216] Countercurrent chromatography has been used in the purification of glycosaminoglycans.[217] In butan-1-ol–aqueous sodium chloride, the fractionation of quaternary ammonium complexes of glycosaminoglycans is dependent primarily upon the chemical composition or charge density, but is independent of molecular weight. Chondroitin 4-sulphate was fractionated according to the degree of sulphation, and was completely resolved from heparin which itself was resolved into three components differing with respect to *N*-sulphate and *O*-sulphate contents.

A method for the isolation of proteoglycans from rat brain by a dissociative technique using 4M guanidine has been re-evaluated and shown to yield neither a significant proportion nor a representative sample of proteoglycans.[218] The sedimentation coefficient distributions for proteoglycans at low solute concentrations have been measured in preformed density gradients and the values were similar to those obtained using conventional analytical centrifugation procedures.[219]

[211] T. G. Sotiroudis, N. G. Dikonomakos, and A. E. Evangelopoulos, *European J. Biochem.*, 1978, **88**, 573.
[212] U. Lindahl and M. Höök, *Ann. Rev. Biochem.*, 1978, **47**, 385.
[213] S. Suzuki, in 'Natural Compounds: Nucleic Acids, Proteins, and Carbohydrates', Vol. 11, pt. 1, ed. F. Korte and M. Goto, Academic Press, New York, 1976, p. 186.
[214] G. P. Diakun, H. E. Edwards, J. C. Allen, and G. O. Phillips, *Biochem. J.*, 1978, **175**, 573.
[215] J. D. Ford and J. R. Baker, *Analyt. Biochem.*, 1978, **84**, 539.
[216] C. S. Sato and F. Gyorkey, *Analyt. Biochem.*, 1978, **87**, 540.
[217] R. E. Hurst, J. Y. Sheng, and Y. Ito, *Analyt. Biochem.*, 1978, **85**, 230.
[218] R. K. Margolis, C. P. Crockett, and R. U. Margolis, *J. Neurochem.*, 1978, **30**, 1177.
[219] J. C. Pita, F. J. Muller, T. Oegema, and V. C. Hascall, *Arch. Biochem. Biophys.*, 1978, **186**, 66.

Vibrational frequencies for important functional groups in chondroitin 4- and 6-sulphate and hyaluronic acid, such as sulphate, glycosidic linkages, and *N*-acetyl groups, have been identified from Raman spectra.[220] Differences between the spectra for the polysaccharides were interpreted in terms of either the geometry of various substituents, or differences in chemical composition.

Infrared spectra of glycosaminoglycans in D_2O and DCl/D_2O solutions have been used in the estimation of hexuronic acid and 2-acetamido-2-deoxyhexose residues of glycosaminoglycans.[221] The values obtained were in agreement with those obtained by colorimetric and conductimetric methods. I.r. absorption spectra and laser-Raman spectra have provided a reliable method for the characterization of the sulphate groups of heparin and related glycosaminoglycans.[222] Although each type of sulphate group has a different absorption and scattering pattern in the S=O stretching region, a given glycosaminoglycan produces vibrational fingerprints, especially in low frequency regions.

The value of electric birefringence in the study of the aggregation of cartilage proteoglycans has been advocated.[223] Transient electric birefringence has provided a rapid means of sizing macromolecules in solution, and results are in close agreement with the estimated length of the extended hyaluronic acid chain which forms the aggregate core of the cartilage proteoglycan.

Proteoglycans and hyaluronic acid have been solubilized from bovine nasal cartilage using lanthanum chloride solutions.[224] The isolated proteoglycans exhibit chemical and physical properties identical to those of similar products isolated by conventional techniques.

Second-order rate constants of periodate oxidation of glycol groups have been compared for glycosaminoglycuronans and the relevant monomers, with a view to assessing perturbations arising from polymer configuration.[225] For a small group of these polysaccharides (chondroitin 4- and 6-sulphates and hyaluronic acid), which are oxidized much more slowly than other glycosaminoglycuronans of analogous compositions, a stable configuration in solution involving carboxy, acetamido, and hydroxy groups in hydrogen-bonded sequences on three adjacent residues, was proposed. It is suggested that this bonding hinders periodate oxidation of D-glucuronic acid in these polymers.

Proteoglycans have been extracted under both associative and disassociative conditions from bovine cornea.[226] It is confirmed that in the cornea different sub-units are present having characteristic aggregating properties and hydrodynamic volumes. Proteoglycans prepared by disassociative extraction of pig laryngeal cartilage undergo self-association at physiological ionic strength.[227] Modification of thiol groups in the polypeptide chains by carboxymethylation

[220] R. Bansil, I. V. Yannas, and H. E. Stanley, *Biochim. Biophys. Acta*, 1978, **541**, 535.
[221] B. Casu, G. Scovenna, A. J. Cifonelli, and A. S. Perlin, *Carbohydrate Res.*, 1978, **63**, 13.
[222] F. Cabassi, B. Casu, and A. S. Perlin, *Carbohydrate Res.*, 1978, **63**, 1.
[223] A. R. Foweraker, M. Isles, B. R. Jennings, T. E. Hardingham, and H. Muir, *Biopolymers*, 1977, **16**, 1367.
[224] J. E. Scott and M. J. Tigwell, *Biochem. J.*, 1978, **173**, 103.
[225] P. J. Roughley, M. Hatt, and R. M. Mason, *Biochim. Biophys. Acta*, 1978, **539**, 445.
[226] P. Speziale, M. S. Speziale, L. Galligani, and C. Balduini, *Biochem. J.*, 1978, **173**, 935.
[227] J. K. Sheehan, I. A. Nieduszynski, C. F. Phelps, H. Muir, and T. E. Hardingham, *Biochem. J.*, 1978, **171**, 109.

caused an inhibition of the association phenomenon. Electron microscopic studies of both intact and partially degraded proteoglycans from bovine nasal cartilage substantiate currently accepted models of the proteoglycan structure.[228]

Sulphated proteoglycans have been isolated from ovine brain.[229] The cerebrum accounts for the major portion of chondroitin sulphate proteoglycan, whereas the brain-stem fraction contains the majority of heparan sulphate proteoglycan.

The sulphated heteroglycans of a number of mammalian cell lines have been characterized.[230] Earlier proposals that heparan sulphate and dermatan sulphate might play a role in cell recognition and adhesiveness and that chondroitin 4- and 6-sulphates might act as a stimulant of cell division have been further discussed.

The glycosaminoglycan and glycoprotein composition of several subcellular fractions of rat brain and liver have been reported.[231] Most of the hyaluronic acid and chondroitin sulphate is closely associated with a low density subfraction of smooth membranes.

In guinea-pig kidney medulla, chondroitin sulphate and dermatan sulphate account for at least half of the hexuronic acid content whereas in the same fraction from rat, heparan sulphates account for 65—70% of the hexuronic acid content.[232]

Hyaluronic acid is the major glycosaminoglycan component of the fat globule of cow's milk.[233] Chondroitin 4- and 6-sulphates, dermatan sulphate, and heparan sulphate are also present.

Resting 3T3 cells from a mouse cell line have been shown to contain glycosaminoglycans, of which the sulphate-containing species predominate.[234] They are also present at a higher level than in growing or SV40 virus transformed cells. The glycosaminoglycans of AH-130 ascites hepatoma cells are composed only of heparan sulphate (93%) and chondroitin sulphate, whilst the ascitic fluid contains heparan sulphate (58%), hyaluronic acid (26%), and chondroitin sulphate (16%).[235]

Significant differences have been observed between the interactions of several sulphate-containing glycosaminoglycans with particulate lysosomal protein, leading to inhibition of a number of lysosomal acid hydrolases.[236] The order of levels of inhibition is heparin > chondroitin 4-sulphate = chondroitin 6-sulphate > dermatan sulphate.

The currently accepted model for the assembly of cartilage proteoglycan has

228 D. Heinegard, S. Lohmander, and J. Thyberg, *Biochem. J.*, 1978, **175**, 913.
229 C. J. B. White, *Experientia*, 1978, **34**, 1036.
230 C. P. Dietrich and H. M. DeOca, *Biochem. Biophys. Res. Comm.*, 1978, **80**, 805.
231 W. L. Kiang, C. P. Crockett, R. K. Margolis, and R. U. Margolis, *Biochemistry*, 1978, **17**, 3841.
232 D. Lis and B. Monis, *Experientia*, 1978, **34**, 693.
233 D. Lis and B. Monis, *J. Supramol. Struct.*, 1978, **8**, 173.
234 S. Vannucchi, M. Del Rosso, C. Cella, P. Urbano, and V. Chiarugi, *Biochem. J.*, 1978, **170**, 185.
235 N. Nakamura, R. E. Hurst, and S. S. West, *Biochim. Biophys. Acta*, 1978, **538**, 445.
236 J. L. Avila, *Biochem. J.*, 1978, **171**, 489.

been reviewed.[237] Three different types of chondroitin sulphate have been isolated from the cartilage of shark, squid, and king crab.[238] In each case, a D-xylosyl-L-serine unit was identified in the glycopeptide linkage region. Chondroitin sulphate from rat rib cartilage has been shown to be heterogeneous with respect to size, degree of sulphation, and contents of 2-acetamido-2-deoxy-D-galactose and sulphate.[239]

Optimal conditions have been established for the β-elimination and reduction of *O*-glycosyl linkages in chondroitin sulphates.[240] Thus whale cartilage chondroitin sulphate contains a D-xylosyl-L-seryl linkage and the linkage region between protein and a chondroitin 4- and 6-sulphate hybrid (1 : 1) from bovine tracheal cartilage is reported to be D-Gal*p*-D-Gal*p*-D-Xyl*p*-*O*-Ser.

The presence of co-polymers of chondroitin 4- and 6-sulphate in chick embryo epiphyseal cartilage has been demonstrated by the isolation of a tetrasaccharide composed of a 4-sulphate- and a 6-sulphate-disaccharide unit.[241]

The stronger binding of Ca^{2+} than Na^{+} by chondroitin sulphate indicates that the association cannot be ascribed solely to electrostatic forces.[242] Sedimentation coefficient and intrinsic viscosity measurements have indicated that chondroitin sulphate in solution can be represented by a semi-flexible coil with a molecular diameter of about 1.2 nm, a value which is consistent with the results of *X*-ray analysis.[243]

A detailed three-dimensional structural *X*-ray analysis has indicated a regular three-fold extended helical conformation for chondroitin 4-sulphate.[244] The polyanion conformation is stabilized by an intramolecular hydrogen bond across each $(1 \rightarrow 4)$-β-D-linkage and two intermolecular hydrogen bonds per disaccharide repeat unit. The two Na^{+} in the crystal asymmetric unit are each assumed to be fully hydrated. Probable differences between the corresponding helical forms of chondroitin 6-sulphate and dermatan sulphate were discussed. Three idealistic models for the organization of chondroitin 4-sulphate molecules have been considered.[245]

Stable interactions between chondroitin sulphate and concanavalin A have not been observed under physiological conditions of salt and pH.[246] The binding, which is observed at low pH, depends upon reversible non-specific electrostatic interactions.

A radioassay method for measurement of the products of proteolytic cleavage of isolated cartilage [^{35}S]proteoglycan has been reported.[247] The method was used to measure the chondroitin sulphate peptides separated from

237 K. Sugahara, *Seikagaku*, 1978, **50**, 582.
238 N. Seno and E. Sekizuka, *J. Biochem.* (*Japan*), 1978, **83**, 953.
239 Y. H. Liau, N. I. Galicki, and M. I. Horowitz, *Biochim. Biophys. Acta*, 1978, **539**, 315.
240 N. Seno and E. Sekizuka, *Carbohydrate Res.*, 1978, **62**, 271.
241 C. R. Faltynek and J. E. Silbert, *J. Biol. Chem.*, 1978, **253**, 7646.
242 K. Tanaka, *J. Biochem.* (*Japan*), 1978, **83**, 647.
243 K. Tanaka, *J. Biochem.* (*Japan*), 1978, **83**, 655.
244 W. T. Winter, S. Arnott, D. H. Isaac, and E. D. T. Atkins, *J. Mol. Biol.*, 1978, **125**, 1.
245 J. J. Carl, W. T. Winter, and S. Arnott, *J. Mol. Biol.*, 1978, **125**, 21.
246 J. E. Morris and S. C. Chan, *Biochim. Biophys. Acta*, 1978, **538**, 571.
247 R. W. Stephens, L. I. Pemberton, P. Ghosh, and T. K. F. Taylor, *Analyt. Biochem.*, 1978, **90**, 726.

digestion products and unchanged proteoglycan by selective dissolution of the cetylpyridinium complexes in trichloracetic acid.

Chemical and enzymatic studies on dermatan sulphate from whale intestine have revealed that sulphate groups are located solely at O-4 of 2-acetamido-2-deoxy-D-galactopyranosyl residues.[248] L-Idopyranosyluronic acid residues are assumed to be distributed uniformly in the backbone of the polysaccharide chain but D-glucopyranosyluronic acid residues are located only in the linkage region of the protein core. ^{23}Na N.m.r. studies on dermatan sulphate have provided information on counterion binding and quadrupole coupling constants.[249]

The proteoglycan of heparin from rat-skin has been proposed to consist of a polypeptide core that is composed of alternate L-seryl and glycyl residues with most of the L-seryl units being substituted with polysaccharide chains through D-xylosidic linkages.[250] This model may represent a common product of heparin biosynthesis in different tissues, with the small molecular weight single chain heparin molecules being formed by limited cleavage of the proteoglycan by *endo*-acting enzymes. Porcine heparin of low molecular weight has been separated into two species, differing in the content of 2-acetamido-2-deoxy-D-glucosyl and 2-deoxy-2-sulphamido-D-glucosyl residues.[251] One of the forms, associated with high anticoagulant activity, contains a tetrasaccharide sequence with 2-deoxy-2-sulphamido-D-glucose at its reducing end as well as equivalent amounts of D-glucuronic acid and L-iduronic acid. Significantly lower levels of this tetrasaccharide sequence were detected in the other form, which is associated with negligible anticoagulant activity. It has been proposed that the anticoagulant activity of heparin is related to the probability of finding in a random distribution of different disaccharides a dodecasaccharide with the sequence required for binding to antithrombin.[252]

Ultraviolet difference spectroscopic and c.d. studies have shown that two forms of heparin having different affinities for solutions of anti-thrombin bind with a 1 : 1 stoicheiometry but with binding constants differing by about two orders of magnitude.[253] Fluorescence studies also corroborated these findings.[254] The binding of either high- or low-affinity heparin to antithrombin results in an increase of the L-tryptophan fluorescence of the antithrombin, and in the case of the high affinity heparin also causes activation of the antithrombin, allowing it to react more rapidly with a number of L-serine-specific proteases of the blood coagulation cascade. The fluorescence enhancement caused by the high affinity heparin is consistent with a conformational change in the antithrombin related to its activation. The results of competition

[248] N. Ototani and Z. Yosizawa, *Carbohydrate Res.*, 1978, **64**, 143.
[249] H. Gustavsson, G. Siegel, B. Lindman, and L. A. Franson, *F.E.B.S. Letters*, 1978, **86**, 127.
[250] H. C. Robinson, A. A. Horner, M. Höök, S. Ogren, and U. Lindahl, *J. Biol. Chem.*, 1978, **253**, 6687.
[251] R. D. Rosenberg, G. Armand, and L. Lam, *Proc. Nat. Acad. Sci., U.S.A.*, 1978, **75**, 3065.
[252] T. C. Laurent, A. Tengblad, L. Thunberg, M. Höök, and U. Lindahl, *Biochem. J.*, 1978, **175**, 691.
[253] B. Nordenman and I. Björk, *Biochemistry*, 1978, **17**, 3339.
[254] B. Nordenman, A. Danielsson, and I. Björk, *European J. Biochem.*, 1978, **90**, 1.

experiments designed to measure the affinity of the two types of heparin to matrix-linked antithrombin also support the proposed existence of a common binding site for the two fractions.[255]

Data obtained from ^{13}C n.m.r. analysis and from potentiometric titration of variously desulphated and periodate-oxidized heparins have supported the interpretation that non-sulphated α-L-iduronate residues may adopt two ring conformations.[256] The non-sulphated L-iduronic acid residues in heparin and heparin sulphate have been shown to be resistant to periodate oxidation at pH 3.0, under conditions which will oxidize a large proportion of the D-glucuronic acid residues.[257]

Absorption and c.d. spectral measurements of Cu^{2+} complexes of heparin and related glycosaminoglycans have provided a method for the determination of stability constants for these complexes, and reveal information regarding the nature of ligand groups and their spatial disposition relative to one another, and thus the conformation of the polymers.[258]

Heparin combines stoicheiometrically with polydiallyldimethyl ammonium chloride (Cat-Floc) to give a flocculent precipitate.[259]

The charge heterogeneity of heparin as observed in isoelectric focusing studies is caused by strong interaction of the polysaccharide with the amphoteric substances used for the generation of the pH gradient.[260]

Three products have been identified from γ-irradiation treatment of the tetrasaccharide β-D-Glc*p*UA-(1 → 3)-β-D-Glc*p*NAc-(1 → 4)-β-D-Glc*p*UA-(1 → 3)-D-Glc*p*NAc, *viz.* the disaccharide (27) arising from direct scission of the central glycosidic bond, D-glucuronic acid from the non-reducing terminus, and the trisaccharide (28).[261]

β-D-Glc*p*UA-(1 → 3)-D-Glc*p*NAc

(27)

β-D-Glc*p*NAc-(1 → 4)-β-D-Glc*p*UA-(1 → 3)-D-Glc*p*NAc

(28)

The depolymerization of hyaluronic acid by lysosomal enzymes from the liver is thought to involve intitial attack by hyaluronidase alone to give an octasaccharide as the product of minimum molecular weight.[262] Hexasaccharides and smaller oligosaccharides are degraded by the action of β-D-glucuronidase and β-D-2-acetamido-2-deoxyglucosidase. Native hyaluronic acid has been hydrolysed by the combined action of β-D-glucuronidase with A and B isoenzymes of β-D-2-acetamido-2-deoxyglucosidase, but with different efficiencies.[263] Trisaccharides from hyaluronic acid and from desulphated

[255] A. Danielsson and I. Björk, *European J. Biochem.*, 1978, **90**, 7.
[256] L-Å. Fransson, T. N. Huckerby, and I. A. Nieduszynski, *Biochem. J.*, 1978, **175**, 299.
[257] L-Å. Fransson, *Carbohydrate Res.*, 1978, **62**, 235.
[258] D. C. Mukherjee, J. W. Park, and B. Chakrabarti, *Arch. Biochem. Biophys.*, 1978, **191**, 393.
[259] T. Katayama, K. I. Takai, R. Kariyama, and Y. Kanemasa, *Analyt. Biochem.*, 1978, **88**, 382.
[260] P. G. Righetti and E. Gianazza, *Biochim. Biophys. Acta*, 1978, **532**, 137.
[261] A. N. Hall, G. O. Phillips, and S. Rabool, *Carbohydrate Res.*, 1978, **62**, 373.
[262] S. Hayashi, *J. Biochem. (Japan)*, 1977, **82**, 1287.
[263] G. Bach and B. Geiger, *Arch. Biochem. Biophys.*, 1978, **189**, 37.

chondroitin 4-sulphate are hydrolysed by the β-D-2-acetamido-2-deoxyglucosidase A only.[264]

Changes in c.d. and o.r.d. spectra of hyaluronic acid in organic–water solvent at various pH values have been ascribed to a major conformational transition.[265] Neither the methyl ester of hyaluronic acid nor of chondroitin sulphate undergoes a transition of this type.[266] Results indicate that hydrogen bonding between the protonated carboxy-group and the carbonyl oxygen of the acetamido group is directly involved in the conformational change.

The electric birefringence technique has been shown to be suitable for investigations of cartilage proteoglycans and their interactions with hyaluronic acid.[267]

A fragment (mol. wt. 6×10^4) of the core protein of the proteoglycans which interact with hyaluronic acid has been isolated.[268] The binding region to hyaluronic acid contains, in addition to accessible L-arginine residues necessary for activity, L-lysine residues near the binding site, which, when substituted with bulky groups such as dansyl groups, sterically block interaction. For interaction, the hyaluronic acid molecule requires a length of at least 4.5 repeat disaccharides, containing three and possibly four unmodified carboxy-groups of D-glucuronic acid residues.

The protein core from keratan sulphate-containing proteoglycan of bovine nasal cartilage has been purified by affinity chromatography on matrix-bound hyaluronic acid.[269]

Biosynthesis.—Each population of fibroblast-like cells from chicken cornea, heart, and skin has been shown to synthesize a pattern of glycosaminoglycans different from the other two, indicating that these cells behave like three distinct and stably differentiated cell types.[270] A comparative study of the uptake of [^{35}S]proteoglycans and [^{35}S]glycosaminoglycans by cultured skin fibroblasts, together with competition experiments, has indicated that during pinocytosis both classes of polysaccharide interact with different receptors on the cell membrane.[271]

Human serum has been reported to stimulate the initial steps in proteoglycan synthesis in chicken cartilage, either at the level of synthesis of the protein core and/or in its D-xylosylation, with the formation of a diffusible product which is rapidly integrated into the tissue matrix.[272]

Increased levels of dermatan sulphate have been observed for rabbit and monkey aorta cells but not for human foreskin fibroblasts grown in the presence of human serum low-density lipoprotein.[273] A mechanism is proposed to explain the increased deposition of lipid in the aorta due to hyperlipidaemia.

264 T. M. Bearpark and J. L. Stirling, *Biochem. J.*, 1978, **173**, 997.
265 J. W. Park and B. Chakrabarti, *Biopolymers*, 1977, **16**, 2807.
266 J. W. Park and B. Chakrabarti, *Biochim. Biophys. Acta*, 1978, **541**, 263.
267 M. Isles, A. R. Foweraker, B. R. Jennings, T. Hardingham, and H. Muir, *Biochem. J.*, 1978, **173**, 237.
268 V. C. Hascall, *J. Supramol. Struct.*, 1977, **7**, 101.
269 J. E. Christner and M. L. Brown, *Analyt. Biochem.*, 1978, **90**, 22.
270 G. W. Conrad, C. Hamilton, and E. Haynes, *J. Biol. Chem.*, 1977, **252**, 6861.
271 W. Truppe and H. Kresse, *European J. Biochem.*, 1978, **85**, 351.
272 Z. Nevo, A. K. Silbergeld, and R. Mamet, *Biochim. Biophys. Acta*, 1978, **541**, 76.
273 K. Ehrlich and M. Murray, *Experientia*, 1978, **34**, 179.

Subcutaneous implantation of demineralized bone-matrix powder in rats results in a sequential chondrogenic–osteogenic development.[274] Some of the characteristics of proteoglycans synthesized at different stages during the sequential process have been characterized. Exogenous proteoglycans have been shown to inhibit proteoglycan synthesis at the level of initiation or elongation of the glycosaminoglycan chains.[275] A lack of L-glutamine, which is required as the amido-group donor in the synthesis of 2-amino-2-deoxy-D-glucose 6-phosphate, causes a decrease in glycosaminoglycan synthesis, which is due to the limitation of synthesis of UDP-2-acetamido-2-deoxy-D-glucose.[276]

Quantitative changes in the chemical composition of, and incorporation of labelled precursors into, sheep cartilage glycosaminoglycans have been reported to occur in response to altered mechanical stress.[277] The observed differences between load-bearing and non-load-bearing joints may be accounted for, in part, by changes in the biosynthesis of cartilage glycosaminoglycans in response to altered functional demand. Vitamin A ,which inhibits the incorporation of 2-amino-2-deoxy-D-glucose into cartilage proteoglycans, stimulates the incorporation of D-mannose into glycopeptides.[278] Chemical and physical changes in proteoglycans during the development of chicken limb bud chondrocytes grown in cell culture have been studied.[279]

Chick embryo epiphyseal microsomal preparations synthesize chondroitin 6-sulphate chains terminated at the non-reducing ends with either D-glucopyranosyluronic acid or 2-acetamido-2-deoxy-D-galactopyranosyl residues.[280] In the presence of 4-nitrophenyl β-D-xylopyranoside, an initiator of chondroitin sulphate synthesis, chicken embryo epiphyseal cartilage preparations synthesize chondroitin sulphate proteoglycan of lower molecular size.[281] Cycloheximide, an inhibitor of protein synthesis, has the reverse effect. Gel chromatography of the chondroitin sulphates from these proteoglycan fractions indicated that the observed changes in molecular size were due to alterations in the average length of the chondroitin sulphate chains attached to the core protein.

Although no adequate explanations exist for the large variations observed in the chain lengths and chemical compositions of cartilage proteoglycans, experimental evidence has indicated that this variability is the result of translational and post-translational mechanisms involved in the biosynthesis of these molecules, rather than the result of the normal catabolic processes that occur after the molecules have been synthesized and secreted.[282] The environmental concentration of sulphate ions is one of the factors which appears to modulate the synthesis of proteoglycans in embryonic cartilage.[283]

[274] A. H. Reddi, V. C. Hascall, and G. A. Hascall, *J. Biol. Chem.*, 1978, **253**, 2429.
[275] C. J. Handley and D. A. Lowther, *Biochim. Biophys. Acta*, 1977, **500**, 132.
[276] G. Speight, C. J. Handley, and D. A. Lowther, *Biochim. Biophys. Acta*, 1978, **540**, 238.
[277] B. Caterson and D. A. Lowther, *Biochim. Biophys. Acta*, 1978, **540**, 412.
[278] J. P. Pennypacker, C. A. Lewis, and J. R. Hassell, *Arch. Biochem. Biophys.*, 1978, **186**, 351.
[279] S. De Luca, D. Heinegard, V. C. Hascall, J. H. Kimura, and A. Kaplan, *J. Biol. Chem.*, 1977, **252**, 6600.
[280] J. E. Silbert, *J. Biol. Chem.*, 1978, **253**, 6888.
[281] Y. Kato, K. Kimata, K. Ito, K. Karasawa, and S. Suzuki, *J. Biol. Chem.*, 1978, **253**, 2784.
[282] S. De Luca, A. I. Caplan, and V. C. Hascall, *J. Biol. Chem.*, 1978, **253**, 4713.
[283] M. Sobue, J. Takeuchi, K. Ito, K. Kimata, and S. Suzuki, *J. Biol. Chem.*, 1978, **253**, 6190.

When chicken chondrocytes are transformed by Rous sarcoma virus, the amount of inorganic sulphate incorporated into proteoglycan is markedly reduced.[284] In addition, these sulphated polysaccharides synthesized by the transformed cells differ qualitatively from those of normal chondrocytes, possibly due to a different core protein being synthesized on transformation.

Retinyl acetate has been reported to cause differences in the levels of sulphated glycosaminoglycans synthesized in dermal and epidermal cells from newborn mice.[285] The levels of glycosaminoglycan sulphotransferases prepared from chicken corneas have been studied throughout embryonic development.[286] Assay conditions have been established using keratan sulphate, heparan sulphate, and chondroitin sulphate as exogenous acceptors of [$^{35}SO_4^-$] from 3′-phosphoadenyl-5′phospho[^{35}S]sulphate. The degree of sulphation of corneal glycosaminoglycans during development is regulated by the availability of 3′-phosphoadenylsulphate rather than by the activity of sulphotransferases. Tunicamycin almost completely abolishes the biosynthesis of the corneal keratan sulphate, only partly inhibits the synthesis of chondroitin sulphate, heparan sulphate, and protein, and has no effect on the synthesis of hyaluronic acid.[287]

The incorporation of amino-sugar and sulphate into low-sulphate content chondroitin 4-sulphate in rat blood shows a high metabolic rate which is quite different from that of tissue glycosaminoglycans.[288]

Chondroitin sulphate fractions in human plasma and urine have similar chemical structures, including low levels of sulphate, and similar molecular weights (8×10^3), making it likely that the plasma low-sulphate content chondroitin sulphate is a precursor of the urinary product in aged subjects.[289]

A glycosaminoglycan with properties resembling keratan sulphate has been isolated from the cerebral cortex proteoglycans of rats.[290] The levels increase during early life then subsequently decrease steadily to negligible levels in the senescent animal.

Amongst the proteoglycans synthesized by the developing embryos of the Japanese sea-urchin *Hemicentrotus pulcherrimus*, dermatan sulphate proteoglycan is the major component.[291] This sulphated polysaccharide seems to have an unusual structure in comparison with the normal mammalian dermatan sulphates in that the latter have only polysaccharide–protein linkages which are resistant to alkaline elimination.

Heparan sulphate produced by bovine endothelial cells of the aorta has a higher *N*-sulphate content than that from the intima and media layers of the aorta of the same animal.[292]

[284] M. Muto, M. Yoshimura, M. Okayama, and A. Kaji, *Proc. Nat. Acad. Sci., U.S.A.*, 1977, **74**, 4713.
[285] S. S. Shapiro and J. P. Poon, *Connective Tissue Res.*, 1978, **6**, 101.
[286] G. W. Hart, *J. Biol. Chem.*, 1978, **253**, 347.
[287] G. W. Hart and W. J. Lennartz, *J. Biol. Chem.*, 1978, **253**, 5795.
[288] R. I. Hata and Y. Nagai, *Biochim. Biophys. Acta*, 1978, **543**, 149.
[289] R. I. Hata, S. I. Ohkawa, and Y. Nagai, *Biochim. Biophys. Acta*, 1978, **543**, 156.
[290] L. Vitello, M. Breen, H. G. Weinstein, R. A. Sittig, and L. J. Blacik, *Biochim. Biophys. Acta*, 1978, **539**, 305.
[291] K. Oguri and T. Yamagata, *Biochim. Biophys. Acta*, 1978, **541**, 385.
[292] G. Gamse, H. G. Fromme, and H. Kresse, *Biochim. Biophys. Acta*, 1978, **544**, 514.

Cell surface heparan sulphate produced by primary cultures of embryonic mouse cells is composed of two forms which can be separated by ion-exchange chromatography.[293] Although these polysaccharides have similar molecular weights, they probably differ in sulphate contents. Endogenous cell-surface heparan sulphate of Chinese hamster cells is released into the medium when cells are treated with heparin.[294] The cell surface localization of heparan sulphate is dependent, at least in part, on cell-surface receptors with binding sites for the oligosaccharide moieties of sulphated glycosaminoglycans.

The UDP-D-glucuronate:glycosaminoglycan β-D-glucuronyl-transferase activity from the media and intima of pig aorta has been determined by the incorporation of D-[^{14}C]glucuronic acid into hyaluronic acid and heparan sulphate.[295]

Sequential extraction of pig arterial wall with guanidine hydrochloride solutions of increasing concentration solubilizes four pools of proteoglycans.[296] An *in vitro* study of the incorporation of $^{35}SO_4^{2-}$ and D-[^{14}C]glucuronic acid into slices of aorta showed the highest levels of incorporation to be in heparan sulphate. The synthesis of glycosaminoglycans by human skin fibroblasts, derived from normal subjects and Hurler and Marfan patients, has been studied before and after transformation of the cells with SV40 virus.[297] Viral transformation results in an increase in the level of hyaluronic acid in normal and Hurler-derived fibroblasts and increases to a lesser extent in Marfan-derived fibroblasts.

Incubation of calf-rib cartilage slices with anti-proteoglycan serum inhibits the biosynthesis of proteoglycan sulphate and total protein.[298]

Degradation.—Rabbit fibroblasts of skin or synovial origin in culture secrete a metal-dependent neutral protease that degrades the protein core of cartilage proteoglycan.[299] The degradation of proteoglycans in rabbit articular cartilage is initiated by a limited proteolysis of the subunit core of the protein, with the production of non-aggregating species which diffuse readily from the tissue.[300] The rapid degradation of the subunits of cartilage proteoglycans into comparatively high molecular weight fragments has been shown to be due to the synthesis and excretion of a metal-dependent neutral protease by rabbit bone marrow macrophages.[301]

Pathology.—Since hyaluronic acid is the only glycosaminoglycan detected in pulmonary secretions of asthma patients, it may be a specific marker for this particular disease.[302]

The role of zinc in the metabolism of glycosaminoglycans has been reported for normal and atheromatous rats.[303] In general, zinc deficiency causes a

293 K. L. Keller, C. B. Underhill, and J. M. Keller, *Biochim. Biophys. Acta*, 1978, **540**, 431.
294 P. M. Kraemer, *Biochem. Biophys. Res. Comm.*, 1977, **78**, 1334.
295 P. Levy and J. Picard, *Internat. J. Biochem.*, 1977, **8**, 789.
296 E. Deudon and J. Picard, *Internat. J. Biochem.*, 1978, **9**, 19.
297 J. J. Hopwood and A. Dorfman, *J. Biol. Chem.*, 1977, **252**, 6536.
298 T. D. Kleine and M. Wunsch, *Experientia*, 1977, **33**, 1414.
299 G. Huybrechts-Godin and G. Vaes, *F.E.B.S. Letters*, 1978, **91**, 242.
300 J. D. Sandy, H. L. G. Brown, and D. A. Lowther, *Biochim. Biophys. Acta*, 1978, **543**, 536.
301 P. Hauser and G. Vaes, *Biochem. J.*, 1978, **172**, 275.
302 S. Sahu and W. S. Lynn, *Biochem. J.*, 1978, **173**, 565.
303 B. Philip and P. A. Kurup, *Indian J. Biochem. Biophys.*, 1977, **14**, 354.

decrease in the concentration of sulphated glycosaminoglycans in the aorta and liver, while the concentration of hyaluronic acid increases in the aorta and not significantly in the liver. High levels of zinc cause increases in the concentration of sulphated glycosaminoglycans in the aorta and liver, with no significant alteration to the levels of hyaluronic acid.

Low levels of keratan sulphate in human femoral heads have been suggested to be one of the factors involved in patients suffering from osteoarthrosis.[304] Observed differences in chain length and in the content of keratan sulphate may be related to the degree of malignancy of human chondrosarcomas.[305]

In two children exhibiting the clinical symptoms of fucosidosis, the diagnosis has been confirmed biochemically by the demonstration of a marked α-L-fucosidase deficiency in cultured skin fibroblasts.[306] Examination of the urinary levels of keratan sulphate and chondroitin sulphate of the two patients showed a five- and eight-fold increase over those of normal children of the same age. Since a number of the terminal non-reducing positions of keratan sulphate molecules are occupied by α-L-fucosyl residues, the increased excretion of keratan sulphate in fucosidosis could result from the impaired degradation of keratan sulphate, due to α-L-fucosidase deficiency.

The pinocytosis by fibroblasts of β-D-2-acetamido-2-deoxyhexosidase excreted by cultured fibroblasts from patients with mucolipidosis II disease is not enhanced by neuraminidase-catalysed removal of neuraminic acid residues from the β-D-2-acetamido-2-deoxyhexosidase.[307]

Proteoglycans isolated from the growth cartilage of vitamin D-deficient chickens have been shown to have the same chemical composition as, but lower molecular weights than, the proteoglycans from normal chickens.[308] These observations have been explained in terms of a depleted proteolytic cleavage.

The enzymic defect of Morquio's disease is attributed to the absence of 2-acetamido-2-deoxy-D-galactose 6-sulphate sulphatase.[309] Extracts of fibroblasts from patients with the disease exhibit normal levels of 2-acetamido-2-deoxy-D-glucose 6-sulphate sulphatose and thus remove sulphate from 2-acetamido-2-deoxy-D-glucose 6-sulphate residues at the non-reducing ends of keratan sulphate.

The characterization of the intracellular glycosaminoglycans from organs of a patient with Sanfillipo A disease shows that both heparan sulphate and the 2-amino-2-deoxy-D-galactosyl-containing glycosaminoglycans (chondroitin sulphate and dermatan sulphate) accumulate in liver, whereas spleen, kidney, heart, cerebellum, and cerebrum have heparan sulphate as the only glycosaminoglycan.[310] The heparan sulphate in all organs is degraded by at least

304 M. Suzuki, A. Boersma, P. Degand, and G. Biserte, *Clinica Chim. Acta*, 1978, **87**, 229.
305 S. Pal, W. Strider, R. Margolis, G. Gallo, S. Lee-Huang, and L. Rosenberg, *J. Biol. Chem.*, 1978, **253**, 1279.
306 H. Greiling, H. W. Stuhlsatz, M. Cantz, and J. Gehler, *J. Clin. Chem. Clin. Biochem.*, 1978, **16**, 329.
307 G. D. Vladutiu and M. C. Rattazzi, *Biochim. Biophys. Acta*, 1978, **539**, 31.
308 I. R. Dickson and P. J. Roughley, *Biochem. J.*, 1978, **171**, 675.
309 A. L. Horwitz and A. Dorfman, *Biochem. Biophys. Res. Comm.*, 1978, **80**, 819.
310 A. Kindler, U. Klein, and K. von Figura, *Z. Physiol. Chem.*, 1977, **358**, 1431.

two *endo*-glycosidases, an *endo*-D-glucuronidase, and an *endo*-2-acetamido-2-deoxy-D-glucosidase, to fragments of low molecular weight (2—6 × 10^3).

Studies using different methods of harvesting cultured human fibroblasts from a patient exhibiting mucopolysaccharidosis, and amniotic fluid, have indicated that reliable prenatal diagnosis of mucopolysaccharidosis can be performed only if the pericellular pool of glycosaminoglycans is removed by pretreatment with trypsin.[311] Thus amniotic fluid cells from a pregnancy carrying a foetus affected by Hurler's disease revealed the expected increase in the level of $^{35}SO_4$-incorporation into fibroblasts only after mild proteolytic treatment.

Elevated levels of glycosaminoglycans including heparan sulphate and dermatan sulphate have been isolated from the hepatic tissue from patients with Hurler syndrome, compared to control levels.[312]

The urinary excretion, in cases of Lowe's syndrome, of undersulphated chondroitin 4-sulphate in the absence of other glycosaminoglycans has led to the suggestion that a disorder in the sulphation of glycosaminoglycans is etiologically implicated in this syndrome.[313]

The increased secretion of urinary hyaluronic acid (mol. wt. 3.6 × 10^5), accounting for 16% of the urinary glycosaminoglycans, has been reported for Werner's syndrome.[314]

Sodium heparinate has been treated sequentially with methanolic hydrogen chloride and nitrous acid to produce *O*-(methyl α-L-idopyranosyluronate 2-sulphate-(1 → 4)-2,5-anhydro-D-mannose.[315] This disaccharide can be used for direct measurement of L-iduronic acid 2-sulphate sulphatase, an enzyme which is defective in Hunter's syndrome (mucopolysaccharidosis II).

Biochemical studies have been presented on two siblings with some features of mucolipidosis III.[316] Levels of β-D-galactosidase, α-D-mannosidase, β-D-glucuronidase, α-L-fucosidase, and 2-acetamido-2-deoxy-D-glucosidase were high but no excessive secretion of glycosaminoglycans was observed.

8 Mammalian Cell and Tissue Glycoproteins

Considerable structural differences in the carbohydrate moieties of glycoproteins present in various soluble and membrane fractions of rat brain have been noted.[317]

An atypical arylsulphatase A (arylsulphate sulphohydrolase, EC 3.1.6.1) purified from chicken brain resembles arylsulphatase A of other animal species in its kinetic properties, such as K_m value, pH optima, and anomalous kinetics.[318] The purified enzyme was shown to be a glycoprotein containing *inter alia* D-glucose, D-mannose, and L-fucose.

311 J. J. H. Fortuin and W. J. Keijer, *Clinica Chim. Acta*, 1978, **82**, 79.
312 J. A. Kofoed, A. A. Tocci, and A. C. Barcelo, *Experientia*, 1978, **34**, 577.
313 M. Akasaki, S. Fukui, T. Sakano, T. Tanaka, T. Usui, and I. Yamashina, *Clinica Chim. Acta*, 1978, **89**, 119.
314 M. Goto and K. Murata, *Clinica Chim. Acta*, 1978, **85**, 101.
315 L. C. Ginsberg, D. T. Di Ferrante, and N. Di Ferrante, *Carbohydrate Res.*, 1978, **64**, 225.
316 R. Sandman, S. Yatsiv, W. van B. Robertson, and P. Erickson, *Clinica Chim. Acta*, 1977, **80**, 471.
317 T. Krusius, J. Finne, R. U. Margolis, and R. K. Margolis, *Biochemistry*, 1968, **17**, 3849.
318 A. Ahmad, A. Surolia, and B. K. Bachhawat, *Indian J. Biochem. Biophys.*, 1977, **14**, 235.

A low molecular weight glycoprotein from bovine brain and liver which has previously been described as a homogeneous sialo-glycoprotein, has been shown to be heterogeneous with respect to charge and to immunological properties and is a mixture of low molecular weight protein and lipid.[319]

A high concentration of D-glucose *in vivo* or an increased concentration of D-glucose or D-glucose 6-phosphate *in vitro* have been found to lead to the glycosylation of ε-amino-groups of lysine residues in the lens crystallins of diabetic rats.[320] The modification of these proteins by D-glucose or D-glucose 6-phosphate occurs non-enzymically in a manner similar to the postsynthetic modification of haemoglobin in the erythrocyte. The glucosylated proteins exhibit an increased susceptibility to disulphide bond formation which results in an increased absorbance of light and an increased formation of high molecular weight aggregates. The addition of reducing agents prevents, as well as reverses, the formation of high molecular weight aggregates and the opalescence of the crystallins. A new interpretation of previous results on cataract formation and a new approach for the development of drugs to prevent cataracts have been proposed.

The biosynthesis of glycoproteins in the retina of chicken embryos involves the partition of D-mannolipids.[321]

The biosynthesis and release of glycoproteins from cultures of breast and prostate cells have been reported.[322] Malignant breast and prostate epithelial cells maintained their capacity to synthesize glycoproteins (mol. wt. 4.8×10^4) as followed by the incorporation of 2-amino-2-deoxy-D-[^{3}H]glucose.

The chemistry of carcinoembryonic antigen has been reviewed.[323] Precipitation by ethanol has been shown to be as efficient as a combination of dialysis, ultra-filtration, and lyophilization for the isolation of carcinoembryonic antigen.[324] The hepatic uptake of circulating carcinoembryonic antigen in the mouse is not inhibited by asialofetuin but is inhibited by the non-ionic detergent Triton WR1339, whilst the uptake of asialocarcinoembryonic antigen is inhibited by asialofetuin but not by the detergent.[325]

The normal crossreacting antigen which crossreacts with carcinoembryonic antigen has been purified from normal spleen tissue and shown to be a glycoprotein (mol. wt. 1×10^5) containing 50% carbohydrate.[326] Striking similarities to carcinoembryonic antigen were demonstrated in the protein portion of the molecule, where the sequences of the first twenty-six amino-acids were identical with one exception.

Digestion of carcinoembryonic antigen with trypsin released seven major glycopeptide fragments; these have been separated and purified by h.p.l.c., and amino-acid sequenced by the phenylthiohydantoin method.[327] Four of the

[319] S. Simonian, J. Heijlman, and G. J. M. Hooghwinkel, *J. Neurochem.*, 1978, **31**, 103.
[320] V. J. Stevens, C. A. Rouzer, V. M. Monnier, and A. Cerami, *Proc. Nat. Acad. Sci., U.S.A.*, 1978, **75**, 2918.
[321] E. L. Kean, *J. Supramol. Struct.*, 1977, **7**, 381.
[322] Z. A. Tokés and G. B. Dermer, *J. Supramol. Struct.*, 1977, **7**, 515.
[323] D. G. Pritchard, C. W. Todd, and M. L. Egan, *Methods Cancer Res.*, 1978, **14**, 55.
[324] D. G. Pritchard and M. L. Egan, *Immunochemistry*, 1978, **15**, 385.
[325] P. Thomas and M. Jones, *Biochem. J.*, 1978, **173**, 981.
[326] E. Engvall, J. E. Shively, and M. Wrann, *Proc. Nat. Acad. Sci. U.S.A.*, 1978, **75**, 1670.
[327] J. E. Shively, M. J. Kessler, and C. W. Todd, *Cancer Res.*, 1978, **38**, 2199.

peptides gave sequences through 20 cycles of Edman degradation, one gave sequences through 13 cycles and two gave sequences through 10 cycles. The abrupt halt in Edman degradation for each peptide was interpreted as the failure to sequence beyond a carbohydrate substitution on the polypeptide chain.

The utilization of 2-acetamido-2-deoxy-D-[^{14}C]galactose and 2-[^{3}H]-acetamido-2-deoxy-D-galactose by perfused rat liver for the biosynthesis of glycoproteins has been described.[328] The ^{14}C and ^{3}H labels were located in a number of glycoproteins, predominantly as residues of neuraminic acid and 2-acetamido-2-deoxy-D-glucose, thus implying that the epimerization step is a major event in the utilization of 2-acetamido-2-deoxy-D-galactose. Although the ratio of labelled neuraminic acid to 2-acetamido-2-deoxy-D-glucose varied for different proteins, the bulk of the 2-acetamido-2-deoxy-D-galactose was incorporated without deacetylation.

The acceptor capacity of liver microsomal proteins for 2-acetamido-2-deoxy-D-glucosyl residues is diminished during the first 48 hours of regeneration after partial hepatectomy,[329] This may be responsible for the decreased transfer of amino-sugar to latent glycoproteins by UDP-2-acetamido-2-deoxy-D-glucose–glycoprotein 2-acetamido-2-deoxy-D-glucosyltransferase (E.C. 2.4.1.51).

Compared with normal cat liver, the liver of an animal affected by G_{M_1} gangliosidosis has been shown to contain increased levels of three, low molecular weight glycopeptides.[330] The storage oligosaccharide appeared to be a hexasaccharide containing residues of D-galactose, D-mannose, and 2-acetamido-2-deoxy-D-glucose (1 : 2 : 3).

The interactions of concanavalin A with acid phosphatase and β-D-glucuronidase from pig liver have been studied.[331]

L-Alanine aminopeptidase has been purified from human kidney.[332] The glycoprotein (mol. wt. 2.36×10^5) contains 21% carbohydrate, including residues of 2-amino-2-deoxy-D-glucose, *N*-acetylneuraminic acid, and hexose.

Hormonal effects on the biosynthesis of sulphated glycoprotein in the endometrium of rabbit uterus have been shown to be confined to the microsomal level.[333] Oestrogen increases the incorporation of L-fucose and 2-acetamido-2-deoxy-D-glucose from their respective nucleotides into a microsomal fraction, while progesterone suppresses the incorporation of the two sugars.

Two forms of sialyltransferase have been isolated from human placental extracts; one (mol. wt. 8×10^4) binds to concanavalin A–Sepharose but the other (mol. wt. 8×10^5) does not bind.[334]

Cell cultures derived from foetal bovine ligamentum nuchae have been used,

[328] A. D. MacNicoll, F. S. Wusteman, G. M. Powell, and C. G. Curtis, *Biochem. J.*, 1978, **174**, 421.
[329] Y. Okamoto, E. Ito, and N. Akamatsu, *Biochim. Biophys. Acta*, 1978, **542**, 21.
[330] E. W. Holmes and J. S. O'Brien, *Biochem. J.*, 1978, **175**, 945.
[331] K. Kimoto, Y. Ogasawara, and M. Kusama, *J. Agric. Chem. Soc.* (*Japan*), 1978, **52**, 61.
[332] Y. J. Kao, W. L. Starnes, and F. J. Behal, *Biochemistry*, 1978, **17**, 2990.
[333] R. Yamamoto, H. Munakata, M. Yamamoto, and Z. Yosizawa, *J. Biochem.* (*Japan*), 1978, **83**, 537.
[334] C. K. Liu, R. Schmied, E. M. Greenspan, and S. Waxman, *Biochim. Biophys. Acta*, 1978, **522**, 375.

in conjunction with a microfibrillar-protein specific antiserum, to identify glycoproteins that are either closely related to or are structural components of elastin-associated microfibrils.[335]

A glycoprotein isolated from rabbit sciatic nerve myelin contained 6.3% carbohydrate in the form of a nonasaccharide unit composed of residues of D-mannose (3), 2-acetamido-2-deoxy-D-glucose (3), *N*-acetylneuraminic acid (1), D-galactose (1), and L-fucose (1).[336] Treatment of sciatic nerve myelin with trypsin releases a glycoprotein fragment (mol. wt. 1.9×10^4) with a higher content of hydrophobic amino-acid residues than the former glycoprotein.[337]

A number of glycoproteins (mol. wt. 1.9—6×10^4) have been detected in peripheral nerve myelin from rabbit and chicken.[338]

Pig thyroid microsomes catalyse the transfer of D-glucose from UDP-D-glucose into glycolipids and glycoproteins.[339] The carbohydrate moiety of the glycoprotein contains oligosaccharide chains composed of an average of nine monosaccharide units and is attached to the protein *N*-glycosidically. The incorporation into the protein of D-glucosyl-containing oligosaccharides formed by calf thyroid slices has been reported to involve an oligosaccharide–lipid intermediate, which contains residues of D-glucose, D-mannose, and 2-acetamido-2-deoxy-D-glucose.[340] If this intermediate acts as a saccharide donor *in vivo*, the transfer of the carbohydrate to protein is followed by a process in which several hexose residues are removed.

Derivatives of cholesterol which inhibit the biosynthesis of cholesterol in aortic smooth muscle, presumably by suppressing the activity of 3-hydroxy-3-methyl-glutaryl-coenzyme A reductase, also effect a corresponding decrease in the levels of glycosylated lipid intermediates with an apparent inhibition of the biosynthesis of glycoproteins.[341]

A method has been proposed for the preparation of viable and relatively homogeneous late spermatocytes from immature rat testes.[342] The capacity of the isolated cells to synthesize glycoproteins and sulphated glycolipid was studied.

Lectins may be used as probes for the study of the organization of specific components of chromatin.[343] Three glycoproteins (mol. wt. 6.9×10^4, 1.25, and 1.35×10^5) have been isolated from the non-histone chromosomal proteins from rat liver nuclei. The H-2 major histocompatibility complex is a complex of genes on murine chromosome 17, and is recognised as controlling the rejection of tissue allografts.[344] Of five major peptides isolated from a

335 C. H. J. Sear, M. A. Kewley, C. J. P. Jones, M. E. Grant, and D. S. Jackson, *Biochem. J.*, 1978, **170**, 715.
336 M. W. Roomi, A. Ishaque, N. R. Khan, and E. H. Eyler, *Biochim. Biophys. Acta*, 1978, **536**, 112.
337 M. W. Roomi and E. H. Eyler, *Biochim. Biophys. Acta*, 1978, **536**, 122.
338 M. W. Roomi, A. Ishaque, N. R. Khan, and E. H. Eyler, *J. Neurochem.*, 1978, **31**, 375.
339 C. Ronin and S. Bouchilloux, *Biochim. Biophys. Acta*, 1978, **539**, 481.
340 R. J. Staneloni and L. F. Leloir, *Proc. Nat. Acad. Sci. U.S.A.*, 1978, **75**, 1162.
341 J. T. Mills and A. M. Adamany, *J. Biol. Chem.*, 1978, **253**, 5270.
342 P. J. Letts, R. C. Hunt, M. A. Shirley, L. Pinteric, and H. Schachter, *Biochim. Biophys. Acta*, 1978, **541**, 59.
343 W. B. Rizzo and M. Bustin, *J. Biol. Chem.*, 1977, **252**, 7062.
344 B. M. Ewenstein, T. Nisizawa, H. Uehara, S. G. Natheason, J. E. Coligan, and T. J. Kindt, *Proc. Nat. Acad. Sci. U.S.A.*, 1978, **75**, 2909.

glycoprotein from this complex, two have been identified as glycopeptides each containing a carbohydrate moiety of 3.3×10^3 daltons.

Cationic polypeptides, such as poly(L-arginine), have been shown to enhance the activity of UDP-D-galactose:glycoprotein galactosyltransferase in cultured skin fibroblasts.[345] The implication of the results in the regulation of glycoprotein synthesis by the fibroblasts and for the pathogenesis of cystic fibrosis is discussed. The single envelope glycoprotein of vesicular stomatitis virus has been used as a specific probe of glycosyltransferase activities in fibroblasts from two cystic fibrosis patients, an obligate heterozygous carrier and a normal individual.[346] No significant differences in the patterns of glycosylation between the different cell cultures were observed. All four cell lines are apparently able to synthesize the D-mannose- and 2-acetamido-2-deoxy-D-glucose-containing core structures and also branch oligosaccharide chains terminating in neuraminic acid which are characteristic of L-asparagine-linked carbohydrate chains in cellular glycoproteins. It was suggested that the basic biochemical defect(s) in cystic fibrosis is not an absence or deficiency in enzymes responsible for the biosynthesis of complex carbohydrate side chains.

The glycosylation of endogenous lipids and proteins by preparations of chicken embryo fibroblasts has been reported.[347] The system synthesizes a heterogeneous mixture of oligosaccharide lipids, in which the oligosaccharide moieties differ in the number of saccharide units and their structure. Glycoproteins immunologically related to oncoviral glycoprotein gp 85 have been detected in uninfected avian cells and in the sera from uninfected birds.[348]

The attachment of D-glucose to a D-manno-oligosaccharide lipid intermediate from Nil 8 fibroblasts seems to serve an important role in the transfer of the oligosaccharide from the lipid carrier to the protein acceptor, since a related D-glucose-free oligosaccharide is transferred from its lipid intermediate to protein at a significantly reduced rate.[349]

The glycocalyx (cell coat) of Ehrlich ascites tumour cells, which is rich in glycoproteins and glycosaminoglycans, contains only one fraction (mol. wt. 1.3×10^5) which interacts with ^{125}I-concanavalin A.[350]

Two glycopeptide fractions of different molecular weights have been isolated from the product of proteolysis of L-fucose-labelled embryonal carcinoma cells.[351] During *in vitro* differentiation of the cells the higher molecular weight L-fucosyl glycopeptides disappear almost completely. The major components of the low molecular weight glycopeptide fraction appear to be different from complex glycopeptides known to occur in adult cells. The results suggest that the carbohydrate profile changes fundamentally during early stages of mammalian development.

Sialoglycopeptides have been isolated from proteolytic digests of the

[345] G. J. S. Rao, D. Chyatte, and H. L. Nadler, *Biochim. Biophys. Acta*, 1978, **541**, 435.
[346] L. A. Hunt and D. F. Summers, *J. Supramol. Struct.*, 1977, 7, 213.
[347] S. S. Krag, *Biochim. Biophys. Acta*, 1978, **538**, 603.
[348] V. Bosch, R. Kurth, and J. E. Smart, *Virology*, 1978, **86**, 226.
[349] S. J. Turco, B. Stetson, and P. W. Robbins, *Proc. Nat. Acad. Sci., U.S.A.*, 1977, **74**, 4411.
[350] H. G. Rittenhouse, J. W. Rittenhouse, and L. Takemoto, *Biochemistry*, 1978, **17**, 829.
[351] T. Muramatsu, G. Gachelin, J. F. Nicolas, H. Condamine, H. Jakob, and F. Jacob, *Proc. Nat. Acad. Sci., U.S.A.*, 1978, **75**, 2315.

glycoproteins produced by cultured B16 mouse melanoma cells.[352] Tetra- and tri-saccharide moieties containing neuraminic acid, D-galactose, and 2-acetamido-2-deoxy-D-galactose are attached *O*-glycosidically, apparently in clusters, to a polypeptide core. An unusual observation is that these sialoglycopeptides, although devoid of 2-acetamido-2-deoxy-D-glucosyl residues, bind to immobilized wheat germ agglutinin and have a strong inhibitory activity on the ability of wheat germ agglutinin to precipitate B16 melanoma cells. Sialoglycopeptides were also isolated by proteolytic digestion of B16 mouse melanoma tumours and appeared identical to those produced by the cultured melanoma cells, suggesting that these glycopeptides are not artefacts of the culture system.[353]

A glycoprotein (mol. wt. 1.1×10^5), capable of dissociation into several components on reduction, has been isolated from T-cell lymphoma, and shown to have F_c receptor properties.[354]

Microsomes from calf pancreas incorporate 2-acetamido-2-deoxy-D-[^{14}C]-glucose from UDP-D-[^{14}C]-2-acetamido-2-deoxy-D-glucose lipid intermediate to form P^1-2-acetamido-2-deoxy-D-glucopyranosyl-P^2-dolichyl pyrophosphate and also P^1-di-*N*-acetyl-α-chitobiosyl-P^2-dolichyl pyrophosphate.[355] This product is similar and possibly identical to a recently prepared synthetic product.[356]

Only slight differences have been detected in the glycoprotein compositions of uninfected mammalian cells and cells infected with herpes simplex virus.[357] Some glycoproteins of high molecular weight and containing residues of 2-amino-2-deoxy-D-glucose were only detected in the non-infected cells.

External surface membrane glycoproteins are synthesized during early sea urchin development.[358] Different rates of glycoprotein synthesis have been demonstrated for three types of blastomeres from 16-cell sea urchin embryos. A glycosyl transferase in the embryos of *Arbacia punctulata*, which synthesizes P^1-2-acetamido-2-deoxy-D-glucosyl-P^2-dolichyl pyrophosphate from UDP-2-acetamido-2-deoxy-D-glucose and dolichyl phosphate, is sensitive to tunicamycin which arrests the development at specific stages.[359] This suggests an essential role for this enzyme in the glycosylation of proteins by a lipid-linked pathway during embryonic development.

9 Cell-surface Glycoproteins

Although the presence of glycosyltransferases on the surfaces of mammalian cells has been reported frequently, and the biological role for these enzymes in cell adhesion and cell recognition postulated, recent workers have found no evidence for the presence of D-galactosyl-, L-fucosyl-, and sialyltransferases on

[352] V. P. Bhavanandan, J. Umemoto, J. R. Banks, and E. A. Davidson, *Biochemistry*, 1977, **16**, 4426.
[353] V. S. Fareed, V. P. Bhavanandan, and E. A. Davidson, *Carbohydrate Res.*, 1978, **65**, 73.
[354] R. Frade and F. M. Kourilsky, *European J. Immunol.*, 1977, **7**, 663.
[355] A. Herscovics, C. D. Warren, B. Burge, and R. W. Jeanloz, *J. Biol. Chem.*, 1978, **253**, 160.
[356] C. D. Warren, A. Herscovics, and R. W. Jeanloz., *Carbohydrate Res.*, 1978, **61**, 181.
[357] S. Olofsson and J. Blomberg, *Arch. Virol.*, 1977, **55**, 293.
[358] A. E. Brown and H. B. Bosmann, *Biochem. Biophys. Res. Comm.*, 1978, **80**, 833.
[359] E. G. Schneider, H. T. Nguyen, and W. J. Lennarz, *J. Biol. Chem.*, 1978, **253**, 2348.

the surface of a variety of cells, including established (normally and virally transformed) as well as non-established cells.[360]

The composition and structure of different classes of membrane glycoproteins and the role played by their carbohydrate moieties in expressing cell-membrane specificity have been surveyed.[361]

A theoretical framework has been proposed for the analysis of adhesion between cells, or of cells to surfaces, when the adhesion is mediated by reversible bonds between specific molecules such as antigen and antibody, lectin and carbohydrate, or enzyme and substrate.[362] Two models have been developed for predicting the rate of bond formation between cells, from the measurements of reaction rates for membrane-bound reactants.

Anti-(microbial carbohydrate) antibodies (directed against the neuraminic acid residues of *Neisseria meningitidis*, and D-mannosyl residues of *Saccharomyces cerevisiae*) have been shown to cross-react with mammalian cell surface glycoproteins and have been used to distinguish between different stages of differentiating mouse cerebellum.[363]

A decrease in sialyltransferase activity in the hearts of cardiomyopathic hamsters has been shown to result in a decrease in the content of neuraminic acid in myocardial cell surface glycoproteins.[364] Since membrane-bound neuraminic acid may control Ca^{2+}-exchangeability in cardiac muscle, the results could provide a biochemical basis for the development of cardiac failure in the hamster.

The insulin receptor in rat liver membranes has been found to interact with another membrane glycoprotein that may represent either a non-recognition portion of the receptor oligomer, or an effector molecule related to the biological action of insulin.[365]

Surface glycopeptides from hepatocytes in an established line of rats have been isolated and characterized.[366] Two types of glycopeptides were recognized; those having *O*-glycosidic linkages, in which residues of D-galactose, 2-acetamido-2-deoxy-D-galactose, and *N*-acetylneuraminic acid are present, and those with *N*-glycosidic-linkages containing residues of D-galactose, D-mannose, L-fucose, 2-acetamido-2-deoxy-D-glucose, and *N*-acetylneuraminic acid.

Evidence has been presented for the existence in hepatocytes of a receptor that binds glycoproteins specifically to molecules containing 3-*O*-α-L-fucosyl-2-acetamido-2-deoxy-D-glucosyl units.[367] Human lactoferrin, which contains this type of linkage, is rapidly cleared from the circulation of mice after intravenous injection, and almost all of the glycoprotein can be recovered in the hepatocytes. The hepatocyte-binding of lactoferrin is mediated through its

[360] W. Deppert and G. Walker, *J. Supramol. Struct.*, 1978, **8**, 19.
[361] J. Järnefelt, J. Finne, T. Krusius, and H. Rauvala, *Trends in Biol. Sciences*, 1978, **3**, 110.
[362] C. I. Bell, *Science*, 1978, **200**, 618.
[363] E. Trenkner and S. Sarkar, *J. Supramol. Struct.*, 1977, **6**, 465.
[364] C. J. Limas and C. Limas, *Biochim. Biophys. Acta*, 1978, **540**, 543.
[365] J. M. Maturo and M. D. Hollenberg, *Proc. Nat. Acad. Sci., U.S.A.*, 1978, **75**, 3070.
[366] H. Debray and J. Montreuil, *Biochimie*, 1978, **60**, 697.
[367] J. P. Prieels, S. V. Pizzo, L. R. Glasgow, J. C. Paulson, and R. L. Hill, *Proc. Nat. Acad. Sci., U.S.A.*, 1978, **75**, 2215.

carbohydrate groups, since its clearance is prolonged after oxidation of the L-fucosyl residues with periodate, or after removal of its oligosaccharide groups with glycosidases. Since the clearance of other glycoproteins, such as human lactoperoxidase, also appears to be mediated through binding to the same hepatocyte receptor as lactoferrin, it is probable that the L-fucosyl-specific receptor may fulfil other functions as well as binding lactoferrin.

Two glycoproteins (mol. wt. 1.8×10^5 and 3.4×10^4) have been identified in both the nuclei and nuclear membranes of rat liver.[368] These glycoproteins, which bind in concanavalin A–ferritin complexes, are located exclusively on the cisternal surface of the membranes. Rough- and smooth-microsomal fractions and Golgi membranes of rat liver have been shown to contain different types and amounts of polyprenyl phosphates and exhibit specificity as sugar acceptors from sugar nucleotides.[369]

The distribution of membrane glycoproteins in hepatoma tissue culture cells has been examined.[370] A set of L-fucosyl-containing glycoproteins are shown to be externally disposed on the plasma membrane of these cells, although they are not unique to the plasma membrane. They are also found associated with a cellular internal membrane(s) system, part of which functions as a direct precursor to the plasma membrane and part may form an intracellular reservoir of preformed plasma membrane. Treatment of the hepatoma tissue culture cells, and of a related epithelial cell which is derived from the rat hepatoma cells, *in situ* with D-galactose oxidase and neuraminidase followed by sodium borotritide reduction was used to label externally-oriented glycoproteins and glycolipids of the plasma membrane.[371] In both cell types, there was a heterogeneity in the turnover of the sugar moieties on some surface glycoproteins. These glycoproteins may represent a class of membrane proteins which are removed from the surface and degraded by a different mechanism from that involved in the turnover of the more abundant plasma membrane proteins.

L-Fucoprotein metabolism in Morris hepatoma 7777 plasma membranes is characteristically different from that in normal liver and host liver membranes.[372] The membrane glycoproteins of Zajdela's hepatoma cells have been shown to contain D-galactosyl- but not 2-acetamido-2-deoxy-D-galactosyl-residues in the non-reducing terminal positions or in positions penultimate to neuraminic acid.[373]

Brush border membranes isolated from the small intestine of rats contain three major glycoproteins which co-migrate, on electrophoresis, with gluco-amylase, *exo*-(1 → 4)-α-D-glucosidase (EC 3.2.1.3), α-D-glucosidase (EC 3.2.1.20), sucrose α-D-glucohydrolase (EC 3.2.1.48), β-D-galactosidase (EC 3.2.1.23), and oligo-1,6-D-glucosidase (EC 3.2.1.10)–α-D-glucosidase–sucrose α-D-glucohydrolase activities.[374] The α-D-glucosidase, sucrose α-D-glucohyrolase, and

[368] I. Virtanen, *Biochem. Biophys. Res. Comm.*, 1977, **78**, 1411.
[369] A. Bergman, T. Mankowski, T. Chojnacki, L. M. de Luca, E. Peterson, and G. Dallner, *Biochem. J.*, 1978, **172**, 123.
[370] D. Doyle, H. Baumann, B. England, E. Friedman, E. Hou, and J. Tweto, *J. Biol. Chem.*, 1978, **253**, 965.
[371] H. Baumann, and D. Doyle, *J. Biol. Chem.*, 1978, **253**, 4408.
[372] P. Vischer and W. Reutter, *European J. Biochem.*, 1978, **84**, 363.
[373] M. Dodeur and R. Bourrillon, *Biochimie*, 1978, **60**, 177.
[374] H. P. Hauri and J. R. Green, *Biochem. J.*, 1978, **174**, 61.

β-D-galactosidase activities were found to be significantly reduced in bacterially contaminated intestinal homogenates, and in particular in the isolated brush borders from the self-filling blind loop.[375]

The large-scale preparation of bovine renal glomerular basement membrane has been reported using conditions to prevent degradation by tissue-associated proteases.[376]

Binding sites having either high or low affinities for concanavalin A have been recognised on the plasma membranes of normal human fibroblasts.[377] The average number of high affinity sites per cell is 8×10^5, representing less than 1% of the total receptor sites for the lectin. A major fibroblast cell surface glycoprotein has been characterized as containing residues of D-galactose, D-mannose, D-glucose, 2-acetamido-2-deoxy-D-glucose, and neuraminic acid.[378]

A protease-resistant, transformation sensitive membrane glycoprotein (subunit mol. wt. 1.7×10^5) and an intermediate filament-forming protein (IFP, mol. wt. 5.6×10^4) have been isolated from hamster embryo fibroblasts.[379] A possible association of the two molecular species is suggested by the fact that the IFP can be co-purified with the glycoprotein using immobilized *Ricinus communis* lectin, although the IFP has no receptor for the lectin.

The transformation of BHK_{21}/C_{13} hamster fibroblasts by Rous sarcoma virus has resulted in the altered glycosylation of nearly every glycoprotein of the cell.[380] The surface D-mannosyl-containing-glycopeptides from human diploid fibroblasts have been compared with those of an SV-40 transformed human cell.[381] Examination of an acidic glycopeptide fraction from the cells using three glycosidases and an *endo*-β-D-2-acetamido-2-deoxy-glucanase showed the presence of a product in the non-growing cells which was absent from the growing cells.

Neuraminidase treatment of wild-type and ricin-resistant strains of baby hamster fibroblasts leads to alterations in their capacity to bind to lectins.[382] A model (29) for the existence of 'exposed' and 'cryptic' oligosaccharide receptors for ricin on the cell membrane glycoproteins of kidney cells has been proposed. Two types of oligosaccharide chains exist, either in the same or different glycoprotein molecules in the membrane. The 'exposed' ricin-binding receptors, R_E, contain a core region of D-mannosyl and 2-acetamido-2-deoxy-D-glucosyl residues linked to an L-asparagine residue, and terminated by sequences containing a ricin-binding D-galactosyl residue. The 'cryptic' ricin-binding receptors, R_C, which may or may not contain a core region, are perhaps more heavily sialylated than the R_E receptors, and are unmasked by neuraminidase

375 A. Jonas, P. R. Flannagan, and G. G. Forsiner, *J. Clin. Invest.*, 1977, **60**, 1321.
376 J. W. Freytag, M. Ohno, and B. G. Hudson, *Prep. Biochem.*, 1978, **8**, 215.
377 M. Feller, C. Richardson, W. D. Behnke, and E. Gruenstein, *Biochem. Biophys. Res. Comm.*, 1977, **76**, 1027.
378 K. M. Yamada, D. H. Schlesinger, D. W. Kennedy, and I. Pastan, *Biochemistry*, 1977, **16**, 5552.
379 W. G. Carter and S. I. Hakomori, *J. Biol. Chem.*, 1978, **253**, 2867.
380 G. P. Tuszynski, S. R. Baker, J. P. Fuhrer, C. A. Buck, and L. Warren, *J. Biol. Chem.*, 1978, **253**, 6092.
381 C. Ceccarini and P. H. Atkinson, *Biochim. Biophys. Acta*, 1978, **500**, 197.
382 S. W. Rosen and R. C. Hughes, *Biochemistry*, 1977, **16**, 4908.

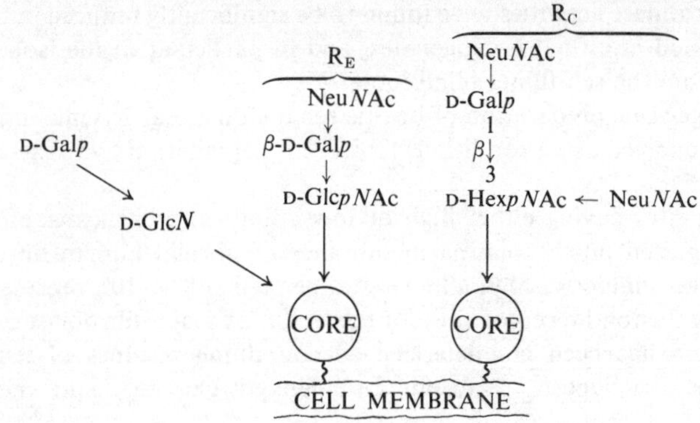

(29)

treatment, when binding to peanut agglutinin becomes possible. In ricin-resistant variants, the assembly or possibly the function of the R_E receptors is altered, while the R_C receptors appear to be unaffected.

Immunological and chemical studies on cell surfaces from normal and transformed BALB/c fibroblasts have shown alterations associated with transformation.[383] The cells studied include normal lines which do not cause tumours when injected into BALB/c mice, viral transformants, and spontaneous transformants. Alterations in glycopeptides derived from several cellular glycoproteins are the only changes which appear to be associated with malignancy.

Normal chicken cells, carrying the chicken-helper factor, express a surface glycoprotein which is antigenically cross-reactive with determinants of the major envelope glycoprotein of avian myeloblastosis virus.[384]

A technique has been developed for the specific labelling of the carbohydrate moieties of membrane-bound glycoproteins that contain 2-acetamido-2-deoxy-D-glucose,[385] but lack residues of D-galactose and neuraminic acid in the nonreducing terminal position. For example, by incubation of retinal disc membranes with rhodopsin, UDP-D-galactose and UDP-D-galactosyltransferase, an enzyme known to catalyse the formation of (1 → 4)-bonds between D-galactose and 2-acetamido-2-deoxy-D-glucose, D-galactosyl residues are incorporated. Oxidation of the incorporated D-galactosyl residues with D-galactose oxidase and coupling of the resulting aldehyde to dansyl hydrazide gave a fluorescent hydrazone which is stable for spectroscopic analysis.

Embryonic chick neutral retina, when grown in culture, releases into the media glycoprotein ligands which bind to homologous cells, and inhibits the lectin-induced redistribution of cell-surface receptors.[386] Material with identical activity was released from neural retina cells that were treated with trypsin.

[383] W. J. Grimes, G. A. Van Nest, and A. R. Kamm, *J. Supramol. Struct.*, 1977, **6**, 51.
[384] J. J. Collins, R. C. Montelaro, T. P. Denny, R. Ishizaki, A. J. Langlois, and D. P. Bolognesi, *Virology*, 1978, **86**, 205.
[385] J. H. Shaper and L. Stryer, *J. Supramol. Struct.*, 1977, **6**, 291.
[386] J. McDonough and J. Lilien, *J. Supramol. Struct.*, 1977, **7**, 409.

A number of glycoprotein receptors for wheat germ agglutinin have been isolated from rat synaptic membranes; each can serve as a substrate for neuraminidase.[387] However, an intrinsic neuraminidase present in the membranes selectively released neuraminic acid residues only from those glycoproteins having an apparent molecular weight $>7 \times 10^4$.

The binding of lectins on insoluble supports has been used to detect the interactions occurring between these molecules and the sites of some receptors of membranes.[388] By use of immobilized peanut lectin, the presence of determinants with terminal D-galactosyl residues on the surface of erythrocytes and lymphocytes has been determined. Equilibrium and kinetic studies of the binding of *Lens culinaris* lectin to rabbit erythrocytes have been reported.[112] The mechanism of agglutination of human erythrocytes by a variety of lectins labelled with gold granules has been studied.[389] Agglutination was found to depend on the size of the gold granules and not upon the number and mobility of the receptors.

Treatment of erythrocytes with neuraminidase increased their antibody affinities for the antigenic receptors.[390] In contrast, the use of proteolytic enzymes, in removing glycopeptide chains, greatly diminishes steric hindrance and brings about higher affinity constants.

A comparative immunochemical study of human erythrocyte membrane glycoprotein which uses two different extraction procedures has been reported.[391] Each preparation contained receptors for influenza virus and was composed of protein (50%) and carbohydrate (30—40%).

Glycophorin, the major sialoglycoprotein of human erythrocyte membrane, has been shown by an immunoferritin technique and with poly-cationized ferritin to be randomly distributed on the cell surface.[392] By use of anti-glycophorin antibodies, it has been established that only bone marrow cells of erythrocyte lineage express this glycoprotein at the cell surface.[393] This occurs in basophilic normoblasts during later stages of erythrocyte differentiation, whereas pronormoblasts do not seem to contain the glycoprotein. Glycophorin has been incorporated into lipid vesicles without the use of detergents; it was demonstrated that the protein spans the lipid bilayer in the vesicles.[394] The glycoprotein is oriented in an asymmetric manner with about 75—85% of the sugar residues being directed towards the outside of the vesicles. Freeze-fracturing electron microscopy has shown that the glycoprotein is present in a slightly aggregated form. Preliminary information on the analysis of human red-cell variants which may lack all or some of the sialoglycopeptides derived

[387] T. F. Cruz and J. W. Gurd, *J. Biol. Chem.*, 1978, **253**, 7314.
[388] M. Caron, M. Felon, A. Faure, and P. Cornillot, *Compt. rend.*, 1978, **286D**, 1407.
[389] M. Horisberger, *Experientia*, 1978, **34**, 721.
[390] C. Doinel, C. Ropars, and C. Salmon, *Immunology*, 1978, **34**, 653.
[391] L. T. Lee, F. A. Janney, J. E. Deas, and C. Howe, *Proc. Soc. Exp. Biol. Med.*, 1978, **158**, 530.
[392] C. G. Gahmberg, G. Tauren, I. Virtanen, and J. Wartiovaara, *J. Supramol. Struct.*, 1978, **8**, 337.
[393] C. G. Gahmberg, M. Jokinen, and L. C. Andersson, *Blood*, 1978, **52**, 379.
[394] E. J. J. Van Zoelen, A. A. Verkleij, R. F. A. Zwaal, and L. L. M. van Deenen, *European J. Biochem.*, 1978, **86**, 539.

from glycophorin A is consistent with the presence in normal cells of a second sialoglycoprotein, glycophorin B.[395]

The plasma glycoprotein, pregnancy-associated α_2-glycoprotein (α_2-PAG) has been demonstrated to reside on the surface of blood leucocytes.[396] By use of an indirect immunofluorescence technique, pre-treatment of leucocytes with anti-α_2-PAG IgG revealed α_2-PAG on peripheral blood B lymphocytes and monocytes but not on T lymphocytes. Earlier work claimed identification of the glycoprotein on peripheral blood T cells but not on B lymphocytes.[397] Surface glycoproteins of normal human blood platelets, granulocytes, monocytes, T and B lymphocytes, and null cells have been labelled using the D-galactose oxidase–sodium borotritide and periodate oxidation–sodium borotritide techniques.[398] All major types of leucocytes showed different and characteristic surface glycoprotein patterns as shown by polyacrylamide gel electrophoresis.

A glycoprotein extract isolated from human platelet plasma membranes exhibits haemagglutination activity and in addition causes agglutination of platelets.[399]

The existence of a membrane form of α_1-acid glycoprotein (apparent mol. wt. 5.2×10^4) on normal human lymphocytes, granulocytes, and monocytes, has been reported.[400] By use of *in vitro* labelling with [^{3}H]leucine, it was shown that the membrane protein is synthesized by the lymphocytes and apparently subsequently cleaved and released as a soluble serum glycoprotein, with a normal molecular weight (4.1×10^4).

Glycopeptides having affinity for *Ricinus sanguineus* lectin have been released from the surface of normal human lymphocytes by digestion with trypsin.[401] Some structural and biochemical characteristics were given and their relationship to HLA determinants and surface immunoglobulins were discussed. All of the covalently linked carbohydrate of the HLA antigens of the major histocompatibility complex from a lymphoblastoid cell line have been accounted for by a single oligosaccharide of 13—16 sugar residues attached by an *N*-glycosidic bond to the peptide backbone.[402] Treatment of the antigens with glycosidases which removed all of the carbohydrate with the exception of single L-fucosyl and a 2-acetamido-2-deoxy-D-glucosyl residues, was not accompanied by any change in HLA antigenic activity. The amino-acid sequences of a number of tryptic glycopeptides showed homology with both the amino-terminal sequences of HLA heavy chains and variable regions of immunoglobulin heavy chains.

The major sialoglycoprotein of human T lymphoblastoid cells of a MOLT-4B cell line, which binds to the lectin from *Ricinis communis*, concanavalin A,

[395] H. Furthmayr, *J. Supramol. Struct.*, 1977, **7**, 121.
[396] C. H. W. Horne, A. W. Thomson, C. M. Towler, F. K. MacMillan, and L. M. Gibb, *Scand. J. Immunol.*, 1978, **8**, 75.
[397] W. H. Stimson, *Clin. Exp. Immunol.*, 1977, **28**, 445.
[398] L. C. Andersson and C. G. Gahmberg, *Blood*, 1978, **52**, 57.
[399] D. J. Bowles and A. Rotman, *F.E.B.S. Letters*, 1978, **90**, 283.
[400] C. G. Gahmberg and L. C. Andersson, *J. Exptl. Med.*, 1978, **148**, 507.
[401] E. Turpin, J. Wantyghem, D. Neel, and Y. Gioussault, *Compt. rend.*, 1978, **286D**, 363.
[402] P. Parham, B. N. Alpert, H. T. Orr, and J. L. Strominger, *J. Biol. Chem.*, 1977, **252**, 7555.

and wheat germ agglutinin, is localized on the cell surface, as indicated by surface labelling using periodate–sodium borotritide and lactoperoxidase–iodination methods.[403] The isolated glycoprotein contains 28% carbohydrate composed of residues of L-fucose, D-galactose, D-mannose, 2-acetamido-2-deoxy-D-glucose, 2-acetamido-2-deoxy-D-galactose, and neuraminic acid. Two types of oligosaccharide chains are probably present since treatment with alkaline sodium borohydride releases both a tri- and a tetra-saccharide.

The structures of three glycopeptides isolated from proteolytic digestion of calf thymocyte plasma membranes have been obtained by a combination of methylation analysis and sequential degradation with glycosidases.[404] Two of the glycopeptides (30, 31) consist of highly branched oligosaccharide chains attached by an *N*-glycosidic bond to peptide, and have the unusual sequence of β-D-Gal*p*-(1 → 3)-β-D-Gal*p*-(1 → 4)-D-Glc*pN*Ac-(1 → as well as the more usual sequence of α-Neu*N*Ac-(2 → 3)-β-D-Gal*p*-(1 → 4)-D-Glc*pN*Ac-(1 →. In addition one glycopeptide also contains disaccharide chains of β-D-Gal*p*-(1 → 3)-D-Gal*pN*Ac-(1 → L-Ser(L-Thr).

The major glycoproteins of the thymocyte plasma membrane which react with concanavalin A do not readily associate with other membrane proteins.[405] It is possible that these concanavalin A-binding glycoproteins directly or indirectly modulate thymocyte function *via* their control of plasma membrane transport and/or permeability.

The binding of tetravalent concanavalin A, and divalent succinylated concanavalin A, to surface glycoproteins of guinea-pig macrophages, causes a marked increase in the rate of oxygen consumption due to activation of the hexose monophosphate shunt.[406]

The presence on alveolar macrophages of a cell surface receptor that binds glycoproteins having terminal D-mannosyl- or D-glucosyl-units has been described.[407] An assay system was developed for the measurement of the binding by cells.

Plasma membranes from rabbit reticulocytes, but not those from erythrocytes, are able to glycosylate endogenous proteins through dolichyl-bound sugar derivatives.[408]

Isolated rough and smooth microsomes and Golgi membranes from the rat are able to incorporate D-mannosyl residues into endogenous protein acceptors, the final products being membrane proteins rather than secretory proteins.[409] Part of the D-mannosyltransferase system is localized on the cytoplasmic side of the membranes.

By use of an indirect immunofluorescence assay, it has been demonstrated

[403] M. Saito, S. Toyoshima, and T. Osawa, *Biochem. J.*, 1978, **175**, 823.
[404] R. Kornfeld, *Biochemistry*, 1978, **17**, 1415.
[405] R. Schmidt-Ullrich, R. B. Mikkelsen, and D. F. H. Wallach, *J. Biol. Chem.*, 1978, **253**, 6973.
[406] T. Yasaka and T. Kambara, *Biochim. Biophys. Acta*, 1978, **508**, 306.
[407] P. D. Stahl, J. S. Rodman, M. J. Miller, and P. H. Schlesinger, *Proc. Nat. Acad. Sci., U.S.A.*, 1978, **75**, 1399.
[408] A. J. Parodi, and J. Martin-Barrientios, *Biochim. Biophys. Acta*, 1977, **500**, 80.
[409] O. S. Nilsson, M. E. De Tomas, E. Peterson, A. Bergman, G. Dallner, and F. W. Hemming, *European J. Biochem.*, 1978, **89**, 619.

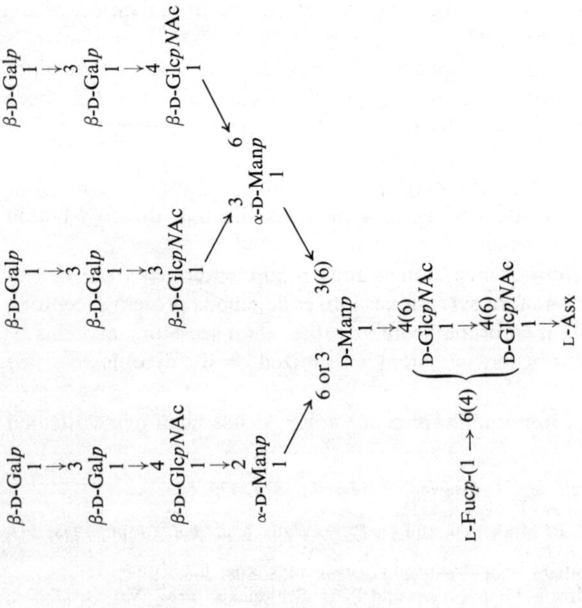
β-D-Galp 1 → 3 β-D-Galp 1 → 4 β-D-GlcpNAc 1 → 2 α-D-Manp 1
β-D-Galp 1 → 3 β-D-Galp 1 → 3 β-D-GlcpNAc 1
β-D-Galp 1 → 3 β-D-Galp 1 → 4 β-D-GlcpNAc 1
3 α-D-Manp 1 6
6 or 3 D-Manp 1 → 4(6) D-GlcpNAc 1 → 4(6) D-GlcpNAc 1 → L-Asx 3(6)
L-Fucp-(1 → 6(4)

(30)

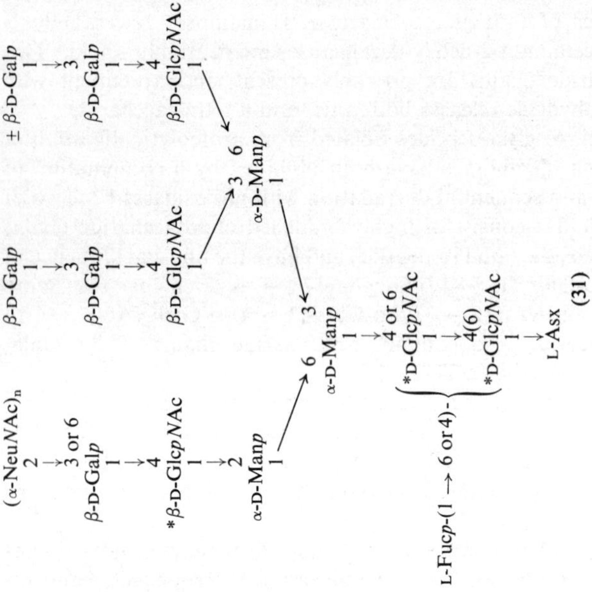
(α-NeuNAc)n 2 → 3 or 6 β-D-Galp 1 → 4 *β-D-GlcpNAc 1 → 2 α-D-Manp 1
β-D-Galp 1 → 3 β-D-Galp 1 → 4 β-D-GlcpNAc 1
± β-D-Galp 1 → 3 β-D-Galp 1 → 4 β-D-GlcpNAc 1
3 6 α-D-Manp 1
6 α-D-Manp 1 → 4 or 6 *D-GlcpNAc 1 → 4(6) *D-GlcpNAc 1 → L-Asx 3
L-Fucp-(1 → 6 or 4)-

(31)

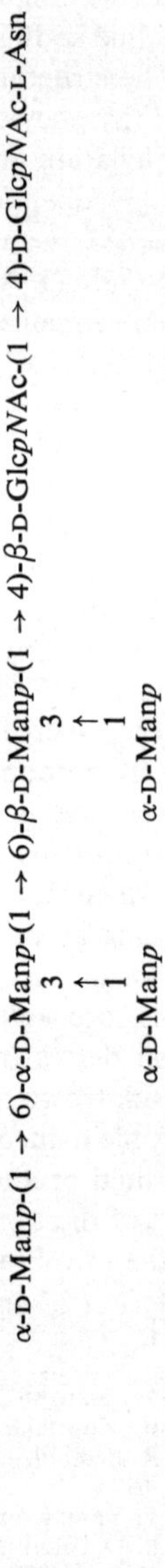
α-D-Manp-(1 → 6)-α-D-Manp-(1 → 6)-β-D-Manp-(1 → 4)-β-D-GlcpNAc-(1 → 4)-D-GlcpNAc-L-Asn
3 ↑ 1 α-D-Manp
3 ↑ 1 α-D-Manp

(32)

that a high molecular weight glycoprotein (LETS glycoprotein) is biosynthesized at a specific stage of preimplantation mouse embryonic development.[410]

The role of fibronectin, an adhesion glycoprotein present on the surface of cells in connective tissue matrices and in extracellular fluids, has been reviewed with reference to malignant transformation, embryonic differentiation, and the function of the reticuloendothelial system.[411] Chemical and physical characteristics of fibronectin and the role of fibronectin and other matrix components in malignant transformation have been surveyed.[412]

Hen oviduct membranes catalyse the biosynthesis of D-glucose-containing oligosaccharides linked to lipid.[413] The entire oligosaccharide may be transferred enzymically from the lipid either to an endogenous membrane protein or to exogenous *S*-carboxymethylated α-lactalbumin. After attachment to the protein, the D-glucose unit is released from the oligosaccharide chains by an enzyme of the oviduct membrane.[414] The residual oligosaccharide chain of the glycoprotein was then identified as $(\alpha\text{-D-Man}p)_4$-β-D-Man*p*-(1 → 4)-β-D-Glc*p*NAc-(1 → 4)-D-Glc*N*Ac.

Two species of glycoprotein are thought to be involved in the atypical cellular behaviour of human trophoblast cells.[415] One glycoprotein is associated with localized cellular growth and contains neuraminic acid in addition to other sugars but not L-fucose, while the other contains both neuraminic acid and L-fucose and is characteristic of neoplastic cells.

Glycoproteins of the Chinese hamster ovary cell membranes have been partially fractionated on the basis of their interactions with concanavalin A and wheat germ agglutinin.[416] The lectin–glycoprotein complexes were precipitated with specific antisera directed against the lectins. Analysis of the immunoprecipitates by polyacrylamide gel electrophoresis indicated that the membranes contain two major classes of 2-amino-2-deoxy-D-glucose-containing glycoproteins of apparent mol. wts. 1×10^5 and 1.3×10^5. Clone 15B cells from Chinese hamster ovary, selected for resistance to ricin toxicity, have abnormal cell surface glycoproteins.[417] One of the glycopeptides (32) isolated from these glycoproteins is present in enhanced amounts in the modified cells. It is proposed that this glycopeptide is the physiological acceptor for the 2-acetamido-2-deoxy-D-glucosyltransferase missing in clone 15B cells, and that transfer of 2-acetamido-2-deoxy-D-glucose is required to facilitate continued processing of the oligosaccharide to form the mature complex sugar chain.

Purified ^{125}I-labelled lectins have been interacted with cells of the mosquito, *Aedes aegyptii*, and with cells of human KB tumour.[418] Relative to human KB cells, mosquito cells bound less of those lectins specific for L-fucose and D-galactose, which are terminal residues in many mammalian glycoproteins,

410 B. R. Zetter and G. R. Martin, *Proc. Nat. Acad. Sci., U.S.A.*, 1978, **75**, 2324.
411 K. M. Yamada and K. Olden, *Nature*, 1978, **275**, 179.
412 A. Vaheri and D. F. Mosher, *Biochim. Biophys. Acta*, 1978, **516**, 1.
413 W. W. Chen and W. J. Lennarz, *J. Biol. Chem.*, 1978, **253**, 5774.
414 W. W. Chen and W. J. Lennarz, *J. Biol. Chem.*, 1978, **253**, 5780.
415 A. Whyte and Y. W. Loke, *J. Exp. Med.*, 1978, **148**, 1087.
416 R. J. Juliano and G. Li, *Biochemistry*, 1978, **17**, 678.
417 E. Li and S. Kornfeld, *J. Biol. Chem.*, 1978, **253**, 6426.
418 T. D. Butters and R. C. Hughes, *Carbohydrate Res.*, 1978, **61**, 159.

whereas the number of binding sites of lectins specific for core-region residues (D-mannose and 2-acetamido-2-deoxy-D-glucose) was similar. Neuraminidase treatment, which enhanced binding of peanut agglutinin or soybean agglutinin to human KB cells, had negligible effects on the binding of these lectins to mosquito cells. The comparative structures of surface oligosaccharides of the two types of cells are discussed in relation to lectin binding studies.

Glycoproteins have been located on the surface of two murine leukaemic cells, as shown by D-galactose oxidase–sodium borotridite and ^{125}I-lacto-peroxidase–iodination techniques.[194] The patterns on polyacrylamide gel electrophoresis of the solubilized glycoproteins from the two sources are indistinguishable but differ from those of the surface glycoproteins of HeLa cells.

Isolation from the sea urchin egg surface of a trypsin-sensitive glycoprotein fraction having species-specific affinity for bindin, the insoluble protein which coats the sperm acrosome process and mediates adhesion of sperm to sea urchin eggs, has been reported.[420] Residues of D-galactose and D-mannose make up the high (34%) carbohydrate content of the glycoprotein fraction. A fraction from the egg jelly of the starfish *Asterias amurensis*, capable of inducing the acrosome process, has been shown to contain glycoproteins with high levels of 6-deoxyhexose (methylpentose).[421]

Two polysaccharides, a D-mannan and an L-arabino D-galactan, have been isolated from the insect trypanosomatid, *Crithidia fasciculata*.[422] Cell agglutination studies as well as immunocytochemical techniques using fluorescein- and ferritin-conjugated antibodies have demonstrated that the L-arabino D-galactan is a cell surface constituent. Receptors for concanavalin A and wheat germ agglutinin have been found on the surface membranes of all parasitic stages of *Schistosoma mansoni*.[423] The major surface coat glycoproteins of *Trypanosoma brucei brucei* has been isolated by affinity chromatography using immobilized lentil lectin.[424]

The biosynthesis of plasma membrane glycoproteins has been the subject of a review.[425]

10 Hormonal Glycoproteins

The processing of placental peptide hormones synthesized in lysates containing membranes derived from tunicamycin-treated ascites tumour cells has been reported.[426] The cleavage of a pre-glycoprotein, pre-human chorionic gonadotrophin, and the sequestration of the processed form within the membranes, occurs in the absence of glycosylation. Thus, in contrast to certain processing

419 K. Y. Chen, R. H. Kramer and E. S. Canellakis, *Biochim. Biophys. Acta*, 1978, **507**, 107.
420 C. G. Glabe and V. D. Vacquier, *Proc. Nat. Acad. Sci., U.S.A.*, 1978, **75**, 881.
421 Y. Uno and M. Hoshi, *Science*, 1978, **200**, 58.
422 M. Gottlieb, *Biochim. Biophys. Acta*, 1978, **541**, 444.
423 K. D. Murrell, D. W. Taylor, W. E. Vannier, and D. A. Dean, *Exp. Parasitol.*, 1978, **46**, 247.
424 J. E. Strickler, P. E. Mancini, and C. L. Patton, *Exp. Parasitol.*, 1978, **46**, 262.
425 G. M. W. Cook, in 'Synthesis, Assembly, and Turnover of Cell Surface Components', Cell Surface Reviews, Vol. 4, ed. G. Poste and G. L. Nicolson, Elsevier–North Holland–Biomed. Press, New York, 1977, p. 85.
426 M. Bielinska, G. A. Grant, and I. Boime, *J. Biol. Chem.*, 1978, **253**, 7117.

steps involved in the maturation of some secretory glycoproteins, glycosylation is not obligatory for pre-protein cleavage.

The biosynthesis of cholesterol from mevalonic acid in sheep ovarian follicles is inhibited in both the follicles and isolated thecal tissues by human chorionic gonadotrophin.[427] The inhibition appears to be subsequent to the formation of lanosterol in the biosynthetic pathway. Although the addition of gangliosides blocks the cholera toxin-mediated stimulation of progesterone biosynthesis in rat ovarian cells, it is ineffectual in blocking gonadotrophin-stimulated steroidogenesis.[428] Sphingolipid did not block the binding of ^{125}I-human chorionic gonadotrophin to either ovarian cells or membranes, implying that gangliosides are not involved in hormone receptor interactions and the subsequent receptor-mediated physiological responses.

Concanavalin A has been reported to stimulate the secretion of human chorionic gonadotrophin by human choriocarcinoma cells.[429] Since the stimulation is thought to occur subsequent to interactions at the membrane level, concanavalin A may be used as a probe for studying membrane-related events in the control of the secretion of this hormone.

Structural analyses of the L-asparagine-linked (33) and L-serine-linked (34) oligosaccharide chains of human chorionic gonadotrophin have been performed.[430]

The L-asparagine-linked oligosaccharide chains of the α- and β-sub-units have similar general structures but in addition the β-unit has one residue of L-fucose, which may be linked to a D-galactosyl or a 2-acetamido-2-deoxy-D-glucosyl unit. Sequential removal of sugar residues from the hormone markedly diminishes its ability to stimulate the accumulation of cyclic AMP by porcine granulosa cells.[431]

Porcine parathyroid cells, on incubation with [^{3}H]- or [^{14}C]-amino-acids, 2-amino-2-deoxy-D-[^{3}H]glucose, and D-[^{3}H]mannose, excreted parathormone and two glycopeptides.[432] One of the glycopeptides, which binds to concanavalin A, appears to be identical to parathyroid secretory protein, which has not been recognized previously as a glycoprotein. The secretion of this glycopeptide is inhibited by Ca^{2+} to a degree greater than or equal to that of the hormone, possibly suggesting that this glycopeptide is involved in the physiological secretion of parathormone by the parathyroid gland.

A common structure (35) has been proposed for the oligosaccharide moieties of glycopeptides isolated from the α-subunits of human follitropin, lutropin, and thyrotropin.[433] Microheterogeneity amongst the glycopeptides is thought to be due to variations in the number of terminal neuraminic acid and L-fucosyl groups.

[427] T. J. Douglas, R. P. Hamilton, and R. F. Seamark, *Austral. J. Biol. Sci.*, 1978, **31**, 405.
[428] S. Azhar, P. Fitzpatrick, and K. M. J. Menon, *Biochem. Biophys. Res. Comm.*, 1978, **83**, 493.
[429] R. Benveniste, K. V. Speeg, A. Long, and D. Rabinowitz, *Biochem. Biophys. Res. Comm.*, 1978, **84**, 1082.
[430] O. P. Bahl, L. Marz, and M. J. Kessler, *Biochem. Biophys. Res. Comm.*, 1978, **84**, 667.
[431] C. P. Channing, C. N. Sakai, and O. P. Bahl, *Endocrinology*, 1978, **103**, 341.
[432] J. J. Morrissey, J. W. Hamilton, and D. V. Cohn, *Biochem. Biophys. Res. Comm.*, 1978, **82**, 1279.
[433] K. Hara, P. Rathnam, and B. B. Saxena, *J. Biol. Chem.*, 1978, **253**, 1582.

α-Neu*f*NAc-(2 → 3)-β-D-Gal*p*-(1 → 4)-β-D-Glc*p*NAc-(1 → 2)-α-D-Man*p*
1
↓
6
β-D-Man*p*-(1 → 4)-β-D-Glc*p*NAc-(1 → 4)-D-Glc*p*NAc-L-Asn
3
↑
1
α-Neu*f*NAc-(2 → 3)-β-D-Gal*p*-(1 → 4)-β-D-Glc*p*NAc-(1 → 2)-α-D-Man*p*

(33)

α-Neu*N*Ac-(2 → 3)-β-D-Gal*p*-(1 → 3)-β-D-Gal*p*NAc-L-Ser
6
↑
2
α-Neu*N*Ac

(34)

α-Neu*f*NAc → β-D-Gal*p* → β-D-Glc*p*NAc → D-Man*p*
↓
α-Neu*f*NAc → β-D-Gal*p* → β-D-Glc*p*NAc → α-D-Man*p*-β-D-Man*p* → D-Gal*p*NAc → D-Glc*p*NAc-L-Asn
↑
α-Neu*f*NAc → β-D-Gal*p* → β-D-Glc*p*NAc → D-Man*p*
α-L-Fuc*p*

(35)

11 Milk Glycoproteins

Six oligosaccharides which are di- and tri-L-fucosyl-derivatives of lacto-*N*-octaose (36—38) and lacto-*N*-neo-octaose (39—41) have been isolated and characterized from the human milk of a non-secretor Lewis ($a^{+}b^{-}$) individual.[434]

```
                  R¹                                          R²
                  1                                           1
                  ↓                                           ↓
                  3                                           3
β-D-Galp-(1 → 4)-β-D-GlcpNAc-(1 → 3)-β-D-Galp-(1 → 4)-β-D-GlcpNAc
                                                              1
                                                              ↓
                                                              6
                                                 β-D-Galp-(1 → 4)-D-Glc
                                                              3
                                                              ↑
                                                              1
                                     β-D-Galp-(1 → 3)-β-D-GlcpNAc
                                                              4
                                                              ↑
                                                              1
                                                              R³
```

(36) $R^1 = R^2 =$ α-L-Fucosyl, $R^3 =$ H.

(37) $R^1 =$ H, $R^2 = R^3 =$ α-L-Fucosyl.

(38) $R^1 = R^2 = R^3 =$ α-L-Fucosyl.

```
                  R¹                                          R²
                  1                                           1
                  ↓                                           ↓
                  4                                           3
β-D-Galp-(1 → 3)-β-D-GlcpNAc-(1 → 3)-β-D-Galp-(1 → 4)-β-D-GlcpNAc
                                                              1
                                                              ↓
                                                              6
                                                 β-D-Galp-(1 → 4)-D-Glc
                                                              3
                                                              ↑
                                                              1
                                     β-D-Galp-(1 → 4)-β-D-GlcpNAc
                                                              3
                                                              ↑
                                                              1
                                                              R³
```

(39) $R^1 = R^2 =$ α-L-Fucosyl, $R^3 =$ H.

(40) $R^1 =$ H, $R^2 = R^3 =$ α-L-Fucosyl.

(41) $R^1 = R^2 = R^3 =$ α-L-Fucosyl.

The milk of marsupials is known to be different from that of other mammals, lactose being the major milk component of almost all eutherians.[435] Oligosaccharides containing between three and ten monosaccharide units, and

[434] Y. Tachibana, K. Yamashita, and A. Kobata, *Arch. Biochem. Biophys.*, 1978, **188**, 83.
[435] M. Messer and G. S. Mossop, *Austral. J. Biol. Sci.*, 1977, **30**, 379.

composed of various proportions of D-galactose, D-glucose, and 2-amino-2-deoxy-D-glucose, have been separated from the milk of the eastern grey kangaroo, *Macropus giganteus*.[435]

Cow colostrum trypsin inhibitor contains a single oligosaccharide moiety which is composed of residues of L-fucose, D-glucose, D-mannose, D-galactose, 2-amino-2-deoxy-D-glucose, 2-amino-2-deoxy-D-galactose, and neuraminic acid, and which is attached to the peptide backbone at L-asparagine-27.[436] A study of the ^{1}H n.m.r. line widths of the intact molecule has indicated that the carbohydrate moiety of the inhibitor causes a pronounced restriction on the overall mobility of the molecule, indicating a sizeable increase in the average radius of gyration as compared to the basic pancreatic trypsin inhibitor.[437]

α-Lactalbumin has been isolated from three different strains of rats and purified by a combination of gel filtration and ion-exchange chromatography.[438] A number of similar, and also different, glycoproteins with α-lactalbumin activity were demonstrated to be present in the milk of lactating rats by polyacrylamide gel electrophoresis. Two distinct genetic variants of α-lactalbumin were shown to exist in two of the strains of rats, and the third strain appeared to be homozygous for one genetic form of α-lactalbumin which may or may not be structurally distinct from the apparently equivalent forms in the other two strains.

Thermodynamic studies on the unfolding of α-lactalbumin caused by lithium chloride, lithium chlorate, and sodium chlorate have been reported.[439] The characterization of the denatured states and the mechanism of the denaturation reactions are discussed and compared with previous results on the unfolding with organic denaturants.

Core glycosylation of newly synthesized rat α-lactalbumin occurs *in vitro* in a microsomal membrane supplemented system.[440] Processing, segregation, and core glycosylation were observed only when membranes were present during translation and not when they were added after translation.

S-Carboxymethylated and *S*-aminoethylated α-lactalbumin, as well as smaller peptides derived from these two forms of the glycoprotein, have been found to serve as substrates for the *in vitro* glycosylation from sugar nucleotides catalysed by hen oviduct membrane preparations.[441] The structural feature common to all of the active carbohydrate acceptors was the tripeptide sequence -L-Asn-L-Gln-L-Ser-, which appears to be sufficient for the *in vitro* synthesis of the *N*-glycosidic bond. Edman degradation of the enzymically glycosylated form of the heptapeptide L-Asn-L-Asn-L-Gln-L-Ser-L-Thr-L-Asp-L-Tyr indicated that the oligosaccharide unit is attached to the second amino-acid residue, which corresponds to L-Asn_{45} in intact α-lactalbumin. Indirect evidence suggests that this specific L-asparaginyl residue is the site of glycosylation for

[436] R. Klauser, D. Čechová, and H. Tschesche, *Z. physiol. Chem.*, 1978, **359**, 173

[437] G. Wagner, K. Wüthrich, and H. Tschesche, *European J. Biochem.*, 1978, **86**, 67.

[438] R. M. McKenzie and B. L. Larson, *J. Dairy Sci.*, 1978, **61**, 714.

[439] S. Maruyama, K. Kuwajima, K. Mitta, and S. Sugai, *Biochim. Biophys. Acta*, 1977, **494**, 343.

[440] V. R. Lingappa, J. R. Lingappa, R. Prasad, K. E. Ebner, and G. Blobel, *Proc. Nat. Acad. Sci., U.S.A.*, 1978, **75**, 2338.

[441] D. K. Struck, W. J. Lennarz, and K. Brew, *J. Biol. Chem.*, 1978, **253**, 5786.

all of the active carbohydrate acceptors used in the study. The results were discussed with respect to the restricted occurrence of tripeptide sequences of the type -L-Asn-X-L-Ser/L-Thr- in eukaryotic secretory proteins.

Membrane preparations from lactating bovine mammary glands have been shown to incorporate D-mannose from GDP-D-mannose into lipid and into metabolically stable acceptors.[442] It is postulated that the sequence GDP-D-mannose ⇆ P^2-D-mannosyl-P^1-pyrophosphorylisoprenol → P^2-D-mannosyl-oligosaccharide-P^1-pyrophosphoryl → D-mannoprotein is followed. The structure of the oligosaccharide–lipid and the endogenous glycoproteins were not established.

Treatment of sialoglycopeptides derived from bovine milk-fat globule membrane with alkaline sodium borohydride released a mixture of oligosaccharides amongst which the structure of a tetrasaccharide (42) and two trisaccharides (43—44) were established.[443]

Neu*N*Ac-(2 → 3)-β-D-Gal*p*-(1 → 3)-D-Glc*N*Ac-ol
6
↑
2
Neu*N*Ac

(42)

Neu*N*Ac-(2 → 3)-β-D-Galp-(1 → 3)-Glc*N*Ac-ol

(43)

β-D-Gal*p*-(1 → 3)-D-Glc*N*Ac-ol
6
↑
2
Neu*N*Ac

(44)

Soluble glycoprotein has been isolated from milk fat globule membrane.[444] Differences in the isolation procedure have accounted for the discrepancies in physicochemical properties of the glycoprotein in earlier reports.

The incorporation of D-mannose and 2-acetamido-2-deoxy-D-glucose into glycoproteins and lipid-linked intermediates of mammary explants of lactating rabbits has been reported.[445] D-Mannose is incorporated largely into glycopeptides (mol. wt. 4—8 × 10^4) and 2-acetamido-2-deoxy-D-glucose is incorporated into a glycopeptide (mol. wt. 3.2 × 10^4) having properties similar to those of rabbit casein. No evidence was obtained to show that a lipid-linked oligosaccharide is involved in the glycosylation of the casein. A microsomal preparation from a lactating rabbit mammary gland has been shown to catalyse the formation of two lipid-linked sugar compounds that appear to be involved in the glycosylation of proteins in the tissue.[446] Variations in the glycosylation pattern of κ-casein with micelle size in bovine milk has been demon-

[442] I. K. Vijay and S. R. Fram, *J. Supramol. Struct.*, 1977, **7**, 251.
[443] G. H. Farrar and R. Harrison, *Biochem. J.*, 1978, **171**, 549.
[444] K. Yamauchi, M. Shimizu, and C. Kannon, *J. Dairy Sci.*, 1978, **61**, 688.
[445] B. K. Speake and D. A. White, *Biochem. J.*, 1978, **170**, 273.
[446] D. A. White, *Biochem. J.*, 1978, **170**, 479.

strated.[447] Glycosylation appears to follow complete micelle formation so that surface or near-surface κ-casein is modified while that incorporated into the interior surface is unchanged. Gel filtration studies revealed that, except for the κ-casein from the largest micelles, the hexose content and the ratio of hexose to glycosylated protein were greater for the larger size polymers of high molecular weight than for those of lower molecular weight.

12 Serum Glycoproteins

A number of highly purified, neuraminidase-treated human serum glycoproteins, have been examined for their cross-reactivity with several lectins having anti-D-galactosyl specificity, in an attempt to characterize these glycoproteins with respect to their slightly differing carbohydrate moieties.[448]

A hepatic receptor, which recognizes and binds specifically to serum glycoproteins bearing terminal nonreducing 2-acetamido-2-deoxy-D-glucosyl residues, has been isolated from chicken liver by chromatography on immobilized agalacto-orosomucoid.[449] In contrast to the corresponding mammalian glycoprotein, the avian one exhibits only minimal binding activity towards asialoglycoproteins and interacts strongly with agalactoglycoproteins. The readily dissociable avian glycoprotein–ligand complex, in contrast to the mammalian one, contains a single high affinity binding site, having K_{diss} 1.4×10^{-9}M.

Evidence has been presented for the existence of a hepatic receptor that specifically binds oligosaccharides containing 2-acetamido-2-deoxy-3-*O*-α-L-fucopyranosyl-D-glucopyranosyl groups.[450] Deglycosylation or periodate oxidation of these glycoproteins enhances the circulation time.

Receptors to asialoglycoproteins have been isolated from rabbit liver by affinity chromatography on immobilized D-galactosyl- and D-glucosyl-derivatives of bovine serum albumin.[451]

Affinity chromatography on immobilized *Ricinus communis* agglutinin has been used for the isolation of a heterogeneous population of human serum inhibitors of the binding of desialylated glycoproteins to rat hepatocyte membranes.[452] The inhibitors are themselves desialylized glycoproteins, which are present in small amounts in normal human serum but in increased levels in serum from patients with cirrhosis.

The clearance from plasma of lysosomal enzymes and glycoproteins terminating in 2-acetamido-2-deoxy-D-glucosyl or D-mannosyl groups has been studied in non-parenchymal cells of the liver.[453]

Clearance of ribonuclease B from rat circulation depends on specific recognition of the α-D-mannosyl residues of the attached oligosaccharide and is

[447] C. W. Slattery, *Biochemistry*, 1978, **17**, 1100.
[448] G. Uhlenbruck, B. A. Baldo, G. Steinhausen, H. G. Schwick, B. P. Chatterjee, V. Hořejši, A. Krajhanzi, and J. Kocourek, *J. Clin. Chem. Clin. Biochem.*, 1978, **16**, 19.
[449] T. Kawasaki, and G. Ashwell, *J. Biol. Chem.*, 1977, **252**, 6536.
[450] L. R. Glasgow, J. C. Paulson, and R. L. Hill, *Proc. Nat. Acad. Sci., U.S.A.*, 1978, **75**, 2215.
[451] C. P. Stowell and Y. C. Lee, *J. Biol. Chem.*, 1978, **253**, 6107.
[452] J. S. Marshall and S. Williams, *Biochim. Biophys. Acta*, 1978, **543**, 41.
[453] P. H. Schlesinger, T. W. Doebber, B. F. Mandell, R. White, C. DeSchryver, J. S. Rodman, M. J. Miller, and P. Stahl, *Biochem. J.*, 1978, **176**, 103.

inhibited by the presence of yeast D-mannan.[454] The clearance from the circulation occurs primarily in liver, spleen, and bone marrow non-parenchymal cells. Quantitative aspects of the hepatic clearance mechanism for the elimination of asialofetuin and asialo-orosomucoid by the rat have been described.[455]

Procedures have been published for the preparation of milligram quantities of four major glycopeptides which are produced when α_1-antitrypsin is cleaved with cyanogen bromide.[456] The four glycopeptides, which account for all the carbohydrate of the glycoprotein, each contain differing amounts of 2-acetamido-2-deoxy-D-glucose, D-mannose, D-galactose, and neuraminic acid. Each glycopeptide, after desialylation and degradation with proteolytic enzymes, serves as an acceptor of neuraminic acid residues in a reaction catalysed by human liver sialyltransferase.[457] The activity of the sialyltransferase required for the addition of CMP-neuraminic acid to any one of the glycopeptide acceptors depends, in part, on the domain of its polypeptide backbone.

Normal and S-variant α_1-antitrypsins have essentially the same carbohydrate composition, indicating that serum deficiency of the glycoprotein in individuals with the Pi S phenotype is not due to diminished carbohydrate content which, in turn, leads to decreased secretion.[458] A Z-variant of α_1-antitrypsin has been isolated from the liver of a patient homozygous for the Z allele of the protease inhibitor locus.[459] This glycoprotein is homogeneous and antigenically competent with the normal protease inhibitor, but it has no protease inhibition capacity. Both neuraminic acid and D-galactosyl residues are absent from the molecule, and there are fewer 2-acetamido-2-deoxy-D-glucosyl residues and an excess of D-mannosyl residues.

The effects of ethanol on the conformation of α_1-antitrypsin, deoxyribonuclease, pepsinogen, soybean trypsin inhibitor, and unfolded ribonucleases have been studied by c.d.[460] In the presence of 50—75 volume % ethanol, the tertiary structure is perturbed and the polypeptide chain is reorganized into new conformations with higher content of α-helix and β-structure than in the native state.

The primary structures of sixteen asialo-oligosaccharides derived from human plasma α_1-acid glycoprotein have been assigned using 360 MHz ^{1}H n.m.r. spectroscopy in combination with partial acetolysis and methylation analysis.[461] The structures of the heteroglycans can be grouped into five classes, and, except for those of two classes of molecules possessing bi- and tri-antennary structures, the structures have not been reported previously. Three glycopeptide fractions containing carbohydrate side chains linked to

[454] T. L. Brown, L. A. Henderson, S. R. Thorpe, and J. W. Baynes, *Arch. Biochem. Biophys.*, 1978, **188**, 418.
[455] E. Regoeczi, M. T. Debanne, M. W. C. Hutton, and A. Koj, *Biochim. Biophys. Acta*, 1978, **541**, 372.
[456] D. E. Roll, J. J. Aguanno, C. J. Coffee, and R. H. Glew, *J. Biol. Chem.*, 1978, **253**, 6992.
[457] J. J. Aguanno, D. E. Roll, and R. H. Glew, *J. Biol. Chem.*, 1978, **253**, 6997.
[458] D. Roll, J. J. Aguanno, C. J. Coffee, R. H. Glew, and R. M. Iammarino, *Biochim. Biophys. Acta*, 1978, **532**, 171.
[459] A. Hercz, E. Katona, E. Cutz, J. R. Wilson, and M. Barton, *Science*, 1978, **201**, 1229.
[460] B. Jirgensons, *Biochim. Biophys. Acta*, 1978, **534**, 123.
[461] B. Fournet, J. Montreuil, G. Strecker, L. Dorland, J. Haverkamp, J. F. G. Vliegenthart, J. P. Binette, and K. Schmid, *Biochemistry*, 1978, **17**, 5206.

L-asparagine have been isolated from human α_1-acid glycoprotein which had been degraded with neuraminidase and proteolytic enzymes.[462] The structure (45) for these glycopeptides was established from a combination of sequential

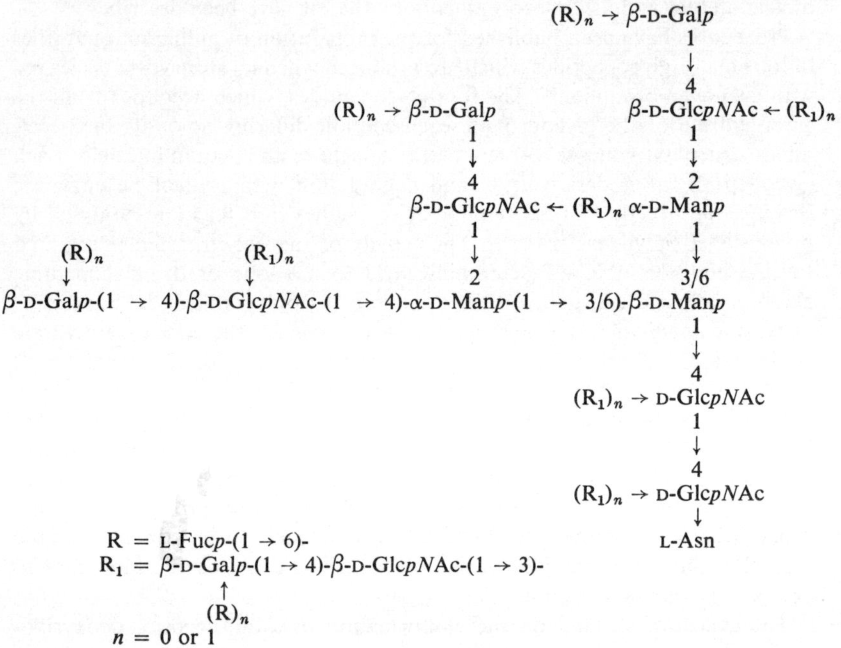

(45)

enzymatic degradation studies and methylation studies. The L-fucosyl residues may be attached to any of the D-galactosyl residues and the sequence β-D-Gal*p*-(1 → 4)-D-Glc*pN*Ac may be linked to any of the other 2-acetamido-2-deoxy-D-glucosyl residues. A glycopeptide which is resistant to attack by *exo*-glycanases has been isolated from α_1-acid glycoprotein, with a view to the isolation of micro-organisms capable of elaborating *endo*-glycanases.[463]

α_1-Acid glycoprotein, which had been isolated from nephrotic urine, has been shown to act as a co-factor in the lipoprotein lipase reaction.[464] The findings suggest that the elevated levels of plasma triglycerides in human nephrotic patients is a result of excessive loss of the glycoprotein from plasma into urine.

A method suitable for the large-scale preparation of α-fetoprotein has been proposed.[465] The amino-acid and carbohydrate composition of the product was reported. Affinity chromatographic procedures have been developed for the

[462] G. O. H. Schwarzmann, V. B. Hatcher, and R. W. Jeanloz, *J. Biol. Chem.*, 1978, **253**, 6983.
[463] A. Amemura, R. H. Shah, O. P. Bahl, and J. M. Merrick, *Carbohydrate Res.*, 1978, **63**, 241.
[464] I. Staprans and J. M. Felts, *Biochem. Biophys. Res. Comm.*, 1977, **79**, 1272.
[465] M. A. Al-Awqati, Y. B. Gordon, and T. Chard, *Clinica Chim. Acta*, 1978, **89**, 173.

separation of α-fetoprotein from the albumin of amniotic fluid.[466, 467] Bovine α-fetoprotein has been shown to exist in two variants on the basis of carbohydrate heterogeneity.[468] The molecular weight and amino-acid composition of the glycoprotein are similar to those of other mammalian α-fetoproteins, and it is related immunochemically to human α-fetoprotein. The corresponding glycoprotein isolated from rat has a molecular weight of 6.5—7.1 × 10^4 and contains 2-acetamido-2-deoxy-D-galactose (1 mole) and 2-acetamido-2-deoxy-D-glucose (8 moles) per mole.[469] The isolation of nine different α-fetoprotein variants from rat by affinity chromatography on immobilized concanavalin A and immobilized *R. communis* lectin suggests the existence of microheterogeneity in the carbohydrate part of the glycoprotein.[470] Heterogeneity of human α-fetoprotein is also implied by the results of isoelectric focusing experiments.[471]

The chemistry of the tumour-associated antigens, α-fetoprotein,[472] carcinoembryonic antigen,[473] and antigens related to carcinoembryonic antigen[474] have been reviewed.

Anti-freeze glycoproteins isolated from Antarctic fish show a high tendency to adopt β-turns which are separated by non-polar sequences, suggesting that the secondary structure of these glycoproteins may play a role in their adsorption on to ice.[475] High resolution ^{1}H n.m.r. and c.d. measurements have been performed on aqueous solutions of anti-freeze glycoproteins.[476] It proved impossible to interpret the peptide contribution to the spectra in terms of α-helix, β-sheet, and random coil. Instead, the c.d. spectrum, which resembles that of β-structure in position, magnitude, and spectral form, is of opposite sign, implying a specific conformation which may possibly be stabilized by non-bonded interactions between the peptide backbone and the carbohydrate side chains. The mechanism of the anti-freeze activity is discussed in terms of the spatial and orientational correlations of the carbohydrate hydroxy-groups and water in the liquid and solid states. The implication of an ordered structure is explained by the comparison of the anti-freeze glycoproteins with synthetic water-soluble polymers, such as polyvinyl pyrrolidone and polyvinyl alcohol, which also show anti-freeze properties. Polar cod, *Boreogadus saida*, normally found in Arctic waters, has been reported to contain anti-freeze glycoproteins similar in structure to those of the Antarctic fish *Trematomus borchgrevinski*, consisting of sequences of L-Ala-L-Ala-L-Thr repeating units with all of the L-threonine residues glycosylated by the disaccharide 2-acet-

[466] J. L. Young and B. A. Webb, *Analyt. Biochem.*, 1978, **88**, 619.
[467] F. K. Khalil, E. Stephen, and S. Guilbaud, *Clinica Chim. Acta*, 1978, **88**, 167.
[468] P. C. W. Lai, E. H. Peters, and F. L. Lorscheider, *Biochim. Biophys. Acta*, 1978, **535**, 138.
[469] V. Versée and A. O. Barel, *Biochem. J.*, 1978, **175**, 73.
[470] B. Bayard and J. P. Kerckaert, *Biochem. Biophys. Res. Comm.*, 1977, **77**, 489.
[471] E. P. Lester, J. B. Miller, and S. Yachnin, *Biochim. Biophys. Acta*, 1978, **536**, 165.
[472] E. Ruoslahti and E. Engvall, *Scand. J. Immunol.*, 1978, **7**, Suppl. 6, 1.
[473] J. E. Shively and C. W. Todd, *Scand. J. Immunol.*, 1978, **7**, Suppl. 6, 19.
[474] S. Hammarström, T. Svenberg, A. Hedin, and G. Sundblad, *Scand. J. Immunol.*, 1978, **7**, Suppl. 6, 33.
[475] M. H. Loucheux-Lefebvre, *Biochem. Biophys. Res. Comm.*, 1978, **81**, 1352.
[476] F. Franks and E. R. Morris, *Biochim. Biophys. Acta*, 1978, **540**, 346.

amido-2-deoxy-3-*O*-β-D-galactopyranosyl-D-galactose.[477] Eight or more anti-freeze glycoproteins from the blood of a number of species of fish have been identified on the basis of molecular weight and electrophoretic mobility. There appears to be a cooperative functioning between the larger and smaller molecular weight components in the depression of the freezing point of the sera.[478]

A sequence of fourteen amino-acids of an L-proline-containing glycopeptide from *T. borchgrevinski* has been determined by m.s.[479] Every L-threonine residue was shown to be linked *O*-glycosidically to the disaccharide, 2-acetamido-2-deoxy-3-*O*-β-D-galactopyranosyl-D-galactose.

Interactions of the 3.8S α_2-histidine-rich glycoprotein of human serum with haeme, metal ions, and organic ligands, have been reported.[480] Although the glycoprotein binds to haeme, it does not appear to compete with serum albumin for haeme, suggesting that the glycoprotein has no major function in serum haeme transport. The glycoprotein also binds bilirubin, diaquocobinamide, Cibacron Blue F3GA, and certain divalent metal ions.

Prostatic acid phosphatase is of considerable biomedical interest as a diagnostic indicator of the development and metastasis of prostatic cancer. Human prostatic acid phosphatases have been purified using affinity chromatography,[481] a combination of affinity chromatography and isoelectric focusing,[482] and affinity chromatography and ion-exchange chromatography.[483, 484] Two forms of acid phosphatase which are inhibited by L-tartrate have been purified from rabbit kidney cortex.[485] Some of the properties of one of the isoenzymes of human prostatic acid phosphatase have been described.[486] Methods for the assay of the enzyme have been developed, based on a double-antibody radioimmunoassay,[487] counterimmuno-electrophoresis,[488] radioimmunoassay,[489] and solid-phase fluorescent immunoassay.[490] Desialylation of human hepatic alkaline phosphatase enhances the sensitivity of the molecule to heat and proteolytic enzymes.[491] Conformational changes, as evidenced by a lower helical content, are also observed.

Chemical and biochemical aspects of pregnancy-specific β-globulin have been reviewed.[492] A comparison has been made of a number of methods for

[477] D. T. Osuga and R. E. Feeney, *J. Biol. Chem.*, 1978, **253**, 5338.
[478] D. T. Osuga, F. C. Ward, Y. Yeh, and R. E. Feeney, *J. Biol. Chem.*, 1978, **253**, 6669.
[479] H. R. Morris, M. R. Thompson, D. T. Osuga, A. I. Ahmed, S. M. Chan, J. R. Vandenheede, and R. E. Feeney, *J. Biol. Chem.*, 1978, **253**, 5155.
[480] W. T. Morgan, *Biochim. Biophys. Acta*, 1978, **535**, 319.
[481] R. L. Van Elten and M. S. Saini, *Clinical Chem.*, 1978, **24**, 1525.
[482] P. Vihko, M. Kontturi, and L. K. Korhonen, *Clinical Chem.*, 1978, **24**, 466.
[483] B. K. Choe, E. J. Pontes, S Bloink, and N. R. Rose, *Arch. Andrology*, 1978, **1**, 221.
[484] B. K. Choe, E. J. Pontes, I. McDonald, and N. R. Rose, *Prep. Biochem.*, 1978, **8**, 73.
[485] J. J. Helwig, A. A. Farooqui, C. Bollack, and P. Mandel, *Biochem. J.*, 1978, **175**, 321.
[486] P. Vihko, *Clinical Chem.*, 1978, **24**, 1783.
[487] B. K. Choe, E. J. Pontes, M. K. Morrison, and N. R. Rose, *Arch. Andrology*, 1978, **1**, 227.
[488] I. McDonald, N. R. Rose, E. J. Pontes, and B. K. Choe, *Arch. Andrology*, 1978, **1**, 235.
[489] P. Vihko, E. Sajanti, O. Janne, L. Peltonen, and R. Vihko, *Clinical Chem.*, 1978, **24**, 1915.
[490] C. Lee, M. C. Wang, G. P. Murphy, and T. M. Chu, *Cancer Res.*, 1978, **38**, 2871.
[491] T. Komoda and Y. Sakagishi, *Biochim. Biophys. Acta*, 1978, **523**, 395.
[492] H. Bohn, *Scand. J. Immunol.*, 1978, **7** Suppl. 6, p. 119.

measuring plasma and urinary levels of this glycoprotein in an attempt to explain discrepancies observed with a small number of plasma samples.[493]

One of the major proteinase inhibitors in human blood plasma, α_2-macroglobulin, has been purified in a one-step procedure on Cibacron Blue–Sepharose.[494] The glycoprotein contains hexose, amino-sugar, and neuraminic acid and is composed of four polypeptide chains each of molecular weight 1.85×10^5.[495]

A serum α_2-globulin which is involved in the transport of vitamin D_3 has been separated into two components with different neuraminic acid contents.[496]

α_2-Opsonic glycoprotein (mol. wt. 2.26×10^5), from rat serum, consists of two sub-units joined by disulphide bonds and contains 19 residues of 2-amino-2-deoxy-D-glucose per molecule.[497] Monospecific antiserum to the human opsonic α_2-serum-binding glycoprotein and to plasma fibronectin demonstrated antigenic identity of the two macromolecules.[498]

Human thyroxine-binding globulin is a glycoprotein containing four carbohydrate side-chains.[499] A homogeneous glycopeptide has been isolated after proteolytic digestion of the glycoprotein, and two other oligosaccharides (A and B) were recovered after alkaline elimination of a mixture of glycopeptides. The structure (46) of the purified glycopeptide was established by periodate oxidation, methylation analysis, and sequential enzymic degradation techniques.[500] The A and B oligosaccharides (47) contain 2-acetamido-2-deoxy-D-glucitol as a consequence of elimination with alkali. Oligosaccharide A has the same structure as (46) except that one half of its branches contain β-D-Glc*p*NAc-(1 → 6)-D-Man*p* rather than β-D-Glc*p*NAc-(1 → 2) D-Man*p*, and oligosaccharide B which also has this modification contains an additional terminal trisaccharide.

Bovine testosterone-binding globulin is a glycoprotein (mol. wt. 8.9×10^4) containing 17% carbohydrate, which includes residues of D-galactose and D-mannose, together with a small amount of neuraminic acid.[501] The levels of carbohydrate in this glycoprotein from different animal species show that the neuraminic acid content decreases in the order human > guinea-pig > bovine whilst the neutral sugar composition varies with guinea-pig < bovine < human.

Human C_{1q}, a subcomponent of the first component of complement, contains six L-asparagine-linked carbohydrate chains, exclusively located at the *C*-terminal end of the molecule.[502] Structural studies indicated that these oligosaccharides are of the typical complex type found in bovine IgG, trans-

[493] C. M. Towler, R. G. Glover, and C. H. W. Thorne, *Clinica Chim Acta*, 1978, **87**, 289.
[494] G. D. Virca, J. Travis, P. K. Hall, and R. C. Roberts, *Analyt. Biochem.* 1978, **89**, 274.
[495] P. K. Hall and R. C. Roberts, *Biochem. J.*, 1978, **173**, 27.
[496] J. Svasti and B. H. Bowman, *J. Biol. Chem.*, 1978, **253**, 4188.
[497] F. A. Blumenstock, P. Weber, and T. M. Saba, *J. Biol. Chem.*, 1977, **252**, 7156.
[498] F. A. Blumenstock, T. M. Saba, P. Weber, and R. Laffin, *J. Biol. Chem.*, 1978, **253**, 4287.
[499] A. B. Zinn, J. S. Marshall, and D. M. Carlson, *J. Biol. Chem.*, 1978, **253**, 6761.
[500] A. B. Zinn, J. S. Marshall, and D. M. Carlson, *J. Biol. Chem.*, 1978, **253**, 6768.
[501] Y. Suzuki, E. Itagaki, H. Mori, and T. Hosoya, *J. Biochem.* (*Japan*), 1977, **81**, 1721.
[502] T. Mizuochi, K. Yonemasu, K. Yamashita, and A. Kobata, *J. Biol. Chem.*, 1978, **253**, 7404.

```
α-NeufNAc-(2 → 6)-β-D-Galp-(1 → 4)-β-D-GlcpNAc-(1 → 2)-α-D-Manp
                                                            1
                                                            ↓
                                                            3
                                                     β-D-Manp-(1 → 4)-β-D-GlcpNAc-(1 → 4)-D-GlcpNAc-L-Asn
                                                            6
                                                            ↑
                                                            1
α-NeufNAc-(2 → 6)-β-D-Galp-(1 → 4)-β-D-GlcpNAc-(1 → 2)-α-D-Manp
```

(46)

```
α-NeufNAc-(2 → 6)-β-D-Galp-(1 → 4)-β-D-GlcpNAc-(1 → 2/6)-α-D-Manp
                                                              1
                                                              ↓
                                                              3
                                                       β-D-Manp-(1 → 4)-β-D-GlcpNAc-(1 → 4)-D-GlcNAc-ol
                                                              6
                                                              ↑
                                                              1
α-NeufNAc-(2 → 6)-β-D-Galp-(1 → 4)-β-D-GlcpNAc-(1 → 2/6)-α-D-Manp
                                                              4
                                                              ↑
                                                              1
                [α-NeufNAc-(2 → 3)-β-D-Galp-(1 → 4)-β-D-GlcpNAc]n
```

A, $n = 0$; B, $n = 1$

(47)

ferrin, and lactotransferrin, although there are subtle structural differences among the oligosaccharide chains of each glycoprotein.

A purified rabbit liver receptor for D-galactosyl residues which are located in non-reducing terminal positions of glycoproteins binds to asialo-human Factor VIII/von Willebrand Factor *via* the non-reducing terminal sequence, 2-acetamido-2-deoxy-4-*O*-β-D-galactopyranosyl-D-glucose.[503] The D-galactosyl residues can be selectively removed with β-D-galactosidase, resulting in the loss of binding to the receptor, a loss of ristocetin-induced platelet aggregating activity, but no change in procoagulant activity.

The role of Ca^{2+} in bovine factor IX structure and activation has been examined.[504] Although several cations effectively interact with factor IX, there appears to be a rigid specificity for the cation in the activation of this pro-enzyme by the intrinsic coagulation system. The lack of co-operative interactions in the binding of Ca^{2+} to factor IX is in contrast to the presence of these interactions in Ca^{2+} binding to other vitamin K-dependent coagulation pro-enzymes.

Human antithrombin III contains residues of D-galactose, D-mannose, 2-acetamido-2-deoxy-D-glucose, and neuraminic acid.[505] The neuraminic acid residues do not appear to be required for the inhibitory action of this glycoprotein on thrombin, and therefore their presence in the glycoprotein may delay the clearance of the glycoprotein from the bloodstream, as has been shown for ceruloplasmin and α_1-acid glycoprotein.

Wheat germ agglutinin has been used in the partial purification of erythropoietin from human serum and urine.[506]

Absorption, c.d., and electron paramagnetic resonance studies have indicated different reduction rates for the two type I Cu^{2+} species in human ceruloplasmin when this metalloprotein is subjected to laser radiations.[507]

The purification and partial characterization of a human plasma glycoprotein (mol. wt. 7×10^4) has been reported.[508]

The previously-stated existence of four carbohydrate chains in dog prothrombin has been challenged.[509] Although four glycopeptides have been isolated by use of a combination of gel filtration and ion-exchange chromatography, it is proposed that the differences are due to microheterogeneity of the glycoprotein.

Purified rabbit serotransferrin has been shown to contain one glycan chain per molecule, in contrast to human serotransferrin which contains two such chains.[510] Heterogeneity of the glycan moiety of the rabbit glycoprotein was observed from the isolation of a disialyl- and a monosialyl-glycopeptide. The

503 J. M. Sodetz, J. C. Paulson, S. V. Pizzo, and P. A. McKee, *J. Biol. Chem.*, 1978, **253**, 7202.
504 G. W. Amphlett, R. Byrne, and F. J. Castellino, *J. Biol. Chem.*, 1978, **253**, 6774.
505 I. Danishesky, A. Zweben, and B. L. Slomiany, *J. Biol. Chem.*, 1978, **253**, 32.
506 J. L. Spivak, D. Small, J. H. Shaper, and M. D. Hollenberg, *Blood*, 1978, **52**, 1178.
507 M. Herve, A. Garnier, L. Tosi, and M. Steinbuch, *Biochem. Biophys. Res. Comm.*, 1978, **80**, 797.
508 K. Schmid, P. F. Garvin, and S. K. Y. Mao, *Experientia*, 1978, **34**, 826.
509 V. N. Timberbaev, G. S. Chenborisova, and D. M. Zubairov, *Biochemistry* (*U.S.S.R.*) 1977, **42**, 649.
510 D. Legler, V. Tordera, G. Spik, L. Dorland, J. Haverkamp, and J. F. G. Vliegenthart, *F.E.B.S. Letters*, 1978, **93**, 255.

construction of molecular models for the human serotransferrin glycan chains shows that they present one compact section linked to the protein in which the pentasaccharide α-D-Man*p*-(1 → 3)-α-D-Man*p*-(1 → 6)- β-D-Man*p*-(1 → 4)-β-D-Glc*p*NAc-(1 → 4)-β-D-Glc*p*NAc-(1 → L-Asn is substituted by two trisaccharides each having the sequence α-Neu*N*Ac-(2 → 6)-β-D-Gal*p*-(1 → 4)-β-D-Glc*p*NAc.[511] The trisaccharide attached to L-asparagine adopts a flat and rigid conformation stabilized by hydrogen bonds whilst the two neuraminic acid-containing oligosaccharides take up helical configurations attached to the pentasaccharide in a Y-shaped or T-shaped configuration.

Some glycopeptides from rabbit serotransferrin have been isolated and characterized.[512]

Two oligosaccharide side-chains (48, 49) have been isolated from apolipoprotein C-III of human plasma low density lipoprotein by treatment with alkaline borohydride.[513]

α-Neu*N*Ac-(2 → 3)-β-D-Gal*p*-(1 → 3)-D-Gal*N*Ac-ol

(48)

α-Neu*N*Ac-(2 → 3)-β-D-Gal*p*-(1 → 3)-D-Gal*N*Ac-ol
6
↑
2
α-Neu*N*Ac

(49)

13 Immunoglobulins

In a review of the dimensions and specificities of recognition sites on lectins and antibodies, a comparison has been made of the specificity of the combining sites of a number of lectins and antibodies with emphasis on those reacting with blood group determinants.[80]

A method has been developed for the pre-treatment of saliva samples to eliminate mucins, thus simplifying and increasing the accuracy of assays for both lysozymes and immunoglobulins.[514]

The elimination of interspecies reactive anti-IgG antibodies has been achieved by chromatography of anti-IgG sera on immobilized heterologous IgG.[515] Using a combination of chromatography on immobilized concanavalin A followed by gel chromatography, IgG has been separated from serum albumin,[516]

Hepatitis *e* antigen has been purified by chromatography on protein A conjugated to Sepharose® CL-4B.[517] The antigen (mol. wt. 3.24×10^5) is probably a dimer of IgG.

511 J. Montreuil, B. Fournet, G. Spik, and G. Strecker, *Comp. rend.*, 1978, **287D**, 837.
512 D. K. Strickland and B. G. Hudson, *Biochemistry*, 1978, **17**, 3411.
513 P. Vaith, G. Assmann, and G. Uhlenbruck, *Biochim. Biophys. Acta*, 1978, **541**, 234.
514 G. Virella, J. Goudswaard, and R. J. Boackle, *Clinical Chem.*, 1978, **24**, 1421.
515 M. E. Johansson and J. A. Espmark, *J. Immunol. Methods*, 1978, **21**, 285.
516 Z. Hrkal and Z. Vodrážka, *J. Chromatog.*, 1977, **135**, 193.
517 H. A. Fields, D. W. Bradley, C. L. Davis, and J. E. Maynard, *Infection and Immunity*, 1978, **20**, 792.

Rotational correlation times (τ values) have been recorded for human and rabbit IgG and their F_c fragments which are spin-labelled in the carbohydrate portion.[518] The data obtained have provided evidence for the existence of an internal lability of the F_c fragment and for rigid fixation of its carbohydrate component in solution.

The e.s.r. spectra of spin-labelled human, rat, and rabbit immunoglobulins illustrate marked freedom of rotation of the bound spin-label.[519] No changes in the spectra were observed on the addition of antigen to the spin-labelled anti-IgG antibodies. Similar results have been obtained with the labelling of oligosaccharides of rabbit IgG.[520]

A three-dimensional model for human IgM, based on small angle *X*-ray scattering results, has been reported.[521]

The complement-activating capacity of the F_c fragment of human IgG increases after selective reduction and carboxymethylation of inter-chain disulphide bonds.[522] Differences observed in the c.d. spectra of the intact and modified fragment suggest that minor conformational differences exist between them.

A partial amino-acid sequence has been reported for the immunoglobulin light-chain isolated from purified plasma membranes of mouse myeloma cells.[523]

A purified human monoclonal IgM (κ) has been characterized as both a cold agglutinin and a cryoglobulin.[524] Removal of neuraminic acid groups does not affect its agglutination properties, but the asialoglycoprotein is then no longer a cryoglobulin. The physical basis for the precipitation of the intact glycoprotein in the cold appears to be the intermolecular immune binding of *N*-acetylneuraminic acid residues.

Anti-(trout immunoglobulin) and anti-(keyhole limpet haemocyanin) sera contain at least two populations of antibodies, one with carbohydrate specificities responsible for cross-reactivity between trout immunoglobulin and keyhole limpet haemacyanin, and another directed towards polypeptide chain determinants.[525]

Anti-carbohydrate antibodies to synthetic glycoconjugates have been elicited in rabbits.[526] A radio-immunoassay procedure was developed to define the specificities of the antibodies towards different carbohydrates.

Thermodynamic parameters for the binding of dansyl hydrazones of oligosaccharides containing 1,2-, 1,3-, 1,4-, and/or 1,6-linked α-D-glucosyl residues

[518] V. P. Timofeev, I. V. Dudich, Y. K. Sykulev, and R. S. Nezlin, *F.E.B.S. Letters*, 1978, **89**, 191.
[519] R. S. Nezlin, V. P. Timofeev, Y. K. Sykulev, and S. E. Zurabyan, *Immunochemistry*, 1978, **15**, 143.
[520] K. J. Willan, B. Golding, D. Givol, and R. A. Dwek, *F.E.B.S. Letters*, 1977, **80**, 133.
[521] P. Wilhelm, I. Pilz, W. Palm, and K. Bauer, *European J. Biochem.*, 1978, **84**, 457.
[522] J. A. L. de Castro, F. Vivanco, and F. Ortiz, *Biochem. Biophys. Res. Comm.*, 1977, **78**, 1319.
[523] O. Wolf, R. Zemell, Y. Burnstein, and I. Schechter, *Biochem. Biophys. Res. Comm.*, 1977, **78**, 1383.
[524] C. M. Tsai, D. A. Zopf, R. K. Yu, R. Wistar, and V. Ginsburg, *Proc. Nat. Acad. Sci., U.S.A.*, 1977, **74**, 4591.
[525] K. M. Yamaga, R. T. Kubo, and H. M. Etlinger, *J. Immunol.*, 1978, **120**, 2068.
[526] D. A. Zopf, C. M. Tsai, and V. Ginsburg, *Arch. Biochem. Biophys.*, 1978, **185**, 61.

to a dextran-specific IgM have been reported.[527] The binding site of the immunoglobulin constitutes a 12 Å deep cavity with sub-sites for oligosaccharides containing a non-reducing nigerosyl group, and a further sub-site located with a more open groove forming interactions of lower specificity.

The binding of dansyl derivatives of D-galacto-oligosaccharides to several immunoglobulins capable of binding (1–6)-β-D-galactans has been reported.[528] Affinity constants which were obtained for dansylated ligands and proteins by fluorescence titration have to be interpreted with care as the derivatizing group can cause additional binding.

The amino-acid sequence of the first 111 residues of the *N*-terminal (variable region) of the light chain of homogeneous antipneumococcal Type III polysaccharide antibody 3368 has been determined.[529] Compared with the sequences of light chains of four other antipneumococcal type III polysacharide antibodies, the 3368 light chain has a unique sequence in those segments of the variable region that contribute to the formation of the antigen binding site.

Ground squirrel IgG contains total hexose (1.8%) and amino-sugar (1.3%) in higher amounts than those found in rabbit and human IgG.[530]

Although thymocytes from rainbow trout react strongly with anti-(trout IgG) antibodies, the reaction appears to be due primarily to cross-reacting specificities for carbohydrate-containing residues of the structure rather than to the IgG-specific antigenic determinants.[531] IgA of human gastric mucosal scrapings has been separated from the mucoproteins by isopycnic density gradient centrifugation, indicating an absence of covalent binding between these two classes of macromolecules.[532]

Optimal conditions for the quantitation of incorporation of [^{3}H]leucine into immunoglobulin by bone marrow cells of patients with multiple myeloma has been reported.[533]

The biosynthesis of the carbohydrate portion of the heavy chains of IgG secreted by LPC1 mouse plasmacytoma cells involves the transfer *en bloc* of a high molecular weight oligosaccharide containing a high proportion of D-mannosyl residues from a lipid carrier to the heavy chain.[59] Within a few minutes of transfer, processing of the oligosaccharide occurs by enzymic removal of an unspecified number of glycosyl residues to leave an oligosaccharide chain of $(\text{Hexose})_7(\text{D-Glc}p\text{NAc})_2$.

D-Mannose and 2-amino-2-deoxy-D-glucose residues become attached to nascent heavy chains of immunoglobulins soon after the site for glycosylation is biosynthesized and therefore prior to the completion of translation.[534] After completion of the polypeptide chain and its release from the ribosome, sequential addition of residues of 2-amino-2-deoxy-D-glucose and D-galactose

[527] G. Schepers, Y. Blatt, K. Himmelspach, and I. Pecht, *Biochemistry*, 1978, **17**, 2239.
[528] A. K. Bhattacharjee and C. P. J. Glaudemans, *Carbohydrate Res.*, 1978, **67**, 536.
[529] L. E. Cannon, M. N. Margolies, and E. Haber, *Biochemistry*, 1978, **17**, 3101.
[530] D. L. Bruce and P. Murtagh, *Immunochemistry*, 1978, **15**, 193.
[531] K. M. Yamaga, R. T. Kubo, and H. M. Etlinger, *J. Immunol.*, 1978, **120**, 2074.
[532] M. Spohn and I. McColl, *Biochem. Biophys. Res. Comm.*, 1977, **79**, 837.
[533] K. Hammerton, D. A. Cooper, M. Duckett, and R. Penny, *J. Immunol.*, 1978, **121**, 409.
[534] L. W. Bergman and W. M. Kuehl, *Biochemistry*, 1977, **16**, 4490.

occurs by the mediation of specific transferases in the smooth membrane. Terminal sugars are added at the time of secretion.

The initial glycosylation of a plastocytoma cell line immunoglobulin heavy chain (γ_{2b}) occurs *in vivo* when the nascent heavy chain reaches a molecular weight of approximately 3.8×10^4.[535] Non-glycosylated, completed heavy chains cannot be glycosylated in these cells. In a different cell line which synthesizes a κ light chain containing an oligosaccharide linked to L-asparagine, the majority of the light chains are glycosylated prior to both chain completion and release from the polysomes. The light chains can be glycosylated only if they are present in a monomeric form, and not in a covalent dimeric form.

14 Erythrocyte Glycoproteins

Sodium deoxycholate dissociates erythrocyte membrane sialoglycoproteins more efficiently than does sodium dodecyl sulphate.[536] Components which are dissociated in sodium deoxycholate aggregate in the presence of sodium dodecyl sulphate.

The physiological mechanism for the elimination of mammalian asialoerythrocytes from the circulation has been shown to be different from that for plasma asialoglycoproteins.[537] Treatment of chicken asialoerythrocytes with D-galactose oxidase results in the oxidation of D-galactosyl or 2-acetamido-2-deoxy-D-galactosyl residues, thereby rendering the erythrocytes nonviable.

A variant of the major transmembrane glycoprotein of the human erythrocyte has been detected following proteolytic cleavage of intact erythrocytes.[538] The glycoprotein has a molecular weight of 6.3×10^4, as opposed to 6.0×10^4 for the normal transmembrane glycoprotein; this is due to the incorporation of an additional segment of peptide and not to increased glycosylation.

Proteolytic digestion of lipid-free human erythrocytes releases 'erythroglycan', a group of glycopeptides (mol. wt. 7×10^3—1.1×10^4) whose carbohydrate moieties comprise chains of (1 → 3)-D-Gal*p*-(1 → 4)-D-Glc*pN*Ac-1 → units.[539] The chains are branched at 50% of the D-galactosyl residues. The glycopeptides themselves may be protein-bound carriers associated with the acroglycolipids which carry blood ABO determinants.[540] The glycopeptides contain 20—70 sugar residues in an alkali-stable chain linked through 2-acetamido-2-deoxy-D-glucose to L-asparagine.[541] In addition to the presence of branch points at C-6 of the D-galactosyl residues, the saccharide chains are terminated in 2-acetamido-2-deoxy-D-glucosyl-, D-galactosyl-, α-L-fucosyl-(1 → 2)-D-galactosyl-, and α-*N*-acetylneuraminyl-(2 → 3)- and -(2 → 6)-D-galactosyl residues. Blood group A and B determinants are present.

[535] L. W. Bergman and W. M. Kuehl, *Biochemistry*, 1978, **17**, 5174.
[536] L. Liljas, *Biochim. Biophys. Acta*, 1978, **532**, 347.
[537] W. C. Bell, G. N. Levy, R. Williams, and D. Aminoff, *Proc. Nat., Acad. Sci., U.S.A.*, 1977, **74**, 4205.
[538] T. J. Mueller and M. Morrison, *J. Biol. Chem.*, 1977, **252**, 6573.
[539] J. Järnfeldt, J. Rush, Y. T. Li, and R. A. Laine, *J. Biol. Chem.*, 1978, **253**, 8006.
[540] M. Dejter-Juszynski, M. Harpaz, H. M. Flowers, and N. Sharon, *European J. Biochem.*, 1978, **83**, 363.
[541] T. Krusius, J. Finne, and H. Rauvala, *European J. Biochem.*, 1978, **92**, 289.

Compared with normal erythrocytes, type M^k and Miltenberger class V erythrocytes are associated with the loss of the major sialoglycoprotein, whereas in M^g erythrocytes there is a modification in the major sialoglycoprotein.[542] The M^g sialoglycoprotein appears to contain fewer sialotetrasaccharides, and this may result from a change in the amino-acid sequence of the protein.

A new chromatographic procedure has been used in the estimation of glycosylated haemoglobin levels.[543] Both the minor phosphorylated and glucosylated haemoglobin components can be identified, in normal and diabetic individuals, providing information on the mechanism for the elevation of the glucosylated products in patients with diabetes mellitus.[544, 545] The glycosylation of the minor haemoglobins is not specific for D-glucose.[546] Incorporation, which is always into the β-chains of the protein, may also be accomplished by D-galactose, D-mannose, and D-glucose 6-phosphate.

15 Salivary, Mucous, and other Mammalian Body Fluid Glycoproteins

A review has described the viscoelastic properties of mucous secretions. Their structural variability may indicate the presence of distinct glycoprotein populations rather than a continuous series of related structures.[547]

Human parotid saliva has been purified using gel- and ion-exchange-chromatographic procedures, to give a family of glycoproteins with a very unusual amino-acid composition, little or no absorbance at 280 nm, and unusual staining characteristics with Coomassie Blue.[548] Differences in charge, size, and total carbohydrate and neuraminic acid contents were noted.

Purified human and monkey glycoproteins containing terminal neuraminic acid residues agglutinate with *Streptococcus sanguis* and *S. mutans* cells.[549] Removal of the sialic acid residues from these glycoproteins results in the loss of agglutination of *S. sanguis* but not *S. mutans*.

Examination of the occurrence of neuraminic acid moieties in glycoproteins from pig submandibular and rat sublingual glands indicates that one in four neuraminic acid residues in the pig glycoprotein occurs as an α-*N*-acylneuraminyl-(2 → 8)-*N*-acylneuraminyl unit.[550] The disaccharide is absent in the corresponding rat glycoprotein.

Alkaline borohydride reductive cleavage of the major glycoprotein fraction of rat sublingual gland results in the release of a series of reduced oligo-

[542] D. J. Anstee and M. J. A. Tanner, *Biochem. J.*, 1978, **175**, 149.
[543] J. Chou, C. A. Robinson, and A. L. Siegel, *Clinical Chem.*, 1978, **24**, 1708.
[544] M. J. McDonald, R. Shapiro, M. Bleichman, J. Solway, and H. F. Bunn, *J. Biol. Chem.*, 1978, **253**, 2327.
[545] H. F. Bunn, K. H. Gabbay, and P. M. Gallop, *Science*, 1978, **200**, 21.
[546] R. Dolhofer and O. H. Wieland, *F.E.B.S. Letters*, 1978, **85**, 86.
[547] J. T. Galagher and A. P. Corfield, *Trends in Biochem. Sciences*, 1978, **3**, 38.
[548] R. L. Henkin, R. E. Lippoldt, J. Bilstad, R. O. Wolf, C. K. L. Lum, and H. Edelhoch, *J. Biol. Chem.*, 1978, **253**, 7556.
[549] M. J. Levine, M. C. Herzberg, M. S. Levine, S. A. Ellison, M. W. Stinson, H. C. Li, and T. Van Dyke, *Infection and Immunity*, 1978, **19**, 107.
[550] B. L. Slomiany, A. Slomiany, and A. Herp, *European J. Biochem.*, 1978, **90**, 255.

saccharides.[551, 552] Five of these neuraminic acid-containing oligosaccharides which have 9, 10, 12, 13, and 15 sugar residues have been characterized. Two types of carbohydrate chains were identified, one type (50) having 2-acetamido-2-deoxy-D-glucose as the non-reducing terminal group and the other (51) having the terminal sequence α-Neu*N*Ac-(2 → 4)-D-Glc*pN*Ac.

A cellular, non-secretory glycoprotein (mol. wt. 2.8×10^4) has been isolated from mouse submandibular glands and partially characterized.[553] A secretory glycoprotein, containing 17% carbohydrate and having 2-amino-2-deoxy-D-glucose as the major carbohydrate component, was isolated from the same source and partially characterized.[554]

A glycoprotein isolated from the submandibular glands of the armadillo gives anomalously high levels of protein when measured by the Lowry method, although its amino-acid composition indicates an absence of chromogenic amino-acids.[555] After desialylation and Smith degradation of the glycoprotein, the Lowry protein value increased by a further 13 and 30% respectively, suggesting that both neuraminic acid and 2-acetamido-2-deoxy-D-hexose residues exert a shielding effect on the reaction.

Treatment of ovine submaxillary mucin with neuraminidase and proteolytic enzymes, releases glycopeptides containing sequences of *O*-α-D-Gal*pN*Ac-L-Ser-L-Glx-L-Pro-Gly-, *O*-α-D-Gal*pN*Ac-L-Ser-, *O*-α-D-Gal*pN*Ac-L-Ser-Gly-Gly-(*O*-α-D-Gal*pN*Ac)-L-Thr-L-Glx-, and Gly-(*O*-α-D-Gal*pN*Ac)-L-Ser-L-Ala-.[556]

Two families of antibodies have been found to be generated after immunization of rabbits with ovine submaxillary mucin.[557] In the case of subcutaneous immunization, the immunodominant determinants are the prosthetic *N*-acetylneuraminosyl-2-acetamido-2-deoxy-D-galactopyranosyl and the 2-acetamido-2-deoxy-D-galactopyranosyl groups, but intravenous immunization of the glycoprotein coupled to red blood cells elicits, in addition, antibodies with specificities for the apoprotein.

Three homogeneous glycoproteins have been isolated from reduced and *S*-carboxymethylated canine tracheal pouch mucous.[558] Alkaline borohydride reduction indicates that the majority of carbohydrate chains are linked to the protein core by *O*-glycosidic bonds involving 2-acetamido-2-deoxy-D-galactose residues and either L-serine or L-threonine.

Products isolated chromatographically from the proteolytic digest of a glycoprotein from bovine cervical mucous contained peptide regions, together with two different sizes of glycopeptide subunit regions.[559]

Alkaline borohydride degradation of rat intestinal glycoprotein releases

[551] B. L. Slomiany and A. Slomiany, *Biochem. Biophys. Res. Comm.*, 1978, **81**, 391.
[552] A. Slomiany and B. L. Slomiany, *J. Biol. Chem.*, 1978, **253**, 7301.
[553] A. V. Nieuw-Amerongen, A. P. Vreugdenhil, and P. A. Roukema, *Biochim. Biophys. Acta*, 1977, **495**, 324.
[554] A. V. Nieuw-Amerongen, M. E. G. Aarsman, A. P. Vreugdenhil, and P. A. Roukema, *Biochim. Biophys. Acta*, 1978, **534**, 26.
[555] A. M. Wu, J. C. Wu and A. Herp, *Biochem. J.*, 1978, **175**, 47.
[556] M. Yamamoto and Z. Yosizawa, *J. Biochem.* (*Japan*), 1978, **83**, 1159.
[557] J. K. Roche, E. D. Day, and H. D. Hill, *Immunochemistry*, 1978, **15**, 339.
[558] G. P. Sachdev, O. F. Fox, G. Wen, T. Schroeder, R. C. Elkins, and R. Carubelli, *Biochim. Biophys. Acta*, 1978, **536**, 184.
[559] G. P. Roberts, *Biochem. J.*, 1978, **173**, 941.

β-D-Glc*pN*Ac-(1 → [3)-β-D-Gal*p*-(1 → 4)-β-D-Glc*pN*Ac-(1]$_n$ → 3)-β-D-Gal*N*Ac-ol
6 ↑ 2 α-Neu*N*Ac (on Gal*p*); 6 ↑ 2 α-Neu*N*Ac (on Gal*N*Ac-ol)

n = 2, 3, or 4
(50)

α-Neu*N*Ac-(2 → 4)-β-D-Glc*pN*Ac-(1 → [3)-β-D-Gal*p*-(1 → 4)-β-D-Glc*pN*Ac-(1]$_n$ → 3)-β-D-Gal*N*Ac-ol
6 ↑ 2 α-Neu*N*Ac (on Gal*p*); 6 ↑ 2 α-Neu*N*Ac (on Gal*N*Ac-ol)

n = 2 or 3
(51)

three oligosaccharide alditols, whose structures (52—54) were established by methylation analysis and ^{1}H n.m.r. spectroscopy.[560] Although the native

β-D-Glc*p*NAc-(1 → 3)-D-Gal*p*NAc-ol

(52)

α-L-Fuc*p*-(1 → 2)-β-D-Gal*p*-(1 → 3)-D-Gal*p*NAc-ol

(53)

α-D-Gal*p*NAc-(1 → 3)-α-L-Fuc*p*-(1 → 2)-β-D-Gal*p*-(1 → 3)-D-Gal*p*NAc-ol

(54)

glycoprotein from pig colonic mucous has a much greater molecular weight (1.5×10^7) than that of the corresponding pig gastric glycoprotein, the two glycoproteins are similar in that they both yield relatively low molecular weight subunits after proteolytic digestion or treatment with mercaptoethanol.[561]

Glycoprotein synthesis, as measured by the incorporation of 2-amino-2-deoxy-D-[^{3}H]glucose into colonic mucosal explants, is significantly decreased in colon cancers compared with that in normal colonic mucosa from the same patient.[562]

Compared with other mucous glycoproteins, a glycoprotein (mol. wt. 9×10^5) derived from rat colonic mucosa has an unusually long average carbohydrate chain length of twenty-two sugar residues.[563]

16 Urinary Glycoproteins

The urine of patients with fucosidosis contains three major L-fucosyl-oligosaccharides each linked to L-asparagine.[564] Their structures are related to other L-fucosyl glycoproteins such as IgG, lactotransferrin, and α_1-acid glycoprotein. The terminal tetrasaccharide sequence of one of the oligosaccharides (55) is unusual for a carbohydrate moiety in a glycoprotein. Three structurally related components, *i.e.* a disaccharide and two L-asparagine-linked oligosaccharides, have been characterized from the urine of patients with fucosidosis type 2.[565] The origin of these metabolites is probably the core region of the carbohydrate chains of glycoproteins.

Mucolipidosis I has been shown to be associated with the excretion of *O*- and *N*-glycopeptides,[566] a finding which is in agreement with a report on a neuraminidase deficiency for this disease.[567] 360 MHz ^{1}H N.m.r. spectra have been recorded for ten urinary sialyl-oligosaccharides isolated from patients with mucolipidosis I and II,[568] 3′-sialyl-lactose has been isolated from normal

560 H. E. Carlsson, G. Sundblad, S. Hammarström, and J. Lönngren, *Carbohydrate Res.*, 1978, **64**, 181.
561 T. Marshall and A. Allen, *Biochem. J.*, 1978, **173**, 569.
562 T. A. O'Gorman, and J. T. La Mont, *Cancer Res.*, 1978, **38**, 2784.
563 V. L. N. Murty, F. J. Downs, and W. Pigman, *Carbohydrate Res.*, 1978, **61**, 139.
564 G. Strecker, B. Fournet, J. Montreuil, L. Dorland, J. Haverkamp, J. F. G. Vliegenthart, and D. Dubesset, *Biochimie*, 1978, **60**, 725.
565 A. Lundslad, J. Lundsten, N. E. Nordén, S. Sjöblad, S. Svensson, P. A. Öckermann, and M. Gehlhoff, *European J. Biochem.*, 1978, **83**, 513.
566 M. Lemonnier, D. Lecat, R. Bourrillon, and P. Maroteaux, *Compt. rend.*, 1978, **287**D, 353.
567 G. Strecker and J. C. Michalski, *F.E.B.S. Letters*, 1978, **85**, 20.
568 L. Dorland, J. Haverkamp, J. F. G. Vliegenthart, G. Strecker, J. C. Michalski, B. Fournet, G. Spik, and J. Montreuil, *European J. Biochem.*, 1978, **87**, 323.

β-D-Gal*p*-(1 → 4)-α-L-Fuc*p*-(1 → 3)-β-D-Glc*pN*Ac-(1 → 2)-α-D-Man*p*-(1 → 6)-β-D-Man*p*-(1 → 4)-β-D-Glc*pN*Ac-(1 → 4)-α-L-Fuc*p*-(1 → 6)-β-D-Glc*pN*Ac-L-Asn

(55)

α-D-Glc*p*-(1 → 6)-α-D-Glc*p*-(1 → 6)-α-D-Glc*p*-(1 → 4)-α-D-Glc*p*-(1 → 4)-α-D-Glc*p*-(1 → 4)-D-Glc

(56)

α-D-Glc*p*-(1 → 4)-α-D-Glc*p*-(1 → 6)-α-D-Glc*p*-(1 → 6)-α-D-Glc*p*-(1 → 4)-α-D-Glc*p*-(1 → 4)-D-Glc

(57)

α-D-Glc*p*-(1 → 6)-α-D-Glc*p*-(1 → 4)-α-D-Glc*p*-(1 → 4)-α-D-Glc*p*-(1 → 4)-α-D-Glc*p*-(1 → 6)-α-D-Glc*p*-(1 → 4)-α-D-Glc*p*-(1 → 4)-α-D-Glc

(58)

human urine by adsorption on charcoal, followed by gel permeation- and ion-exchange chromatography.[569]

Two serologically active urinary glycoproteins have been isolated from the urine of a patient suffering from tubular proteinuria.[570] One of the glycoproteins had an *N*-terminal amino-acid sequence with many similarities to that of another urinary glycoprotein (HC), in addition to analogous antigenic determinants.

A homogeneous glycoprotein (mol. wt. 2.7×10^4) has been recovered in significant concentrations from the urine of patients with plasma cell leukaemia.[571]

Patients with glycogen storage disease type II have been shown to excrete three urinary oligosaccharides (56—58), and patients with type III disease excrete an additional branched heptasaccharide.[572] All four of these oligosaccharides are considered to originate from glycogen. Structural analysis suggests that the number of (1 → 6)-α-D-linkages in the oligosaccharides is higher than expected and this may indicate the existence of a shorter distance between branches in glycogen than is generally assumed.

17 Miscellaneous Glycoproteins and Chitin

Additional heterogeneity has been described amongst the asparaginyl-carbohydrates from ovalbumin.[573] Digestion of fractions with *exo*-glycosidases is discussed in relation to the use of these glycopeptides for studies of glycosyltransferase enzymes.

The structures of glycopeptides I (59) and II from ovalbumin have been established.[574] Glycopeptide II is now recognized as being composed of two glycopeptides, IIA (60) and IIB (61).

The complete structures of the carbohydrate units of three other glycopeptides from ovalbumin (62—64) have been established, using a combination of sequential digestion with glycosidases, Smith degradation, selective acetolysis, and methylation analysis.[575]

Glycopeptide III from ovalbumin is recognized as being a mixture of three glycopeptides.[576] Subtle differences in their structures have been identified by hydrolysis with *endo*-β-D-2-acetamido-2-deoxyglucanases. L-Asparaginyl glycopeptides having a low content of D-mannosyl residues are hydrolysed by an *endo*-β-D-2-acetamido-2-deoxyglucanase H.[577]

Ovomucin has been separated into two fractions (A and B) having high and

[569] M. Koseki and K. Tsurumi, *J. Biochem.* (*Japan*), 1977, **82**, 1785.
[570] A. Dautigny, I. Bernier, J. Jollès, J. Colombani, and P. Jollès, *Comp. rend.*, 1977, **285**D, 1535.
[571] B. K. Seon and D. Pressman, *Biochemistry*, 1978, **17**, 2815.
[572] G. Lennartson, A. Lundblad, J. Lundsten, S. Svensson, and A. Häger, *European J. Biochem.*, 1978, **83**, 325.
[573] V. Shepherd and R. Montgomery, *Carbohydrate Res.*, 1978, **61**, 147.
[574] K. Yamashita, Y. Tachibana, and A. Kobata, *J. Biol. Chem.*, 1978, **253**, 3862.
[575] J. Conchie and I. Strachan, *Carbohydrate Res.*, 1978, **63**, 193.
[576] T. Tai, K. Yamashita, S. Ito, and A. Kobata, *J. Biol. Chem.*, 1977, **252**, 6687.
[577] R. B. Trimble, A. L. Tarentino, T. H. Plummer, and F. Maley, *J. Biol. Chem.*, 1978, **253**, 4508.

```
      α-D-Manf                     R¹
           1                        1
           ↓                        ↓
           3                        4
R³-(1 → 6)-α-D-Manp-(1 → 6)-β-D-Manp-(1 → 4)-β-D-GlcpNAc-(1 → 4)-β-D-GlcpNAc-Asn
           3                        3
         ↗                          ↑
       1                            1
  α-D-Manp    R⁴-(1 → 4)-α-D-Manp
                                    2
                                    ↑
                                    1
                                    R₂
```

(59) R^1 = β-D-Glc*p*NAc, R^2 = β-D-Glc*p*NAc, R^3 = α-D-Man*p*, R^4 = β-D-Gal*p*-(1 → 4)-β-D-Glc*p*NAc.

(60) R^1 = β-D-Glc*p*NAc, R^2 = β-D-Glc*p*NAc, R^3 = H, R^4 = β-D-Gal*p*-(1 → 4)-β-D-Glc*p*NAc.

(61) R^1 = β-D-Glc*p*NAc, R^2 = β-D-Glc*p*NAc, R^3 = α-D-Man*p*, R^4 = β-D-Glc*p*NAc.

(62) R^1 = R^2 = H, R^3 = α-D-Man*p*, R^4 = H.

(63) R^1 = α-D-Man*p*, R^2 = H, R^3 = α-D-Man*p*, R^4 = H.

(64) R^1 = α-D-Man*p*, R^2 = β-D-Glc*p*NAc, R^3 = α-D-Man*p*, R^4 = H.

low levels respectively of carbohydrate.[578] Fraction A corresponds to β-ovomucin, containing carbohydrate (53%) made up of residues of D-galactose, 2-acetamido-2-deoxy-D-glucose, and 2-acetamido-2-deoxy-D-galactose, while in fraction B, which contains 15% carbohydrate, residues of 2-acetamido-2-deoxy-D-galactose are replaced by D-mannose.

Two sulphated oligosaccharides have been isolated from ovomucin by alkaline borohydride treatment of glycopeptides.[579] The structure (65) of one of these components has been reported.[580]

$$\text{Neu}f\text{NAc-}(2 \rightarrow 3)\text{-D-Gal}p\text{-}(1 \rightarrow 3)\text{-D-Gal}N\text{Ac-ol 6-SO}_4$$

(65)

Methylation analysis and periodate oxidation studies have confirmed earlier structural concepts of mammalian lung D-galactan.[581]

Homogeneous D-galactan sulphates have been isolated from the testes of the tunicates *Halocynthia roretzi* and *Styela plicata*.[582] The results of optical rotation, i.r.spectroscopy, and acid hydrolysis studies suggest that the polysaccharides have repeating sequences of α-L-galactopyranosyl 4-sulphate and α-L-galactopyranosyl 6(or 3)-sulphate.

A multi-volume treatise on the glycoconjugates has been published. Volume 1 deals with methods of analysis and structural determination, mucous glycoproteins, blood glycoproteins, and glycoproteins of other specific systems.[583] In Volume 2 appear chapters on proteoglycans and structural glycoproteins, glycoprotein catabolism and immunological aspects of lectins.[584]

Aspects of the chemistry and biochemistry of glycopeptides have been surveyed.[585]

Chitin has been isolated in a highly *N*-acetylated form by treatment of *Sordaria* and *Ascaris* species with hydrogen peroxide–H_4 e.d.t.a. solutions and ozone.[586]

The action of lysozyme on the hydrolysis of chitin and chito-oligosaccharides which contain 2-amino-2-deoxy-D-glucosyl residues has been reported.[587]

A rapid and sensitive assay for chitinase using [^{3}H]chitin has been developed.[588]

X-ray diffraction studies have confirmed the antiparallel chain structure for α-chitin.[589] The structure contains two types of amide groups which differ in

578 A. Kato, K. Ogino, N. Matsudomi, and K. Kobayashi, *Agric. and Biol. Chem. (Japan)*, 1977, **41**, 1925.
579 A. Kato, S. Hirata, H. Sato, and K. Kobayashi, *Agric. and Biol. Chem. (Japan)*, 1978, **42**, 835.
580 A. Kato, S. Hirata, and K. Kobayashi, *Agric. and Biol. Chem. (Japan)*, 1978, **42**, 1025.
581 N. Roy, and C. P. J. Glaudemans, *Carbohydrate Res.*, 1978, **63**, 318.
582 N. Toda, T. Horikawa, K. Anno, and N. Seno, *Carbohydrate Res.*, 1978, **62**, 389.
583 'The Glycoconjugates', Vol. 1, ed. M. I. Horowitz and W. Pigman, Academic Press, New York, 1977.
584 'The Glycoconjugates', Vol. 2, ed. M. I. Horowitz and W. Pigman, Academic Press, New York, 1978.
585 M. Isemura, *Seikagaku*, 1978, **50**, 291.
586 C. J. Hackett and K. C. Chen, *Analyt. Biochem.*, 1978, **89**, 487.
587 K. Amano and E. Ito, *European J. Biochem.*, 1978, **85**, 97.
588 J. Molano, A. Durán, and E. Cabib, *Analyt. Biochem.*, 1977, **83**, 648.
589 R. Minke and J. Blackwell, *J. Mol. Biol.*, 1978, **120**, 167.

their hydrogen bonding and account for the observed splitting of the amide I band in the i.r. spectrum. The inability of this form of chitin, and not β-chitin, to swell in water is explained by extensive intermolecular hydrogen bonding.

Deacetylation of chitin with alkali under heterogeneous conditions has been proposed to proceed preferentially from the amorphous region to produce block-type co-polymers of 2-acetamido-2-deoxy-D-glucose and 2-amino-2-deoxy-D-glucose.[590] Under homogeneous conditions, the deacetylation proceeds indiscriminately to produce random-type co-polymers of the two units.

Chitin, but not chitosan, exhibits an e.s.r. spectrum due to the presence of the acetamido group.[591]

18 Biosynthesis of Glycoproteins

Sialyl- and L-fucosyl-transferases have been shown to transfer their respective sugars to the terminal β-D-Gal*p*-(1 → 4)-D-Glc*p*NAc sequence of L-asparagine-linked oligosaccharides forming α-Neu*f*NAc-(2 → 6)-β-D-Gal*p*-(1 → 4)-D-Glc*p*NAc- and β-D-Gal*p*-(1 → 4)-[α-L-Fuc*p*-(1 → 3)]-D-Glc*p*NAc- oligosaccharides respectively.[592] When the terminal D-galactosyl residue from asparaginyl-linked oligosaccharides containing the sequence β-D-Gal*p*-(1 → 4)-D-Glc*p*NAc is cleaved by the β-D-galactosidase of *Streptococcus pneumoniae* the resulting oligosaccharides are not substrates for the two glycosyl transferases. Thus the substrate requirements for the three enzymes make them suited to the study of the biosynthesis of the L-asparaginyl-linked oligosaccharides in glycoproteins.

Sialyltransferases are localized in the smooth membranes and the Golgi apparatus of rat brain.[593] The sialyltransferase responsible for the incorporation of *N*-acetylneuraminic acid from CMP-*N*-acetylneuraminic acid into endogenous glycoprotein in rat brain appears to differ from glycolipid sialyltransferase since exogenously added asialofetuin does not compete with both the exogenous and endogenous glycolipid acceptors.[594]

Thyroid rough microsomal preparations catalyse the transfer of D-[^{14}C]-manno-oligosaccharide–lipid to a synthetic dinitrophenyl heptapeptide (DNP-L-Ala-L-Leu-L-Glu-L-Asn-L-Ala-L-Thr-L-Arg) containing the tripeptide sequence L-Asn-L-Ala-L-Thr.[595] It has been anticipated from the structural examination of numerous glycoproteins that this tripeptide sequence is a necessary but not a sufficient requirement for *N*-glycosylation. The oligosaccharide was tentatively identified as being attached *N*-glycosidically to the L-asparaginyl residue. In the presence of thyroid rough microsomes, D-manno-oligosaccharyl-*N*,*N'*-diacetylchitobiosyl units are transferred to protein after being preassembled on lipids.[596]

The requirements of the hen oviduct enzymes involved in the transfer of

590 K. Kurita, T. Sannan, and Y. Iwakura, *Makromol. Chem.*, 1977, **178**, 3197.
591 F. N. Bruscato, M. Holdiness, and A. G. Danti, *J. Polymer. Sci., Polymer Chem.*, 1978, **16**, 1163.
592 J. C. Paulson, J. P. Prieels, L. R. Glasgow, and R. L. Hill, *J. Biol. Chem.*, 1978, **253**, 5617.
593 S. S. Ng and J. A. Dain, *J. Neurochem.*, 1977, **29**, 1085.
594 S. G. Ng and J. A. Dain, *J. Neurochem.*, 1977, **29**, 1075.
595 C. Ronin, C. Granier, J. van Rietschoten, and S. Bouchilloux, *Biochem. Biophys. Res. Comm.*, 1978, **81**, 772.
596 C. Ronin and S. Bouchilloux, *Biochim. Biophys. Acta*, 1978, **539**, 470.

GDP-D-mannose to exogenous acceptor proteins *via* oligosaccharide–lipid complexes have been characterized.[597] A variety of proteins of known amino-acid sequences and which contain one or more of the appropriate L-asparagine glycosylation sites (the tripeptide sequence L-Asn-X-L-Ser/L-Thr), but which do not normally exist in a glycosylated form, have been tested as substrates. Rabbit muscle triosephosphate isomerase and ovine prolactin accept oligosaccharide chains from oligosaccharide–lipid complexes after the tertiary structure of the protein is disrupted. The mixture of polypeptide fragments arising from cyanogen bromide cleavage of catalase and concanavalin A, but not the intact proteins themselves, serve as oligosaccharide acceptors.

19 Analysis of Glycoproteins

The monosaccharide content of glycoproteins containing 6—20% carbohydrates has been determined by high voltage electrophoresis of [^{3}H]-labelled sugar alditols.[598] Methods for the carbohydrate analysis of glycoproteins have been reported.[599–601] Picomolar quantities of monosaccharides in the form of trifluoroacetate derivatives of methyl glycosides are separated and detected by g.l.c. with an electron capture detector.

Recoveries of monosaccharides, amino-sugars, and amino-acids arising from the hydrolysis of glycoproteins have been improved by the addition of glycerol prior to concentration by freeze-drying or rotary evaporation.[602] Neutral and amino-sugars have been separated on an auto-analyser using anion- and cation-exchange resins respectively.[603]

N-Acetylneuraminic acid, released from glycoproteins with acid or neuraminidase, has been estimated by g.c.–m.s. of the derived *O*-trimethylsilyl-ether.[604]

A rapid analytical system sensitive in the nanomole range and based on anion-exchange chromatography has been developed for the determination of amino-acids without interference from neutral sugars or amino-sugars.[605]

Hexitols and aminohexitols, derived from the monosaccharide components of glycoproteins by reduction with sodium borotritide, have been determined after separation using high voltage electrophoresis in borate buffer.[606]

A method based on mass spectrometric sequence study of the oligosaccharides has been described for micro-scale fingerprinting of mixtures of glycopeptides released from glycoproteins after proteolysis.[607] The type of branching in complex L-asparagine-linked glycan chains of glycoproteins has been

597 K. E. Kronquist and W. Lennarz, *J. Supramol. Struct.*, 1978, **8**, 51.
598 M. Tomona, W. Niedermeier, C. Spivey, and J. Gerard, *Microchem. J.*, 1978, **23**, 93.
599 D. G. Pritchard and W. Niedermeier, *J. Chromatog.*, 1978, **152**, 487.
600 M. Tomana, W. Niedermeier, and C. Spivey, *Analyt. Biochem.*, 1978, **89**, 110.
601 M. M. Wrann and C. W. Todd, *J. Chromatog.*, 1978, **147**, 309.
602 R. Dawson and K. Mopper, *Analyt. Biochem.*, 1978, **84**, 186.
603 D. Bladier, G. Perret, J. Pre, and P. Cornillot, *J. Chromatog.*, 1978, **154**, 330.
604 J. Roboz, R. Suzuki, and J. G. Bekesi, *Analyt. Biochem.*, 1978, **87**, 195.
605 R. Dawson and K. Mopper, *Analyt. Biochem.*, 1978, **84**, 191.
606 M. Tomana, W. Niedermeier, C. Spivey, and J. Gerard, *Microchem. J.*, 1978, **23**, 93.
607 K. A. Karlsson, I. Pascher, B. E. Samuelson, J. Finne, T. Krusius, and H. Rauvala, *F.E.B.S. Letters*, 1978, **94**, 413.

studied by 360 MHz ^{1}H n.m.r. spectroscopy.[608] The bi-, tri-, and tetra-antennary structures of asialo- and afuco-glycan chains can be distinguished on the basis of the chemical shifts of protons of the branched D-manno-oligosaccharide core and of the 2-acetamido protons of the 2-acetamido-2-deoxy-D-glucosyl residues.

4-Deoxy-α-L-*threo*-hex-4-enopyranosyluronic acid-containing disaccharides, formed by the action of chondroitin ABC lyase and chondroitin AC lyase have been determined by a periodate–thiobarbituric acid method.[609]

The *O*-trimethylsilyl derivatives of five reference disaccharide alditols, composed of 2-acetamido-2-deoxyhexitol substituted at C-3, C-4, or C-6 with a hexose grouping, have been studied by g.l.c.–m.s.[610] The data were used in the determination of the linkage in disaccharide alditols derived from rat brain glycoproteins and gangliosides.

The terminal D-galactosyl residues of asialoglycopeptides have been oxidized to yield the corresponding uronic acid in a two-step sequence using D-galactose oxidase followed by treatment with Tollen's reagent.[611] When applied to a glycopeptide derived from α_1-acid glycoprotein, 2-amino-2-deoxy-4-*O*-β-D-galactopyranosyluronic acid-D-glucose was isolated. The method has been used to distinguish between 1 → 3-, 1 → 4-, and 1 → 6-linked isomers of the corresponding aldobiouronic acids because the colour yield of various *O*-substituted 2-acetamido-2-deoxyhexoses in the Morgan-Elson reaction is indicative of the position of the substitution.

An *endo*-β-D-2-acetamido-2-deoxyglucanase from *Streptomyces plicatus* has been used in the determination of the molecular weight of the protein moiety of glycoproteins and their inherent number of carbohydrate chains.[612] Carboxypeptidase Y is confirmed as having a molecular weight of 5.1×10^4 after removal of carbohydrate. Molecular weight changes were also demonstrated on the removal of oligosaccharide side-chains from mung bean nuclease I and *Penicillium* nuclease P_1.

β-D-2-Acetamido-2-deoxyhexosidases A and B from human liver are able to hydrolyse the terminal 2-acetamido-2-deoxy-β-D-glucosyl groups from a variety of oligosaccharides derived *via* acetolysis of ovomucoid and from an asialogalactoglycopeptide from human serotransferrin.[613] The enzymes are less active towards branched oligosaccharides.

Conjugates of alchohol dehydrogenase and lactate dehydrogenase, when bound covalently to concanavalin A, have been reported to retain their abilities to bind with glycoproteins and to catalyze oxidation–reduction reactions with NAD in the presence of appropriate substrates.[614] This dual capacity renders these protein conjugates useful reagents for the location of glycoprotein bands

608 L. Orland, J. Haverkamp, J. F. G. Vliegenthart, B. Fournet, G. Strecker, G. Spik, J. Montreuil, K. Schmid, and J. F. Binette, *F.E.B.S. Letters*, 1978, **89**, 149.
609 M. Koseki, A. Kimura and K. Tsumuri, *J. Biochem.* (*Japan*), 1978, **83**, 553.
610 J. Finne, I. Mononen, and J. Karkkainen, *Biomed. Mass Spec.*, 1977, **4**, 281.
611 J. S. Myers and O. Gabriel, *Carbohydrate Res.*, 1978, **67**, 223.
612 R. B. Trimble and F. Maley, *Biochem. Biophys. Res. Comm.*, 1977, **78**, 935.
613 T. Bearpark, S. Boughelet, B. Fournet, J. Montreuil, G. Spik, J. Stirling, and G. Strecker, *F.E.B.S. Letters*, 1978, **84**, 379.
614 G. Avigad, *Analyt. Biochem.*, 1978, **86**, 443.

on polyacrylamide gels; precipitation of formazan dye occurs as a result of the dehydrogenase activity.

A general procedure, involving a combination of cation- and anion-exchange chromatography and charcoal–celite column chromatography, has been used in the isolation of acidic oligosaccharides and glycopeptides from normal and pathological urines.[615]

Problems encountered in estimating the molecular weights of glycoproteins by gel filtration have been highlighted.[616] For sialoglycoproteins of relatively low molecular weight, such as fetuin and α_1-acid glycoprotein, gel filtration gives apparent molecular weights much larger than the known values. The corresponding asialo-glycoproteins have apparent molecular weights lower than expected from sole removal of neuraminic acid.

[615] G. Strecker and A. Lemaire-Poitau, *J. Chromatog.*, 1977, **143**, 553.
[616] J. A. Alhadeff, *Biochem. J.*, 1978, **173**, 315.

6
Enzymes

BY J. F. KENNEDY

1 Introduction

General Aspects, Nomenclature, and Composite Works.—Recommendations of the IUPAC–IUB Commission on Biochemical Nomenclature for the nomenclature of multiple forms of enzymes have been published.[1]

New volumes of 'Methods in Enzymology' cover the preparation and use of affinity labelled enzymes,[2] the structures of enzymes,[3–5] and purine and pyrimidine nucleotide metabolism.[6] A further volume [7] entitled 'Complex Carbohydrates, Part C' includes sections dealing with glycoside hydrolases. The forty-eighth volume in the series 'Advances in Enzymology and Related Areas of Molecular Biology' has been published.[8]

The third volume in the edited series 'Topics in Enzyme and Fermentation Biotechnology' has been published.[9] 'Fermentation and Enzyme Technology' outlines the fundamental microbiological, biochemical, genetic, and engineering aspects of fermentation, presenting advanced methods of fermentation and control.[10] It covers the isolation of enzymes, especially those found in intracellular contents of micro-organisms. Enzyme immobilization and the factors influencing the use of enzymes in reactors are examined.

An edited volume entitled 'Molecular Movements and Chemical Reactivity as Conditioned by Membranes, Enzymes, and Other Macromolecules' includes sections on synzymes (synthetic polymers with enzyme-like catalytic activities), diffusion-reaction in structured media and membranes bearing enzymes, and structures and energetics of proteins and their active sites.[11]

[1] Nomenclature of Multiple Forms of Enzymes, *Mol. Cell. Biochem.*, 1978, **19**, 61.
[2] 'Methods in Enzymology', Vol. 46, ed. W. B. Jakoby and M. Wilchek, Academic Press, New York, 1977.
[3] 'Enzyme Structure, Part E', in 'Methods in Enzymology', Vol. 47, ed. C. H. W. Hirs and S. N. Timasheff, Academic Press, New York, 1977.
[4] 'Enzyme Structure, Part F', in 'Methods in Enzymology', Vol. 48, ed. C. H. W. Hirs and S. N. Timasheff, Academic Press, New York, 1978.
[5] 'Enzyme Structure, Part G', in 'Methods in Enzymology', Vol. 49, ed. C. H. W. Hirs and S. N. Timasheff, Academic Press, New York, 1978.
[6] 'Methods in Enzymology', Vol. 51, ed. P. A. Hoffee and M. E. Jones, Academic Press, New York, 1978.
[7] 'Methods in Enzymology', Vol. 50, ed. V. Ginsburg, Academic Press, New York, 1978.
[8] 'Advances in Enzymology and Related Areas of Molecular Biology', Vol. 48, ed. A. Meister, J. Wiley & Sons Ltd., Chichester, 1978.
[9] 'Topics in Enzyme and Fermentation Biotechnology', Vol. 3, ed. A. Wiseman, Ellis Horwood Ltd., Chichester, 1979.
[10] D. I. C. Wang, C. L. Cooney, A. L. Demain, P. Dunnill, A. E. Humphrey, and M. D. Lilley, in 'Fermentation and Enzyme Technology', J. Wiley & Sons Ltd., Chichester, 1978.
[11] 'Molecular Movements and Chemical Reactivity as Conditioned by Membranes, Enzymes, and other Macromolecules', ed. R. Lefever and A. Goldbeter, J. Wiley & Sons Ltd., Chichester, 1978.

The subcellular distribution of enzymes involved in α-D-glucan utilization in *Klebsiella pneumoniae* has been investigated.[12] The eleventh volume in the series 'Current Topics in Membranes and Transport' and entitled 'Cell Surface Glycoproteins: Structure, Biosynthesis, and Biological Functions' includes a chapter on the subject of glycoprotein membrane enzymes.[13]

A book on cellular and biochemical aspects in diabetic retinopathy refers to glycoside hydrolases.[14]

Methods of Assay.—'Principles of Enzymatic Analysis' is the subject of an edited volume.[15]

Properties.—The properties of glycoside hydrolases have been reviewed.[16] As part of a series on surface carbohydrates of the prokaryotic cell, a comprehensive chapter deals with enzymes acting on bacterial surface carbohydrates.[17] Enzymes hydrolysing capsular and slime polysaccharides, lipopolysaccharides, and teichoic acids are treated in detail.

Kinetics.—A book dealing with the kinetics of enzyme-catalysed reactions has been designed for teaching purposes.[18]

A procedure, based on the Gauss-Newton method for non-linear regression, has been developed to obtain constants from the analysis of progress curve data.[19] Rules are presented which greatly simplify the derivation of the necessary equations. Values for kinetic parameters from the method agreed well with those obtained from actual steady-state rate measurements.

Mechanisms of Action.—'Chemistry and Control of Enzyme Reaction' is the subject of a recently published book.[20]

Structures.—Mechanisms of the activation and inhibition of enzymes by antibodies and the use of immunochemical methods for studying the structure of the active site, quaternary structure, and conformational changes in enzymes during the interaction with substrates and other ligands have been discussed.[21]

Isolation.—Discussions on the extraction and purification of enzymes from plant tissue,[22] purification of enzymes by affinity chromatography,[23] and sub-

[12] G. Wöhner and G. Wöber, *Arch. Mikrobiol.*, 1978, **116**, 311.

[13] J. R. Riordan and G. G. Forstner, 'Cell Surface Glycoproteins: Structure, Biosynthesis and Biological Structures,' in 'Current Topics in Membranes and Transport', Vol. 11, Academic Press, New York, 1978, p. 145.

[14] 'Cellular and Biochemical Aspects in Diabetic Retinopathy', eds. F. Regnault and T. Duhault, North Holland Publ. Co., Amsterdam, 1978.

[15] 'Principles of Enzymatic Analysis', ed. H. U. Bergmeyer, Verlag Chimie, Weinheim, 1978.

[16] H. M. Flowers and N. Sharon in 'Advances in Enzymology and Related Areas of Molecular Biology', ed. A. Meister, J. Wiley and Sons Ltd., Chichester, 1978. p. 29.

[17] 'Surface Carbohydrates of the Prokraryotic Cell', ed. I. W. Sutherland, Academic Press, London, 1977.

[18] D. Piszkiewicz, 'Kinetics of Chemical and Enzyme Catalyzed Reactions', Oxford University Press, New York, 1977.

[19] R. G. Duggleby and J. F. Morrison, *Biochim. Biophys. Acta*, 1977, **481**, 297.

[20] K. G. Scrimgeour, 'Chemistry and Control of Enzyme Reaction', Academic Press, London, 1977.

[21] N. K. Nagradova and I. D. Grozdova, *Biochemistry* (*U.S.S.R.*), 1977, **42**, 897.

[22] M. J. C. Rhodes, in 'Regulation of Enzyme Synthesis and Activity in Higher Plants', Phytochemical Society Symposia Series No. 14, ed. H. Smith, Academic Press, London, 1977. p. 245.

[23] P. D. G. Dean and M. J. Harvey, ref. 22, p. 271.

cellular fractionation techniques in enzyme distribution studies [24] form sections in a book on the 'Regulation of Enzyme Synthesis and Activity in Higher Plants'.

A procedure for enzyme purification by affinity chromatography using a non-covalently bound absorbent has been described with particular reference to lipophilic interactions.[25]

Biosynthesis.—Various aspects of the regulation of enzyme synthesis and activity in higher plants are covered by an edited volume.[26]

Applications.—A new approach to preparative organic synthesis in aqueous–organic systems is based on the idea that the relevant enzymic process is carried out in a biphasic 'water–water-immiscible organic solvent'.[27] Thereby the enzyme is localized in the aqueous phase; this eliminates the traditional problem of stabilizing the enzyme against inactivation by a non-aqueous solvent. Hence, in contrast to the commonly used combinations 'water–water-miscible organic solvent', it was suggested that in this system the content of water may be infinitely low. This allows a dramatic shift of the equilibrium of reactions forming water as a product, towards products. The fact that the system consists of two phases provides another important source for an equilibrium shift, *i.e.* free energies of the transfer of a reagent from one phase to another. Equations are derived describing the dependence of the equilibrium constant in a biphasic system on the ratio of the volumes of the aqueous and nonaqueous phases and the partition coefficients of the reagents between the phases.

The application of glycoside hydrolases to the study of complex carbohydrates and cell surfaces has been reviewed.[16]

'Biotechnological Applications of Proteins and Enzymes' is the title of the published proceedings of a conference which covered production of enzymes, exploitation of soluble and insolubilized enzymes, medical applications of proteins and enzymes, and proteins as food.[28]

In the second volume in the series 'Topics in Enzyme and Fermentation Biotechnology' attention is given to various aspects of enzyme biotechnology.[29] An interdisciplinary approach is favoured and subjects covered include the use of immobilized enzymes, the stability problem of enzymes in use and in storage, and the biological treatment of industrial waste water viewed as a fermentation process. It relates also to the utilization of enzymes in many industries, including pharmaceutics, food, textiles, and paper.

The innovative series 'Enzyme Engineering' is intended to fill the need for

[24] R. M. Leech, ref. 22, p. 290.
[25] A. G. Lacko and T. F. Chen, *J. Chromatog.*, 1977, **130**, 446.
[26] 'Regulation of Enzyme Synthesis and Activity in Higher Plants', ed. H. Smith, Academic Press, New York, 1977.
[27] A. M. Klibanov, G. P. Samokhin, K. Martinek, and I. V. Berezin, *Biotechnol. and Bioeng.*, 1977, **19**, 1351.
[28] 'Biotechnological Applications of Proteins and Enzymes', ed. Z. Bohak and N. Sharon, Academic Press, New York, 1977.
[29] 'Topics in Enzyme and Fermentation Biotechnology', Vol. 2, ed. A. Wiseman, Ellis Horwood Ltd., Chichester, 1978.

concise reports on theoretical and experimental applications of enzymes.[30, 31] Recent volumes include papers on innovation in enzyme engineering, new methods of stabilizing and immobilizing enzymes and whole cells, enzyme engineering and enzyme purification, affinity chromatography and enzyme co-factor immobilization, application and regenerations, analytical and biomedical application of immobilized enzymes, and industrial applications of immobilized enzymes.

A survey of immobilized enzymes constitutes part of the first volume (entitled 'Immobilized Enzyme Principles') of a new series 'Applied Biochemistry and Bioengineering'.[32] 'Immobilized Enzymes' is the title of an edited book which considers immobilization methods for enzymes and microbial cells in detail and then reviews general properties.[33] Kinetic constants and stability as well as other properties are examined, and the effectiveness of various methods in relation to particular applications, and applications to industrial processes based on the author's experience are discussed.

The fundamental principles of operation of an immobilized enzyme flow analyser have been studied.[34] Models were developed to explain the dependence of peak area, height, and half-width on sample concentration, sample volume, system flow rate, and column size. The dependencies were analogous to the behaviour of a concentration-sensitive chromatographic detector.

An edited series of papers on the subject of immobilized enzymes for industrial reactors cover the following aspects: basic enzymology; carriers including controlled-pore glasses for immobilization; immobilization by covalent attachment, entrapment, adsorption, and inorganic bridge formation; comparative characteristics of free and immobilized enzymes; immobilized co-enzymes; design and operation of immobilized enzyme reactors; and applications of immobilized enzymes.[35]

A two volume edited work critically surveys the current state of research on possible uses of immobilized enzymes and proteins in diagnosis, in prevention, and in therapy in such areas as the construction of artificial kidneys and livers, detoxification, fuel cells, and enzyme therapy.[36, 37] The first volume was devoted to the multiplicity of current therapeutic applications as well as future prospects, including the feasibility of immobilizing various enzymes and pro-

[30] 'Enzyme Engineering', Vol. 4, ed. G. B. Broun, G. Manecke, and L. B. Wingard, jun., Plenum Press, New York, 1978.

[31] 'Enzyme Engineering', Vol. 3, ed. E. K. Pye and H. W. Weetall, Plenum Press, New York, 1978.

[32] 'Immobilized Enzyme Principles', in 'Applied Biochemistry and Bioengineering', Vol. 1, ed. L. B. Wingard, jun., E. Katchalski-Katzir, and L. Goldstein, Academic Press, New York, 1976.

[33] 'Immobilized Enzymes: Research and Development', ed. I. Chibata, Kodansha Scientific Ltd. and J. Wiley & Sons Ltd., Chichester, 1979.

[34] R. S. Schilfreen, D. A. Hanna, L. D. Bowers, and P. W. Carr, *Analyt. Chem.*, 1977, **49**, 1929.

[35] 'Immobilized Enzymes for Industrial Reactors', ed. R. A. Messing, Academic Press, New York, 1975.

[36] 'Biomedical Applications of Immobilized Enzymes and Proteins', Vol. 1, ed. T. M. S. Chang, Plenum Press, New York, 1977.

[37] 'Biomedical Applications of Immobilized Enzymes and Proteins', Vol. 2. ed. T. M. S. Chang, Plenum Press, New York, 1977.

teins for use in specific treatment regimes. The volume also offers a brief introduction to the subject for those in biomedical sciences who wish to use immobilized enzymes but require a review of their chemical properties. Volume 2 focuses on the applications of these substances in diagnosis and public health, detailing the various techniques that are now available, or could be developed and employed in clinical and preventive medicine. This is followed by a perspectives section in which a number of authors suggest possible areas for future exploration.

A review on the subject of immobilized enzymes in biochemical analysis covers: preparation and properties of immobilized enzymes; assay methods using immobilized enzymes (spectrophotomatic assays, automated methods in biochemical analysis, enzyme electrodes, and colorimetric analyses using immobilized enzymes); and applications of immobilized enzymes in biochemical analysis.[38]

A further volume in the series 'Advances in Enzymology and Related Areas of Molecular Biology' includes a chapter on the subject of immobilized co-enzymes in general ligand affinity chromatography and their use as active co-enzymes.[39]

2 α- and β-D-2-Acetamido-2-deoxygalactosidases, α- and β-D-Acetamido-2-deoxyglucosidases, and β-D-2-Acetamido-2-deoxyhexosidases

Various alkylated derivations of agarose, differing in the alkyl chain length and the presence of different functional groups, have been synthesized to provide materials for hydrophobic chromatography for the isolation and purification of β-D-2-acetamido-2-deoxyhexosidase.[40] It was found that an increase in alkyl chain length results in stronger binding of the enzyme, whereas protonated amino-groups inhibit hydrophobic binding.

Significant differences occur between the interactions of glycoaminoglycans with a particular lysosomal protein leading to inhibition of lysosomal enzymes including β-D-2-acetamido-2-deoxy-glucosidase and -galactosidase.[41]

The screening and prevention of Tay–Sachs disease, which involves a deficiency of β-D-2-acetamido-2-deoxyhexosidase A, has been reviewed as part of the 'Progress in Clinical and Biological Research' series.[42]

It has been shown that the α-D-2-acetamido-2-deoxygalactosidase and α-D-galactosidase B activities of human tissues are manifested by the same protein containing a single catalytic site (p. 392).[43]

An immunoadsorbent column has been prepared using a specific antibody to human β-D-2-acetamido-2-deoxyglucosidase, and agarose cyclic imidocarbonate.[44] Although no precipitating activity of the antisera was found with

[38] J. Everse, C. L. Ginsburgh, and N. O. Kaplan, *Methods Biochem. Analysis*, 1979, **25**, 135.
[39] 'Adv. Enzymol. and Related Areas of Molec. Biol.', Vol. 46, ed. A. Meister, J. Wiley & Sons Ltd., New York, 1978.
[40] A. A. Artyukov and N. V. Molodtsov, *J. Chromatog.*, 1977, **130**, 451.
[41] J. L. Avila, *Biochem. J.*, 1978, **171**, 489.
[42] 'Tay-Sachs disease. Screening and Prevention.' 'Progress in Clinical and Biological Research', Vol. 18, ed. M. M. Kayback, Alan R. Liss Inc., New York, 1977.
[43] A. W. Schram, M. N. Hamers, and J. M. Tager, *Biochim. Biophys. Acta*, 1977, **482**, 138.
[44] R. P. Erickson and R. Sandman, *Experientia*, 1977, **33**, 14.

mouse or rat liver extracts, enzyme was easily eluted from the column which provided about 50-fold, single-step purifications of the heterologous enzymes.

To examine the possibility in infantile G_{M_2} gangliosidosis (AB variant) of a mutant β-D-2-acetamido-2-deoxyhexosidase A, well capable of hydrolysing the fluorogenic synthetic substrates, but unable to attack the relevant ganglioside, the enzyme was isolated from patient tissue and characterized biochemically and immunologically in comparison with an enzyme preparation from normal tissue.[45] The lack of differences between the two enzyme preparations indicated that the defect involved in the disease is not at the genetic level of production of either α or β chains of β-D-2-acetamido-2-deoxyhexosidase A.

Levels of lysosomal β-D-2-acetamido-2-deoxyhexosidase have been determined in normal human hair roots and in hair roots obtained from a patient with mannosidosis.[46] The significance of the results was discussed in relation to the levels of other glycoside hydrolases present, and to the detection of lysosomal storage disease. The β-D-2-acetamido-2-deoxyhexosidase was found to be the most active glycoside hydrolase in normal hair roots.

As part of a study on the neurochemistry of genetic hyperglycosaminoglycanuria (mucopolysaccharidosis), lysosomal enzymes including α-D-2-acetamido-2-deoxyglucosidase and β-D-2-acetamido-2-deoxyglucosidase in normals and four types of patients have been investigated in brain and liver.[47]

A study of the glycoside hydrolases of human lung showed that the β-D-2-acetamido-D-deoxyglucosidase is highly active.[48] Its electrophoretic mobility on polyacrylamide gels indicated that the enzyme is different from that in the serum. The secretory mechanism of the lung enzyme was discussed.

β-D-2-Acetamido-2-deoxyhexosidase isoenzymes A and B have been isolated from human kidney, in a homogeneous state as demonstrated by electrophoretic and enzymic criteria.[49] The enzymes (K_m's 0.8 mM for 4-nitrophenyl 2-acetamido-2-deoxy-β-D-glucopyranoside, double pH optima for both, 4.5 and 4.8) were stable for at least 18 months at -20 °C, pH 6.5. On sodium dodecyl sulphate–polyacrylamide gel electrophoresis isoenzyme A (mol. wt. $1.110 \pm 0.015 \times 10^5$ by gel filtration) dissociated into one subunit type (mol. wt. 6.8×10^4) and isoenzyme B (mol. wt. $1.140 \pm 0.016 \times 10^5$) into three subunits (mol. wts. 1.00×10^5, 0.68×10^5, and 0.37×10^5) and one protein band (mol. wt. 1.40×10^5). The effects of carboxymethylation on subunit phenomena and the activating/inactivating effects of metal ions were examined.

It has been shown that the α-D-galactosidase B from human liver also exhibits α-D-2-acetamido-2-deoxygalactosidase activity (p. 392).[43] A highly purified fraction of α-D-galactosidase B from human liver was found to contain α-D-2-acetamido-2-deoxygalactosidase activity (K_m 2-nitrophenyl 2-acetamido-2-deoxy-α-D-galactopyranoside 1.3 mM, K_m 4-methylumbelliferyl α-D-galactopyranoside 6.8 mM).[50] The hydrolysis of the latter substrate was completely

[45] E. Couzelmann, K. Sandhoff, N. Nehrkorn, B. Geiger, and R. Arnon, *European J. Biochem.*, 1978, **84** 27.
[46] N. C. Phillips and R. Thorpe, *Clinica Chim. Acta*, 1978, **83**, 93.
[47] G. Constantopoulos and A. S. Dekaban, *J. Neurochem.*, 1978, **30**, 965.
[48] G. E. R. Hook, *Biochemistry*, 1978, **17**, 520.
[49] D. V. Marinkovic and J. N. Marinkovic, *Biochem. J.*, 1977, **163**, 133.
[50] K. J. Dean, S.-S. J. Sung, and C. C. Sweeley, *Biochem. Biophys. Res. Comm.*, 1977, **77**, 1411.

inhibited by the former substrate (K_i 2.7 mM). The enzyme also uniquely catalysed the hydrolysis of Forssman hapten, a glycolipid containing a terminal 2-acetamido-2-deoxy-α-D-galactopryanosyl group (K_m 0.59 mM).

The isoenzymes of β-D-2-acetamido-2-deoxyhexosidase A from human liver and serum have been separated by isoelectric focusing in thin layers of polyacrylamide gel.[51] β-D-Acetamido-2-deoxyhexosidases A and B have been purified from human liver and tested for their ability to hydrolyse terminal 2-acetamido-2-deoxy-β-D-glucopyranosyl groups from a variety of oligosaccharides derived from acetolysis of ovomucoid and from an asialo-agalacto-glycopeptide of human serotransferrin.[52] The role of the two forms of the enzyme in the degradation of glycoproteins was discussed.

Two forms of α-D-2-acetamido-2-deoxygalactosidase, which also display α-D-galactosidase activity, have been isolated from human liver *via* affinity chromatography on immobilized concanavalin A.[53] In fresh normal human liver, most of the α-D-2-acetamido-2-deoxygalactosidase activity was found in the fractions containing α-D-galactosidase B and a minor portion in the fractions containing α-D-galactosidase A. However, the minor peak of α-D-2-acetamido-2-deoxygalactosidase activity did not coincide with the peak of α-D-galactosidase A activity. Both peaks of α-D-2-acetamido-2-deoxygalactosidase activity were precipitated by antiserum raised against α-D-galactosidase B but not by antiserum raised against α-D-galactosidase A. In fresh Fabry liver (which contains no α-D-galactosidase A) more than 98% of the α-D-2-acetamido-2-deoxygalactosidase activity was recovered in the fractions containing α-D-galactosidase B. Less than 2% was found in the fractions containing A-like α-D-galactosidase. Upon ageing of the liver, the yield of the latter form increased (to almost 25% after 17 months' storage at −20 °C). The conversion of α-D-2-acetamido-2-deoxygalactosidase to the second form upon ageing was accompanied by a decrease in apparent molecular weight from $1.10 \pm 0.05 \times 10^5$ to $0.99 \pm 0.03 \times 10^5$, as measured by gel filtration. The kinetic properties of the enzyme changed slightly upon ageing. In contrast, there was no effect of ageing on the pH optimum of the enzyme. It was concluded that only one lysosomal α-D-2-acetamido-2-deoxygalactosidase is present in human liver and that one of the two forms of α-D-2-acetamido-2-deoxygalactosidase observed in extracts of liver is formed by chemical modification of the enzyme during storage of the liver and during storage of the purified preparation.

Those proteins (mol. wts. 7.0—20 × 10^4) of human liver that cross-reacted with antibodies raised to apparently homogeneous β-D-2-acetamido-2-deoxyhexosidases A and B were detected by immunodiffusion.[54] The possible structure and functional relationships between the enzymes and the cross-reacting proteins were discussed.

The activities of purified human placental β-D-2-acetamido-2-deoxyhexo-

[51] G. Pott, D. Chi-Boesler, and U. Gerlach, *J. Clin. Chem. Clin. Biochem.*, 1978, **16**, 15.
[52] T. Bearpark, S. Bouquelet, B. Fournet, J. Montreuil, G. Spik, J. Stirling, and G. Strecker, *F.E.B.S. Letters*, 1977, **84**, 379.
[53] A. W. Schram, P. G. de Groot, M. N. Hamers, B. Brouwer-Kelder, W. E. Donker-Koopman, and J. M. Tager, *Biochem. Biophys. Acta*, 1978, **525**, 410.
[54] M. Carroll, *Biochem. J.*, 1978, **173**, 191.

sidases A and B against hyaluronic acid in the presence of β-D-glucuronidase have been investigated.[55] Without the latter, the β-D-2-acetamido-2-deoxy-hexosidase cleaved only the terminal 2-acetamido-2-deoxy-β-D-glucopyranosyl group. Although there were some differences in the rates of reaction of the A and B isoenzymes in the mixed system, the data indicated that hyaluronic acid can be hydrolysed by the combined activities of β-D-glucuronidase with either the isoenzyme A or the isoenzyme B.

The β-D-2-acetamido-2-deoxyglucosidase activity in hydatidiform mole has been shown to be two-fold higher than that in full-term human placenta.[56] Qualitatively, the enzymes from the two tissues were similar with respect to K_m values and pH optima. The molar enzyme was more susceptible to heat denaturation, presumably due to the presence of a higher level of the heat-labile isoenzyme form A in this tissue. Data were also presented indicating that the placenta is not the source of the β-D-2-acetamido-2-deoxyglucosidase activity in maternal serum.

An automated and reliable method has been described for the identification of heterozygous carriers of Tay–Sachs disease.[57] The method was based on determination of β-D-2-acetamido-2-deoxyhexosidase A activity in tears using 4-methylumbelliferyl 2-acetamido-2-deoxy-β-D-glucopyranoside as substrate.

A method of separating the isoenzymes of β-D-2-acetamido-2-deoxy-glucosidase using electrofocusing in thin layers of polyacrylamide gel has been demonstrated.[58] Application to blood enzymes from normals, patients with liver disease, and pregnant women showed the simplicity and potential wide application of the method in clinical fields.

With a view to the biochemical detection of homo- and hetero-zygous carriers and G_{M_2}-gangliosidosis, serum β-D-2-acetamido-2-deoxyhexosidase activities have been investigated in patients with Tay–Sachs and Sandhoff diseases and in normals.[59] Two related methods for the differential determination of the A and B enzymes were possible, but more conclusive results for heterozygous carriers were attained by the heat inactivation method.

The isoenzymes of β-D-2-acetamido-2-deoxyhexosidase in amniotic fluid and maternal serum samples collected during gestation have been quantitated *via* ion-exchange chromatography.[60] Isoenzymes A and B constituted the major components of most aminiotic fluids but some samples characterized by high enzyme activity also contained a high proportion of an isoenzyme apparently identical to isoenzyme P of maternal serum. It was pointed out that this passage of maternal serum enzyme into the amniotic cavity could lead to false negative diagnosis of type B and type O G_{M_2}-gangliosidosis, if the diagnoses rely solely upon isoenzyme analysis in amniotic fluid.

The specific activity and distribution of the isoenzymes of β-D-2-acetamido-2-deoxyhexosidase have been studied in platelets, granulocytes, monocytes, and

[55] G. Bach and B. Geiger, *Arch. Biochem. Biophys.*, 1978, **189**, 37.
[56] K. C. Leung, K. P. Fung, C. C. Yu, Y. M. Choy, and C. Y. Lee, *Clinica Chim. Acta*, 1977, **74**, 43.
[57] J. D. Goldberg, J. H. Truex, and R. J. Desmick, *Clinica Chim. Acta*, 1977, **77**, 43.
[58] G. Pott, M. Schneider, and U. Gerlach, *Science Tools*, 1978, **25**, 67.
[59] B. Molzer and H. Bernheimer, *Clinica Chim. Acta*, 1976, **73**, 163.
[60] M. Potier, G. Boire, L. Dallaire, and S. B. Melancon, *Clinica Chim. Acta*, 1977, **76**, 309.

lymphocytes isolated from human venous blood.[61] Since there are differences in the contents of various isoenzymes in the cell types, the relative activity of the A enzyme in a sample is dependent on the proportions of the individual cell types present, and therefore activity measurements do not accurately affect the genotype of the blood donor. Accordingly, chromatographic separation and assay of the platelet extracts for activity of the isoenzymes was considered a more convenient and satisfactory method of discriminating carriers and non-carriers of the Tay–Sachs mutant gene.

The distributions of glycoside hydrolases including β-D-2-acetamido-2-deoxy-galacto- and gluco-sidases in human polymorphonuclear leucocytes have been determined.[62]

A method described for the continuous kinetic measurement of serum β-D-2-acetamido-2-deoxyhexosidase activity has been applied to detection of heterozygotes for Tay–Sachs gene.[63] In contrast to single-point methods, a pH of 4.5, which is optimal for the enzyme activity on 4-methylumbelliferyl 2-acetamido-2-deoxy-β-D-glucopyranoside, is maintained whilst the increase in fluorescence for the reaction product, 4-methylumbelliferone, is determined. Under the conditions described, the reaction followed zero-order kinetics, and activity was linearly related to serum dilution.

β-D-2-Acetamido-2-deoxyhexosidase P, the main isoenzyme of this activity in human pregnancy serum, has been isolated and purified by a two-step procedure involving both affinity chromatography, on a 2-acetamido-2-deoxy-D-glucopyranosyl derivative of agarose, and ion-exchange chromatography.[64] Whereas the kinetic behaviour of the isoenzyme P was analogous to that of isoenzymes A and B, it differed from these in its electrophoretic mobility and in its molecular weight, (1.5×10^5, *cf.* 1.0×10^5 and 1.0×10^5). Immunochemical analysis indicated that the P isoenzyme is antigenically cross-reactive with isoenzymes A and B, but it does not contain the A-specific antigenic determinants. Possible structures for isoenzyme P are a β-D-2-acetamido-2-deoxyhexosidase B-like structure with greater glycosylation or a hexamer of the β chain, possibly arranged as three β_2 subunits.

α-D-2-Acetamido-2-deoxyglucosidase has been purified from human urine to a state of apparent homogeneity.[65] The enzyme (mol. wt. 3.07×10^5 by gel filtration) is a glycoprotein and polyacrylamide gel electrophoresis in the absence and presence of sodium dodecyl sulphate indicated molecular weight heterogeneity of isocharged forms. The enzyme exhibited the following characteristics: pH optimum 4.5, K_m 0.14—0.74 mM, and V_{max} 1.04—3.68 μmol mg^{-1} min^{-1} for three aryl and UDP 2-acetamido-2-deoxy-α-D-glucopyranosides. Heparan sulphate, heparin, and dermatan sulphate are competitive inhibitors and although Hg^{2+} and Cu^{2+} are inhibitors, thiol-protective agents had no affect on the enzyme activity. The enzyme has been shown to act upon heparan sulphate and heparin (pH optimum 4.4) by an *exo* pattern.[66]

[61] S. Nakagawa, S. Kumin, D. Fox, and H. M. Nitowsky, *J. Lab. Clinica Med.*, 1978, **91**, 922.
[62] R. F. Rest, M. H. Cooney, and J. K. Spitznagel, *Biochem. J.*, 1978, **174**, 53.
[63] L. Zinterhofer and G. Schuttringer, *Clinical Chem.*, 1976, **22**, 1631.
[64] B. Geiger, E. Calef, and R. Arnon, *Biochemistry*, 1978, **17**, 1713.
[65] K. von Figura, *European J. Biochem.*, 1977, **80**, 525.
[66] K. von Figura, *European J. Biochem.*, 1977, **80**, 535.

Isoelectric focusing showed multiple forms of the enzymes (p*I* 3.3—6.0) which differed in their recognition and endocytosis by cultivated skin fibroblasts. Sodium periodate treatment reduced the α-D-2-acetamido-2-deoxyglucosidase recognition by fibroblasts and suggested that the recognition sites on the enzyme are associated with its carbohydrate moiety.

Secretion of β-D-2-acetamido-2-deoxyglucosidase by normal human fibroblast cultures was linear with respect to time up to 96 h.[67] Two forms of the A isoenzyme were found in the culture medium. One form was similar to the isoenzyme found in other extracellular fluids such as plasma and tears, and the other resembled the intracellular (lysosomal) enzyme. The presence of the two isoenzymes in the culture medium appeared to represent two distinct secretory processes.

The levels of β-D-2-acetamido-deoxyglucosidase activity in skin fibroblasts from cases of cystic fibrosis and controls have been compared.[68] In a study of I-cell disease (mucolipidosis II) a model has been presented for the structure of the carbohydrate recognition site of fibroblast-derived β-D-2-acetamido-2-deoxyhexosidase that may extend to the other affected hydrolases and that is responsible for specific uptake of the enzyme by fibroblasts.[69]

The pinocytosis by fibroblasts of β-D-2-acetamido-2-deoxyhexosidase excreted by cultural skin fibroblasts from a patient with I-cell disease was not enhanced by neuraminidase treatment of the enzyme.[70] The uptake of the normal plasma enzyme was minimal and this was not appreciably enhanced by neuraminidase treatment. In contrast, normal seminal fluid β-D-2-acetamido-deoxyhexosidase was readily pinocytosed regardless of neuraminidase treatment. Thus it was concluded that a neuraminidase deficiency in I-cell disease may not be directly responsible for excessive extracellular enzyme.

The K_m values and relative catalytic constants (k_{cat}^{rel}) have been determined for the bovine liver β-D-2-acetamido-2-deoxyglucosidase-catalysed hydrolysis of a series of substituted-phenyl 2-acetamido-2-deoxy-β-D-glucopyranosides.[71] Plots of log K_m with respect to Hammett substituent constant and Hantsch substituent constant showed the enzyme–substrate affinity to be dependent on the electronic nature of the substituents but not on the hydrophobic nature. Methanol competed with water for the glycosyl-enzyme to some extent, but did not increase k_{cat}^{rel} indicating that the deglycosylation is not rate limiting.

Two isoenzymes (A and B) of β-D-2-acetamido-2-deoxyglucosidase (mol. wts. 2.0×10^5 and 1.9×10^5 by gel filtration, p*I* values 5.31 and 6.78) have been purified to homogeneity on isoelectric focusing from bull seminal plasma.[72] Two identical subunits (mol. wts. 5.3×10^4 and 1.34×10^4) were obtained from the A and B isoenzymes when subjected to polyacrylamide gel electrophoresis in the presence of sodium dodecyl sulphate or other dissociating agents. The β-D-2-acetamido-2-deoxy-glucosidase and -galactosidase activities of the two

[67] P. Willcox, *Biochem. J.*, 1978, **173**, 433.
[68] T. F. Scanlin, S. S. Matacic, M. Pace, U. V. Santer, and M. C. Glick, *Biochem. Biophys. Res. Comm.*, 1977, **79**, 869.
[69] G. D. Vladutiu, *Biochem. J.*, 1978, **171**, 509.
[70] G. D. Vladutiu and M. C. Rattazzi, *Biochim. Biophys. Acta*, 1978, **539**, 31.
[71] M. Tanaka, S. Kyosaka, and Y. Sekiguchi, *Chem. and Pharm. Bull. (Japan)*, 1976, **24**, 3144.
[72] A. Khar and S. R. Anand, *Biochim. Biophys. Acta*, 1977, **483**, 152.

isoenzymes were strongly inhibited by Hg^{2+} and Ag^{+}, indicative of the presence of thiol groups in the proteins.

β-D-2-Acetamido-2-deoxyglucosidase has been purified from bull sperm acrosomal extracts *via* ion-exchange chromatograpy and isoelectric focusing.[73] The enzyme (p*I* 7.96, mol. wt. 1.9×10^5, K_m's 0.52 mM and 0.27 mM for 4-nitrophenyl 2-acetamido-2-deoxy-β-D-glucopyranoside and -galactopyranoside respectively, pH optima 4.5 and 3.4 respectively) was inhibited by a number of compounds, *etc.*, the most potent being Hg^{2+} and lactones.

The β-D-2-acetamido-2-deoxyhexosidase of porcine brain fractions has been investigated.[74]

The contribution of β-D-2-acetamido-2-deoxyhexosidase to the degradation of hyaluronic acid by canine liver lysosomes has been investigated (see p. 453).[75]

The effect of cortisone and thyroxine on β-D-2-acetamido-2-deoxy-glucosidase levels in rat forebrain and cerebellum have been examined during early postnatal development.[76]

In normal and regenerating rat liver tissue, β-D-2-acetamido-2-deoxyglucosidase has been shown to be localized primarily in the fraction of light mitochondria, enriched with lysosomes.[77] The interaction of rat liver β-D-2-acetamido-2-deoxyhexosidase with agarose-immobilized ovalbumin has been investigated.[78] The use of the latter as an affinity chromatography material was also studied.

Rat colonic β-D-2-acetamido-2-deoxyhexosidase has been separated into three forms by ion-exchange chromatography.[79] Although the ratio of β-D-2-acetamido-2-deoxy-2-gluco- and galacto-sidase activities varied during purification, it was not possible to separate them by any of a variety of chromatographic procedures. The pH optima were identical for both activities and all three forms. Data from mixed substrate experiments were consistent with an active site model wherein the two activities reside in separate active sites, each able to be inhibited by the substrate of the other site. The effects of acetate and thiol reagents confirmed the two-site model. The relationship between the enzyme forms was compared with that of similar enzyme forms from other species.

The specific activity of lysosomal β-D-2-acetamido-2-deoxyhexosidase in the caput and cauda of rat epididymis has been determined as a function of age.[80] The results indicated a relationship between changes occurring in sperm in the epididymis and the lysosomal enzymes.

High β-D-2-acetamido-2-deoxyhexosidase activity has been found in the

[73] A. Khar and S. R. Anand, *Biochim. Biophys. Acta*, 1977, **483**, 141.
[74] B. Venerando, A. Preti, A. Lombardo, B. Cestaro, and G. Tettamanti, *Biochim. Biophys. Acta*, 1978, **527**, 17.
[75] S. Hayashi, *J. Biochem. (Japan)*, 1977, **82**, 1287.
[76] H. C. Lau, C. Horowitz, J. Jumawan, and O. Koldovsky, *J. Neurochem.*, 1978, **31**, 261.
[77] T. I. Valyakina, N. D. Gabriélyan, R. L. Komaleva, and A. Ya. Khorlin, *Biochemistry (U.S.S.R.)*, 1977, **42**, 393.
[78] R. T. Dean, *J. Chromatog.*, 1978, **150**, 279.
[79] N. Mian, D. G. Herries, and E. A. Batte, *Biochim. Biophys. Acta*, 1978, **523**, 454.
[80] A. F. Milne, J. N. Williams IV, and S. L. Goodwin, *Comp. Biochem. Physiol. Pt. B.*, 1978, **59**, 199.

plasma of hereditary diabetic XA line hamsters.[81] Kidney fractions from XA line and non-diabetic M line animals and many other tissues also displayed this activity. Two isoenzymes (B_1 and B_2) separated by gel filtration from the plasma showed identical properties (pH optima 3.5—4.0, K_m's 0.18 mM for 4-nitrophenyl 2-acetamido-2-deoxy-β-D-glucopyranoside) whereas three isoenzymes, A, B_1, and B_2, were obtained from liver extracts (pH optima 4.0—4.5, 3.5, and 3.5—4.0 respectively, K_m's 0.48, 0.19, and 0.19 respectively). The occurrence of the various isoenzymes in various tissues was compared.

Infection of mice with an avirulent strain of Semliki forest virus caused an increase in brain β-D-2-acetamido-2-deoxyglucosidase activity.[82] In a study of the secretion of lysosomal hydrolases by stimulated and non-stimulated macrophages from mice, β-D-2-acetamido-2-deoxyglucosidase was found to be secreted in considerable amounts by a mechanism independent of phagocytosis.[83]

The purification and properties of a β-D-2-acetamido-2-deoxyhexosidase from the mollusc *Chamelea gallina* have been reported.[84] The enzyme was shown to be active against the 4-nitrophenyl glycosides of both 2-acetamido-2-deoxy-β-D-glucopyranose and -galactopyranose.

An affinity column prepared from 4-aminophenyl 1-thio-β-L-fucopyranoside residues attached to agarose through succinylated 3,3′-diaminodipropylamine bridges has been shown to bind β-D-2-acetamido-2-deoxyglucosidase from the serum of horseshoe crab (*Limulus polyphemus*).[85] The enzyme was eluted by 2-acetamido-2-deoxy-D-glucose at a concentration of 2 mg ml^{-1} and by other monosaccharides at higher concentrations in a form free from other glycoside hydrolases.

Polyacrylamide gel electrophoresis of gut preparations had permitted the separation of two isoenzymes of β-D-2-acetamido-2-deoxyglucosidase during larval development and metamorphosis of *Discoglossus pictus*.[86] The ratio of the two activities was followed through the various stages.

An analysis of the pH dependence of the K_m and V_{max} of the reaction catalysed by β-D-2-acetamido-2-deoxyhexosidase in electrophoretically pure form from *Halocynthia roretzi*, a tunicate ascidian, gave evidence of two fractionally important ionogenic groups of the enzyme.[87] One of these controls the binding of the substrate, and in the free enzyme, has a pK_a of 5.5. The other group (pK_a 5.6) is essential for the catalytic step and evidently does not change its pK_a value when substrate is added.

A β-D-2-acetamido-2-deoxyhexosidase from the digestive gland of a terrestial mollusc (*Helicella ericetorum*) exhibited both β-D-2-acetamido-2-deoxy-gluco- and galacto-sidase activities (pH optima 4.5 and 4.5, stability range pH 3.8—4.6, p*I* 4.83, mol. wt. 1.20—1.45 × 10^5, K_m and V_{max} 0.43 mM and

[81] A. Y. Chang, *Biochim. Biophys. Acta*, 1978, **522**, 503.
[82] A. J. Suckling, S. Jagelman, and H. E. Webb, *Infection and Immunity*, 1977, **15**, 386.
[83] J. Schnyder and M. Baggiolini, *J. Exp. Med.*, 1978, **148**, 435.
[84] N. Perez and J. A. Cabezas, *Biochimie*, 1977, **59**, 729.
[85] R. S. Jain, R. L. Binder, C. Walz, C. A. Buck, and L. Warren, *J. Chromatog.*, 1977, **136**, 141.
[86] J. Hourdry, *Compt. rend.* 1977, **284D**, 2023.
[87] E. V. Sundukova and A. A. Baikov, *Biochemistry* (*U.S.S.R.*), 1976, **41**, 1577.

30.1 μmol min^{-1} and 0.19 mM and 8.6 μmol min^{-1} for 4-nitrophenyl 2-acetamido-2-deoxy-β-D-gluco- and galaco-pyranosides respectively).[88] The enzyme was inhibited by Hg^{2+}, Fe^{3+}, acetate, D-mannose, 2-acetamido-2-deoxy-D-glucose, and -galactose, and exhibited *endo*-β-D-2-acetamido-2-deoxyglycanase activity on ovalbumin, ovomucoid, and glycosaminoglycans. Mixed substrate analysis indicated that the two activities are catalysed at the same active site.

A β-D-2-acetamido-2-deoxyglucosidase (pH optimum 3.4, p*I* 4.95, mol. wt. 1.6 × 10^5) has been purified from *Helix pomatia* and separated from endogenous chitinases.[89]

α-D-2-Acetamido-2-deoxygalactosidase, devoid of β-D-2-acetamido-2-deoxyhexosidase activity, has been purified from the limpet (*Patella vulgata*).[90] The final preparation gave one protein band on disc gel electrophoresis in the presence and absence of sodium dodecyl sulphate, and on disc gel isoelectric focusing. By gel filtration, the molecular weight of the enzyme (pH optimum 3.8, p*I* 5.5, K_m 0.6 mM for 4-nitrophenyl 2-acetamido-2-deoxy-α-D-galactoside) was 2.0 × 10^5 at pH 4.2 (active) and 0.45 × 10^5 at pH 7.0 (inactive) indicating that the enzyme exists as an active oligomer at acidic pH and as an inactive monomer at neutral and alkaline pH. The minor amount of associated α-D-galactosidase activity may be an intrinsic property of the α-D-2-acetamido-2-deoxygalactosidase.

Two forms of β-D-2-acetamido-2-deoxyhexosidase from the hepatopancreas of *Mytilus edulis* have been separated and purified by ammonium sulphate fractionation, ion-exchange chromatography, and gel filtration.[91] The two forms I and II had identical molecular weights, were constituted by two subunits, and exhibited similar kinetic properties (K_m 4-nitrophenyl 2-acetamido-2-deoxy-β-D-glucopyranoside: form I 0.44 mM, form II 0.4 mM; K_m 4-nitrophenyl 2-acetamido-2-deoxy-β-D-galactopyranoside: form I 0.108 mM, form II 0.3 mM).

Four molecular forms of β-D-2-acetamido-2-deoxyglucosidase located in different parts of germinating fenugreek (*Trigonella foenum graecum*) seeds have been characterized and isolated.[92] The pH and the temperature optima and K_m values towards synthetic substrates were found to be different for the β-D-2-acetamidodeoxy-glucosidase and -galactosidase activities of each molecular form. The activities of the forms against various inter-saccharide linkages were measured and inhibition studies were tried with a range of reagents. The four forms (mol. wt. 8.40 × 10^4, 7.20 × 10^4, 18.0 × 10^4, and 15.0 × 10^4; p*I* values 6.8, 6.3, 4.9, and 4.7, respectively) appeared, according to dissociation studies, to be polymers formed by association of between three and six subunits of identical molecular weight.

The identification and purification of a surface-bound β-D-2-acetamido-2-

[88] P. Calvo, A. Reglero, and J. A. Cabezas, *Biochem. J.*, 1978, **175**, 743.
[89] G. Lundblad, M. Elander, and J. Lind, *Acta Chem. Scand. Ser. B.*, 1976, **30**, 889.
[90] Y. Uda, S.-C. Li, and Y.-T. Li, *J. Biol. Chem.*, 1977, **252**, 5194.
[91] P. S. Mozo, M. F. Rama, R. V. Pernas, and M. R. Amil, *Comp. Biochem. Physiol.*, 1977, **58B**, 29.
[92] S. Bouquelet and G. Spik, *European J. Biochem.*, 1978, **84**, 551.

deoxyhexosidase from spores of *Bacillus cereus* have been reported.[93] The formation of β-D-2-acetamido-2-deoxyglucosidase by batch and continuous cultures of *Bacteroides fragilis* under a variety of conditions has been investigated.[94]

Two distinct 2-acetamido-2-deoxy-α-D-galactosidases have been separated from filtrates of cultured *Clostridium perfringens* by electrophoresis in polyacrylamide gels.[95] One of the enzymes was identified as a α-D-2-acetamido-2-deoxygalactosidase. The second enzyme was probably an *endo*-α-D-2-acetamido-2-deoxygalactanase. The A-specificities of human type-A erythrocytes and of water-soluble glycoproteins having A-activity were both destroyed by incubation with the α-D-2-acetamido-2-deoxygalactosidase, but not the *endo*-α-D-2-acetamido-2-deoxygalactanase. A concomitant rise in blood-group O(H) activity, as indicated by the use of a lectin from *Ulex europeus*, occurred in the A-erythrocytes treated with the α-D-2-acetamido-2-deoxygalactosidase.

A mutation in the post-translational modification of the glycoside hydrolases of *Dictyostelium discoideum* affected the β-D-2-acetamido-2-deoxyglucosidase molecules (see p. 418).[96, 97] β-D-2-Acetamido-2-deoxyglucosidase has been shown to be released during autolysis of *Neurospora crassa* under certain culture conditions.[98]

Vegetative microplasmodia of the slime mould, *Physarum polycephalum*, have been shown to introduce an intracellular β-D-2-acetamido-2-deoxyhexosidase.[99] Under certain conditions extracellular activity was also produced. Evidence demonstrating dissimilarities between the two enzymes included marked differences in specificities for natural and synthetic substrates, kinetic parameters, relative net charges, carbohydrate contents, subunit polypeptide molecular weights (extracellular 2.53×10^4, intracellular 4.05×10^4), amino-acid compositions, and immunological properties. The apparent distinct nature of the structural genes coding for the enzymes was discussed.

3 α-D- and α-L-Arabino(furano)sidases

Significant differences occur between the interactions of glycosaminoglycans with a particular lysosomal protein leading to inhibition of lysosomal enzymes including an arabinosidase.[41]

Chromatographies, polyacrylamide gel electrophoresis, and isoelectric focusing have been used to study the α-L-arabinosidase activity of a partially purified preparation of almond emulsin.[100]

Extraction of germinating-lupin cotyledon, followed by ion-exchange and

[93] W. C. Brown, D. Vellom, E. Schnepf, and C. Greer, *F.E.M.S., Microbiol. Letters*, 1978, **3**, 247.
[94] J. O. Berg, C. E. Nord, and T. Wadstrom, *Appl. Environ. Microbiol.*, 1978, **35**, 269.
[95] W. C. Bell, N. Pomato, and D. Aminoff, *Carbohydrate Res.*, 1978, **61**, 447.
[96] S. J. Free, R. T. Schimke, H. Freeze, and W. F. Loomis, *J. Biol. Chem.*, 1978, **253**, 4102.
[97] S. J. Free and R. T. Schimke, *J. Biol. Chem.*, 1978, **253**, 4107.
[98] F. Reyes, R. Lahoz, and P. Cornago, *Trans. Brit. Mycol. Soc.*, 1977, **68**, 357.
[99] G. D. Kuehn, R. C. Williams, M. K. Cox, and J. C. Jackson, *Arch. Biochem. Biophys.*, 1977, **183**, 686.
[100] Yu. A. Zhdanov, R. M. Kessler, N. R. Yakubova, and K. B. Sherstnev, *Biochemistry (U.S.S.R.)* 1977, **42**, 20.

gel chromatography, gave two α-L-arabinofuranosidases.[101] Some fractions were further purified by using affinity chromatography on agarose-immobilized concanavalin A. The changes in the activities of the enzymes during germination have been determined. Some kinetic and physical properties of these enzymes were described (pH optima 4.4 and 3.5, mol. wt. 7.0×10^4 and 12.0×10^4 respectively) and their role in the modification of cell-wall polysaccharides was discussed.

α-L-Arabinofuranosidase has been obtained from the culture medium of *Scopolia japonica* calluses *via* gel filtration, ion-exchange chromatography, and isoelectric focusing.[102] Various properties of the enzyme were studied, and the effect of lactones on its activity were determined.

The taxonomy of the brown rot fungi (*Monilinia* spp.) has been related to their extracellular cell wall-degrading enzymes.[103] Patterns of α-L-arabinofuranosidase secreted by *M. fructigena*, *M. fructicola*, and *M. laxa* were studied by column isoelectric focusing (p*I* values 5.1; 3.0 and 7.0; and 6.7 and 7.2 respectively).

4 β-D-Fructofuranosidases

The levels of β-D-fructofuranosidase in seeds of a pasture legume (*Stylosanthes humilis*) have been followed during germination.[104]

An increase in β-D-fructofuranosidase content of tomato fruit during the ripening process has been demonstrated by a radioimmunoassay process.[105] The amount of tomato fruit β-D-fructofuranosidase extractable from the cell walls during ripening paralleled the changes in activity of the enzyme.[106] Using the techniques of radioimmunoassay, double immunodiffusion analysis, and immunotitration, no differences in immunological properties of the β-D-fructofuranosidase between the various stages of fruit ripening were detected.

Hyperproduction of β-D-fructofuranosidase resulted from a cross between a hyperderepressed strain and a normal strain of *Neurospora crassa*, and was due to the gene for hyperderepression.[107]

The β-D-fructofuranosidases of various races of *Phytophthora megasperma* have been purified until they gave a single band, whether stained for protein or carbohydrate, on isoelectric focusing in flat bed gels.[108] The enzymes contained 25% carbohydrate consisting of 97% of D-mannose and 2-amino-2-deoxy-D-glucose although the ratio of the two residue types differed from race to race. The existence of race-specific carbohydrate structures in the different races suggested that the carbohydrates may be involved in determining the specificity of host–protein interactions.

101 N. K. Matheson and H. S. Saini, *Carbohydrate Res.*, 1977, **57**, 103.
102 M. Tanaka and T. Uchida, *Biochim. Biophys. Acta*, 1978, **522**, 531.
103 H. J. Willetts, R. J. W. Byrde, A. H. Fielding, and A. L. Wong, *J. Gen. Microbiol.*, 1977, **103**, 77.
104 A. Bacic, G. N. Richards, and D. Yellowlees, *Phytochemistry*, 1977, **16**, 845.
105 T. Iwatsubo, K. Sekiguchi, K. Kurata, T. Tada, K. Iki, H. Nakagawa, N. Ogura, and H. Takehana, *Agric. & Biol. Chem. (Japan)*, 1976, **40**, 1243.
106 K. Iki, K. Sekiguchi, K. Kurata, T. Tada, H. Nakagawa, N. Ogura, and H. Takehana, *Phytochemistry*, 1978, **17**, 311.
107 D. Sternberg and A. S. Sussman, *Arch. Mikrobiol.*, 1974, **101**, 303.
108 E. Ziegler and P. Albersheim, *Plant Physiol.*, 1977, **59**, 1104.

An enzymic test which allows the localization of yeast β-D-fructofuranosidase directly on sodium dodecyl sulphate gels is dependent upon the appearance of an activity band (mol. wt. 1.9×10^5) arising from membrane fractions and additional to those of the external and internal β-D-fructofuranosidases.[109] The band showed kinetic characteristics of a precursor of the external enzyme, and when the increase in the external enzyme slowed down, this membrane-associated form almost disappeared completely. Tunicamycin-initiated inhibition studies indicated that the transient form contained already those carbohydrate parts which contain 2-acetamido-2-deoxy-D-glucose and are transferred *via* dolichyl phosphate-bound intermediates.

Evidence for the involvement of salt links in the stabilization of bakers' yeast β-D-fructofuranosidase has been obtained from studies of immobilization of the enzyme on cellulose, derivatives of Sephadex® and insolubilized concanavalin A, and modification with citraconic anhydride, methyl acetimidate, or ethylene diamine.[110] The general stability to such treatments was offered as the evidence, since modification of either amino- or carboxy-groups would result in loss of conformational stability, unless the stabilizing salt linkages were present.

External yeast β-D-fructofuranosidase has been labelled by [^{3}H]conduritol-B-epoxide, an active-site directed inhibitor of this enzyme.[111] During the inactivation, 1—2 mol of inhibitor were bound per mol. Only the 1-L-1,2-anhydro-*myo*-inositol and not the 1-D-1,2-anhydro-*myo*-inositol was the reactive component of the racemic epoxide. It was concluded that an ester linkage had been formed and that a carboxy-group is situated in the active site of the enzyme.

β-D-Fructofuranosidase has been immobilized, with retention of enzymic activity, by ionic reaction with DEAE- and CM-Sephadexes® and by interaction with immobilized concanavalin A.[110]

5 α-D-, β-D-, and α-L-Fucosidases

Significant differences occur between the interactions of glycosaminoglycans with a particular lysosomal protein leading to inhibition of lysosomal enzymes including α-D-fucosidase.[41]

Levels of lysosomal α-L-fucosidase have been determined in normal hair roots and in hair roots obtained from a patient with mannosidosis.[46] The significance of the results was discussed in relation to the levels of other glycoside hydrolyses present and to the detection of lysosomal storage diseases.

Human α-L-fucosidase is a soluble lysosomal enzyme which hydrolyses α-L-fucosyl residues linked to O-2 of galactose or O-3, 4, or 6 of 2-acetamido-2-deoxy-D-glucose.[112] Demonstration of activity towards natural oligosaccharide or glucosphingolipid substrates was achieved by measuring liberated L-fucose by coupling to L-fucose hydrogenase and NAD and measuring NADH

[109] P. Babczinski and W. Tanner, *Biochim. Biophys. Acta*, 1978, **538**, 426.
[110] J. Woodward and A. Wiseman, *Biochim. Biophys. Acta*, 1978, **527**, 8.
[111] H. Braun, *Biochim. Biophys. Acta*, 1977, **485**, 141.
[112] G. Dawson and G. Tsay, *Arch. Biochem. Biophys.*, 1977, **184**, 12.

production spectrophotomatically. Activity of purified human spleen, brain, and cultured skin fibroblast, or crude cell extracts, towards 4-methylumbelliferyl α-L-fucopyranoside had a pH optimum of 4.5—5.5 and was unaffected by the presence of neutral detergents such as Triton X-100. Various substrate specificity and kinetic studies were also reported together with comparative data for samples from cases of fucosidosis.

α-D-Fucosidase activity has been demonstrated in human and pig kidneys.[113] During purification the two forms of the enzyme co-chromatographed with α-D-galactosidase activity. Similar pH optima and thermostabilities of the two activities and their inhibition by D-fucose, D-galactose-, and D-galactono-1,4-lactone suggested that the α-D-fucosidase and α-D-galactosidase activities are manifested by one molecule.

Human liver α-L-fucosidase, after purification to homogeneity, was found to contain carbohydrate (7% by weight) consisting of D-mannose (8.3), 2-amino-deoxy-D-glucose (4.3), neuraminic acid (1.6), D-glucose (1.6), D-galactose (0.8), and L-fucose (1.8 residues per 5×10^4 molecular weight subunit).[114] The presence of D-galactose and L-fucose was possibly due to artifacts of the purification procedure.

Human liver α-L-fucosidase has been separated into four components by gel filtration and ion-exchange chromatography.[115] These components differed in their relative stabilities to heat and acid and in their response to neuraminidase. The enzymes were able to liberate L-fucose from 2′-L-fucosyllactose. The serum enzyme was also investigated.

The chemical relationship of the seven forms of human liver α-L-fucosidase has been studied by isoelectric focusing of neuraminidase and sialyltransferase-treated preparations of α-L-fucosidase.[116] Neuraminidase treatment leads to a decrease in the activity of the more-acidic forms, and a concomitant increase in the activity of the more-neutral forms. Incubation of the neuraminidase-treated enzyme forms with a radiolabelled CMP-neuraminic acid and sialyltransferase led to generation of more-acidic forms of the enzyme. The seven isoenzymes were separated by preparative isoelectric focusing and were characterized kinetically and immunochemically.

Human serum α-L-fucosidase has been purified 240 000 fold with 35% yield by an affinity chromatographic procedure utilizing agarose-immobilized 2-amino-2-deoxy-L-fucose.[117] Isoelectric focusing of the purified enzyme indicated the presence of several forms (major form p*I* 5.0). Sodium dodecyl sulphate–polyacrylamide gel electrophoresis of the enzyme (mol. wt. $2.96 \pm 0.30 \times 10^5$ by gel filtration) indicated that the presence of two subunits with similar molecular weights (5.65×10^4 and 5.40×10^4). Using the 4-nitrophenyl substrate (K_m value 0.52 mM) the enzyme displayed a broad pH optimum centred round pH 4.8 with a second minor optimum at pH 6.1. The enzyme was shown to contain 1.7% neuraminic acid; quantitative immunoprecipitation studies were also conducted on the enzyme.

[113] G. Y. Wiederschain and E. M. Beyer, *Arch. Biochem. Biophys.*, 1977, **182**, 335.
[114] J. A. Alhadeff and H. Freeze, *Mol. Cell. Biochem.*, 1977, **18**, 33.
[115] M. A. Chester, B. Hultberg, and S. Sjöblab, *Biochim. Biophys. Acta*, 1977, **485**, 147.
[116] J. A. Alhadeff, G. Cimino, and A. J. Janowsky, *Mol. Cell. Biochem.*, 1978, **19**, 171.
[117] J. A. Alhadeff and A. J. Janowsky *Clinica Chim. Acta*, 1978, **82**, 133.

β-D-Fucosidase activity in fibroblasts from a case of juvenile (type 3) Gaucher disease has been found to occur at the normal level.[118] α-L-Fucosidase activity has been shown to be elevated in skin fibroblasts from cystic fibrosis patients.[68] The activities of nine other 'acid' hydrolases in cystic fibrosis and control fibroblasts are, however, similar. The relationship of these results to the finding that a decreased activity of α-L-fucosidase exists in the serum of cystic fibrosis patients was discussed. It was proposed that an abnormal distribution of the enzyme is involved in the pathogenesis of this disease.

The distributions of glycoside hydrolases including α-L-fucosidase in human polymorphonuclear leucocytes have been determined.[62]

Peripheral blood serum α-L-fucosidases from various mammalian species: pig (*Sus scropha*), goat (*Capra hincus*), bull (*Bos taurus*), horse (*Equus caballus*), and donkey (*Equus asinus*) have been studied *via* fluorimetric and spectrophotometric procedures.[119] The pH optima were; 5.5 for all species using fluorescent substrates, 5.0, 4.0, 4.5, 4.5—5.0, and 4.5—5.0 respectively using chromogenic substrates. The serum α-L-fucosidases were all found to be heat labile.

A study of the effect of porcine kidney α-L-fucosidase on human and rabbit immunoglobulins, ceruloplasmin and ovine submaxillary gland glycoprotein showed that free L-fucose was not released.[120] Isoelectric focusing demonstrated the existence of multiple forms of the enzyme.

α-L-Fucosidase has been purified to apparent homogeneity from monkey brain.[121] The enzyme (pH optimum 5.0, K_m 0.22 mM) was shown to be a tetramer (mol. wt. 2.85×10^5) of identical subunits (mol. wt. 7.35×10^4) held together by non-covalent forces. Tetramer, dimer, and monomer were all present and active and gave rise to a triphasic heat inactivation profile and distinctly different pH activity profiles of the differentially heat-inactivated enzymes. L-Fucose was a competitive inhibitor. Partial immunological identities with human brain and liver α-L-fucosidase were demonstrated.

The effect of cortisone and thyroxine on α-D-fucosidase levels in rat forebrain and cerebellum have been examined during early postnatal development.[76]

An affinity column for the purification of α-L-fucosidases has been constructed by linking 4-aminophenyl 1-thio-α-L-fucopyranoside to agarose through succinylated 3,3′-diaminodipropylamine bridges.[122] Excellent purification of α-L-fucosidases from rat epididymus, horseshoe crab (*Limulus polyphemus*), and *Clostridium perfringens* could be effected in one step with good yield. The enzymes purified thereby were free from other glycoside hydrolases and from proteases. The column material was stable and reusable over two years.

Two glucosidases which hydrolysed *inter alia* β-D-fucosidases have been isolated from the digestive juice of *Achatina balteata*.[123]

[118] Y. B. Chiao, G. M. Hoyson, S. P. Peters, R. E. Lee, W. Diven, J. V. Murphy, and R. H. Glen, *Proc. Nat. Acad. Sci. U.S.A.*, 1978, **75**, 2448.
[119] E. Villar P. Calvo, and J. A. Cabezas, *Comp. Biochem. Physiol.*, 1978, **60B** 459.
[120] E. M. Beier and G. Ya. Vidershain, *Biochemistry* (*U.S.S.R.*), 1977, **42**, 684.
[121] T. Alam and A. S. Balasubramanian, *Biochim. Biophys. Acta*, 1978, **524** 373.
[122] R. S. Jain, R. L. Binder, A. Levy-Benshimol, C. A. Buck, and L. Warren *J. Chromatog.*, 1977, **139**, 283.
[123] B. Colas and J. Attias, *Biochimie*, 1977, **59**, 577.

The α-L-fucosidase from the marine mollusc *Littorina littorea* has been purified; only one form was detectable by isoelectric focusing at pH 5.4.[124] L-Fucose was a competitive inhibitor and Hg^{2+} ion was a non-competitive inhibitor of the enzyme (mol. wt. 2.2×10^5, pH optimum 6.5, K_m 3.0×10^{-4} M and V_{max} 2.6 μmol mg^{-1} min^{-1} for 4-nitrophenyl α-L-fucopyranoside). The enzyme proved to be a glycoprotein containing D-glucose (7.4), D-galactose (2.2), D-mannose (1.8), and 2-amino-2-deoxy-D-hexose (2.2%).

A glycoside hydrolase isoenzyme, which is both a β-D-glucosidase and a β-D-galactosidase and was isolated from almond emulsin, has been shown to possess β-D-fucosidase activity.[125] It has been concluded that all three activities reside in a single catalytic site for the following reasons; D-glucosylamine, D-galatosylamine, and D-fucosylamine each act competitively against all three of the corresponding 4-nitrophenyl glycoside substrates; any given inhibitor exhibits the same K_i value when tested in the presence of any of the three substrates; when the enzyme is incubated with any two of the substrates, at or above their respective K_m values, the rate of 4-nitrophenol formation is not additive, but rather is equal to the value calculated on the basis of the individual K_m values and relative maximum velocities.

The α-L-fucosidase activity of culture filtrates of *Scopolia japonica* calluses have been investigated by isoelectric focusing.[102]

α-L-fucosidase has been immobilized with retention of enzymic activity by affinity binding to agarose-immobilized concanavalin A.[120]

6 α- and β-D-Galactosidases and D-Galactolipid-oriented Galactosidases

The conversion of β-D-galactosidase (lactase)-hydrolysed acid whey to alcohol has been reported.[126]

Purified α-D-galactosidase B from normal human tissues also contained α-D-glucosidase, α-D-xylosidase, and α-D-2-acetamido-2-deoxygalactosidase activities, as measured with 4-nitrophenyl glycoside substrates.[43] Treatment of the α-D-galactosidase B with homologous antisera reduced the α-D-galactosidase, α-D-xylosidase, and α-D-2-acetamido-2-deoxygalactosidase activities but not the α-D-glucosidase activity. When the B isoenzyme was incubated with 4-nitrophenyl α-D-galactopyranoside and 4-nitrophenyl 2-acetamido-2-deoxy-α-D-galactopyranoside together, the rate of 4-nitrophenol formation was considerably less than the sum of the rates observed when the substrates were added singly. It was concluded that the α-D-galactosidase B and α-D-2-acetamido-2-deoxygalactosidase activities are due to the same protein containing a single catalytic site.

Significant differences occur between the interactions of glycosaminoglycans with a particular lysosomal protein leading to inhibition of lysosomal enzymes including α- and β-D-galactosidases.[41]

Tissues from a case of juvenile (type 3) Gaucher disease exhibited a deficiency

[124] M. A. de Pedro, A. Reglero, and J. A. Cabezas, *Comp. Biochem. Physiol.*, 1978 **60B** 379.
[125] D. E. Walker and B. Axelrod, *Arch. Biochem. Biophys.*, 1978, **187**, 102.
[126] V. S. O'Leary, R. Green, B. C. Sullivan, and V. H. Holsinger, *Biotechnol. and Bioeng.*, 1977, **17**, 1019.

of β-D-galactosidase, and additionally β-D-glucosidase and β-D-xylosidase activities. Evidence for the manifestation of all three activities by a single enzyme species was provided.[118]

An assay procedure has been developed for accurate estimation of lactosylceramidase II in the presence of a relatively high activity of lactosylceramidase I.[127] The procedure involved determination of lactosylceramidase-cleaving activities under two different assay conditions. Applicability of the procedure was evaluated with separated soluble fractions of the two β-D-galactosidases from normal human brains, and with whole homogenates of grey and white matter, liver and cultured fibroblasts from control individuals, and from patients with globoid cell leukodystrophy or G_{M_1}-gangliosidosis. Assay for the I and II enzymes permitted accurate diagnosis of both globoid cell leukodystrophy and G_{M_1}-gangliosidosis with one natural substrate, lactosylceramide, irrespective of the relative proportion of the two β-D-galactosidases in the tissue.

As part of a study on the neurochemistry of the genetic hyperglycoaminoglycanuria conditions (mucopolysaccharidoses), lysosomal enzymes including α- and β-D-galactosidases in patients of four types and normals have been investigated in brain and liver.[47]

α-D-Galactosidase activity has been demonstrated in human and porcine kidneys.[113] During purification the two forms of the enzyme co-chromatographed with α-D-fucosidase activity. Similar pH optima, thermostabilities, and inhibition by D-fucose, D-galactose, and D-galactono-1,4-lactone suggested that the α-D-galactosidase and α-D-glucosidase activities are manifested by one molecular type.

Previously reported differences in the substrate specificities of α-D-galactosidases A and B from human liver have been further investigated.[50] A highly purified fraction of the B enzyme was found to contain α-D-2-acetamido-2-deoxygalactosidase activity (K_m 2-nitrophenyl 2-acetamido-2-deoxy-α-D-galactopyranoside 1.3 mM, K_m 4-methylumbelliferyl α-D-galactopyranoside 6.8 mM) (see also p. 379).

The association of α-D-galactosidase B and A-like activities with α-D-2-acetamido-2-deoxygalactosidase activities from human liver has been investigated (p. 380).[53]

A method has been described for the rapid isolation of α-D-galactosidases A and B from normal human liver.[128] Interconversion of the B form to an A-like form occurred during storage. Application of the method to liver from cases of Fabry's disease showed a majority of B-like and little A-like enzyme although the kinetic properties of both isoenzymes from the pathological tissue were normal. From these and immunological studies it was concluded that the B-like α-D-galactosidase in Fabry tissues is identical with normal α-D-galactosidase B and the small amount of A-like activity in Fabry material is due to a modified form of α-D-galactosidase B.

[127] H. Tanaka and K. Suzuki, *Clinica Chim. Acta*, 1977, **75**, 267.
[128] A. W. Schram, M. N. Hamers, B. Brouwer-Kelder, W. E. Donker-Koopman, and J. M. Tager, *Biochim. Biophys. Acta*, 1977, **482**, 125.

Two acidic β-D-galactosidase isoenzymes (designated A_1 and A_2) were separated by isoelectric focusing from β-D-galactosidase A of human liver.[129] Kinetic studies with 4-methylumbelliferyl β-D-galactopyranoside substrate revealed similar parameters for both (pH optima 4.2 and 4.5, K_m values 0.32 and 0.30 mmol l^{-1} respectively). The A_1 enzyme form was shown to be more heat labile than the A_2. Significant differences were observed with antibody preparations against the two enzyme forms. Using the anti-A_1 antibodies two precipitin arcs with residual enzymatic activity were obtained by immunoelectrophoresis of β-D-galactosidase A whereas only one was obtained with anti-A_2 antibodies. Anti-A_1 precipitated 85% of the original activity present in β-D-galactosidase A and only 56% could be precipitated by anti-A_2. The possibility of common structural components was suggested.

Human liver 'acid' β-D-galactosidases A_2 and A_3 have been isolated *via* chromatography on agarose-immobilized concanavalin A, and a 6-aminohexyl 1-thio-β-D-galactopyranoside derivative of concanavalin A.[130] Form A_2 (mol. wt. $1.50 \pm 0.15 \times 10^5$ by gel filtration) appeared as a single band (mol. wt. $6.5 \pm 0.1 \times 10^4$) on electrophoresis in the presence of sodium dodecyl sulphate. Form A_3 (mol. wt. $6.60 \pm 0.66 \times 10^5$) appeared under similar conditions as a major band (mol. wt. $6.5 \pm 0.1 \times 10^4$) and minor bands (mol. wt. $4.4 \pm 0.1 \times 10^4$, $2.6 \pm 0.1 \times 10^4$ and $2.2 \pm 0.1 \times 10^4$). Gel filtration of purified β-D-galactosidase A_3 generated approximately equal amounts of forms A_3 and A_2. β-D-Galactosidase A_1 was not detected by gel filtration of partially or highly purified preparations of forms A_2 and A_3. Both A_2 and A_3 forms had identical isoelectric points (4.42 ± 0.02). It was concluded that the forms A_2 and A_3 are dimeric and multimeric forms of β-D-galactosidase A_1. β-D-Galactosidase A_2 contained 7.5% carbohydrate by weight consisting of neuraminic acid, D-galactose, 2-amino-2-deoxy-D-glucose, and D-mannose (molar ratio 1.1 : 1.0 : 1.7 : 2.7).

The activities of human liver 'acid' β-D-galactosidase against 4-methylumbelliferyl β-D-galactopyranoside, G_{M1}-ganglioside and lactosylceramide appear to be manifested by a single molecular species.[131] Evidence for this included copurification of all activities by affinity chromatography on immobilized concanavalin A and on a β-D-galactopyranosyl derivative of agarose to yield a single polyacrylamide gel electrophoretic band.

Two 'neutral' β-D-galactosidases have been purified from human liver *via* affinity chromatography on immobilized concanavalin A, to remove 'acid' β-D-galactosidases, and affinity chromatography on a β-D-galactopyranosyl derivation of agarose.[132] The purified isoenzymes (mol. wts. 3.75—3.95 $\times 10^4$, pH optima 6.2, similar K_m and V_{max} values) also exhibited β-D-glucosidase and D-galactosylglucosylceramidase (pH optimum 5.5) activities. No immunological crossreactivity was observed between the 'neutral' and 'acid' forms.

129 Y. Ben-Yoseph, E. Shapira, and H. L. Nadler, *Enzyme*, 1977, **22**, 276.
130 R. G. Frost, E. W. Holmes, A. G. W. Norden, and J. S. O'Brien, *Biochem. J.* 1978 ,**175**, 181.
131 A. L. Miller, R. G. Frost, and J. S. O'Brien, *Biochem. J.*, 1977, **165**, 591.
132 Y. Ben-Yoseph, E. Shapira, D. Edelman, B. K. Burton, and H. L. Nadler, *Arch. Biochem. Biophys.*, 1977, **184** 373.

The type A or 'acid' and type B or 'neutral' β-D-galactosidase activities have been measured in post-mortem liver samples from cases of non-genetic disease or ganglioside storage diseases other than G_{M_1}-gangliosidosis.[133] The type A activities fall within the established normal range in all samples. The type B activities showed a bimodal distribution suggestive of the occurrence of two distinct populations of human individuals. No clinical symptoms were associated with the much lower type B β-D-galactosidase activities and it appeared that this β-D-galactosidase deficiency could be found in the original tissues. Methods of screening for type B β-D-galactosidase deficiency were described and the significance of the enzyme deficiency was discussed.

α-D-Galactosidase A has been purified from human placenta.[134] Polyacrylamide gel electrophoresis of the purified, *S*-carboxymethylated enzyme in sodium dodecyl sulphate showed one component (mol. wt. 6.5×10^4), but electrophoresis of the non-derivatized enzyme showed a major (mol. wt. 6.75×10^4) and a minor (mol. wt. 4.70×10^4) component, the latter being a degradation product. The native enzyme was suggested to be a dimer (mol. wt. 1.5×10^5). Homologous antibody did not precipitate the α-D-galactosidase in Fabry's disease fibroblasts. The instability of the normal enzyme to heat and pH imposes a serious obstacle to its use in the treatment of Fabry's disease.

Two α-D-galactosidase isoenzymes (mol. wts. 1.03×10^5 and 1.17×10^5, K_m's 1.5 mM and 13.1 mM for 4-methylumbelliferyl α-D-galactopyranoside, pH optima 4.4 and 4.4, p*I*'s 4.7 and 4.4 respectively) have been purified from human placental tissue to homogeneity on disc gel electrophoresis.[135] According to their affinity for immobilized concanavalin A and their positive reaction in the periodate–Schiff stain, they both contained carbohydrate. Minimum molecular weights of 5.77×10^4 and 4.77×10^4 in the presence of sodium dodecyl sulphate indicated that both isoenzymes are dimers. The isoenzymes exhibited different patterns of inhibition by various carbohydrates. It was concluded that they are not structurally related and that they do not share a precursor–product relationship.

The human plasma form of α-D-galactosidase A has been purified.[136] The enzyme (mol. wt. 9.66×10^4, p*I* 3.7, K_m 1.9 mM for 4-methylumbelliferyl α-D-galactopyranoside, 0.23 mM for globotriglycosylceramide) was inhibited by *myo*-inositol (K_i 0.29 M) similarly to α-D-galactosidase from other tissues. Comparison with the liver enzyme, and the increased p*I* value obtained after neuraminidase treatment suggested that the plasma form is an isoenzyme with a higher neuraminic acid content that the tissue α-D-galactosidase isoenzymes.

Human plasma α-D-galactosidase A has been purified 7000 fold from Cohn Fraction IV–I on a pilot scale.[137] The ratio of activities against 4-methylumbelliferyl α-D-galactopyranoside and ceramide trihexoside was 6.2 : 1; both activities were heat labile and exhibited the same mobilities on polyacrylamide gel electrophoresis.

[133] P. S. J. Cheetham, N. E. Dance, and D. Robinson, *Clinica Chim. Acta*, 1978, **83**, 67.
[134] J. S. Mayes and E. Beutler, *Biochim. Biophys. Acta*, 1977, **484**, 408.
[135] J. W. Kusiak, J. M. Quirk, and R. O. Brady, *J. Biol. Chem.*, 1978, **253**, 184.
[136] D. F. Bishop and C. C. Sweeley, *Biochim. Biophys. Acta*, 1978, **525**, 399.
[137] D. F. Bishop, D. E. Wampler, J. T. Sgouris, R. J. Bonefeld, D. K. Anderson, M. C. Hawley, and C. C. Sweeley, *Biochim. Biophys. Acta*, 1978, **524**, 109.

'Acid' β-D-galactosidases from human urinary samples have been characterized using G_{M_1}-ganglioside, asialofetuin, and 4-methylumbelliferyl β-D-galactopyranoside as substrates.[138] Gel filtration revealed the presence of three forms of the enzyme (mol. wts. 5—7 × 10^5, 9—12 × 10^4 and 2.0—2.7 × 10^4). This report was claimed to represent the first characterization of urinary 'acid' β-D-galactosidase with respect to naturally occurring glycolipid and glycoprotein substrates. In addition, data were presented to indicate that the enzyme may be composed of an enzymically active form (apparent mol. wt. 2.0—2.7 × 10^4) which is also capable of hydrolysing glycolipid and glycoprotein.

The physicochemical and kinetic properties (pH optimum 4.7, K_m 2.6 mM for 4-methylumbelliferyl α-D-glucopyranoside) of α-D-galactosidase A activity in normal cultured endothelial cells have been determined.[139] In association with studies on the molecular pathology of Fabry's disease, a mechanism was derived for the accumulation of trihexosylceramide due to the defective α-D-galactosidase A activity.

The levels of β-D-galactosidase activity in skin fibroblasts from cases of cystic fibrosis and controls have been compared.[68] Complementation studies have been conducted with enucleated fibroblasts from different variants of β-D-galactosidase deficiency.[140]

A patient was previously characterized as having a varient form of G_{M_1}-gangliosidosis, based on severe deficiencies in β-D-galactosidase activity in both fibroblasts and leucocytes using 4-methylumbelliferyl β-D-galactopyranoside and G_{M_1}-ganglioside as substrates.[141] Re-examination of the cultured fibroblasts showed a severe deficiency in neuraminidase activity against certain, but not all substrates. The presence of normal levels of β-D-galactosidase activity in leukocytes from the mother of the patient indicated that the β-D-galactosidase deficiency is not the primary enzyme defect.

The distribution of glycoside hydrolases including α- and β-D-galactosidases in human polymorphonuclear leucocytes has been determined.[62]

A method has been developed for the isolation of β-D-galactosidase from bovine intestinal juice.[142] The properties of the enzyme were investigated using lactose and 2-nitrophenyl β-D-galactopyranoside as substrates. The β-D-galactosidase of porcine brain fractions has been investigated.[74]

The purification of two forms of β-D-galactosidase from monkey brain has been described.[143] The enzyme forms (pH optima 3.8—4.0, K_m's 0.22 and 0.22 mM for 4-nitrophenyl β-D-galactopyranoside, mol. wts. 1.2 × 10^5 and 1.2 × 10^6) exhibited subcomponent phenomena on polyacrylamide gel electrophoresis.

G_{M_1}-ganglioside β-D-galactosidase has been purified from feline brain (2000 fold, 36% yield) and liver (30 000 fold, 40% yield) by a combination of

[138] B. C. Kress and A. L. Miller, *Clinica Chim. Acta*, 1978, **85**, 23.
[139] D. L. Johnson and R. J. Desnick, *Biochim. Biophys. Acta*, 1978, **538** 195.
[140] H. A. De Wit-Verbeek, A. Hoogeveen, and H. Galjaard, *Experimental Cell Res.*, 1978, **113**, 215.
[141] D. A. Wenger, T. J. Tarby, and C. Wharton, *Biochem. Biophys. Res. Comm.*, 1978, **82**, 589.
[142] G. V. Vikha, V. I. Tyapko, N. A. Zhuravskaya, P. E. Pavlovskii, and E. D. Kaverzneva, *Biochemistry* (*U.S.S.R.*), 1977, **42**, 1066.
[143] T. Alam and A. S. Balasubramanian, *J. Neurochem.*, 1978, **20**, 1199.

classical and affinity procedures. Disc gel electrophoresis showed multiple and single components for the brain and liver preparations respectively.[144] The brain and liver enzymes (mol. wts. 2.50×10^5 and 2.50×10^5, pH optima *ca.* 4.2 and 4.2, K_m's for 4-methylumbelliferyl β-D-galactopyranoside 0.173 mM and 0.325 mM respectively) were inhibited by mercuric chloride. This and their sensitivities to hydrogen peroxide and persulphate suggested the involvement of a thiol group in the catalysis mechanism.

The β-D-galactosidase activity of guinea-pig brush border membrane has been investigated.[145]

The effect of cortisone and thyroxine on β-D-galactosidase levels in rat forebrain and cerebellum have been examined during early postnatal development.[76] In normal and regenerating rat liver tissue, β-D-galactosidase has been shown to be localized primarily in the fraction of light mitochondria, enriched with lysosomes.[77] The interaction of rat liver β-D-galactosidase with agarose-immobilized ovalbumin has been investigated.[78] The use of the latter as an affinity chromatography material was also studied.

β-D-Galactosidase has been purified from rat mammary gland *via* gel filtration and affinity chromatography on agarose-immobilized 4-aminophenyl 1-thio-β-D-galactopyranoside.[146] The purified enzyme moved electrophoretically as a single band. At pH's 7.0 and 5.0 different molecular sizes (1.1×10^5 and 2.0×10^5 respectively) were observed by gel filtration. The enzyme exhibited both hydrolytic and transferase activities, the latter accounting for the biosynthesis of 6β-*O*-D-galactopyranosyl *myo*-inositol. The β-D-galactosidase was strongly inhibited by heavy-metal atoms, D-galactono-1,4-lactone, phenyl 1-thio-β-D-galactopyranoside, and 4-chloromercuribenzoate.

The activities of α- and β-D-galactosidases were shown to be significantly lower in the kidneys, spleen, liver, leg muscle, cheek pouch, and spinal cord of diabetic XA line that those in the non-diabetic M line Chinese hamsters.[147] Both the β-D-galactosidase isoenzymes separable by ion-exchange chromatography were shown to be depressed. The M and XA line α-D-galactosidase (mol. wt. 2.4×10^5) and β-D-galactosidases (mol. wts. 1.6×10^5 and 1.9×10^5) exhibited similar pH optima and K_m values for 4-nitrophenyl glycosides.

β-D-Galactosidase activity, in foetal mice, was shown to appear first at 16 days gestation and to have a pH optimum of 4.[148] In postnatal development the enzyme activity of cell homogenates tended to show bimodal pH optima at 4 and 5.6. There were two molecular forms of the enzyme, separable both by molecular-sieve chromatography and electrophoresis. One of the molecular forms of the enzyme was active over a wider range of pH (3.2—6.2) and had half as much activity at pH 5.6 as it did at pH 4. This isoenzyme was continuously present in both foetal and postnatal stages. The second isoenzyme first appeared at birth, was active over a narrower range of pH (4.6—6.2) and inactive at pH 4. The bimodal pH optima, observed in postnatal stages in the cell

[144] J. K. Anderson, J. E. Mole, and H. J. Baker, *Biochemistry*, 1978, **17**, 467.
[145] J. R. Green and B. Hadorn, *Biochim. Biophys. Acta*, 1977, **467**, 86.
[146] C. H. Kwo and W. W. Wells, *J. Biol. Chem.*, 1978, **253**, 3550.
[147] A. Y. Chang, *Biochim. Biophys. Acta*, 1978, **522**, 491.
[148] K. I. Krishna and P. R. V. Nayudu, *Enzyme*, 1978, **23**, 145.

homogenates, appeared to be due to the combined activity of the two molecular forms. In isolated brush border membranes, isoenzyme 2 was the only one present. The other organelles (mitochondria, microsomes, lysosomes, nuclei, and cytoplasm) had variable proportions of both isoenzymes, as indicated by the activity ratio at pH 4/5.6.

Two glycosidases have been purified to homogeneity on polyacrylamide gel electrophoresis, from the digestive juice of *Achatina balteata* by acetone fractionation, ion-exchange chromatography, ammonium sulphate fractionation, and gel filtration.[123] Both enzymes were highly specific for the β-D-anomeric configuration of the glycosidic linkage. They hydrolysed cellobiose, lactose, and synthetic β-D-galactosides, -glucosides, and -fucosides at a pH optimum of 5.2—5.6. The substrates were recognized by the same catalytic site as shown by mutual competition studies between substrates and competitive inhibition observed with aldonolactones and glycopyranoses such as D-galacto-, D-gluco, and D-fuco-pyranoses. The kinetic parameters and physicochemical properties of the two enzymes differed.

Some evidence has been offered for the manifestation of an α-D-galactosidase and an α-D-2-acetamido-2-deoxygalactosidase activity of the limpet (*Patella vulgata*) by a single molecular species.[90]

The β-D-galactosidase activity of partially purified almond emulsin has been studied using chromatographic analyses, polyacrylamide gel electrophoresis, and isoelectric focusing.[100] A β-D-galactosidase isoenzyme (mol. wt. 1.358×10^5, protomer mol. wt. 6.515×10^4) isolated in homogeneous form from almond emulsin was found to be a single glycoprotein.[149] The β-D-galacto- and gluco-sidase activities comigrated under a variety of chromatographic and electrophoretic examinations.

A purified β-D-galactosidase isoenzyme purified from almond emulsin was found to catalyse the hydrolysis of both β-D-galacto- and β-D-gluco-pyranosides.[150] The different inhibitions of the enzyme activities by various compounds was investigated in detail. Analysis of the kinetic data indicated that the catalytic site of the enzyme responsible for the β-D-galactosidase activity is kinetically different from the β-D-glucosidase site. Evidence has been provided for a single catalytic site for β-D-fucosidase, β-D-glucosidase, and β-D-galactosidase activities of almond emulsin (see p. 392).[125]

Five α-D-galactosidases have been identified by chromatography and polyacrylamide disc-gel electrophoresis in the germinated seeds of white clover *Trifolium repens*).[151] One of the enzymes (pH optimum 4.0, mol. wt. 4.1—4.3×10^4 by sodium dodecyl sulphate–polyacrylamide gel electrophoresis) was studied in further detail and was shown to possess D-galactosyltransferase activities.

The properties of two forms (A and B) of α-D-galactosidase from coconut kernel has been described.[152] They were interconvertible, and in the absence

[149] A. K. Grover, D. D. MacMurchie, and R. J. Cushley, *Biochim. Biophys. Acta*, 1977, **482**, 98.
[150] A. K. Grover and R. J. Cushley, *Biochim. Biophys. Acta*, 1977, **482**, 109.
[151] J. Williams, H. Villarroya, and F. Petek, *Biochem. J.*, 1977, **161**, 509.
[152] K. Balasubramanian, P. M. Dey, and J. B. Pridham, *Phytochemistry*, 1976, **15**, 1445.

of KCl, the lower molecular weight, B form, was favoured. Combination of B with an endogenous enzymically inactive protein yielded almost complete conversion to A.

The location and level of β-D-galactosidase activity in *Hippeastrum* flower petals have been examined.[153]

A simple procedure developed for the purification of Jack bean β-D-galactosidase employed affinity chromatography on agarose-immobilized 6-*N*-β-(4-aminophenyl)-ethylamino-3-*O*-β-D-galactopyranosyl-6-deoxy-L-gulitol.[154] The purified enzyme (K_m 0.59 mM for 4-nitrophenyl β-D-galactopyranoside) was obtained in 64% yield. Possible application of this new type of affinity ligand to purification of other glycoside hydrolases was discussed.

Extraction of germinating lupin cotyledons, followed by ion-exchange and gel chromatography, gave three β-D-galacto(pyrano)sidases.[101] Some fractions were further purified by using affinity chromatography on agarose-immobilized concanavalin A. The changes in the activities of the enzymes during germination have been determined. Some kinetic and physical properties of these enzymes were described (pH optima 3.2, 3.9, and 3.1; mol. wt. 7.4×10^4, 5.4×10^4, and 6.5×10^4 respectively), and their role in the modification of cell-wall polysaccharides was discussed.

The levels of α- and β-D-galactosidases have been followed in the cotyledons and embryos of germinating seeds of cultivated and wild peas (*Pisum sativum* and *P. elatius*).[155] α-D-Galactosidase is preformed in the cotyledons but its activity increases during germination in the embryos. β-D-Galactosidase activity in embryos increased during germination but showed little change in cotyledons. The possible functions of the enzymes were discussed.

The α- and β-D-galactosidases of the culture medium of *Scopolia japonica* calluses have been investigated by isoelectric focusing.[102]

β-D-Galactosidase isolated from a partially purified preparation obtained from cultured cells of a particular strain of *Aspergillus oryzae* gave a single band on polyacrylamide gel electrophoresis and was free from other glycoside hydrolase activities.[156] The enzyme was stable in aqueous solution, retaining full activity at 4 °C for several months. At pH 4.5, the optimum pH for activity, and 37 °C full activity was maintained for several days.

The formation of β-D-galactosidase by batch and continuous cultures of *Bacteroides fragilis* under a variety of conditions has been investigated.[94]

The activity and stability of *Escherichia coli* β-D-galactosidase are markedly increased by an acid-resistant dialysable factor that accompanies commercial preparations of albumin.[157] The thermal stability of the enzyme has been found to be *in situ* greater than *in vitro* but smaller than *in vivo*.

The effect of pH upon the *Escherichia coli* β-D-galactosidase-catalysed hydrolyses of aryl β-D-galactosides has been shown to be essentially similar for each

[153] H. C. Butcher, G. J. Wagner, and H. W. Siegelman, *Plant Physiol.*, 1977, **59**, 1098.
[154] P. V. Wagh, *Biochim. Biophys. Acta*, 1978, **522**, 515.
[155] I. Marbach, A. M. Mayer, and R. Maron, *Phytochemistry*, 1978, **17**, 655.
[156] M. Akasaki, M. Suzuki, I. Funakoshi, and I. Yamashina, *J. Biochem.* (*Japan*), 1976, **80**, 1195.
[157] J. Moreno and A. Sols, *Mol. Cell. Biochem.*, 1975, **9**, 59.

of the three steps of their hydrolysis.[158] It differs markedly from that on the hydrolysis of D-galactosyl pyridinium salts; these proceed through a 'non-bottleneck' pathway. While pH increase abolishes the rate of every step of the reaction for aryl β-D-galactosides, it favours the first step of hydrolysis of the D-galactosyl pyridinium salts, which supports the hypothesis that catalysis of these compounds originates largely in non-covalent interactions.

The binding of a series of alkyl 1-thio-β-D-galactopyranosides to β-D-galactosidase from *E. coli* has been investigated.[159] The inhibition constants were compared to the partition coefficients for the transfer of these substrate-analogues from water to octan-1-ol. The relationships between the observed binding constants and the partition coefficients indicate that part of the agylcon group binds to a hydrophobic area that is limited in relation to the length of hydrocarbon chain that can be accommodated. Outside this area, the hydrocarbon chain is only partially desolvated. The main driving-force for binding of the agylcon group is the increase in entropy resulting from the return of water molecules from the more-organized layer around the solute molecule to the bulk-water phase.

The binding of a series of substituted phenyl 1-thio-β-D-galactopyranosides to the active site of β-D-galactosidase from *E. coli* has been investigated.[160] The inhibition constants and the standard free energy of transfer $\Delta G^0(K_i)$ were compared with the duration coefficients and ΔG^0 (octan-1-ol) values of these inhibitor molecules in a model octan-1-ol–water system. All of the evidence available suggested that the agylcon group of the inhibitor was bound to the active site by unspecific, hydrophobic forces. Correlations between ΔG^0 (K_i) and ΔG^0 (octan-1-ol) could not be found for the whole series but only for simpler sub-series. Probably, the octanol system was too incomplete a model for the hydrophobic region in the active site of β-D-galactosidase.

The site of attachment to *E. coli* β-D-galactosidase of the active site-directed inhibitor methyl 4-nitrophenyltriazine β-D-galactopyranoside has been determined.[161] When the enzyme was completely inactivated, 1 mole of the methyl D-galactopyranoside group was bound per mole of monomer with retention of the tetrameric structure. After reactions with the [^{14}C]methyl version of the reagent, labelled peptides were isolated and analysed. The radioactive label was found to be covalently bound to L-methionine residue 500.

The action of different effectors, glycosides, and alcohols on the reactions catalysed by *E. coli* β-D-galactosidase has been studied with a view to determining the conformation adaptability of the enzyme active site.[162] Effectors as large as tri- and tetra-saccharides had no effect on the enzyme activity, suggesting a small size for the binding site; most of the β-D-galactosides produced a competitive inhibition; other compounds acting as uncompetitive inhibitors with a better affinity for the reaction intermediate than the free

[158] S. G. Withers, M. Jullien, M. L. Sinnott, O. M. Viratelle, and J. M. Yon, *European J. Biochem.*, 1978, **87**, 249.
[159] C. K. De Bruyne, and M. Yde, *Carbohydrate Res.*, 1977, **56**, 153.
[160] M. Yde and C. K. De Bruyne, *Carbohydrate Res.*, 1978, **60**, 155.
[161] A. V. Fowler, I. Zabin, M. L. Sinnott, and P. J. Smith, *J. Biol. Chem.*, 1978, **253**, 5283.
[162] P. J. Deschavanne, O. M. Viratelle, and J. M. Yon, *J. Biol. Chem.*, 1978, **253**, 833.

enzyme-transfer products were frequently encountered. It was concluded that the active site is made up of two subsites, a D-galactose and a D-glucose subsite.

Removal of the C-5 hydroxymethyl group of the D-galactopyranosyl substrates of *E. coli* β-D-galactosidase has only a slight effect, usually a decrease, on the rate of the conformation step with aryl glycosides, on the rate of cleavage of the C–N bond of the glycosylpyridium ion, or on the rate of hydrolysis of the glycosyl–enzyme.[163] Values of K_m were all substantially increased. It was concluded that substrate destabilization of the particular type commonly envisaged for lysosyme catalysis therefore plays no part in catalysis by this particular glycosidase, and that the C-5 hydroxymethyl group plays a role only on binding.

On incubation with *E. coli* β-D-galactosidase, D-[2-^{2}H]galactal (1) plus glycerol yielded 1-deoxyglycerol-1-yl 2-deoxy-β-D-*lyxo*-(*S*)-[2-^{2}H]hexopyranoside. By ^{1}H n.m.r. analysis, it was shown that the hydrogen atom introduced on C-2 is *trans*-related to the agylcon moiety.[164] In contrast to this stereospecific, enzyme-catalysed addition, the reaction of phenol with peracetylated (1), catalysed by 4-toluenesulphonic acid, which yields phenyl 3,4,6-tri-*O*-acetyl-2-deoxy-α-D-*lyxo*-[2-^{2}H]hexopyranoside, was shown to entail both a *trans* and *cis* addition.

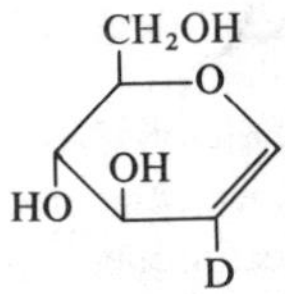

(1)

The inactivation of *E. coli* β-D-galactosidase by *N*-bromacetyl β-D-glucopyranosylamine has been tested and compared with the action of *N*-bromoacetyl β-D-galactopyranosylamine with a view to investigation of the possibility of the existence of two subsites at the active centre of the enzyme.[165]

The amino-acid sequence of 72 chymotryptic peptides isolated from [^{14}C], [^{3}H]-labelled carboxymethyl-β-D-galactosidase of *E. coli* has been determined.[166] A variety of techniques were used in isolation procedures including separation by solubility, size, and chromatography. These peptides contain approximately 500 amino-acids. Peptides from this study and those reported earlier account together for the sequence of about 600 of the 1021 residues in the subunit.

Hydrolysis with trypsin of citraconyl-carboxymethyl-β-D-galactosidase from *E. coli* has been carried out under limiting conditions.[167] L-Aspartic acid–L-arginine–X sequences were cleaved and many large peptides were produced. From this digest and two earlier preparations, all 80 theoretically possible tryptic fragments have been isolated and their structures determined.

All the 24 cyanogen bromide peptides of β-D-galactosidase from *E. coli* have

[163] P. Marshall, C. G. Reed, M. L. Sinnott, and I. J. L. Souchard, *J. C. S. Perkin II*, 1977, 1198.

[164] J. Lehmann and B. Ziegler, *Carbohydrate Res.*, 1977, **58**, 73.

[165] O. M. Viratelle, J. M. Yon, and J. Yariv, *F.E.B.S. Letters*, 1977, **79**, 109.

[166] A. V. Fowler, *J. Biol. Chem.*, 1978, **253**, 5484.

[167] A. V. Fowler, A. J. Brake, and I. Zabin, *J. Biol. Chem.*, 1978, **253**, 5490.

been isolated in pure form.[168] This study has been carried out as part of the determination of the amino-acid sequence of the enzyme. The amino-acids in 9 of these peptides have been sequenced and high proportions of L-glutamic acid and L-aspartic acid residues and low proportions of L-lysine were discovered in some of these peptides.[169] The amino-acid sequences in 8 of the cyanogen bromide peptides comprising the central segment of the β-D-galactosidase have been presented.[170] This portion of the molecule, about 27% of the protein, was shown to contain over 40% of the L-lysine and L-tyrosine residues and had a slight excess of basic amino-acids.

The whole amino-acid sequence of *E. coli* β-D-galactosidase has been assembled.[171] The monomer contained 1021 amino-acid residues in a single polypeptide chain (mol. wt. 1.16349×10^5). The sequence determination was aided by analysis of cyanogen bromide peptides obtained from a polypeptide fragment produced by a *lacZ* termination mutant strain.

The sequence of the *C*-terminal third of the β-D-galactosidase has been presented.[172] L-Tyrosine, L-phenylalanine, and L-valine were low in this segment whereas L-alanine and L-lysine were high. This segment has a slight excess of basic groups.

The amino-acid sequence of β-D-galactosidase from *E. coli* has been compared with that of other proteins.[173] Identities with the dihydrofolate reductase of the same organism were highlighted.

Fragments from carboxymethylated β-D-galactosidase from *E. coli* have been shown to be capable of restoring enzymic activity to other fractions of the enzyme by a complementation technique.[174]

The functional properties of C_{ZP} protein, a mutant deriving from wild-type *E. coli* β-D-galactosidase by a point mutation, have been investigated.[175] A large decrease of the specificity, as evaluated by k_{cat}/K_m ratio, was observed and principally originated by a weaker binding of the substrates. Conclusive evidence was given for the occurrence of a two-step mechanism, with a sequential release of the products.

The regulation of the *in vitro* biosynthesis of the *N*-terminal portion of the β-D-galactosidase molecule (α-peptide) of *E. coli* has been investigated using DNA fragments of the lactose operon as template.[176]

Two forms of initiation factor 2 (mol. wt. 9.0×10^4) have been isolated from *E. coli* extracts and tested for their ability to support β-D-galactosidase biosynthesis in a phage DNA-directed *in vitro* protein biosynthesis system.[177] Only the IF-2α form was found to function in β-D-galactosidase biosynthesis.

168 A. V. Fowler, *J. Biol. Chem.*, 1978, **253**, 5499.
169 A. V. Fowler and I. Zabin, *J. Biol. Chem.*, 1978, **253**, 5505.
170 A. V. Fowler and I. Zabin, *J. Biol. Chem.*, 1978, **253**, 5510.
171 A. V. Fowler and I. Zabin, *J. Biol. Chem.*, 1978, **253**, 5521.
172 A. V. Fowler, A. J. Brake, and I. Zabin, *J. Biol. Chem.*, 1978, **253**, 5515.
173 J. M. Hood, A. V. Fowler, and I. Zabin, *Proc. Nat. Acad. Sci., U.S.A.*, 1978, **75**, 113
174 D. V. Marinkovic and J. N. Marinkovic, *Biochem. J.*, 1977, **165**, 417.
175 P. J. Deschavanne, O. M. Viratelle, and J. M. Yon, *Proc. Nat. Acad. Sci., U.S.A.*, 1978, **75**, 1892.
176 H.-F. Kung, M. Tainsky, and H. Weissbach, *Biochem Biophys. Res. Comm.*, 1978, **81**, 1000.
177 B. Eskin, B. Treadwell, B. Redfield, C. Spears, H.-F. Kung and H. Weissbach, *Arch. Biochem. Biophys.*, 1978, **189**, 531.

Studies of the lactose transport activity of cytoplasmic membrane vesicles from *E. coli* have demonstrated the functional symmetry of the β-D-galactosidase carrier in the organism.[178]

The properties of α-D-galactosidase (mol. wt. 3.0×10^5) from *Saccharomyces carlsbergensis* have been shown to be similar to those of other extracellular enzymes in yeast.[179] The carbohydrate content (57%) consists of D-mannose (*ca.* 92%), D-glucose (7%), and 2-amino-2-deoxy-D-glucose (1%); 35% of its amino-acid residues can be accounted for by L-threonine, L-serine, and L-aspartic acid. Although the enzyme remains active after treatment with sodium dodecyl sulphate, it is denatured by urea and guanidinium chloride. With urea or 60 °C heat denaturation, the enzyme dissociates into two types of subunits and was therefore considered to be the first reported external enzyme from yeast with an oligomeric structure.

The properties of *Saccharomyces lactis* β-D-galactosidase which occurs in milk have been reported.[180] The existence and characteristics of the enzyme as carried through into milk products were also investigated.

The production of β-D-galactosidase by *Streptococcus thermophilus* has been studied using milk whey as substrate.

β-D-Galactosidase has been immobilized with retention of enzymic activity by modification with Brilliant Orange KKh and adsorption of the product on to DEAE-Sephadex®;[182] and by reaction with agarose cyclic imidocarbonate and with co-immobilization with lysozyme on the same support.[183]

7 α- and β-D-Glucosidases and D-Glucolipid-oriented D-Glucosidases

β-D-Glucosidase activity has been determined using an automated system with a cyanide electrode under flow-stream conditions.[184] The effects of solution concentrations, pH, and temperature on the reaction with amygdalin as substrate were investigated.

A rapid method reported for the direct observation of β-D-glucosidases after separation on polyacrylamide gel was based on pre-incorporation of 2-nitrophenyl β-D-glucopyranoside into the gel and incubation at 37 °C until sufficient 2-nitrophenol had been liberated by the separated enzymes to give visible yellow bands.[185]

Significant differences occur between the interactions of glycosaminoglycans with a particular lysosomal protein leading to inhibition of lysosomal enzymes including α- and β-D-glucosidases.[41]

Human D-glucocerebrosidase has been shown to be cleared from rat

178 R. M. Teather, O. Hamelin, H. Schwarz, and P. Overath, *Biochim. Biophys. Acta*, 1977, **467**, 386.
179 P. S. Lazo, A. G. Ochoa, and S. Gascón, *Arch. Biochem. Biophys.*, 1978, **191**, 316.
180 E. J. Guy and E. W. Bingham, *J. Dairy Sci.*, 1978, **61**, 147.
181 M. V. R. Rao and S. M. Dutta, *Appl. Environ. Microbiol.*, 1977, **34**, 185
182 B. Mattiasson, *Biotechnol. Bioeng.*, 1977, **19**, 777.
183 V. S. Baeva, L. V. Kozlov, V. K. Antonov, and A. S. Tikhomirova, *Biochemistry* (*U.S.S.R.*,) 1977, **42**, 544.
184 G. Gambaretto and A. E. Pedler, *Ann. Chim.* (*Italy*), 1975, **64**, 717.
185 H. Verachtert, K. Ramasamy, P. de Schuttere, and J. Bevers, *J. Chromatog.*, 1978, **147**, 443.

circulation with a half-life of 21 minutes after intravenous administration.[186] The half-life of the desialylized enzyme is 1.2 minutes. Desialylized orosomucoid inhibits the disappearance of the modified enzyme from the blood. The uptake of the native and desialylized enzyme by hepatocytes and Kupffer cells was also investigated.

Tissues from a case of juvenile (type 3) Gaucher disease exhibited a deficiency of β-D-glucosidase activity.[118] Evidence was presented for the manifestation of β-D-glucosidase, β-D-galactosidase, and β-D-xylosidase activities by a single enzyme species.

Four cases of atypical forms of glycogenosis with α-D-glucosidase ('acid' maltase) deficiency have been described. Studies of the enzyme activity in muscle, leucocytes, and fibroblasts emphasized the heterogeneity of α-D-glucosidase deficiency in man.[187] The properties of the enzyme purified from various tissues were studied. The pH optimum of the placental enzyme was found to be *ca.* pH 4.9.

An 'acid' α-D-glucosidase having both (1 → 4)-α-D-glucosidase and (1 → 6)-α-D-glucosidase activities has been purified from human liver.[188] The enzyme (mol. wt. 1.1×10^5 by polyacrylamide gel electrophoresis; K_m maltose 8.9 mM, K_m glycogen 9.62 mM outer chain end groups, K_m isomaltose 33 mM) existed in several active charge isomer forms (p*I* values 4.4—4.7) which did not differ in their specific activities. Some evidence for the presence of carbohydrate in some forms of the enzyme was offered, and subunit phenomena (mol. wt. 8.8×10^4 and 7.6×10^4) were shown to exist. Using antibodies to the enzyme, it was shown that liver and skeletal muscle from cases of Type II glycogen storage disease (Pompé's disease), all lacking in lysosomal α-D-glucosidase, do not have altered, enzymically inactive proteins cross-reactive with the normal enzyme.

The activation of human liver 'acid' α-D-glucosidase by albumin, according to Arrhenius plots, pH dependence, and Lineweaver–Burk plots, is due not only to stabilization of the enzyme but also to the actual activation influencing the enzymic activity.[189] It was proposed that for optional functioning, the enzyme needs a protein environment. Glycogen had a competitive inhibitory affect on the hydrolysis of 4-methylumbelliferyl α-D-glycopyranoside, in contrast to maltose which exhibited non-competitive inhibition. It was concluded that two catalytic sites exist, one for glycogen and one for maltose, while both sites influence each other. Antibodies to the enzyme did not react with any material from the liver of a patient with Type II glycogen storage disease.

During a study of the properties of two β-D-galactosidases from human liver, it emerged that the isoenzymes also possessed β-D-glucosidase activity (pH optima 6.2) (see p. 394).[132]

Studies of the purification and properties of a heat-stable D-glucocerebrosidase-activating factor from spleens from controls and cases of Gaucher's

[186] F. S. Furbish, C. J. Steer, J. A. Barranger, E. A. Jones, and R. O. Brady, *Biochem. Biophys. Res. Comm.*, 1978, **81**, 1047.
[187] J. Bienvenu, H. Carrier, F. Freycon, and M. Mathieu, *Clinica Chim. Acta*, 1978, **84**, 277.
[188] A. K. Murray, B. I. Brown, and D. H. Brown, *Arch. Biochem. Biophys.*, 1978, **185**, 511.
[189] J. F. Koster and R. G. Slee, *Biochim. Biophys. Acta*, 1977, **482**, 89.

disease showed that the proteinaceous molecules differed in amino-acid composition and contained carbohydrate only in the case of the disease.[190] The activating capacities of detergent, phospholipid, *etc.* and the endogenous materials were compared. It appeared that the latter is/are probably unimportant in the catabolism of D-glucocerebroside in normal tissues.

The subcellular localization of a heat-stable D-glucocerebrosidase activator substance in spleen from a case of Gaucher's disease has been investigated.[191] The substance was shown to be an activator of the enzyme in normal human tissues, and the observations reported indicate that the activator is associated with the storage deposits contained in lysosomes of the Gaucher cell.

α-D-Glucosidase activity associated with various glycoside hydrolase activities, particularly α-D-galactosidase B from human liver, has been investigated.[43] Enzyme replacement therapy for the alleviation of Gaucher's disease has been impeded because of the difficulty in preparing large amounts of β-D-glucocerebrosidase, the enzyme that is deficient in patients with this disorder.[192] A large-scale procedure for the purification of human placental β-D-glucocerebrosidase has been reported, and is based on cholate extraction, fractional precipitation, and hydrophobic chromatography. The contamination of enzyme preparations intended for human infusion, prepared by isolation procedures involving immobilized concanavalin A, was investigated. An analogous report for the purification of the enzyme describes *inter alia* stages in the procedure involving affinity chromatography on agarose-immobilized concanavalin A and agarose-immobilized phosphatidylserine.[193] Intravenous administration to patients of the enzyme in free form and entrapped in resealed erythrocytes was investigated.

A D-glucocerebroside β-D-glucosidase-rich detergent-free preparation has been obtained from human placentae by a rapid method combining affinity chromatography on concanavalin A–agarose and organic-solvent precipitation.[194]

Human seminal plasma α-D-glucosidase activity has been determined using 4-nitrophenyl α-D-glucopyranoside as substrate.[195] The level of activity was found to be decreased in patients with infertility disorders. Tris(hydroxymethyl)aminomethane exerted a non-competitive inhibition of the enzyme.

The levels of α-D- and β-D-glucosidase activities in skin fibroblasts from cases of cystic fibrosis and controls have been compared.[68] Cultured skin fibroblasts from normal and glucosylcereamidotic subjects have been found to contain one β-D-glucosidase in contrast to the occurrence of multiple enzymes in other tissues.[196] The fibroblast enzyme (mol. wt. 1.5×10^5 under isotonic conditions, by gel filtration; pH optimum 4.2) occurs as a large aggregate at

[190] S. P. Peters, P. Coyle, C. J. Coffee, R. H. Glew, M. S. Kuhlenschmidt, L. Rosenfeld, and Y. C. Lee, *J. Biol. Chem.*, 1977, **252**, 563.
[191] Y. B. Chiao, J. P. Chambers, R. H. Glew, R. E. Lee, and D. A. Wenger, *Arch. Biochem. Biophys.*, 1978, **186**, 42.
[192] F. S. Furbish, H. E. Blair, J. Shiloach, P. G. Pentchev, and R. O. Brady, *Proc. Nat. Acad. Sci., U.S.A.*, 1977, **74**, 3560.
[193] G. L. Dale and E. Beutler, *Proc. Nat. Acad. Sci., U.S.A.*, 1976, **73**, 4672.
[194] I. P. Braidman and G. Gregoriadis, *Biochem. J.*, 1977, **164**, 439.
[195] P. Chapdelaine, R. R. Tremblay, and J. Y. Dubé, *Clinical Chem.*, 1978, **24**, 208.
[196] O. T. Mueller and A. Rosenberg, *J. Biol. Chem.*, 1977, **252**, 825.

low ionic strength. In whole cell homogenates it is membrane-bound and is solubilized by Triton X-100 and sodium taurocholate. The β-D-glucosidase displayed close similarity to one of the forms of β-D-glucosidase in human spleen, specifically that form which is affected in Gaucher's disease. The fibroblast levels of the enzyme in that disease were found to be considerably reduced, but the enzyme retained its normal identity.

The distribution of glycoside hydrolases including α-D- and β-D-glucosidases in human polymorphonuclear leucocytes has been determined.[62]

The properties of the α-D-glucosidase which has been shown to occur in the reproductive tract of the ewe have been investigated.[197]

The results of a detailed comparison of the relative efficiencies of different procedures for solubilizing the membrane-bound α-D-glucosidases of porcine intestinal mucosa have been reported.[198] Procedures for the selective solubilization of certain members of the complex group of α-D-glucosidases have been developed. The selective solubilization led to the conclusion that separate α-D-glucosidases (isomaltase and limit dextrinase respectively) are involved in the hydrolysis of isomaltose and the α-limit dextrins formed from starch by α-amylase.

The effects of fasting on the rythmic changes in the activities of α-D-glucosidase, *etc.* in the small intestine of rats have been studied.[199] The rhythms were found to persist when the animals were starved.

α-D-Glucosidase (maltase, glucoamylase) has been solubilized from rat intestinal mucosa by digestion with papain and was subsequently purified to immunological homogeneity.[200] The enzyme (mol. wt. 5.0×10^5) was shown to be an acidic glycoprotein (20% carbohydrate) which contained low amounts of L-cysteine and no neuraminic acid. Sodium dodecyl sulphate caused the conversion of the enzyme into five new species (mol. wts. $1.34—4.80 \times 10^5$) and the possibility that these five represented stable aggregates of a common monomer of the enzyme was discussed.

The separation by polyacrylamide gel electrophoresis and subsequent analysis of the components of guinea-pig intestinal brush border membrane revealed the presence of three enzyme complexes: α-D-glucosidase–glucoamylase, α-D-glucosidase–sucrose α-D-glucohydrolase–glucoamylase, and α-D-glucosidase–sucrose α-D-glucohydrolase.[145]

Under *in vitro* inhibition by D-glucose of α-D-glucosidase activity in haemolymph of bee (*Apis mellifica*) prenymphs, the reaction order (*n*) predetermined according to the initial natural glycemia, decreased with increasing inhibitor concentration, and the affinity constant, between enzyme and substrate, underwent lower variations than in other cases where *n* did not change.[201] The study of inhibited molecular forms suggested the possibility of a gradual negative-co-operativity mechanism concerning a dimeric form of the enzyme.

197 T. O'Shea and B. E. Murdoch, *Austral. J. Biol. Sci.*, 1978, **31**, 363.
198 J. J. Marshall, C. M. Sturgeon, and W. J. Whelan, *Analyt. Biochem.*, 1977, **82**, 435.
199 M. Saito, E. Murakami, T. Nishida, Y. Fujisawa, and M. Suda, *J. Biochem.* (*Japan*), 1976 **80**, 563.
200 P. R. Flanagan and G. G. Forstner, *Biochem. J.*, 1978, **173**, 553.
201 M. Bounias, *Compt. rend.*, 1977, **285D**, 1263.

Two glycosidases which are active against *inter alia* β-D-glucosides have been purified from the digestive juice of *Achatina balteata*.[123]

The β-D-glucosidase activity of a partially purified preparation of almond emulsin has been studied by chromatographic analysis, polyacrylamide gel electrophoresis, and isoelectric focusing.[100] The enzyme occurred as two components of equal molecular weight the aggregation of which with inactive proteins probably leads to the appearance of multiple forms of the enzyme with similar specific activities. The β-D-glucosidase activity was not separated from the accompanying β-D-xylosidase, β-D-galactosidase, and α-L-arabinosidase activities; this fact may be evidence that all the activities are manifested by a single species which does not exhibit strict specificity with respect to the stereochemistry of either hexosides at C-4 and C-6, or pentosides at C-4 and C-5.

The hydrolysis of salicin by almond β-D-glucosidase in a hollow fibre reactor has been studied.[202] The application of a simple theoretical model to the data obtained was demonstrated.

A β-D-glucosidase isoenzyme (mol. wt. 1.3518×10^5, protomer mol. wt. 6.515×10^4) isolated in homogeneous form from almond emulsin was shown to be a single glycoprotein.[149] The β-D-galacto- and gluco-sidase activities comigated under a variety of chromatographic and electrophoretic examinations. A purified β-D-glucosidase isoenzyme purified from almond emulsin was found to catalyse the hydrolysis of β-D-gluco- and galacto-pyranosides.[150] The different inhibitions of the enzyme activities of various compounds were investigated in detail. Analysis of the kinetic data indicated that the catalytic site of the enzyme responsible for the β-D-glucosidase activity is kinetically distinct from the β-D-galactosidase site. Evidence has been provided for a single catalytic site for β-D-fucosidase, β-D-glucosidase, and β-D-galactosidase activities of almond emulsin (see p. 392).[125]

The β-D-glucosidase isoenzymes B from sweet and bitter almonds showed considerable differences in their kinetic and inhibition parameters, but both were inhibited much more strongly by basic β-D-glucosyl derivatives than by their neutral analogues.[203] The additional interaction energy apparently due to the basic character ranged from 4.3 kcal mol^{-1} for β-D-glucopyranosylamine compared to β-D-glucose, to 6.9 kcal mol^{-1} for *N*-benzyl β-D-glucopyranosylamine compared to *N*-(4-toluidinyl)-β-D-glucopyranosylamine. Similar compounds, which could not be protonated at the glycosidic nitrogen were very weak inhibitors. The additional binding energy was assumed to be due to the electrostatic interaction of the inhibitor cation with a carboxy-group in an environment of low polarity.

β-D-Glucosidase A from bitter almonds was shown to be inhibited by the substrate analogue 6-bromo-3,4,5-trihydroxycyclo[2-^{3}H]hex-1-ene oxide.[204] Incorporation of 2 mol inhibitor per mol of dimeric enzyme resulted in total loss of activity. Tryptic peptides containing the label (*i.e.* active site regions)

[202] D. B. Van Dongen and D. O. Cooney, *Biotechnol. Bioeng.*, 1977, **19**, 1253.
[203] G. Legler, *Biochim. Biophys. Acta*, 1978, **524**, 94.
[204] G. Legler and A. Harder, *Biochim. Biophys. Acta*, 1978, **524**, 102.

were sequenced. The specificity of the reaction of β-D-glucosidases with substrate-related epoxides indicated that an L-aspartyl residue participates in the hydrolytic action.

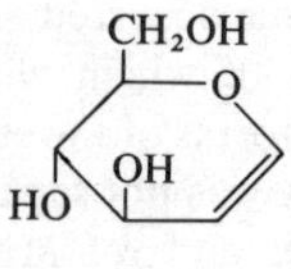

(2)

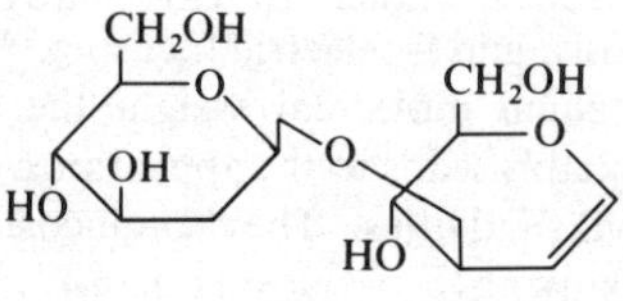

(3)

The by-product (3) in the hydration of D-glucal (2) catalysed by almond emulsin β-D-glucosidase has been identified as 1,5-anhydro-2-deoxy-3-*O*-(2-deoxy-β-D-*arabino*-hexopyranosyl)-D-*arabino*-hex-1-enitol.[205] Two models for the formation of (3) are discussed, involving transfer of a 2-deoxy-D-*arabino*-hexopyranosyl cation to HO-3 of D-glucal (glycon transfer) and transfer of an allylic D-pseudoglucal cation to HO-1 of 2-deoxy-D-*arabino*-hexopyranose (aglycon transfer). The enzymic production of (3) is highly regiospecific, which lends support to the second model and implies the presence of a specific binding-site for the aglycon moiety.

An α-D-glucosidase purified from flint corn by precipitation with ammonium sulphate, ion-exchange chromatography, and gel-filtrations was homogeneous in ultracentrifugal and disc electrophoretic analysis.[207] The enzyme (6.5 *S*, mol. wt. 6.5×10^4 by gel filtration) showed a pH optimum of 3.6 for both maltose and soluble starch. The ratio of velocities of hydrolysis for maltose, phenyl α-D-glucopyranoside, and soluble starch was estimated to be 100 : 14.3 : 6.1. The α-D-glucosidase hydrolysed soluble starch by an *exo* action.

Crude cell-wall preparation from chick pea (*Cicer arietinum*) cell suspension cultures showed high activity for the hydrolysis of coniferyl alcohol β-D-glucopyranoside (coniferin).[208] Conventional purification of the salt-extracted β-D-glucosidase activities gave a preparation, isoelectric focusing of which indicated the occurrence of isoenzymes (p*I*'s 8.5—10.0, mol. wts. 1.10×10^5 with two subunits of 6.3×10^4 and 4.3×10^4, pH optima 5) with identical catalytic activity. Coniferin was found to be one of the best substrates for these β-D-glucosidase isoenzymes, and the data suggested that a β-D-glucosidase-catalysed reaction might be involved in lignification of these plant cell cultures.

Four different β-D-glucosidases, active against isoflavone 7-β-glucosides, have been separated from leaves of chick pea plants (*Cicer arietinum*).[209] Two of these enzymes were purified to pure single homogeneous bands on disc electrophoresis and isoelectric focusing (p*I* values 4.35 and 4.45, mol. wts. 1.2—1.4×10^5, subunit mol. wts. 6.5×10^4). Both enzymes preferentially catalysed hydrolysis of diglycosides such as biochanin A 7-(2-*O*-apiosyl-β-D-glucopyranoside). The disaccharide unit was liberated intact. The enzymes also

[205] J. Lehmann and E. Schröter, *Carbohydrate Res.*, 1977, **58**, 65.
[206] I. Maeda, Z. Nikuni, H. Taniguchi, and M. Takamura, *Carbohydrate Res.*, 1978, **61**, 309.
[207] S. Chiba and T. Shimomura, *Agric. & Biol. Chem.* (*Japan*), 1975, **39**, 1033.
[208] W. Hösel, E. Surholt, and E. Borgmann, *European J. Biochem.*, 1978, **84**, 487.
[209] W. Hösel, *Z. physiol. Chem.*, 1976, **357**, 1673.

hydrolysed biochanin A 7-β-D-glucopyranoside and 2-nitrophenyl β-D-glucopyranoside. The enzymes (pH optima 5.5), which showed transferase activity, were strongly inhibited by 4-hydroxymercuribenzoate and bromocondurite; D-glucono-1,5-lactone, Ag^{2+} and Hg^{2+} showed only weak inhibition.

The locations and levels of α-D- and β-D-glucosidase activities in *Hippeastrum* flower petals have been examined.[153]

Potato tuber α-D-glucosidase (p*I* 4.7, mol. wt. 1.2×10^5, pH optimum 6.5—7.0, K_m 0.21 mM for 4-nitrophenyl α-D-glucopyranoside) has been shown to act on maltose and malto-oligosaccharides.[210] The enzyme exhibited transglucosidase activity.

The transglucosylation reaction of buckwheat α-D-glucosidase has been examined under the co-existence of 2-deoxy-D-glucose and maltose.[211] As the transglucosylation products, two kinds of new disaccharide were chromatographically isolated in a crystalline form (hemihydrate). It was confirmed that these disaccharides were 3-*O*-α-D-glucopyranosyl-2-deoxy-D-glucose and 4-*O*-α-D-glucopyranosyl-2-deoxy-D-glucose. The principal product formed in the enzyme reaction was 3-*O*-α-D-glucopyranosyl-2-deoxy-D-glucose.

The α-D- and β-D-glucosidases of the culture medium of *Scopolia japonica* calluses have been investigaged by isoelectric focusing.[102]

The amounts of α-D-glucosidase in cells of *Bacillus amyloliquefaciens* grown on amylaceous polysaccharides were found to be higher than in cells grown on non-carbohydrate carbon sources.[212] By subcellular fractionation and solubilization studies, the α-D-glucosidase was exclusively found to be associated with membranes from ruptured spheroplasts. The purified enzyme (pH optimum 6.8, mol. wt. $2.7 \pm 0.1 \times 10^4$, temperature optimum 39 °C) hydrolysed maltose, maltotriose, isomaltose, and isomaltotriose at rates slower than for 4-nitrophenyl α-D-glucopyranoside and sucrose.

The formation of α-D-glucosidase by batch and continuous cultures of *Bacteroides fragilis* under a variety of conditions has been investigated.[94]

The hydrolysis of 2-nitrophenyl β-D-glucopyranoside by β-D-glucosidase from *Botryodiplodia theobromae* has been studied in the presence of dioxan.[213] At donor saturation, the maximum hydrolysis rate in the presence of up to 50% dioxan was at pH 4.3—4.5. The hydrolytic reaction was found to proceed with retention of configuration at the anomeric site. The kinetics of the enzyme catalysis in the presence of added acceptors indicated that water was necessary for the maintenance of the active enzyme conformation apart from its acceptor function.

One of the three, synergistically related, enzymes in a cellulose preparation from *Chaetornium thermophile* was shown to be a β-D-glucosidase.[214]

A mutation in the post-translational modification of the glycoside hydrolases of *Dictyostelium discoideum* affected the α-D-glucosidase and β-D-glucosidases 1 and 2 molecules (see p. 418 ref. manase).[96, 97].

[210] S. D. Killilea and M. J. Clancy, *Phytochemistry*, 1978, **17**, 1429.
[211] S. Chiba, T. Shimomura, and K. Hatakeyama, *Agric. Biol. Chem.* (*Japan*), 1975, **39**, 591.
[212] H. Urlaub and G. Wöber, *Biochim. Biophys. Acta*, 1978, **522**, 161.
[213] G. M. Umezurike, *Biochem. J.*, 1978, **175**, 455.
[214] J. Eriksen and J. Goksøyr, *European J. Biochem.*, 1977, **77**, 445.

The α-D-glucosidase activities in extracts of antennal segments and in the haemolymph of the cotton stainer (*Dysdercus intermedius*) have been measured quantitatively using αα-trehalose, 4-nitrophenyl α-D-glucopyranoside, sucrose, turanose, maltose, isomaltose, and palatinose as substrates.[215] Three active species were isolated by gel filtration from the segments, but only two from the haemolymph. The results of the isolation were discussed with reference to the hypothesized presence of sugar reception processes in insect gustatory cells.

β-D-Glucosidase from a preparation of cellulase of *Geotrichum candidum* was freed from C_1 enzymes and (1 → 4)-β-D-glucanases by ion-exchange chromatography, gel filtration, and isoelectric focusing (forms A—F: p*I* values 3.8, 4.2, 4.6, 5.1, 5.6, and 6.2, molecular weights A 3.5×10^4, B 12.3×10^4, C 18.8×10^4, F 22.3×10^4).[216] Components A, C, E, and F hydrolysed cellobiose and did not act on 4-nitrophenyl β-D-glucopyranoside whereas the converse held for B and D. Other substrate specificities of the isoenzymes were also investigated.

The purification and properties of an extracellular β-D-glucosidase from *Lengites trabea* have been reported.[217]

The β-D-glucosidase of *Mucor racemosus* was shown to be biosynthesized when the organism was grown in the presence of such diverse carbon sources as glycerol, lactate, D-xylose, D-ribose, methyl and phenyl α-D-glucopyranosides, maltose, and cellobiose.[218] Enzyme biosynthesis was strongly repressed in the presence of hexoses and of adenosine 3′,5′-monophosphate. The role of adenosine 3′,5′-monophosphate in the control of β-D-glucosidase synthesis in *M. racemosus* was discussed.

Due to the gene for hyperderepression, a cross between a hyperderepressed strain and a normal strain of *Neurospora crassa* exhibited hyperproduction of β-D-glucosidase (cellobiase).[107]

The carbohydrate and amino-acid compositions of α-D-glucosidase from *Saccharomyces logos* have been examined.[219] The enzyme was shown to contain D-galactose and D-mannose (molar ratio 1 : 2) and 2-amino-2-deoxy-D-hexose.

Cellobiose or cellulose were necessary to induce the production of β-D-glucosidase activity by *Sporotrichum pulverulentum*.[220] Purification *via* preparative slab gel isoelectric focusing and affinity chromatography on a phenyl derivative of agarose gave five active fractions (p*I*'s 4.5—5.2, mol. wts. 1.65—1.82×10^5 by sodium dodecyl sulphate–polyacrylamide gel electrophoresis).

An induced β-D-glucosidase from *Stachybotrys atra* has been purified and was found to be different from the constititive β-D-glucosidase.[221] The enzyme (mol. wt. 6.55—6.90×10^4, pH optimum 6.7 p*I* 4.8) contained carbohydrate

[215] K. P. Gaffal, *Comp. Biochem. Physiol.*, 1977, **58B**, 71.
[216] N. A. Rodionova, N. G. Rumyantseva, N. A. Tiunova, L. I. Martinovich, and L. N. Bakhtadze, *Biochemistry* (*U.S.S.R.*), 1977, **42**, 34.
[217] D. Herr, F. Baumer, and H. Dellweg, *European J. App. Microbiol.*, 1978, **5**, 29.
[218] P. Borgia and P. S. Sypherd, *J. Bacteriol.*, 1977, **130**, 812.
[219] T. Saeki, S. Chiba, and T. Shimomura, *Agric. & Biol. Chem.* (*Japan*), 1975, **39**, 551.
[220] V. Deshpande, K.-E. Eriksson, and B. Pettersson, *European J. Biochem.*, 1978, **90**, 191.
[221] R. L. de Gussem, G. M. Aerts, M. Claeyssens, and C. K. De Bruyne, *Biochim. Biophys. Acta*, 1978, **525**, 142.

and also displayed β-D-xylosidase and β-D-thioglucosidase activities, the three activities appearing to be manifested by the one molecular species. The enzyme was strongly inhibited by D-glucono-1,5-lactone and nojirimycin and was very sensitive to thiol reagents.

Evidence has been obtained for the presence of a β-D-glucosidase in a yeast cell-wall lytic system obtained from a *Streptomyces* species.[222]

The production of mutants of *Trichoderma reesei* with enhanced β-D-glucosidase production has been reported. In some mutant strains of *Trichoderma viride* induced and isolated with the aid of mystatin selection and devoid of cellulose activity, the β-D-glucosidase activity was missing.[223] Three distinct β-D-glucosidase (cellobiase) components have been isolated from a commercial *T. viride* cellulose preparation by repeated chromatography on a cellulosic ion-exchanger using a salt eluent gradient.[224] The kinetics of the enzyme showed that cellobiase hydrolyses cellobiose by a non-competitive mechanism and that the product, D-glucose, inhibits the enzyme. Product inhibition significantly retarded the rate of cellobiose hydrolysis.

β-D-Glucosidase has been immobilized and insolubilized with retention of enzymic activity by covalent reaction with cellulose cyclic imidocarbonate[225] and chitosan,[226] by attachment to collagen membranes,[227] and by entrapment in liposomes,[194] and in resealed eyrthocytes.[193]

8 β-D-Glucuronidases

The preparations of D-glucaro-1-[*N*-(6-aminohexyl)amide (4) and *N*-1-β-D-glucopyranosyl(6-aminohexanamide)uronic acid (5), two inhibitors of β-D-glucuronidase, for possible use in affinity sorbents have been reported.[228]

Several differences occur between the interactions of glycosaminoglycans with a particular lysosomal protein leading to inhibition of lysosomal enzymes including β-D-glucuronidase.[41]

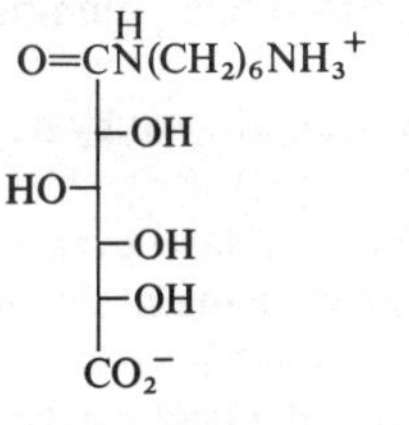

(4)

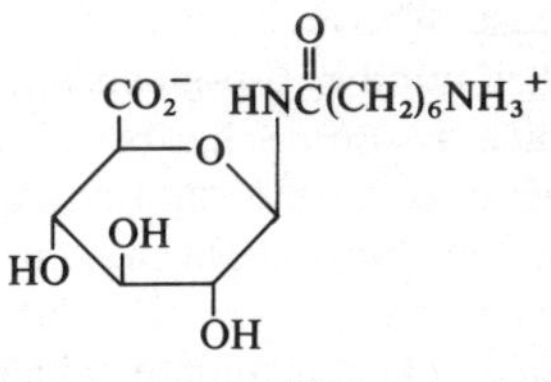

(5)

As part of a study on the neurochemistry of genetic hyperglycosaminoglycanuria (mucopolysaccharidosis), lysosomal enzymes including β-D-glucuronidase in four types of patients and in normals have been investigated in brain and liver.[47]

222 K. H. Rademacher and G. Reuter, *Z. Allgem. Mikrobiol.*, 1978, **18**, 107.
223 B. S. Montencourt and D. E. Eveleigh, *Appl. Environ. Microbiol.*, 1977 **34**, 777.
224 C.-S. Gong, M. R. Ladisch, and G. T. Tsao, *Biotechnol. Bioeng.*, 1977, **19**, 959.
225 D. Venardos, *Biotechnol. Bioeng.*, 1977, **19**, 1405.
226 F. Bissett and D. Sternberg, *Appl. Environ. Microbiol.*, 1978, **35**, 750.
227 M. Sugiura and H. Toyoda, *J. Agric. Chem. Soc. Japan*, 1977, **51**, 703.
228 R. T. Lee and Y. C. Lee, *Carbohydrate Res.*, 1978, **64**, 302.

A study of the glycoside hydrolases of human lung showed that the level of β-D-glucuronidase activity is low. The electrophoretic mobility of the enzyme on polyacrylamide gels indicated that it is different from serum β-D-glucuronidase. The secretory mechanism of the lung enzyme was discussed.[48]

Human placental β-D-glucuronidase has been purified to homogeneity in three steps: immunoadsorption, ion-exchange chromatography, and reduction and alkylation followed by ion-exchange chromatography.[229] Homogeneity was demonstrated by polyacrylamide gel electrophoresis, gel filtration (mol. wt. $3.10 \pm 0.10 \times 10^5$), and sedimentation equilibrium (mol. wt. $2.86 \pm 0.10 \times 10^5$). The enzyme (pH optimum 3.8) contained carbohydrate: D-mannose (68), 2-acetamido-2-deoxy-D-glucose (19), D-galactose (12), D-glucose (13), and L-fucose (15 residues for 2.86×10^5 mol. wt.) but no neuraminic acid. Polyacrylamide gel electrophoresis in the presence of sodium dodecyl sulphate revealed the presence of three entities (mol. wts. 7.7×10^4, 6.0×10^4, and 1.8×10^4). This is compatible with the enzyme being a tetramer of 7.7×10^4 mol. wt. subunits some of which may consist of 6.0×10^4 and 1.8×10^4 polypeptides.

The combined action of human placental β-D-glucuronidase and β-D-2-acetamido-2-deoxyhexosidase isoenzymes A and B on hyaluronic acid has been investigated (see p. 380).[55]

The levels of β-D-glucuronidase activity in skin fibroblasts from cases of cystic fibrosis and controls have been compared.[68] The purification (470 fold, 46% yield, 6 stages) of the saline extractable β-D-glucuronidase from human leucocytes and some of its catalytic and molecular properties have been reported.[230] Comparisons of these properties were made with those of other human and non-human β-D-glucuronidases. The subcellular distribution of glycoside hydrolases including β-D-glucuronidase in human polymorphonuclear leucocytes has been determined.[62]

Evidence has been provided for the receptor-mediated binding of β-D-glucuronidase by alveolar macrophages.[231]

The contribution of β-D-glucuronidase to the degradation of hyaluronic acid by canine liver lysosomes has been investigated (see p. 453).[75]

The effect of cortisone and thyroxine on β-D-glucuronidase levels in rat forebrain and cerebellum have been examined during early postnatal development.[76]

To investigate the chemical relationships between rat liver lysosomal and microsomal β-D-glucuronidase, which are essentially identical catalytically and in reactivity with antibody and are similar in molecular weight, the two enzymes were isolated by procedures in which modifications of the proteins were avoided.[232] The enzymes were found to differ in both carbohydrate and amino-acid compositions. The microsomal enzyme contained much more D-mannose and, in contrast to the lysosomal enzyme, contained neuraminic acid but no D-glucose.

[229] F. E. Brot, C. E. Bell, jun., and W. S. Sly, *Biochemistry*, 1978, **18**, 385.
[230] J. P. Périn and P. Jollés, *Biochimie*, 1978, **60**, 543.
[231] P. D. Stahl, J. S. Rodman, M. J. Miller, and P. H. Schlesinger, *Proc. Nat. Acad. Sci., U.S.A.*, 1978, **75**, 1399.
[232] D. R. P. Tulsiani, H. Six, and O. Touster, *Proc. Nat. Acad. Sci., U.S.A.*, 1978, **75**, 3080.

The microsomal (p*I* 6.9—7.6) and lysosomal (p*I* 5.8—6.6) β-D-glucuronidases purified from rat liver have been subjected to chemical analyses.[233] The enzymes had very similar amino-acid compositions and did not contain sulphur-containing amino-acids. The microsomal β-D-glucuronidase contained 6.5% carbohydrate, consisting of D-mannose, D-glucose, L-fucose, D-galactose, and 2-amino-2-deoxy-D-glucose in a ratio of 69 : 12 : 3 : 3 : 25. Neuraminic acid was not detected in this enzyme. The lysosomal enzyme contained 5.9% carbohydrate, consisting of D-mannose, D-glucose, L-fucose, D-galactose, and 2-amino-2-deoxy-D-glucose in a ratio of 57 : 14 : 3 : 4 : 23, with a trace of neuraminic acid. To confirm the presence of neuraminic acid in the lysosomal enzyme, more-acidic forms of lysosomal β-D-glucuronidase with p*I* values ranging from pH 5.4 to 6.1 were also purified. They contained 0.11% neuraminic acid and were neuraminidase-sensitive. This suggested that a very small part of the lysosomal enzyme may be a sialoglycoprotein. Since it has been suggested that neuraminic acid in the lysosomal glycoproteins is cleaved rapidly by autolytic degradation [A. Goldstone and H. Koenig, *Biochem. J.*, 1974, **141**, 527], the sugar acceptor activity of the purified lysosomal β-D-glucuronidase was determined using sialyltransferase in the Golgi fraction of rat liver. As the lysosomal β-D-glucuronidase was not a substrate for the sialyltransferase, the possibility of cleavage of terminal neuraminic acid by autolytic degradation was probably ruled out. The microsomal β-D-glucuronidase was also not a substrate for the sialyltransferase. This suggested that sialylization may occur very seldom in β-D-glucuronidase in the Golgi region. The relationship between the carbohydrate moieties of the microsomal and lysosomal β-D-glucuronidases was discussed in connection with intracellular transport of β-D-glucuronidase.

β-D-Glucuronidase found in the Golgi fraction of rat liver lysosomes has been proved to be indigenous to the Golgi complex by hypertonic-shock treatment of various subcellular fractions.[234] Deoxycholate treatment indicated that 65% of the enzyme was in the Golgi contents, and the remainder bound to the Golgi membrane. Detergent-dependent release of β-D-glucuronidase from the microsomal fraction also indicated that 60% of the activity was in the cisternal lumen and that the remainder was bound to the endoplasmic reticulum.

In order to trace the rate of intracellular transport of lysosomal enzymes, β-D-glucuronidase in the endoplasmic reticulum and the lysosomes was purified from rat liver and the sequence of appearance of the newly biosynthesized enzyme was followed with respect to these organelles after labelling the enzyme with L-[^{3}H]leucine.[235] The time-course of the appearance of labelled β-D-glucuronidase in these organelles after administration of the labelled amino-acid suggested that the newly biosynthesized enzyme is transported from the endoplasmic reticulum to the lysosomes, probably through the Golgi complex.

Infection of mice with an avirulent strain of Semliki forest virus caused an increase in brain β-D-glucuronidase activity.[82] In a study of the secretion of lysosomal hydrolases by stimulated and non-stimulated macrophages from

[233] M. Himeno, Y. Nishimura, K. Takahashi, and K. Kato, *J. Biochem.* (*Japan*), 1978, **83**, 511.
[234] H. Tsuji, N. Hattori, T. Yamamoto, and K. Kato, *J. Biochem.* (*Japan*), 1977, **82**, 619.
[235] H. Tsuji and K. Kato, *J. Biochem.* (*Japan*), 1977, **82**, 637.

mice, β-D-glucuronidase was found to be secreted in considerable amounts by a mechanism independent of phagocytosis.[83]

β-D-Glucuronidase has been purified from the urine of androgen-treated female mice; the enzyme had the same physiological properties as the lysosomal form of the enzyme in mouse kidney.[236] The effects of testostrone treatment of the animals upon the rate of β-D-glucuronidase biosynthesis and excretion were considered.

9 α-L-Iduronidases

A simple assay for α-L-iduronidase activity in leucocytes has been reported for the detection of Hurler syndrome and its carriers.[237] The method avoids time-consuming stages and tedious extraction procedures, but is based on phenyl α-L-idopyranosyluronic acid as substrate.

As part of a study on the neurochemistry of genetic hyperglycosaminoglycanuria (mucopolysaccharidoses), lysosomal enzymes, including α-L-iduronidase, in normals and four types of patients have been investigated in brain and liver.[47]

α-L-Iduronidase has been purified from human kidney *via* affinity chromatography on agarose-immobilized heparin.[238] The enzyme (mol. wt. $6.00 \pm 0.65 \times 10^4$ by sedimentation equilibrium), dissociable into subunits (mol. wt. $3.10 \pm 0.65 \times 10^4$), contained 2-amino-2-deoxy-D-glucose and exhibited binding to concanavalin A. The α-L-iduronidase (pH optima 4.0, 4.5; K_m 1.5 mM and 9 mM; V_{max} 16 and 270 μmol mg^{-1} protein min^{-1} for phenyl and mannitol glycosides of α-L-idopyranosyluronic acid) was of the low uptake non-corrective form with respect to skin fibroblasts cultured from cases of Hurler syndrome. The enzyme was inhibited by 4-chloromercuribenzoate and Cu^{2+} but not significantly by other divalent cations, H_4edta or thiol compounds.

10 α-D- and β-D-Mannosidases

Several differences occur between the interactions of glycosaminoglycans with a particular lysosomal protein leading to inhibition of lysosomal enzymes including α-D-mannosidase.[41] The mitogenic effect of α-D-mannosidase on lymphocytes has been studied.[239]

Levels of lysosomal α-D-mannosidase have been determined in normal hair roots and in hair roots obtained from a patient with mannosidosis.[46] The significance of the results was discussed in relation to the levels of other glycoside hydrolases present, and to the detection of lysosomal storage diseases.

A study of the glycoside hydrolases of human lung showed that the level of α-D-mannosidase activity is high.[48] The electrophoretic mobility of the enzyme on polyacrylamide gels indicated that it is different from serum α-D-mannosidase. The secretory mechanism of the lung enzyme was discussed.

The ability of acidic α-D-mannosidases from human placenta and urine,

[236] N. C. Mills, C. Gupta, and C. W. Bardin, *Arch. Biochem. Biophys.*, 1978, **185**, 100.
[237] T. Momoi, M. Sudo, K. I. Tanioka, and H. Kushida, *Clinica Chim. Acta*, 1977, **81**, 311.
[238] L. H. Rome, A. J. Garvin, and E. F. Neufeld, *Arch. Biochem. Biophys.*, 1978, **189**, 344.
[239] E. Paus, *Nature*, 1978, **272**, 452.

porcine kidney, bovine liver, and jack bean to correct the deranged metabolism of cultured skin fibroblasts from mannosidosis patients has been studied.[240] One cause for the different corrective abilities was the different rates of endocytosis of the enzyme forms into the cells. It was also demonstrated that there are differences in the intracellular corrective activity of the internalized enzymes, which can be explained by different specificities of the enzymes against the storage material.

Normal human serum contains at least three forms of α-D-mannosidase: an acidic form which has a pH optimum of 4.25, is inhibited by Co^{2+} and is thermostable; an intermediate form, which has a pH optimum of 5.6—5.7, is stimulated by Co^{2+} and is heat labile at 50 °C; and a neutral form (pH optimum 6.0—6.5).[241] In cases of mucolipidosis II and III the serum acidic activity persists while the intermediate activity is absent or altered. Heating the serum does not affect the pH-activity curve, the electrofocusing profile or the response to Co^{2+} of the enzyme. The absent or altered intermediate α-D-mannosidase may be responsible for aberrant biochemical properties reported for other glycoside hydrolases in mucolipidosis II and III.

A differential assay based on their differences in thermal stability has been used to measure the acidic and true intermediate α-D-mannosidases in the plasma of controls and individuals homozygous or heterozygous for mannosidosis.[242] The intermediate activity was found to be independent of age, sex, or mannosidosis genotype. The acidic α-D-mannosidase did not vary significantly with age or between sexes for groups of the same age. The concentrations of acidic and intermediate α-D-mannosidase showed a positive correlation for adults but not for children. It was suggested that the ratio of acidic to true intermediate α-D-mannosidase activities might be a useful secondary test for the detection of adult heterozygotes for mannosidosis.

The levels of α-D-mannosidase activity in skin fibroblasts from cases of cystic fibrosis and controls have been compared.[68] The effects of pH, metal ions, and heat have been studied on α-D-mannosidase from normal and mannosidosis fibroblasts and in culture medium derived from these.[243] The results showed that enzymes secreted from mannosidosis fibroblasts behaved identically to the enzymes present in culture medium from normal fibroblasts. The implication of this finding was discussed. Two acidic forms of α-D-mannosidase activity, A and B have been separated and characterized from fibroblasts from controls and patients with mannosidosis.[244] In normal cells, A and B differ in adsorption on DEAE-cellulose, electrophoretic mobility, stability at 70 °C, and resistance to freezing at −20 °C. In five of six mutant cell lines, A and B both exhibited altered K_m's for artificial substrates, and decreased thermal stability. Zn^{2+} had no effect on A or B in mutant or normal cells. In contrast, Co^{2+} slightly inhibited B in normal cells but markedly enhanced the activity of B in mutant cells. It was concluded that the mutation in mannosidosis results

[240] G. Mersmann, F.-J. Wilde, and E. Buddecke, *Z. physiol. Chem.*, 1978, **359**, 421.
[241] B. C. Kress and A. L. Miller, *Biochem. Biophys. Res. Comm.*, 1978, **81**, 756.
[242] S. Hirani, B. G. Winchester, and A. D. Patrick, *Clinica Chim. Acta*, 1977, **81**, 135.
[243] B. Hultberg and P. K. Masson, *Biochim. Biophys. Acta*, 1977, **481**, 573.
[244] B. K. Burton and H. L. Nadler, *Enzyme*, 1978, **23**, 29.

in an alteration in the biochemical characteristics of both major acidic forms of α-D-mannosidase. Evidence has been provided that the mutant enzyme in fibroblasts of a patient with mannosidosis does not cross-react with antiserum raised against normal acidic α-D-mannosidase.[245]

The subcellular distribution of glycoside hydrolases including α-D-mannosidase in human polymorphonuclear leucocytes has been determined.[62]

α-D-Mannosidase from porcine kidney has been shown to exist in multiple forms differing in their capability to be endocytosed by α-D-mannosidase-deficient cultured cells.[246] 'Uptake' and 'non-uptake' forms of the enzyme were recognized. The content of 'uptake' forms in a preparation was identical with that of enzyme molecules bearing a phosphorylated carbohydrate group necessary for the recognition of cell surface receptors.

The effect of cortisone and thyroxine on α-D-mannosidase levels in rat forebrain and cerebellum have been examined during early postnatal development.[76]

Rat liver has been shown to contain α-D-mannosidase in lysosomes, Golgi membranes, and cytosol.[247] The lysosomal enzyme (mol. wt. 3.35×10^5 by molecular sieve, 2.00×10^5 by sucrose density centrifugation, pH optimum 4.6) was a glycoprotein as evidenced by its binding to immobilized concanavalin A, *etc.* Prolonged exposure of the enzyme to H_4edta caused loss of activity which could be restored by addition of zinc sulphate. Substrate specificities and immunological reactions were investigated for the α-D-mannosidases from the three sources.

In a study of the secretion of lysosomal hydrolases by stimulated and non-stimulated macrophages from mice, α-D-mannosidase was found to be secreted in a considerable amount by a mechanism independent of phagocytosis.[83]

α-D-Mannosidase solubilized by 3M-LiCl from the cell walls of the coleoptiles of *Avena sativa* has been purified *via* fractional precipitation, gel filtration, ion-exchange chromatography, and isoelectric focusing to homogeneity on disc gel electrophoresis.[248] The enzyme (mol. wt. 6.3×10^5, pH optimum 4.5, K_m 3.2 mM for 4-nitrophenyl α-D-mannopyranoside) had no metal ion cofactor requirement and was insensitive to 4-chloromercuribenzoate. According to the periodate–Schiff reaction the enzyme contained carbohydrate.

A simple procedure reported for the isolation of jack-bean α-D-mannosidase was based on affinity chromatography on agarose-immobilized benzidine.[249]

The influence of substituents on the hydrolysis of substituted phenyl α-D-mannopyranosides by α-D-mannosidase from *Medicago sativa* seeds has been investigated.[250] As indicated by structure–activity relations, the electronic effect of the substituent has an influence on the rate of formation of the intermediate D-mannosyl–enzyme complex. This effect depends not only on the nature of the substituent, but also on its position (*meta* or *para*) and on the temperature of the experiment. Hammett-type linear free energy relationships show that the reaction constant ρ changes its sign at *ca.* 27 °C. Substrates

245 L. J. Burditt, K. A. Chotai, and B. G. Winchester, *F.E.B.S. Letters*, 1978, **91**, 186.
246 G. Mersmann, R. Becker, and E. Buddecke, *Biochim. Biophys. Acta*, 1978, **525**, 154.
247 D. J. Opheim and O. Touster, *J. Biol. Chem.*, 1978, **253**, 1017.
248 L. C. Greve and L. Ordin, *Plant Physiol.*, 1977, **60**, 478.
249 P. V. Wagh, *J. Chromatog.*, 1978, **152**, 565.
250 J. De Prijcker, A. De Bock, and C. K. De Bruyne, *Carbohydrate Res.*, 1978, **60**, 141.

with strong electron-withdrawing groups show values of log *V* that are linearly related to 1/T, whereas the Arrhenius plots for other substrates are severely curved. This complex behaviour was tentatively explained by assuming that some *meta*-substituents have an unusual, temperature and substituent dependent influence on the formation of the Michaelis–Menten complex.

When the pK_m of α-D-mannosidase from *Phaseolus vulgaris* was determined at different pH values, the results indicated that ionizable groups with p*K* values of 3.8 and 5.7 could be essential to activity.[251] Modification with carbodi-imide or Woodward's Reagent K abolished the activity, but the substrate analogue, methyl α-D-mannopyranoside, protected the enzyme against inactivation. Incorporation of a ^{14}C-labelled nucleophile in the presence or absence of the analogue suggested that 2—4 carboxy-groups were protected. Exchange studies indicated that the essential Zn^{2+} could be bound to such groups.

β-D-Mannosidase has been purified from the cell extract of an edible mushroom (*Tremella fuciformis*) to a single protein band (mol. wt. 1.4×10^5 by gel filtration) on disc gel electrophoresis.[252] The enzyme (pH optimum 5.0, K_m 2.1 mM for 4-nitrophenyl β-D-mannopyranoside) readily hydrolysed 4-*O*-β-D-mannopyranosyl-D-mannose, but β-(1 → 4)-linked D-mannotriose and D-mannotetraose were hydrolysed much more slowly. Unlike other β-D-mannosidases, its activity was not inhibited by D-mannono-1,3-lactone but it was strongly inhibited by 4-chloromercuribenzoate. The enzyme tended to be inactivated in the presence of oxygen, but could be reactivated by dithiothreitol.

The α- and β-D-mannosidases of the culture medium of *Scopolia japonica* calluses have been investigated by isoelectric focusing.[102]

The β-D-mannosidase (mol. wt. 1.2×10^5) purified to homogeneity from the culture filtrate of *Aspergillus niger* was shown to be a glycoprotein.[253] Analyses of ^{14}C-glycopeptides derived from the enzyme *via* culture in the presence of [^{14}C]-D-fructose and alkaline borohydride cleavage of the intact enzyme indicated the presence of carbohydrate chains containing mannose and 2-acetamido-2-deoxyglucose linked *via* an N^4-(2-acetamido-2-deoxy-β-D-glucopyranosyl) hydrogen L-asparaginate linkage to the protein chain. The activity of the enzyme (pH optimum 3.5—4.0) upon various glycosides and manno-oligosaccharide was investigated.

β-D-Mannosidase from culture filtrates of *Aspergillus niger* has been purified in large amounts by fractional precipitation and ion-exchange chromatography, the removal of traces of α-D-galactosidase being achieved by affinity chromatography.[254] The enzyme (mol. wt. $1.30 \pm 0.05 \times 10^5$, pH optimum 3.5, temperature optimum 55 °C, p*I* 4.7) was characterized by a high proportion of acidic amino-acids but absence of L-cysteine residues. The presence of 80% of β-sheet structure, 20.8% of carbohydrate (D-galactose 21, D-mannose 111, 2-acetamido-2-deoxy-D-glucose 16 moles mol^{-1}) possibly explained the high heat stability of the β-D-mannosidase, which was inhibited by D-mannose (K_i 7.85 mM).

[251] E. Paus, *Biochim. Biophys. Acta*, 1978, **526**, 507.
[252] Y. Sone and A. Misaki, *J. Biochem.*, 1978, **83**, 1135.
[253] A. D. Elbein, S. Adya, and Y. C. Lee, *J. Biol. Chem.*, 1977, **252**, 2026.
[254] S. Bouquelet, G. Spik, and J. Montreuil, *Biochim. Biophys. Acta*, 1978, **522**, 521.

The isolation is described of a mutant of *Dictyostelium discoideum* M31 which produces a reduced number of α-D-mannosidase-1 molecules per cell during the developmental programme of the organism.[96] Several glycosidases, including α-D-glucosidase, β-D-glucosidases-1 and -2 and β-D-2-acetamido-2-deoxyglucosidase, were shown to be altered in strain M31. These enzymes do not share a common protein subunit but are known to share a common antigenic determinant which is, in part, carbohydrate in nature. α-D-Mannosidase-1 was found to contain phospho-L-serine/-L-threonine residues. The glycoside hydrolases isolated from M31 were less negative than those from wild type cells. It was concluded that the mutation present in M31, *Mod A*, affects post-translational modification. The presence of the *Mod A* mutation affected the intracellular level of the multimeric glycoside hydrolases differently.[97] The mutant cell contains only a small fraction of wild-type β-D-glucosidase-1 activity. The alteration renders β-D-glucosidase-1 holoenzyme thermolabile and susceptible to degradation *in vivo*. α-D-Mannosidase and β-D-2-acetamido-2-deoxyglucosidase were found at *ca.* 33% wild-type level in the mutant. Degradation of the holoenzyme did not appear to be responsible for the low level of these activities. It was proposed that α-D-mannosidase-1 and β-D-2-acetamido-2-deoxyglucosidase subunits are being degraded prior to subunit assembly. It was concluded that the modification bestows different properties upon the various glycoside hydrolases.

A unique α-D-mannosidase activity has been isolated from *Saccharomyces cerevisiae*.[255] The cellular concentration of the enzyme was inversely proportional to growth rate of the organism, suggesting that the enzyme is under catabolite repression. The α-D-mannosidase was localized in a crude particulate fraction of the cell extract and was not solubilized by treatment with detergents or solutions of high ionic strength. The enzyme (pH optimum 6.3, K_m 50 μM for 4-nitrophenyl α-D-mannopyranoside) was competitively inhibited by D-mannose (K_i 20mM), but was not affected by H_4edta or thiol reagents. Its inhibition by 4-chloromercuriphenylsulphonic acid was prevented by addition of substrate. Evidence has been obtained for the presence of an α-D-mannosidase in a guest cell-wall lytic system obtained from a *Streptomyces* species.[222]

11 Neuraminidases (Sialidases)

Neuraminidase activity in fibroblast homogenates has been measured by a rapid and simple assay with a synthetic substrate.[256] The 2-(3-methoxyphenyl) glycoside of *N*-acetylneuraminic acid (6) liberates methoxyphenol on attack by the neuraminidase (Scheme 1); addition of aminoantipyrine (7) in the presence of an oxidizing agent, potassium ferricyanide, yields a coloured quinone (8). The neuraminidase activity in virus-transformed hamster fibroblasts was found to be increased over the normal counterpart. In addition, the differential activity seen using the synthetic substrate and fetuin made it possible to detect an enzyme activity not hitherto described. The advantages of this assay system for metabolic screening were discussed.

[255] D. J. Opheim, *Biochim. Biophys. Acta*, 1978, **524**, 121.
[256] U. V. Santer, J. Yee-Foon, and M. C. Glick, *Biochim. Biophys. Acta*, 1978, **523**, 435.

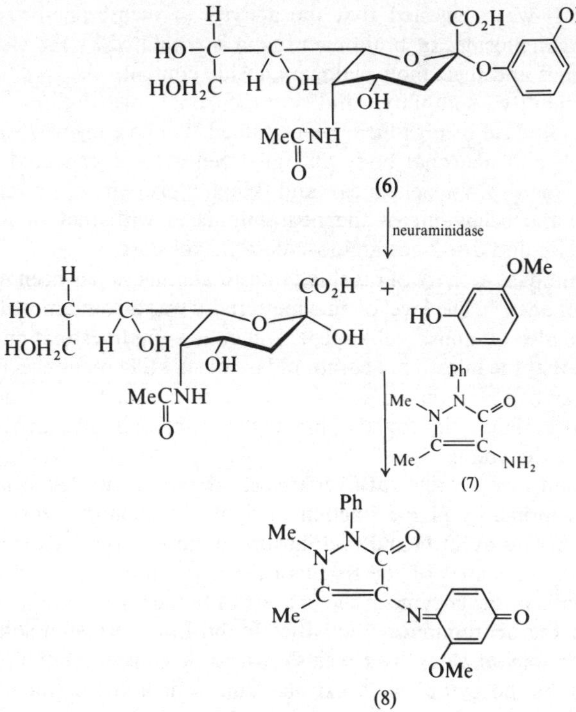

Scheme 1

The levels of neuraminidase activity in skin fibroblasts from controls and cases of cystic fibrosis have been compared.[68] Re-examination of the cultured β-D-galactosidase-deficient fibroblasts from a case of G_{M_1} gangliosidosis revealed a severe deficiency in neuraminidase activity using neuraminlactose, fetuin, and the 2-(3-methoxyphenyl) glycoside of *N*-acetylneuraminic acid as substrates, but normal neuraminidase using G_{M_3} ganglioside as a substrate.[141]

Studies on pinocytosis by human fibroblasts of β-D-2-acetamido-2-deoxyhexosidase have led to the conclusion that a neuraminidase deficiency in I-cell disease may not be directly responsible for excessive extracellular β-D-2-acetamido-2-deoxyhexosidase.[70] The proposed neuraminidase deficiency in I-cell disease has been discussed in the light of its significance in influencing the final monosaccharide sequence in the carbohydrate structure of the recognition site of the fibroblasts-derived β-D-2-acetamido-2-deoxyhexosidase.[69]

The degradation of ganglioside G_{D1a} and sialyl-lactitol by membrane-bound neuraminidase from calf brain has been studied as a function of the lipophilic/hydrophilic character of the substrate.[257] Availability and enzymic breakdown of the ganglioside were enhanced by general anaesthetics such as nitrous oxide and halothane, whereas the degradation of sialyl-lactitol was not affected by

[257] K. Sandhoff and B. Pallmann, *Proc. Nat. Acad. Sci., U.S.A.*, 1978, **75**, 122

these agents. It was suggested that the activity of membrane-bound neuraminidase on gangliosides of brain membrane is regulated by the viscosity of these membranes and their monosialoganglioside content.

The extractability, solubility, and intraneuronal distribution of brain cytosol neuraminidase of pigs have been studied.[74] The enzyme from nerve-ending cytosol, and neuronal body and glial cell cytosol appeared to differ somewhat in their physicochemical and kinetic properties. Comparisons were made of the behaviour of the neuraminidases with that of lysosomal β-D-galactosidase and β-D-2-acetamido-2-deoxyhexosidase.

The neuraminidase activity of rat small intestinal mucosa has been measured as a function of age.[258] The level of enzyme activity was also examined in other rat tissues, and also in mouse, rabbit, cat, and guinea-pig intestinal mucosa. It was suggested that the intestinal neuraminidase of suckling mammals functions primarily to remove neuraminic acid from various components of milk, thus providing neuraminic acid for the biosyntheses of sialoglycoproteins and gangliosides by the young.

In normal and regenerating rat liver tissue, neuraminidase has been shown to be localized primarily in the fraction of light mitochondria, enriched with lysosomes.[77] Actinomycin D and cycloheximide did not lower the neuraminidase activity of lysosomes of the regenerating liver, indicating absence of *de novo* biosynthesis of the enzyme in the process of tissue regeneration.

The sum of the neuraminidase activities found in the isolated soluble and particulate fractions of chick liver was shown to be considerably higher than that observed in the cytoplasmic extract from which these fractions were obtained.[259] Addition of increasing amounts of particulate neuraminidase to a constant amount of the soluble preparation resulted in a progressive loss of enzyme activity.

High neuraminidase activities have been shown to be produced by *Arthrobacter ureafaciens*, *A. oxydans*, and *A. ausescens*.[260] Both colominic acid and *N*-acetylneuraminic acid itself were shown to be effective as carbon sources. Affinity chromatographic purification on colominic acid cross-linked to starch by epichlorohydrin gave a product free from phospholipase C, glycoside hydrolases, and *N*-acetylneuraminic acid aldolase.

A neuraminidase purified from the culture filtrate of *Arthrobacter ureafaciens* has been shown to act upon G_{M1} ganglioside and the oligosaccharide derived therefrom.[261] Addition of detergents, especially bile salts, resulted in a marked increase in the hydrolysis of the intact ganglioside but not of the free oligosaccharide. The actions of neuraminidases from *Clostridium perfringens* and *Vibrio cholerae* in the same circumstances were also investigated.

Neuraminidase from *Arthrobacter* species has been purified by conventional procedures until judged to be homogeneous on gel electrophoresis in sodium dodecyl sulphate.[262] The enzyme (mol. wt. 8.8×10^4, K_m 0.78 mM for *N*-acetylneuraminlactose) was subfractionated by isoelectric focusing (major

[258] J. J. Dickson and M. Messer, *Biochem. J.*, 1978, **170**, 407.
[259] D. R. P. Tulsiani, V. P. Bhavanandan, and R. Carubelli, *Experientia*, 1976, **32**, 1501.
[260] Y. Uchida, Y. Tsukada, and T. Sugimori, *J. Biochem. (Japan)*, 1977, **82**, 1425.
[261] K. Sugano, M. Saito, and Y. Nagai, *F.E.B.S. Letters*, 1978, **89**, 321.
[262] P. Wang, S. W. Tanenbaum, and M. Flashner, *Biochim. Biophys. Acta*, 1978, **523**, 170.

band: p*I* 5.35, 6 minor bands: p*I*'s 5.25—5.70), was not inhibited by *N*-acetylneuraminic acid, and contained 2-amino-2-deoxy-glucose and -galactose and neutral carbohydrate. Inhibition and modification experiments suggested that an L-tryptophan residue was essential for catalysis.

A fluorogenic substrate for the assay of neuraminidase, 4-methylumbelliferone *N*-acetylneuraminic acid ketoside, has been synthesized.[263] A detailed examination of the K_m values for this substrate of neuraminidase from *Clostridium perfringens* and strains of A- and B-type influenza neuraminidases from chicken red blood cell influenza virus was conducted.

Purification of neuraminidase (mol. wt. 4.8×10^4, pH optimum 6.2) from *Erysipelothrix rhusiophatiae* has been reported.[264] The enzyme exhibited highest activity against oligosaccharide (9). The rate of cleavage of (10) and of the α-(2 → 8)-linkage of (11) were 20% of this. Gangliosides and colominic

α-Neu*p*NAc-(2 → 3)-β-D-Gal*p*-(1 → 4)-D-Glc
(9)

α-Neu*p*NAc-(2 → 6)-β-D-Gal*p*-(1 → 4)-D-Glc
(10)

α-Neu*p*NAc-(2 → 8)-α-Neu*p*NAc-(2 → 3)-β-D-Gal*p*-(1 → 4)-D-Glc
(11)

acid were poor substrates. In general the enzyme showed a specificity pattern similar to that found in other bacteria of low pathogenicity towards man.

The purification of neuraminidase from *Streptomyces griseus via* affinity chromatography on agarose-immobilized fetuin has been reported.[265] The enzyme was found to be a glycoprotein and selective chemical modication demonstrated that L-arginine residues are essential for full enzymic activity.

The alteration of cell shape in the early chick embryo mesoderm by neuraminidase from *Vibrio cholera* has been reported.[266]

Neuraminidase of a non-agglutinating *Vibrio* species has been isolated *via* affinity chromatography on a 4-aminophenylhydroxamic acid-peptide derivative of agarose.[267] The three molecular forms (mol. wts. 9.0×10^4, 5.8×10^4, and 5.2×10^4) separated were found to be immunochemically related.

Neuraminidase activity has been demonstrated in highly concentrated preparations of human para-influenza 1 virus (HA2 virus).[268] The enzyme (pH optimum 5.0—5.4, temperature optimum 37—40 °C, K_m 5 mM for *N*-acetylneuraminlactose) was found to be thermolabile, and exhibited some characteristics similar to those of neuraminidases from other paramyxoviruses (Sendai, NDV, mumps, human para-influenza 2 virus).

The neuraminidase of B/Lee/40 influenza virus has been isolated following detergent or tryptic digestion of virus particles.[269] Analytical results indicated

[263] J. J. Thomas, E. C. Folger, D. L. Nist, B. J. Thomas, and R. H. Jones, *Analyt. Biochem.*, 1978, **88**, 461.
[264] H. von Nicolai, H. E. Müller, and F. Zilliken, *Z. physiol. Chem.*, 1978, **359**, 393.
[265] J. P. Kabayo and D. W. Hutchinson, *F.E.B.S. Letters*, 1977, **78**, 221.
[266] J. Wakely and M. A. England, *Experientia*, 1978, **34**, 243.
[267] Yu. V. Vertiev and Yu. V. Ezepchuk, *Biochemistry* (*U.S.S.R.*), 1977, **42**, 1273.
[268] N. Kessler, G. Bardeletti, and M. Aymard, *J. Gen. Virol.*, 1977, **37**, 547.
[269] A. K. Allen, J. J. Skehel, and V. Yuferov, *J. Gen. Virol.*, 1977, **37**, 625

that the carbohydrate side-chains of the enzyme molecule contain D-galactose, D-mannose, L-fucose, and 2-acetamido-2-deoxy-D-glucose and that they are attached through N^4-(2-acetamido-2-deoxy-β-D-glucopyranosyl) hydrogen L-asparaginate linkages, and that the intact molecule (mol. wt. 2.8×10^5) consists of four subunits (mol. wt. 7.0×10^4).

The neuraminidase of the influenza virus recombinant strain (HON2) has been solubilized by detergents and isolated by affinity chromatography on a 4-aminophenylhydroxamic acid–peptide derivative of agarose.[270] The neuraminidase was purified to a single high molecular weight glycoprotein when assayed under non-reducing conditions on sodium dodecyl sulphate–polyacrylamide gel electrophoresis.

Neuraminidase (mol. wt. 1.525×10^5 by sedimentation diffusion, 1.470×10^5 by sedimentation equilibrium) has been isolated from the X7-(F1) (HON2) strain of influenza virus by proteolysis and purified by gel filtration.[271] In 6M guanidinium chloride the molecular weight was halved (6.6×10^4) and after irreversible reduction and blocking of thiol groups was found to be halved again (3.35×10^4, both by sedimentation equilibrium). The enzyme was therefore considered to be a tetramer composed of two disulphide-linked dimers, themselves associated by non-covalent linkages. Theoretical considerations based on this model predict that assembly of the molecule must be accompanied by allosteric conformational changes in the subunits. The high carbohydrate content (46%) was thought to explain the exceptional resistance of the molecule to proteolytic enzymes.

Neuraminidase has been immobilized to give a water-insoluble derivative with retention of enzymic activity by covalent reaction with agarose cyclic imidocarbonate,[272, 273] a hydrazide derivative of agarose cyclic imidocarbonate,[273] and periodate-oxidized controlled-pore glass (CPG)-glycophase.[273]

12 β-D-Thioglucosidases

An induced molecular species from *Stachybotrys atra* displayed β-D-thioglucosidase, β-D-glucosidase, and β-D-xylosidase activities.[221]

13 α-D- and β-D-Xylosidases

Tissues from a case of juvenile (type 3) Gaucher tissue exhibited a deficiency of β-D-xylosidase activity.[118] Evidence for the manifestation of β-D-xylosidase, β-D-glucuronidase, and β-D-galacosidase activities by a single enzyme molecule was provided.

The α-D-xylosidase activity associated with various glycoside hydrolase activities, particularly α-D-galactosidase, from human liver has been investigated.[43]

Physiochemical studies of the β-D-xylosidase activity of partially purified

[270] D. J. Bucher, *Biochim. Biophys. Acta*. 1977, **482**, 393.
[271] N. P. Groome, G. Belyavin, A. Lansdell, and D. Ashford, *Biochim. Biophys. Acta*, 1977, **495**, 58.
[272] Yu. V. Vertiev, G. K. Belyanskaya, and Yu. V. Ezepchuk, *Biochemistry, U.S.S.R.*, 1977, **42**, 1365.
[273] T. L. Parker, A. P. Corfield, R. W. Veh, and R. Schauer, *Z. physiol. Chem.*, 1977, **358**, 789.

almond emulsin has led to the conclusion that the activity is manifest by the same enzyme species as the β-D-glucosidase activity.[100]

β-D-Xylosidase from *Bacillus pumilus* has been shown to contain one intramolecular disulphide bridge and four L-cysteinyl residues per monomer (mol. wt. 6.0 × 10^4).[274] Under native conditions, one L-cysteinyl residue per polypeptide chain reacted with 5,5′-dithiobis(2-nitrobenzoic acid) or 4-chloromercuribenzoate resulting in complete loss of enzymic activity. Reductive treatment with dithiothreitol completely reversed this loss. Kinetic analysis of the inactivation reaction pointed to a saturation effect, occurring when the concentration of the —SH-directed reagent was increased. A reaction model, which included prior association of the reagent with the enzyme, was proposed.

The influence of substituents on the binding and hydrolysis of several substituted β-D-xylopyranosides by β-D-xylosidase from *Bacillus pumilus* has been investigated.[275] From a comparison of the inhibition constants of the substrates, 1-thio-β-D-xylopyranosides, with the apparent Michaelis–Menten constants of the substrates, it followed that the latter constants are good approximations of the true equilibrium constants. The influence of the substituent on the rate and activation parameters is small. The results are in agreement with, but do not prove, a one-step mechanism without the formation of a glycosyl–enzyme intermediate.

The determination of the binding parameters of 4-methylumbelliferyl β-D-ribopyranoside, a fluorescent ligand of β-D-xylosidase from *Bacillus pumilis*, has been reported.[276] A single binding-site per polypeptide chain (mol. wt. 6.0 × 10^4) was found, and the homogeneity of the binding sites in the dimeric and tetrameric forms of the enzyme were shown. The association constants, as a function of temperature and ionic strength, were obtained from equilibrium binding experiments, and compared with the inhibition constants determined kinetically. The apparent discrepancies were attributed to a temperature, ionic strength, and concentration-dependent shift in the dimer–tetramer equilibrium of the enzyme and different affinities of the ligand for both oligomeric sites.

An induced molecular species from *Stachybotrys atra* displayed β-D-xylosidase, β-D-glucosidase, and β-D-thioglucosidase activities.[221]

14 *endo*-α-D-2-Acetamido-2-deoxygalactanases

The *endo*-α-D-2-acetamido-2-deoxygalactanase and α-D-2-acetamido-2-deoxygalactosidase of *Clostridium perfringens* culture filtrates have been separated by polyacrylamide gel electrophoresis.[95] The substrate specificities of the two enzymes were compared (see p. 387).

15 *endo*-β-D-2-Acetamido-2-deoxyglucanases

The purification procedure for *endo*-β-D-2-acetamido-2-deoxyglucanase has been improved to yield an enzyme preparation which was homogeneous upon

[274] E. Saman, M. Claeyssens, and C. K. De Bruyne, *European J. Biochem.*, 1978, **85**, 301.
[275] H. Kersters-Hilderson, E. Van Doorslaer, and C. K. De Bruyne, *Carbohydrate Res.*, 1978, **65**, 219.
[276] M. Claeyssens and C. K. De Bruyne, *Biochim. Biophys. Acta*, 1978, **533**, 98.

gel electrophoresis.[277] The molecular weight of the enzyme as estimated by gel filtration was 2.8×10^5 while sodium dodecyl sulphate gel electrophoresis after reduction with 2-mercaptoethanol gave a value of 1.5×10^5. The purified enzyme did not show any chitinase, hyaluronidase, or lysozyme activity. In the presence of *exo*-glycoside hydrolases removing peripheral sugars, the *endo*-glycanase acted on serum glycoproteins such as transferrin and fetuin. The enzyme also hydrolysed an oligosaccharide (12) indicating that the peptide portion of substrates does not have much effect on susceptibility to the enzyme.

(D-Man*p*)$_5$-(Glc*p*NAc)$_2$

(12)

Commercially available fig latex contains several *endo*-β-D-2-acetamido-2-deoxyglucanases which catalyse the reaction shown in Scheme 2.[278]

(D-Man*p*)$_n$-β-D-Glc*p*NAc-(1 → 4)-D-Glc*p*NAc-L-Asn →

(D-Man*p*)$_n$-D-Glc*N*Ac + D-Glc*pN*Ac-L-Asn

Scheme 2

By fractional precipitation, gel filtration, and ion-exchange chromatography two distinct types of the enzyme have been partially purified and characterized. The two forms exhibited differences in the rates of cleavage of, and specificity for, glycosidic linkages in analogous glycopeptides.

A simple procedure described for the detection of *endo*-β-D-2-acetamido-2-deoxyglucanase H activity is based on the use of *N*-[^{14}C]methylribonuclease B as substrate.[279] Because the carbohydrate moiety of ribonuclease B has α-D-mannopyranosyl residues as terminal non-reducing groups, but such residues are lost as part of an oligosaccharide on action of the *endo*-β-D-2-acetamido-2-deoxyglucanase, agarose immobilized concanavalin A was used to separate radioactive substrate (absorbed) from radioactive product (unadsorbed). The assay was used to follow the preparation of the enzyme from *Streptomyces plicatus*.

endo-β-D-2-Acetamido-2-deoxyglucanase C_{II} (from *Clostridium perfringens*) was found to hydrolyse 75% of an ovalbumin glycopeptide, whereas *endo*-β-D-2-acetamido-2-deoxyglucanase H (from *Streptomyces griseus*) cleaved it completely.[280] For the purpose of clarifying the difference of substrate specificities of the C_{II} and H enzymes, their action upon various ovalbumin glycopeptides was determined (Table 1).

endo-β-D-2-Acetamido-2-deoxyglucananse from *Streptomyces plicatus* can be useful in determining both the molecular weight of the protein moiety of glycoproteins and their inherent number of oligosaccharide chains.[281] Although

[277] T. Muramatsu, N. Koide, and K. Maeyama, *J. Biochem. (Japan)*, 1978, **83**, 363.
[278] S.-F. Chien, R. Weinburg, S.-C. Li, and Y.-T. Li, *Biochem. Biophys. Res. Comm.*, 1977, **76**, 317.
[279] J. S. Tkacz, *Analyt. Biochem.*, 1978, **84**, 49.
[280] T. Tai, K. Yamashita, S. Ito, and A. Kobata, *J. Biol. Chem.*, 1977, **252**, 6687.
[281] R. B. Trimble and F. Maley, *Biochem. Biophys. Res. Comm.*, 1977, **78**, 935.

Table 1 *Specificity differences of* endo-β-D-2-*acetamido*-2-*deoxyglucanases* C_{II} *and* H

Ovalbumin glycopeptide structures	endo-β-D-2-*Acetamido*-2-*deoxyglucanase activity* C_{II}	H
α-D-Man*p*-(1 → 3)-α-D-Man*p*-(1 → 6)-β-D-Man*p*-(1 → R 6 (on the second α-D-Man*p*) ↑ 1 α-D-Man*p* 3 (on β-D-Man*p*) ↑ 1 α-D-Man*p*	+	+
α-D-Man*p*-(1 → 3)-α-D-Man*p*-(1 → 6)-β-D-Man*p*-(1 → R 6 (on the second α-D-Man*p*) ↑ 1 α-D-Man*p* 3 (on β-D-Man*p*) ↑ 1 α-D-Man*p* 2 ↑ 1 α-D-Man*p*	+	+
β-D-Glc*pN*Ac 1 ↓ 4 α-D-Man*p*-(1 → 3)-α-D-Man*p*-(1 → 6)-β-D-Man*p*-(1 → R 6 (on the second α-D-Man*p*) ↑ 1 α-D-Man*p* 3 (on β-D-Man*p*) ↑ 1 α-D-Man*p* 2 ↑ 1 β-D-Glc*pN*Ac	+	+
β-D-Glc*pN*Ac 1 ↓ 4 α-D-Man*p*-(1 → 3)-α-D-Man*p*-(1 → 6)-β-D-Man*p*-(1 → R 3 (on β-D-Man*p*) ↑ 1 β-D-Glc*pN*Ac-(1 → 4)-α-D-Man*p* 2 ↑ 1 β-D-Glc*pN*Ac	−	+

R = remainder of glycopeptide.

most of the carbohydrate was removed by the *endo*-glycanase from each of the native glycoproteins assayed, denaturation of the glycoprotein was necessary to effect complete removal. Enzyme activity of the substrate was not affected by carbohydrate depletion.

16 *endo*-β-D-2-Acetamido-2-deoxyglycanases

A β-D-2-acetamido-2-deoxyhexosidase (mol. wt. 1.20—1.45 × 10^5) purified from the digestive gland of a terrestial mollusc *Helicella ericetorum* has been shown to exhibit *endo*-β-D-acetamido-2-deoxyglycanase activity on natural substrates such as ovalbumin, ovomucoid, chondroitin-4-sulphate, chitin, and hyaluronic acid.[88]

17 Agarases

Agarase, prepared by affinity chromatography of a culture filtrate of an agar-degrading *Pseudomonas*-like bacterium, solubilized 72% of beaded agarose gel after incubation for 48 hours.[282] The gel structure was maintained but, in combination with mechanical disintegration, 95% of the gel could be solubilized. The molecular-sieving properties of the partially degraded agarose gels were changed. The molecular weight selectivity curve for dextran was not parallel to that of the original gel. The molecular weight distribution of the carbohydrate chains in agarose and partially degraded agarose was measured by gel filtration of melted samples at 60 °C on Sepharose® CL-4B. It was found that degradation of agarose gel is a two-step reaction. The first step is a rapid hydrolysis of very long polysaccharide chains to chains having a molecular weight of 10^4 to 10^5, indicating restrictions due to the gel structure. The second step is a slower hydrolysis of these chains into soluble oligosaccharides. Independently of the degree of degradation, some very large polysaccharide chains are left, and these chains may be of importance for maintaining gel structure. Enzymic degradation of melted agarose resulted in a rapid depolymerization to soluble oligosaccharides, mainly neoagarotetraose and neoagarohexaose.

18 Alginases and Alginate Lyases

The alginate lyase associated with bacteriophage infection of *Azotobacter vinelandii* has been used in the analysis of sodium alginate.[283] The enzyme degraded the polysaccharide to a series of oligosaccharides each containing a terminal 4-deoxy-α-L-*erythro*-hex-4-enepyranosyluronic acid group. Analysis of these oligosaccharides, together with kinetic information, indicated that the enzyme was specific for D-mannuronic acid-containing regions; this allowed primary structural identification of the substrate. The phage-induced enzyme was shown to be distinct from the alginate lyase elaborated by the host organisms by its pH optimum, molecular weight, K_m value, and stability.

19 α-Amylases

A strong stimulating effect by thiocyanate ions has been observed during the estimation of α-amylase by the amyloclastic method.[284] Thiocyanate ions themselves were shown to catalyse the decolorization of the starch–iodine complex and therefore may lead to wrong values during determination of

[282] M. Malmqvist, *Carbohydrate Res.*, 1978, **62**, 337.
[283] I. W. Davidson, C. J. Lawson, and I. W. Sutherland, *J. Gen. Microbiol.*, 1977, **98**, 223.
[284] D. Lathia and M. Brenderbach, *Clinica Chim. Acta*, 1978, **82**, 209.

α-amylase by this method. The rate of thiocyanate–iodine reactions depended upon the hydrogen ion concentration of the iodine reagent.

A study of some factors affecting the Phadebas α-amylase test has been reported.[285] Anomalous results previously reported for the estimation of urinary α-amylase levels, and dependent on the protein contained, were investigated in detail.

A method reported for the quantitative determination of α- and β-amylase activities present together in amylolytic preparations has been based on the use of non-modified amylase and cross-linked amylose substrates.[286] The former is cleaved by both enzymes whereas the latter is cleaved by α-amylase and not by β-amylase (see Scheme 3).

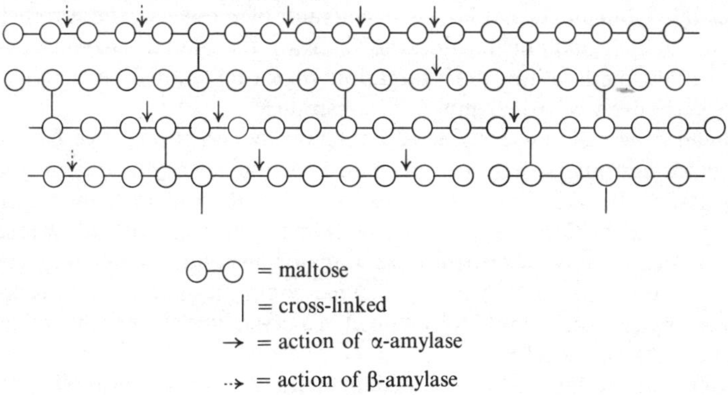

Scheme 3

A rapid and simple method for determination of α-amylase activity, by using a fraction of an amylo-oligosaccharide (amylodextrin) as substrate, has been described.[287] The sample solution is incubated with a solution of the amylo-oligosaccharide fraction and the substrate consumed is estimated by measuring the difference in absorbance at 460 nm. The K_m value of the amylo-oligosaccharide fraction is about half that of starch. The homogeneity of the amylo-oligosaccharide fraction in its chemical structure and molecular weight facilitated the specification of α-amylase units according to the international definition. This procedure was applied for assay of human-serum α-amylase with excellent reproducibility. This method did not require a large dilution factor, as the standard curve showed a linear relationship over a wide range of α-amylase concentrations.

Mathematical treatments have been developed to describe the action of α-amylase on amylose.[288] The treatments are based on the unique properties of the exponential (or most-probable) distribution of molecular weights of the substrate, namely, that (a) the principal averages are invarient to chain-end attack if the product molecules are ignored, and (b) the ratio of the principal

[285] K. Wiener and C. H. Foot, *Clinica Chim. Acta*, 1977, **75**, 177.
[286] M. A. Mateescu, *Biochimie*, 1978, **60**, 535.
[287] S. Kikumoto, N. Nimura, Y. Higara, and T. Kinoshita, *Carbohydrate Res.*, 1978, **61**, 369.
[288] W. Banks and C. T. Greenwood, *Carbohydrate Res.*, 1977, **57**, 301.

averages is invariant to random attack. The relations so developed allow published, qualitative data for the α-amylolysis of amylose to be interpreted in a quantitative manner. As a result, it appears that multiple attack is of little or no significance in the action patterns of α-amylases, with the exception of those derived from the pancreas.

A method described for the determination of (human) salivary α-amylase is based on the hydrolysis of 4-nitrophenyl α-maltoside to give 4-nitrophenol.[289] Precision estimates for the method were considered in detail. The production of oligosaccharide substrates for α-amylase and the actions of the human parotid and human and porcine pancreatic α-amylases upon them were compared.[290]

The effects of various ions in conjunction with *tris* or amino-acids upon the activity of human salivary α-amylase have been investigated.[291] Though it has been claimed previously that Na^+ is an activator of α-amylase, the current study reconfirmed that sodium ion does not activate the human enzyme. A continuous flow method for the estimation of serum α-amylase activity uses amylopectin dyed with DyAmyl-L as substrate.[292]

Human pancreatic juice, collected antiseptically and strictly free from other body fluids and juices, has been examined for the isoenzyme composition of α-amylase.[293] Unlike those of the extract of and exudate from pancreatic tissue, the composition of α-amylase isoenzyme in the secreted juice was simple and consisted of only two isoenzymes, a major one and a minor one, and the reason for the difference in the isoenzyme composition was discussed. The major α-amylase isoenzyme was obtained in a crystalline state with a molecular weight of $5.4—5.5 \times 10^4$.

A study of the kinetics of the Beckman Enzymatic Amylase Method, for determination of human serum and urinary amylase, by use of the centrifugal analyser showed that the rate of reaction is not linear.[294] It showed a pronounced initial lag period of 10 min, after which the lag was diminished but never completely abolished. Furthermore, the function relating rate of reaction to enzyme activity was not linear but increased disproportionately with increase in enzyme activity. When glycogen was substituted for soluble starch in the Beckman reagent, the initial lag period was less than 3 minutes and subsequently the reaction became linear. When mixed soluble starch–glycogen substrate was used, the α-amylase activity of serum and urine specimens correlated well with the activity determined by the Phadebas method. A dialysable factor present in a urine specimen from a kidney transplant patient inhibited the Beckman assay but not the Phadebas assay. This inhibition was simulated by the addition of pyruvate and reversed by the addition of NADH.

Three gelatinized tapioca starches [unmodified (A), hydroxypropyl distarch

289 B. K. Gillard, H. C. Markman, and S. A. Feig, *Clinical Chem.*, 1977, **23**, 2279.
290 K. Wallenfels, P. Földi, H. Niermann, H. Bender, and D. Linder, *Carbohydrate Res.*, 1978, **61**, 359.
291 S. K. Meur and K. B. De, *Experientia*, 1976, **32**, 1133.
292 J. A. Carney, S. A. Osibeluwo, J. A. Clements, and R. F. Jay, *Ann. Clin. Biochem.*, 1977, **14**, 350.
293 K. Matsuura, M. Ogawa, G. Kosaki, N. Minamiura, and T. Yamamoto, *J. Biochem. (Japan)*, 1978, **83**, 329.
294 N. Q. Hanson and W. G. Yasmineh, *Clinical Chem.*, 1978, **24**, 762.

phosphate (B), and hydroxypropyl starch (C)] were hydrolysed by porcine pancreatic α-amylase.[295] Oligosaccharide constituents of the hydrolysates were separated by t.l.c. and quantified by spectrodensitometry and were shown to be D-glucose, maltose, maltotriose, and maltotetraose. Hydrolysates from starches B and C had lesser amounts of the lower molecular weight oligosaccharides (G_1—G_4) than did those from starch A. The products of hydrolysis varied quantitatively with the degree of hydrolysis.

The kinetics of inhibition of porcine pancreatic α-amylase have been studied by use of D-glucono-1,5-lactone and maltobiono-1,5-lactone as transition-state analogues.[296] With D-glucono-1,5-lactone, α-amylase can be inhibited, to a degree, non-competitively ($K_i = 0.81$ mM, β *ca.* 0.2), whereas with maltobionolactone, the inhibition is competitive ($K_i = 0.31$ mM). The ratio of the binding affinity of the lactones, products, and substrates, permits the conclusion that ring distortion takes place in the transition state. Comparisons were made with the results of inhibition studies with glucoamylase and β-amylase (see p. 438). A study on the crystallization of porcine pancreatic α-amylase and the preparation of a new crystalline variety suitable for high resolution *X*-ray analysis has been reported.[297] Preliminary *X*-ray data were also recorded.

The effects of puromycin and actinomycin D on the serum and liver α-amylase levels in the rabbit, mouse, and rat have been investigated.[298] The report includes a commentary on the validity of the arguments in previous reports that question the biosynthesis of α-amylase in the liver.

Amylase activity has been measured in rat submaxillary salivary gland during heat acclimation (34 °C). It was suggested that the observed decrease in enzyme activity is due to liver atrophy which occurs during prolonged exposure to heat.[299]

Shellfish glycogen was cross-linked by treatment with cyanogen bromide followed by 1,6-diaminohexane.[300] The resulting, insoluble product efficiently adsorbed *Helix pomatia* α-amylase from crude solutions of the enzyme at 0 °C, but only poorly at higher temperatures. A method was developed for the purification of *Helix pomatia* α-amylase involving formation of an enzyme–adsorbent complex in the cold and recovery of the α-amylase by suspending the washed complex in buffer at 37 °C. After gel chromatography of the desorbed α-amylase, the enzyme was homogeneous as judged by polyacrylamide gel electrophoresis. An overall purification of 360-fold was achieved with a recovery of 35%.

Thermal stabilization resulting from protein–protein association between two inhibitor molecules from wheat flour and α-amylase from yellow mealworm (*Tenebrio molitor*) larvae has been investigated by differential scanning calorimetry.[301] Thermograms (plots of heat flow *versus* temperature) for the two

[295] R. L. Conway and L. F. Hood, *Stärke*, 1976, **28**, 341.
[296] E. Lásló, J. Holló, A. Hoschke, and G. Sárosi, *Carbohydrate Res.*, 1978, **61**, 387.
[297] M. Pierrot, J. P. Astier, B. Abadie, and G. Marchis-Mouren, *F.E.B.S. Letters*, 1977, **79**, 105.
[298] R. L. McGeachin, W. M. Abshier, and K. O'Leary, *Carbohydrate Res.*, 1978, **61**, 425.
[299] M. Horowitz, Y. Oron, and E. Atias, *Comp. Biochem. Physiol*, 1978, **60B**, 351.
[300] J. J. Marshall and W. Woloszczuk, *Carbohydrate Res.*, 1978, **61**, 407.
[301] V. Silano and J. C. Zahnley, *Biochim. Biophys. Acta*, 1978, **533**, 181.

inhibitors showed broad endothermic peaks with the same extrema (denaturation temperatures) at 93 °C, and equal, small enthalpies of denaturation (2 cal g^{-1}). The α-amylase produced a sharp endotherm at 70.5 °C, but a larger enthalpy change on denaturation (6 cal g^{-1}). The α-amylase–inhibitor complexes differed in thermal stability, but both showed significant stabilization relative to free enzyme.

α-Amylase has been purified from de-embryonated seeds of barley (*Hordeum vulgare*) which had been gibberellic acid-stimulated.[302] Treatment of the seeds with 5-fluorouracil had no effect on the thermal stability, molecular weight, isoenzyme composition, or specific activity of the enzyme or the amount synthesized. Contrary to the work of others, it was therefore concluded that 5-fluorouracil cannot be used as a probe to examine the timing of α-amylase in mRNA synthesis.

Gibberellic acid enhances ultimately α-amylase production in isolated barley aleurone layers and a method of assessing density shift of protein biosynthesis as a function of substrate actively available was devised.[303] The density shifts of α-amylase and of newly biosynthesized proteins (radiolabelled) were determined by isopycnic centrifugation in CsCl density gradients, and apparently all the α-amylase molecules are *de novo* biosynthesized. Thus it was concluded that there is no accumulation of an inactive α-amylase precursor in barley aleurone layers between the time of application of gibberellic acid and the time of the rapid increase in α-amylase activity. α-Amylases secreted by the aleurone layers of whole barley grains were relatively rich in L-histidine and relatively poor in L-glutamic acid/L-glutamine and L-serine residues compared with other eukaryotic proteins.[304] The α-amylase (mol. wt. *ca.* 4.1×10^4) contained D-glucose, D-mannose, and 2-acetamido-2-deoxy-D-glucose (0.5 moles $mole^{-1}$) and therefore appeared to be heterogeneous with respect to glycosylation.

Myoinositol hexaphosphate strongly inhibited α-amylase from barley, chick peas, wheat, maize, peanut, and *Bacillus subtilis*.[305] The inhibition of the wheat enzyme was non-competitive (K_i^{app} 1 mM), pH dependent, and markedly increased by preincubation with the inhibitor prior to substrate addition. Addition of Ca^{2+} did not reverse the inhibition, indicating that its inhibition was not due to Ca^{2+} binding by the inhibitor.

Changes in the isoenzyme pattern of α-amylase have been shown to coincide with the beginning of the starch degradation period in the pods of field beans (*Vicia faba*).[306] Mesophyll cells of the pods contained the same α-amylase isoenzymes as the endocarp, but exhibited a higher α-amylase activity that parallelled the much higher starch content of this tissue in comparison to the endocarp. Regulation of the starch breakdown was considered to be mediated in part by the formation of a special α-amylase isoenzyme.

Three α-amylases, which could be subdivided according to their heat stabilities, have been found in lentil (*Lens culinaris*) cotyledons during germina-

[302] S. J. Rodaway and H. Kende, *Plant Physiol.*, 1978, **61**, 1.
[303] T.-H. D. Ho and J. E. Varner, *Arch. Biochem. Biophys.*, 1978, **187**, 441.
[304] S. J. Rodaway, *Phytochemistry*, 1978, **17**, 385.
[305] C. B. Sharma, M. Goel, and M. Irshad, *Phytochemistry*, 1978, **17**, 201.
[306] K. Müntz, *Phytochemistry*, 1977, **16**, 1491,

tion.[307] Polyacrylamide gel electrophoresis provided an appraisal of all the amylolytic enzymes present, and their levels were determined as a function of germination time.

The levels of amylase activity in seeds of a pasture legume (*Stylosanthes humilis*) have been followed during germination.[104] Partial purification of the activity was achieved by ion-exchange chromatography.

α-Amylase has been found in the axis portion of ungerminated pea (*Pisum sativum*) seeds; its presence was demonstrated by the hydrolysis of β-limit dextrin, the change in adsorption spectrum for the starch–iodine complex, and the correlation between starch level decrease and reducing sugar level increase.[308] Changes occurring in the α-amylase levels during germination were quantified.

The α-amylase of the cell wall of *Pichia burtonii* has been investigated.[309]

Evidence has been reported for the presence of inhibitors of porcine pancreatic α-amylase in rye and wheat.[310] The inhibitors were isolated by affinity chromatography on the agarose derivative-immobilized enzyme. The location in the seed, molecular properties, and biological role of proteinaceous inhibitors of α-amylase in wheat have been reviewed.[311] Inhibition specificity of albumin inhibitors and structural features essential for interaction with inhibited amylases were also examined. The possible significance of these naturally occurring inhibitors in relation to their presence in foods in active form was described.

An immunochemical characterization of wheat α-amylase at different ontogenical steps has been reported.[312]

The α-amylase of mycelial cells of *Aspergillus oryzae* has been shown to exist in a particular form in 8×10^3 g pellets.[313] The purification of lysozomes by discontinuous sucrose gradient in D_2O centrifugation showed that the α-amylase activity is bound to these particles.

The action of *Aspergillus oryzae* α-amylase on reducing-end, and uniformly radiolabelled maltotriose through maltodecaose has been studied.[314] The enzyme was found to hydrolyse more than a single glycosidic bond during some enzyme–substrate encounters (Scheme 4). The extent of this repetitive attack was quantitated.

Aspergillus oryzae α-amylase catalyses degradation of oligosaccharides by a variety of pathways; a quantitative study of the degradation of maltotriose by this enzyme has been reported.[315] The results lead to a scheme (Schemes 5 and 6) involving multiple transglycosylation reactions and shifted binding due to simultaneous binding of two substrate molecules. The effect of substrate concentration on the products of maltotriose degradation is correctly predicted

[307] J. F. Tárrago and G. Nicolás, *Plant Physiol.*, 1976, **58**, 618.
[308] B. D. Davis, *Plant Physiol.*, 1977, **60**, 513.
[309] G. Moulin and P. Galzy, *Z. Allgem. Mikrobiol.*, 1978, **18**, 269.
[310] J. J. Marshall, *Carbohydrate Res.*, 1977, **57**, C27.
[311] V. Buonocore, T. Petrucci, and V. Silano, *Phytochemistry*, 1977, **16**, 811.
[312] J. Daussant, *Ann. Immunol.*, 1978, **129C**, 215.
[313] J. Bata, P. Vallier, and L. Colobert, *Experientia*, 1978, **34**, 572.
[314] J. D. Allen and J. A. Thoma, *Carbohydrate Res.*, 1978, **61**, 377.
[315] J. D. Allen and J. A. Thoma, *Biochemistry*, 1978, **17**, 2345.

(*A*) Bond-cleavage frequencies measured from reducing-end labelled maltoheptaose

0.00 0.34 0.17 0.41 0.08 0.00

○—○—○—○—○—○—ø

(*B*) Predicted products of hydrolysis of 100 uniformly labelled maltoheptaose molecules in the absence of repetitive attack

Substrate	*Number of molecules*	*Mole fraction*
●—●—●—●—●—●—ø	100 (before hydrolysis)	
Products		
ø	0	0.00
●—ø	42	0.21
●—●—ø	58	0.29
●—●—●—ø	58	0.29
●—●—●—●—ø	42	0.21
●—●—●—●—●—ø	0	0.00

Scheme 4 *Expected results for hydrolysis of uniformly labelled maltoheptaose in the absence of repetitive attack.* ○, *α-D-Glucopyranosyl unit;* ● *radiolabelled α-D-glucopyranosyl unit;* ø, *reducing radiolabelled D-glucose residue;* ——, (1 → 4) *bond. The bond-cleavage frequencies calculated from degradation of reducing-end labelled maltoheptaose were used to predict the products of hydrolysis of* 100 *molecules of uniformly labelled maltoheptaose in the absence of repetitive attack.*

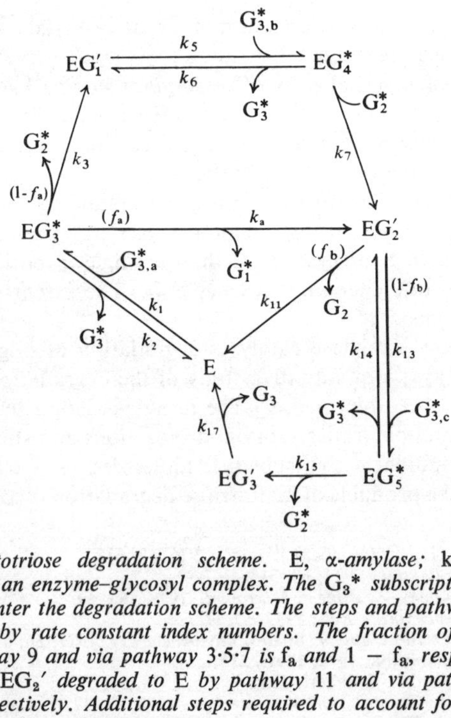

Scheme 5 *Maltotriose degradation scheme.* E, *α-amylase;* k, *rate constant, a prime indicates an enzyme–glycosyl complex. The* G_3* *subscripts indicate the point at which they enter the degradation scheme. The steps and pathways in this scheme are referred to by rate constant index numbers. The fraction of* EG_3* *degraded to* EG_2' *via pathway 9 and via pathway 3·5·7 is* f_a *and* 1 − f_a, *respectively. Similarly, the fraction of* EG_2' *degraded to* E *by pathway* 11 *and via pathway* 13·15·17 *is* f_b *and* 1 − f_b, *respectively. Additional steps required to account for all of the experimental data are shown in Scheme 6*

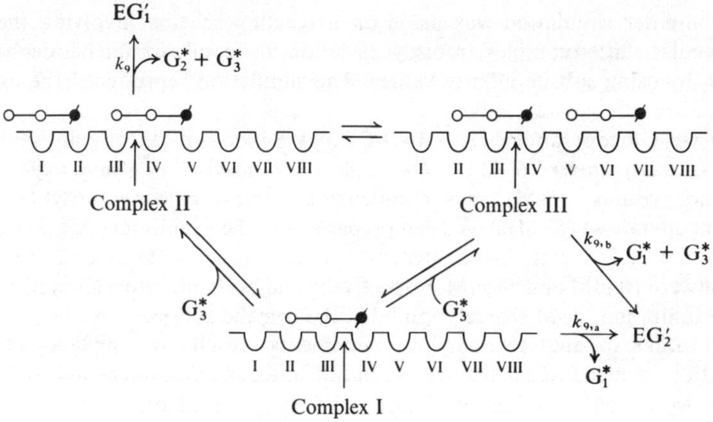

Scheme 6 *Two–one binding in the degradation of maltotriose.* ○—○—●, *reducing-end* ^{14}C *labelled maltotriose;* ∪, *subsite on the enzyme corresponding to a glucopyranoside unit binding position;* ↑, *bond cleavages site. This mechanism is an integral part of the model in Scheme 5. The* EG_3^* *complex in Scheme 5 is actually a composite of complexes* I, II, *and* III. EG_1', EG_2', *and* k_3 *are identical with their corresponding symbols in Scheme 5. The rate constant* k_9 *in Scheme 5 is now proportioned between* $k_{9,a}$ *and* $k_{9,b}$. *The enzyme is composed of eight subsites as shown* (*Allen & Thoma*, 1976). *Maltotriose molecules that are bound so as to expose a susceptible bond to the catalytic amino-acids are referred to as productively bound*

over a 10^4 fold concentration range, and the time course for maltotriose degradation was closely approximated by this scheme. The scheme was proposed as a general model of carbohydrate hydrolase action.

The utilization of pathways other than simple glycosidic bond scission by *A. oryzae* α-amylase increases with the concentration of substrate, implicating a multimolecular substrate mechanism.[316] Reducing-end labelled and uniformly labelled malto-oligosaccharides were used to elucidate these alternative degradation mechanisms. Condensation followed by hydrolysis is not a significant pathway. Transglycosylation is concluded to occur but no one single mechanism accounts for the degradation of maltotriose. Explanatory schemes were offered.

A depolymerization computer model, which uses a minimization routine to establish a subsite map for a depolymerase, has been applied to the experimental data for two α-amylases.[317] An eight-subsite map was arrived at from the K_m values and bond-cleavage frequencies for oligosaccharide substrates for *Aspergillus oryzae* α-amylase. Similarly ten subsites have been identified for *Bacillus amyloliquefaciens* α-amylase.

The action of Taka-amylase A, an α-amylase from *Aspergillus oryzae*, has been studied quantitatively by a product analysis method using unlabelled maltotriose and maltotriose labelled at the reducing end as substrates.[318]

[316] J. D. Allen and J. A. Thoma, *Biochemistry*, 1978, **17**, 2338.
[317] J. D. Allen and J. A. Thoma, *Biochem. J.*, 1976, **159**, 121.
[318] T. Suganuma, M. Ohnishi, R. Matsuno, and K. Hiromi, *J. Biochem.* (*Japan*), 1976, **80**, 645.

A computer simulation was made on a reaction scheme involving the termolecular-shifted complex, transglycosylation, and condensation besides hydrolysis, by using subsite affinity values. The simulation reproduced the experimental results satisfactorily.

O-α-D-Glucopyranosyl-(1 → 4)-*O*-6-deoxy-6-iodo-α-D-glucopyranosyl- (1 → 4)-D-glucopyranose (6′-MT), *O*-α-D-glucopyranosyl-(1 → 4)-6-deoxy-6-iodo-D-glucopyranose (6-M), and *O*-6-deoxy-6-iodo-α-D-glucopyranosyl-(1 → 4)-D-glucopyranose (6′-M) have been prepared and their inhibitory action against Taka-amylase A was investigated.[319] The inhibitor constants of 6′-MT and 6′-M were 10 mM and 54 mM, respectively, and both inhibitors showed mixed-type inhibition. 6-M scarcely inhibited the enzyme action.

In order to analyse complex experimental results of amylase-catalysed reactions, which include transglycosylation and condensation as well as hydrolysis, with amylose and linear oligosaccharides as substrates, a general equation governing the concentrations of substrate and product species in a reaction mixture has been derived based on the subsite theory.[320] The equation was devised so as to follow the location in products of radioactive D-glucose which was originally located at a certain position in a substrate. The reaction scheme assumed is as follows: (1) enzyme can bind any substrate reversibly to form an enzyme–substrate complex, ES; (2) From the ES complex, a D-glucosyl bond or C^1–OH bond at the reducing end of the substrate is cleaved to form a reactive intermediate, releasing a product of shorter chain or H_2O; (3) the reactive intermediate binds an acceptor molecule reversibly to form a ternary complex from which the product of transglycosylation or condensation is produced. The binding constants between enzyme and substrate, and between reactive intermediate and acceptor can be estimated from the subsite affinities. The rate constants of catalytic processes such as cleavage of a D-glucosyl bond and C^1–OH bond, hydrolysis, and transglycosylation are assumed to be independent of the degree of polymerization as well as of the binding mode of the substrate. The applicability of this scheme and rate equation was tested successfully with the Taka-amylase A-catalysed reaction of maltohexaose labelled with ^{14}C at the reducing end.

Bacillus acidocaldarius has been shown to produce an inducible thermoacidophilic α-amylase during the stationary phase of growth in the presence of (1 → 4)-linked α-D-glucans.[321] The enzyme fractions purified from both the culture medium and the cells were identical. The α-amylase (mol. wt. 6.8×10^4, pH optimum 3.5, temperature optimum 75 °C) was shown to comprise a single polypeptide chain and to be stable at acidic pH.

The α-amylases from five strains of *Bacillus amyloliquefaciens* have been compared to determine whether differences in primary structure are responsible for variations in catalytic properties among the enzymes.[322] Amino-acid analysis established virtually identical compositions for the proteins; reaction with

[319] K. Omichi, K. Fujii, T. Mizukami, and Y. Matsushima, *J. Biochem. (Japan)*, 1978, **83**, 1443.
[320] R. Matsuno, T. Suganuma, H. Fujimori, K. Nakanishi, K. Hiromi, and T. Kamikubo, *J. Biochem. (Japan)*, 1978, **83**, 385.
[321] V. Buonocore, C. Caporale, M. De Rosa, and A. Gambacorta, *J. Bacteriol.* 1976, **128**, 515.
[322] P. T. Borgia and L. L. Campbell, *J. Bacteriol.*, 1978, **134**, 389.

dimethylaminonaphthylene sulphonylchloride indicated an *N*-terminal L-valine for each α-amylase. Carboxypeptidase A analysis revealed an -L-leucyl-L-tyrosine-OH *C*-terminus. The enzymes (p*I* range 5.09—5.18) showed identical tryptic digests. It was concluded that the primary structures of the five enzymes are identical or exhibit only conservative substitutions.

The nature of basic differential equations governing the concentration of substrate and product species in a reaction of maltose catalysed by saccharifying α-amylase from *Bacillus subtilis* has been investigated by analysing their behaviour near zero time.[323] Near zero time, the differential equations could be approximated by simultaneous linear differential equations with respect to time. A characteristic equation was formulated from the coefficient matrix, and the eigen values, which are functions of initial maltose concentration and kinetic constants, were calculated. The eigen values were found to be negative real numbers except for one, which had a negative real value at low maltose concentrations and a positive one at high concentrations. Threshold concentration was defined as the initial maltose concentration at which the eigen value had zero value. The concentration was estimated to be 38.2 mM. Above the threshold concentration, the concentrations of oligomers having chains longer than maltose increased explosively at the initial stage of reaction and the concentrations at the steady state increased sharply with increase of substrate concentration, showing a typical threshold phenomenon. This type of threshold phenomenon was considered not necessarily characteristic of this enzyme.

The hydrolytic reaction of phenyl β-maltoside catalysed by *Bacillus subtilis* α-amylase has been studied at 25 °C and pH 5.4 by measuring the total reducing power, the amount of phenol liberated, and by t.l.c.[324] The enzyme hydrolysed phenyl β-maltoside at the D-glucosidic linkage between the two D-glucose residues, to give D-glucose and phenyl β-D-glucopyranoside; in addition maltose, maltotriose, and phenyl β-maltotrioside were produced. The enzymic reaction has been studied kinetically by a polarimetric method; the time-course of the reaction with purified maltose showed a distinct lag-phase in the initial stage of the reaction, followed by a linear, steady-state phase.[325] The results of product analysis showed that the reaction involves glycosyl transfer and/or condensation in addition to hydrolysis, and that the maltotriose and maltotetraose formed strongly accelerate the reaction. To interpret these phenomena, a reaction scheme involving hydrolysis, glycosyl transfer, and condensation was presented along with a computer simulation. The major factors responsible for the lag and sigmoidicity were: (1) The D-glucosidic linkage in maltose is difficult to cleave either for hydrolysis or for glycosyl transfer. (2) The rate constant of glycosyl transfer is much larger than that of hydrolysis. Maltotriose and maltotetraose are much better substrates than maltose in hydrolysis and/or in glycosyl transfer. (3) Once these malto-oligosaccharides are formed from maltose, even in a small amount, by transglycosylation or

[323] R. Matsuno, K. Nakanishi, M. Ohnishi, K. Hiromi, and T. Kamikubo, *J. Biochem.* (*Japan*), 1978, **83**, 859.
[324] K. Ishikura, Y. Nitta, and T. Watanabe, *J. Biochem.* (*Japan*), 1977, **81**, 1187.
[325] H. Fujimori, M. Ohnishi, M. Sakoda, R. Matsuno, and K. Hiromi, *J. Biochem.* (*Japan*), 1977, **82**, 417.

condensation, a reaction cycle is formed by which glycosyl transfer and hydrolysis are greatly accelerated. In this cycle maltose is consumed as an acceptor and D-glucose is formed as a product split off either in the hydrolysis or in the glycosyl transfer. This mechanism, together with the subsite structure, also explains why this enzyme yields D-glucose in much higher amount that other α-amylases.

Four L-tryptophan residues of saccharifying α-amylase from *Bacillus subtilis* out of eleven in total are reactive towards *N*-bromosuccinimide (NBS), suggesting that they are on the surface of the enzyme.[326] This is consistent with the results of solvent perturbation difference spectrophotometry with ethylene glycol. One of four L-tryptophan residues was clearly distinguished from the other three in reactivity with NBS by the stopped-flow method. This most reactive L-tryptophan residue was not protected from modification by substrates or analogues, indicating that it is not located in the substrate binding site. One of the other three L-tryptophan residues, probably the second most reactive one, was considered to be related in some way to the glycosyl transfer in the reaction of the enzyme with maltose as a substrate.

Conditions for the maximum yield of extracellular α-amylase from *Bacteroides amylophilus* have been reported.[327] Isoelectric focusing and polyacrylamide gel electrophoresis demonstrated the presence of six isoenzymes; one of these was purified by ion-exchange chromatography and gel filtration. This enzyme (pH optimum 6.3, temperature optimum 43 °C, pH stability range 5.8—7.5, p*I* 4.6, mol. wt. 9.2×10^4 by sodium dodecyl sulphate–polyacrylamide gel electrophoresis) was inhibited by cyclo-hexa and hepta-amyloses, phenyl α-D-glucopyranoside, and Hg^{2+} whereas Ca^{2+} and Co^{2+} were strong activators. The relative rates of hydrolysis of amylose, soluble starch, amylopectin, and dextrin were 100, 97, 92, and 60%, respectively (K_m values 18.2, 18.7, 18.2, and 16.7 μmol D-glucosidic bonds l^{-1}, respectively).

Due to the gene for hyperderepression, a cross between a hyperderepressed strain and a normal strain of *Neurospora crassa* exhibited hyperproduction of amylase activity.[107]

An α-amylase preparation from *Thermoactinomyces vulgaris* has been used to hydrolyse pullulan to panose.[328] The digest was applied to a carbon–Celite column and eluted with a linear gradient of propan-1-ol from 0 to 5%. From the trisaccharide fractions eluted, panose was prepared in about 70% yield. The production of amylase activity by *Thermomonospora curvata* has been investigated.[329]

Various enzymically active immobilized forms of α-amylase have been reported,[330] including reaction of the enzyme with hydrous titanium(IV) oxide coated on magnetic iron oxide, and with diazotized 1,3-diaminobenzene coated on magnetic iron oxide which was itself either uncoated or coated with

[326] H. Fujimori, M. Ohnishi, and K. Hiromi, *J. Biochem.* (*Japan*), 1978, **83**, 1503.
[327] S. J. McWethy and P. A. Hartman, *J. Bacteriol.*, 1977, **129**, 1537.
[328] Y. Sakano, M. Kogure, T. Kobayashi, M. Tamura, and M. Suekane, *Carbohydrate Res.*, 1978, **61**, 175.
[329] F. Stutzenberger and R. Carnell, *Appl. Environ. Microbiol.*, 1977, **34**, 234.
[330] I. P. Ilalych, O. S. Cyperovych, L. O. Kolesnyk, and V. D. Cesarska, *Ukrainskü Biokhimichnii Zhur.*, 1976, **48**, 480.

hydrous titanium(IV) oxide,[331] and reaction with a dextran derivate to give a water-soluble product.[332]

20 β-Amylases

A method reported for differential determination of β- and α-amylase activities in mixtures of the two enzymes relies on the non-susceptibility of cross-linked amylase to β-amylase (see p. 427).[286]

β-Amylase has been shown to occur in ungerminated seeds of a pasture legume (*Stylosanthes humilis*).[104] The level of activity was followed during germination of the seeds.

A β-amylase has been found in lentil (*Lens culinaris*) cotyledons during germination.[307] Polyacrylamide gel electrophoresis provided an appraisal of all the amylolytic enzymes present, and their levels were determined as a function of germination time.

The increased levels of β-amylase activity in pea (*Pisum sativum*) seeds which occur on germination have been investigated.[308]

The specificity of sweet-potato β-amylase in the vicinity of the phosphate ester groups was studied by determining the structures of the phosphorylated oligosaccharides (α-phosphodextrin and β-limit-α-phosphodextrin) formed by its action on potato starch.[333] The β-limit-α-phosphodextrin was separated by ion-exchange chromatography into three fractions that were distinguishable by the d.p. and by the ratio of D-glucose 6-phosphate residues to total organic phosphate. Each fraction contained linear molecules having one phosphate ester group that was not located at the reducing or non-reducing terminals. The smallest phosphodextrin was 6^2-phosphorylmaltotriose. It was deduced that β-amylase hydrolysed (1 → 4)-α-D-linkages from the non-reducing end until one or two D-glucosyl residues remained attached to the phosphorylated residue, depending on whether there was originally an odd or even number of D-glucosyl residues on the non-reducing side of the phosphorylated residue.

Crystalline β-amylase catalysed essentially complete hydrolysis of α-maltosyl fluoride to give maltose and hydrogen fluoride.[334] The rates of hydrolysis catalysed by β-amylase preparations from sweet potatoes and soybeans acting on a range of concentrations of the substrate produced linear curves for the relationship, $1/v$ *vs.* $1/S$; reaction constants for crystalline, sweet-potato enzyme were K_m 3.6 mM and V_{max} *ca.* 2 μmol min^{-1} mg^{-1}. The finding that α-maltosyl fluoride is hydrolysed 30—60 times faster than maltotriose demonstrates for the first time that β-amylase is capable of effecting hydrolysis at an appreciable rate of a substrate having only two D-glucose residues.

Sweet potato β-amylase-catalysed hydrolyses of aryl β-maltotriosides with substituents, NO_2-, Cl-, and Br- at the 2-, 3-, and 4-positions in the phenyl ring have been studied at pH 4.8 and 25 °C.[335] The hydrolyses of a few of the maltotriosides by soybean β-amylase were also studied at pH 5.4 and 25 °C.

[331] J. F. Kennedy, S. A. Barker. and C. A. White. *Stärke*, 1977, **29**, 240.
[332] J. J. Marshall and J. D. Humphreys, *J. Chromatog.*, 1977, **137**, 468.
[333] S. Tabata, S. Hizukuri, and K. Nagata, *Carbohydrate Res.*, 1978, **67**, 189.
[334] D. S. Genghof, C. F. Brewer, and E. J. Hehre, *Carbohydrate Res.*, 1978, **61**, 291.
[335] N. Suetsugu, K. Takeo, Y. Sanai, and T. Kuge, *J. Biochem.* (*Japan*), 1978, **83**, 473.

It was found that the aryl β-maltotriosides were preferentially hydrolysed into maltose and aryl β-D-glucosides by both β-amylases. The Michaelis constant K_m and the molecular activity k_0 were determined for the hydrolyses of these maltotriosides and compared with those of maltotriose and maltotetraose. Aryl β-maltotriosides were more rapidly hydrolysed than maltotriose by a factor of 30—80, and more slowly hydrolysed than maltotetraose by a factor of 10—30, depending on the kinds of substituents. The rapid hydrolysis of aryl β-maltotrioside as compared with maltotriose was considered due to the interaction of an aryl group with the subsite of β-amylase. This is in contrast with *Rhizopus niveus* glucoamylase-catalysed hydrolysis of phenyl β-maltoside, whose phenyl group does not interact so much with the subsite of the enzyme.

The kinetics of inhibition of porcine pancreatic α-amylase, sweet-potato β-amylase, and *Aspergillus niger* glucoamylase enzymes have been studied by use of D-glucono-1,5-lactone and maltobiono-1,5-lactone as transition-state analogues.[296] With D-glucono-1,5-lactone, α-amylase can be inhibited, to a degree, non-competitively (K_i = 0.81 mM, β *ca.* 0.2), whereas with maltobionolactone, the inhibition is competitive (K_i = 0.31 mM). The effect of β-amylase can be inhibited with maltobionolactone in a completely competitive way (K_i = 0.11 mM), whereas with D-gluconolactone the inhibition is very poor (K_i = 21 mM). Glucoamylase cannot be inhibited with maltobionolactone, whereas, with D-gluconic acid, a completely mixed inhibition may be observed (K_i = 1.3 mM). The ratio of the binding affinity of the lactones, products, and substrates permits the conclusion that ring distortion takes place in the transition state with all three enzymes.

Bacillus polymyxa β-amylase is reported to be much less susceptible to inhibition by cyclodextrins, and by reagents which react with thiol groups, than is sweet-potato β-amylase.[336]

21 Amylo-1,6-D-glucosidases

[^{14}C]Glycogen has been synthesized *in vitro* by incubating D-[^{14}C]glucose with rabbit-liver glycogen in the presence of a pure preparation of the glycogen-debranching enzyme (amylo-1,6-glucosidase-4-α-glucanotransferase)[337] The course of the reaction has been monitored and ^{14}C-products isolated after 30 minutes and 5 hours. The distribution of D-[^{14}C]glucose groups in the polysaccharides has been determined by debranching the molecules with a crystalline isoamylase from a *Pseudomonas* species. The quantities of unlabelled and ^{14}C-linear unit chains containing D-[^{14}C]glucose at their reducing ends have been determined by paper chromatography followed by enzymic degradation and analysis. In the 30 minute product, between 65 and 85% of the D-[^{14}C]-glucose groups were covered by unlabelled groups because of transferase action. In the 5 hour product, the extent of covering approached 100%. Extensive redistribution of unlabelled groups also was found to have occurred, even in the early stages of the reaction. It is concluded that the D-[^{14}C]glucose incorporation assay for amylo-1,6-D-glucosidase, as ordinarily carried out, is

[336] J. J. Marshall, *Mol. Cell. Biochem.*, 1975, **7**, 127.
[337] D. H. Brown, D. J. Sprinkle, and B. J. Brown, *Carbohydrate Res.*, 1978, **61**, 265.

probably not specific just for the hydrolytic action of this enzyme, but that it depends indirectly on its transferase activity as well.

22 Cellulases

The enzymic hydrolysis of common cellulosic wastes by cellulase has been investigated further.[338]

Cellulase activity has been measured polarographically by coupling the liberation of D-glucose with oxygen consumption.[339] Cellulase is coupled with an excess mixture of '(1 → 4)-α-D-glucosidase', D-glucose oxidase, mutarotase, and catalase. The assay is capable of detecting D-glucose concentrations of 5 $\mu g\ ml^{-1}$.

An automated method reported for the determination of cellulase activity is based on stopping the enzyme reaction at alkaline pH, dialysis, and determination of the reducing power of the diffusate by the spectrophotometric ferricyanide–molybdoarsenic acid reaction.[340]

The role of acquired digestive enzymes in the digestion of cellulase in the mid-gut of the fungus-growing termite *Macrotermes matalensis* has been studied.[341]

Two cellulases purified from growing regions of auxin-treated peas hydrolysed cellulose powder, carboxymethylcellulose, cellodextrins and mixed-linkage glucans, such as barley β-D-glucan, at rates comparable to those reported for the most active fungal cellulases, and with kinetics and product formation characteristic of *endo*-hydrolase action.[342] They were unable to cleave (1 → 3)-linkages in β-D-glucans, or (1 → 4)-linkages in dextrins containing excessive substitution at C6, α-D-configuration, alternating (1 → 3)- and (1 → 4)-β-D-linkages or residues other than D-glucose. Despite marked differences in their physical and immunochemical properties, close similarities between the two enzymes suggested identity of their active sites.

Solid media containing carboxymethylcellulose have been developed to detect C_x cellulase-producing micro-organisms.[343] Hydrolysis of the carboxymethylcellulose was seen as a clear zone around colonies after flooding plates with 1% aqueous hexadecyltrimethyl ammonium bromide. Tests with ten bacterial and four fungal species have shown that the degree of substitution of the carboxymethylcellulose affected both growth and enzyme production. The solid media substrates provided a more rapid assay for C_x cellulase production than did a medium containing native cellulose.

Factors affecting the cellulolytic activity of rumen contents have been examined.[344] The levels and nature of cellulase activity during growth of *Achlya bisexualis* on D-glucose, cellulose, and selected polysaccharides have been studied.[345]

[338] S. Dhawan and J. K. Gupta, *J. Gen. Appl. Microbiol.*, 1977, **23**, 155.
[339] T. R. Green, Y. W. Han. and A. W. Anderson, *Analyt. Biochem.*, 1977, **82**, 404.
[340] K. A. Holm, *Analyt. Biochem.*, 1978, **84**, 522.
[341] M. M. Martin and J. S. Martin, *Science*, 1978, **199**, 1453.
[342] Y. Wong, G. B. Fincher, and G. A. Maclachlan, *J. Biol. Chem.*, 1977, **252**, 1402.
[343] L. Hankin and S. L. Anagnostakis, *J. Gen. Microbiol.*, 1977, **98**, 109.
[344] C. S. Stewart, *Appl. Environ. Microbiol.*, 1977, **33**, 497.
[345] W. H. Miele and A. E. Linkins, *Canad. J. Botany*, 1978, **56**, 1974.

Aspergillus clavatus when grown on synthetic liquid media with soluble or insoluble cellulose as the sole carbon source released cellulolytic enzymes into the media.[346] These enzymes were separable by gel filtration and ion-exchange chromatography into four components. When assayed for hydrolysis of insoluble cellulose, the combined components released more D-glucose than did the individual, separate components.

A celluloytic enzyme isolated from a commercial cellulase preparation from *Aspergillus niger*, and which contained no carbohydrate, was found to be active towards carboxymethylcellulose but inactive towards cellobiose and 4-nitrophenyl β-D-glucopyranoside.[347] The enzyme (pH optimum 3.9, mol. wt. 2.6×10^4 by sodium dodecyl sulphate–polyacrylamide gel electrophoresis) according to kinetic studies had groups with pK values between 4.2 and 5.3 involved in the enzyme–substrate complex.

A *Cellulomonas* species which hydrolysed cellulose, as shown by clear-zone formation on a cellulose-agar medium, exhibited catabolite repression of cellulase biosynthesis on addition of D-glucose to the culture medium.[348] A stable mutant that no longer exhibited catabolite repression was produced through treatment of the wild-type organism with *N*-methyl-N^1-nitro-*N*-nitrosoguanidine. Cellobiose, cellulose, and sophorose induced cellulase biosynthesis. Cellulase regulation was concluded to depend upon a complex relationship involving catabolite repression, inhibition, and induction.

A *Cellulomonas* species produced two classes of cellulases when grown in the presence of cellulose.[349] One class comprised enzymes which were tightly bound to cellulose, and the other comprised enzymes which were found free in the culture supernatant liquid. The cellulose-bound activity reached a maximum at the beginning of the stationary phase and then decreased, whereas the soluble activity increased to a plateau late in the stationary phase. The production of both classes of enzyme was repressed by adding cellobiose or glycerol to the cellulose.

Three extracellular cellulases have been purified from the cultures of a *Cellulomonas* species.[350] One was found in solution in the cell-free supernatant and two others were found to be bound to the cellulose added as a carbon source. The free enzyme and one of the cellulose-bound enzymes exhibited binding for Sephadex®. The two cellulose-bound enzymes are glycoproteins. All three enzymes behaved as *endo*-cellulases towards soluble carboxymethylcellulose and had little activity on cellulose powder.

When cultivated in a liquid medium with cellulose, *Chaetomium thermophile* produced extracellular cellulolytic enzymes.[214] Ion-exchange chromatography and gel filtration gave three electrophoretically pure components; a typical C_x *endo*-enzyme (pI 4.6, mol. wt. 4.1×10^4) active on carboxymethylcellulose in solution but of low effect on cellulose; a C_1 *exo*-enzyme (pI 4.6, mol. wt. 6.7×10^4) active on cellulose but hardly on carboxymethylcellulose; and a

346 P. O. Olutiola, *J. Gen. Microbiol.*, 1977, **102**, 27.
347 P. L. Hurst, J. Nielsen, P. A. Sullivan, and M. G. Shepherd, *Biochem. J.*, 1977, **165**, 33.
348 B. J. Stewart and J. M. Leatherwood, *J. Bacteriol.*, 1976, **128**, 609.
349 P. Béguin, H. Eisen, and A. Roupas, *J. Gen. Microbiol.*, 1977, **101**, 191.
350 P. Béguin and H. Eisen, *European J. Biochem.*, 1978, **87**, 525.

β-D-glucosidase, inactive upon cotton or carboxymethylcellulose. The three components in admixture showed synergistic effects on highly ordered cellulose. The cellulase system of a *Cytophaga* species has been investigated.[351]

The C_1 component from *Fusarium solani* cellulase has been purified extensively by molecular-sieve chromatography on Ultrogel® and ion-exchange chromatography.[352] The purified component showed little capacity for hydrolysing highly ordered substrates (*e.g.* cotton fibre), but poorly ordered substrates (*e.g.* phosphoric acid-swollen cellulose), and the soluble cello-oligosaccharides cellotetraose and cellohexaose, were readily hydrolysed; cellobiose was the principal product in each case. Attack on carboxymethylcellulose, a substrate widely used for measuring the activity of the randomly acting enzymes (C_x enzymes) of the cellulase complex, was minimal, and ceased after the removal of a few unsubstituted residues from the end of the chain. These observations, and the fact that the rate of change of degree of polymerization of H_3PO_4-swollen cellulose was very slow compared with that effected by the randomly acting *endo*-D-glucanases (C_x, CM-cellulases), indicated that C_1 is a cellobiohydrolase. Fractionation by a variety of methods gave no evidence for the non-identity of the cellobiohydrolase and the component that acted in synergism with the randomly acting C_x enzyme when solubilizing cotton fibre.

Mycelia and ungerminated conidia of *Neurospora crassa* have been shown to secrete an *endo*-cellulase.[353] A simple induction system of phosphate buffer plus inducer relied on the internal metabolic reserves of conidia or mycelia to provide energy and substrates for protein synthesis. Cellobiose was the best inducer. In deionized water, the cellulase remained mostly associated with the cell, but a variety of salts stimulated the release of the enzyme into the medium.

The induction of β-D-glucosidase activities in *Sporotrichum pulverulentum* by cellulose and cellobiose has been investigated (see p. 410).[220] A new source of cellulase used by Macrotermitinae has been described. Ingested conidiophores of the symbiotic fungus *Termitomyces* species provide this cellulase.[354]

Four principal *endo*-β-D-glucanase components of *Trichoderma homingii* cellulase have been separated by gel filtration, ion-exchange chromatography, and isoelectric focusing.[355] All four enzymes hydrolysed carboxymethylcellulose, phosphoric acid-swollen cellulose, cellotetraose, and cellopentaose, but differed in the rate and mode of attack. All four enzymes acted synergistically with the *exo*-β-D-glucanase (cellobiohydrolase) of the same organism. The results of these studies appeared compatible with the interpretation that the 'crystalline' areas of cotton cellulose are hydrolysed only by those *endo*-β-D-glucanases capable of forming an enzyme–enzyme complex with cellobiohydrolase on the surface of the cellulose chains.

The development of an agar plate screening technique has allowed the isolation of a range of mutants of *Trichoderma reesei* capable of biosynthesizing

[351] W. T. H. Chang and D. W. Thayer, *Canad. J. Microbiol.*, 1977, **23**, 1285.
[352] T. M. Wood and S. I. McRae, *Carbohydrate Res.*, 1977, **57**, 117.
[353] B. M. Eberhart, R. S. Beck, and K. M. Goolsby, *J. Bacteriol.*, 1977, **130**, 181.
[354] N. Abokhatwa *Experientia*, 1978, **34**, 559.
[355] T. M. Wood and S. I. McCrae, *Biochem. J.*, 1978, **171**, 61.

cellulase under conditions of high catabolite repression.[223] Commercial aspects of the production and isolation of hyper-cellulase producing mutants of *T. reesei* were discussed.

Continuous culture studies for the production of cellulase have been carried out growing *Trichoderma viride* on a medium containing commercial D-glucose as the carbon source.[356] Some data on cellulase biosynthesis were examined and correlated in terms of a maturation time model. Specific oxygen uptake rates were correlated with specific production rates. Mutant strains not producing cellulases have been induced and isolated from *Trichoderma viride*.[357] Enrichment of the mutants was carried out by nystatin selection.

Differential speed two roll milling was found to be an effective pretreatment for increasing the susceptibility of cellulose to enzymic hydrolysis, as exemplified using cellulase from *Trichoderma viride*.[358] A further advantage of the roll mill was the increased wet density of the product permitting higher slurry concentrations during hydrolysis. Since the first step in the enzymic hydrolysis of cellulase is the adsorption of the enzyme(s) on to the surface of the cellulose fibre, the process of adsorption of *Trichoderma viride* cellulase system on three different preparations of pure cellulose (cellulose pulp, milled pulp, and microcrystalline cellulose) has been investigated.[359] The adsorption of protein and enzyme from solution was largely independent of pH, but dependent upon temperature and the type of cellulose used.

A cellulase component has been purified from a crude cellulase preparation of *Trichoderma viride* until homogeneous on polyacrylamide disc gel electrophoresis.[360] The enzyme (mol. wt. 4.5×10^4 by gel filtration, pH optimum 4.5—5.0, temperature optimum 50 °C) was thermostable but was inactivated by Hg^{2+}, Ag^+, and Cu^{2+}. The K_m values for the enzyme decreased in parallel with increase of chain length of the substrates, and the enzyme produced cellobiose and D-glucose from various cellulosic substrates. The enzyme was also found to catalyse the rapid synthesis of cellotetraose from cellobiose.

A low molecular weight cellulase (mol. wt. 2.0×10^4, p*I* 7.52) has been isolated from culture filtrates of *Trichoderma viride* by a two-step procedure of gel filtration and ion-exchange chromatography.[361] Subjection of the preparation to polyacrylamide gel electrophoresis, isoelectric focusing, sedimentation equilibrium analysis, and chromatography of the reduced and alkylated form indicated it to be homogeneous.

Cellulase has been immobilized with retention of enzymic activity by reaction with polyacrylonitrile resin[362] and by glutaraldehyde coupling to collagen fibrils in suspension.[363]

[356] D. E. Brown and M. A. Zainudeen, *Biotechnol. Bioeng.*, 1977, **19**, 941.
[357] K. M. H. Nevalainen and E. T. Palva, *Appl. Environ. Microbiol.*, 1978, **35**, 11.
[358] T. Tassinari and C. Macy, *Biotechnol. Bioeng.*, 1977 **19**, 1321.
[359] N. Peitersen, J. Medeiros, and M. Mandels, *Biotechnol. Bioeng.*, 1977, **19**, 1091.
[360] G. Okada, *J. Biochem. (Japan)*, 1976, **80**, 913.
[361] K. Håkansson, L. Fägerstam, G. Pettersson, and L. Andersson, *Biochim. Biophys. Acta*, 1978, **524**, 385.
[362] T. Handa, S. Goto, T. Shindo, and H. Hukai, *J. Jap. Soc. Starch Sci.*, 1976, **23**, 71.
[363] I. Karube, S. Tanaka, T. Shirai, and S. Suzuki, *Biotechnol. Bioeng.*, 1977, **19**, 1183.

23 Chitinases

Radioactive chitin (reacetylated chitosan) has been used as the substrate for the sensitive assay of chitinase.[364] The procedure is based on the insolubility of chitin and solubility of the reaction product *N,N'*-diacetylchitobiose. Reacetylated chitosan proved to be a more satisfactory adsorbent during the purification of chitinase than did colloidal chitin.

The display of weak chitinase activities by hen and turtle egg lysosomes has been reported.[365]

Two chitinases (pH optima 4.2 and 4.3, mol. wts. 2.6×10^4 and 1.3×10^4 respectively) have been purified from *Helix pomatia* digestive juice and the endogenous β-D-2-acetamido-2-deoxyglucosidase.[89] The larger chitinase molecule gave two chitinase-active regions on isoelectric focussing (p*I*'s 5.7 and 3.8) whereas the smaller gave one active region (p*I* 7.3).

During the moult, chitin in the old cuticle of the insect *Manduca sexta* was shown to be digestive by the chitinase taken up from moulting fluid, but the chitin in intact (*i.e.* premoult) cuticle was not accessible to chitinase.[366] As a prerequisite of digestion, old cuticle was rendered chitinase-labile by a reaction attributable to a trypsin-like activity in the moulting fluid.

Chitinase activity has been shown to be released during autolysis of *Neurospora crassa* under certain culture conditions.[98]

24 Chondroitin ABC and Chondroitin AC Lyases

An 'improved' method described for the micro determination of glycosaminoglycans is based on digestion with chondroitin ABC and AC lyases.[367] The method has, however, been published previously. The determination is based on the colour production of 4-deoxy-α-L-*threo*-4-enepyranosyluronic acid-containing disaccharides produced by the action of chondroitin ABC and AC lyases when the periodate–thiobarbituric acid method is applied to the α,β-unsaturated disaccharides. Suitable conditions for the quantitative assay were described.

The modes of action of two chondroitinases AC from *Arthrobacter aurescens* and *Flavobacterium heparinum* have been examined.[368] By comparison of the change of viscocity and products of digests, it was shown that the *Arthrobacter* enzyme degrades the substrate by a stepwise attack, though the first attack of both enzymes is of *endo* type. Further studies were carried out on the initial rate and final extent of the enzymic degradation of various glycosaminoglycans and modified forms thereof.

To decide upon the action patterns of these enzymes, a new method for the calculation of the degree of multiple attack, based on a concept established by J. F. Rohyt and D. French (*Arch. Biochem. Biophys.*, 1967, **122,** 8), was employed.[369] It was shown that the degree of multiple attack was represented by

[364] J. Molano, A. Durán, and E. Cabib, *Analyt. Biochem.*, 1977, **83**, 648.
[365] S. K. Gayen, S. Som, N. K. Sinha, and A. Sen, *Arch. Biochem. Biophys.*, 1977, **183**, 432.
[366] M. L. Bade and A. Stinson, *Biochem. Biophys. Res. Comm.*, 1978, **84**, 381.
[367] M. Koseki, A. Kimura, and K. Tsurumi, *J. Biochem.* (*Japan*), 1978, **83**, 553.
[368] K. Hiyama and S. Okada, *J. Biochem.* (*Japan*), 1976, **80**, 1201.
[369] K. Hiyama, *J. Biochem.* (*Japan*), 1976, **80,** 1209.

the ratio of the initial velocity of the number-average degree of scission to that of the viscocity-average degree of scission. The degree of multiple attack was estimated to be 3.03 for the *Arthrobacter* chondroitinase AC and 1.31 for the *Flavobacterium* chondroitinase AC.

The actions of the two chondroitinases AC have been compared *via* steady-state kinetics carried out at pH 6.0.[370] The *Arthrobacter* enzyme was susceptible to salt activation/regression and bivalent metal ions such as Ca^{2+}, Co^{2+}, Mg^{2+}, and Ba^{2+} were more effective for activation than virulent ions such as Na^{+} and K^{+}. Various kinetic parameters of the actions of the enzymes was determined and thermodynamic parameters were calculated. The results obtained suggested that the driving force of enzyme–substrate complex formation could be attributed to ionic interaction for *Arthrobacter* chondroitinase AC and to hydrophobic bonding for *Flavobacterium* chondroitinase AC.

25 Dextranases

A bacterium previously shown to produce a novel type of dextranase (isomaltose-releasing dextranase) has been identified as *Arthrobacter globiformis*.[371]

Two distinct dextranase activities have been isolated from the culture supernatant of a soil bacterium, *Leuconostoc mesenteroides*.[372] Among the various α-D-glucans tested, only dextrans containing (1 → 2)-α-D-linkages at the branch points were susceptible to one of the activities. The hydrolysate of such a dextran was composed of D-glucose as sole low molecular weight product and limit dextrin. It was therefore suggested that the enzyme has a specificity for (1 → 2)-α-linkage-branched dextrans.

An *endo*-dextranase isolated from the culture fluid of *Penicillium purpurogenum* did not act on isomaltose, hydrolyses isomaltotriose slowly, and hydrolysed isomaltotetraose rapidly with the formation of two molecules of isomaltose.[373] The action of the enzyme upon dextrans with various degrees of branching and upon isomaltopentaitol was investigated.

The mode of action of dextranase D_2 from a *Pseudomonas* species has been investigated using a series of isomalto-oligosaccharides as substrates.[374] The scission of reduced isomaltodecaose was represented diagrammatically (Scheme 7).

The action pattern and subsite map of a dextranase from *Streptococcus mutans* have been reported.[375] The experimental and calculated bond-cleavage frequencies for the enzyme and its action on isomalto-oligosaccharides were compared (Scheme 8).

Dextranase has been immobilized with retention of enzyme activity by reaction with 1,3-diaminobenzene-coated porous titanium(IV) oxide particles [376]

[370] K. Hiyama and S. Okada, *J. Biochem.* (*Japan*), 1977, **82**, 429.
[371] T. Sawai, Y. Ukigai, and A. Nawa, *Agric. Biol. Chem.* (*Japan*), 1976, **40**, 1249.
[372] M. Kobayashi, Y. Mitsuishi, and K. Matsuda, *Biochem. Biophys. Res. Comm.*, 1978, **80**, 306.
[373] A. L. Minakova and M. E. Preobrazhenskaya, *Biochemistry* (*U.S.S.R.*), 1977, **42**, 1264.
[374] G. N. Richards and M. Streamer, *Carbohydrate Res.*, 1978, **62**, 191.
[375] A. Pulkownik, J. A. Thoma, and G. J. Walker, *Carbohydrate Res.*, 1978, **61**, 493.
[376] J. F. Kennedy and I. M. Kay, *Carbohydrate Res*,. 1977, **59**, 553.

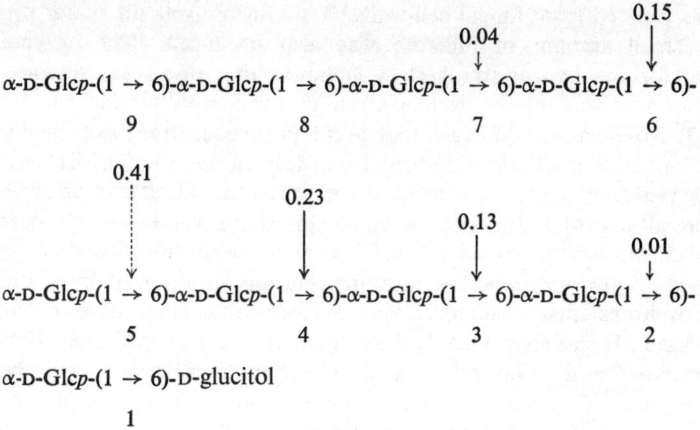

Scheme 7

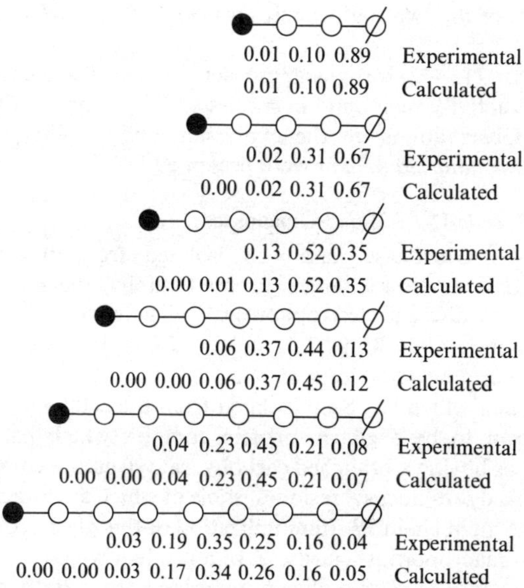

○, D-Glucose residue; ∅, "reducing" D-glucose residue; ●, D-[^{14}C]glucose residue; —, (1 → 6)-α-D-glucosidic linkage.

Scheme 8

and with a coating of aminobenzoic acid–formaldehyde resin on nickel, tin, cobalt, iron, and aluminium.[377]

26 *endo*-(1 → 3)-α-D-Glucanases

Cladosporium resinae (1 → 3)-α-D-glucanase has been characterized as an *endo*-D-glucanase capable of completely hydrolysing insoluble (1 → 3)-α-D-

[377] M. F. Chaplin, and J. F. Kennedy, *Carbohydrate Res.*, 1976, **50**, 267.

glucans isolated from fungal cell-walls.[378] D-Glucose was the major product, but a small amount of nigerose was also produced. The enzyme was specific for the hydrolysis of (1 → 3) bonds that occur in sequence, and nigerotetraose was the smallest substrate that was rapidly attacked. Isolated (1 → 3)-α-D-glucosidic linkages that occur in mycodextran, isolichenin, dextrans, and oligosaccharides derived from dextran were not hydrolysed. Insoluble D-glucans synthesized from sucrose by culture filtrates of *Streptococcus sp.* were all hydrolysed to various limits; the range was 11—61%. A soluble D-glucan, synthesized by an extracellular D-glucosyltransferase of *S. mutans* was not a substrate, whereas insoluble glucans synthesized by a different D-glucosyltransferase, isolated from *S. mutans* strains, were extensively hydrolysed. It was suggested that dextranase-CB, a bacterial *endo*-(1 → 6)-α-D-glucanase that does not release D-glucose from any substrate, could be used together with *C. resinae* (1 → 3)-α-D-glucanase to determine the relative proportions of (1 → 6)-linked to (1 → 3)-linked sequences of D-glucose residues in the insoluble D-glucans produced by oral streptococci. The simultaneous action of the two D-glucanases was highly effective in solubilizing the D-glucans.

A survey for (1 → 3)-α-D-glucanase activity has been conducted among yeasts.[379] The activity was found in species of *Cryptococcus*, *Rhodotorula*, and *Endomyces*. Observations on the expression and stability of (1 → 3)-α-D-glucanase in *Rhodatorula minuta* were presented.

27 *endo*-(1 → 3)-β-D-Glucanases

endo-(1 → 3)-β-D-Glucanase has been isolated from the marine bivalve *Chlamys abbidus via* ion-exchange chromatography, the final fraction being homogeneous on disc gel electrophoresis.[380] The enzyme (pH optimum 4.4, temperature optimum 35 °C, p*I* 7.5, mol. wt. 2.0×10^4) was stable in the pH range 3.5—7.5 and heat stable to 45 °C.

A glycoprotein which has been isolated from the cell walls of baker's yeast has been shown to be a glucan–protein complex which has been partially characterized as having a branched carbohydrate structure composed of chains of (1 → 3)-linked β-D-glucosyl residues, some of which are attached by (1 → 6)-linkages to the main chain.[381] Immobilization of this glycoprotein by covalent attachment to macroporous agarose gave a product which was used to isolate a number of *endo*-(1 → 3)-β-D-glucan hydrolases from *Helix pomatia*, malted barley, and a *Basidiomycete* species.

The levels of *endo*- plus *exo*-(1 → 3)-β-D-glucanase activities in seeds of a pasture legume (*Stylosanthes humilis*) have been followed during germination.[104] The two activities were separated by ion-exchange chromatography.

In *Phaseolus aureus* hypocotyl segments, more than 80% of the *endo*-(1 → 3)-β-D-glucanase activity has been shown to be located in particular

[378] G. J. Walker and M. D. Hare, *Carbohydrate Res.*, 1977, **58**, 415.
[379] M. T. Meyer and H. J. Phaff, *J. Bacteriol.*, 1977, **131**, 702.
[380] N. M. Privalova and L. A. Elyakova, *Comp. Biochem. Physiol.*, 1978, **60B**, 225.
[381] M. Edward and R. J. Sturgeon, *Carbohydrate Res.*, 1977, **57**, 3.

fractions.[382] The amounts of soluble and particular activities were increased by incubation of the hypocotyl segments with β-indolyl acetic acid.

An *endo*-(1 → 3)-β-D-glucanase from germinated rye has been purified *via* ion-exchange chromatography and gel filtration.[383] The enzyme (pH optima 5.0 for laminaran, 5.7 for carboxymethylpachyman, mol. wt. 2.43×10^4) was stable up to 45 °C, and its activity could be increased by the addition of bovine serum albumin and certain chlorides. Isoelectric focusing indicated the *endo*-(1 → 3)-β-D-glucanase to be a basic protein. The properties of the enzyme were compared with those of similar enzymes from other higher plants.

The isolation and some properties of the two components of an *Arthrobacter* species (1 → 3)-β-D-glucanase (glucanase II-1 and glucanase II-2) have been described.[384] Both of these enzymes hydrolysed various (1 → 3)-β-D-glucans, accumulating as final products, D-glucose and laminaribiose. They resembled each other in instability at acidic pH values, lability at temperatures above 55 °C, an optimal pH range of 5.5—6.5, and the requirement of the presence of glucanase I for the effective hydrolysis of yeast glucan. Glucanase II-1 showed a high lytic activity for isolated yeast cell wall, while the activity of glucanase II-2 was low. Culture fluids of early cultures contained glucanase II-1 predominantly, while those of late cultures contained glucanase II-2 predominantly.

An *endo*-(1 → 3)-β-D-glucanase which is able to disrupt the cells of living yeast has been purified in homogeneous state from the culture filtrate of *Flavobacterium dormitator* var. *glucanolyticae*.[385] The mode of enzyme action has been suggested to be a 'random' type of (1 → 3)-β-D-glucanase. The enzyme (mol. wt. 1.7—2.2×10^4) preferred larger chain saccharides as substrate for its action; however, smaller oligosaccharides such as laminaritriose and laminaribiose were also decomposed by the enzyme. The K_m values of the enzyme for laminarin, laminarihexaose, and laminaritetraose were determined. The ability of this enzyme to disrupt the cells of living yeast is remarkable, since the *endo*-D-glucanase of a smaller oligosaccharide-producing type from most sources has been recognized to be inactive (or very weakly active) on living yeast cells.

An *endo*-(1 → 3)-β-D-glucanase (EC 3.2.1.6), isolated from the culture filtrate of the fungus *Mucor hiemalis*, was purified by ammonium sulphate fractionation, gel filtration, and ion-exchange chromatography.[386] The optimum pH, optimum temperature, and K_m value of the enzyme were 5.0, 50 °C, and 0.048%, respectively. The enzyme (mol. wt. 3.0×10^4 by gel filtration) was strongly inactivated by Pb^{2+}, Cu^{2+}, and Hg^{2+} ions and also inhibited by Zn^{2+} and Fe^{3+} ions. The enzyme was specific for laminaran and the action pattern of the enzyme was of the *endo*-type.

Penicillium italicum has been shown to produce three (1 → 3)-β-D-glucanases

[382] R. Goldberg, *Compt. rend.*, 1977, **284D**, 1409.
[383] G. M. Ballance and D. J. Manners, *Phytochemistry*, 1978, **17**, 1539.
[384] K. Doi, A. Doi, T. Ozaki, and T. Fukui, *Agric. Biol. Chem.* (*Japan*), 1976, **40**, 1355.
[385] S. Nagasaki, Y. Nishioka, H. Mori, and S. Yamamoto, *Agric. Biol. Chem.* (*Japan*), 1976, **40**, 1059.
[386] T. Miyazaki, T. Yadomae, H. Yamada, and H. Oikawa, *Carbohydrate Res.*, 1977, **55**, 65.

separable by ion-exchange chromatography.[387] One displayed *endo* activity, the second *exo* activity, and the third probably both modes of action. The repressive action of D-glucose on the production of these enzymes was examined, and the results were discussed in the context of possible functions of the enzymes in the fungal cells.

The extracellular *endo*-(1 → 3)-β-D-glucanase produced by *Rhizopus arrhizus* has been purified until homogeneous by ultracentrifugation (sedimentation velocity and equilibrium studies), by electrophoresis on polyacrylamide gel with normal, sodium dodecyl sulphate, and urea–acetic acid gels, and upon isoelectric focusing.[388] The amino-acid composition of the enzyme has been determined and the enzyme possesses a carbohydrate moiety composed of D-mannose and D-galactose (in the ratio *ca.* 5 : 1) that is linked to the protein through a 2-acetamido-2-deoxyglucose residue. The molecular weight, as determined by equilibrium sedimentation, is 2.88×10^4 and this number was confirmed by electrophoresis on gels of sodium dodecyl sulphate. The enzyme does not possess subunit structure. It hydrolysed its substrates with retention of configuration and possessed transglycosylating ability. The rates of hydrolysis of a wide variety of substrates were determined, and its action pattern on a series of oligosaccharides containing mixed 1,3-, 1,4-, and 1,6- linked β-D-glucopyranosyl residues was investigated. The enzyme favoured stretches of (1 → 3)-β-D-linkages, but it could hydrolyse (1 → 4)-β-D-linkages that are flanked on the non-reducing side with stretches of (1 → 3)-β-D-links. The enzyme did not act either on (1 → 6)-β-D-glucosyl linkages located in stretches of (1 → 3)-β-D linkages or on (1 → 3)-β-D-glycosidic linkages involving sugars other than D-glucose.

The role of (1 → 3)-β-D-glucanase in cell wall biosynthesis and its distribution in the mycelium of *Sclerotium rolfsii* have been studied.[389] Specific zones of immunofluorescence appeared in the hyphal tips, clamp connections, new septa, and lateral branching when a specific antiserum was used with the direct method of fluorescent antibody staining. Enzyme activity in the cell wall preparation was inactivated by diethylpyrocarbonate but a majority of the activity was in a latent form which was unaffected by the ester.

A survey for (1 → 3)-β-D-glucanase activity has been conducted among yeasts.[379]

28 *exo*-(1 → 3)-β-D-Glucanases

The levels of *exo*- plus *endo*-(1 → 3)-β-D-glucanase activities in seeds of a pasture legume (*Stylosanthes humilis*) have been followed during germination.[104] The *exo*- and *endo*-activities were separated by ion-exchange chromatography.

An intracellular-periplasmic *exo*-(1 → 3)-β-D-glucanase has been extracted from *Kluyveromyces aestuarii* and purified to immunoelectrophoretic homogeneity by ion-exchange chromatography and gel permeation.[390] The enzyme consisted of a single chain (mol. wt. 4.3×10^4) and contained *ca.* 24% neutral

[387] T. Santos, M. Sanchez, J. R. Villanueva, and C. Nombela, *J. Bacteriol.*, 1978, **133**, 465.
[388] D. R. Clark, J. Johnson, K. H. Chung, and S. Kirkwood, *Carbohydrate Res.*, 1978, **61**, 457.
[389] G. Kritzman, I. Chet, and Y. Henis, *J. Bacteriol.*, 1978, **134**, 470.
[390] M.-A. Lachance, T. G. Villa, and H. J. Phaff, *Canad. J. Biochem.*, 1977, **55**, 1001.

carbohydrate. Kinetic constants and activation energies were determined for laminaran, pustulan, and 4-nitrophenyl β-D-glucopyranoside. (K_m 0.23, 0.37, and 5.0; E_a 3.1, 1.9, and 18.0 kcal $mole^{-1}$, respectively).

The production, and repression by D-glucose, of *exo*-(1 → 3)-β-D-glucanases by *Penicillium italicum* have been studied.[387] The functions of the enzymes, which were separable by ion-exchange chromatography, in the fungal cells were discussed.

29 *endo*-(1 → 4)-β-D-Glucanases

The production of mutants of *Trichoderma reesei* with enhanced *endo*-(1 → 4)-β-D-glucanase production has been reported.[223]

Regulation of *endo*-(1 → 4)-β-D-glucanases in the white rot fungus *Sporotrichum pulverulentium* has been investigated *via* the viscocity-lowering effect of the enzymes on carboxymethylcellulose solutions.[391] The effect of inducers (*e.g.* cellobiose) and repressors could determine whether the enzymes were localized on the cell wall surfaces or were released into the surrounding medium. In the case of *Trichoderma viride* sophorose, not cellobiose, was a good inducer of *endo*-(1 → 4)-β-D-glucanases. Comparisons, based on the enzyme data, were made of the ways that the two fungi in their natural environments assure their nutrition.

Four electrophoretically distinct *endo*-(1 → 4)-β-D-glucanases from *Trichoderma viride* have been identified.[392] Three of them (mol. wts. 3.72×10^4, 5.20×10^4, and 4.95×10^4 by sedimentation equilibrium analysis) contained D-mannose, D-glucose, D-galactose, and 2-amino-2-deoxy-D-glucose (total 4.5, 15.0, and 15.2% respectively) and were high in acidic- and hydroxy-amino-acids and glycine, but low in basic amino-acids.

30 *endo*-(1 → 6)-α-D-Glucanases

It has been suggested that dextranase-CB, a bacterial (*endo*)-(1 → 6)-α-D-glucanase that does not release D-glucose from any substrate, could be used together with *endo*-(1 → 3)-α-D-glucanase to determine the relative proportions of (1 → 6)-linked to (1 → 3)-linked sequences of D-glucose residues in the insoluble glucans produced by oral streptococci.[378] The simultaneous action of the two D-glucanases was highly effective in solubilizing the D-glucans.

31 *exo*-(1 → 6)-α-D-Glucanase

A Gram-positive and pleomorphic bacterium isolated from soil as a producer of *exo*-(1 → 6)-α-D-glucanase [EC 3.2.1.70] was identified as *Arthrobacter globiformis*.[393] This enzyme, inducible by dextran extracellularly, was partially purified from a cell-free culture supernatant. It was found most active at pH around 6.0 and most stable at pH 6.0—6.5. The enzyme was proved, by several experiments, to attack dextran in the *exo*-wise fashion to release only D-glucose leaving a macromolecular limit dextrandextrin. Transglucosylation from dextran to accumulating or added D-glucose was not observed.

[391] K.-E. Eriksson and S. G. Hamp, *European J. Biochem.*, 1978, **90**, 183.
[392] S. P. Shoemaker and R. D. Brown, *Biochim. Biophys. Acta*, 1978, **523**, 147.
[393] T. Sawai, T. Yamaki, and T. Ohya, *Agric. Biol. Chem.* (*Japan*), 1976, **40**, 1293.

32 *endo*-(1 → 6)-β-D-Glucanase

Two types of extracellular (1 → 6)-β-D-glucanases produced by *Bacillus circulans* have been differentiated by their different abilities to lyse yeast cell-walls.[394] The non-lytic (1 → 6)-β-D-glucanase was isolated by a combination of gel filtration and ion-exchange chromatography. The purified enzyme (mol. wt. 5.2×10^4) was electrophoretically homogeneous. For the substrate pustulan, the enzyme exhibited the following kinetic properties: pH optimum 5.0; K_m 0.83 mg of pustulan ml^{-1}; V_{max} 104 microequivalents of D-glucose released min^{-1} mg^{-1} of protein. Pustulan was hydrolysed by an *endo*-mechanism, producing D-glucose and gentiobiose as preponderant final products. The non-lytic enzyme was specific for the (1 → 6)-β-D-glucosidic linkage and did not hydrolyse branched, (1 → 3)-β-D-linked glucans containing (1 → 6)-interchain linkages. In contrast, the lytic (1 → 6)-β-D-glucanase produced D-glucose, gentiobiose, and gentiotriose as the final products of pustulan hydrolysis, and exhibited significant activity on branched (1 → 3)-D-glucans having (1 → 6)-interchain linkages. In these cases, the major products were gentiobiose and D-glucose, suggesting an ability of the lytic enzyme to cleave some (1 → 3)-linkages surrounding a (1 → 6)-branch point. This latter property may explain the ability of this enzyme to lyse yeast cell-walls weakly.

A survey for (1 → 6)-β-D-glucanase activity has been conducted among yeasts.[379]

33 D-Glucanases (Miscellaneous)

A β-D-glucanase highly specific for β-D-glucans containing a linkage sequence (13) has been isolated from several commercial preparations of *Bacillus subtilis* α-amylase including one purified by repeated crystallization.[395] The

..... β-D-Glc*p*-(1 → 4)-β-D-Glc*p*-(1 → 3)-β-D-Glc*p*-(1 → 4)-β-D-Glc*p* .....

(13)

enzyme was separated from the α-amylase by gel filtration, did not hydrolyse cellulose or laminaran, and was of the *endo* type since changes in substrate viscosity occurred long before the appearance of detectable reducing sugars. Various kinetic aspects of the properties of the enzyme (pH optimum 6.5) which was active on lichenan were reported.

34 Glucoamylases

The separation by polyacrylamide gel electrophoresis and subsequent analysis of the components of guinea pig intestinal brush border membrane revealed the presence of *inter alia* enzyme complexes: α-D-glucosidase–glucoamylase, α-D-glucosidase–sucrose, and α-D-glucohydrolase–glucoamylase.[145]

Glucoamylase activity has been solubilized from rat intestinal mucosa (see p. 406).[200]

[394] F. M. Rombouts, G. Fleet, D. J. Manners, and H. J. Phaff, *Carbohydrate Res.*, 1978, **64**, 237.
[395] D. J. Huber and D. J. Nevins, *Plant Physiol.*, 1977, **60**, 300.

A change in the fluorescence intensity of a 1-anilinonaphthalene-8-sulphonate probe, bonded to wheat seed β-amylase, after addition of various inorganic nitrates has been established.[396] The parallel change in the β-amylase activity as considered to be due to changes in the conformation of the enzyme protein under the influence of salts.

Maltose has been used frequently as a substrate in the assay of glucoamylase because of the simplicity of analysis, a practice which has caused no difficulty since the enzyme is reported to follow Michaelis–Menten kinetics.[397] However, tests of glucoamylase from *Aspergillus niger* over an extended concentration range exhibited a decided deviation from Michaelis–Menten kinetics. Explanations as to why this may be the rule rather than the exception were offered.

The hydrolysis of starch in a membrane reactor has been shown to suffer from disadvantageous formation of a concentrated gel layer at the membrane surface caused by the effect of concentration polarization and to some extent by the ageing of the starch solution.[398] It was demonstrated that glucoamylase, from *Aspergillus niger*, reduces the formation of a gel layer and gives higher and more-stable performances of the membrane reactor.

The action patterns of *Aspergillus niger* glucoamylase and glucosyltransferase (transglucosylase) on D-[1-^{14}C]glucose, [1-^{14}C]maltose, and [1-^{14}C]malto-oligosaccharides (labelled at position 1 of the D-glucose residue at the reducing end) have been compared by oligosaccharide-mapping techniques.[399] The extent of conversion of D-glucose and of maltose into new oligosaccharides was 2.2 and 1.9% with glucoamylase, and 5.7 and 33% with glucosyltransferase. The major oligosaccharides produced by both enzymes were isomaltose (6-*O*-α-D-glucopyranosyl-α-D-glucose), panose [*O*-α-D-glucopyranosyl-(1 → 6)-*O*-α-D-glucopyranosyl-(1 → 4)-α-D-glucose], and nigerose (3-*O*-α-D-glucopyranosyl-α-D-glucose). The glucosyltransferase also synthesized oligosaccharides from malto-oligosaccharides of higher molecular weight to yield compounds having α-(1 → 6)-linked D-glucosyl units at the non-reducing ends. Glucoamylase exhibited little, if any, such activity on malto-oligosaccharides.

The kinetics of inhibition of *Aspergillus niger* glucoamylase has been studied by use of D-glucono-1,5-lactone and maltobiono-1,5-lactone as transition-state analogues.[296] Glucoamylase could not be inhibited with maltobiono-lactone, whereas with D-gluconic acid, it was found that a completely mixed inhibition may be observed (K_i = 1.3 nM). The ratio of the binding affinity of the lactones, products, and substrates permitted the conclusion that ring distortion takes place in the transition state. Comparisons were made with the results of similar inhibition studies with α-amylase and β-amylase (see p. 429).

Glucoamylase from an *Endomycopsis* species has been purified to homogeneity on disc electrophoresis in polyacrylamide gel, isoelectric focusing, and ultra-

[396] A. A. Anisimov, I. F. Aleksandrova, K. A. Iudina, Sh. Kh. Abseev, M. O. Kondrat'eva, and G. G. Beresneva, *Biochemistry* (*U.S.S.R.*), 1977, **42**, 1748.
[397] S. J. Swanson, A. Emery, and H. C. Lim, *Biotechnol. Bioeng.*, 1977, **19**, 1715.
[398] A. M. Madgavkar, Y. T. Shah, and J. T. Cobb, *Biotechnol. Bioeng.*, 1977, **19**, 1719.
[399] J. H. Pazur, A. Cepure, S. Okada, and L. S. Forsberg, *Carbohydrate Res.*, 1977, **58**, 193.

centrifugation.[400] The enzyme (mol. wt. 5.3 × 10^4, pH optimum 4.35, p*I* 3.8) was found to be a glycoprotein.

The thermophilic fungus *Humicola lanuginosa* has been shown to elaborate two glucoamylases. Some properties of one of these enzymes (glucoamylase II) were investigated.[401] The enzyme was found to have a higher pH optimum than reported for other fungal glucoamylases, and to be very thermostable. In particular, it displayed unusual resistance to heat in the presence of its substrate. It appeared to be capable of completely degrading starch. However, the possibility that traces of contaminating α-amylase assist in degradation could not be ruled out.

In a series of studies on the genus *Monascus*, the purification and properties of two forms of glycoamylase from *Monascus kaoliang* have been reported.[402] The substrate specificities of the two enzymes were invesitgated.[403]

The hydrolysis of phenyl β-maltoside by *Rhizopus niveus* glucoamylase has been compared with that by β-amylases (see p. 437).[335] Maltodextrin (high-d.p. malto-oligosaccharides) was found to produce a trough at 303 nm in the difference spectrum of glucoamylase from *R. niveus* upon binding with the enzyme; this trough disappeared upon hydrolysis.[404] The trough, which was ascribed to a change in the electrostatic environment of an L-tryptophan residue at the terminal subsite of the enzyme, was found to be closely related to the formation of the enzyme–substrate complex. The kinetics of binding of malto-dextrin and maltotriose to the enzyme were studied at pH 4.5 and 5 °C, by monitoring the trough by the stopped-flow method. The result was consistent with a two-step mechanism, in which a fast, bimolecular association is followed by a slower, unimolecular isomerization process. The latter process involves an environmental change of the L-trytophan residue, and is considered to be closely connected to the formation of the productive complex essential for the catalysis.

35 α-L-Guluronan Lyase

Klebsiella aerogenes has been shown to produce an extracellular α-L-guluronan lyase (alginolyase); this has been partly purified.[405] The enzyme is specific for the α-L-guluronosyl linkages in whole alginate and fractions therefrom. The end products of its action on poly-D-guluronic acid blocks are mainly the unsaturated di- and tri-saccharides, with a smaller proportion of the homologous tetrasaccharide. Some general properties of the enzyme were reported.

36 Heparin Hydrolases

An *endo*-glycosidase present in rat liver lysosomes has been shown to hydrolyse heparin.[406] Inorganic sulphate equivalent to approximately one sulphamide

[400] I. M. Gracheva, T. A. Lushchik, Yu. A. Tyrsin, and E. E. Pinchukova, *Biochemistry (U.S.S.R.)*, 1977, **42**, 1258.
[401] P. M. Taylor, E. J. Napier, and I. D. Fleming, *Carbohydrate Res.*, 1978, **61**, 301.
[402] H. Iizuka and S. Mineki, *J. Gen. Appl. Microbiol.*, 1977, **23**, 217.
[403] H. Iizuka and S. Mineki, *J. Gen. Appl. Microbiol.*, 1978, **24**, 185.
[404] M. Ohnishi and K. Hiromi, *Carbohydrate Res.*, 1978, **61**, 335.
[405] J. Boyd and J. R. Turvey, *Carbohydrate Res.*, 1977, **57**, 163.
[406] B. Arbogast, J. J. Hopwood, and A. Dorfman, *Biochem. Biophys. Res. Comm.*, 1977, **75**, 610.

bond cleaved per heparin chain was also released in incubations of *N*-[^{35}S]-sulphate heparin with crude lysosomal preparations. There was no evidence of *exo*-glycosidase or further sulphamidase activity although oligosaccharides approaching the size of tetrasaccharide were produced. The *endo*-glycosidase has a broad pH-dependence (pH optimum 4.4).

37 Hyaluronate Lyases and Hyaluronidases

In a study of chondroitin 4-sulphate and chondroitin sulphate throughout digestion with lysosomes under acidic conditions, the contributions of hyaluronidase, β-D-glucuronidase, β-D-2-acetamido-2-deoxyhexosidase, and sulphatase [EC 3.1.6.4] to the degradation of chondroitin 4-sulphate were examined on the basis of characterization and identification of the digestion products.[407] Chondroitin 4-sulphate was first degraded only by hyaluronidase, and sulphatase did not significantly contribute to the initial degradation of polymeric chrondroitin 4-sulphate. In further degradation, the main oligosaccharides produced were normally-sulphated odd-numbered oligosaccharides such as tri-, penta-, hepta-, and nona-saccharides. In contrast to the results of digestion studies of hyaluronic acid, octa- and deca-saccharides of chondroitin 4-sulphate were not digested by hyaluronidase but by the concerted action of glycoside hydrolases and sulphatase. It was concluded that the dodecasaccharide derived from chondroitin 4-sulphate serves as the lowest-molecular-weight substrate for hyaluronidase and the decasaccharide as the largest-size substrate for β-D-glucuronidase in the degradation of chondroitin 4-sulphate by the enzymes of lysosomes. The degradation of chondroitin by the lysosomes occurred in a similar way. Therefore the difference in the pathways of degradation between hyaluronic acid and chondroitin 4-sulphate may be attributed to their 2-amino-2-deoxy-hexose compositions. β-D-Glucuronidase cleaved D-glucuronic acid in the terminal non-reducing position of even-numbered oligosaccharides which were derived from chondroitin 4-sulphate regardless of the presence of the sulphate group on the adjacent 2-acetamido-2-deoxy-D-galactose residue. The sulphated odd-numbered oligosaccharides were resistant to β-D-2-acetamido-2-deoxyhexosidase. The affinity of β-D-2-acetamido-2-deoxyhexosidase for the odd-numbered oligosaccharides from chondroitin was lower than that of β-D-glucuronidase for the even-numbered oligosaccharides. No contribution of sulphatase which cleaved the sulphate group of 2-acetamido-2-deoxy-D-galactose at the non-reducing end was found.

Hyaluronic acid and its oligosaccharides (tetra-, hexa-, and octa-saccharides) were digested by canine liver lysosomes at pH 4.0.[75] The degradation products were separated by gel filtration and ion-exchange chromatography. The contribution of hyaluronidase and *exo*-glycosidases (β-D-glucuronidase and β-D-2-acetamido-2-deoxyhexosidase) on the degradation of hyaluronic acid was inferred from the distribution of the oligosaccharides, D-glucuronic acid, and 2-acetamido-2-deoxy-D-glucose in the digestion mixture. Hyaluronic acid was degraded into a large amount of even-numbered oligosaccharides and to a trace amount of free D-glucuronic acid in the initial stage of the diges-

[407] S. Hayashi, *J. Biochem. (Japan)*, 1978, **83**, 149.

tion. On the other hand, the prolonged digestion study showed an increase of odd-numbered oligosaccharides such as the tri- and penta-saccharides in proportion to an increase of free D-glucuronic acid. On the degradation of the hyaluronic acid oligosaccharides, tetra- and hexa-saccharides gave tri- and penta-saccharides as the main oligosaccharides in the degradation products respectively. Octasaccharide was not degraded into hepta-saccharide by the action of β-D-glucuronidase but into a mixture of di-, tetra-, and hexa-saccharides by hyaluronidase. The activity of β-D-glucuronidase for tetra- and hexa-saccharides was higher than that of β-D-2-acetamido-2-deoxyhexosidase for tri- and penta-saccharides. On the basis of these findings, it was concluded that hyaluronic acid is first depolymerized by hyaluronidase alone and octasaccharide is the minimum substrate size in the degradation by hyaluronidase. Hexasaccharide and other smaller oligosaccharides are degraded by the action of glycoside hydrolases. The level of β-D-2-acetamido-2-deoxyhexosidase activity is a limiting factor in the degradation by the concerted action of glycosidase hydrolases.

Primary cultures of fibroblasts isolated from embryonic chick skin contain hyaluronidase activity, both associated with the cells and secreted into the medium.[408] The enzyme has a pH optimum at 3.7 and is most active against hyaluronate. The oligosaccharide products of the action of the enzyme on this substrate are similar in size to those from the action of testicular hyaluronidase, ruling out significant degradation by *exo*-enzymes.

The isolation and partial characterization of hyaluronidase which occurs in rat carrageenan granuloma has been reported.[409]

Urea treatment of *Streptococcus pyogenes* followed by fractional precipitation, ion-exchange chromatography, and affinity chromatography on agarose-immobilized lentil lectin yielded a hyaluronidase (hyaluronate lyase?) (mol. wt. 7.5×10^4 by gel filtration, 7.1×10^4 by polyacrylamide gel electrophoresis in sodium dodecyl sulphate, pH optimum 5.5, partial specific volume 0.725 $cm^3\ g^{-1}$).[410] Calcium and magnesium were found to be effectors of the enzyme which was a glycoprotein (7.25% carbohydrate) containing D-galactose, D-glucose, and 2-amino-2-deoxy-D-glucose.

38 Inulinases

Culture conditions for inulinase production by an *Aspergillus* species have been reported.[411] The crystallization and general properties of an extracellular inulinase obtained from an *Aspergillus* species have been described.[412]

Studies reported on microbial inulinases include a description of the general properties of extracellular inulinase from a *Penicillium* species.[413]

[408] R. W. Orkin, G. Jackson, and B. P. Toole, *Biochem. Biophys. Res. Comm.*, 1977, **77**, 132.
[409] A. Ito, K. Kitamura, and Y. Mori, *Chem. Pharm. Bull.* (*Japan*), 1977, **25**, 1010.
[410] L. C. Benchetrit, E. D. Gray, R. D. Edstrom, and L. W. Wannamaker, *J. Bacteriol.*, 1978, **134**, 221.
[411] T. Nakamura, S. Hoashi, and S. Nakatsu, *J. Agric. Chem. Soc. Japan*, 1978, **52**, 105.
[412] T. Nakamura, T. Kurokawa, S. Nakatsu, and S. Ueda, *J. Agric. Chem. Soc. Japan*, 1978, **52**, 159.
[413] T. Nakamura and S. Nakatsu, *J. Agric. Chem. Soc. Japan*, 1977, **51**, 681.

39 Isoamylases

Pseudomonas species isoamylase has been shown to have little action on starch granules.[206] *Pseudomonas* isoamylase has been shown to hydrolyse (1 → 6)-α-D-glucosidic linkages of amylopectin, glycogen, and various branched dextrins and oligosaccharides.[414] The detailed structural requirements for the substrate have been examined qualitatively and quantitatively, in comparison with the pullulanase of *Klebsiella aerogenes*. As with pullulanase, isoamylase is unable to cleave D-glucosyl stubs from branched saccharides. Isoamylase differs from pullulanase in the following characteristics: (1) the favoured substrates for isoamylase are higher-molecular-weight polysaccharides; most of the branched oligosaccharides examined were hydrolysed at a lower rate, 10% or less of the rate of hydrolysis of amylopectin; (2) maltosyl branches are hydrolysed off by isoamylase very slowly in comparison with maltotriosyl branches; (3) isoamylase requires a minimum of three D-glucose residues in the B- or C-chain.

40 Isopullulanases

Isopullulanase from *Aspergillus niger* has been used to prepare isopanose [*O*-α-D-glucopyranosyl-(1 → 4)-*O*-α-D-glucopyranosyl-(1 → 6)-*O*-α-D-glucopyranose] from pullulan.[328] The digest was applied to a carbon–celite column and eluted with a linear gradient of propan-1-ol from 0 to 5%. From the trisaccharide fractions eluted, isopanose was obtained in about 90% yield.

41 Limit Dextrinases

The selective solubilization of α-D-glucosidases from the complex mixture of such enzymes in porcine intestinal mucosa led to the conclusion that separate enzymes are involved in the hydrolysis of isomaltose and the α-limit dextrins formed from starch by α-amylase ('isomaltase' and limit dextrinase respectively).[198] Various aspects of the enzymology of oligo-1,6-D-glucosidase (limit dextrinase?) from peas have been reported.[415]

42 Lysozymes

The use of optical rotatory dispersion in quantitative analysis has been applied to the microdetermination of lysozyme.[416] A spin-label assay for lysozyme which is based on the enzymic hydrolysis of spin-labelled peptidoglycan has been described.[417] Hydrolysis of this polymer by lysozyme results in sharpening of the e.s.r. spectrum. According to this method, human lysozyme is 3.5 times as active as hen egg-white lysozyme. The assay is suitable for measuring lysozyme levels in biological fluids.

Human and hen egg-white lysozymes have been shown to convert the *Micrococcus luteus* peptidoglycan into the disaccharide and tetrasaccharide of glycopeptide (14) and the peptide itself. With goose egg-white lysozyme,

[414] K. Kainuma, S. Kobayashi, and T. Harada, *Carbohydrate Res.*, 1978, **61**, 345.
[415] D. Yellowlees, *Proc. Austral. Biochem. Soc.*, 1978, **11**, 26.
[416] S. S. M. Hassan, *Z. Analyt. Chem.*, 1975, **274**, 375.
[417] L. S. Johnston and F. C. Neuhaus, *Analyt. Biochem.*, 1978, **85**, 56.

disaccharide and (14) were formed in equimolar amounts.[418] Papaya lysozyme

β-D-GlcpNAc-(1 → 4)-MurNAc
|
L-Ala-D-isoGlu(Gly)-L-Lys-D-Ala-D-Ala

(14)

afforded mainly products without peptide substitution and T4 phage lysozyme gave only (14). This pattern of specificity was discussed in terms which correlate it with the differences in structures of their presumed natural substrates.

Lysozyme has been purified from human tears, serum, and urine of acute monocytic leukaemia patients, renal disease patients, and residents in cadmium-polluted areas of Tsushima Island, using affinity chromatography on an adsorbent containing lysozyme-lysate of *Micrococcus lysodeikticus* cell walls.[419] The specific activities and heterogeneities of the preparations on disc gel electrophoresis were found to differ from one source of enzyme to another.

Elevated levels of serum lysozyme have been observed in cases of Gaucher disease; this high level was considered to reflect an increased body mass of reticuloendothelial cells.[420] It was suggested that the enzyme elevation might be of use in diagnosis of Gaucher disease. In cases of monocytic leukaemia, myelomonocytic leukaemia, and myeloma no evidence has been found for immunoglobulin (IgG)–lysozyme complexes.[421] The mobility of the serum lysozyme was identical to that of free urinary lysozyme.

A method of two-dimensional electrophoresis has been derived to allow a study of the electrophoretic mobility of human urinary lysozyme.[422] Three different phenotypes have been defined in a study of 13 purified lysozymes obtained from different patients with monocytic leukaemia.

Lysozyme of mink spleen, kidneys, and liver has been isolated by affinity chromatography on deaminated chitin.[423] The L-histidine content of the enzyme was unusually high. The lysozyme preparations contained an unknown, firmly bound component which absorbed at 400—420 nm.

An autoradiographic study has been made of the distribution of ^{125}I-labelled lysozyme after its intravenous injection in rats.[424] As part of a study on structure–activity relationships of mammalian tissue lytic enzymes, the location and characterization of multiple forms of rat kidney lysozyme have been reported.[425] Iron-deficient rats have been shown to have higher levels of kidney lysozyme activity than control animals.[426] There were no significant changes in the serum and spleen lysozyme activity levels in iron deficiency.

[418] M. Schindler, D. Mirelman, and N. Sharon, *Biochim. Biophys. Acta*, 1977, **482**, 386.
[419] T. Yoshimoto, M. Tobiishi, and D. Tsuru, *J. Biochem.* (*Japan*), 1976, **80**, 703.
[420] E. Silverstein and J. Friedland, *Clinica Chim. Acta*, 1977, **74**, 21.
[421] G. Virella, W. A. Hipp, and W. Hood, *Clinical Chem.*, 1978, **24**, 514.
[422] G. Virella, J. Goudswaard, F. W. Tischendorf, and H. H. Fudenberg, *Clinica Chim. Acta*, 1977, **79**, 503.
[423] G. M. Malinina, I. A. Cherkasov, N. A. Kravchenko, and V. A. Berestov, *Biochemistry* (*U.S.S.R.*), 1977, **42**, 680.
[424] T. Yuzuriha, K. Katayama, and S. Ohtake, *Chem. Pharm. Bull* (*Japan*), 1977, **25**, 836.
[425] R. Raghunathan and S. Gurnana, *Indian J. Biochem. Biophys.*, 1977, **14**, 229.
[426] A. R. Sherman and I. Wolinsky, *Experientia*, 1978, **34**, 367.

Tortoise (*Trionyx gangeticus*) lysozyme has been purified and the crystallized form was found to be homogeneous on polyacrylamide gel electrophoresis in sodium dodecyl sulphate, on immunological tests, and on sedimentation.[365] The physiochemical and enzymic properties of the enzyme (average mol. wt. 1.54×10^4 by sedimentation and diffusion) were strikingly similar to those of the hen enzyme. The tortoise molecule underwent pH-dependent expansion at pH 2 and dimerization above pH 5.7. Further comparisons were made with hen and turtle lysozymes.

The purification of ostrich (*Strathio camelus*) egg-white lysozyme has been reported.[427] The quantitative amino-acid compositions, molecular weight (2.05×10^4), *N*-terminal sequence, and kinetic properties allowed classification of the enzyme among the goose-type lysozymes.

The binding constants of 2-acetamido-2-deoxy-α and β-D-glucopyranoses to hen and turkey egg-white lysozymes have been determined at various pH values.[428] The pH dependence of the binding of 2-acetamido-2-deoxy-β-D-glucopyranose to both enzymes was essentially the same. Binding-constant magnitudes were interpreted in terms of interactions with certain amino-acid residues, and the following conclusions were reached. Above pH 4, 2-acetamido-2-deoxy-α-D-glucopyranose binds to hen enzyme in both α- and β-modes, which correspond to the binding orientation of the α- and β-monosaccharides, respectively, but below pH 4 it binds predominantly in the β-mode. 2-Acetamido-2-deoxy-β-D-glucopyranose binds to hen and turkey lysozymes predominantly in the β-mode above pH 4 and in both α- and β-modes below pH 4. 2-Acetamido-2-deoxy-α-D-glucopyranose binds to turkey lysozyme predominantly in the β-mode over a wide pH range.

The crystal structure of turkey egg-white lysozyme, determined by the molecular replacement method at 5 Å resolution, has now been refined to 2.8 Å resolution and a model has been built to fit the electron density.[429] A comparison of the co-ordinates with those of the hen enzyme was made; a significant difference was observed in the region of residues 98–115 of the structure. The interactions between various saccharides bound at subsites E and F′ and the enzyme molecules were discussed.

A rapid method devised for the purification of hen egg-white lysozyme was based on the finding that gel filtration on agarose can be used selectively, due to the fact that lysozyme interacts with the agarose matrix and elutes later than the corresponding total volume for the column.[430]

The effect of *N*-unsubstitution in chitin and chitin oligosaccharides on the hydrolysis of these carbohydrates by hen egg-white lysozyme has been studied.[431] A series of oligosaccharide substrates was derived from partially deacetylated chitin by the action of the enzyme. The results of digestion of the isolated oligosaccharides with lysozyme, together with the structural features

[427] J. Jollès, J. P. Périn, and P. Jollès, *Mol. Cell. Biochem.*, 1977, **17**, 39.
[428] Y. Yang, S. Kuramitsu, Y. Nakae, K. Ikeda, and K. Hamaguchi, *J. Biochem.* (*Japan*), 1976, **80**, 425.
[429] R. Sarma and R. Bott, *J. Mol. Biol.*, 1977, **113**, 555.
[430] J. M. Fernandez-Sousa, R. Perez-Castells, and R. Rodriguez, *Biochim. Biophys. Acta*, 1978, **523**, 430.
[431] K.-I. Amano and E. Ito, *European J. Biochem.*, 1978, **85**, 97.

of the saccharides were accounted for by the importance of the *N*-acetyl group of the 2-acetamido-2-deoxy-D-glucose residues located on subsites C and E in the enzyme–substrate complexes.

Lytic activity of hen egg-white lysozyme towards bacterial cells of *Micrococcus lysodeitkticus* was found to be pH dependent and inhibited by several aminoglycosidic antibodies the structures of which are related to the saccharidic substrates of the enzyme.[432] The extents of inhibition suggested a role of the positive charges of this type of antibiotic in the mechanism of lysozyme activity inhibition.

The rate of catalysis by hen egg-white lysozyme has been determined in the presence of various metal ions (Co^{2+}, Zn^{2+}, and eight of the trivalent lanthanide ions).[433] The inhibition data were fitted to several models of the interactions of the metal ion with the enzyme, and the degrees of fit were compared. The models ranged in complexity from a single inhibitory total binding site on the enzyme to the presence of two non-independent and non-equivalent inhibitory metal binding sites, but the simplest models appeared to be the most valid.

The binding of methyl 2-acetamido-2-deoxy-β-D-glucopyranoside to hen egg-white lysozyme in the tetragonal crystal form has been studied by *X*-ray diffraction techniques to a resolution of 0.25 nm.[434] The binding of the β-glycoside was almost identical to that of 2-acetamido-2-deoxy-β-D-glucopyranose. Real-space refinement of the lysozyme complexes of these two carbohydrates allowed preliminary analysis of the conformational changes observed on binding monosaccharide inhibitors, especially in the region involving L-tryptophan-62 and residues 70—76. The close similarity of the lysozyme structures in the presence and absence of acetate was contrary to expectations from previous n.m.r. spectroscopic studies.

The interactions of 2-acetamido-2,6-dideoxy-D-glucose and 2-acetamido-3,6-dideoxy-D-glucose with hen egg-white lysozyme have been studied at various pH values by measurement of the changes in the c.d. band at 295 nm.[435] It was shown that both compounds bind at subsite C of the lysozyme and compete with 2-acetamido-2-deoxy-D-glucose. Whereas *X*-ray crystallographic studies show that O-6 and O-3 of 2-acetamido-2-deoxy-D-glucose at subsite C are hydrogen-bonded to the indole NH's of L-trytophans 62 and 63, studies of the binding constants for the monosaccharides indicated that L-tryptophan 63 and not 62 is important for the interaction of 2-acetamido-2-deoxy-D-glucose with the lysozyme.

The interactions of the substrate analogues methyl 2-acetamido-2-deoxy-β-D-glucopyranoside, di-*N*-acetylchitobiose, and tri-*N*-acetylchitotriose with hen egg-white lysozyme in which an ester linkage had been formed between L-glutamic acid 35 and L-tryptophan 108 were studied by c.d. and fluorescence techniques and were compared with those for the native enzyme.[436] The binding

[432] J. M. Fernandez-Sousa, J. G. Gavilanes, A. M. Munico, and R. Rodriguez, *Biochem. Biophys. Res. Comm.*, 1977, **75**, 895.
[433] F. Ostroy, R. A. Gams, J. D. Glickson, and R. E. Lenkinski, *Biochim. Biophys. Acta*, 1978, **527**, 56.
[434] S. J. Perkins, L. N. Johnson, P. A. Machin, and D. C. Phillips, *Biochem. J.*, 1978, **173**, 607.
[435] S. Kuramitsu, Y. Yang, K. Ikeda, and K. Hamaguchi, *J. Biochem. (Japan)*, 1976, **80**, 417.
[436] Y. Nakae, K. Ikeda, and K. Hamaguchi, *J. Biochem. (Japan)*, 1976, **80**, 435,

constants of the methyl glycoside and disaccharide were essentially the same as those for intact lysozyme in the pH range 1—5. Above pH5, the binding constants of these saccharides to modified enzyme did not change with pH, whereas the binding constants to intact lysozyme decreased, indicating that L-glutamic acid 35 participates in the binding of these saccharides. The participation of L-aspartic acids 101 and 66 in the binding of disaccharides was also implicated.

The ovomucin–lysozyme (hen egg-white) aggregation was found to be remarkably affected by pH or ionic strength.[437] The extent of interaction of F-ovomucin with lysozyme was much larger than that of S-ovomucin. The ovomucin–lysozyme interaction decreased correspondingly at a rate depending on the time at which ovomucin was modified by neuraminidase. On the other hand, the ovomucin–lysozyme interaction disappeared completely by acetylation, succinylation, or carbamylation of L-lysyl ε-amino groups in lysozyme, but it was not greatly affected by guanidination of L-lysyl ϵ-amino groups in lysozyme. From these results, it was confirmed that the electrostatic interaction between the negative charges of the terminal neuraminic acid group in ovomucin and the positive charges of L-lysyl ϵ-amino groups in lysozyme is essential for the ovomucin–lysozyme interaction.

Lysozyme (hen egg-white presumed) has been shown to complex with *Pseudomonas aeruginosa* lipopolysaccharide in two distinct stages.[438] The initial stage did not produce turbidity detectable by nephelometry but did permit sedimentation of the complex. The association, which was considered to provide a model for investigation of *in vivo* protein–lipopolysaccharide interactions, was disrupted 100% by the action of Mg^{2+}.

The effects of dioxan on the conformation of hen egg-white lysozyme at low pH have been studied by c.d. spectroscopy and viscosimetry.[439] The results indicated that the tertiary structure reflected by the near-ultraviolet c.d. bands was initially disrupted at 20—30% dioxan with possible slight disorganization of the ordered secondary structure, and then a marked increase in helical structure was induced by high dioxan concentrations.

Ultraviolet difference absorption spectra of lysozyme at pH 2.1 produced by dioxan were measured as a function of dioxan concentration at 25 °C.[440] At 20—30% dioxan an abrupt decrease in molar difference absorption coefficients ($\Delta\epsilon$) was observed in plots of $\Delta\epsilon_{292}$ and $\Delta\epsilon_{285}$ *vs.* dioxan concentration; this reflected an unfolding process of lysozyme by dioxan. The magnitude of the decrease in $\Delta\epsilon_{292}$ indicated the exposure of at least one buried L-tryptophyl residue to the solvent environment in this unfolding process. On the other hand, two linear portions were observed at 0—12% and 40—70% dioxan in the plots, corresponding to solvent effects of dioxan on native and denatured lysozyme, respectively. The number of exposed L-tryptophyl and L-tyrosyl residues in these dioxan concentration regions was estimated. The results showed that 3.9 L-tryptophyl and 1.5 L-tyrosyl residues were exposed at

[437] A. Kato, T. Imoto, and K. Yagishita, *Agric. Biol. Chem. (Japan)*, 1975, **39**, 541.
[438] D. F. Day, M. L. Marceau-Day, and J. M. Ingram, *Canad. J. Microbiol.*, 1978, **24**, 196.
[439] T. Izumi and H. Inoue, *J. Biochem. (Japan)*, 1977, **81**, 1769.
[440] T. Izumi and H. Inoue, *J. Biochem. (Japan)*, 1978, **83**, 1555.

0—12% dioxan while 3.1 L-tryptophyl and 2.5 L-tyrosyl residues were exposed at 40—70% dioxan. L-Tryptophyl and L-tyrosyl exposures of denatured lysozyme at dioxan concentrations above 40% were also estimated on the basis of difference spectra obtained using a lysozyme solution containing 40% dioxan as a reference. In good agreement with the above results, this modified method gave an exposure of 3.4 L-tryptophyl and 2.3 L-tyrosyl residues for denatured lysozyme and some advantages of its use for denatured proteins were discussed. These results indicate that the tryptophyl behaviour in denaturation is different from that of L-tyrosyl residues: the L-tyrosyl exposure is simply increased upon denaturation, while the L-tryptophyl exposure is increased in the unfolding process at 20—30% dioxan, and then decreased even below that in the native state above 40% dioxan. The decrease in L-tryptophyl exposure was attributed to intramolecular reorganization of lysozyme (with the disulphide bridges situated near the L-tryptophyl residues remaining intact) in the direction of a more helical conformation at higher dioxan concentrations. Fluorescence spectra were also measured for lysozyme in aqueous dioxan mixtures at apparent pH 2.1—2.3 and 25 °C. The results were consistent with the unfolding process observed by difference spectral measurements, but at high dioxan concentrations the fluorescence spectra of lysozyme behaved in a complicated manner. This appeared to reflect the state of the L-tryptophyl residues in relation to the conformational changes in lysozyme. Preliminary results on the sedimentation velocity behaviour of lysozyme were also reported.

The pH difference spectra of intact hen egg-white lysozyme and ester derivatives thereof have been measured at various ionic strengths and temperatures.[441] The pH dependence of the extinction difference at 301.5 nm was well interpreted in terms of the ionization of the catalytic groups of L-aspartic acid 52 and L-glutamic acid 35 which interact electrostatically. The ionizations of these residues and the associated thermodynamic parameters were studied in detail.

Studies of individual carbon sites of hen egg-white lysozymes by ^{13}C n.m.r. spectroscopy have been reported.[442] Chemical shift considerations, the effect of pH, and partially relaxed Fourier transform n.m.r. spectra were used to assign each resonance to one of the seven types of non-protonated aromatic carbons of amino-acid residues. Spectra of chemically modified lysozyme samples yielded various assignments to specific residues in the sequence.

The ionization constant of L-histidine 15 of hen egg-white lysozyme, determined by ^{1}H n.m.r. spectroscopy, exhibited a dependence upon concentration, whereas the pK_a values of this residue and of ester 35, 108 lysozyme were observed to be almost independent of protein concentration.[443] The L-tryptophan residues of lysozyme were examined by ^{13}C n.m.r. spectroscopy as functions of pH, protein concentration, and temperature. On the basis of the results of these investigations, it was proposed that the portion of the lysozyme surface involved in head-to-tail association includes L-histidine 15

[441] S. Kuramitsu, K. Ikeda, and K. Hamaguchi, *J. Biochem. (Japan)*, 1977, **82**, 585.
[442] A. Allerhand, R. S. Norton, and R. F. Childers, *J. Biol. Chem.*, 1977, **252**, 1786.
[443] H. Shindo, J. S. Cohen, and J. A. Rupley, *Biochemistry*, 1977, **16**, 3879.

and possibly one or more adjacent positively charged residues. The other spatially distinct region involved in the self-association comprises L-glutamic acid 35, L-tryptophan 108, and L-tryptophan 63. There is an interaction between L-histidine 15 and L-glutamic acid 35 within the intermolecular interface.

Various lanthanide complexes of hen egg-white lysozyme have been used to test the validity of the assumption of the magnetic susceptibility tensor of the metals being axially symmetric in the use of trivalent lanthanide ions as probes in n.m.r. studies of macromolecules.[444] For Gd^{3+}–lysozyme, the broadening data place the metal ions in close proximity to the reported *X*-ray position. Contrary to current assumptions, the shifts induced by Nd^{3+} and Ce^{3+} exhibit considerable non-axial character. On the basis of a statistical hypothesis test, the axially symmetric model was rejected with >97% confidence.

A general method for analysing dipolar shifts induced by paramagnetic ions in the n.m.r. spectra of ligand molecules has been applied to the shift perturbation induced by Co^{2+} in the spectrum of hen egg-white lysozyme.[445] The results presented are in accordance with the non-validity of the assumption of axial symmetry. However, it was suggested that the shift data furnished a means of determining the site of metal complexation in macromolecules whose structure has been determined by *X*-ray crystallography.

N.m.r. relaxation times have been measured for water protons in cross-linked hen egg-white lysozyme crystals below the freezing point as a function of the mole fraction of protons in the water phase.[446] Data were analysed using a cross relaxation model that eliminated the necessity of postulating long residence times for water molecules in the domain of the protein.

In a study of the extension of the transition phenomenon in hen egg-white lysozyme crystals the case of monoclinic rather than tetragonal crystals as the starting materials was considered.[447] The rotational and transitional parameters relating to the known co-ordinates of the crystalline form of hen egg-white lysozyme molecule have been related to those of the turkey enzyme which differs in its primary structure in 7 of the 129 residues.[448] An electron density map has been calculated at 5 Å resolution; the packing of the molecules in this form appears to present the entire length of the active cleft in the vicinity of the crystallographic six-fold axis and does not appear to be blocked by neighbouring molecules.

A method reported for the measurement of water strongly held by (hen egg-white) lysozyme was based on vacuum drying and titration with Karl Fischer reagent.[449] Investigations were conducted over the temperature range −20 °C—+20 °C. The 23 or 20 water molecules held per lysozyme molecule fell into several classes. Three to four of them were quite strongly held and may correspond to the three buried water molecules observed during *X*-ray

[444] D. G. Agresti, R. E. Lenkinski, and J. D. Glickson, *Biochem. Biophys. Res. Comm.*, 1977, **76**, 711.
[445] R. E. Lenkinski, D. G. Agresti, D. M. Chen, and J. D. Glickson, *Biochemistry*, 1978, **17**, 1463.
[446] E. Hsi and R. G. Bryant, *Arch. Biochem. Biophys.*, 1977, **183**, 588.
[447] J. Berthou and P. Jollès, *Biochimie*, 1978, **60**, 209.
[448] R. Bott and R. Sarma, *J. Mol. Biol.*, 1976, **106**, 1037.
[449] P. B. Rao and W. P. Bryan, *Biopolymers*, 1978, **17**, 291.

structural analysis. The presence of carbohydrate in the lysozyme active cleft had no effect on drying at 0 °C.

The exact boundary, residue, conformational, and directional definitions of the three antigenic sites of hen egg-white lysozyme have been described.[450] The results revealed that the three antigenic sites account quantitatively for the total antigenic reactivity of the protein. Thus the entire antigenic structure of the enzyme has now been determined precisely. Its nature was discussed together with the power of the surface-simulation synthetic concept.

Several surface-simulation synthetic peptides have been synthesized in order to investigate the conformation restrictions of the antigenic site of hen egg-white lysozyme and to determine if the spatially constructed antigenic site has a preferred 'direction'.[451] The results from interaction direct and *via* immuno-adsorption showed that the peptide (15) possessed the maximum inhibitory

H-L-Phe-Gly-L-Lys-L-Lys-L-Asn-L-Thr-L-Asp-OH

(15)

activity towards the lysozyme immune reaction, and accounted quantitatively for the maximum expected reactivity of the site. When the peptide sequence was reversed, the immunochemical reactivity of the peptide was drastically reduced with goat antisera but was unchanged with rabbit antisera; it was concluded that the antigenic site exhibits a preferred 'direction' towards certain antisera.

Several peptides have been synthesized to simulate the surface of the hen egg-white lysozyme antigenic site.[452] These enabled the precise definition of the site as well as an investigation of its conformational and directional requirements. The results showed that the antigenic site is made up of the spatially contiguous surface residues L-arginine 125, L-arginine 5, L-glutamic acid 7, L-arginine 14, and L-lysine 13. The antigenic site exhibited restricted conformational freedom.

Self-association of hen egg-white lysozyme in solution has been examined by ^{13}C n.m.r. spectroscopy.[453] The effect of pH on the resonances of the non-protonated aromatic carbons of 9 mM lysozyme, and the effect of protein concentration on these resonances, both indicated that self-association significantly affects the chemical shift of C^{γ} of L-tryptophan 62, but not the chemical shifts of the other non-protonated aromatic carbons. Therefore the resonance of C^{γ} of L-tryptophan 62 was advocated as a convenient monitor of lysozyme self-association.

The effect of hydration on the thermal denaturation of lysozyme as measured by differential scanning calorimetry has been examined.[454] The Raman spectrum of chemically denatured (dimethyl sulphoxide, urea, LiBr, guanidinium chloride, sodium dodecyl sulphate) hen egg-white lysozyme has been studied.[455]

[450] M. Z. Atassi and C. L. Lee, *Biochem. J.*, 1978, **171**, 429.
[451] C. L. Lee and M. Z. Atassi, *Biochim. Biophys. Acta*, 1977, **495**, 354.
[452] M. Z. Atassi and C. L. Lee, *Biochem. J.*, 1978, **171**, 419.
[453] R. S. Norton and A. Allerhand, *J. Biol. Chem.*, 1977, **252**, 1795.
[454] Y. Fujita and Y. Noda, *Bull. Chem. Soc. Japan*, 1978, **51**, 1567.
[455] R. S. Porubcan, K. L. Watters, and J. T. McFarland, *Arch. Biochem. Biophys.*, 1978, **186**, 255.

The correlation between a Raman band intensity parameter and denaturation as evidenced by enzyme activity and viscosity measurements was good, leading to the conclusion that the amide III Raman spectrum is useful for assessing the degree of denaturation.

A new technique for the rapid measurement of ultrasonic absorption with a sampling interval of 5 msec has been developed and applied to a kinetic study of denaturation and subsequent reduction of hen egg-white lysozyme.[456] The rate constant of denaturation by guanidinium chloride obtained from the time variance of ultrasonic absorption agreed well with that from u.v. absorption and optical rotation. Analysis of the constants as functions of reagent concentrations indicated that the intermediate state between native and reduced state is not necessarily the completely denatured state.

Kinetic analysis of the protease digestion of several derivatives of hen egg-white lysozyme showed that only the denatured (D) state of the protein is digested and that the reaction velocity is proportional to the equilibrium constant of the native (N)–denatured transition of the enzyme.[457] Alteration of the net charge of lysozyme by acetylation caused a shift of the N $\rightleftharpoons$ D transition to the right. Both the formation of a lysozyme–inhibitor complex and the introduction of a covalent bond in the lysozyme molecule restricted the transition. The magnitude of the N $\rightleftharpoons$ D transition was related to the susceptibility of lysozyme to protease digestion. Measurement of the enzymic activity and fluorescence properties showed that the gross conformation of acetylated lysozyme is very similar to that of the native enzyme.[458] Further studies on N $\rightleftharpoons$ D transition led to the conclusion that charge balance plays an important role in the maintenance of conformation.

Methodology for the regeneration to the native form of hen egg-white lysozyme from its forms protected by derivatization has been described.[459]

Natural-abundance ^{13}C n.m.r. spectroscopy has been used to investigate the effect of chemical modification of L-tryptophan 108 of hen egg-white lysozyme on the lanthanide ion-binding properties of the enzyme.[460] Unlike the intact enzyme, some modified forms did not show evidence of aggregation at neutral pH.

The modification of L-tryptophan residues of hen egg-white lysozyme by *N*-bromosuccinimide has been studied kinetically by the stopped-flow method, changes in absorbance and fluorescence being monitored.[461] One L-tryptophan (62?) was distinguished from others in terms of modification rate, and was protected by modified chitin, 2-acetamido-2-deoxy-D-glucose, and tri-*N*-acetylchitotriose.

The products of air-oxidation in the presence of 2-mercaptoethanol of reduced partially carboxymethylated hen egg-white lysozyme have been fractionated and physiochemical properties of them were determined.[462] The

[456] K. Yamanaka, H. Nakajima, and Y. Wada, *Biopolymers*, 1978, **17**, 2159.
[457] T. Imoto, K. Fukuda, and K. Yagishita, *J. Biochem.* (*Japan*), 1976, **80**, 1313.
[458] T. Imoto, S. Moriyama, and K. Yagishita, *J. Biochem.* (*Japan*), 1976, **80**, 1319.
[459] S. Aimoto and Y. Shimonishi, *Bull. Chem. Soc. Japan*, 1978, **51**, 205.
[460] K. Dill and A. Allerhand, *Biochemistry*, 1977, **16**, 5711.
[461] K. Hiromi, T. Kawagishi, and M. Ohnishi, *J. Biochem.* (*Japan*), 1977, **81**, 1583.
[462] A. S. Acharya and H. Taniuchi, *Proc. Nat. Acad. Sci., U.S.A.*, 1977, **74**, 2362.

results indicated that eight isomers, containing three presumably native disulphide bonds and one free and one carboxymethylated thiol group, are formed during the air oxidation. Since in each of these isomers the formation of one of the four native disulphide bonds is permanently blocked, it was argued that no one of the four disulphide bonds of native lysozyme is obligatory in the formation of the other three native disulphide bonds.

Conditions for the selective cleavage of the kyneurenyl bond in oxidized hen egg-white lysozyme in aqueous hydrazine solution have been established.[325] Further derivatization and peptide analysis showed that the *N'*-formylkynurenine is present at position 62 in place of L-tryptophan in the native protein.

The structure of phage T4 lysozyme in solution has been shown to differ noticeably from its crystalline structure.[463]

Lysozyme has been immobilized with retention of enzymic activity by reaction with agarose cyclic imidocarbonate and has been co-immobilized with β-D-galactosidase on the same support.[182]

43 Mannanases (Miscellaneous)

A simple assay procedure for β-D-mannanase enzyme which has been developed employs carob D-galacto-D-mannan dyed with Remazolbrilliant Blue.[464] Additionally, the procedure is quantitative, relatively sensitive, and highly specific for β-D-mannanase enzyme. It can be readily used for the determination of β-D-mannanase activity in crude enzyme preparations and column-chromatography eluates.

The modes of action of β-D-mannanase enzymes of diverse origin on legume seed D-galacto-D-mannans have been studied.[465]

Two components of konjac β-D-mannanases (I and II) were purified by ion-exchange chromatography and gel filtration.[466] The enzyme preparations (mol. wts. 3.2×10^4 and 2.9×10^4 respectively) behaved as homogeneous proteins in ultracentrifugal analysis. Kinetic studies of these β-D-mannanases indicated that the hydrolysis of konjac D-gluco-D-mannan, D-manno-oligosaccharides, and D-gluco-D-manno-oligosaccharides proceeds in a random or an *endo* mechanism, but the randomness of enzyme I was less than that of enzyme II. Both β-D-mannanases I and II hydrolysed (1 → 4)-β-D-manno-oligosaccharides and D-gluco-D-manno-oligosaccharides, but the D-mannosidic bonds linked to the D-glucose residues were attacked more easily than those between D-mannose residues.

A technique which employs substrate affinity chromatography on D-gluco-D-mannan or D-mannan derivatives of agarose has been developed for the purification of legume seed β-D-mannanases.[467] *Via* this technique, β-D-mannanase B has been purified from lucerne seeds.

[463] A. A. Timchenko, D. B. Ptitsyn, A. V. Troitsky, and A. I. Denesyuk, *F.E.B.S. Letters*, 1978, **88**, 109.
[464] B. V. McCleary, *Carbohydrate Res.*, 1978, **67**, 213.
[465] B. V. McCleary, *Proc. Austral. Biochem. Soc.*, 1978, **11**, 33.
[466] H. Shimahara, H. Suzuki, N. Sugiyama, and K. Nisizawa, *Agric. Biol. Chem. (Japan)*, 1975, **39**, 301.
[467] B. V. McCleary, *Phytochemistry*, 1978, **17**, 651.

Mutant strains of *Trichoderma viride* not producing D-mannanase have been induced and isolated with the aid of nystatin interaction.[357]

44 β-Neoagarotetraose Hydrolases

A quantitative thin-layer chromatographic assay has been reported for the β-neoagarotetraose hydrolase of *Pseudomonas atlantica*, an enzyme which hydrolyses *O*-3,6-anhydro-α-L-galactopyranosyl-(1 → 3)-*O*-β-D-galacto-pyranosyl-(1 → 4)-*O*-3,6-anhydro-α-L-galactopyranosyl-(1 → 3)-D-galactose by cleaving the central (1 → 4)-β-D-linkage to yield two molecules of neo-agarobiose.[468]

45 Oligo-(1 → 6)-D-glucosidases

Various aspects of the enzymology of oligo-(1 → 6)-D-glucosidase (limit dextrinase?) from peas have been reported.[415]

46 Pectate, Pectin, and Poly-D-galacturonate Lyases

Pectin lyase from the culture medium of *Aspergillus japonicus* has been purified 122-fold by ammonium sulphate fractionation, gel filtration, ion-exchange chromatography, and gel filtration.[469] The purified enzyme (mol. wt. 3.2×10^4 by gel filtration) was free from pectinesterase and *endo*-polygalacturonase, and was homogeneous on ultracentrifugal analysis. The isoelectric point and the sedimentation coefficient of the enzyme were at pH 7.7. and 3.4S, respectively. The pH optimum of the enzyme for pectin was 6.0, but was 7.0 for polymethyl polygalacturonate methyl glycoside. The addition of calcium ion stimulated the activity and shifted the pH optimum to 7.0 when pectin was used as substrate. A marked inhibition of the activity was observed by oxidizing reagents such as iodine and *N*-bromosuccinimide, although hydrogen peroxide did not inhibit the activity. Metal chelating reagents, reducing reagents, and thiol inhibitors did not affect the activity. The purified enzyme was able to macerate parenchymateous tissues, of which onion tissue was the most susceptible.

The pectic enzymes of six strains of pigmented, pectolytic clostridia isolated from potatoes have been examined and compared with those of strains of *Clostridium felsineum* and *C. aurantibutyricum* using cup-plate assays.[470] All the organisms were found to produce pectate lyase activity.

4-*O*-(4-Deoxy-β-L-*threo*-hex-4-enepyranosyluronic acid)-D-galacturonic acid has been proposed as the inducer of pectic acid (poly-D-galacturonate) lyase in *Erwinia carotovora*.[471] No induction was observed for D-galacturonic acid.

The taxonomy of the brown rot fungi (*Monilinia* species) has been related to their extracellular cell wall-degrading enzyme. Patterns of pectin lyase secreted by *M. fructigena*, *M. fructicola*, and *M. laxa* were studied by isoelectric focusing.[103]

A pectate lyase has been purified from the culture filtrate of *Streptomyces fradiae* by precipitation with ammonium sulphate, followed by ion-exchange

[468] D. Groleau and W. Yaphe, *J. Chromatog.*, 1977, **137**, 234.
[469] S. Ishii and T. Yokotsuka, *Agric. Biol. Chem.* (*Japan*), 1975, **39**, 313.
[470] B. M. Lund and T. F. Brocklehurst, *J. Gen. Microbiol.*, 1978, **104**, 59.
[471] S. Tsuyumu, *Nature*, 1977, **269**, 237.

chromatography and gel filtration.[472] The enzyme was an *endo*-pectate lyase (pH optimum 9.0—9.2) and required only calcium ion for the activity. The enzyme had macerating activity and was active on both native pectin and D-galacturonan, but the activity toward D-galacturonic acid oligosaccharides was small.

47 Poly-D-galacturonases

The role of the secondary alcoholic groups of D-galacturonan in its degradation by *endo*-D-galacturonanase has been investigated.[473]

The *in vitro* degradation of green tomato (*Lycopersicon esculentum*) cell walls by an extract of ripe fruits has been shown to be catalysed solely by an *endo*-poly-D-galacturonase.[474] However, the action of the enzyme *in vitro* did not lead to the substantial decrease in cell wall D-galactose which was observed *in situ*.

Ungerminated pollen of *Nicotiana tabacum* has been shown to contain a pectolytic enzyme (pH optimum *ca.* 6.0) which was devoid of pectin lyase activity.[475]

The pectin enzymes of six strains of pigmented, pectolytic clostridia isolated from potatoes have been examined and compared with those of strains of *Clostridium felsineum* and *C. aurantibutyricum* using cup-plate assays.[470] The *C. felsineum* strains only produced poly-D-galacturonase activity.

The taxonomy of the brown rot fungi (*Monilinia* species) has been related to their extracellular cell wall-degrading enzymes.[103] Patterns of poly-D-galacturonases secreted by *M. fructigena* and *M. laxa* were studied by column isoelectric focusing.

An electrophoretically homogeneous preparation of *endo*-poly-D-galacturonase from culture filtrates of *Rhizoctonia fragariae*, a pathogenic agent in strawberry plants, has been resolved into two isoenzymes (p*I* values 6.76 and 7.08, mol. wts. both 3.6×10^4 by gel filtration) by narrow pH range isoelectric focusing.[476] The two enzymes each consisted of a single polypeptide chain and exhibited similar substrate specificities.

Two inducible poly-D-galacturonases from culture filtrates of *Trichoderma koningii* have been purified to homogeneity by ion-exchange chromatography and narrow pH range isoelectric focusing.[477] Both enzymes consisted of a single polypeptide chain (mol. wt. 3.2×10^4 by gel filtration) and contained carbohydrate.

48 *exo*-Poly-D-galacturonases

Exo-poly-D-galacturonase has been extracted from apple cortical tissue.[478] The enzyme (pH optimum *ca.* 4.8, mol. wt. 5.8×10^4) was found to be in-

[472] M. Sato and A. Kaji, *Agric. Biol. Chem. (Japan)*, 1975, **39**, 819.
[473] L. Rexová-Benková, M. Mračková, O. Luknár, and R. Kohn, *Coll. Czech. Chem. Comm.*, 1977, **42**, 3204.
[474] S. J. Wallner and H. L. Bloom, *Plant Physiol.*, 1977, **60**, 207.
[475] M. Kroh and M. W. Loewus, *Phytochemistry*, 1978, **17**, 797.
[476] F. Cervone, A. Scala, M. Foresti, M. G. Cacace, and C. Noviello, *Biochim. Biophys. Acta*, 1977, **482**, 379.
[477] C. Fanelli, M. G. Cacace, and F. Cervone, *J. Gen. Microbiol.*, 1978, **104**, 305.
[478] I. M. Bartley, *Phytochemistry*, 1978, **17**, 213.

hibited by H_4edta and citrate, and was activated by Ca^{2+} and to a lesser extent by Sn^{2+}.

Poly-D-galacturonase activity has been detected in a number of plants including seeds of *Avena sativa*, *Phaseolus vulgaris*, *Pisum sativum*, and *Zea mays*.[479] The enzyme from oat seedlings (pH optimum 5.0—5.5) was activated by Ca^{2+} and to a lesser extent by Cd^{2+}. The rate of attack of the enzyme was maximal for substrates of DP = 2 and above; the enzyme was shown to act by an *exo* pattern.

The isolation, characterization, and mode of action of *exo*-D-galacturonanase from carrot have been described.[480]

In an attempt to characterize the groups essential for the catalytic action of *endo*-D-galacturonanase from *Aspergillus niger*, the behaviour of the kinetic parameters as a function of pH has been examined.[481] It was concluded that a carboxy-group, and a protonated imidazole group of L-histidine, participate in the catalytic reaction.

49 Pullulanases

An assay reported for pullulanase levels in the presence of other carbohydrases is dependent upon the three-fold increase in release of reducing groups by the enzyme acting upon pullulan when glucoamylase is present in excess.[482]

A debranching enzyme (R-enzyme), free from contaminating carbohydrases and homogeneous by polyacrylamide disc-gel electrophoresis, has been purified from malted barley.[206] A partially purified preparation of this enzyme accelerated the rate of digestion of barley-starch granules by the action of purified α- and β-amylases to the same extent as was affected by the dialysed, crude extract from malted barley. Contrary to expectation, the debranching enzyme, purified to homogeneity, had very little accelerating affect. These results indicate that a factor or factors, which may be maltase or α-D-glucosidase and were lost during the purification of the debranching enzyme, may play a role in the digestion of starch granules by the dialysed, crude extract from malted barley *in vitro* and by enzymes in the endosperm of germinating barley seeds *in vivo*. The debranching enzymes, including barley-malt R-enzyme, *Aerobacter* pullulanase, and *Pseudomonas* isoamylase, did not digest starch granules to a detectable extent.

The purification of pullulanase from *Aerobacter aerogenes via* affinity chromatography on an agarose-immobilized cyanogen bromide-activated cyclohexa-amylose has been reported.[483] Elution from the affinity column was achieved using cyclohepta-amylose. Crossed immunoelectrophoresis was used to follow the various steps of the purification.

Some properties of crystalline extracellular pullulanase from *Aerobacter aerogenes* were compared with those of crystalline intracellular pullulanase.[484] The amino-acid compositions of these enzymes were found to be alike on the

[479] R. Pressey and J. K. Avants, *Plant Physiol.*, 1977, **60**, 548.
[480] K. Heinrichová, *Coll. Czech. Chem. Comm.*, 1977, **42**, 3214.
[481] L. Rexová-Benková and M. Mračková, *Biochim. Biophys. Acta*, 1978, **523**, 162.
[482] B. J. Catley, *Carbohydrate Res.*, 1978, **61**, 419.
[483] B. S. Enevoldsen, L. Reimann, and N. L. Jansen, *F.E.B.S. Letters*, 1977, **79**, 121.
[484] R. Ohba and S. Ueda, *Agric. Biol. Chem.* (*Japan*), 1975, **39**, 967.

whole. Both enzymes also gave three different isoelectric points. The viscosity of the pullulan solution treated with pullulanase decreased rapidly at the early stage of the reaction. It was found that pullulanase cleaved at random (1 → 6)-α-D-glucosidic linkages in pullulan. Ca^{2+} stimulated the activity of pullulanases, while heavy metal ions except Mn^{2+} and Ag^{+} considerably inactivated the enzyme and particularly Hg^{2+} and Al^{3+} severely inactivated. Metal chelating agents and thiol reagents did not inhibit. *N*-Bromosuccinimide completely inhibited both pullulanases. It was suggested that L-tryptophan residues might possibly concern the enzymatic action of pullulanases.

Pullulanase activity in a *Klebsiella* species, and which is of course an enzyme of starch catabolism, has been shown to be associated with the outer membrane of the cells.[485]

The structural requirements for oligosaccharides to act as substrates for *Klebsiella aerogenes* pullulanase have been compared with those for isoamylase (see p. 455).[414]

50 Sucrose α-D-Glucohydrolases

The separation by polyacrylamide gel electrophoresis and subsequent analysis of the components of guinea-pig intestinal brush border membrane revealed the presence of *inter alia* enzyme complexes: α-D-glucosidase–sucrose, α-D-glucohydrolase–glucoamylase, and α-D-glucosidase–sucrose α-D-glucohydrolase.[145]

51 αα-Trehalases

The αα-trehalase activity of guinea-pig brush border membrane has been investigated.[145]

Rat intestinal αα-trehalase solubilized from microvillous membrane by Triton X-100 plus sodium deoxycholate has been purified by conventional techniques.[486] The enzyme (mol. wt. 2.4×10^5 by gel filtration, pH optimum 5.7 or 4.7 according to buffer type, K_m 5.5—11 mM for αα-trehalase according to buffer) was sensitive to thiol agents, and evidence was obtained for its undergoing conformational changes in acetate. The enzyme, which was slightly stimulated by Ca^{2+}, Mg^{2+}, and Zn^{2+} ions, was suggested to play a role in sugar transport in intestinal mucosal cells.

The midguts from *Rhynchosciara americana* larvae display an αα-trehalase activity which is soluble.[487] The enzyme (mol. wt. 1.22×10^5, pH optimum 6.0, p*I* 4.6, K_m 0.67 mM for αα-trehalose, E_{act} 16.7 kcal mol^{-1}) was not inhibited by thiol reagents. It was proposed that two carboxy-groups exist in the active site, one of them having a very high (8.3) pK_a value. The increase of the pK_a values of the essential groups of the free enzyme in the presence of increasing concentrations of dioxan supported this hypothesis.

αα-Trehalase purified after solubilization from *Humicola lanuginosa* was highly specific for αα-trehalose (K_m 0.4 mM).[488] The enzyme (mol. wt. 1.7 ×

[485] G. Wöhner and G. Wöber, *Arch. Mikrobiol.*, 1978, **116**, 303.
[486] M. Nakano, Y. Sumi, and M. Miyakawa, *J. Biochem.* (*Japan*), 1977, **81,** 1041.
[487] W. R. Terra, C. Ferreira, and A. G. de Bianchi, *Biochim. Biophys. Acta*, 1978, **524**, 131.
[488] A. R. S. Prasad and R. Maheshwari, *Biochim. Biophys. Acta*, 1978, **525**, 162.

10^5) was afforded protection against heat inactivation by protease-susceptible factors present in mycelial extracts. Bovine serum albumin and casein also protected the enzyme from heat inactivation.

Due to the hyperderepressive gene, a cross between a hyperderepressed strain and a normal strain of *Neurospera crassa* exhibited hyperproduction of αα-trehalase activity.[107]

52 *endo*-(1 → 4)-β-D-Xylanases

Two *endo*-(1 → 4)-β-D-xylanases (mol. wts. 2.4×10^4 and 4.1×10^4) have been shown to be produced by *Aspergillus niger*.[489] The smaller species in electrophoretically homogeneous form (p*I* 4.2, pH optimum 4.2 for carboxymethylxylan) split arabinoglucuronoxylan to xylo-oligosaccharides but did not hydrolyse xylobiose.

53 *exo*-(1 → 4)-β-D-Xylanases

Six *exo*-(1 → 4)-β-D-xylanases have been shown to occur in *Aspergillus niger*.[489] One of these, in electrophoretically homogeneous form (p*I* 4.6, mol. wt. 3.0×10^4, pH optimum 3 for methyl β-D-xylopyranoside), split off D-xylose from the ends of the D-xylan chains of xylotriose, *etc*., and exhibited high transglycosylase activity.

54 Xylanases (Miscellaneous)

The gibberellic acid-induced biosynthesis and release of cell-wall degrading *endo*-D-xylanase by isolated aleurone layers of barley have been studied.[490]

Studies on the D-xylanase system of *Streptomyces* species have included a study of the action of the enzyme on various D-xylans and D-xylo-oligosaccharides.[491] The isolation and structural identification of arabino-D-xylo-oligosaccharides arising from action of the enzyme on corn cob arabino-D-xylan were reported.[492]

Mutant strains of *Trichoderma viride* not producing D-xylanase have been induced and isolated with the aid of nystatin selection.[357]

55 Carbohydrate Isomerases

D-Glucose Isomerases (D-Xylose isomerases).—A discussion of enzyme chemistry in the D-glucose isomerase field demonstrates its role in the displacement of enzyme equilibria by oxyanions (including germanate) to obtain high yields of D-fructose from D-glucose.[493]

D-Glucose isomerase has continued to be immobilized with retention of activity; the large-scale commercial form is available as immobilized 'Sweetzyme'.[494] In another instance, the enzyme has been fixed to an organism mycelium and as such immobilized by spraying a mixture with gelatin into

[489] N. A. Rodionova, I. V. Gorbacheva, and V. A. Buivid, *Biochemistry* (*U.S.S.R.*), 1977, **42**, 505.
[490] W. V. Dashek and M. J. Chrispeels, *Planta*, 1977, **134**, 251.
[491] I. Kusakabe, T. Yasui, and T. Kobayashi, *J. Agric. Chem. Soc.* (*Japan*), 1977, **51**, 439.
[492] I. Kusakabe, T. Yasui, and T. Kobayashi, *J. Agric. Chem. Soc.* (*Japan*), 1977, **51**, 669.
[493] 'Topics in Enzyme and Fermentation Biotechnology', Vol. 3, ed. A. Wiseman, J. Wiley and Sons, Ltd., Chichester, 1979.
[494] J. Oestergaard and S. L. Knudsen, *Stärke*, 1976, **28**, 350.

water-immiscible solvent and then glutaraldehyde crosslinking within the resultant particles.[495]

56 Carbohydrate Oxidases

Cellobiose Oxidases.—An extracellular enzyme which utilizes molecular oxygen to oxidise cellodextrins to the corresponding aldonic acids has been isolated from culture filtrates of the white rot fungus *Sporotrichum pulverulentum.*[496] This enzyme, named cellobiose oxidase, was purified by classical techniques and was shown to be a glycoprotein (mol. wt. 9.3×10^4). The u.v. spectrum of the enzyme was characteristic of a haemoprotein, there being approximately one flavin component per enzyme molecule. The possible role of this complex enzyme in cellulose degradation was discussed.

D-Galactose Oxidases.—A D-galactose oxidase–NADH–peroxidase coupled assay system has been used for the estimation of D-galactose oxidase activity.[497] Spectrophotometric measurement of NADH consumption yields a direct quantitative value of enzymic activity.

D-Glucose Oxidases.—The effects of periodate oxidation upon the D-glucose oxidase of *Aspergillus niger* include a reduction in mass of up to 14%.[498] This was considered in the light of changes in sedimentation and diffusion coefficients and the Stokes radius. Gel filtration studies indicated that the oxidized enzyme exists as a complex oligomerizing system.

Comparative c.d. studies for *Aspergillus niger* D-glucose oxidase and the corresponding apoenzyme suggested that the FAD co-enzyme is not a gross structural determinant of the enzyme.[499]

D-Glucose oxidase has been immobilized, with retention of activity, by glutaraldehyde crosslinking on to an albumin-treated carbon paste electrode surface,[500] and in conjunction with peroxidase by reaction with activated cellulose.[501] A soluble, immobilized but active form of the enzyme has been produced by glutaraldehyde crosslinking.[499]

57 Glycopeptide-linkage Hydrolases

4-L-Aspartyl-β-D-glucosylamine Amidohydrolases.—The effect of various compounds on 4-L-aspartyl-β-D-glucosylamine amidohydrolase (mol. wt. 6.3×10^4) from human liver has been studied.[502] *N*-Acetyl-L-cysteine inhibited the enzyme non-competitively (K_i 3.2 mM) whereas 3-hydroxybutanone inhibited competitively (K_i 4.1 mM). L-Methionine, L-isoleucine, and cystathione enhanced enzyme activity. The differences between the characteristics of the enzymes from human and non-human sources were discussed.

Whereas normally 4-L-aspartyl-β-D-glucosylamine amidohydrolase is specific for substrates which contain both the free α-amino- and α-carboxy-groups of

[495] J. V. Hupkes and R. van Tilburg, *Stärke*, 1976, **28**, 356.
[496] A. R. Ayers, S. B. Ayers, K. E. Eriksson, *European J. Biochem.*, 1978, **90**, 171.
[497] G. Avigad, *Analyt. Biochem.*, 1978, **86**, 470.
[498] M. Gerber, O. Bodmann, and G. V. Schulz, *F.E.B.S. Letters*, 1977, **80**, 294.
[499] B. Solomon, N. Lotan, and E. Katchalski-Katzir, *Biopolymers*, 1977, **16**, 1837.
[500] F. R. Shu and G. S. Wilson, *Analyt. Chem.*, 1976, **48**, 1679.
[501] P. Cremonesi, R. Bovara, and G. Mazzola, *Cellulose Chem. Technol.*, 1976, **10**, 567.
[502] B. Dugal, *Biochem. J.*, 1978, **171**, 799.

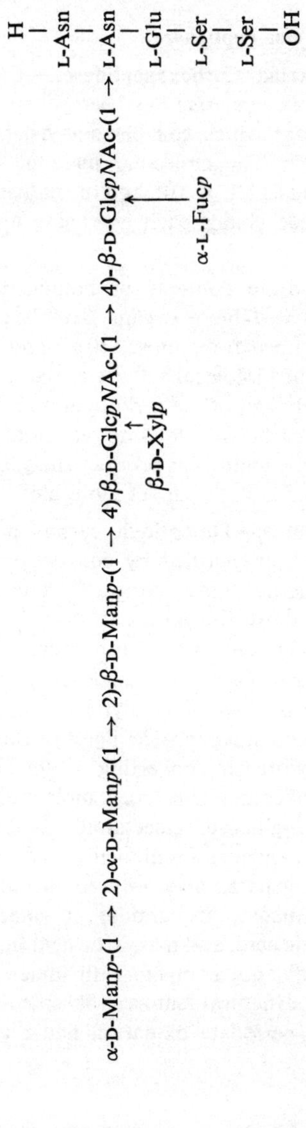
α-D-Manp-(1 → 2)-α-D-Manp-(1 → 2)-β-D-Manp-(1 → 4)-β-D-GlcpNAc-(1 → 4)-β-D-GlcpNAc-(1 → L-Asn
β-D-Xylp
α-L-Fucp
H
L-Asn
L-Glu
L-Ser
L-Ser
OH

(16)

L-aspartic acid involved in the glycopeptide linkage, *i.e.* cannot act on glycopeptides with several amino-acid residues, a new amidase which overrides these requirements has been discovered in almond emulsin.[503] The enzyme was shown to cleave the glycopeptide linkage in the glycopeptide (16) from pineapple-stem bromelain.

58 Proteinases and Peptidases

Carboxypeptidases and Serine Carboxypeptidases.—Carboxypeptidase from baker's yeast *Saccharomyces cerevisiae* has been shown to be a glycoprotein, the carbohydrate moieties of which can be removed by an *endo*-β-D-2-acetamido-2-deoxy-glucanase.[281] The molecular mass of the carbohydrate-free protein chain was confirmed as 5.1×10^4, and the native enzyme was shown to contain four oligosaccharide chains each averaging approximately fourteen D-mannose residues.

Cathepsins.—The carbohydrate contents of homogeneous preparations of cathepsin D from various vertebrate organs have been determined.[504] The chicken liver enzyme contained D-mannose (10%) and 2-acetamido-2-deoxy-D-glucose (11.6%). Since the biological activity of the enzyme did not decrease noticeably upon removal of most of the carbohydrate by glycoside hydrolases, it was concluded that the carbohydrate is not essential to the catalytic functions of cathepsin D. Tryptic hydrolysis product analysis indicate the presence of several carbohydrate chains in the intact molecule.

Dipeptidyl Carboxypeptidases.—Dipeptidyl carboxypeptidase (angiotensin I-converting enzyme) has been purified by conventional techniques from a particulate fraction of porcine kidney cortex.[505] The purified enzyme gave two protein bands on standard disc-gel electrophoresis, but showed a single protein component after treatment with neuraminidase. It was concluded that the preparation was a mixture of sialo- and asialo-enzymes and that neuraminic acid residues did not contribute to the catalytic activity.

Two glycopeptide fractions in a pronasic digest of rabbit pulmonary dipeptidyl carboxypeptidase (angiotensin-converting enzyme) have been resolved by gel filtration.[506] The minor component (*ca.* 1 mole mol^{-1} enzyme) contained D-mannose, D-galactose, D-glucose, 2-acetamido-2-deoxy-D-glucose, 2-acetamido-2-deoxy-D-galactose, and neuraminic acid (molar ratio 1 : 5 : 3 : 4 : 1 : 2) and molar equivalents of L-aspartic acid, L-threonine, and L-serine; the major component contained L-fucose, D-mannose, D-galactose, 2-acetamido-2-deoxy-D-glucose, neuraminic acid, and L-aspartic acid (molar ratio 1 : 4 : 4 : 4 : 1 : 1). This glycopeptide did not compete with intact glycoprotein for anti-enzyme antibodies. Some structural features (Scheme 9) for the major glycopeptide were deduced *via* periodate oxidation and digestion with glycoside hydrolases.

503 N. Takahashi, *Biochem. Biophys. Res. Comm.*, 1977, **76**, 1194.
504 O. V. Kazakova, V. N. Orekhovich, and G. V. Vikha, *Biochemistry* (*U.S.S.R.*), 1977, **42**, 1728.
505 G. Oshima, K. Nagasawa, and J. Kato, *J. Biochem.* (*Japan*), 1976, **80**, 477.
506 J. L. Hartley and R. L. Soffer, *Biochem. Biophys. Res. Comm.*, 1978, **83**, 1545.

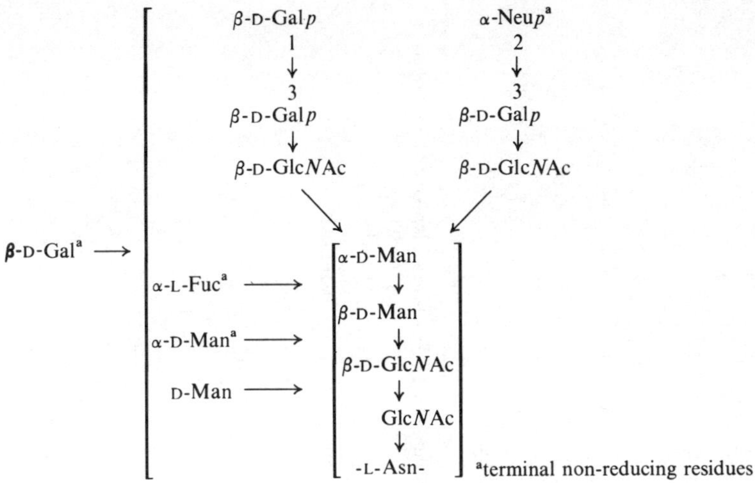

Scheme 9

59 Deoxyribonucleases and Ribonucleases

Nucleases.—On treatment of mung bean nuclease I, a glycoproteinaceous enzyme, with β-D-2-acetamido-2-deoxyglucanase the molecular mass decreased from 3.9×10^4 to 3.1×10^4.[281] The peptides produced on reduction of this enzyme with a thiol were of molecular mass 1.87×10^4 and 1.25×10^4, indicating that carbohydrate had been attached to both. Nuclease P_1 from a *Penicillium* species, also a glycoproteinaceous enzyme, was decreased in size from 4.0×10^4 to 3.0×10^4 by the action of the enzyme.

Ribonucleases.—The primary structure of the main glycopeptide obtained by pronasic digestion of horse pancreatic ribonuclease has been investigated by 360 MHz ^{1}H n.m.r. spectroscopy and methylation analysis.[507] Results demonstrated that the glycopeptide had the structure (17), this being the first reported occurrence of the presence of both (2 → 3)- and (2 → 6)-linkages from neuraminic acid groups in a glycopeptide.

60 Phosphatases

Alkaline Phosphatases.—Various degrees of specific interaction between concanavalin A and human alkaline phosphatases, extracted by n-butanol from various sources, have been demonstrated by a simple electrophoretic retardation technique and affinity chromatography.[508] The lectin-binding abilities of the enzymes demonstrated the glycoproteinaceous characters of the latter.

As part of the studies on the structure–function relationships of human tissue alkaline phosphatases, changes in physicochemical properties, activity, affinity for various lectins and blood group antisera, carbohydrate composition,

[507] B. L. Schut, L. Dorland, J. Haverkamp, J. F. G. Vliegenthart, and B. Fournet, *Biochem. Biophys. Res. Comm.*, 1978, **82**, 1223.
[508] C.-H. Chang, D. Angellis, and W. H. Fishman, *Mol. Cell. Biochem.*, 1975, **9**, 55.

```
α-NeuNAc-(2 → 6)-β-D-Galp-(1 → 4)-β-D-GlcpNAc-(1 → 2)-α-D-Manp
                                                          1
                                                          ↓
                                                          6
                                                      β-D-Manp-(1 → 4)-β-D-GlcpNAc-(1 → 4)-β-D-GlcpNAc-(1 → L-Asn
                                                          3                                      6
                                                          ↑                                      ↑
                                                          1                                      1
α-NeuNAc-(2 → 3)-β-D-Galp (1 → 4)-β-D-GlcpNAc-(1 → 2)-α-D-Manp                                 α-L-Fucp
```

(17)

and biological half-life of the enzymes have been investigated.[509] The desialylized hepatic enzyme was shown to be more heat labile and more sensitive to protease digestion in the presence of sodium dodecyl sulphate than the native hepatic enzyme. Helix contents of these two forms of the enzyme (mol. wts. 1.65×10^5 and 1.75×10^5, respectively) were calculated to be 30.8 and 39.0% respectively. Intestinal neuraminidase preparations treated with α-D-mannosidase, β-D-2-acetamido-2-deoxyglucosidase, or *endo*-β-D-2-acetamido-2-deoxyglucanase displayed a decrease in enzymic activity.

61 Sulphamidases

Endo-glycosidase activity present in rat liver lysosomes, during its degradation of heparin, also released inorganic sulphate equivalent to one sulphamide bond cleaved per heparin chain.[406]

62 Sulphatases

2-Acetamido-2-deoxy-D-galactose 6-sulphate Sulphatases.—Extracts of fibroblasts from cases of Morquio syndrome lack 2-acetamido-2-deoxy-D-galactose 6-sulphate sulphatase activity, but exhibit normal levels of 2-acetamido-2-deoxy-D-glucose 6-sulphate sulphatase activity.[510] Thus the defective enzyme in the disease is a sulphatase specific for the 6-sulphate group linked to sugars with the D-*galacto* configuration. Hydrolysis of ester sulphate by this enzyme is limited to 6-sulphate groups occurring at the non-reducing terminal.

Arylsulphatases A and B.—Arylsulphatase A (cerebroside sulphatase) from human liver did not form a stable complex when chromatographed with a 12-fold excess of its physiological activator protein.[511] The activator was shown to complex with substrates (sulphatides) and products (cerebrosides) of the enzyme and its action.

The distribution of soluble arylsulphatases in human tissues has been examined *via* ion-exchange chromatography. All tissues contained arylsulphatase A and B.[512] In addition, brain singularly contained significant quantities of a minor anionic form (B_m); only trace amounts of B_m were found in liver, kidney, testis, and placenta. Arylsulphatases B and B_m had equal activity towards methylumbelliferyl sulphate, nitrocatechol sulphate, and UDP-2-acetamido-2-deoxy-D-galactose 4-sulphate, but both forms were inactive toward the arylsulphatase A substrates cerebroside sulphate and L-ascorbic acid 2-sulphate. The physicochemical properties of arylsulphatases B and B_m differed with respect to thermal stability, ionicity, and behaviour on polyacrylamide gel electrophoresis and isoelectric focusing.

A study of extracellular hydrolases of human lung showed that the level of arylsulphatase activity is low.[48] The electrophoretic mobility of the enzymes on polyacrylamide gels indicated that they differed from those occurring in serum. A secretory mechanism of the lung enzymes was discussed.

[509] T. Komoda and Y. Sakagishi, *Biochim. Biophys. Acta*, 1978, **523**, 395.
[510] A. L. Horwitz and A. Dorfman, *Biochem. Biophys. Res. Comm.*, 1978, **80**, 819.
[511] G. Fischer and H. Jatzkewitz, *Biochim. Biophys. Acta*, 1977, **481**, 561.
[512] R. L. Stevens, A. L. Fluharty, A. R. Killgrove, and H. Kihara, *Biochim. Biophys. Acta*, 1977, **481**, 549.

The levels of arylsulphatase activity in skin fibroblasts from cases of cystic fibrosis and controls have been compared.[68]

The arylsulphatases A and B of human urine, rat liver, rat brain, and sheep brain have been separated by a simple and rapid affinity chromatographic method on agarose-immobilized Cibacron Blue F3GA.[513]

A theoretical basis has been given for methods of determining the apparent velocity constant ($k^{\star}$) for the substrate-induced inactivation of bovine liver (aryl)sulphatase A, and the initial velocity (v_0) of the catalytic reaction.[514] The method has been applied to the characterization of the inactivation occurring during the hydrolysis of a number of substrates, and it was suggested that the inactivation of the enzyme does not proceed through an enzyme–substrate complex but through the enzyme–SO_4^{2-} complex produced during the catalytic reaction.

A purification procedure has been described for an atypical arylsulphatase A which exists in chicken brain.[515] Investigation of the properties of the enzyme revealed *inter alia* that the molecule is a glycoprotein.

Four protein peaks showing arylsulphatase activity have been isolated from the gut of the giant African snail *Achatina achatina via* ion-exchange chromatography and gel filtration.[516] The existence of multiple forms of the enzyme (pH optima 6.0, 5.8, 6.5, 6.3; mol. wts. 11.0×10^4, 5.0×10^4, 9.3×10^4, and 11.0×10^4 respectively) was discussed in relation to their possible involvements in the various metabolic cycles of the organism.

Cerebroside Sulphatases.—Cerebroside sulphatase (arylsulphatase A) from human liver did not form a stable complex when chromatographed with a 12-fold excess of its physiological activator protein.[511] The activator was shown to complex with substrates (sulphatides) of the enzyme and with products (cerebrosides) of the enzyme action.

Acidic forms of human cerebroside sulphatase have been partially purified from invertebrates; *Tethya aurantium* (Porifera), *Patella vulgata* (mollusca), *Maja squinado* (Arthropoda), *Martlasterias glacialis* (Echinodermata), and *Microcosmus sulcatus* (Tunicata).[517] Enzyme preparations thus obtained cleaved cerebroside sulphates only in the presence of either specific detergents, *e.g.* taurodeoxycholate, or an activator protein isolated from human liver. On a molar basis, less of the activator protein was required to achieve the same activation as detergent.

L-Iduronic acid 2-sulphate Sulphatase.—Commercially available sodium heparinate has been sequentially treated with methanolic 0.06 molar hydrogen chloride and nitrous acid.[518] The non-degraded material was separated by gel filtration from the nonsulphated and monosulphated disaccharides produced. The latter ones, obtained in 10% yield, have been used as a substrate for the

[513] A. Ahmad, A. Surolia, B. K. Bachhawat, *Biochim. Biophys. Acta*, 1977, **481**, 542.
[514] A. B. Roy, *Biochim. Biophys. Acta*, 1978, **526**, 489.
[515] A. Ahmad, A. Surolia, and B. K. Bachhawat, *Indian J. Biochem. Biophys.*, 1977, **14**, 235.
[516] S. I. O. Agogbua, E. O. Anosike, and E. N. Ugochukwu, *Comp. Biochem. Physiol. Pt. B.*, 1978, **59**, 169.
[517] W. Mraz, G. Fischer, and H. Jatzkewitz, *Z. physiol. Chem.*, 1976, **357**, 301.
[518] L. C. Ginsberg, D. T. Di Ferrante, and N. Di Ferrante, *Carbohydrate Res.*, 1978, **64**, 225.

direct measurement of the enzyme L-iduronic acid 2-sulphate sulphatase present in human plasma and fibroblast homogenates. Studies of the kinetics and pH optimum of the enzyme, by use of plasma of a patient with mucolipidosis II, indicated an apparent K_m of 2.5 mM and a pH optimum of 4.6—4.8. The levels of activity in normal plasma and plasma of a patient with Hunter's disease were found to be 20.4 ± 1.22 units (μmol sulphate 24 h^{-1} g $protein^{-1}$) and 3.25 ± 0.36 units, respectively. In homogenates of cultured skin fibroblasts, the levels were 137.6 ± 10.7 units for normal controls and 6.4 ± 5.1 units for patients with Hunter's disease. The plasma of two obligated heterozygotes gave intermediate levels of activity, whereas the plasma of two possible heterozygotes gave either intermediate levels or entirely normal levels of activity.

63 Miscellaneous Enzymes

Adenosinetriphosphatase.—The effects of wheat germ agglutinin and concanavalin A on a variety of parameters of two purified ATP phosphohydrolases from the rectal salt gland of *Squalus acanthias* and from the electroplax of *Electrophorus electricus* have been examined.[519] Neuraminidase treatment of the rectal gland enzyme has no effect on wheat germ agglutinin inhibition, but that inhibition was reversible by di-*N*-acetylchitobiose. The inhibition of the rectal gland adenosinetriphosphatase by concanavalin A was less severe.

Cell Lytic Enzyme.—A lytic enzyme isolated from the lysate of *Pseudomonas aeruginosa* infected with a new strain of bacteriophage, phage 93, was purified by ion-exchange chromatography and by gel filtration.[520] The molecular weight was estimated as 2.1×10^4. The optimal conditions for the hydrolysis of acetone-dried cells of *P. aeruginosa* were determined (optimal pH 6.5—7.0, temperature optimum 70 °C). The enzyme was strongly inhibited by Ag^+, Hg^{2+}, Ni^{2+}, Fe^{2+}, and Cu^{2+} ions. When peptidoglycan obtained from *P. aeruginosa* was digested, free amino-groups were liberated without release of reducing sugars. The enzyme was suggested to be an amidase or peptidase.

Dextransucrases.—Dextransucrase has been purified from *Leuconostoc mesenteroides* and examined.[521] The structures of dextrans enzymatically synthesized therefrom were studied by periodate oxidation. It was found that the content of (1 → 6)-α-D-glucosidic linkages in synthesized dextran was raised considerably as the purification of enzyme proceeded, and that the content of branch links was increased by adding to the purified enzyme a fraction obtained from the culture broth of the organism by chromatography.

Reactions of dextransucrase of *Leuconostoc mesenteroides* and sucrose in the presence of sugars (acceptors) of low molecular weight have been observed to give a dextran of low molecular weight and a series of oligosaccharides.[522] The acceptor reaction of dextransucrase was examined in the absence and presence of sucrose by using D-[^{14}C]glucose, D-[^{14}C]fructose, and ^{14}C-reducing-end labelled maltose as acceptors. A purified dextransucrase was pre-incubated with sucrose, and the resulting D-fructose and unreacted sucrose

[519] J. R. Perrone and L. E. Hokin, *Biochim. Biophys. Acta*, 1978, **525**, 446.
[520] K. Arima, A. Yanai, and T. Beppu, *Agric. Biol. Chem. (Japan)*, 1976, **40**, 1301.
[521] D. Suzuki and T. Kobayashi, *Agric. Biol. Chem. (Japan)*, 1975, **39**, 557.
[522] J. F. Robyt and T. F. Walseth, *Carbohydrate Res.*, 1978, **61**, 433.

were removed from the enzyme by gel filtration. The enzyme, which migrated at the void volume, was collected and referred to as 'charged enzyme'. The charged enzyme was incubated with ^{14}C-acceptor in the absence of sucrose. Each of three acceptors gave two fractions of labelled products, a high molecular weight product, identified as dextran, and a product of low molecular weight that was an oligosaccharide. It was found that all three of the acceptors were incorporated into the products at the reducing end. Similar results were obtained when the reactions were performed in the presence of sucrose, but higher yields of labelled products were obtained and a series of homologous oligosaccharides was produced when D-glucose or maltose was the acceptor. It was proposed that the acceptor reaction proceeds by nucleophilic displacement of D-glucosyl and dextranosyl groups from a covalent enzyme–complex by a specific, acceptor hydroxy-group, and that this reaction effects a glycosidic linkage between the D-glucosyl and dextranosyl groups and the acceptor. It was concluded that the acceptor reactions serve to terminate polymerization of dextran by displacing the growing dextran chain from the active site of the enzyme; the acceptors, thus, do not initiate dextran polymerization by acting as primers.

α-Lactalbumins.—α-Lactalbumin from human milk showed a heterogenous behaviour when subjected to ion-exchange chromatography with DEAE-Sephadex®.[523] The two components separated showed identical patterns in the following studies; amino-acid compositions, fluorescence and c.d. spectra, transition temperature of denaturation, antigenicity, lactose synthase specifying activity, and hydrodynamic properties. After rechromatography of either peak, these two components appeared to be in equilibrium; this equilibrium varies with the temperature and the pH of chromatography. The two peaks were concluded to be the result of some conformational change induced upon the binding of the protein to the solid matrix.

The environment of L-tryptophan residues in α-lactalbumin was determined by a comparative study of bovine, human, and guinea-pig α-lactalbumins in terms of solvent perturbation and denaturation.[524] The results were interpreted on the basis of the fact that L-tryptophans-26 and -60 of bovine α-lactalbumin are replaced by other amino-acids in human and guinea-pig α-lactalbumins, respectively, and on the assumption that the three α-lactalbumins have similar conformations. Solvent perturbation measurements using ethylene glycol, glycerol, and dimethylsulphoxide as perturbants showed almost the same degree of exposure of L-tryptophan residues among the three α-lactalbumins at pH 7, and 25 °C and 2 °C, suggesting that L-tryptophan-26 is buried in the interior of the molecule and is inaccessible to solvent, and that L-tryptophan-60 is almost buried. A cationic detergent, cetyldimethylbenzylammonium chloride (CDBA) caused an abnormal difference spectrum of α-lactalbumin, which has a major positive peak at about 302 nm characteristic of the native structure of α-lactalbumin. The similarity of the spectrum of α-lactalbumin to that of lysozyme was considered to suggest that α-lactalbumin has a cleft-like region

[523] J.-P. Prieels and J. Schlusselberg, *Biochim. Biophys. Acta*, 1977, **491**, 76.
[524] K. Takase, R. Niki, and S. Arima, *J. Biochem.* (*Japan*), 1978, **83**, 371.

homologous with the active site cleft of lysozyme. The denaturations induced by guanidinium chloride, acid, and heat were followed by difference spectroscopy. The results of the studies are consistent with the proposed conformations of α-lactalbumin based on its similarity to hen egg-white lysozyme, and support the validity of the proposed molecular models.

Effects of deuteriation on the Raman spectrum of an L-tryptophan residue of bovine α-lactalbumin have been studied.[525] Deductions were made concerning the chance of exposure to the solvent by the L-tryptophan residues.

It has been shown that α-lactalbumin undergoes a three-state denaturation involving a helical intermediate state, on treatment with guanidinium chloride.[526] The unfolding of the molecule and the characteristics of the intermediate state were examined by means of c.d. spectroscopy, difference spectra, and pH-jump measurements to investigate the temperature dependence and kinetic properties of the unfolding and refolding, the pH dependence of the transition between the intermediate and fully unfolded states, and the effect of disulphide bond reduction on the stabilization of the intermediate. The mechanism of folding of the protein was discussed, and the folding process from the fully unfolded to the native state is apparently divided into at least three main steps: (1) the formation of incipient helical structures dictated by local interactions; (2) the packing of the helical segments accompanied by hydrophobic interactions; (3) the final stabilization by the electrostatic interactions.

α-Lactalbumin has been purified to homogeneity from rat milk.[527] In contrast to other α-lactalbumins, it is a glycoprotein and exhibits an abnormally high molecular weight when obtained by gel filtration or electrophoresis in sodium dodecyl sulphate. The molecular weight by sedimentation equilibrium is 1.54×10^4; the molecules have a low α-helical content. At least three major charge forms, all containing carbohydrate and active in the lactose synthase reaction, were demonstrated.

Lactose Synthases.—Lactose synthase has been immobilized with retention of enzymic activity by reaction as D-galactosyltransferase with an activated mixed disulphide: macroporous agarose-glutathione-2(5-nitropyridyl)-disulphide.[528]

NAD$^+$ Nucleosidase.—NAD$^+$ Nucleosidase (NAD$^+$ glycohydrolases) of isolated fractions of microsomes and plasma membranes obtained from rat liver has been solubilized with lipase, and purified by ion-exchange chromatography, gel filtration, and isoelectric focusing; the enzymes had similar molecular weights, pH-activity profiles, and K_m values.[529] Marked charge heterogeneity was observed for the microsomal enzyme, but not the plasma membrane enzyme, on isoelectric focusing (*ca.* p*I* 8.0 and 6.7 respectively). Treatment of either enzyme with neuraminidase resulted in a single peak of activity (p*I* 9.2). The microsomal enzyme was shown to contain hexose, 2-amino-2-deoxyhexose,

[525] H. Takesada, M. Nakanishi, A. Y. Hirakawa, and M. Tsuboi, *Biopolymers*, 1976, **15**, 1929.
[526] K. Kuwajima, *J. Mol. Biol.*, 1977, **114**, 241.
[527] R. C. Brown, W. W. Fish, B. G. Hudson, and K. E. Ebner, *Biochim. Biophys. Acta*, 1977, **491**, 82.
[528] J. Grunwald and L. J. Berliner, *Biochim. Biophys. Acta*, 1978, **523**, 53.

and neuraminic acid. It was suggested that differences in carbohydrate composition might be important in determining the subcellular distribution of the NAD glycohydrolases.

Pectinesterases.—A rapid assay for inhibitors of pectinesterase has been developed to facilitate large-scale screening.[530] Pectin and an indicator are incorporated into an agar plate, and the hydrolysis of pectin to pectic acid causes a colour change in the indicator on account of the lowering of the pH. The area of colour change is proportional to the amount of pectinesterase activity present. The pectic enzymes of six strains of pigmented, pectolytic clostridia isolated from potatoes have been examined and compared with those of strains of *Clostridium felsineum* and *C. aurantibutyricum* using cup-plate assays.[470] The *C. felsineum* strains only were found to be devoid of pectinesterase activity.

The taxonomy of the brown rot fungi (*Monilinia* species) has been related to their extracellular cell wall-degrading enzymes.[103] Patterns of pectinesterase secreted by *M. fructigena*, *M. fructicola*, and *M. laxa* were studied by column isoelectric focusing.

Two forms of pectinesterase, apparently similar in molecular weight but differing in charge and response to cation concentration, have been demonstrated in limes and oranges.[531] In the orange, one form was located in the peel while the other was located within the segment covers and juice sacs.

Three multiple forms of pectinesterase have been found in extracts and purified products from carrot.[532] The enzymes (mol. wts. 2.7×10^4) were found to possess properties similar to those of tomato pectinesterase.

Index of Enzymes Referred to in Chapter 6[a]

Trivial name and name used in this volume	*Systematic name*	*E.C. No.*	*Page*
endo-α-D-2-Acetamido-2-deoxygalactanase	D-Galactosyl-*N*-acetamidodeoxy-α-D-galactoside D-galactosyl-*N*-acetamidodeoxy-D-galactohydrolase	3.2.1.97	423
2-Acetamido-2-deoxy-D-galactose 6-sulphate sulphatase			475
α-D-2-Acetamido-2-deoxy-galactosidase	2-Acetamido-2-deoxy-α-D-galactoside acetamido-deoxygalactohydrolase	3.2.1.49	378
β-D-2-Acetamido-2-deoxy-galactosidase	2-Acetamido-2-deoxy-β-D-galactoside acetamidodeoxy-galactohydrolase	3.2.1.53	378
endo-β-D-2-Acetamido-2-deoxyglucanase	D-Mannosyl-glycoprotein 1,4-*N*-acetamidodeoxy-β-D-glycohydrolase (see also Chitinase, EC 3.2.1.14)	3.2.1.96	423

[529] R. P. Diaugustine, T. Abe, and P. Voyter, *Biochim. Biophys. Acta*, 1978, **526**, 518.
[530] R. E. Zimmerman, *Analyt. Biochem.*, 1978, **85**, 219.
[531] R. Evans and D. McHale, *Phytochemistry*, 1978, **17**, 1073.
[532] O. Markovic, *Experientia*, 1978, **34**, 561.

[a] See Introduction (Part II, Chapter 1, p. 215).

Trivial name and name used in this volume	*Systematic name*	*E.C. No.*	*Page*
D-Galactose oxidase	D-Galactose: oxygen 6-oxidoreductase	1.1.3.9	470
α-D-Galactosidase	α-D-Galactoside galactohydrolase	3.2.1.22	392
β-D-Galactosidase	β-D-Galactoside galactohydrolase	3.2.1.23	392
endo-(1 → 3)-α-D-Glucanase	1,3-(1,3;1,4)-α-D-Glucan 3-glucanohydrolase	3.2.1.59	445
endo-(1 → 3)-β-D-Glucanase	1,3-β-D-Glucan glucanohydrolase	3.2.1.39	446
endo-(1 → 4)-β-D-Glucanase		(3.2.1.6?)	449
endo-(1 → 6)-α-D-Glucanase			449
exo-(1 → 6)-α-D-Glucanase	1,6-α-D-Glucan glucohydrolase	3.2.1.70	449
endo-(1 → 6)-β-D-Glucanase	1,6-β-D-Glucan glucanohydrolase	3.2.1.75	450
exo-(1 → 3)-β-D-Glucanase	1,3-β-D-Glucan glucohydrolase	3.2.1.58	448
D-Glucanases (miscellaneous)			450
Glucoamylase	1,4-α-D-Glucan glucohydrolase	3.2.1.3.	450
β-D-Glucocerebrosidase			403
D-Glucolipid oriented D-glucosicases			403
D-Glucose isomerase	*see* D-Xylose isomerase E.C. 5.3.1.5.	[5.3.1.18]	469
D-Glucose oxidase	β-D-Glucose:oxygen 1-oxidereductase	1.1.3.4	470
α-D-Glucosidase	α-D-Glucoside glucohydrolase	3.2.1.20	403
β-D-Glucosidase	β-D-Glucoside glucohydrolase	3.2.1.21	403
β-D-Glucuronidase	β-D-Glucuronide glucuronosohydrolase	3.2.1.31	411
Glycopeptide-linkage hydrolases			470
α-L-Guluronan lyase			452
Heparin hydrolase		[3.2.1.19]	452
Hyaluronate lyase	Hyaluronate lyase	4.2.2.1	453
Hyaluronidase	Hyaluronate 4-glycanohydrolase		453
L-Iduronic acid 2-sulphate sulphatase			476
α-L-Iduronidase	α-L-Iduronide iduronohydrolase	3.2.1.76	414
Inulinase	2,1-β-D-Fructan fructanohydrolase	3.2.1.7	454
Isoamylase	Glycogen 6-glucanohydrolase	3.2.1.68	455
Isopullulanase	Pullulan 4-glucanohydrolase	3.2.1.57	455
α-Lactalbumin			478
Lactose synthase	UDPgalactose: D-glucose 4-β-D-galactosyltransferase	2.4.1.22	479
Lactosylceramidase			392
Limit dextrinase		(3.2.1.10, and 3.2.1.41?)	455
Lysozyme	Mucopeptide-*N*-acetyl-muramoylhydrolase	3.2.1.17	455
Maltase	*see* α-D-Glucosidase		403
D-Mannanases (miscellaneous)		[3.2.1.22]	464
α-D-Mannosidase	α-D-Mannoside mannohydrolase	3.2.1.24	414
β-D-Mannosidase	β-D-Mannoside mannohydrolase	3.2.1.25	414
NAD^+ Nucleosidase	NAD^+ Glycohydrolase	3.2.2.5	479
β-Neoagarotetraose hydrolase			465
Neuraminidase	Acylneuraminyl hydrolase	3.2.1.18	418
Nuclease			473
Oligo-(1 → 6)-D-glucosidase	Dextrin 6-α-D-glucanohydrolase	3.2.1.10	465

Trivial name and name used in this volume	*Systematic name*	*E.C. No.*	*Page*
Pectate lyase	Poly(1,4-α-D-galacturonide) lyase	4.2.2.2	465
Pectin lyase	Poly(methoxygalacturonide) lyase	4.2.2.10	465
Pectinesterase	Pectin pectylhydrolase	3.1.1.11	480
Peptidases			472
Phosphatases			473
Poly-D-galacturonase	Poly(1,4-α-D-galacturonide) glycanohydrolase	3.2.1.15	466
exo-Poly-D-galacturonase	Poly(1,4-α-D-galacturonide) galacturonohydrolase	3.2.1.67	466
Poly-D-Galacturonate lyase			465
Proteinases			472
Pullulanase	Pullulan 6-glucanohydrolase	3.2.1.41	467
Ribonuclease			473
Serine carboxypeptidase	Peptidyl-L-amino-acid hydrolase	3.4.16.1	472
Sialidase[b]			418
Sucrose α-D-glucohydrolase	Sucrose α-D-glucohydrolase	3.2.1.48	468
Sulphamidase			475
Sulphatase			475
β-D-Thioglucosidase	Thioglucoside glucohydrolase	3.2.3.1	422
αα-Trehalase	αα-Trehalose glucohydrolase	3.2.1.28	468
endo-(1 → 4)-β-D-Xylanase	1,4-β-D-Xylan xylanohydrolase	3.2.1.8	469
exo-(1 → 4)-β-D-Xylanase	1,4-β-D-Xylan xylohydrolase	3.2.1.37	469
Xylanase (miscellaneous)			469
D-Xylose isomerase	D-Xylose ketol-isomerase	5.3.1.5	469
α-D-Xylosidase			422
β-D-Xylosidase			422

[b] Sialidase is a term generally used to include neuraminidase and unspecified acylneuraminyl hydrolases. Since there is little evidence for the existence of acylneuraminyl hydrolases that do not cleave 5-acetamido-3,5-dideoxy-D-*glycero*-D-*galacto*-2-nonulosonyl residues, use of the term neuraminidase is preferred.

7

Glycolipids and Gangliosides

BY I. M. MORRISON

1 Introduction and General Methods

During the year a number of book chapters and review articles on various aspects of glycolipids have appeared. Such chapters have reviewed the enzymology of the sphingolipidoses,[1] cell surface carbohydrates and biological recognition,[2] and the biosynthesis and assembly of bacterial cell walls.[3] A review of fucolipids has discussed the chemical nature of these compounds, the methods and techniques for their isolation and identification, and the biosynthesis and biology of this group of glycolipids.[4] A review of sphingolipidoses has considered chemical, biochemical, and therapeutic studies on these disorders.[5] Some of the past accomplishments in the treatment of these disorders have been considered as well as the future directions in research. On the analytical side, the application of mass spectrometry to the structural analysis of glycosphingolipids has been successful in determining the positions of linkage in the sugar component of sphingolipids.[6] The trimethylsilyl ethers of glycosphingolipids have been shown to be the most suitable derivatives for the separation and characterization of these compounds by g.c.–m.s.[7]

It has been recommended that IR45 (OH^-) resin be used instead of silver carbonate for the neutralization of methanolysates of glycoproteins and glycolipids prior to the derivatization and analysis of the carbohydrates by g.l.c.[8] Glycolipids have been separated from neutral lipids and phospholipids by chromatography on phenylboronic acid covalently bound to a polystyrene matrix.[9] A new chromatographic approach to the resolution of individual gangliosides, called ganglioside mapping, has been developed.[10] The method involves an initial separation of the ganglioside mixture on a DEAE–Sepharose column followed by t.l.c. on silica gel of the successive fractions from the column. Between 30 and 40 individual gangliosides could be separated by this method.

[1] 'Laboratory and Research Methods in Biology and Medicine', Vol. 1, ed. R. H. Glew and S. P. Peters, Alan R. Liss Inc., New York, 1977.
[2] 'Progress in Clinical and Biological Research', Vol. 23, ed. V. T. Marchesi, V. Ginsburg, P. W. Robbins, and C. F. Fox, Alan R. Liss, Inc., New York, 1978.
[3] J. M. Ghuysen, in 'Synthesis, Assembly, and Turnover of Cell Surface Components', Cell Surface Reviews Vol. 4, ed. G. Poste and G. L. Nicolson, Elsevier North-Holland, New York, 1977, p. 463.
[4] J. M. McKibbin, *J. Lipid Res.*, 1978, **19**, 131.
[5] P. G. Pentchev and J. A. Barranger, *J. Lipid Res.*, 1978, **19**, 401.
[6] H. Egge, *Chem. and Phys. Lipids*, 1978, **21**, 349.
[7] C. J. W. Brooks, C. G. Edmonds, S. J. Gaskell, and A. G. Smith, *Chem. and Phys. Lipids*, 1978, **21**, 403.
[8] S. J. Rickert and C. C. Sweeley, *J. Chromatog.*, 1978, **147**, 317.
[9] K. Krohn, K. Eberlein, and G. Gercken, *J. Chromatog.*, 1978, **153**, 550.
[10] M. Iwamori and Y. Nagai, *Biochim. Biophys. Acta*, 1978, **528**, 257.

Neutral glycosphingolipids have been separated by h.p.l.c. of their perbenzoylated derivatives and detected by their absorbance at 230 nm.[11] Using a Zipak column and a dioxane–hexane gradient, complete separations were achieved when up to four sugar residues were present in the glycolipid. Glycerylphosphorylinositol mannosides were separated by t.l.c. on silica gel using propan-2-ol : ammonia (1 : 1) as eluant.[12] H.p.t.l.c. of brain gangliosides has proved so sensitive that 90 pmol of lipid-bound sialic acid could be detected by direct densitometric scanning of resorcinol-positive spots on the plate.[13] Different amounts of individual gangliosides were found in a wide variety of species.

Field desorption mass spectrometry of oligosaccharides and glycolipids by the cationization method has been studied.[14] A synthetic cord factor, dimycoloyl-trehalose gave an intense peak at the expected m/e value. Every carbon nucleus in the bovine brain ganglioside G_{M1} has been assigned by ^{13}C n.m.r. spectroscopy.[15] Confirmation of these assignments has been gained from a comparison with authentic standards. It was proposed that the carboxy and side chain glyceryl groups of the sialic acid residues, along with additional ligands donated by the 2-acetamido-2-deoxy-β-D-galactosyl and β-D-galactopyranosyl residues, were intimately involved in cation binding. The use of radiolabelled acetic anhydride for the acetylation of glycosphingolipids has meant that 10^{-9} g of these lipids could be detected and tentatively identified.[16] Tritium has been specifically introduced into the ceramide portion of gangliosides from potassium borotritide using palladium as the catalyst.[17] The yield of tritiated compounds in virtually quantitative and no hydrolysis of even the very labile bonds in trisialosyl-gangliotetraosylceramide G_{T1} was observed.

It has been found that dialysis bags require to be prewashed with sodium carbonate and H_4 edta to reduce the contamination of glycolipids during dialysis.[18] Mixtures of trichloroacetic and phosphotungstic acids did not cause the production of artifacts or hydrolysis of gangliosides as previously reported; the odd effects could be removed by treatment with H_4 edta followed by dialysis.[19]

Fluorescence polarization of 1,6-diphenyl-1,3,5-hexatriene has been used to demonstrate the interaction between vesicles containing gangliosides and the lectin from *Limulus polyphemus*.[20]

A new class of model glycolipids has been synthesized, the structures of which are shown in (1).[21] Sucrose octa-esters have been synthesized using as acidic groups the even chain fatty acids 8—18 C as well as the unsaturated acids

[11] M. D. Ullman and R. H. McCluer, *J. Lipid Res.*, 1978, **19**, 910.
[12] G. H. Khuller and B. Banerjee, *J. Chromatog.*, 1978, **150**, 518.
[13] S. Ando, N. C. Chang, and R. K. Yu, *Analyt. Biochem.*, 1978, **89**, 437.
[14] J. C. Prome and G. Puzo, *Israel J. Chem.*, 1978, **17**, 172.
[15] L. O. Sillerud, J. H. Prestegard, R. K. Yu, D. E. Schafer, and W. H. Konigsberg, *Biochemistry*, 1978, **17**, 2619.
[16] J. Blomberg, *Analyt. Biochem.*, 1978, **88**, 302.
[17] G. Schwarzmann, *Biochim. Biophys. Acta*, 1978, **529**, 106.
[18] M. T. Coleman and A. J. Yates, *J. Chromatog.*, 1978, **166**, 611.
[19] T. N. Seyfried, E. J. Webber, and R. K. Yu, *Lipids*, 1977, **12**, 979.
[20] A.-C. Roche, R. Maget-Dana, A. Obrenovitch, K. Hildenbrand, C. Nicolau, and M. Monsigny, *F.E.B.S. Letters*, 1978, **93**, 91.
[21] T. J. Williams, N. R. Plessas, and I. J. Goldstein, *Carbohydrate Res.*, 1978, **67**, C1.

$$\left.\begin{array}{l}\alpha\text{-D-Gal}p\ 1\rightarrow\\ \beta\text{-D-Gal}p\ 1\rightarrow\\ \alpha\text{-D-Glc}p\ 1\rightarrow\\ \beta\text{-D-Glc}p\ 1\rightarrow\end{array}\right\}\ \mathrm{OCH_2(CHOH)_4\text{-}\overset{\displaystyle O}{\overset{\|}{C}}\text{-}NH\text{-}(CH_2)_{\mathit{n}}Me}$$

n = 7, 9, 11, 13, or 15

(1)

oleic, elaidic, and linoleic.[22] All of the esters gave stable phases similar in their *X*-ray diffraction pattern to that given by the α-phase of the corresponding triglyceride. No β-phase was obtained from these esters either from the melt or by solvent evaporation. Poly {1-[10-(1-*O*-β-D-glucopyranosyl-oxycarbonyl) decyloxycarbonyl]-1-methylethylene} has been synthesized by conventional methods for use as a polymeric glycolipid analogue.[23]

Mono-, di-, tri-, and tetrahexosylceramides have force-area isotherms with similar limiting molecular areas when measured at the same pH value and in salt solutions of the same molar concentration.[24] Increasing numbers of negatively charged sialosyl residues in sialogangliosides induced a progressive increase in the liquid expander characteristics of the lipid films and the limiting area occupied per molecule due to electrostatic repulsion. Attempts to calculate the dipolar contribution to the surface potential by each carbohydrate residue suggested that the non-reducing terminal sialosyl groups in the di- and trisialogangliosides contributed to the potential with a vertical dipole moment opposite to that of the interior sialosyl group. Mixed monolayers of any of the different gangliosides showed immiscible behaviour at high surface pressures probably due to steric hindrance and electrostatic repulsion between the polar head groups.[25] Di- and trisialogangliosides could be differentiated from neutral sphingolipids and monosialogangliosides by their interactions with phospholipids which were correlated to the perpendicular electrical field at the interface contributed by the carbohydrate groups. Phosphocholine head groups were important in establishing an interaction with di- and trisialogangliosides.

Evidence has been presented to show that the hydrophilic portion of ganglioside G_{T1}, when included in a lipid bilayer, can react as a specific acceptor for thyrotropin.[26] At low surface pressures (15 dynes cm^{-1}), concanavalin A is able to penetrate dipalmitoylphosphatidylcholine monolayers but the penetration is inhibited by the presence of methyl α-D-mannopyranoside in the sub-phase.[27] At higher surface pressures, the lectin can interact with mono-D-glucosyldiglyceride or di-glucosylglyceride and also with synthetic glycolipids provided they contain two or three α-1,4-linked D-glucose residues in the head group, but it will not interact with gangliosides containing *N*-acetylneuraminic acid. Cholera toxin will react with the sugar residues in the G_{M1} molecules

22 R. J. Jandacek and M. R. Webb, *Chem. and Phys. Lipids*, 1978, **22**, 163.
23 K. Nishio, T. Nakaya, and M. Imoto, *Makromol. Chem.*, 1978, **179**, 1117.
24 B. Maggio, F. A. Cumar, and R. Caputto, *Biochem. J.*, 1978, **171**, 559.
25 B. Maggio, F. A. Cumar, and R. Caputto, *Biochem. J.*, 1978, **175**, 1113.
26 A. Poss, M. Deelers, and J. M. Ruysschaert, *F.E.B.S. Letters*, 1978, **86**, 160.
27 B. D. Read, R. A. Demel, H. Wiegandt, and L. L. M. van Deenen, *Biochim. Biophys. Acta*, 1977, **470**, 325.

when these are present at the membrane surface of lipid bilayers.[28] It is suggested that a complex is formed which results in the appearance of ion-conducting channels.

2 Animal Glycolipids and Gangliosides

Sulpho-D-galactosylglycerolipid was barely detectable in brain tissue of rats up to 10 days of age but accumulated rapidly during the next 15 days and then remained at a fairly constant level.[29] Most of the sulpho-D-galactosylglycerolipid is in the myelin fraction and is present in a 1 : 1 mixture of the diacyl and alkylacyl forms. This lipid is synthesized from D-galactosyldiacylglycerol and adenosine 3′-phosphate 5′-sulphatophosphate.[30]

A 40% reduction in the levels of mono-D-galactosyldiacylglycerides in brain tissue occurred when rats were made hypothyroid at birth.[31] Tri-iodothyronine could restore the D-galactosyltransferase activity and this hormone, when added neonatally, increased the level of mono-D-galactosyldiacylglycerol in the brain tissue. D-Gluco- and D-galacto-cerebrosides, sulphatides, and sphingomyelins from rat brain have been separated on borate-coated silicic acid.[32] The glycocerebroside accounted for 54.1 nmol per g of fresh brain tissue. Sulphated glyceroglycolipids isolated from rat brain were found to consist of a mixture of the 3′-sulphate derivatives of 1,2-di-*O*-acyl- and 1-*O*-alkyl-2-*O*-acyl-3-*O*-β-D-galactosylglycerol.[33] Similar sulpholipids were isolated from fish brain but not from human brain. The disialoganglioside G_{D1a} was the largest component of the total ganglioside fraction isolated from rat brain subcellular particles.[34] The molar percentage of G_{D1a}, relative to G_{M1}, G_{D1b}, and G_{T1} increased with increasing age and the level persisted in mature rat brain tissue when the levels of the other gangliosides were declining. The sialosylation activity in neuronal perikarya fractions of rat brain were 3—12 fold higher than that of the whole homogenate when Tay–Sachs ganglioside, lactosylceramide, or haematoside were used as acceptors.[35] The sialosylation activity in the synaptosomal fractions was negligible.

A higher concentration of ganglioside was found in the cerebrum and brain stem tissue of mice susceptible to audiogenic seizures than was found in resistant strains, but a lower concentration was found in the cerebellum.[36] Of the individual gangliosides, a higher concentration of G_{M1} was found in all three regions of the brain of susceptible mice.

The G_{M1} ganglioside β-D-galactosidases from feline liver and brain have been purified and characterized.[37] The preparation from brain tissue contained several isoenzymes whilst that from liver appeared to be homogeneous. The

[28] M. T. Tosteson and D. C. Tosteson, *Nature*, 1978, **275**, 142.
[29] J. Pieringer, G. Rao, P. Mandel, and R. A. Pieringer, *Biochem. J.*, 1977, **166**, 421.
[30] G. S. Rao, L. N. Norcia, J. Pieringer, and R. A. Pieringer, *Biochem. J.*, 1977, **166**, 429.
[31] T. J. Flynn, D. S. Deshmukh, and R. A. Pieringer, *J. Biol. Chem.*, 1977, **252**, 5864.
[32] E. K. Korniat and L. Hof, *J. Neurochem.*, 1978, **30**, 557.
[33] I. Ishizuka, M. Inomata, K. Ueno, and T. Yamakawa, *J. Biol. Chem.*, 1978, **253**, 898.
[34] H. K. M. Yosuf and J. W. T. Dickerson, *Biochem. J.*, 1978, **174**, 655.
[35] H. J. F. Maccioni, S. S. Defilipo, C. A. Landa, and R. Caputto, *Biochem. J.*, 1978, **174**, 673.
[36] T. N. Seyfried, G. H. Glaser, and R. K. Yu, *J. Neurochem.*, 1978, **31**, 21.
[37] J. K. Anderson, J. E. Mole, and H. J. Baker, *Biochemistry*, 1978, **17**, 467.

degradation of the lipophilic ganglioside G_{D1a} by membrane-bound neuraminidase is faster than when water-soluble substrates are used since G_{D1a} is taken up by cell membranes of calf brain.[38] A model presented suggests that the activity of membrane-bound neuraminidase on gangliosides of brain membranes is regulated by the viscosity of these membranes and their monosialoganglioside content. D-Glucosylphosphoryldolichol has been shown to act as a D-glucosyl donor in the biosynthesis of oligosaccharide lipids in calf brain membranes.[39]

The structure of a novel L-fucose-containing ganglioside from pig cerebellum has been established (2).[40] 9-*O*-Acetyl-*N*-acetylneuraminic acid accounts for up to 20% of the brain ganglioside fractions from man, cow, horse, pig, sheep, cat, rabbit, rat, chicken, and codfish.[41] In the pig and cow, traces of the *N*-glycolyl derivative were also found. The novel trisialoganglioside, G_{T1a} (3), has been isolated and characterized and is responsible for 0.6% of the total ganglioside fraction of human brain.[42]

```
β-D-Galp-(1 → 3)-β-D-GalpNAc
       2                1
       ↑                ↓
       1                4
α-L-Fucp           β-D-Galp-(1 → 4)-β-D-Glcp-(1 → 1)-ceramide
                        3
                        ↑
                        2
α-NeupNAc-(2 → 8)-α-NeupNAc
```

(2)

```
α-NeupNAc-(2 → 8)-α-NeupNAc-(2 → 3)-β-D-Galp-(1 → 3)-β-D-GalpNAc-(1 → 4)-
                                         β-D-Galp-(1 → 4)-β-D-Glcp-(1 → 1)-ceramide
                                              3
                                              ↑
                                              2
                                         α-NeupNAc
```

(3)

The D-glucocerebrosidase from the spleens of patients with Gaucher disease had a lower specific activity than that from the spleens of normal humans although the enzymes appeared to be similar from other parameters.[43] The data strongly suggested that the genetic basis of Gaucher disease is a structural mutation of D-glucocerebrosidase. Gel electrophoretic data for the enzymes suggested that there are other differences between the two types of enzyme. The heat-stable D-glucocerebrosidase activator substance in Gaucher spleen is

[38] K. Sandhoff and B. Pallmann, *Proc. Nat. Acad. Sci., U.S.A.*, 1978, **75**, 122.
[39] C. J. Waechter and M. G. Scher, *Arch. Biochem. Biophys.* 1978, **188**, 385.
[40] S. Sonnino, R. G. Hidoni, G. Galli, and G. Tetamanti, *J. Neurochem.*, 1978, **31**, 947.
[41] J. Haverkamp, R. W. Veh, M. Sander, R. Schauer, J. P. Kamerling, and J. F. G. Vliegenthart, *Z. physiol. Chem.*, 1977, **358**, 1609.
[42] S. Ando and R. K. Yu, *J. Biol. Chem.*, 1977, **252**, 6247.
[43] P. G. Pentchev, R. O. Brady, H. E. Blair, D. E. Britton, and S. H. Sorrell, *Proc. Nat. Acad. Sci., U.S.A.*, 1978, **75**, 3970.

also found in the storage deposits contained in the lysosomes of the Gaucher cell.[44]

$$\beta\text{-D-Gal}p\text{-}(1 \rightarrow 4)\text{-}\beta\text{-D-Glc}pN\text{Ac-}(1 \rightarrow 2)\text{-}\alpha\text{-D-Man}p\text{-}(1 \rightarrow 6)\text{-}\beta\text{-D-Man}p\text{-}(1 \rightarrow 4)\text{-D-Glc}pN\text{Ac}$$

$$\begin{array}{c} 3 \\ \uparrow \\ 1 \\ \text{R} \end{array}$$

(4) R = H
(5) R = β-D-Gal*p*-(1 → 4)-β-D-Glc*pN*Ac-(1 → 2)-α-D-Man*p*

Two structurally related oligosaccharides, a pentasaccharide (4) and an octasaccharide (5), have been isolated from the urine of a patient with juvenile G_{M1} gangliosidosis.[45] Both are probably degradation products derived from the core of the carbohydrate chains of glycoproteins. Extracts from human kidneys, devoid of β-D-2-acetamido-2-deoxyhexosidase activity, were able to stimulate the β-D-2-acetamido-2-deoxyhexosidase A-catalysed degradation of ganglioside G_{M2} and glycolipid G_{A2}.[46] The stimulant factor was also found in extracts from the kidneys of patients with infantile G_{M2} gangliosidosis, variants B and O, but was absent from the kidneys of patients with the AB-variant. The β-D-2-acetamido-2-deoxyhexosidase A from patients with the AB-variant has been purified and characterized and it was shown that the defect is not at the genetic level in the production of either the α- or the β-chains of the enzyme.[47] The structures of minor oligosaccharides isolated from the liver of a patient with G_{M2} gangliosidosis, variant O (Sandhoff–Jatzkewitz disease) have been established by ^{13}C n.m.r. spectroscopy.

The ceramide trihexoside (6) accumulating in Fabry's disease has been synthesized by a conventional Koenigs–Knorr procedure.[49] A novel ganglioside, disialosylparagloboside (7), has been isolated from human kidneys and its structure determined.[50]

α-D-Gal*p*-(1 → 4)-β-D-Gal*p*-(1 → 4)-β-D-Glc*p*-(1 → 1)-ceramide

(6)

α-Neu*pN*Ac-(2 → 8)-α-Neu*pN*Ac-(2 → 3)-β-D-Gal*p*-(1 → 4)-β-D-Glc*pN*Ac-(1 → 3)-
β-D-Gal*p*-(1 → 4)-β-D-Glc*p*-(1 → 1)-ceramide

(7)

The genetics and biochemistry of some human blood-groups have been comprehensively reviewed.[51] The structures of several neutral glycolipids and gangliosides in bovine erythrocyte membranes have been established by the results of acid hydrolysis, methylation, periodate and chromium(VI) oxide

44 Y.-B. Chiao, J. P. Chambers, R. H. Glew, R. E. Lee, and D. A. Wenger, *Arch. Biochem. Biophys.*, 1978, **186**, 42.
45 A. Lundblad, S. Sjöblad, and S. Svensson, *Arch. Biochem. Biophys.*, 1978, **188**, 130.
46 E. Conzelmann and K. Sandhoff, *Proc. Nat. Acad. Sci., U.S.A.*, 1978, **75**, 3979.
47 E. Conzelmann, K. Sandhoff, H. Nehrkorn, B. Geiger, and R. Arnon, *European J. Biochem.*, 1978, **84**, 27.
48 N. M. K. N. Y. Kin and L. S. Wolfe, *Carbohydrate Res.*, 1978, **67**, 522.
49 D. Shapiro and A. J. Acher, *Chem. and Phys. Lipids*, 1978, **22**, 197.
50 H. Rauvala, T. Krusius, and J. Finne, *Biochim. Biophys. Acta*, 1978, **531**, 266.

oxidation analyses.[51] The two major glycolipids were characterized as lactosylceramide and (8); the two major gangliosides had the structures (9) and (10). Other workers have isolated even more complex glycolipids (11) and (12) from the same type of tissue.[52] A number of other lipids, of minor importance, were found in both of these studies.

α-D-Gal*p*-(1 → 3)-β-D-Gal*p*-(1 → 4)-β-D-Glc*p*NAc-(1 → 3)-β-D-Gal*p*-(1 → 4)-
β-D-Glc*p*-(1 → 1)-ceramide
(8)

Neu*p*NGl-(2 → 3)-β-D-Gal*p*-(1 → 4)-β-D-Glc*p*-(1 → 1)-ceramide
(9)

Neu*p*NGl-(2 → 3)-β-D-Gal*p*-(1 → 4)-β-D-Glc*p*NAc-(1 → 3)-β-D-Gal*p*-(1 → 4)-
β-D-Glc*p*-(1 → 1)-ceramide
(10)

α-Neu*p*NR-(2 → 3)-β-D-Gal*p*-(1 → 4)-β-D-Glc*p*NAc-(1 → 3)-β-D-Gal*p*-(1 → 4)-
β-D-Glc*p*NAc-(1 → 3)-β-D-Gal*p*-(1 → 4)-β-D-Glc*p*-(1 → 1)-ceramide

(11) R = Acetyl
(12) R = Glycolyl

Ten new types of gangliosides, in addition to haematoside and paragloboside, have been isolated from human erythrocyte membranes.[54] Very strong blood-group Ii and moderate blood-group H activities were demonstrated in some of the ganglioside fractions. The fraction with the strongest blood-group I activity behaved like sialosyldeca- to sialosyldodeca-glycosylceramides on t.l.c. Certain of the blood-group I and most of the i determinants were in partially or completely cryptic forms and could be unmasked by treatment with neuraminidase. Thus the I- and i-antigens, which are known to occur on the internal structures of blood-group ABH-active glycoproteins, also occur in interior chains of gangliosides from erythrocyte membranes.

A blood-group A-active glycolipid, which contains 23 sugar residues, has been isolated from human erythrocyte membranes.[55] As a result of Smith degradation and methylation analysis, the structure (13) has been proposed for the non-reducing end of the oligosaccharide chain. A method for the determination of glycosphingolipids in the erythrocyte membranes of various mammalian species has been developed in which, after purification, the sphingosine has been released by acid hydrolysis and estimated by the formation of a complex with Methyl Orange.[56]

Nine separate blood-group type glycolipids, some previously unknown, were isolated from feline small intestine; their structures were tentatively assigned

[51] W. M. Watkins, *Proc. Roy. Soc.*, 1978, **202**, 31.
[52] K. Uemura, M. Yuzawa, and T. Taketomi, *J. Biochem.* (*Japan*), 1978, **83**, 463.
[53] J. L. Chien, S. C. Li, R. A. Laine, and Y. T. Li, *J. Biol. Chem.*, 1978, **253**, 4031.
[54] T. Feizi, R. A. Childs, S. I. Hakomori, and M. C. Powell, *Biochem. J.*, 1978, **173**, 245.
[55] A. Gardas, *European J. Biochem.*, 1978, **89**, 471.
[56] J. Oulevey, E. Bodden, and O. W. Thiele, *European J. Biochem.*, 1977, **79**, 265.

```
L-Fucp-(1 → 2)-D-Galp-(1 → 4)-D-GlcpNAc
                                    1
                                    ↓
                                    6
                              D-Galp-(1 → 4)-[D-GlcpNAc-(1 → 3)-D-Galp]7-
                                    3             (1 → 4)-D-Glcp-(1 → 1)-ceramide
                                    ↑
                                    1
L-Fucp-(1 → 2)-D-Galp-(1 → 4)-D-GlcpNAc
                      3
                      ↑
                      1
                 D-GalpNAc
```

(13)

by the use of mass fragmentography.[57] Two types of sequence, namely β-D-galactosyl-(1 → 4)-2-acetamido-2-deoxy-β-D-glucosyl-(1 → 6)-D-galactosyl-(1 → R_1) and β-D-galactosyl-(1 → 4)-2-acetamido-2-deoxy-β-D-glucosyl-(1 → 4)-2-acetamido-2-deoxy-D-glucosyl-(1 → R_2), have been identified in poly(glycosyl)ceramides with blood-group I-activity which have been isolated from human erythrocyte membranes.[58] The anti-I serum specificity was directed against the non-reducing end of the first type of chain.

A new group of complex glycosphingolipids has been isolated from hog gastric mucosa.[59] Four compounds, each containing 12—14 sugar residues but of similar composition, were isolated and were found to be fucolipids with blood-group (A + H) activities. The structures of these four compounds have been established by conventional methods (14)—(17).[60]

```
α-L-Fucp-(1 → 2)-β-D-Galp-(1 → 3/4)-β-D-GlcpNAc
                 2                   1
                 ↑                   ↓
                 1                   3
           α-D-GalpNAc         β-D-Galp-(1 → 4)-β-D-GlcpNAc-(1 → R
                                     6
                                     ↑
                                     1
α-L-Fucp-(1 → 2)-β-D-Galp-(1 → 3/4)-β-D-GlcpNAc
(14) R = 3)-β-D-Galp-(1 → 4)-β-D-Galp-(1 → 4)-β-D-Glcp-(1 → 1)-ceramide
(15) R = 4)-β-D-GlcpNAc-(1 → 3)-β-D-Galp-(1 → 4)-β-D-Glcp-(1 → 1)-ceramide
(16) R = 3)-β-D-Galp-(1 → 4)-β-D-Galp-(1 → 4)-β-D-Glcp-(1 → 1)-ceramide
             6
             ↑
             1
         β-D-GlcpNAc-(4 ← 1)-β-D-Galp
(17) R = 4)-β-D-GlcpNAc-(1 → 3)-β-D-Galp-(1 → 4)-β-D-Glcp-(1 → 1)-ceramide
                                     6
                                     ↑
                                     1
           β-D-Galp-(1 → 4)-β-D-GlcpNAc
```

[57] M. E. Breimer, G. C. Hansson, K.-A. Karlsson, H. Leffler, W. Pimlott, and B. E. Samuelsson, *F.E.B.S. Letters*, 1978, **89**, 42.
[58] E. Zdebska and J. Koscielak, *European J. Biochem.*, 1978, **91**, 517.
[59] A. Slomiany and B. L. Slomiany, *F.E.B.S. Letters*, 1978, **90**, 293.
[60] B. L. Slomiany and A. Slomiany, *European J. Biochem.*, 1978, **90**, 39.

Blood-group ABH-active antigens and glycolipids have been isolated from human saliva.[61] Since the activity was still present in delipidated samples but was absent from alkaline-degraded samples, it was concluded that the lipid portion was inactive in the ABH system. In all, eight glycolipids were characterized and each contained D-glucose, glyceryl ethers, and fatty acids. Blood-group ABH-specific macroglycolipids were isolated in high yield from a crude membrane glycoprotein fraction of human erythrocytes.[62] These compounds are glycosphingolipids and it was concluded that these compounds are the predominant ABH-specific components in human erythrocyte membranes. On average, these glycolipids had about 31 sugar residues in the carbohydrate fraction of the molecule.

Eight glycolipids, each containing 4—6 sugar residues per molecule, were isolated from human plasma of blood-group O Le^b secretors.[63] These comprised two blood-group Le^a-active ceramide pentasaccharides, three blood-group Le^b-active ceramide hexasaccharides, two different globosides, and a new ceramide tetrasaccharide containing residues of D-glucose, D-galactose, and 2-acetamido-2-deoxy-D-glucose (1 : 2 : 1). Glycolipids have been isolated from human plasma of blood-group A Le^a-active individuals.[64] Three different Le^a-active ceramide pentasaccharides were found and each contained residues of D-glucose, D-galactose, 2-acetamido-2-deoxy-D-glucose, and L-fucose (1 : 2 : 1 : 1). In addition, an Le^a-active ceramide hexasaccharide, two globosides, two haematosides, and a new 2-acetamido-2-deoxy-D-glucose-containing ceramide tetrasaccharide were characterized.

Meconium was found to be a rich source of epithelial blood-group glycolipids; about fifty times the concentration of fucolipid was found compared with that found in adult blood.[65]

The five major neutral glycosphingolipids of human plasma have been characterized and each has a structure which has been identified previously.[66] One of these glycolipids, D-glucosylceramide, and possibly a second, lactosyl ceramide, exchange between plasma and erythrocyte pools. The conclusion is drawn in terms of the relative roles of carbohydrate and lipid moieties of the glycosphingolipids in maintaining that association with erythrocyte membranes.

A comparison has been made of the glycosphingolipids of human lymphocytes.[67] Thymocytes and peripheral blood lymphocytes did not have the same distribution of the simpler lipids as tonsil lymphocytes and contained larger amounts of more complex glycosphingolipids. The peripheral blood lymphocytes contained more bound sialic acid than did either tonsil lymphocytes or thymocytes.

A comparative study of the glycolipid composition of plasma membranes from two types of rat ascites hepatomas and normal rat liver has been

[61] B. L. Slomiany and A. Slomiany, *European J. Biochem.*, 1978, **85**, 249.
[62] M. Dejter-Juszynski, N. Harpaz, H. M. Flowers, and N. Sharon, *European J. Biochem.*, 1978, **83**, 363.
[63] P. Hanfland, *European J. Biochem.*, 1978, **87**, 161.
[64] P. Hanfland, R.-G. Kladetzky, and H. Egli, *Chem. and Phys. Lipids*, 1978, **22**, 141.
[65] K. A. Karlsson and G. Larson, *F.E.B.S. Letters*, 1978, **87**, 283.
[66] J. Kościelak, W. Maśliński, J. Zieleński, E. Zdebska, T. Brudzyński, H. Miller-Podraza, and B. Cedergren, *Biochim. Biophys. Acta*, 1978, **530**, 385.
[67] K. E. Stein and D. M. Marcus, *Biochemistry*, 1977, **16**, 5285.

reported.[68] Normal liver membranes contained ceramide monohexose (CMH), ceramide dihexoside (CDH), and haematoside (G_{M3}). Island-type hepatoma membranes contained ceramide trihexoside (CTH) and globoside as well as CMH, CDH, and G_{M3}, whereas free-type hepatoma membranes were characterized by the presence of asialogangliosides but not G_{M3}. A blood-group H-active fucolipid was a major component of free-type hepatoma membranes. It was noted that an increase in the glycolipid concentration was accompanied by a decrease in cell adhesiveness. Further compositional studies have tentatively characterized some of the ceramide fractions.[69]

Seven glycero-D-glucolipids have been isolated from human saliva.[70] The carbohydrate chains are exclusively linear oligosaccharides formed from up to eight residues of D-glucose linked by α-(1 → 6) bonds.

A Lewis-precursor glycolipid has been isolated from an Le-negative individual.[71] This glycolipid, in the presence of GDP-L-fucose and L-fucosyltransferase, gave an Le(a-b-) precursor which could be transferred to give both Le^{a}- and Le^{b}-active glycolipids. Furthermore, the Le^{b}-active glycolipid could be synthesized from the Le^{a}-active one.

Two different glycolipid L-fucosyltransferases, involved in the biosynthesis of blood-group related glycosphingolipids, were detected in a membrane preparation isolated from a human neuroblastoma-derived clonal cell line.[72] One catalysed the transfer of L-fucose to neolactotetraosyl- or pentaosylceramide to form H-1 and B-1 glycolipids respectively. The second catalysed the transfer of L-fucose to lactotriaosylceramide to form a tetraosylceramide which was an intermediate in the synthesis of the Le^{a}-type glycolipid.

Monosialoganglioside haematoside, G_{M3}, a precursor of more complex gangliosides, has been isolated from human liver and characterized.[73] The haematoside gave twin bands on t.l.c. but each band had similar base and sugar compositions although there were striking differences in the fatty acid contents.

In studies on human anti-thrombin, β-D-glucosylceramide was found to complex firmly with anti-thrombin III.[74] It was suggested that the anti-thrombin might act as a carrier for glycolipids.

A D-mannosylphosphoryldolichol, synthesized by the outer membrane of mouse liver mitochondria, has been characterized.[75] The factors affecting the transfer of D-glucosyl and D-mannosyl residues to dolichylmonophosphate by cell-free preparations from both rat and pig liver have been determined.[76] The presence of GDP- and UDP-D-mannose markedly inhibited the transfer of D-glycosyl residues from UDP-D-glucose. It was suggested that two D-glucosyltransferases are present but that only one of these is inhibited. UDP-D-

[68] T. Taki, Y. Hirabayashi, Y. Suzuki, M. Matsumoto, and K. Kojima, *J. Biochem.* (*Japan*), 1978, **83**, 1517.
[69] Y. Hirabayashi, T. Taki, M. Matsumoto, and K. Kojima, *Biochim. Biophys. Acta*, 1978, **529**, 96.
[70] B. L. Slomiany, A. Slominay, and G. B. J. Glass, *European J. Biochem.*, 1978, **84**, 53.
[71] R. Prohaska, H. Schenkel-Brunner, and H. Tuppy, *European J. Biochem.*, 1978, **84**, 161.
[72] K. A. Presper, M. Basu, and S. Basu, *Proc. Nat. Acad. Sci., U.S.A.*, 1978, **75**, 289.
[73] T. N. Seyfried, S. Ando, and R. K. Yu, *J. Lipid Res.*, 1978, **19**, 538.
[74] I. Danishefsky, A. Zweben, and B. L. Slomiany, *J. Biol. Chem.*, 1978, **253**, 32.
[75] O. Gateau, R. Morelis, and P. Louisot, *Compt. rend.*, 1977, **285**, 1523.
[76] A. K. A. Kerr and F. W. Hemming, *European J. Biochem.*, 1978, **83**, 581.

glucose did not inhibit the transfer of D-mannosyl residues and it was suggested that the pool of dolichylmonophosphate used by the D-mannosyltransferase is not available to the D-glucosyltransferase. The incorporation of D-mannose into D-mannosylretinylphosphate, D-mannosyldolichylphosphate, and specific glycolipids of rat liver was stimulated two-fold by the presence of vitamin A.[77] The same vitamin had no effect on the incorporation of D-galactose into D-galactolipids or glycoproteins. A particulate fraction from rat liver catalysed the synthesis of trisialoganglioside from disialohaematoside.[78]

About 20% of the lipid-bound sialic acid in bovine mammary tissue was present in three gangliosides.[79] They had the structures, D-Gal*p* → D-Gal*p*NAc → Neu*p*NAc → D-Gal*p* → D-Glc*p* → ceramide, D-Gal*p* → D-Gal*p*NAc → Neu*p*NAc → Neu*p*NAc → D-Gal*p* → D-Glc*p* → ceramide, and D-Gal*p*NAc → Neu*p*NAc → Neu*p*NAc → D-Gal*p* → D-Glc*p* → ceramide. The major ganglioside from cultured cells of a rat mammary gland carcinoma cell line had the structure Neu*p*NAc → D-Gal*p* → D-Gal*p*NAc → Neu*p*NAc → D-Gal*p* → D-Glc*p* → ceramide.[80] In total, the ganglioside level was four times that of normal cells.

Four neutral glycosylceramides were found in the lipid fraction from bovine thyroid gland.[81] One of these contained only D-glucose residues, two others contained residues of D-glucose and D-galactose in the ratios 1 : 1 and 1 : 2, and the fourth contained residues of 2-acetamido-2-deoxy-D-glucose as well as D-glucose and D-galactose. Of the lipid-bound sialic acid in rabbit thymus 92% was present in the di- and trisialoganglioside fraction.[82] From this fraction the G_{D3} ganglioside (18), which has a novel disialosyl sequence, was isolated. The gangliosides from human thyroid tissue were found to have a high affinity for thyrotropin and were thus considered to have a potential role in the regulation of the function of that gland.[83]

α-Neu*p*NGl-(2 → 8)-α-Neu*p*NAc-(2 → 3)-β-D-Gal*p*-(1 → 4)-β-D-Glc*p*-(1 → 1)-ceramide

(18)

Lactosylceramide is a precursor of the gangliosides G_{D1a} and G_{T1b} in the bovine thyroid gland.[84] In studies on the labelling of lipid- and protein-linked oligosaccharides in pig thyroid rough microsomes, the D-mannosylated residues in the lipid-linked saccharides were characterized as poly-D-mannosylated *N,N'*-diacetylchitobioses which, on average, contained nine monosaccharide residues.[85] In similar studies on the D-glucosylated residues, the

77 J. R. Hassell, C. S. Silverman-Jones, and L. M. De Luca, *J. Biol. Chem.*, 1978, **253**, 1627.
78 C. L. Richardson, W. D. Merritt, T. W Keenan, and D. J. Morre, *Experientia*, 1978, **34**, 571.
79 A. A. Bushway and T. W. Keenan, *Lipids*, 1978, **13**, 59.
80 T. W. Keenan, E. Schmid, and W. W. Franke, *Lipids*, 1978, **13**, 451.
81 G. van Dessel, A. Lagrou, M. J. Hilderson, W. Dierick, and G. Dacremont, *Biochimie*, 1977, **59**, 839.
82 M. Iwamori and Y. Nagai, *J. Biol. Chem.*, 1978, **253**, 8328.
83 B. R. Mullin, T. Pacuszka, G. Lee, L. D. Kohn, R. O. Brady, and P. H. Fishman, *Science*, 1978, **199**, 77.
84 T. Pacuszka, R. O. Duffard, R. N. Nishimura, R. O. Brady, and P. H. Fishman, *J. Biol. Chem.*, 1978, **253**, 5839.
85 C. Ronin and S. Bouchilloux, *Biochim. Biophys. Acta*, 1978, **539**, 470.

poly-D-glucosyl sequences also contained, on average, nine D-glucose residues but there were differences between the two oligosaccharide species in their molecular size distribution.[86]

Both glycosphingolipids and ceramides have been isolated from human and pig epidermal tissue.[87] The least polar of four chromatographically distinct D-glucosylceramides from pig epidermal tissue, and probably from human epidermal tissue due to similar properties, was characterized as 1-(3′-*O*-octadecadienoyl)-*β*-D-glucosyl-*N*-dihydroxypentatriacontadienoyl-sphingosine.[88]

β-D-Gal*p*-(1 → 4)-*β*-D-Glc*pN*Ac-6-S-(1 → 3)-*β*-D-Gal*p*-(1 → 4)-*β*-D-Glc*p*-(1 → 1)-ceramide

(19)

A sulphated neolactotetraosylceramide isolated from hog gastric mucosa was characterized as (19) by conventional procedures.[89] A unique D-glucosphingolipid pattern which has not been previously encountered in any type of tissue or cell has been seen in chicken egg-yolk.[90] The structures were determined and the major ones were found to be Neu*pN*Ac → D-Gal*p* → ceramide, Neu*pN*Ac → lactosyl → ceramide, and Neu*pN*Ac → Neu*pN*Ac → lactosyl → ceramide. The only other neutral glycosphingolipid was D-galactosylceramide.

Six bands were present in the ganglioside fraction isolated from peripheral nerve cells undergoing Walleman degeneration.[91] One of these bands contained three sialic acid residues per molecule, two bands contained two sialic acid residues per molecule while three bands contained only one sialic acid residue per molecule.

Glycosphingolipids have been isolated from adenocarcinoma cells from the human colon.[92] Compared with those from normal tissue, the lactosylceramide content was between two- and five-fold higher but the D-galactosylceramide content was lower. G_{M3} G_{M1}, G_{D1a}, G_{D3}, and lacto-*N*-neotetraosylceramide are abundant in both types of tissue and there is no change in the D-glucosylceramide, triaosyl ceramide, globoside, or paragloboside content. There was a marked decrease, and in some instances a total loss, of blood-group ABH activity in cancerous tissue. A hexosylceramide with Le^b-activity was present in all tumour tissues, the carbohydrate portion of which contained residues of L-fucose, D-galactose, 2-acetamido-2-deoxy-D-glucose, and D-glucose (2 : 2 : 1 : 1).

In a comparative study of the ganglioside fraction of normal and quaking mice, an increase in the gangliosides G_T and G_Q was found in the quaking mice but a 50% decrease in G_{M1} was noted.[93] It was suggested that the decreased concentration of G_{M1} could be linked to an enzymic deficiency in the latter stages of the myelination process. A novel ganglioside, ALG, was also found

[86] C. Ronin and S. Bouchilloux, *Biochim. Biophys. Acta*, 1978, **539**, 481.
[87] G. M. Gray and R. J. White, *J. Invest. Dermatol.*, 1978, **70**, 336.
[88] G. M. Gray, R. J. White, and J. R. Majer, *Biochim. Biophys. Acta*, 1978, **528**, 127.
[89] B. L. Slomiany and A. Slomiany, *J. Biol. Chem.*, 1978, **253**, 3517.
[90] S. C. Li, J. L. Chien, C. C. Wan, and Y. T. Li, *Biochem. J.*, 1978, **173**, 697.
[91] A. J. Yates and D. K. Thompson, *J. Neurochem.*, 1978, **30**, 1649.
[92] B. Siddiqui, J. S. Whitehead, and Y. S. Kim, *J. Biol. Chem.*, 1978, **253**, 2168.
[93] N. Baumann, S. Sonnino, R. Ghidoni, M.-L. Harpin, and G. Tettamanti, *Compt. rend.*, 1978, **286**, 1807.

which contained an alkali-labile constituent and gave the ganglioside G_{T1b} as an alkaline degradation product. Another comparison between the ganglioside fractions from human, cat, and rabbit spinal cord myelins showed that the human spinal cord contained only 10% of the ganglioside level of cerebral grey matter,[94] and there was a very low level of ganglioside in cat and rabbit tissue. D-Galactocerebroside is the major glycolipid in myelin; this lipid can thus be used as a specific cell surface antigen marker for oligodendrocytes, certainly in rat and probably in mouse and in human tissue, since these cells are responsible for synthesizing myelin in the central nervous system.[95]

The structure of the saccharide moiety of the D-glucose-containing dolichyl-diphosphate oligosaccharide which is a key intermediate in protein glycosylation has been investigated by the use of various glycosidases.[96] Jack bean α-D-mannosidase released some residues of D-mannose but complete removal of all the D-mannose residues was only achieved after the action of an α-D-glucosidase. From the results, a tentative structure was proposed in which the oligosaccharide chain is branched and some side chains terminate in D-glucose residues.

An *endo*-β-D-galactosidase has been used to release oligosaccharides from various glycosphingolipids.[97] The lacto-*N*-glycosylceramide series, which has the general structure R → 2-acetamido-2-deoxy-β-D-glucosyl-(1 → 3)-β-D-galactosyl-(1 → 4)-D-glucose, was hydrolysed at the β-D-galactosyl-(1 → 4)-D-glucose bond. The β-D-galactosyl residues in the 'globo-' or 'ganglio-' series were not susceptible. The presence of sialosyl-groups at the non-reducing termini of chains increased the enzyme-susceptibility of the linkage. In glycolipids with more than one such linkage, the other inner linkages were more susceptible than the one adjacent to the ceramide moiety. The ganglioside G_{M1} was thought to be resistant to the action of neuraminidase due to the steric hindrance of the D-galactosyl → 2-acetamido-2-deoxy-D-galactose unit.[98] However, in the presence of bile salts, the seemingly resistant neuraminic acid linkage was hydrolysed by the action of the neuraminidase of *Clostridium perfringens*. The neuraminidase of *Arthrobacter ureafaciens* is even more active towards the substrate since the linkage is hydrolysed in the absence of bile salts.

Concanavalin A is able to stimulate the synthesis of glycolipid in mouse thymocytes.[99] The rate of biosynthesis during stimulation has been determined; the glycolipids biosynthesized have been identified the major product being D-glucosyl-ceramide.

Only small amounts of globoside and Forssman hapten were found in the glycolipid fraction from hamster cell lines and their polyoma transformant.[100]

94 K. Ueno, S. Ando, and R. K. Yu, *J. Lipid Res.*, 1978, **19**, 863.
95 M. C. Raff, R. Mirsky, K. L. Fields, R. P. Lisak, S. H. Dorfman, D. H. Silberberg, N. A. Gregson, S. Leibowitz, and M. C. Kennedy, *Nature*, 1978, **274**, 813.
96 R. A. Ugalde, R. J. Staneloni, and L. F. Leloir, *F.E.B.S. Letters*, 1978, **91**, 209.
97 M. N. Fukuda, K. Watanabe, and S. I. Hakomori, *J. Biol. Chem.*, 1978, **253**, 6814.
98 K. Sugano, M. Saito, and Y. Nagai, *F.E.B.S. Letters*, 1978, **89**, 321.
99 G. Rosenfelder, R. V. W. van Eijk, D. A. Monner, and P. F. Mühlradt, *European J. Biochem.*, 1978, **83**, 571.
100 N. Yokosawa, S. Kijomoto-Ochiai, and A. Makita, *Biochim. Biophys. Acta*, 1978, **528**, 138.

The BHK-polyoma cells showed little 2-acetamido-2-deoxy-β-D-galactosyltransferase activity in contrast to a greatly elevated 2-acetamido-2-deoxy-α-D-galactosyltransferase activity in the parent BHK cells.

Preparations from chicken embryo fibroblasts were able to transfer sugar residues from GDP-D-mannose and UDP-2-acetamido-2-deoxy-D-glucose to endogenous lipid.[101] These glycolipids became intermediates in glycoprotein biosynthesis. The antibiotic bacitracin inhibited the incorporation of D-mannose and 2-acetamido-2-deoxy-D-glucose from their corresponding sugar nucleotides into lipid-linked saccharides in pig aorta tissue.[102] The inhibition of the incorporation of 2-acetamido-2-deoxy-D-glucose was more sensitive than that of D-mannose. High concentrations of dolicholphosphate overcame the inhibition of D-mannose but not that of 2-acetamido-2-deoxy-D-glucose.

The level of glycosphingolipids in the urine of patients with proteinuria of various etiologies was markedly elevated.[103] In no case was there any evidence of a glycosphingolipid storage disorder. Thus glycosphingolipiduria may occur in proteinuria as well as in glycosphingolipid storage disorders.

A model has been postulated for the behaviour of gangliosides in cell membranes.[104] In phosphatidylcholine bilayers, oligosaccharide chains are quite mobile and show a tendency to interact with each other, probably due to hydrogen bonding. Evidence was presented to show that physiological concentrations of Ca^{2+} and Mg^{2+} ions lead to cross-linking and condensing of the terminal non-reducing sialosyl groups in the gangliosides. There was no tendency for these non-reducing terminal sialosyl groups to penetrate the hydrophobic region of the lipid bilayers. The interaction of gangliosides with glycophorin and intact BHK cells was investigated at the same time. It was suggested that these interactions could contribute to the glycocalyx of the eucaryotic cell. The authors predict that laterally mobile carbohydrate-bearing components of cell surfaces will show a tendency to cluster about glycoprotein arrays, especially if the species involved bear accessible carboxylic acid functions.

Choleragen interacts with the oligosaccharides of ganglioside G_{M1} in a way which suggests that multiple binding sites occur on the toxin.[105] Each toxin molecule appears to be capable of binding 5–6 oligosaccharide molecules, confirmation of which was obtained by determining the molecular weight of the toxin before and after binding.

The protein kinase activity which was stimulated in rats by cholera enterotoxin was blocked by the addition of a ganglioside mixture.[106] The same gangliosides were, however, not directly involved in the binding of gonadotrophin to ovarian cell membranes.

The ganglioside G_{D3} caused crenation in chicken erythrocytes followed by the appearance of rounded and oddly shaped erythrocytes that were 'pinched

[101] S. S. Krag, *Biochim. Biophys. Acta*, 1978, **538**, 603.
[102] J. P. Spencer, M. S. Kang, and A. D. Elbein, *Arch. Biochem. Biophys.*, 1978, **190**, 829.
[103] R. R. Townsend, R. M. Órth, C. M. Clawson, S. C. Li, and Y. T. Li, *J. Clin. Invest.*, 1978, **62**, 119.
[104] F. J. S. Harom and C. M. W. Grant, *Biochim. Biophys. Acta*, 1978, **507**, 280.
[105] P. H. Fishman, J. Moss, and J. C. Osborne, jun., *Biochemistry*, 1978, **17**, 711.
[106] S. Azhar and K. M. J. Menon, *Biochim. Biophys. Res. Comm.*, 1978, **81**, 205.

off' by fusion of the cell membrane.[107] The gangliosides G_{D1c} and G_{T1} caused similar changes except that there was no 'pinching off' of the membrane. Other gangliosides, such as G_{M1} and G_{M3} as well as glycosphingolipids, had no effect.

A novel method for the diagnosis of G_{M1} gangliosidosis has been reported.[108] Normal tears contain a β-D-galactosidase which has a pH optimum of 4.2. Tears from patients with G_{M1} gangliosidosis contain only about 20% of the control value.

Membrane preparations from immature stages of the fruit fly *Ceratitis capitata* catalyse the transfer of D-mannose into lipid-linked oligosaccharides.[109] The lipid-linked trisaccharide has been partially characterized and tentatively assigned the structure D-mannosyl → *NN'* diacetylchitobiosyl-poly(prenyl-pyrophosphate). These insect poly(prenyl)-oligosaccharides may participate in glycoprotein biosynthesis.

The glycosphingolipids of the starfish, *Asterina pectinifera*, have been isolated and the ceramide mono- and dihexoside fractions characterized.[110] D-Glucose and lactose were the major sugar residues in these fractions. The lactose sequence has not previously been reported in starfish glycolipids. The fatty acids and long-chain bases were also unusual in having high concentrations of 2-hydroxy-acids and branched-chain bases.

The aminoalkylphosphorylcerebrosides in the muscle tissue of the marine snail, *Turbo cornutus*, have been characterized by g.c.–m.s. The major component was 1-*O*-[*O'*, *O'*(*N*-methylaminoethylphosphonyl)-D-galactosyl]-ceramide while the minor component lacked the *N*-methyl group.[111]

6-*O*-mycoloyl-*N*-acetylmuramyl-L-alanyl-D-isoglutamine, a lipoglycopeptide with anti-tumour activity, has been synthesized.[112]

3 Plant and Microbial Glycolipids

The changes which occur in the composition of glycolipids and phospholipids in the cotyledons of soybeans during germination have been investigated.[113] The major components were the glycolipids found in the chloroplast and these appeared at an early stage of growth. The non-polar lipids were metabolized at a faster rate than the polar lipids. The mono- and di-D-galactosyl-diglyceride concentration increased in seeds grown under illumination and the changes appeared to come from both the formation of photosynthetic tissues and the interconversion of phospholipids.

A major glycophosphoceramide from tobacco leaves, named PSL-I, has been characterized from the results of methylation, and periodate and chromium(VI) oxide oxidation analyses to have the structure (20).[114] Six other novel

[107] B. Maggio, F. A. Cumar, and R. Caputto, *F.E.B.S. Letters*, 1978, **90**, 149.
[108] A. Tsuboyama, F. Miki, M. Yoshida, Y. Ogura, and T. Miyatake, *Clinica Chim. Acta*, 1977, **80**, 237.
[109] L. A. Quesada Allue and E. Belocopitow, *European J. Biochem.*, 1978, **88**, 529.
[110] M. Sugita, *J. Biochem.* (*Japan*), 1977, **82**, 1307.
[111] A. Hayashi and F. Matsuura, *Chem. and Phys. Lipids*, 1978, **22**, 9.
[112] T. Shiba, S. Okada, S. Kusumoto, I. Azuma, and Y. Mamamura, *Bull. Chem. Soc. Japan*, 1978, **51**, 3307.
[113] H. Yoshida and G. Kajimoto, *Agric. and Biol. Chem.* (*Japan*), 1977, **41**, 1857.
[114] T. C.-Y. Hsieh, K. Kaul, R. A. Laine, and R. L. Lester, *Biochemistry*, 1978, **17**, 3575.

α-D-Glc*p*NAc-(1 → 4)-α-D-Glc*p*-(1 → 2)-myo-inositol-1-*O*-phosphoceramide

(20)

phosphoinositol-containing sphingolipids have also been isolated from tobacco leaves and were shown to be related to the previously characterized lipids PSL-I (20) and PSL-II, where PSL-II has the same structure as PSL-I except that the amino-sugar does not carry an *N*-acetyl group.[115] The precise structures of these lipids has not been determined but their composition was PSL-I-[Ara_2 Gal_2], PSL-I-[Ara_3 Gal_2], PSL-I-[Ara_4 Gal_2], PSL-II-[Ara_3 Gal], PSL-II-[Ara_{2-3} Gal_2], and PSL-II-[Ara_2 Gal_2 Man]. The presence of 2-amino-2-deoxy-D-glucose in the PSL-II series was shown not to be an artifact.

^{14}C-Labelled sodium bicarbonate or carbon dioxide could be incorporated into the mono-D-galactosyldiacylglycerol fraction of spinach leaves and chloroplasts but the labelled compounds were only incorporated into the fatty acids.[116]

Lipopurothionins, petroleum ether-soluble lipid–protein complexes from the *Gramineae* sp., were converted to a petroleum ether-insoluble form on extraction with acetone.[117] The only component of the acetone extract which was able to restore the solubility in petroleum ether was di-D-galactosyldiglyceride.

Low levels (5 μg ml^{-1}) of the antibiotic tunicamycin inhibited the incorporation of 2-acetamido-2-deoxy-D-glucose into polyprenol sugars in cotton bolls and mung bean seedlings.[118] Conversely, even high concentrations (500 μg ml^{-1}) of the antibiotic had no effect on the incorporation of D-mannose residues in similar tissue.

Two different enzymes for D-galactosylation occur in isolated chloroplast envelopes of spinach leaves.[119] The UDP-D-galactose-diglyceride-D-galactosyltransferase is responsible for the biosynthesis of mono-D-galactosyldiglyceride while the second enzyme, D-galactolipid-D-galactolipid-D-galactosyltransferase, is responsible for the biosynthesis of di-D-galactosyldiglycerides and higher homologues. The latter enzyme responded to variations in the concentration of UDP-D-galactose but the reaction could proceed in the absence of the sugar nucleotide. The rates of synthesis of mono- and di-D-galactosyldiglycerides in *Vicia faba* leaves were found to be similar even when grown in two very different concentrations of CO_2 (0.03% and 1%).[120] Changes in light intensity also cause similar changes in these rates of synthesis. The glycosyltransferases involved in the transfer of sugar residues from UDP-sugars to lipids in membranes have been identified and their specificities have been determined.[121] The involvement of lipid-linked sugars in the synthesis of cellulose precursors has been studied.[122] Preparations from the Chlorophyta *Protetheca zopfii* cata-

[115] K. Kaul and R. L. Lester, *Biochemistry*, 1978, **17**, 3569.
[116] J. W. A. McKee and J. C. Hawke, *F.E.B.S. Letters*, 1978, **94**, 273.
[117] C. Hernandez-Lucas, F. Fernandez de Caleya, P. Carbonero, and F. Garcia-Olmedo, *J. Agric. Food Chem.*, 1977, **25**, 1287.
[118] M. C. Ericson, J. T. Gafford, and A. D. Elbein, *J. Biol. Chem.*, 1977, **252**, 7431.
[119] A. Van Besouw and J. F. G. Wintermans, *Biochim. Biophys. Acta*, 1978, **529**, 44.
[120] N. W. Lem and J. P. Williams, *Phytochemistry*, 1978, **17**, 1893.
[121] M. Ledig and C. Peaud-Lenoël, *Biochimie*, 1978, **60**, 35.
[122] H. E. Hopp, P. A. Romero, G. R. Daleo, and R. P. Lezica, *European J. Biochem.*, 1978, **84**, 561.

lysed the incorporation of D-glucose into lipid-P-D-glucose, lipid-PP-D-glucose, and lipid-PP-oligosaccharides. These oligosachacrides had DPs ranging from 2 to 12 and all of them contained β-(1 → 4) linkages. When GDP-D-glucose was used as the nucleotide sugar, an alkali-insoluble polymer was formed which had similar properties to cellulose.

Six glycolipids have been isolated from the polar lipid fraction of *Lactobacillus casei*.[123] Each one was a derivative of 1,2-di-*O*-acylglycerol and the oligosaccharide fragments were characterized as (21)—(26). The structure of

α-D-Glc*p*-diacylglycerol

(21)

α-D-Gal*p*-(1 → 2)-α-D-Glc*p*-diacylglycerol

(22)

α-D-Gal*p*-(1 → 2)-α-D-Glc*p*-6-acylate-diacylglycerol

(23)

β-D-Glc*p*-(1 → 6)-α-D-Gal*p*-(1 → 2)-α-D-Glc*p*-diacylglycerol

(24)

β-D-Glc*p*-(1 → 6)-α-D-Gal*p*-(1 → 2)-α-D-Glc*p*-6-acylate-diacylglycerol

(25)

β-D-Glc*p*-(1 → 6)-β-D-Glc*p*-(1 → 6)-α-D-Gal*p*-(1 → 2)-α-D-Glc*p*-diacylglycerol

(26)

α-D-Glc*p*-(1 → 2)-α-D-Glc*p*-diacylglycerol
6
↑
R

R = hexadecanoic acid

(27)

a glycolipid from *Streptococcus lactis* was characterized as (27).[124] The acyl group, which was linked to C-6 of the inner D-glucose residue, was mainly comprised of hexadecanoic acid. The same fatty acid was found linked to C-3 of an analogue found in *S. diacetilactis*. An acyl-di-D-glucosyldiacylglycerol-containing lipoteichoic acid with a poly(3-*O*-D-galactobiosyl-2-*O*-D-galactosyl-*sn*-glycero-1-phosphate) main chain has been isolated from a mutant of the same organism.[125] The repeating units are joined together by phosphodiester bonds at C-6 of the non-reducing terminal D-galactose group of the D-galactobiosyl unit.

The occurrence and concentration of phosphoglycolipids in various families of Gram-positive bacteria have been determined.[126] Phosphatidylglycolipids occur exclusively in group D *streptococci* and *S. haemolyticus*. The prevalent species, phosphatidyl-α-kojibiosyldiacylglycerol, accounted for 28% of the

[123] M. Nakano and W. Fischer, *Z. physiol. Chem.*, 1977, **358**, 1439.
[124] W. Fischer, R. A. Laine, M. Nakano, D. Schuster, and H. Egge, *Chem. and Phys. Lipids*, 1978, **21**, 103.
[125] H. U. Koch and W. Fischer, *Biochemistry*, 1978, **17**, 5275.
[126] W. Fischer, M. Nakano, R. A. Laine, and W. Bohrer, *Biochim. Biophys. Acta*, 1978, **528**, 288.

[6-(*sn*-glycero-1-phospho)-α-D-Glc*p*-(1 → 2)-α-D-Glc*p*-6-acetylate-diacylglycerol

(28)

[6-(*sn*-glycero-1-phospho)-β-D-Glc*p*-(1 → 6)-α-D-Gal*p*-(1 → 2)-α-D-Glc*p*]-diacylglycerol

(29)

[6-(*sn*-glycero-1-phospho)-β-D-Glc*p*-(1 → 6)-α-D-Gal*p*-(1 → 2)-α-D-Glc*p*-6-acylate]-diacylglycerol

(30)

polar lipid in these organisms. The related glycerophospho-phosphatidyl-α-kojibiosyldiacylglycerol was only found in *S. faecalis*. Except for *Micrococcus lysodeiktus*, all of the bacteria which were investigated contained one or more glycerophosphoglycolipids. Eight of these were isolated from six different Gram-positive bacteria.[127] As well as *sn*-glycero-1-phospho-β-gentiobiosyldiacylglycerol and *sn*-glycero-1-phospho-α-kojibiosyldiacylglycerol, three novel structures (28—30) were established. Identical structural features of these glycerophosphoglycolipids were found in the corresponding lipoteichoic acids.

S. lactis has been found to contain three unusual polar glycerophosphoglycolipids.[128] The structure of one of them was established as 1,2-di-*O*-acyl-3-*O*-[6-(*sn*-glycero-1-phospho-3-*sn*-glycero-1-phospho)-α-D-glucopyranosyl-(1 → 2)-(6-*O*-acyl-α-D-glucopyranosyl)]-glycerol. The second was similar to this except that it lacked the 6-*O*-acyl group on the D-glucose residue. The third, too, was similar to the first except that it contained an extra α-D-galactosyl group at C-2 of the inner glycolipid-linked glycerophosphate moiety. These lipids are considered to be the missing link between simple glycerophosphoglycolipids and lipoteichoic acids of Gram-negative bacteria.

Lipoteichoic acids from *Lactobacillus casei* have been degraded to give glycolipid and poly(glycerophosphate) fractions.[129] The lipid fraction contained trihexosyldiacylglycerol and acyl trihexosyldiacylglycerol which were considered to be 'lipid anchors' and had the structures (31) and (32) respectively.

1, 2-di-*O*-acyl-3-*O*-[β-D-Glc*p*-(1 → 6)-*O*-α-D-Gal*p*-(1 → 2)-α-D-Glc*p*]-glycerol

(31)

1, 2-di-*O*-acyl-3-*O*-[β-D-Glc*p*-(1 → 6)-*O*-β-D-Gal*p*-(1 → 2)-α-D-Glc*p*-6-acylate]-glycerol

(32)

N-Acetylneuraminic acid was shown to be a component of the lipopolysaccharide of *Salmonella touera* along with residues of D-glucose, D-galactose, 2-acetamido-2-deoxy-D-glucose, 2-keto-3-deoxyoctonate, and an unidentified amino sugar.[130]

A novel D-glucuronic acid-containing sphingoglycolipid has been found in *Flavobacterium devorans*.[131] The compound is probably *N*-2-hydroxymyristyl-dihydrosphingosine-(1 ← 1)-D-glucuronic acid.

The structure of the cord factor from *Mycobacterium smegmatis* has been confirmed as α,α trehalose 6,6′-dimycolate ($C_{166}H_{322}O_{15}$).[132] The α,α trehalose was identified by paper chromatography and the positions of ester linkages at O-6 and O-6′ was established by ^{13}C n.m.r. spectroscopy.

An unusual glycolipid was isolated from *Rhodospirillium tenue* and its

[127] W. Fischer, R. A. Laine, and M. Nakano, *Biochim. Biophys. Acta*, 1978, **528**, 298.
[128] R. A. Laine and W. Fischer, *Biochim. Biophys. Acta*, 1978, **529**, 250.
[129] M. Nakano and W. Fischer, *Z. physiol Chem.*, 1978, **359**, 1.
[130] B. Kedzierska, *European J. Biochem.*, 1978, **91**, 545.
[131] A. Yamamoto, I. Yano, M. Masui, and E. Yabuuchi, *J. Biochem. (Japan)*, 1978, **83**, 1213.
[132] B. Mompon, C. Federici, R. Toubiana, and E. Lederer, *Chem. and Phys. Lipids*, 1978, **21**, 97.

Fatty acids

L-Ara*N*-1-P → D-Glc*pN*-(1 → 6)-D-Glc*pN*-1-P ← 1-D-Ara*f*

4

↑

1

D-Glc*pN*

Fatty acids = 3-hydroxydecanoic acid (amide-linked)
palmitic and myristic acids (ester-linked)

(33)

partial structure was reported (33).[133] None of the peripheral sugars, 4-amino-4-deoxy-L-arabinose 1-phosphate, 2-amino-2-deoxy-D-glucose, or D-arabinofuranose-1-phosphate are esterified but the internal disaccharide is esterified with palmitic and myristic acids and 3-hydroxy-decanoic acid is also present in amide-linked form. The precise points of linkage of these acids awaits clarification.

A cell-free particulate fraction from a halotolerant, moderately halophilic unnamed bacterium, could incorporate D-glucose and phosphatidylglycerol into D-glucophosphatidylglycerol.[134] Exogenously added phosphatidylglycerol was only available to the enzyme when dispersed in a detergent. Mg^{2+} and Ca^{2+} ions were also required for enzymic activity.

Yeast membranes are able to catalyse the transfer of *N,N′*-diacetylchitobiose and *β*-D-mannosyl-*N,N′*-diacetylchitobiose from their dolichylphosphate-bound forms into glycoproteins.[135] Similarly linked residues of 2-acetamido-2-deoxy-D-glucose were not transferred. The results were considered to be the first proof of the involvement of dolichylphosphate-activated sugars in the glycosylation of L-serine/L-threonine residues in extracellular glycoproteins. Glycosylation of the surface glycoprotein of *Halobacterium salinarium* took place on the cytoplasmic side of the bacterial membrane by transfer of a sugar residue from a glycolipid.[136] The probable precursor was isoprenylpyrophospho-2-acetamido-2-deoxy-D-glucose.

The lipid intermediates in the biosynthesis of the linkage unit between teichoic acids and peptidoglycans have been investigated in *Staphylococcus aureus* and *Micrococcus sp*2102.[137] Prenyl-pyrophospho-2-acetamido-2-deoxyglucose was found in the lipid pool. In studies on the oligosaccharide-linked lipid from *Streptococcus sanguis*, which contained residues of D-glucose, L-rhamnose, and 2-acetamido-2-deoxy-D-glucose, it was found that the 2-acetamido-2-deoxy-D-glucose residue was linked closer to the lipid group than the L-rhamnose residue which in turn was closer than the D-glucose residue.[138] The oligosaccharide fragment was probably linked to the lipid by phosphodiester bonds. Apart from dolichylmonophospho-D-mannose, which has previously been characterized as a lipid-linked saccharide from *Saccharomyces*

[133] R. N. Tharanathan, J. Weckesser, and H. Mayer, *European J. Biochem.*, 1978, **84**, 385.
[134] N. Stern and A. Tietz, *Biochim. Biophys. Acta*, 1978, **530**, 357.
[135] L. Lehle and W. Tanner, *European J. Biochem.*, 1978, **83**, 563.
[136] M. F. Mescher and J. L. Strominger, *F.E.B.S. Letters*, 1978, **89**, 37.
[137] H. A. I. McArthur, F. M. Roberts, I. C. Hancock, and J. Baddiley, *F.E.B.S. Letters*, 1978, **86**, 193.
[138] T. H. Chiu and C. Saralkar, *J. Bacteriol.*, 1978, **133**, 185.

cerevisiae, the only other glycolipid present which was not sensitive to alkaline hydrolysis was a polyprenyldiphospho-oligosaccharide in which the sugar moiety had at least twelve hexose residues.[139] This glycolipid is synthesized by the yeast from dolichylpyrophospho-*N*,*N′*-diacetyl-chitobiose and GDP-D-mannose and the saccharide moiety can be transferred intact to glycoprotein.

[139] L. Lehle and W. Tanner, *Biochim. Biophys. Acta*, 1978, **539**, 218.

8

Chemical Synthesis and Modification of Oligosaccharides, Polysaccharides, Glycoproteins, Enzymes, and Glycolipids

BY C. M. STURGEON

1 Synthesis of Polysaccharides, Oligosaccharides, Glycoproteins, Glycopeptides, and Glycolipids

Polysaccharides.—The coupling of D-glucose and malto-oligosaccharides by hydrazone linkage to a number of polysaccharide carriers by means of acid hydrazide groups has been investigated.[1] Malto-oligomeric grafts of degree of polymerization ≥4 could be extended enzymatically by phosphorolytic synthesis to amylose chains of various lengths. Carbohydrate side-chains (D-glucose, maltose, and malto-oligosaccharides with degrees of polymerization 6—8) have been coupled on to polymeric carriers *via* amide bonds (Scheme 1).[2] The

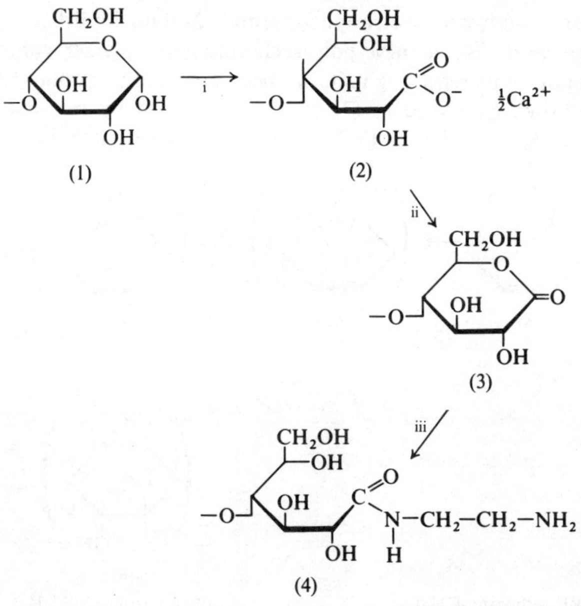

Reagents: i, $CaBr_2$–$CaCO_3$, electrolytic oxidation; ii, Amberlite IR 120 (H^+); iii, 1,2-Diaminoethane

Scheme 1

[1] H. Andresz, G. C. Richter, and B. Pfannemuller, *Makromol. Chem.*, 1978, **179**, 301.
[2] W. N. Emmerling and B. Pfannemuller, *Makromol. Chem.*, 1978, **179**, 1627.

sugar (1) is activated by electrolytic oxidation to yield the calcium salt of the aldonic acid (2) which is then converted to the corresponding aldonolactone (3) by treatment with an anion-exchange resin. The activated cyclic ester allows a direct coupling to carriers with amino groups, as shown for 1,2-diaminoethane (4). Carriers with carboxylic groups are first activated by transesterification with the 4-nitrophenyl ester of trifluoroacetic acid and then treated with (4).

Direct addition polymerization of 2-amino-2-deoxy-D-glucose and a di-isocyanate (4,4′-methylene-diphenyl-1,1′-di-isocyanate, phenyl-1,4-di-isocyanate, toluene-2,4-di-isocyanate, or butyl-1,4-di-isocyanate) yields a linear sugar-containing poly(urea-urethane) (5) having good solubility in organic solvents, unlike chitin or chitosan.[3]

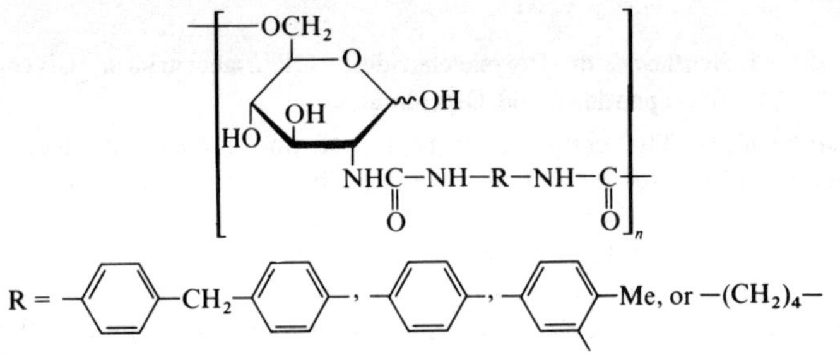

(5)

Poly(*trans*-3,4-dihydroxytetrahydropyran-6,2-diyloxymethene)[(1→6)-(4-deoxy-α-D-glucan)] (6), a new polysaccharide-type polymer containing two hydroxy-groups on its repeating unit, has been prepared from poly(5,6-dihydro-2*H*-pyran-6,2-diyloxymethylene) (7) which is obtained by ring-opening polymerization of 6,8-dioxabicyclo[3.2.1]-oct-3-ene (8) (Scheme 2).[4] The water

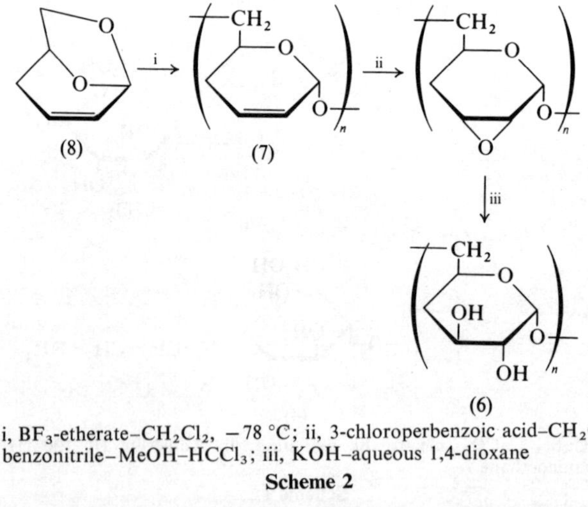

Reagents: i, BF_3-etherate–CH_2Cl_2, −78 °C; ii, 3-chloroperbenzoic acid–CH_2Cl_2, or H_2O_2–benzonitrile–MeOH–$HCCl_3$; iii, KOH–aqueous 1,4-dioxane

Scheme 2

3 K. Kurita, N. Hirakawa, and Y. Iwakura, *Makromol. Chem.*, 1977, **178**, 2939.
4 M. Okada, H. Sumitomo, and H. Komada, *Makromol. Chem.*, 1978, **179**, 949.

sorption isotherm of the polymer shows a marked resemblance to that of some natural dextrans, which have a related backbone structure.

1,6-Anhydro-2,3,4-tri-*O*-(4-methylbenzyl)-β-D-glucopyranose has been co-polymerized with 1,6-anhydro-2,3,4-tri-*O*-benzyl-β-D-glucopyranose and the mechanism of reaction analysed.[5] Physical properties of the co-polymer indicate a high stereoregularity of structure.

The cationic polymerization of 5,6-anhydro-1,2-*O*-isopropylidene-α-D-glucofuranose (9) yields a polymer whose structure (10) was found to be poly(5,6-anhydro-1,2-*O*-isopropylidene-α-D-glucofuranose).[6] Optimal conditions for the polymerization were investigated and the mechanism discussed.

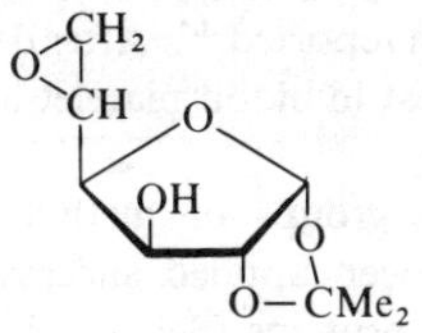

(9)

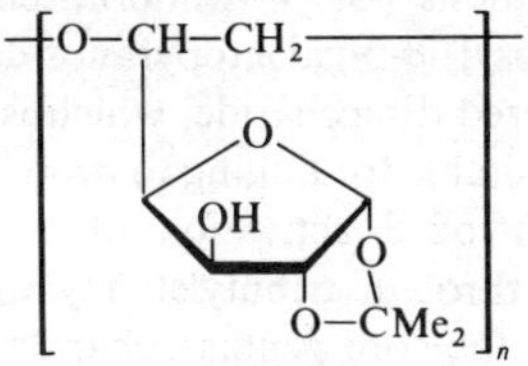

(10)

The cationic polymerization of 1,2,4-orthoacetyl-3-*O*-acetyl-α-D-xylopyranose, initiated by various electrophilic reagents, has been studied and optimum reaction conditions determined.[7]

Oligosaccharides.—Reaction pathways utilizing the Koenigs–Knorr reaction for the synthesis of glycopyranosides and other glycosides, particularly furanosides and heptanosides, have been reviewed.[8] The Koenigs–Knorr reaction has been applied to the synthesis of 2-*O*-α-L-rhamnopyranosyl-L-rhamnose, which is part of the immunodeterminant grouping in a streptococcal polysaccharide.[9] Methyl and benzyl 3-*O*-β-D-xylopyranosyl-α-D-mannopyranosides have been prepared by D-xylosylation (Koenigs–Knorr) of methyl and benzyl 4,6-*O*-benzylidene-α-D-mannopyranosides.[10]

The reaction of 2,3,4,6-tetra-*O*-acetyl-α-D-glucopyranosyl bromide with methyl (methyl 2-*O*-methyl-α-D-galactopyranosid)uronate in the presence of silver carbonate has yielded the crystalline pseudoaldobiouronic acid derivatives methyl [methyl 2-*O*-methyl-3-*O*- and 4-*O*-(2,3,4,6-tetra-*O*-acetyl-β-D-glucopyranosyl)-α-D-galactopyranosid]uronate.[11] Deacetylation of the compounds followed by Kuhn methylation afforded mainly crystalline, fully methylated products.

The synthesis of *O*-β-D-mannopyranosyl-(1 → 4)-*O*-α-L-rhamnopyranosyl-(1 → 3)-[*O*-α-D-glucopyranosyl-(1 → 6)]-D-galactose, the repeating unit of the *O*-specific polysaccharide chain of the lipopolysaccharide from *Salmonella*

[5] H. Ito and C. Schuerch, *J. Polymer Sci., Polymer Chem.*, 1978, **16**, 2217.
[6] T. Uryu, K. Kitano, H. Tachikawa, K. Ito, and K. Matsuzaki, *Makromol. Chem.*, 1978, **179**, 1773.
[7] A. F. Bochkov and A. V. Rodionov, *Bull. Acad. Sci. U.S.S.R.*, 1976, **25**, 2599.
[8] K. Igarashi, in 'Advances in Carbohydrate Chemistry and Biochemistry', Vol. 34, ed. R. S. Tipson and D. Horton, Academic Press, New York, 1977, p. 243.
[9] W. Schalch, W. Hochstrasser, and D. G. Braun, *Tetrahedron Letters*, 1978, 4153.
[10] R. T. Lee and Y. C. Lee, *Carbohydrate Res.*, 1978, **67**, 389.
[11] J. Hirsch, P. Kováč, and V. Kováčík, *Carbohydrate Res.*, 1977, **56**, 391.

senftenberg, has been reported.[12] The oligosaccharide was obtained by glycosylation of benzyl 2,4-di-*O*-benzyl-6-*O*-(2,3,4-tri-*O*-benzyl-6-*O*-4-nitrobenzoyl-α-D-glucopyranosyl)-β-D-galactopyranoside or benzyl 2-*O*-acetyl-6-*O*-(2,3,4-tri-*O*-benzyl-6-*O*-4-nitrobenzoyl-α-D-glucopyranosyl)-β-D-galactopyranoside with 3-*O*-acetyl-4-*O*-(2,3,4,6-tetra-*O*-acetyl-β-D-mannopyranosyl)-β-L-rhamnopyranose 1,2-(methyl orthoacetate) followed by removal of the protecting groups. The tetrasaccharide *O*-β-D-glucopyranosyl-(1 → 4)-*O*-α-L-rhamnopyranosyl-(1 → 3)-[*O*-α-D-glucopyranosyl-(1 → 6)]-D-galactose which is an analogue of the repeating unit of the *S. senftenberg* *O*-antigenic polysaccharide, has also been synthesized.[13]

The synthesis of 4-trifluoroacetamidophenyl 3,4,6-tri-*O*-acetyl-2-*O*-(α-D-glucopyranosyl)-β-D-galactopyranoside has been reported.[14] After deacetylation, this derivatized disaccharide, which is of interest in blood platelet agglutination studies, is suitable for linking to proteins.

Regio-controlled activation of the hydroxy-groups of methyl α-D-mannopyranoside through tributylstannylation has been applied successfully to the synthesis of the core pentasaccharides of glycoproteins (Scheme 3).[15] The key intermediate (11) could be transformed into a core saccharide containing either disaccharide (β-D-galactopyranosyl-(1 → 4)-2-acetamido-2-deoxy-D-glucopyranose) units (12) or oligo-D-mannosidic units (13).

Synthesis of one of the core oligosaccharides of human blood group substances, the reduced form of which is reportedly present among the degradation products of soluble Le[a]-active glycoprotein from ovarian cysts, has been reported.[16] Condensation of the per-*O*-acetyloxazoline derivative (14) with the trisaccharide (15), followed by removal of the protecting groups, yielded the branched pentasaccharide (16).

Synthesis of part of a proposed composite structure for carbohydrate chains in a human Le[b] substance has been reported (Scheme 4).[17] Two sequential bromide ion-catalysed condensations with 2,3,4-tri-*O*-benzyl-α-L-fucopyranosyl bromide are followed by catalytic hydrogenation to yield the tetrasaccharide *O*-α-L-fucopyranosyl-(1 → 2)-*O*-β-D-galactopyranosyl-(1 → 4)-*O*-[α-L-fucopyranosyl-(1 → 3)]-2-acetamido-2-deoxy-α-D-glucopyranose. Blood group substance oligosaccharides with terminal 2-acetamido-2-deoxy-D-galactosyl groups have been synthesized.[18]

Three 6-deoxy-6-iodomalto-oligosaccharides, *O*-α-D-glucopyranosyl-(1 → 4)-6-deoxy-6-iodo-*O*-α-D-glucopyranosyl-(1 → 4)-D-glucopyranose, *O*-α-D-glucopyranosyl-(1 → 4)-6-deoxy-6-iodo-D-glucopyranose, and 6-deoxy-6-iodo-*O*-α-D-glucopyranosyl-(1 → 4)-D-glucopyranose, have been synthesized and their inhibitory action against one of the α-amylases (Taka-amylase A) of *Aspergillus*

[12] N. K. Kochetkov, N. N. Malysheva, V. I. Torgov, and E. M. Klimov, *Carbohydrate Res.*, 1977, **54**, 269.
[13] N. K. Kochetkov, N. N. Malysheva, V. I. Torgov, and E. M. Klimov, *Bull. Acad. Sci. U.S.S.R.*, 1977, **26**, 589.
[14] P. J. Garegg, I. J. Goldstein, and T. Iversen, *Acta Chem. Scand.*, 1976, **30B**, 876.
[15] T. Ogawa, K. Katano, and M. Matsui, *Carbohydrate Res.*, 1978, **64**, C3.
[16] C. Augè, S. David, and A. Veyriéres, *J.C.S. Chem. Comm.*, 1977, 449.
[17] J.-C. Jacquinet and P. Sinay, *J. Org. Chem.*, 1977, **42**, 720.
[18] H. Paulsen, W. Stenzel, and C. Kólař, *Tetrahedron Letters*, 1977, 2785.

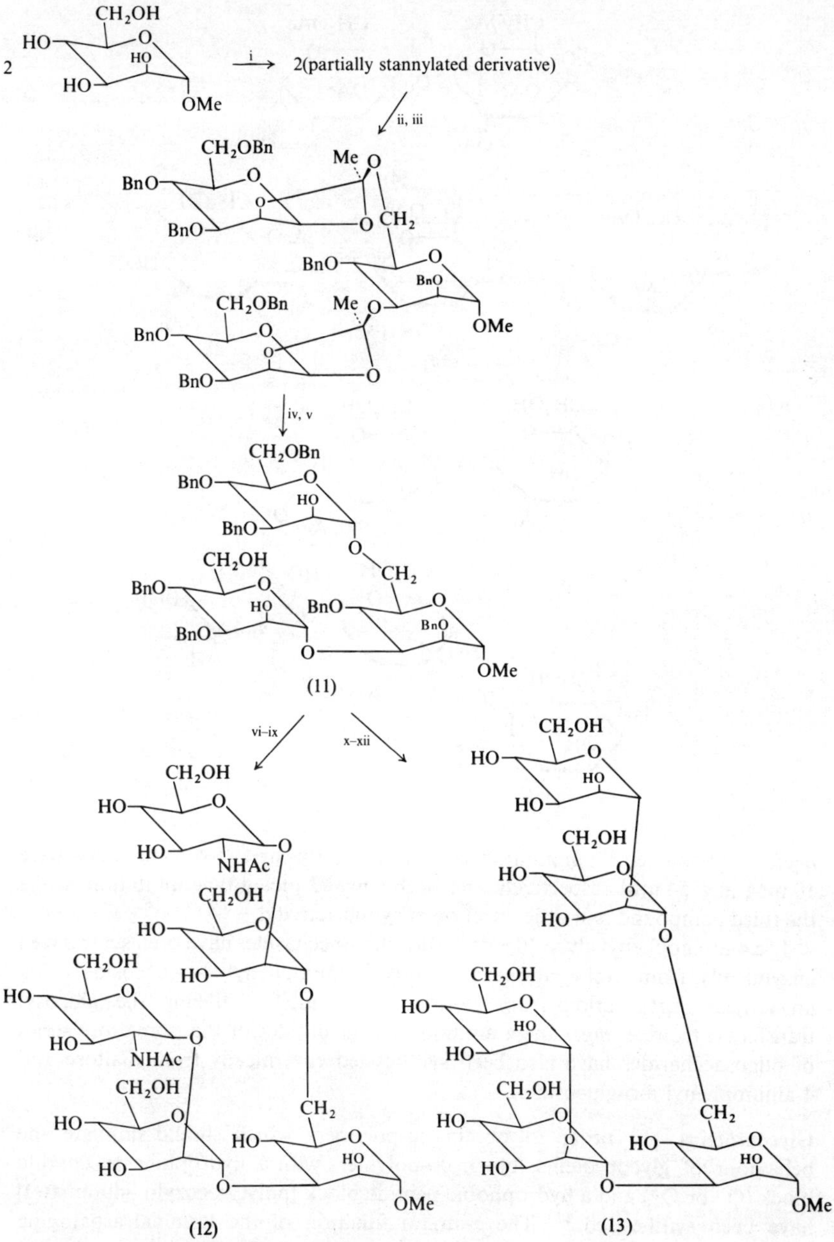

Reagents: i, $(Bu_3Sn)_2O$; ii, 2-*O*-acetyl-3,4,6-tri-*O*-benzyl-α-D-mannopyranosyl chloride; iii, NaH–2-bromotoluene–DMF; iv, $HgBr_2$–N_2, 120 °C; v, NaOMe–MeOH–THF; vi, silver triflate–*s*-collidine–CH_3NO_2–3,4,6-tri-*O*-acetyl-2-deoxy-2-phthalimido-D-glucopyranosyl chloride; vii, boiling ethanolic hydrazine hydrate, under argon; viii, Ac_2O–MeOH; ix, H_2–10% Pd–C–aq. EtOH; x, silver triflate–tetramethylurea; ii, CH_2Cl_2–N_2; xi, NaOMe–MeOH–THF; xii, H_2–10% Pd–C–aq. EtOH

Scheme 3

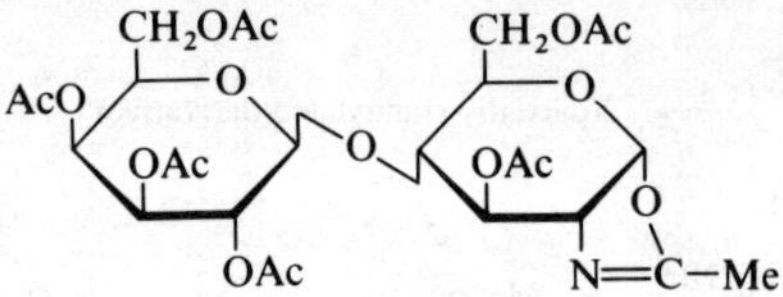

(14)

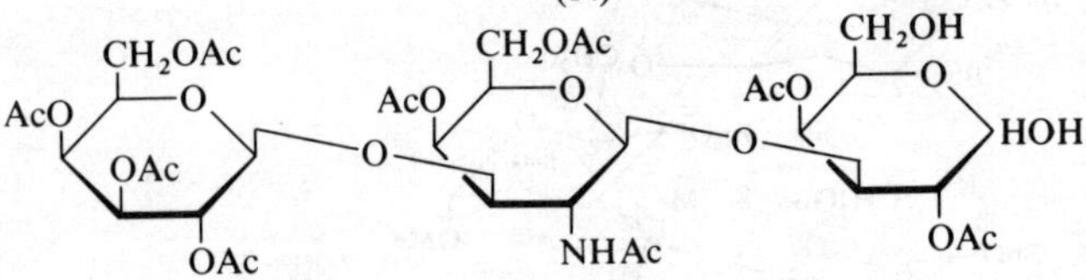

(15)

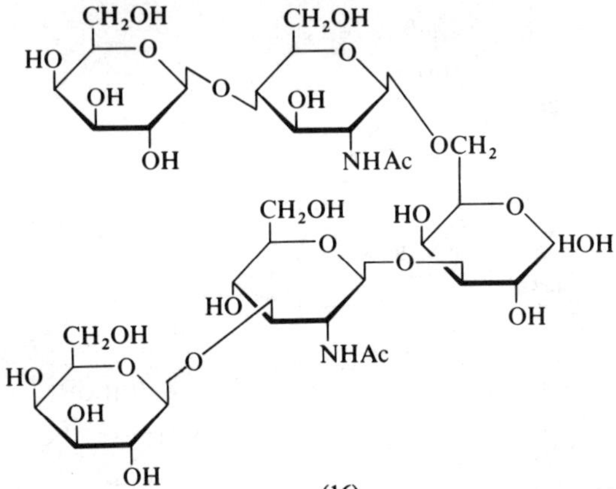

(16)

oryzae investigated.[19] Inhibition constants for the first two compounds were 10 mM and 54 mM respectively and both showed mixed-type inhibition, while the third compound had little effect on enzymic activity.

The 4-aminophenyl glycosides of malto-oligosaccharides have been synthesized enzymically from cyclomaltohexaose and 4-aminophenyl β-D-glucoside using an amylase preparation from *Bacillus macerans*.[20] Utilizing the glucosyl transferase from *A. niger* the 4-aminophenyl glycosides of the isomaltose series of oligosaccharides have also been synthesized enzymically from maltose and 4-aminophenyl β-D-glucoside.

Glycoproteins.—In order to obtain co-polymers which should simulate the behaviour of glycoproteins, block co-polymers with a hydrophilic saccharide block (O^{α} or O^{β}) and a hydrophobic peptide block [poly(γ-benzyl-L-glutamate)] have been synthesized.[21] The α-amino function of the terminal asparagine

[19] K. Omichi, K. Fujii, T. Mizukami, and Y. Matsushima, *J. Biochem. (Japan)*, 1978, **83**, 1443.

[20] J. H. Pazur, Y. Tominaga, K. L. Dreher, L. S. Forsberg, and B. M. Romanic, *J. Carbohydrates, Nucleosides, Nucleotides*, 1978, **5**, 1.

[21] A. Douy and B. Gallot, *Makromol. Chem.*, 1977, **178**, 1595.

residue of glycopeptide fragments (O^{α}, mol. wt. 1.85×10^3, and O^{β}, mol. wt. 3.2×10^3) obtained by pronase digestion of ovomucoid were used to initiate polymerization of the *N*-carboxyanhydride derivative of γ-benzyl glutamate. Saccharide–peptide block co-polymers containing 14–86% polypeptide were prepared.

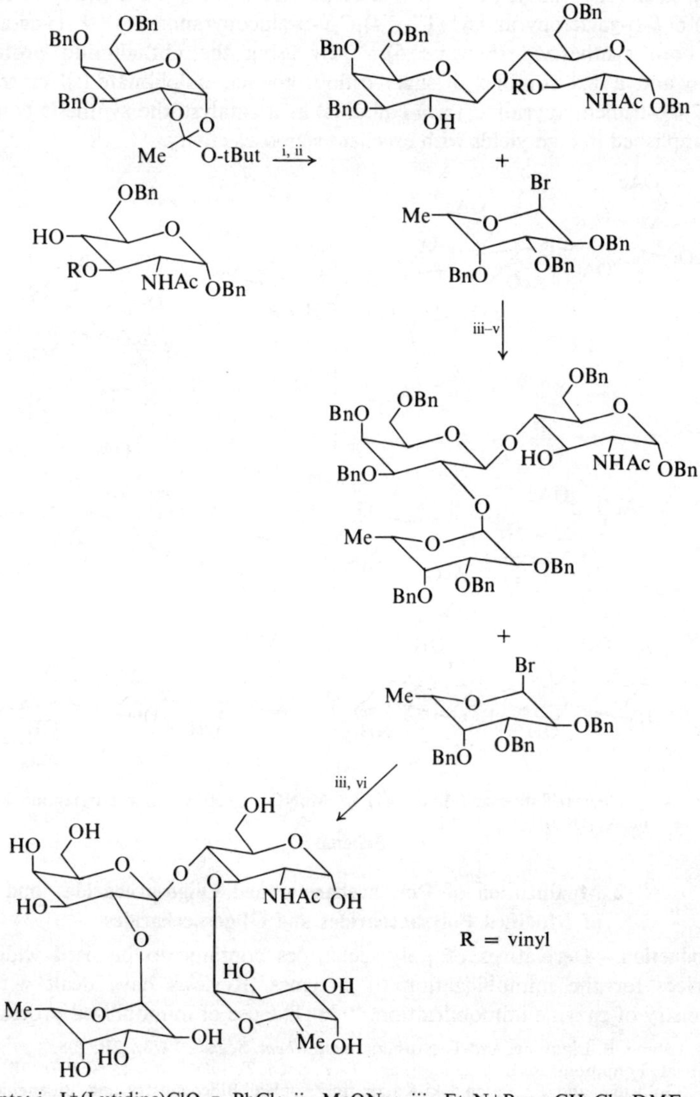

Reagents: i, I^+(Lutidine)ClO_4^-–PhCl; ii, MeONa; iii, $Et_4N^+Br^-$, CH_2Cl_2–DMF; iv, t-BuOK–DMSO; v, $HgCl_2$–HgO; vi, H_2–Pd/C.

Scheme 4

Glycopeptides.—The synthesis of O^3-[O-β-D-galactopyranosyl-(1→4)-O-β-D-xylopyranosyl-L-serine, which forms part of the linkage region between polysaccharide and protein chains in heparin and related glycosaminoglycuronans, has been reported.[22]

Glycolipids.—A benzyl analogue (17) of the tetrasaccharide paragloboside, O-β-D-galactopyranosyl-(1 → 4)-O-2-acetamido-2-deoxy-β-D-glucopyranosyl (1 → 3)-O-β-D-galactopyranosyl-(1 → 4)-O-β-D-glucopyranosyl-(1 → 1)-ceramide, has been synthesized (Scheme 5).[23] By using the phthalimido protecting group and a 1:1 complex of silver trifluoromethanesulphonate (silver triflate) and 2,4,6-trimethylpyridine (*sym*-collidine) as a catalyst, the synthesis could be accomplished in high yields with excellent stereoselectivity.

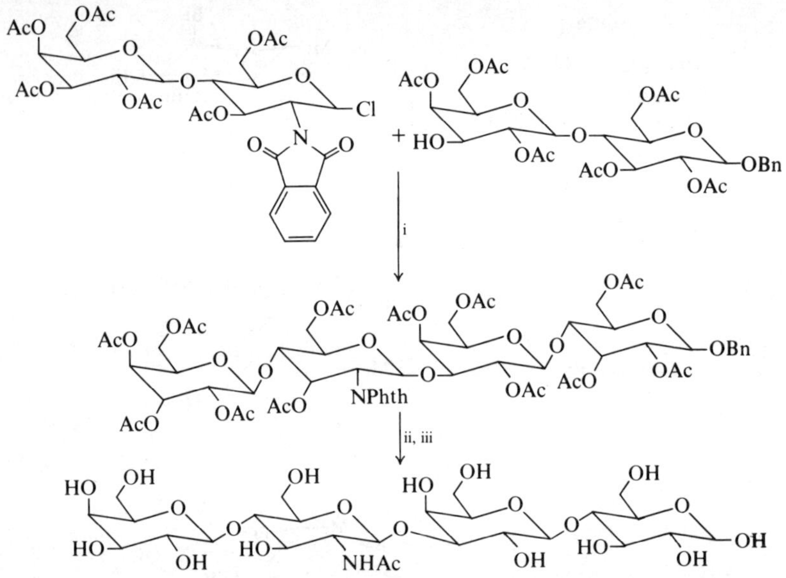

(17)

Reagents: i, silver triflate–*s*-collidine (1/1) in $MeNO_2$, −30 °C; ii, n-butylamine–MeOH; iii, Ac_2O–py

Scheme 5

2 Modification of Polysaccharides and Oligosaccharides and Uses of Modified Polysaccharides and Oligosaccharides

Introduction.—Derivatives of polysaccharides continue to be used widely as matrices for the immobilization of enzymes. Reviews have dealt with the chemistry of enzyme immobilization.[24,25] The use of immobilized proteases as

22 B. Erbing, B. Lindberg, and T. Norberg, *Acta Chem. Scand.*, 1978, **32B**, 308.

23 M. M. Ponpipom, R. L. Bugianesi, and T. Y. Shen, *Tetrahedron Letters*, 1978, 1717.

24 L. Goldstein and E. Katchalski-Katzir, in 'Applied Biochemistry and Bioengineering, Vol. 1, Immobilized Enzyme Principles', ed. L. B. Wingard Jr., E. Katchalski-Katzir, and L. Goldstein, Academic Press, New York, 1976, p. 1.

25 G. Manecke and J. Schlunsen, in 'Polymeric Drugs', ed. L. G. Donaruma and O. Vogl, Academic Press, New York, 1978, p. 39.

anti-thrombogenic drugs has been discussed,[26] and the preparation of soluble and immobilized macromolecular adducts for controlled chemotherapy described.[27] Industrial and research applications of immobilized enzymes and antibodies,[28] and of immobilized enzymes and microbial cells[29] have been reviewed. An article discussing the application of immobilized enzymes in affinity chromatography has appeared.[30] Considerable use of polysaccharide derivatives has been made in the field of affinity chromatography, which has been extensively reviewed.[31] Basic principles of affinity chromatography have been described by several authors.[32,33]

The preparation of immobilized co-enzymes and their applications in general affinity chromatography and as active co-enzymes have been reviewed.[34,35] The most significant methods of immobilization of polynucleotides have been described and their applications discussed.[36,37] Another article reviews affinity electrophoresis, a new separation technique involving electrophoresis on a carrier containing immobilized ligands capable of specific interaction with some component(s) of the mixture to be separated.[38]

Polysaccharide carriers have been widely used for the preparation of immunoadsorbents. Techniques and instrumentation for preparative immunoadsorptive separations have been discussed.[39] The application of 'reversed immunosorbents' to the immobilization of specific antibodies has been outlined,[40] and a review of the use of affinity chromatography for the isolation of immunologically reactive cells has appeared.[41]

Agaroses.—Agarose gel chromatography of a number of proteins and glycoproteins has been investigated in media of different ionic strengths and after application of various amounts of samples.[42] K_{av} was negligibly affected by chromatography conditions, but when K_{av} was plotted against Stokes radius, molecules with different frictional ratios were found to follow different relationships, which were discussed with regard to effects of molecular asymmetry.

A new affinity technique involving 'salting-out' chromatography on columns of

26 N. A. Platé, in Ref. 25, p. 63.
27 A. Zaffaroni and P. Bonsen, in Ref. 25, p. 1.
28 J. F. Kennedy, *Chem. in Britain*, 1978, **14**, 436.
29 I. Chibata and T. Tosa, in Ref. 24, p. 329.
30 'Biotechnological Applications of Proteins and Enzymes', ed. Z. Bohak and N. Sharon, Academic Press, New York, 1977.
31 J. Turková, 'Affinity Chromatography: Journal of Chromatography Library, Vol. 12', Elsevier, Amsterdam, 1978.
32 P. D. G. Dean and M. J. Harvey, in 'Regulation of Enzyme Synthesis and Activity in Higher Plants', ed. H. Smith, Academic Press, New York, 1977, p. 271.
33 P. C. Allen, E. A. Hill, and A. M. Stokes, in 'Plasma Proteins: Analytical and Preparative Techniques', Blackwell Scientific Publications, 1977, p. 145.
34 K. Mosbach, *Adv. Enzymol. and Mol. Biol.*, 1978, **46**, 205.
35 C. R. Lowe, *Trends in Biochemical Sciences*, 1978, **3**, 134.
36 H. Potuzak and P. D. G. Dean, *F.E.B.S. Letters*, 1978, **88**, 161.
37 I. P. Trayer and M. A. Winstanley, *Internat. J. Biochem.*, 1978, **9**, 449.
38 V. Hořejší, M. Tichá, and J. Kocourek, *Trends in Biochemical Sciences*, 1979, **4**, N6.
39 J. W. Eveleigh, *J. Chromatog.*, 1978, **159**, 129.
40 D. E. Levy and J. W. Eveleigh, *J. Immunol. Methods*, 1978, **22**, 131.
41 L. Hudson, *J. Chromatog.*, 1978, **159**, 123.
42 I. Axelsson, *J. Chromatog.*, 1978, **152**, 21.

unsubstituted agarose has been demonstrated with phenylalanyl- and isoleucyl-tRNA synthetases from baker's yeast.[43] After complexation of the crude enzyme mixture with the respective t-RNA the solution is passed through an agarose column packed in 50% saturated ammonium sulphate. Specific elution of the enzyme is achieved using a decreasing salt gradient (50% → 25% w/v).

Agarose itself has also been used as an affinity support for the purification of a lectin from Chinook salmon ova, the action of which is inhibited by D-galactose.[44]

The γ-irradiation-induced coupling of trypsin, pepsin, and ribonuclease to agarose has been described.[45] In the presence of agarose, less aggregation of ribonuclease occurred than in free solution, although some radiation damage was observed.

Acid treatment of agarose increases the number of D-galactose end-groups in the molecule. The hydrolysed matrix has been used successfully for the affinity chromatographic separation of two polypeptide chains from ricin E,[46] for the purification of a galactan-reactive agglutinin from *Tridacna maxima*,[47] and for the purification of a lectin from *Viscum album*.[48]

Co-polymerization of agarose with acrylamide, *N*,*N'*-methylene-*bis*-acrylamide, and acrylamido-6-hexanoic acid or methacryloyl glycylglycine has produced rigid gels suitable for affinity chromatography.[49] Using a carbodi-imide reaction mediator, 4-aminophenyl-lactic acid and 4-aminobenzyl 1-thio-2-acetamido-2-deoxy-β-D-glucopyranoside have been treated with the new supports and the products used for the purification of lactate dehydrogenase and wheat germ agglutinin respectively.

Alkyl derivatives obtained by reacting pentyl and decyl glycidyl ethers with cross-linked agarose have been used for hydrophobic interaction chromatography of human serum albumin, ovalbumin, and phycoerythrin.[50] The effects of neutral salts on the adsorption was investigated and results correlated with the results of c.d. spectroscopic studies. Hydrophobic chromatography using a number of 6-aminoalkyl agaroses has been applied to the purification of acetate kinase from *Escherichia coli*.[51] Supports compared included 1,6-diamino-hexyl-, 1-(*N*-chlorotriazinyl)-1,6-diaminohexyl-, 1,8-diamino-octyl-, carboxy-propionylaminohexyl-, and dimethylaminohexyl- derivatives of agarose. Binding was examined as a function of pH, ionic strength, and other conditions of adsorption–desorption, and the relative merits of each adsorbent were discussed.

Cyclic imidocarbonate derivatives of agarose continue to be used extensively

[43] F. von der Haar, *F.E.B.S. Letters*, 1978, **94**, 371.
[44] E. W. Voss, J. L. Fryer, and G. M. Banowetz, *Arch. Biochem. Biophys.*, 1978, **186**, 25.
[45] J. Brandt and L.-O. Andersson, *Acta Chem. Scand.*, 1976, **30B**, 815.
[46] G. Funatsu, T. Mise, H. Matsuda, and M. Funatsu, *Agric. and Biol. Chem. (Japan)*, 1978, **42**, 851.
[47] B. A. Baldo, W. H. Sawyer, R. V. Stick, and G. Uhlenbruck, *Biochem. J.*, 1978, **175**, 467.
[48] P. Ziska, H. Franz, and A. Kindt, *Experientia*, 1978, **34**, 123.
[49] E. Brown, R. Joyeau, E. Boschetti, and Y. Moroux, *J. Chromatog.*, 1978, **150**, 111.
[50] S. Påhlman, J. Rosengren, and S. Hjertén, *J. Chromatog.*, 1977, **131**, 99.
[51] D. P. Karpavichene, V. V. Kulene, Y. Y. Kulis, and B. S. Kurtinaitene, *Biochemistry (U.S.S.R.)*, 1978, **43**, 353.
[52] C. Fischer, S. P. Rothenberg, and M. Da Costa, *Analyt. Biochem.*, 1978, **85**, 15.
[53] A. Goździcka-Józefiak and J. Augustyniak, *J. Chromatog.*, 1977, **131**, 91.

for the immobilization of biologically active macromolecules and for affinity chromatography.[52,53] Whereas the cyclic imidocarbonate group may not be the only product from the reaction of a polysaccharide with cyanogen bromide, the reaction of a cyclic imidocarbonate group with a protein presumably occurs by nucleophilic attack of an amino-group on the *trans*-imidocarbonate ring (16*a*) to give either an isourea or an *N*-substituted imidocarbonate, or an *N*-substituted carbonate (see vol. 7, p. 511; Vol. 10, p. 433).

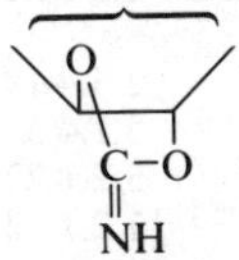

(16a)

Recent references to the preparation of active immobilized forms of enzymes (for use as immobilized enzymes) and immunologically active macromolecules (for use as immunoadsorbents) and of various affinants (for use as affinity chromatography matrices) by such methods are summarized in Tables 1,[54–78] 2,[79–110] and 3,[111–183] respectively.

54 Ma. Naoi, Mi. Naoi, and K. Yagi, *Biochim. Biophys. Acta*, 1978, **523**, 19.
55 R. Horowitz and G. M. Whitesides, *J. Amer. Chem. Soc.*, 1978, **100**, 4632.
56 J. Tramper, F. Müller, and H. C. van der Plas, *Biotechnol. and Bioeng.*, 1978, **20**, 1507.
57 V. G. Shore, A. S.-W. Sae, and B. Shore, *Biochim. Biophys. Acta*, 1978, **529**, 319.
58 A. Spiro-Kern and P. S. Chen, *Insect Biochem.*, 1977, **7**, 453.
59 J. S. Patton, P. A. Albertsson, C. Erlanson, and B. Borgström, *J. Biol. Chem.*, 1978, **253**, 4195.
60 S. Sakamoto, M. Sakamoto, P. Goldhaber, and M. J. Glimcher, *Arch. Biochem. Biophys.*, 1978, **188**, 438.
61 D. Dhermy, O. Bournier, and P. Boivin, *Biochem. Biophys. Res. Comm.*, 1978, **85**, 906.
62 A. W. Schram, M. N. Hamers, E. Oldenbroek-Haverkamp, A. Strijland, A. de Jonge, F. A. J. T. M. van den Bergh, and J. M. Tager, *Biochim. Biophys. Acta*, 1978, **527**, 456.
63 V. I. Muronetz, R. A. Asryants, and N. K. Nagradova, *Biochim. Biophys. Acta*, 1978, **525**, 291.
64 T. O. Golovina, T. V. Cherednikova, A. T. Mevkh, and N. K. Nagradova, *Analyt. Biochem.*, 1977, **83**, 778.
65 G. Spies-Karotkin and S. M. Constantinides, *Mol. and Cell. Biochem.*, 1978, **21**, 153.
66 R. Toftgård, T. Anfält, and A. Granéli, *Analyt. Chim. Acta*, 1978, **99**, 383.
67 A. Kilara, K. M. Shahani, and F. W. Wagner, *Biotechnol. and Bioeng.*, 1977, **19**, 1703.
68 B. Mattiasson, *Biotechnol. and Bioeng.*, 1977, **19**, 777.
69 A. Kato, K. Ogino, N. Matsudomi, and K. Kobayashi, *Agric. and Biol. Chem.* (*Japan*), 1977, **41**, 1925.
70 Yu. V. Vertiev, G. K. Belyanskaya, and Yu. V. Ezepchuk, *Biochemistry* (*U.S.S.R.*), 1977, **42**, 1365.
71 A. P. Corfield, J.-M. Beau, and R. Schauer, *Z. physiol. Chem.*, 1978, **359**, 1335.
72 F. Gubenšek, D. Zunic, and J. Babnik, *Periodicum Biologorum*, 1978, **80**, 97.
73 Ch. Flemming, H. Göbel, H. Wand, A. Gabert, and W. Bock, *Acta Biol. Med. Germanica*, 1978, **37**, 179.
74 J. E. Brown, C. L. Carton, A. S. Dietz, R. F. Baugh, and C. Hougie, *Thrombosis and Haemostasis*, 1978, **40**, 144.
75 H. Y. K. Chuang, S. F. Mohammad, and R. G. Mason, *Res. Comm. Chem. Pathol. Pharm.*, 1978, **21**, 379.
76 G. A. Kochetov and O. N. Solov'eva, *Biochemistry* (*U.S.S.R.*), 1977, **42**, 1477.
77 D. B. Johnson and M. P. Coughlan, *Biotechnol. and Bioeng.*, 1978, **20**, 1085.
78 S. Pallante, M. A. Lyle, and C. Fenselau, *Drug Metabolism and Disposition*, 1978, **6**, 389.
79 R. P. Erickson and R. Sandman, *Experientia*, 1977, **37**, 14.
80 B. Mattiasson, *F.E.B.S. Letters*, 1977, **77**, 107.

Table 1 *Use of agarose cyclic imidocarbonates for the preparation of active immobilized enzymes*

Enzyme coupled to agarose cyclic imidocarbonate	*E.C. No.*	*Agarose intermediate*[a]	*Ref.*
D-Amino acid oxidase	1.4.3.3	(18), (19)	54
Carbonate dehydratase	4.2.1.1	—	55
Catalase	1.11.1.6	—	56
Chymotrypsin	3.4.21.1	—	57, 58
Colipase		—	59
Collagenase	3.4.4.19	—	60
Deoxyribonuclease I	3.1.4.5	—	61
α-D-Galactosidase	3.2.1.22	—	62
Glyceraldehyde 3-phosphate dehydrogenase	1.2.1.12	—	63, 64, 65
Lactate dehydrogenase	1.1.1.27	—	66
Lipase	3.1.1.3	(20)	67
		—	59
Lysozyme	3.2.1.17	—	68, 69
Neuraminidase	3.2.1.18	—	70, 71
Papain	3.4.22.2	(20)	67
Phospholipase	3.1.1.4	—	72
Polygalacturonase	3.2.1.15	—	73
Superoxide dismutase	1.15.1.1	—	56
Thrombin	3.4.21.5	—	74, 75
Transketolase	2.2.1.1	—	76
Trypsin	3.4.21.4	—	57
Urate oxidase	1.7.3.3	—	77
UDP-D-glucuronosyl transferase	2.4.1.17	—	78
Xanthine oxidase	1.2.3.2	—	56, 77

[a] See text p. 527 and Scheme 6.

[81] B. Mattiasson, C. Borrebaeck, B. Sanfridson, and K. Mosbach, *Biochim. Biophys.* 1977, **483**, 221.
[82] F. K. Khalil, E. Stephen, and S. Guilbaud, *Clinica. Chim. Acta*, 1978, **88**, 167.
[83] J. L. Spivak, D. Small, J. H. Shaper, and M. D. Hollenberg, *Blood*, 1978, **52**, 1178.
[84] H. J. van Leeuwen, *Agric. and Biol. Chem.* (*Japan*), 1978, **42**, 1375.
[85] C.-H. Wong, C.-L. Ho, and K.-T. Wang, *J. Chromatog.*, 1978, **154**, 25.
[86] G. Marcoullis, E. M. Salonen, and R. Gräsbeck, *Biochim. Biophys. Acta*, 1977, **495**, 336.
[87] D. H. Conrad and A. Froese, *Immunochemistry*, 1978, **15**, 283.
[88] N. S. Taylor and M. Pollack, *Infection and Immunity*, 1978, **19**, 66.
[89] M. R. A. Morgan, P. J. Brown, M. J. Leyland, and P. D. G. Dean, *F.E.B.S. Letters*, 1978, **87**, 239.
[90] E. Ruoslahti, *J. Immunol.*, 1978, **121**, 1687.
[91] V. Vila, E. Reganon, and J. Aznar, *Clinica. Chim. Acta*, 1978, **87**, 245.
[92] R. T. Proffitt, L. Sankaran, and B. M. Pogell, *Analyt. Biochem.*, 1978, **88**, 236.
[93] D. C. Rueger, D. Dahl, and A. Bignami, *Analyt. Biochem.*, 1978, **89**, 360.
[94] L. T. Clement, A. M. Kask, and E. M. Shevach, *Immunochemistry*, 1978, **15**, 393.
[95] G. L. Manderino, G. T. Gooch, and A. B. Stavitsky, *Cellular Immunol.*, 1978, **41**, 264.
[96] K. Berg, R. Alacam, R. D. Hamilton, and I. Heron, *Texas Rep. Biol. Med.*, 1977, **35**, 187.
[97] J. E. McEntire, *J. Immunol. Methods*, 1978, **24**, 39.
[98] J. S. Welch and C. Dobson, *Trans. Roy. Soc. Trop. Med. Hygiene*, 1978, **72**, 282.
[99] G. V. Hillyer and M. Cervoni, *J. Immunol. Methods*, 1978, **20**, 385.
[100] J. Folkersen, B. Teisner, J. Westergaard, and S. E. Svehag, *J. Immunol. Methods*, 1978, **23**, 137.
[101] J. Folkersen, B. Teisner, S. Ahrons, and S. E. Svehag, *J. Immunol. Methods*, 1978, **23**, 117.
[102] R. Walsh, D. T. Mason, M. J. Tonkon, and J. Wickman-Coffelt, *Prep. Biochem.*, 1976, **6**, 177.
[103] F. Shimizu, H. Mossmann, K. Himmelspach, H. Takamiya, and A. Vogt, *Res. Expt. Med.*, 1978, **173**, 165.

Table 2 *Use of agarose cyclic imidocarbonates for the preparation of immunoadsorbents*

Immunologically active component coupled to agarose cyclic imidocarbonate	*Use of product*	*Ref.*
Anti-(β-D-2-acetamido-2-deoxyglucosidase) antibody	Purification of β-D-2-acetamido-2-deoxyglucosidase	79
Anti-(serum albumin) antibody	Development of an enzyme thermistor for measurement of a variety of substrates using different immobilized enzymes	80
	Development of a thermometric enzyme linked immunoadsorbent assay for determination of serum albumin	81
	Removal of serum albumin during purification of α-fetoprotein.	82
	Removal of albumin during purification of erythropoietin	83
Anti-(bitter peptide from tryptic hydrolysate of casein) antibody	Removal of bitter peptide from tryptic casein hydrolysates	84
Anti-cardiotoxin antibody	Purification of synthetic cardiotoxin	85
Anti-(pig cobalophilin) antibody	Purification of vitamin B-12 binding proteins	86
Anti-(2,4-dinitrophenyl) antibody	Isolation of immunoglobulin E receptor complexes from rat basophilic leukaemia cells	87
Anti-(*Pseudomonas aeruginosa* exotoxin) antibody	Purification of *Pseudomonas aeruginosa* exotoxin	88
Anti-ferritin antibody	Purification of ferritin by electrophoretic desorption from the immobilized antibody	89
Anti-(human α-fetoprotein) antibody	Purification of bovine serum α-fetoprotein	90
Anti-fibrinogen antibody	Isolation of fibrin(ogen) degradation products from normal plasma	91
Anti-(D-fructose 1,6-diphosphatase) antibody	Quantitative immunochemical determination of radioactive proteins synthesized by chick embryo hepatocyte cultures	92
Anti-(bovine glial fibrillary acidic protein) antibody	Purification of bovine glial fibrillary acidic protein from brain extracts	93
Anti-(Ia antigen from strain 2 and 13 guinea-pig lymphocytes) antibody	Purification of detergent-solubilized Ia antigens	94
Anti-(rabbit immunoglobulin G) antibody	Purification of goat anti-rabbit Fab′ fragments	95
Anti-(human leucocyte interferon) antibody	One-step purification of human leucocyte, lymphoblastoid, and fibroblast interferon	96
Anti-(human intrinsic factor) antibody	Purification of vitamin B-12 binding proteins	86
Anti-(human α-2-macroglobulin) antibody	Purification of α-2-macroglobulin	97

Table 2 (*cont.*)

Immunologically active component coupled to agarose cyclic imidocarbonate	*Use of product*	*Ref.*
Anti-(oestriol 16 α-D-glucopyranosyluronic acid) antibody	Purification of oestriol 16 α-D-glucopyranosyluronic acid by electrophoretic desorption from the immobilized antibody	89
Anti-[parasite antigen (*Dirofilaria immitis*, *Toxocara canis*, *Angiostrongyeus cantonensis* and *Ascaris lumbricoides*)] antibodies	Purification of parasite antigens for immunodiagnosis of parasitic zoonoses	98
Anti-[parasite antigen (*Fasciola hepatica* trematode)] antibody	Isolation of specific parasite antigen from crude extracts	99
Anti-phospholipase A_2 antibody	Purification of phospholipase A_2 from *Vipera ammodytes*	68
Anti-(pregnancy zone protein) antibody	Purification of pregnancy zone protein	100, 101
Anti-renin antibody	Purification of canine plasma renin	102
Anti-transferrin antibody	Removal of contaminant transferrin during purification of α-fetoprotein	82
2,4-Dinitrophenyl$_{33}$-human serum albumin	Purification of specific anti-(2,4-dinitrophenyl) antibodies	103
Fibrinogen	Purification of anti-fibrinogen antibodies	104
Haptoglobin	Isolation of anti-haptoglobin antibodies	105
Heparin	Isolation of specific anti-heparin antibodies	60
Immunoglobulin G (heterologous) (rabbit or human)	Removal of interspecies reactivity from anti-immunoglobulin G sera (human or rabbit)	106
unc 54-Myosin	Preparation of specific anti-(unc 54-myosin) antibodies	107
Poly(L-Ala)-poly(L-Lys)-poly-(L-Tyr-L-Glu)	Purification of antibodies raised against the synthetic antigen	108
Protein A (ex. *Staphylococcus aureus*)	Purification of antigen–antibody complexes from incubates of viral polypeptides and virus-specific antisera	109
Staphylococcus aureus cells	Purification of mouse spleen B lymphocytes	110

[104] A. W. Stemberger and H. Hormann, *Thrombosis and Haemostasis*, 1978, **39**, 574.
[105] J. Osada and W. Dobryszycka, *Immunochemistry*, 1978, **15**, 591.
[106] M. E. Johansson and J. Å. Espmark, *J. Immunol. Methods*, 1978, **21**, 285.
[107] F. Schachat, R. L. Garcea, and H. F. Epstein, *Cell*, 1978, **15**, 406.
[108] H. Ungar-Waron, A. Gluckman, R. Meirom, and Z. Trainin, *J. Immunol. Methods*, 1978, **23**, 213.
[109] E. C. Hayes, L. L. Wright, and H. J. Zweerink, *Analyt. Biochem.*, 1978, **91**, 276.
[110] V. Ghetie, G. Mota, and J. Sjöquist, *J. Immunol. Methods*, 1978, **21**, 133.
[111] T. Ishibashi, S. Gasa, I. Ohkubo, and A. Makita, *Biochim. Biophys. Acta*, 1978, **525**, 265.
[112] I. Emöd and B. Keil, *F.E.B.S. Letters*, 1977, **77**, 51.
[113] C. R. Brown and E. F. Hartree, *Biochem. J.*, 1978, **175**, 227.
[114] J. A. Swack, G. L. Zander, and M. F. Utter, *Analyt. Biochem.*, 1978, **87**, 114.
[115] V. M. Stepanov, G. N. Rudenskaya, V. V. Yanonis, V. I. Ostoslavskaya, M. V. Gonchar, E. K. Kotlova, and A. Ya. Strongin, *Bioorg. Khim.*, 1978, **4**, 1256.
[116] P. V. Wagh, *J. Chromatog.*, 1978, **152**, 565.
[117] N. Elango, S. Janaki, and A. R. Rao, *Biochem. Bicphys. Res. Comm.*, 1978, **83**, 1388.

Table 3 *Use of agarose cyclic imidocarbonates for the preparation of affinity chromatography materials without linkage extension*

Ligand or affinant coupled to agarose cyclic imidocarbonate	*Use of product*	*Ref.*
L-Alanine	Purification of nucleosidase diphosphatase	111
Albumin	Stabilization of co-immobilized enzymes	56
4-Aminophenyl mercuric acetate	Affinity chromatography and purification of *Clostridium* collagenase and clostripapain	112
4-(4′-Aminophenoxypropoxy)-benzamidine	Affinity chromatography of acrosin	113
L-Arginine monochlorohydrate	Affinity chromatography of *Clostridium* collagenase and clostripapain	112
Avidin	Purification of inactive biotin enzymes from crude cellular extracts	114
Bacitracin	Affinity chromatography and purification of carboxylic proteinases	115
Benzidine	Affinity chromatography purification of jack bean α-D-mannosidase	116
Blue dextran	Isolation of glyoxylase I	117
	Affinity chromatography and purification of mammalian fibroblast interferons	118
Bovine submaxillary mucin	Affinity chromatography of lectins of Japanese horseshoe crab (*Tachypleus tridentatus*)	119
Casein	Affinity chromatography of rat liver nuclear protein kinases	120
Casein hydrolysate	Purification of acid protease (*Aspergillus niger*)	121
Collagen	Purification of D-galactosyl-hydroxy-L-lysyl-D-glucosyl transferase	122
Concanavalin A	Affinity chromatography of proteases of germinating peanut seeds	123
	Removal of albumin and immunoglobulins from human serum	124
	Affinity chromatography of allergens from bovine hair and dander	125
	Purification of anti-picryl T suppressor factor	126
	Purification of arylsulphatase A	127
	Affinity chromatography of collagen glucosyltransferase	128
	Purification of placental α-fetoprotein	78

[118] E. Bollin, K. Vastola, D. Oleszek, and E. Sulkowski, *Prep. Biochem.*, 1978, **8**, 259.
[119] S. Shimizu, M. Ito, and M. Niwa, *Biochim. Biophys. Acta*, 1977, **500**, 71.
[120] W. Thornburg, R. Kieras, and T. J. Lindell, *Prep. Biochem.*, 1978, **8**, 133.
[121] A. Shinmyo, I. K. Davis, F. Nomoto, T. Tahara, and T. Enatsu, *European J. Appl. Microbiol. Biotechnol.*, 1978, **5**, 59.
[122] H. Anttinen, R. Myllylä, and K. I. Kivirikko, *Biochem. J.*, 1978, **175**, 737.
[123] R. B. Vanhuystee, *Z. Pflanzenphysiol.*, 1978, **89**, 51.
[124] Z. Hrkal and Z. Vodrážka, *J. Chromatog.*, 1977, **135**, 193.
[125] T. C. Bøg-Hansen, P. Prahl, and H. Løwenstein, *J. Immunol. Methods*, 1978, **22**, 293.
[126] G. L. Asherson, M. Zembala, and J. Noworolski, *Immunology*, 1978, **35**, 1051.
[127] A. Ahmad, A. Surolia, and B. K. Bachhawat, *Indian J. Biochem. Biophys.*, 1977, **14**, 235.
[128] H. Anttinen, R. Myllylä, and K. I. Kivirikko, *European J. Biochem.*, 1977, **78**, 11.

Table 3 (*cont.*)

Ligand or affinant coupled to agarose cyclic imidocarbonate	*Use of product*	*Ref.*
	Affinity chromatography and purification of galactosyl-hydroxylysyl glucosyltransferase	122
	Affinity chromatography of glycopeptides obtained by proteolytic digestion of rat-brain glycoproteins	129
	Affinity chromatography of phytotoxic glycopeptide mixture from *Ceratocystis ulmi*	130
	Purification of pituitary lutrotropin and thyrotropin	131
	Purification of human fibroblast interferon	132
	Affinity chromatography of lymphocyte plasma membrane subfractions	133
	Affinity chromatography of sialyltransferases	134
	Affinity chromatography and purification of *Adenia digitata* phytohaemagglutinin ('modeccin')	135
	Identification of a breast-tumour-associated orosomucoid	136
	Purification of an oviducal enzyme from *Bufo arenarum*	137
	Purification of tartrate-sensitive acid phosphatase	138
	Removal of unreacted enzyme from synthetic enzyme–dextran conjugates after glycosylation	139
Cytochrome *c*	Isolation of a multiprotein complex containing cytochromes *b* and *c* from *Neurospora crassa* mitochondria	140
Deoxyribonucleic acid (double- and single-stranded)	Purification of non-histone chromatin proteins with preferential affinity for single-stranded deoxyribonucleic acid	141
Dextran sulphate	Purification of hepatitis B surface antigen	142

[129] T. Krusius and J. Finne, *European J. Biochem.*, 1978, **84**, 395.
[130] G. Strobel, N. van Alfren, K. D. Hapner, M. McNeil, and P. Albersheim, *Biochim. Biophys. Acta*, 1978, **538**, 60.
[131] G. A. Bloomfield, M. R. Faith, and J. G. Pierce, *Biochim. Biophys. Acta*, 1978, **533**, 371.
[132] J. W. Heine, A. J. Mikulski, E. Sulkowski, and W. A. Carter, *Arch. Virol.*, 1978, **57**, 185.
[133] K. Resch, A. Loracher, B. Mahler, M. Stoeck, and H. N. Rode, *Biochim. Biophys. Acta*, 1978, **511**, 176.
[134] C.-K. Liu, R. Schmied, E. M. Greenspan, and S. Waxman, *Biochim. Biophys. Acta*, 1978, **522**, 375.
[135] S. Olsnes, T. Haylett, and K. Refsnes, *J. Biol. Chem.*, 1978, **253**, 5069.
[136] L. D. Papsidero, T. Nemoto, and T. M. Chu, *Cancer Biochem. Biophys.*, 1978, **3**, 15.
[137] D. C. Miceli, S. N. Fernández, and E. J. del Pino, *Biochim. Biophys. Acta*, 1978, **526**, 289.
[138] J.-J. Helwig, A. A. Farooqui, C. Bollack, and P. Mandel, *Biochem. J.*, 1978, **175**, 321.
[139] J. J. Marshall, J. D. Humphreys, and S. L. Abramson, *F.E.B.S. Letters*, 1977, **83**, 249.
[140] H. Weiss and B. Juchs, *European J. Biochem.*, 1978, **88**, 17.
[141] D. L. Bidney and G. R. Reeck, *Biochem. Biophys. Res. Comm.*, 1978, **85**, 1211.
[142] M. Einarsson, L. Kaplan, and G. Utter, *Vox Sanguinis*, 1978, **35**, 224.

Table 3 (*cont.*)

Ligand or affinant coupled to agarose cyclic imidocarbonate	*Use of product*	*Ref.*
1,2-Diaminoethane	Purification of L-fucosyltransferase	143
3,9-Diamino-7-ethoxyacridine lactate ('Rivanol')	Purification of *Clostridium* sp. collagenase and clostripapain	112
1,6-Diaminohexane	Affinity chromatography of aminoacyl-tRNA synthetases	144
Dodecylamine	Affinity chromatography and purification of lecithin : cholesterol acyltransferases	145
Ferredoxin	Purification of spinach ferredoxin-$NADP^+$ reductase	146
	Purification of spinach ferredoxin nitrite reductase	147
Ferredoxin, dithionite-reduced	Purification of *Anacystis nidulans* ferredoxin-nitrate reductase and NADP reductase	148
Fetuin	Purification of *Datura stramonium* phytohaemagglutinin	149
	Affinity chromatography of *Phaseolus vulgaris* phytohaemagglutinin	150
Fetuin, neuraminidase-treated	Purification of *Adenia digitata* phytohaemagglutinin ('modeccin')	135
Fibrin	Affinity chromatography and purification of fibrinogens from plasma of normal and cobalt-treated rabbits	151
Fibrinogen	Determination of soluble fibrin monomer complexes by affinity chromatography.	104
Fibrinogen, fibrinogen fragment D	Affinity chromatography and quantitation of soluble fibrin in human plasma	152
Fibrinogen, fibrinogen fragment D, crosslinked fibrin, noncrosslinked fibrin	Affinity chromatography of soluble fibrin in human plasma	153
D-Glucan–protein complex (*Saccharomyces cerevisiae*)	Isolation of *exo*-(1 → 3)-*β*-D-glucanase	154
Glutathione	Affinity chromatography of proteases of germinating peanut seeds	123
Gramicidin S	Affinity chromatography and purification of *Bacillus subtilis* metalloendoproteinase	155

[143] M. N. Pérez-Gonzalez, P. Broquet, and P. Louisot, *Compt. rend.*, 1977, **285D**, 729.
[144] H. Jakubowski, *J. Chromatog.*, 1977, **139**, 331.
[145] A. G. Lacko and T. F. Chen, *J. Chromatog.*, 1977, **130**, 446.
[146] M. Shin and R. Oshino, *J. Biochem.* (*Japan*), 1978, **83**, 357.
[147] S. Ida, *J. Biochem.* (*Japan*), 1977, **82**, 915.
[148] C. Manzano, P. Candau, and M. G. Guerrero, *Analyt. Biochem.*, 1978, **90**, 408.
[149] D. C. Kilpatrick and M. M. Yeoman, *Biochem. J.*, 1978, **175**, 1151.
[150] A. Pusztai and J. C. Stewart, *Biochim. Biophys. Acta*, 1978, **536**, 38.
[151] S. Krantz and M. Lober, *Acta Biol. Med. Germanica*, 1978, **37**, 545.
[152] N. Semeraro and D. Collen, *Thrombosis Res.*, 1978, **12**, 1059.
[153] A. Bini, N. Semeraro, and D. Collen, *Thrombosis Res.*, 1978, **13**, 443.
[154] M. Edward and R. J. Sturgeon, *Carbohydrate Res.*, 1977, **57**, 3.
[155] T. I. Vaganova, L. V. Lastovetskaya, A. Ya. Strongin, L. A. Lyublinskaya, and V. M. Stepanov, *Biochemistry* (*U.S.S.R.*), 1976, **41**, 1805.

Table 3 (*cont.*)

Ligand or affinant coupled to agarose cyclic imidocarbonate	*Use of product*	*Ref.*
Guanosine diphosphate : EF-Tu (protein synthesis elongation factor) complex	Purification of EF-Ts, a protein synthesis elongation factor	156
Haemoglobin	Affinity chromatography separation of renin and pseudo-renin	157
	Purification of proteolytic enzymes in green wheat leaves	158
Heparin	Affinity chromatography separation of ribosomes from reticulocyte postmitochondrial supernatants	159, 160
	Purification of hepatitis B surface antigen	142
	Affinity chromatography of rat high density lipoproteins	161
	Purification of oestradiol receptor of calf uterus	162
	Purification of collagenase	60
Histones	Affinity chromatography of rat liver nuclear protein kinases	120
Kunitz bovine trypsin inhibitor	Purification of trypsin	163
Lens culinaris lectin	Purification of detergent-solubilized Ia antigens	94
L-Lysine	Isolation of plasminogen activator	164
Micrococcus lysodeikticus cells (lysozyme lysate)	Affinity chromatography purification of human lysozyme	165
D-Mannan–protein complex (*Saccharomyces cerevisiae*)	Isolation of α-D-mannosidase from jack bean	154
Modulator protein (rabbit skeletal muscle)	Purification of modulator-deficient myosin light chain kinase	166
	Isolation of heat-stable inhibitor protein of Ca^{2+}-activated cyclic nucleotide phosphodiesterase	167
Non-intrinsic factor cobalamin-binding protein	Isolation of cobalamin and cobalamin analogues	168
Ovalbumin	Affinity chromatography of glycosidases	169

[156] G. R. Jacobson and J. P. Rosenbusch, *F.E.B.S. Letters*, 1977, **79**, 8.
[157] H. J. Chou, J. H. Shaper, and R. I. Gregerman, *Biochim. Biophys. Acta*, 1978, **524**, 183.
[158] G. J. T. Frith, L. B. Swinden, and M. J. Dalling, *Plant Cell Physiol.*, 1978, **19**, 1029.
[159] J. Hradec and O. Kříž, *Biochem. J.*, 1978, **173**, 349.
[160] J. Hradec and Z. Dušek, *Biochem. J.*, 1978, **172**, 1.
[161] C. Hay, J. A. Rooke, and E. R. Skinner, *F.E.B.S. Letters*, 1978, **91**, 30.
[162] A. M. Molinari, N. Medici, B. Moncharmont, and G. A. Puca, *Proc. Nat. Acad. Sci., U.S.A.*, 1977, **74**, 4886.
[163] R. S. Temler and J. H. R. Kägi, *Enzyme*, 1977, **22**, 249.
[164] R. Radcliffe and T. Heinze, *Arch. Biochem. Biophys.*, 1978, **189**, 185.
[165] T. Yoshimoto, M. Tobiishi, and D. Tsuru, *J. Biochem.* (*Japan*), 1976, **80**, 703.
[166] H. Yazawa and K. Yagi, *J. Biochem.* (*Japan*), 1978, **84**, 1259.
[167] R. K. Sharma, R. Desai, T. R. Thompson, and J. H. Wang, *Canad. J. Biochem.*, 1978, **56**, 598.
[168] J. F. Kolhouse and R. H. Allen, *Analyt. Biochem.*, 1978, **84**, 486.
[169] R. T. Dean, *J. Chromatog.*, 1978, **150**, 279.

Table 3 (*cont.*)

Ligand or affinant coupled to agarose cyclic imidocarbonate	*Use of product*	*Ref.*
Phaseolus vulgaris globulin	Isolation of large amounts of the *P. vulgaris* globulin	170
L-Phenylalanine	Purification of phenylalanine ammonia lyase	171
Phosvitin	Affinity chromatography of rat liver nuclear protein kinases	120
Phytohaemagglutinin	Purification of erythropoietin	83
Phytolacca americana mitogens Pa1 and Pa2	Purification by affinity chromatography of membrane receptors for Pa1 and Pa2	172
Picrylalbumin	Purification of anti-(picryl T) suppressor factor	126
Poly(adenylic acid)	Affinity chromatography of poly(uridylic acid) sequences and negative strands of Sindbis virus-specific RNA's	173
Poly(L-arginine)	Affinity chromatography of *Clostridium* collagenase and clostripapain	112
Poly(uridylic acid)	Purification of mouse interferon	174
Prealbumin	Purification by affinity chromatography of retinol-binding protein	175
Ribonucleic acid	Isolation of a proviral coding strand of avian myeloblastosis virus from leukaemic chicken myeloblast	176
Soybean trypsin inhibitor	Purification by affinity chromatography of trypsin	177
	Purification by affinity chromatography of chymotrypsin	178
	Affinity chromatography of proteases of germinating peanut seeds	123
Streptomyces albogriseolus protease inhibitor (S-SI)	Affinity chromatography and purification of chymotrypsin and microbial alkaline proteases	179
α-Thrombin	Purification by affinity chromatography of heparin	180
Transferrin	Affinity chromatographic separation of reticulocytes from erythrocytes and leucocytes	181

[170] D. R. Stockman, T. C. Hall, and D. S. Ryan, *Plant Physiol.*, 1976, **58**, 272.
[171] S. Gupta, *Proc. Austral. Biochem. Soc.*, 1978, **11**, 22.
[172] K. Yokoyama, T. Terao, and T. Osawa, *Biochim. Biophys. Acta*, 1978, **538**, 384.
[173] A. Margotat, J. Laplane, M. R. Pisano, and J. Nicoli, *Ann. Microbiol.*, 1976, **127B**, 243.
[174] J. de Maeyer-Guignard, M. G. Tovey, I. Gressner, and E. de Maeyer, *Nature*, 1978, **271**, 622.
[175] G. Fex and B. Hansson, *Biochim. Biophys. Acta*, 1978, **537**, 358.
[176] K. Som and D. P. Nayak, *Analyt. Biochem.*, 1978, **87**, 527.
[177] S. Katoh, T. Kambayashi, R. Deguchi, and F. Yoshida, *Biotechnol. and Bioeng.*, 1978, **20**, 267.
[178] D. Gabel, V. Kasche, H. Amnéus, and H. Lundqvist, in 'Affinity Chromatography', ed. O. Hoffmann-Ostenhof, Pergamon Press, 1978, p. 99.
[179] S. Sato, T. Kimura, and S. Murao, *Agric. and Biol. Chem.* (*Japan*), 1975, **39**, 415.
[180] M. J. Griffith, H. S. Kingdon, and R. L. Lundblad, *Biochem. Biophys. Res. Comm.*, 1978, **83**, 1198.
[181] N. D. Light and M. J. A. Tanner, *Analyt. Biochem.*, 1978, **87**, 263.

Table 3 (*cont.*)

Ligand or affinant coupled to agarose cyclic imidocarbonate	*Use of product*	*Ref.*
Troponin I	Isolation by affinity chromatography of troponin C	182
Wheat germ haemagglutinin	Affinity chromatography of glycopeptides obtained by proteolytic digestion of rat-brain glycoproteins	129
	Affinity chromatographic separation of mouse bone marrow cells bearing different amounts of surface immunoglobulin	183
	Purification by affinity chromatography of erythropoietin	83

These derivatives are listed according to the uses prescribed since, for example, the insoluble form of an enzyme used for affinity chromatography may not be enzymically active.

Agarose derivatives have been used in a variety of novel procedures applying the principles of affinity chromatography. A new preparative technique involves use of immobilized antibodies to ferritin or oestriol-16 α-D-glucopyranosyluronic acid as affinity gels which are placed on top of polyacrylamide gels after pre-electrophoresis.[89] By including an affinity ligand in the intermediate gel when performing crossed immunoelectrophoresis with a mixture, different heights of 'rockets' are obtained in the final gel; this allows prediction of optimal use of affinity chromatography for the system under consideration. This has been successfully demonstrated using concanavalin A for the fractionation of allergens from bovine hair and dander.[125] On the basis of predictions made from immunoelectrophoresis results, optimal conditions were chosen for affinity chromatography of small amounts of allergen.

The temperature at which affinity chromatography is performed has been shown to be critical for the isolation of ribosomes on heparin–agarose derivatives; the ribosomes are retained at 4 °C but do not bind at 22 °C.[159]

Cross-reactive heterologous antibodies have been used to prepare immunoadsorbents possessing very low affinity for bovine α-fetoprotein, allowing elution of the bound protein without resort to chaotropic elution buffers.[90]

Using an agarose derivative of anti-(pregnancy zone protein) antibody,[101] the use of an automated immunospecific affinity chromatography system for continuous use has been demonstrated for the purification of pregnancy zone protein.[100] The automated system includes time-controlled sample application, washing and elution steps, and automatic dialysis of eluted fractions.

Immobilized antibodies to human serum albumin have been employed in a general enzyme thermistor which can be used for measurement of a number of substrates.[80] The appropriate enzyme is coupled to human serum albumin and mixed with the immobilized antibodies in the thermistor unit. The heat produced

[182] J. F. Head, R. A. Weeks, and S. V. Perry, *Biochem. J.*, 1977, **161**, 465.
[183] N. A. Nicola, A. W. Burgess, D. Metcalf, and F. L. Battye, *Austral. J. Exp. Biol. Med. Sci.*, 1978, **56**, 663.

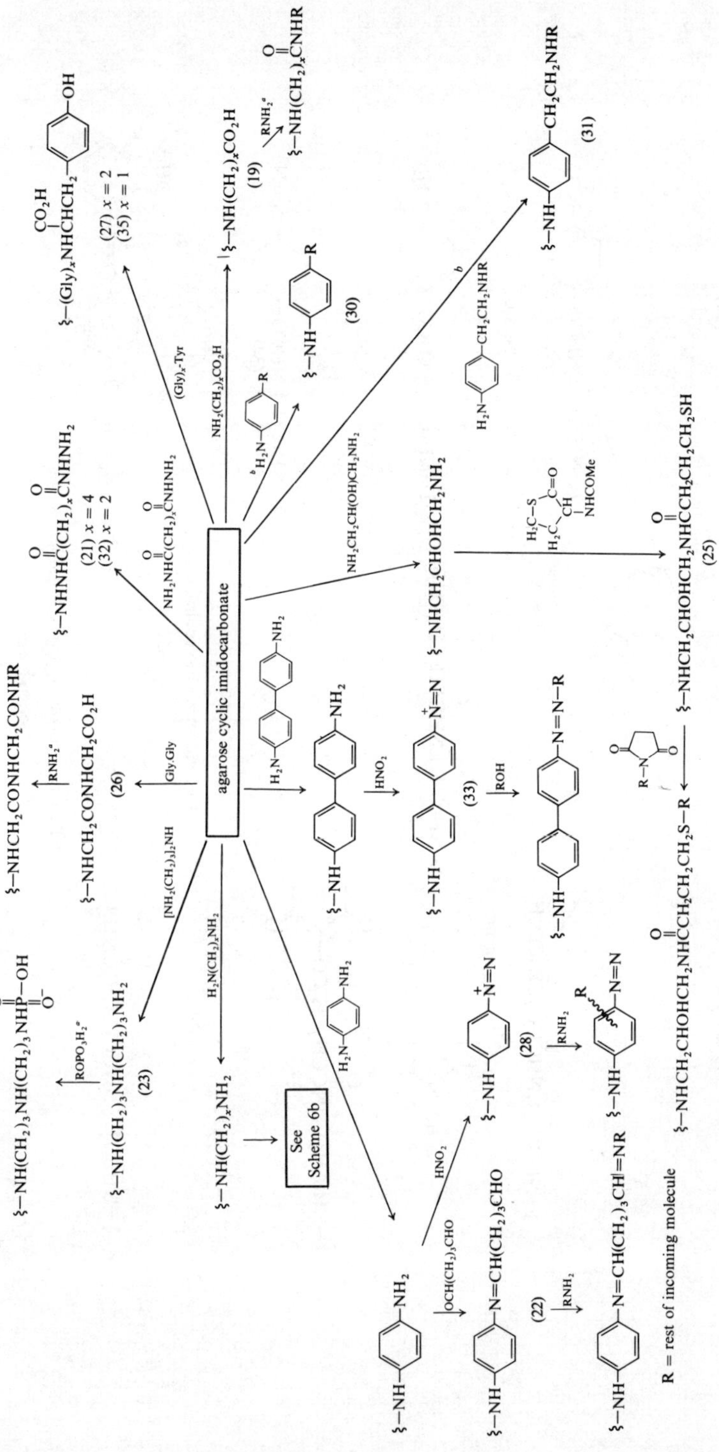

[a] Reaction involving carbodi-imide.
[b] Indicated linkage extension attached to ligand prior to reaction with support.
[c] In subsequent references to this scheme, the suffix A denotes the use of $H_2N(CH_2)_3$-$NH(CH_2)_3NH_2$ instead of $NH_2(CHN_2)_xH_2$ for the production of (18), *etc.*

Scheme 6a

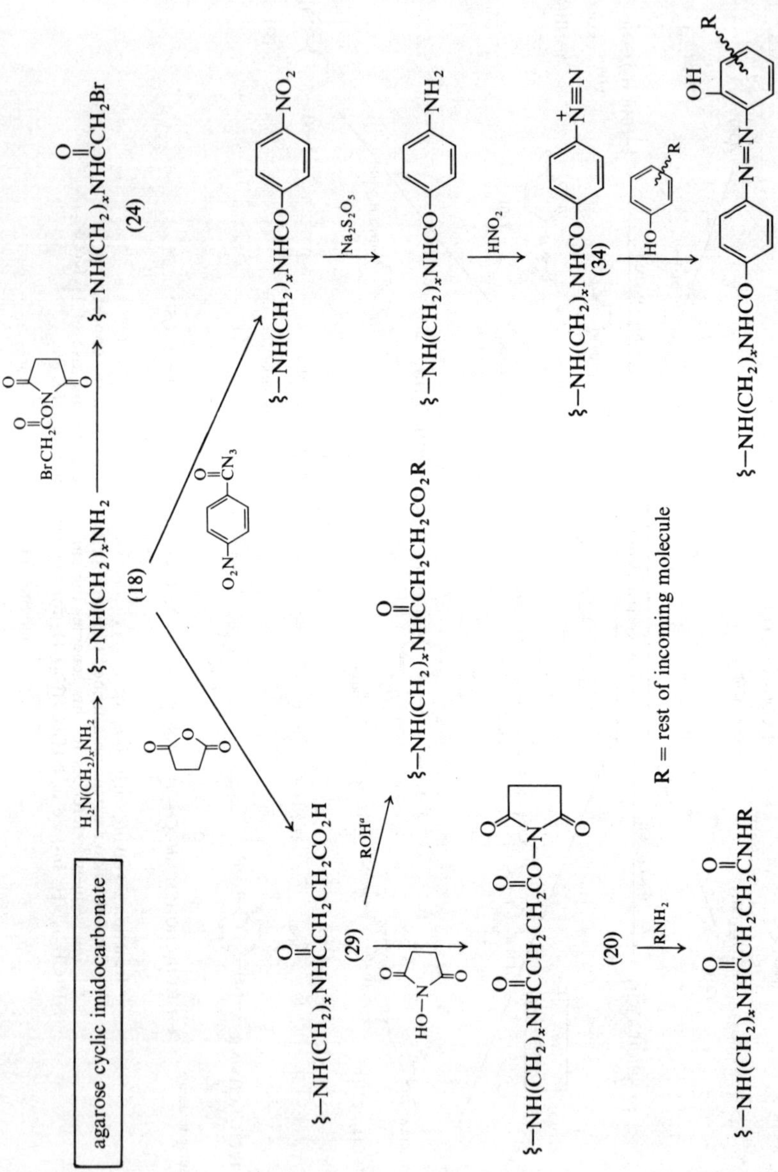

[a] Reaction involving carbodi-imide.

Scheme 6b

by the enzyme-catalysed reaction during the passage of the substrate pulse is registered by means of a thermistor immersed in the top of the bed.

Direct attachment of an affinity ligand or enzyme to agarose cyclic imidocarbonate has often proved to be unrewarding since the approach of the counter molecule to the ligand or enzyme may be sterically impeded. This difficulty can be overcome by inserting a covalent bridge between the matrix and the ligand, and bridging is also used when functional groups on the intended ligand do not react with those on the matrix, since a new type of functionality can be introduced. Modifications used recently to provide bridges in the preparation of affinity-chromatography matrices, *etc.*, from agarose cyclic imidocarbonate are shown in Scheme 6 and Table 4.[184—256] When bridging has been used to immobilize enzymes, details are given in Table 1. Other derivatives of agarose cyclic imidocarbonate used for miscellaneous purposes are listed in Table 5.[257—280]

Table 4 *Use of agarose cyclic imidocarbonate for the preparation of affinity chromatography materials* via *linkage extension*

Species coupled to agarose cyclic imidocarbonate	*Agarose intermediate*[a]	*Use of product*	*Ref.*
2-Acetamido-2-deoxy-β-D-glucopyranoside	(18)	Isolation of D-galactosyl-transferases by affinity chromatography	184
N-Acetylhomocysteine thiolactone	(18)[b]	Purification of specific DNA sequences of mercurated polynucleotides	185
N-Acetyl-L-tryptophan	(20)	Affinity purification of indolyl-3-alkane α-hydroxylase	186
2-Acylamido-2-deoxy-D-glucopyranose	[c]	Affinity chromatographic purification of hexokinases; investigation of effects of different spacer arms (propionyl, butyryl, hexanoyl, or octanoyl) in hexokinase purification	187
Adenosine diphosphate	(21)	Affinity fractionation and study of heavy meromyosin	188
	(18)	Affinity chromatography of H^+-translocating adenosine triphosphatase	189
	[d]	Purification by affinity chromatography of yeast glutathione reductase	190
Adenosine monophosphate	[d]	Purification of human lactate dehydrogenase isoenzyme X	191

184 C. A. Smith and K. Brew, *J. Biol. Chem.*, 1977, **252**, 7294.
185 S. S. Longacre and B. Mach, *J. Biol. Chem.*, 1978, **253**, 7500.
186 G. Schmer and J. Roberts, *Biochim. Biophys. Acta*, 1978, **527**, 264.
187 C. L. Wright, A. S. Warsy, M. J. Holroyde, and I. P. Trayer, *Biochem. J.*, 1978, **88**, 183.
188 P. D. Wagner, *F.E.B.S. Letters*, 1977, **81**, 81.
189 G. D. Webster and J. B. Jackson, *Biochim. Biophys. Acta*, 1978, **503**, 135.
190 I. Carlberg and B. Mannervik, *Biochim. Biophys. Acta*, 1977, **484**, 268.
191 A. H. J. Kolk, L. van Kuyk, and B. Boettcher, *Biochem. J.*, 1978, **173**, 767.

Table 4 (*cont.*)

Species coupled to agarose cyclic imidocarbonate	*Agarose inter-mediate*[a]	*Use of product*	*Ref.*
	(18)[c]	Affinity chromatographic purification of adenosine triphosphatase from rat liver mitochondria	192
	(18)	Affinity chromatographic purification of adenosine deaminase	193
	[d]	Affinity chromatographic purification of histidinol dehydrogenase	194
	[d]	Purification of dehydrogenase enzymes	195
Adenosine triphosphate	(22)	Preparation of affinity chromatography media	196
	(21)	Affinity chromatography purification of *Escherichia coli* dnaB protein	197
	(21)	Purification of ciliary Ca^{2+}-adenosine triphosphatase	198
	(21)	Affinity chromatographic purification of poly(adenylic acid) polymerase	199
S-Adenosyl-L-homocysteine	(23)	Affinity chromatographic purification of indolethylamine *N*-methyl transferases	200
	(18)	Purification of furano-coumarin *O*-methyl transferase by affinity chromatography	201
	(24)	Purification of *S*-adenosyl methionine : protein-carboxyl *O*-methyltransferase by affinity chromatography	202
Agalacto-orosomucoid	(19)	Isolation from chicken liver of a hepatic receptor specific for non-reducing 2-acetamido-2-deoxy-D-glucose residues	203
Alprenolol	(25)	Affinity chromatographic purification of β-adrenergic receptors	204
4-Aminobenzamidine	(26)	Affinity chromatography of human enterokinase	205

192 J.-Y. Le Deaut, J. M. Egly, M. Ledig, and P. Mandel, *Biochim. Biophys. Acta*, 1978, **525**, 438.
193 H. Rosemeyer and F. Seela, *Carbohydrate Res.*, 1978, **62**, 155.
194 E. H. Creaser, G. J. Hughes, and J. A. Lindsay, *Proc. Austral. Biochem. Soc.*, 1978, **11**, 23.
195 C. R. Lowe and M. G. Gore, *F.E.B.S. Letters*, 1977, **77**, 247.
196 P. Gacesa and W. J. D. Whish, *Biochem. J.*, 1978, **175**, 349.
197 E. Lanka, C. Edelbluth, M. Schlicht, and H. Schuster, *J. Biol. Chem.*, 1978, **253**, 5847.
198 M. J. Doughty, *Comp. Biochem. and Physiol.*, 1978, **60**, 339.
199 M. Grez and J. Niessing, *F.E.B.S. Letters*, 1977, **77**, 57.
200 J. P. G. Mack and M. B. Slaytor, *J. Chromatog.*, 1978, **157**, 153.
201 S. K. Sharma and S. A. Brown, *J. Chromatog.*, 1978, **157**, 427.
202 S. Kim, S. Nochumson, W. Chin, and W. K. Paik, *Analyt. Biochem.*, 1978, **84**, 415.
203 T. Kawasaki and G. Ashwell, *J. Biol. Chem.*, 1977, **252**, 6536.
204 G. Vauquelin, P. Geynet, J. Hanoune, and A. D. Strosberg, *Life Sciences*, 1978, **23**, 1791.
205 D. A. W. Grant, A. I. Magee, and J. Hermon-Taylor, *European J. Biochem.*, 1978, **88**, 183.

Table 4 (*cont.*)

Species coupled to agarose cyclic imidocarbonate	*Agarose intermediate*[a]	*Use of product*	*Ref.*
4-Aminobenzyl 1-thio-2-acetamido-2-deoxy-β-D-glucopyranoside	(19)	Purification of wheat germ phytohaemagglutinin	206
4-Aminophenyl 2-acetamido-2-deoxy-β-D-galactopyranoside	(19)	Affinity chromatographic separation of β-D-2-acetamido-2-deoxyhexosidases A and B from C	207
4-Aminophenylhydroxamic acid	(27)	Isolation of neuraminidase by affinity chromatography	208
4-Aminophenyl lactic acid	(19)	Affinity chromatographic purification of lactate dehydrogenase	206
4-Aminophenyl mercuric acetate	(19)	Affinity chromatography of *Clostridium* collagenase and clostripapain	112
3-Aminophenyl propionate	(28)	Affinity chromatographic purification of carboxypeptidase A	209
4-Aminophenyl β-D-thiogalactopyranoside	(19)	Affinity chromatographic purification of type A botulinum neurotoxin	210
N-(4-Androsten-3-on-17β-oxycarbonyl)-6-aminohexanoic acid	(18), (21)	Affinity chromatographic purification of the 4-en-3-oxosteroid: (acceptor)-1-en-oxidoreductase from *Nocardia opaca*	211, 212
Aniline	(26)	Affinity chromatography of human enterokinase	205
L-Arginine monochlorhydrate	(19)	Affinity chromatography of *Clostridium* collagenase and clostripapain	112
L-Arginine peptides: L-Arg-L-Ala-Gly, L-Arg-Gly-Gly, L-Arg-Gly-Gly-Gly-Gly, L-Arg-L-Val, L-Arg-L-Val-Gly, L-Arg-L-Val-Gly-Gly, L-Arg-L-Val-Gly-Gly-Gly	(29)	Quantitative study of the mechanism of substrate binding of trypsin; affinity chromatography of trypsin	213, 214
Aspartic acid	(18)[e]	Photocontrolled affinity chromatography of asparaginase	215
17-α-carboxyethynyl-17-hydroxy-4-androsten-3-one	(18)	Isolation of testosterone-binding globulin from bovine serum by affinity chromatography	216

[206] E. Brown, R. Joyeau, E. Boschetti, and Y. Moroux, *J. Chromatog.*, 1978, **150**, 111.
[207] J. J. W. Lisman and B. Overdijk, *Z. physiol. Chem.*, 1978, **359**, 1019.
[208] Y. V. Vertiev and Y. V. Ezepchuk, *Biochemistry* (*U.S.S.R.*), 1977, **42**, 1273.
[209] G. Oshima and K. Nagasawa, *J. Biochem.* (*Japan*), 1977, **81**, 1285.
[210] L. J. Moberg and H. Sugiyama, *Appl. Envir. Microbiol.*, 1978, **35**, 878.
[211] P. Atrat, V. Deppmeyer, and C. Hörhold, *J. Steroid Biochem.*, 1978, **9**, 579.
[212] P. Atrat, V. Deppmeyer, C. Hörhold, L. Möller, and R. Knöll, *J. Steroid Biochem.*, 1978, **9**, 691.
[213] M. Nishikata, K. Kasai, and S. Ishii, *J. Biochem.* (*Japan*), 1977, **82**, 1475.
[214] K. Kasai and S. Ishii, *J. Biochem.* (*Japan*), 1978, **84**, 1051.
[215] I. Karube, Y. Ishimori, S. Suzuki, and T. Sato, *Biotechnol. and Bioeng.*, 1978, **20**, 1775.
[216] Y. Suzuki, E. Itagaki, H. Mori, and T. Hosoya, *J. Biochem.* (*Japan*), 1977, **81**, 1721.

Table 4 (*cont.*)

Species coupled to agarose cyclic imidocarbonate	*Agarose intermediate*[a]	*Use of product*	*Ref.*
4-Chloromercuribenzoate	(18)	Affinity purification of L-glutathione : D-fructose 6-phosphate aminotransferase	217
Cortisol, 21-deoxycortisol	(29)	Evaluation of affinity matrices for purification of transcortin; affinity chromatographic purification of transcortin	218
Cyanocobalamin, carboxylic acid derivative	(18)	Affinity chromatographic purification of human intrinsic factor and pig cobalophilin	86
Daunomycin	(19)	Fractionation of non-histone proteins from rat leukaemia cells	219
Dehydroepiandrosterone-17-(*O*-carboxymethyl)oxime	(18)	Isolation of dehydroepiandrosterone sulphotransferase by affinity chromatography	220
2,4-Dinitrophenylhydrazine	(18)	Affinity chromatography of *Bacillus subtilis* metalloendoproteinase	155
Flavins: *N*(3)-carboxymethylriboflavin; *N*(10)-6-carboxybutyl-7,8-dimethylisoalloxazine; *N*(10)-6-carboxypentyl-7,8-dimethylisoalloxazine; 8α-[*S*-(*N*-acetyl)-L-cysteinyl] riboflavin; 8α-[*S*-Mercaptopropionic acid] riboflavin; 8α-(*N*-β-alanyl)riboflavin	(18)	Establishment of general methods for purification of proteins that bind riboflavin; purification of riboflavin carrier protein, flavokinase and flavodoxin	221
5′-Fluoro-2′-deoxy-uridine 5′-(4-aminophenyl phosphate)	(29)	Isolation by affinity chromatography of thymidylate synthetase	222
β-L-Fucopyranosylamine	(19)	Affinity chromatographic purification of *Ulex europeus* lectin I	223
	(19)	Purification of carbohydrate-binding proteins from *Dictyostelium discoideum* by affinity chromatography	224
α-D-Galactopyranose	(30)	Affinity chromatographic purification of haemagglutinin from *Maclura pomifera* seeds	225

[217] K. Hosoi, S. Kobayashi, and T. Ueha, *Biochem. Biophys. Res. Comm.*, 1978, **85**, 558.
[218] D. W. Chan, M. Sharma, and W. R. Slaunwhite, *Arch. Biochem. Biophys.*, 1977, **182**, 197.
[219] H. Kikuchi and S. Sato, *Biochim. Biophys. Acta*, 1978, **532**, 113.
[220] J. B. Adams, and D. McDonald, *Proc. Austral. Biochem. Soc.*, 1978, **11**, 22.
[221] A. H. Merrill and D. B. McCormick, *Analyt. Biochem.*, 1978, **89**, 87.
[222] J. M. Whiteley, *Methods Enzymol.*, 1978, **51**, 98.
[223] M. E. A. Pereira, E. C. Kisailus, F. Gruezo, and E. A. Kabat, *Arch. Biochem. Biophys.*, 1978, **185**, 108.
[224] D. C. Kilpatrick, J. A. Schmidt, and J. L. Stirling, *F.E.M.S. Microbiol. Letters*, 1978, **4**, 67.
[225] J. N. Bausch and R. D. Portez, *Biochemistry*, 1977, **16**, 5790.

Table 4 (*cont.*)

Species coupled to agarose cyclic imidocarbonate	*Agarose intermediate*[a]	*Use of product*	*Ref.*
3-*O*-β-D-Galactopyranosyl-6-deoxy-L-gulitol	(31)	Purification of β-D-galactosidase by affinity chromatography	226
Glycogen[f]	(32)	Isolation of glycogen phosphorylase B by affinity chromatography	227
Guanine, 4-aminoethoxyphenyl derivative	[c]	Purification of guanine aminohydrolase by affinity chromatography	228
Guanosine diphosphate[f]	(18)[g]	Isolation of protein synthesis factor Tu from *Escherichia coli*	156
Guanosine monophosphate	(18)	Affinity chromatography of inosinic acid dehydrogenase from *Aerobacter aerogenes*	229
Hyaluronic acid	(21)	Purification by affinity chromatography of protein-keratan sulphate core of bovine nasal cartilage proteoglycan after digestion with chondroitin ABC lyase	230
17-Hydroxy-5α-androstan-3-one	(18)	Isolation of testosterone-binding globulin from bovine serum	216
2-Hydroxyphenylacetate	(33)	Affinity chromatographic purification of homogentisate-1,2-dioxygenase	231
Indole-3-carboxaldehyde	(18)[g]	Affinity chromatographic purification of indolyl-3-alkane α-hydroxylase	186
L-Lysine	(18)	Affinity chromatography of rat liver aminoacyl-tRNA synthetase	232
L-Lysine, N^6-[pteroyltetra-(γ-glutamyl)]-derivative	[c]	Affinity chromatographic purification of thymidylate synthetase	233
Methotrexate	(18)	Isolation of sheep and rat liver dihydropteridine reductases by affinity chromatography	234
1-Methyl-2-aminopropylamino acridinium bromide hydrobromide	(19)	Purification of acetylcholinesterase by affinity chromatography	235
10-Methyl-2-aminopropylamino acridinium dibromide	(19)	Selective purification of either 11S or 18S plus 14S forms of acetyl cholinesterase from mixtures by differential affinity chromatography	236
N-Methyl-*N*-prop-1-yne benzylamine-4-carboxylic acid	(18)	Affinity chromatographic purification of monoamine oxidase	237

226 P. V. Wagh, *Biochim. Biophys. Acta*, 1978, **522**, 515.
227 L. N. Viktorova, B. A. Klyashchitskii, and E. Ramensky, *F.E.B.S. Letters*, 1978, **91**, 194.
228 J. D. Bergstrom and A. L. Bieber, *Prep. Biochem.*, 1978, **8**, 275.
229 P. E. Brodelius, R. A. Lannom, and N. O. Kaplan, *Arch. Biochem. Biophys.*, 1978, **188**, 228.
230 J. E. Christner, M. L. Brown, and D. D. Dziewiatkowski, *Anal. Biochem.*, 1978, **90**, 22.
231 M. Sugumaran and C. S. Vaidyanethan, *F.E.M.S. Microbiol. Letters*, 1978, **4**, 343.
232 C. V. Dang and D. C. H. Yang, *Biochem. Biophys. Res. Comm.*, 1978, **80**, 709.
233 L. T. Plante, Y. Gaumont, and R. L. Kisliuk, *Prep. Biochem.*, 1978, **8**, 91.
234 S. Webber, T. L. Detts, W. R. Snyder, and J. M. Whiteley, *Analyt. Biochem.*, 1978, **84**, 491.
235 C. A. Reavill and D. T. Plummer, *J. Chromatog.*, 1978, **157**, 141.
236 G. Webb and D. G. Clark, *Arch. Biochem. Biophys.*, 1978, **191**, 278.
237 C. M. Buess, J. K. Price, B. D. Roberts, and W. R. Carper, *Experientia*, 1977, **33**, 163.

Table 4 (*cont.*)

Species coupled to agarose cyclic imidocarbonate	*Agarose intermediate*[a]	*Use of product*	*Ref.*
Nicotinamide adenine dinucleotide		Isolation of *myo*-inositol-1-phosphate synthase by affinity chromatography	238
	(34)	Affinity chromatographic purification of alcohol dehydrogenase	239
	(19)	Affinity chromatographic purification of sorbitol dehydrogenase	240
Nicotinamide adenine dinucleotide phosphate	(18)	Purification by affinity chromatography of D-glutamate dehydrogenase and of D-glucose-6-phosphate dehydrogenase	241
	(21)	Affinity chromatography of $NADP^+$-specific isocitrate dehydrogenase	242
Oxamate	(18)	Affinity chromatographic purification of lactic acid dehydrogenase	243
	(18)	Isolation of lactate dehydrogenase isoenzyme X by affinity chromatography	191
	(18)	Affinity chromatography of bacterial lactate dehydrogenases; separation of D-and L-specific enzymes by alteration of elution conditions	244
Pepstatin	(18)	Purification of cathepsin D by affinity chromatography	245
Phenylacetate, *N*-succinimide derivative	(21)[c]	Affinity chromatographic purification of lysozymes	246
D-Phenylalanine methyl ester	(19)	Affinity chromatographic purification of elastase	247
Phosphorylcholine	(35)	Affinity chromatographic purification of C-reactive protein from ascites fluid	248
Retinoic acid	(18)	Purification of retinol-binding protein by affinity chromatography	175
Saligenin β-D-glucopyranoside	(21)	Affinity chromatographic purification of β-D-glucosidase	249

238 E. Olugbenga Ogunyemi, F. Pittner, and O. Hoffmann-Ostenhof, *Z. physiol. Chem.*, 1978, **359**, 613.
239 M. J. Runswick and J. I. Harris, *F.E.B.S. Letters*, 1978, **92**, 365.
240 N. Leissing and E. T. McGuinness, *Biochim. Biophys. Acta*, 1978, **524**, 254.
241 W. Leicht, *European J. Biochem.*, 1978, **84**, 133.
242 F. Miake, T. Torikata, K. Koga, and K. Hayashi, *Agric. and Biol. Chem.* (*Japan*), 1976, **40**, 1269.
243 A. R. Place, D. A. Powers, and Y. C. Lee, *Analyt. Biochem.*, 1977, **83**, 636.
244 N. Kelly, M. Delaney, and P. O'Carra, *Biochem. J.*, 1978, **171**, 543.
245 A. Linde and B. Persliden, *Prep. Biochem.*, 1978, **8**, 231.
246 P. Bailon and A. H. Nishikawa, *Prep. Biochem.*, 1977, **7**, 61.
247 K. Morihara and H. Tsuzuki, *Agric. and Biol. Chem.* (*Japan*), 1975, **39**, 1123.
248 J. E. Volanakis, W. L. Clements, and R. E. Schrohenloher, *J. Immunol. Methods*, 1978, **23**, 285.
249 W. W. Roth and V. R. Srinivasan, *Prep. Biochem.*, 1978, **8**, 57.

Table 4 (*cont.*)

Species coupled to agarose cyclic imidocarbonate	*Agarose intermediate*[a]	*Use of product*	*Ref.*
Soybean trypsin inhibitor	(19)	Affinity chromatographic purification of *Clostridium* collagenase and clostripapain	112
Spermine	(19)	Isolation of spermine synthase by affinity chromatography	250
Tartaric acid	(18)	Affinity purification of human prostatic acid phosphatase	251
Tartaric acid derivative	(18)	Purification of tartrate-inhibitable acid phosphatases by affinity chromatography	252
Thiamine monophosphate	(18)	Affinity chromatographic isolation of thiamine pyrophosphokinase	253
Thiamine pyrophosphate	(18)	Affinity chromatography of the pyruvate dehydrogenase complex of wild type *Escherichia coli*	254
Tri-L-alanine	(19)	Affinity chromatographic purification of elastase	255
L-Tryptophan	(18)[h]	Purification by affinity chromatography of indolyl-3-alkane α-hydroxylase	186
Uridine diphosphate	(18)	Affinity chromatographic purification of D-galactosyltransferase	184
UDP-D-glucuronic acid	(18)	Hydrophobic affinity chromatography of collagen D-glucosyltransferases	122, 128
5′-Uridine triphosphate	(18)	Affinity chromatographic purification of ribonuclease A and catalase	256

[a] See Schemes 6a and 6b, pp. 525, 526. [b] Commercially available crosslinked agarose used for coupling. [c] Bridge indicated initially coupled to affinant molecule and then treated with agarose cyclic imidocarbonate. [d] Commercially available derivative. [e] Followed by reaction with spiropyran. [f] Pre-oxidized with $NaIO_4$. [g] Reduced with $NaBH_4$ after coupling. [h] Followed by activation with glutaraldehyde and reaction with affinant.

Table 5 *Use of agarose cyclic imidocarbonates for the preparation of miscellaneous derivatives*

Species coupled to agarose cyclic imidocarbonate	*Agarose intermediate*[a]	*Use of product*	*Ref.*
Abrin, abrus phytohaemagglutinins	—	Study of lectin interactions with cell-surface receptors	257
Adenosine triphosphate	(28)	Development of affinity chromatography media	258

[250] R. L. Pajula, A. Raina, and J. Kekoni, *F.E.B.S. Letters*, 1978, **90**, 153.
[251] P. Vihko, *Clinical Chem.*, 1978, **24**, 1783.
[252] R. L. Van Elten and M. S. Saini, *Clinical Chem.*, 1978, **24**, 1525.
[253] Y. Wakabayashi, A. Iwashima, and Y. Nose, *Biochim. Biophys. Acta*, 1976, **429**, 1085.
[254] J. Visser, M. Strating, and W. Van Dongen, *Biochim. Biophys. Acta*, 1978, **524**, 37.
[255] K. Katagiri, T. Takeuchi, K. Taniguchi, and M. Sasaki, *Analyt. Biochem.*, 1978, **86**, 159.
[256] G. K. Smith, K. J. Schray, and S. W. Schaffer, *Analyt. Biochem.*, 1978, **84**, 406.
[257] K. Sandvig, S. Olsnes, and A. Pihl, *European J. Biochem.*, 1978, **88**, 307.
[258] P. Gacesa and W. J. D. Whish, *Biochem. J.*, 1978, **175**, 349.

Table 5 (*cont.*)

Species coupled to agarose cyclic imidocarbonate	*Agarose intermediate*[a]	*Use of product*	*Ref.*
Albumin	(19)	Quantitation of salicylic acid and other drugs; study of drug–protein binding	259
Aminoalkyl derivatives: 1,6-Diaminohexane, Chlorotriazine-1,6-diaminohexane, 1,8-Diamino-octane, Carboxypropionylaminohexane, *NN*-Dimethylaminohexane	(18)	Hydrophobic chromatography of acetate kinase; comparison of different sidechains	51
L-Arginine-peptides (from salmine)	—	Assessment of performance of affinity chromatography support for purification of trypsin	177
Bandeiraea simplicifolia lectin I	—	Study of specificity of the phytohaemagglutinin by competitive-binding assay with oligosaccharides	260
Blue Dextran	—	Study of interaction of Blue Dextran with calf spleen NAD glycohydrolase	261
bis-(Carboxymethyl)amine		Isolation of lactoferrin using metal chelate chromatography with copper ions	262
Concanavalin A	—	Electrophoretic desorption of bound immunoglobulins	263
	—	Structural studies of oligosaccharides from *endo*-β-D-2-acetamido-2-deoxyglucanase digests of ovalbumin	264
	—	Binding of malaria schizonts which when matured release merozoites which can be readily isolated	265
	—	Concentration of 'acid' glycosidase and arylsulphatase enzymes from cerebrospinal fluid	266
	—	Binding of β-D-fructofuranosidase and investigation of the involvement of salt links in the stabilization of the enzyme	267

259 N. I. Nakano, T. Oshio, Y. Fujimoto, and T. Amiya, *J. Pharm. Sci.*, 1978, **67**, 1005.
260 E. C. Kisailus and E. A. Kabat, *Carbohydrate Res.*, 1978, **67**, 243.
261 F. Schuber and M. Pascal, *Biochimie*, 1977, **59**, 735.
262 B. Lönnerdal, J. Carlsson, and J. Porath, *F.E.B.S. Letters*, 1977, **75**, 89.
263 M. R. A. Morgan, P. M. Johnson, and P. D. G. Dean, *J. Immunol. Methods*, 1978, **23**, 381.
264 K. Yamashita, Y. Tachibana, and A. Kobata, *J. Biol. Chem.*, 1978, **253**, 3862.
265 P. H. David, M. Hommel, J. C. Benichou, H. A. Eisen, and L. H. Pereiradasilva, *Proc. Nat. Acad. Sci.* (*U.S.A.*), 1978, **75**, 5081.
266 R. Cherian and A. S. Balasubramanian, *Clinica Chim. Acta*, 1978, **89**, 411.
267 J. Woodward and A. Wiseman, *Biochim. Biophys. Acta*, 1978, **527**, 8.

Table 5 (*cont.*)

Species coupled to agarose cyclic imidocarbonate	*Agarose intermediate*[a]	*Use of product*	*Ref.*
	—	Comparison of binding affinities of bovine γ-glutamyltransferases from different organs	268
	—	Identification of a breast tumour-associated orosomucoid	136
	—	Characterization of the isotoxins of modeccin	135
D-Galactose, 2-imino-2-methoxyethyl 1-thioglycoside	[b]	Study of binding of D-galactosyl neoglycoproteins to the hepatic asialoglycoprotein receptor	269
D-Glucose, 2-imino-2-methoxyethyl 1-thioglycoside	[b]	Study of binding of D-glucosyl neoglycoproteins to the hepatic asialoglycoprotein receptor	269
Haemoglobin	—	Study of properties of immobilized haemoglobin	270
Haptoglobin	—	Investigation of haptoglobin binding with α- and β-chain tryptic peptides of haemoglobin	271
Mitomycin C	—	*In vitro* and *in vivo* comparison of anti-tumour activity of the immobilized and soluble drugs	272
Nicotinamide adenine dinucletide	(18)	ADP-ribosylation of reticulocyte elongation factor EF-2	273
NADH, carbamoyl-methyl derivative	(18)	Study of interaction between alcohol dehydrogenase and the NADH analogue	274
Octylamine	[c]	Hydrophobic adsorption of phospholipids and their application for the characterization of purified phospholipase	275
L-Phenylalanine-4-nitroanilide	—; (19)	Kinetic studies of α-chymotrypsin catalysed hydrolysis of the immobilized substrate; investigation of the effect of different spacer arm lengths	276
Phenylbutylamine	—	Hydrophobic interaction chromatography of chymotrypsin at sub-zero temperatures	277

[268] K. Iwami, K. Yasumoto, T. Fushiki, Y. Takigawa, and H. Mitsuda, *J. Biochem.* (*Japan*), 1978, **84**, 1237.
[269] C. P. Stowell and Y. C. Lee, *J. Biol. Chem.*, 1978, **253**, 6107.
[270] M. R. R. Fanelli, G. Amiconi and E. Antonini, *European J. Biochem.*, 1978, **92**, 253.
[271] A. Tsapis, C. Mihaesco, A. Alfsen, Y. Beuzard, and J. Rosa, *Biochem. Biophys. Res. Comm.*, 1978, **80**, 319.
[272] T. Kojima, M. Hashida, S. Muranishi, and H. Sezaki, *Chem. and Pharm. Bull.* (*Japan*), 1978, **26**, 1818.
[273] K. S. Lam and R. L. Heintz, *European J. Biochem.*, 1978, **88**, 459.
[274] L. Anderson, P. O. Larsson, and K. Mosbach, *F.E.B.S. Letters*, 1978, **88**, 167.
[275] M. Malmqvist, T. Malmqvist, and R. Möllby, *F.E.B.S. Letters*, 1978, **90**, 243.
[276] J. Fischer, L. Lange, S. Koch, and H. D. Jakubke, *European J. Biochem.*, 1978, **88**, 445.
[277] C. Le Peuch and C. Balny, *F.E.B.S. Letters*, 1978, **87**, 232.

Table 5 (*cont.*)

Species coupled to agarose cyclic imidocarbonate	*Agarose inter-mediate*[a]	*Use of product*	*Ref.*
Protein A (*Staphylococcus aureus*)	[c]	Removal of contaminant ribonuclease activity from immunoglobulin G preparations	278
	[c]	Electrophoretic desorption of bound immunoglobulins	263
	[c]	Characterization of hepatitis e antigen	279
Ricinus communis phytohaemagglutinin I	—	Study of microheterogeneity of rat α-fetoprotein	280

[a] See Schemes 6a and 6b, pp. 525, 526. [b] Agarose cyclic imidocarbonate first treated with bovine serum albumin and then treated with the appropriate 2-imino-2-methoxyethyl 1-thioglycoside. [c] Cross-linked agarose used for coupling.

The use of immobilized albumin for the quantitation of salicylic acid is a procedure which may be extended to the determination of other drugs.[259] The method is simple and precise and the immobilized albumin derivative may be used for at least eight months without alteration of the binding capacity for salicylate.

Frontal chromatography on affinity adsorbents may provide information about the dissociation constant (K_d) of enzyme-immobilized ligand complexes and is useful as a general method of investigating the interaction of a protein and a specific ligand.[214] Results obtained by this method were in good agreement with those of kinetic experiments when trypsin was used as the model enzyme.

Photocontrol of affinity chromatography provides a novel means of regulating binding and release of the desired molecule from an affinant, as has been demonstrated for L-asparaginase using an L-aspartic acid–spiropyran derivative of agarose (36).[215] The enzyme is bound in visible light and released in the dark.

Commercially available 'Affi-gel' supports, agarose beaded gel included in a polyacrylamide support having different spacer arm lengths, have been used for the preparation of affinity matrices. The spacers are anchored by extremely stable ether linkages. Antibodies to mouse interferon have been immobilized on 'Affi-gel 10' and used for the purification of mouse interferon.[174] Histones, histone fractions, poly(L-arginine) and poly(L-lysine) have also been immobilized on the support and competitive elution studies performed with histone-binding proteins from mouse liver.[281] Organomercurial agarose derivatives ('Affi-gel 501') have been used for the immobilization of pyridoxamine 5′-phosphate oxidase.[282] L-Ornithine decarboxylase has been purified successfully on columns of 'Affigel-10' coupled with pyridoxamine 5′-phosphate.[283]

Activation of agarose itself by treatment with an ether derivative has been carried out using butane-1,4-diol-glycidyl ether.[254] The activated agarose was

278 T. J. Miller and H. O. Stone, *J. Immunol. Methods*, 1978, **24**, 111.
279 H. A. Fields, D. W. Bradley, C. L. Davis, and J. E. Maynard, *Infection and Immunity*, 1978, **20**, 792.
280 B. Bayard and J. P. Kerckaert, *Biochem. Biophys. Res. Comm.*, 1977, **77**, 489.
281 B. J. Conner and D. E. Comings, *Biochim. Biophys. Acta*, 1978, **532**, 122.
282 H. Tsuge, K. Sen-Maru, and K. Ohashi, *F.E.B.S. Letters*, 1978, **93**, 331.
283 R. J. Boucek and K. J. Lembach, *Arch. Biochem. Biophys.*, 1977, **184**, 408.

then treated with thiochrome and used to study the pyruvate dehydrogenase complex of wild-type *Escherichia coli.* Agarose cross-linked with 2,3-dibromopropanol has been treated with bromosulphophthalein and used for the affinity chromatographic purification of ligandin.[284] Divinylsulphone activation of agarose has been used for preparation of matrices containing L-fucose.[285,286] These products can be successfully applied to the isolation of L-fucose-binding lectins.

Epichlorohydrin-activated agarose has been used as a matrix for affinity chromatography with or without additional activation with cyanogen bromide. Immobilized RNA and DNA have been prepared by reaction with desulphated and epichlorohydrin cross-linked agarose cyclic imidocarbonate.[287] Insulin has been coupled to a succinyldiaminodipropylamino-derivative of epichlorohydrin-crosslinked agarose.[288] Thiol containing agarose matrices of improved performance have been prepared using epichlorohydrin-activated agarose by reacting it with sodium thiosulphate.[289] The *S*-alkylthiosulphate gel was then reduced with DL-dithiothreitol and used to recover mercurated nucleic acid hybrids.

Agarose into which epoxy-groups have been introduced by reaction with compounds such as diglycidyl ether or 1,4-*bis*-2,3-epoxypropoxybutane have been used to prepare affinity chromatography matrices. Lectins have been isolated using affinant supports containing 2-acetamido-2-deoxy-D-galactose,[290] and 2-amino-2-deoxy-D-galactose.[47]

Epoxy-activated agarose has been treated with 8-amino-xanthine and used to isolate urate oxidase.[291] *Escherichia coli* lipopolysaccharide immobilized on epoxy-activated agarose provides an immunoadsorbent for isolation of antisera,[292] and L-fucose immobilized on the same support may be used for the purification of L-fucose-specific lectins.[293] Glutathione–agarose matrices prepared after epoxy-activation have been successfully applied to the purification of glutathione *S*-transferases.[294] An *S*-hexyl glutathione derivative similarly prepared has been used for purification of glyoxalase I from rabbit liver.[117]

Hydrocarbon ligands of different lengths have been treated with epoxy-activated agarose and the binding of tubulin, the major protein constituent of rat brain microtubules, investigated.[295] Epoxy-activated agarose has been reacted with iminodiacetic acid to produce *bis*-carboxy-methylamino-agarose.[296] Columns of this support have been charged with metal ions (Cu^{2+}, Zn^{2+}) and

284 A. Grahnen and I. Sjoholm, *European J. Biochem.*, 1977, **80**, 573.
285 H. J. Allen and E. A. Z. Johnson, *Carbohydrate Res.*, 1977, **58**, 253.
286 H. J. Allen, E. A. Z. Johnson, and K. L. Matta, *Immunol. Comm.*, 1977, **6**, 585.
287 J. Kempf, N. Pfleger, and J. M. Egly, *J. Chromatog.*, 1978, **147**, 195.
288 S. Jacobs, Y. Schechter, K. Bissell, and P. Cuatrecasas, *Biochem. Biophys. Res. Comm.*, 1977, **77**, 981.
289 A. R. Thomason and F. M. Rottman, *Analyt. Biochem.*, 1978, **89**, 501.
290 L. J. Blacik, M. Breen, H. G. Weinstein, R. A. Sittig, and M. Cole, *Biochim. Biophys. Acta*, 1978, **532**, 225.
291 T. Watanabe and T. Suga, *Analyt. Biochem.*, 1978, **86**, 357.
292 J. Fox and K. Hechemy, *Infection and Immunity*, 1978, **20**, 867.
293 L. G. Gürtler, *Biochim. Biophys. Acta*, 1978, **544**, 593.
294 P. C. Simons and D. L. Van Der Jagt, *Analyt. Biochem.*, 1977, **82**, 334.
295 K. Hemminki, *Acta Chem. Scand.*, 1976, **30B**, 794.
296 E. Bollin and E. Sulkowski, *Arch. Virology*, 1978, **58**, 149.

used for the purification of hamster interferon by metal chelate affinity chromatography.

A thioether linkage stable between pH 4—8 has been prepared by reaction of thiol-containing agarose with 2-hydroxy-5-nitrobenzyl bromide.[297] Chymotrypsin and β-amylase are adsorbed strongly to the matrix and retain high enzymic activity.

By incorporating methylmercuric hydroxide into agarose gel, a gel electrophoresis medium effectively containing a denaturing agent has been prepared and applied to the study of RNA's from avian myeloblastosis virus.[298]

Triazinyl dyes (usually Cibacron Blue F-3GA) which react with agarose *via* the chlorine atoms on the triazine ring have been successfully applied to a number of purifications,[299] including those of α-fetoprotein,[300] α-2-macroglobulin,[301] deoxycytidine kinase,[302] erythropoietin,[83] mammalian interferons,[118,303] lactate dehydrogenase,[304] NADP$^+$-dependent dehydrogenase,[305] and sex hormone-binding globulin,[306] and to the removal of albumin from plasma.[307] Immobilized Cibacron Blue has been used in the purification of human serum albumin using an electrophoretic desorption technique.[89] Cibacron Blue immobilized by reaction with epichlorohydrin-activated agarose has been applied to the separation of arylsulphatases A and B.[308]

A number of new methods for the preparation of biospecific agarose adsorbents utilizing 2,4,6-trichloro-*sym*-triazine as linking agent have been described.[309] Trichloro-*sym*-triazine-activated agarose has been coupled with 2-aminoethanol, 1,2-diaminoethane, 1,4-diaminobutane, and 1,6-diaminohexane for the hydrophobic chromatography of β-D-2-acetamido-2-deoxyhexosidases.[310]

Agarose treated with phenylacetyl hydrazide has been used for the isolation of lysozyme.[246] Actin filaments, with phalloidin present as stabilizer, have been entrapped on agarose beads by using suitable conditions for actin polymerization.[311] Myosin, heavy meromyosin and heavy meromyosin subfragment 1 were purified by affinity chromatography on the support.

'Ultrogel', a commercially available polyacrylamide–agarose medium, has been used as support matrix for the coupling of D-tryptophan methyl ester in a comparative study of the efficacy of two condensation agents, 1-cyclohexyl-3-(2-morpholinoethyl)-carbodi-imide-metho-4-toluene sulphonate and *N*-ethoxy-

297 K. Caldwell, R. Axén, and A. T. Mevkh, *Bioorg. Khim.*, 1978, **4**, 707.
298 J. S. Smith, J. C. Ostenburg, and J. M. Bailey, *Analyt. Biochem.*, 1978, **89**, 309.
299 R. S. Beissner and F. B. Rudolph, *J. Chromatog.*, 1978, **161**, 127.
300 J. L. Young and B. A. Webb, *Analyt. Biochem.*, 1978. **88**, 619.
301 G. D. Virca, J. Travis, P. K. Hall, and R. C. Roberts, *Analyt. Biochem.*, 1978, **89**, 274.
302 A. Baxter, L. M. Currie, and J. P. Durham, *Biochem. J.*, 1978, **173**, 1005.
303 A. Mizrahi, J. A. O'Malley, W. A. Carter, A. Takatsuki, G. Tamura, and E. Sulkowski, *J. Biol. Chem.*, 1978, **253**, 7612.
304 G. M. Rothe, A. Cörper, and W. Bacher, *Phytochemistry*, 1978, **17**, 1701.
305 D. H. Watson, M. J. Harvey, and P. D. G. Dean, *Biochem. J.*, 1978, **173**, 591.
306 M. J. Iqbal, P. Ford, and M. W. Johnson, *F.E.B.S. Letters*, 1978, **87**, 235.
307 S. Angal and P. D. G. Dean, *F.E.B.S. Letters*, 1978, **96**, 346.
308 A. Ahmad, A. Surolia, and B. K. Bachhawat, *Biochim. Biophys. Acta*, 1977, **481**, 542.
309 T. H. Finlay, V. Troll, M. Levy, A. J. Johnson, and L. T. Hodgins, *Analyt. Biochem.*, 1978, **87**, 77.
310 A. A. Artyukov and N. V. Molodtsov, *J. Chromatog.*, 1977, **130**, 451.
311 A. Grandmont-Leblanc and J. Gruda, *Canad. J. Biochem.*, 1977, **55**, 949.

carbonyl-2-ethoxy-1,2-dihydroquinoline.[312] The second generally proved advantageous, and chymotrypsin was purified using the affinity matrix. Ultrogel activated with glutaraldehyde has been treated with soybean trypsin inhibitor and applied to the purification of trypsin.[313]

An agar derivative, an ester of 4-β-hydroxymethyl-sulphonyl-2-amino-anisole, has been diazotized and treated with phosphorylase B.[314] The enzyme lost enzymic activity but retained immunological activity on immobilization.

Amyloses.—The diffusion of helium, oxygen, nitrogen, and carbon dioxide in amylose acetate has been studied in the temperature range 30—60 °C.[315] Arrhenius parameters were evaluated and the solubility of the gases in the polymer discussed in terms of the condensing tendency of the gases studied.

Light scattering and viscometric techniques have been applied to the study of the solution properties of sodium amylose sulphate in salt solutions of different ionic strengths and at several temperatures.[316] The molecular weight is correlated with the radius of gyration and the second virial coefficient. The stiffness parameters of the polymer chains were determined and compared with those of other polymers.

A study has been made of the thermal stability of amylose and some of its derivatives, *viz.*, amylose palmitate, amylose succinate, and sodium carboxymethylamylose.[317] Results indicated that except for amylose succinate, esterification increased the thermal stability of all the amylosic derivatives compared to amylose. The order of stability noted was amylose palmitate > amylose acetate > amylose propionate > amylose > sodium carboxymethylamylose > amylose succinate.

Crystals of amylose propionate have been grown from dioxane and aniline at different temperatures.[318] Study of the morphology of various growth structures revealed that generation of vertical lamellae is very common in polymer crystals, and orientation of these lamellae on further growth determines the final morphology of resulting growth structures. Probable mechanisms of crystal growth were discussed.

Per(trimethylsilyl)ated amylose has been selectively acylated at primary positions using a mixture of pyridine, acetic anhydride, and acetic acid.[319]

Comb-like derivatives of amylose have been prepared by condensation of 1,2-(t-butyl orthoacetate) and 1,2-(ethyl orthoacetate) derivatives of 3,4,6-tri-*O*-acetyl-α-D-glucopyranose with 2,3-di-*O*-phenylcarbamoyl derivatives of amylose in boiling chlorobenzene in the presence of 2,6-lutidinium perchlorate as catalyst.[320] The distance between the branch points, attached mainly by (1→6)-β-D- with a small proportion of (1→6)-α-D- linkages as indicated by n.m.r.

312 E. Boschetti, M. Corgier, and R. Garelle, *Biochimie*, 1978, **60**, 425.
313 E. Boschetti and M. Delay, *Science Tools*, 1978, **25**, 18.
314 N. B. Kozlova, L. V. Roze, and P. L. Vul'fson, *Biochemistry* (*U.S.S.R.*), 1978, **43**, 403.
315 V. M. Patel, K. C. Patel, and R. D. Patel, *Stärke*, 1977, **29**, 339.
316 K. S. Patel, K. C. Patel, and R. D. Patel, *Stärke*, 1977, **29**, 181.
317 D. H. Desai, K. C. Patel, and R. D. Patel, *Stärke*, 1976, **28**, 377.
318 R. M. Gohil and R. D. Patel, *Stärke*, 1977, **29**, 9.
319 D. Horton and J Lehmann, *Carbohydrate Res.*, 1978, **61**, 553.
320 B. Pfanemüller, G. C. Richter, and E. Husemann, *Carbohydrate Res.*, 1977, **56**, 139.

and i.r. spectroscopy, was 2—3 D-glucose residues when 1 molar equivalent was used for each hydroxy-group in the polysaccharide. Derivatives with oligomeric side-chains were similarly prepared using peracetylated 1,2-(ethyl orthoacetate) derivatives of maltose, maltotetraose, and maltohexaose.[321] The distances between the introduced branch points were 3—4, 10, and approximately 17 D-glucose residues respectively, representing considerably higher branch frequency than has been previously obtained. After saponification of the modified polysaccharides, the side-chains of d.p. 4 and 6 attached to position 6 in the D-glucose residues of amylose could be extended readily using potato phosphorylase.

The interactions of carboxymethyl- and diethylaminoethyl-amyloses of low degree of substitution with iodine in the presence of iodide ion have been spectrophotometrically investigated as a function of iodide ion concentration, temperature, and polymeric charge.[322] The interaction of amylose with a hydrophobic fluorescent probe, 2-(4-toluidinyl)-naphthalene-6-sulphonate (TNS) has been investigated.[323]

Factors affecting the 'Phadebas' amylase test as performed on samples of plasma, urine, and duodenal juice have been discussed and the importance of performing suitable blank determinations emphasized.[324] A new method for the quantitative determination of α- and β-amylase activities in amylolytic preparations involves the use of unmodified amylose as substrate for the determination of total ($\alpha + \beta$) activity, and the use of amylose cross-linked with epichlorohydrin for the determination of α-amylolytic activity alone.[325] In each case the maltose produced is measured using 2,4-dinitrosalicylate reagent.

Epichlorohydrin-crosslinked amylose has been used as an affinity matrix for the isolation of the extracellular isoamylase of *Pseudomonas amyloderamosa*,[326] and for the purification of the periplasmic maltose-binding protein of *Escherichia coli*.[327]

Celluloses.—The characterization and properties of cellulose solutions of the 'network' rather than 'particle' type have been reviewed with particular regard for their application in chemical engineering.[328]

By utilizing commercially available antibiotic filter discs, which are slightly more resistant to cellulase activity and have a higher degree of crystallinity than conventional filter paper, an improved assay for cellulase activity has been developed.[329]

The carbonyl compound spectrum of cellulose cigarette smoke condensate has

321 B. Pfanemüller, G. C. Richter, and E. Husemann, *Carbohydrate Res.*, 1977, **56**, 147.
322 A. Cesàro and D. A. Brant, *Biopolymers*, 1977, **16**, 983.
323 H. Nakatani, K.-I. Shibata, H. Kondo, and K. Hiromi, *Biopolymers*, 1977, **16**, 2363.
324 K. Wiener and C. H. Foot, *Clinica Chim. Acta*, 1977, **75**, 177.
325 M. A. Mateescu, *Biochimie*, 1978, **60**, 535.
326 K. Kato, Y. Konishi, A. Amemura, and T. Harada, *Agric. and Biol. Chem. (Japan)*, 1977, **41**, 2077.
327 T. Ferenci and U. Klotz, *F.E.B.S. Letters*, 1978, **94**, 213.
328 J. Schurz, *Cellulose Chem. Technol.*, 1977, **11**, 3.
329 B. S. Montenecourt, D. R. Eveleigh, G. K. Elmund, and J. Parcells, *Biotechnol. and Bioeng.*, 1978, **20**, 297.

been investigated; 63 such compounds were positively or tentatively identified.[330]

When aqueous solutions of D-glucose and vanillyl alcohol are treated under mild conditions an ether linkage is formed between the benzyl group and O-6 of D-glucose; this has been presented as a possible model for bonding between lignin and cellulose.[331]

I.r. spectra of graft co-polymers of poly(acrylic acid) and poly(methacrylic acid) on cellulose have been studied.[332] It was concluded that strong interchain hydrogen bonds are formed between poly(acrylic acid) and the polysaccharide backbone but that grafted poly(methacrylic acid) is less likely to form such bonds. Cotton cellulose has been grafted with acrylic acid and the sodium salt of the derivatives formed was coupled with streptomycin in an ion-exchange reaction.[333] The advantages of using such cellulose derivatives, and those obtained with tetracyclines *via* a pathway involving diazonium salts of some aromatic derivatives of cellulose, as medical bandages with anti-microbial properties were discussed.

Reversible redox-systems have been used in a method for the synthesis of graft co-polymers of cellulose and α,α-dihydroperfluorheptilacrylate.[334]

Graft co-polymerization of different synthetic polymers (*e.g.* polyacrylonitrile, polystyrene, polymethylvinylpyridine) with cellulose has been performed at different stages of synthetic fibre production.[335] The distribution of grafted chains was dependent on the amount of graft polymer present and on the nature of the graft monomer. A connection between the morphology of grafted fibres and their properties was demonstrated, and methods of regulating the morphological structures of grafted fibres were proposed.

An acylpentapeptide inhibitor of pepsin and acid proteases has been activated with ethyl chloroformate and treated with 4-aminobenzyl cellulose.[336] The derivative is useful for the affinity chromatographic purification of acid proteases. 6-Aminohexylcellulose treated with cyanogen bromide-activated gallotannin provides a suitable matrix for the immobilization of naringinase.[337] Phenolic and sulphonic groups have been introduced into 4-aminophenylcellulose by diazotization and reaction with 1,2-dihydroxybenzene-3,5-disulphonic acid.[338] The metal ion binding properties of the support were determined.

A new method of reversible immobilization of mono- and oligo-nucleotides

330 H. Sakuma, N. Shimojima, and S. Sugawara, *Agric. and Biol. Chem.* (*Japan*), 1978, **42**, 359.
331 J. A. Hemmingson, G. J. Leary, I. J. Miller, W. A. Thomas, and A. D. Woodhouse, *J. C. S. Chem. Comm.*, 1978, 92.
332 E. Gruber, I. Mohammed, S. Alloush, and J. Scurz, *Cellulose Chem. Technol.*, 1976, **10**, 577.
333 C. Simionescu, D. Dumitriu, and R. Butnaru, *Cellulose Chem. Technol.*, 1974, **8**, 393.
334 Y. P. Zhdanova, M. S. Topelson, L. S. Sletkina, I. Y. Tzarevskaya, and Z. A. Rogovin, *Cellulose Chem. Technol.*, 1976, **10**, 315.
335 B. P. Morin, M. P. Bereza, I. Y. Tzarevskaya, G. I. Stanchenko, and Z. A. Rogovin, *Cellulose Chem. Technol.*, 1977, **11**, 293.
336 M. Hoshino, R. Miyajima, S. Murao, and K. Mitsugi, *Agric. and Biol. Chem.* (*Japan*), 1977, **41**, 709.
337 M. Ono, T. Tosa, and I. Chibata, *Agric. and Biol. Chem.* (*Japan*), 1978, **42**, 1847.
338 P. Burba, M. Griesbach, and K. H. Lieser, *Z. analyt. Chem.*, 1977, **284**, 257.

using amino-oxybutylcellulose as support has been described.[339] The nucleotides were activated with mesitoyl chloride prior to coupling and could be decoupled by mild acid hydrolysis. The same support has been used for the immobilization of periodate-oxidized nucleosides, their mono-, di-, and tri-phosphates, and tRNA, poly(adenylic acid), poly(uridylic acid), and *S*-adenoxyl-L-homocysteine.[340]

Complexes formed between cellulose pretreated with liquid ammonia and the sterically hindered amines, 2-aminobutane, 2-amino-2-methylpropane, and 1,1,4,4-tetramethyl-1,4-diaminobutane have been studied.[341] The bulk of the amine group, rather than its length, may be a factor in the degree of distension observed in diffractometer tracings of the complexes.

Chlorination of cellulose with methane sulphonyl chloride in *N,N*-dimethylformamide and chloral was rapid and selective, initially at C-6.[342] When most of the primary hydroxy- groups had been replaced reaction occurred at O-3 to yield 3,6-dichloro-3,6-dideoxy-D-allose residues. Gel permeation chromatography was used to investigate cellulose degradation during chlorination.

Carboxymethylation and methylation of cellulose have been achieved in a dimethyl sulphoxide–paraformaldehyde solvent system.[343] The resulting carboxymethylcellulose possessed a limiting low degree of substitution of about 0.20 while methylcellulose samples could be prepared over a wide range of degrees of substitution. Substitution patterns were investigated and it was shown that by careful choice of reaction conditions a significant degree of control over the substituent distribution produced in cellulose etherification could be achieved.

Carboxymethylcellulose has been used as a solid substrate for the qualitative measurement of cellulase activity of micro-organisms.[344] The substrate allows more rapid assay than does native cellulose.

Affinity elution chromatography, using 3-phosphoglycerate as eluant and carboxymethylcellulose as support matrix, has been applied to the purification of 3-phosphoglycerate kinase from a variety of sources.[345]

A new 'molecular-associative' mechanism has been proposed for the formation of cyanoethylcellulose by means of reactive complexes having hydroxy- and anhydro-D-glucose hydroxy-groups and including polarized molecules of acrylonitrile in the sphere of elementary interaction.[346] Experimental evidence against the previously accepted ionic mechanisms for cyanoethylation is presented.

The preparation of dialdehydocellulose from microcrystalline cellulose and cotton linters has been compared for a variety of reaction conditions.[347] Microcrystalline cellulose was shown to be more reactive, and under optimal

339 V. V. Shumyantseva and R. M. Khomutov, *Bioorg. Khim.*, 1978, **4**, 1477.
340 A. A. Nedospasov and R. M. Khomutov, *Bioorg. Khim.*, 1978, **4**, 645.
341 J. J. Creely and R. H. Wade, *J. Polymer Sci., Polymer Letters*, 1978, **16**, 477.
342 T. Ishii, A. Ishizu, and J. Nakano, *Carbohydrate Res.*, 1977, **59**, 155.
343 M. D. Nicholson and D. C. Johnson, *Cellulose Chem. Technol.*, 1977, **11**, 349.
344 L. Hankin and S. L. Anagnostakis, *J. Gen. Microbiol.*, 1977, **98**, 109.
345 T. Fifis and R. K. Scopes, *Biochem. J.*, 1978, **175**, 311.
346 V. N. Syutkin, B. Y. Dudrov, V. V. Gromova, and S. N. Danilov, *Cellulose Chem. Technol.*, 1976, **10**, 133.
347 N. E. Kotelnikova and G. A. Petropavlovski, *Cellulose Chem. Technol.*, 1974, **8**, 203.

conditions formed a water-soluble material in 90% yield. Comb-like derivatives of cellulose have been prepared as for amylose (*cf.* p. 539).[320,321]

Diethylaminoethylcellulose has been used for the immobilization of methanol oxidase[348] and D-glucose isomerase[349] by physical adsorption.

N,N-Diethyl-*N*-2-hydroxypropylamino-cellulose has been used to adsorb adenosine-3,5-phosphate and other adenine nucleotides from an extract of human platelets previously incubated with radioactively-labelled adenine.[350] This forms the basis of a simple assay for adenylate cyclase in the intact platelet cells as adenosine-3,5-phosphate can be eluted specifically from the support with an adenosine-3,5-phosphate-dependent protein kinase, and the radioactivity of the eluate determined.

The effect of epichlorohydrin- or formaldehyde-crosslinking on the fine structure of cellulose has been investigated.[351] Crosslinking was shown to increase the number of accessible hydroxy-groups, thus improving cellulose reactivity in some esterification reactions.

The substitution of hydroxy-groups in cellulose II during the process of partial acetylation has been investigated.[352] A dielectric method for the quantitative estimation of the degree of substitution at O-6 was developed.

Three different reagents (glacial acetic acid, acetyl chloride, and acetic anhydride) for esterifying solutions of cellulose in mixtures of dimethyl sulphoxide and paraformaldehyde have been evaluated.[353] The thermoplastic resins obtained softened between 80 and 110 °C, and had thermal stabilities similar to those of native cellulose and degrees of acetyl-substitution ranging from 0 to 2.0 depending on the acetylation time. The mechanism of formation and the structural characteristics of cellulose acetate membranes suitable for reverse osmosis have been investigated.[354] Scanning electron microscopy studies showed the membranes to be high molecular weight condensation structures of the cellular type.

Rheological properties of solutions of cellulose acetates, butyrates, and acetobutyrates have been investigated for a wide range of applied stresses and shear rate gradients and related to the nature of the esters and their average degree of polymerization.[355] The latter most markedly affected the values of the initial Newtonian viscosities of the solutions. Cellulose molecules containing acetyl and butyryl groups in different proportions were shown to have different flexibilities and hence difficient rheological properties.

The effect of heating on structural transitions and crystallization of cellulose acetates, butyrates, and valerates has been investigated.[356] The glass transition

348 J. Baratti, R. Couderc, C. L. Cooney, and D. I. C. Wang, *Biotechnol. and Bioeng.*, 1978, **20**, 333.
349 G. Huitron and J. Limon-Lason, *Biotechnol. and Bioeng.*, 1978, **20**, 1377.
350 A. K. Sinha and R. W. Colman, *Biochem. J.*, 1978, **174**, 699.
351 Ľ. Kuniak, *Cellulose Chem. Technol.*, 1974, **8**, 247.
352 O. Y. Bufetchikova, A. J. Artjukhov, T. I. Borisova, and E. A. Abramova, *Cellulose Chem. Technol.*, 1976, **10**, 453.
353 R. B. Seymour and E. L. Johnson, *J. Polymer Sci., Polymer Chemistry*, 1978, **16**, 1.
354 I. N. Vlodavets, G. Z. Nefedova, N. E. Kozhevnikova, S. A. Tverskaya, O. L. Obolonskaya, and A. N. Khodan, *Bull. Acad. Sci. U.S.S.R.*, 1977, **26**, 94.
355 N. K. Klenkova and E. N. Khlebosolova, *Cellulose Chem. Technol.*, 1977, **11**, 191.
356 N. K. Klenkova, L. A. Volkova, N. D. Tsimara, O. M. Kulakova, and N. A. Dolotova, *Cellulose Chem. Technol.*, 1974, **8**, 463.

point and the temperature of transition into the viscous state decrease regularly with the increasing length of the acyl group of the ester.

A method for the qualitative and quantitative determination of ether groups in cellulose derivatives with one or more alkoxy and hydroxyalkyl groups has been reported.[357] The groups are degraded with hydriodic or hydrobromic acid and the resulting monoiodoalkanes and alkenes or monobromo- and dibromoalkanes separated and identified by g.l.c. analysis.

A systematic study has been made of the influence of the specific water structure and diphylic nature of some cellulose ethers (methyl-, methylhydroxypropyl-, hydroxypropyl-, hydroxyethyl-, and hydroxyethylhydroxypropyl-) on the properties of their aqueous solutions.[358] The presence of nonpolar groups in the cellulose ether macromolecule and of specific water structures account for the solution properties given by hydrophobic hydration and hydrophilic bonds.

Characteristics of the ^{13}C n.m.r. spectra of methyl-, carboxymethyl-, and hydroxyethyl-cellulose ethers have been examined at 22.6 MHz.[359] Partial pre-depolymerization with acid or cellulase proved to be a requisite step. The spectra permitted an assessment of the degree of substitution at individual positions of the D-glucose residues and afforded other information in agreement with data obtained by other methods.

The distribution of substituents in ethylcellulose has been determined by hydrolysis and conversion to the alditol acetates which were determined quantitatively by g.l.c.[360] Reference ethyl ethers of D-glucose were synthesized and isolated from a hydrolysate of ethylcellulose of high degree of substitution, and the conformation of these derivatives determined by mass spectrometry. The ratio of ethylation on O-2, O-3, and O-6 was found to be 7.45 : 1 : 1.5.

Hydroxypropylcellulose has been used for the development of a peroral controlled-release dosage form of isoproterenol hydrochloride.[361]

Dry films of methylcellulose prepared in the presence of dimethylol urea (0.05—0.2% w/w) have been shown to exhibit a faint maximum of tensile strength and elongation at break, reflecting formation of a very widely spaced network.[362] Further increasing the concentration of dimethylol urea resulted in a decrease in tensile strength.

Cellulose cyclic imidocarbonate has been used for the preparation of active immobilized β-D-glucosidase[363] and alcohol dehydrogenase.[364]

The production of a strong ionite by graft co-polymerization of cellulose with poly(2-methyl-5-vinylpyridine) using epichlorohydrin has been studied.[365] The basic regularity of the reaction as well as the composition of its products was investigated.

[357] G. Bartelmus and R. Ketterer, *Z. analyt. Chem.*, 1977, **286**, 161.
[358] N. V. Petryuk and A. M. Wasserman, *Cellulose Chem. Technol.*, 1977, **11**, 173.
[359] A. Parfondry and A. S. Perlin, *Carbohydrate Res.*, 1977, **57**, 39.
[360] L. B. Ivannikova, I. A. Mikhailopulo, M. I. Prigozhin, V. A. Tropina, A. I. Fanar, and M. V. Prokofieva, *Cellulose Chem. Technol.*, 1976, **10**, 673.
[361] Y. Machida and T. Nagai, *Chem. and Pharm. Bull. Japan*, 1978, **26**, 1652.
[362] G. A. Petropavlovski, L. K. Yakovenko, and G. G. Vasilieva, *Cellulose Chem. Technol.*, 1976, **10**, 335.
[363] D. Venardos, *Biotechnol. and Bioeng.*, 1977, **19**, 1405.
[364] J.-C. W. Kuan, S. S. Kuan, and G. G. Guilbault, *Analyt. Chim. Acta*, 1978, **100**, 229.
[365] M. Y. Mazov and M. A. Tyuganova, *Cellulose Chem. Technol.*, 1976, **10**, 185.

Nitrocellulose has been modified by graft co-polymerization with methyl methacrylate.[366] For the first time, complete separation of the graft co-polymer and the two individual homopolymers was effected, by means of Soxhlet extraction in different solvent systems.

Nitrocellulose has been used for the isolation of mouse globin mRNA.[367] Denatured mouse liver DNA immobilized on nitrocellulose by adsorption provides a template with which RNA molecules of about 30 nucleotides in length could be synthesized by RNA polymerase from *Escherichia coli*.[368]

The synthesis, antimicrobial activity, and biodeterioration resistance of cellulose 5-nitrofuran derivatives and furan derivatives of model saccharides have been studied, and the relationship between their biological properties and chemical structures was discussed.[369]

Oligo-(dT) cellulose derivatives have been applied to the fractionation of mouse globin m-RNA,[367] and to the purification of terminal deoxynucleotidyl transferase.[370]

The preparation and periodate oxidation of C-6 oxycellulose have been studied and a conformational interpretation of hemiacetal stability presented.[371] Cellulose acylated with palmitoyl chloride provides a suitable affinity matrix for the purification of *Chromobacterium viscosum* lipase.[372]

Affinity elution techniques, involving desorption of the required molecule from a general affinant support using an eluant with special affinity for the molecule, have been applied very successfully to purifications by utilizing phosphocellulose. Glucose-6-phosphate dehydrogenase, 6-phosphogluconate dehydrogenase, glutathione reductase, pyruvate kinase,[373] pyruvate kinase iso-enzymes,[374] L-threonyl-tRNA synthetase,[375] and aminoacyl-tRNA synthetases (L-tyrosyl-, L-threonyl-, L-valyl-, and L-isoleucyl-)[376] have all been purified rapidly by this method.

The molecular masses of samples of sulphoethylcellulose with different degrees of substitution have been determined by ultracentrifugal analysis.[377]

Nucleophilic substitution reactions of the secondary sulphonyloxy groups of sulphonate derivatives of cellulose have been investigated in hexamethyl phosphorus triamide using acetoxy-ions as the nucleophile.[378]

Cellulose beads activated with tolylene-2,4-di-isocyanate have been used to

366 D. Sudhakar, K. S. V. Srinivasan, K. T. Joseph, and M. Santappa, *J. Polymer Sci., Polymer Letters*. 1978, **16**, 457.
367 A. A. Karavanov, S. N. Khil'ko, and E. E. Makarovskaya, *Biochemistry* (*U.S.S.R.*), 1977, **42**, 1324.
368 A. A. Akhundova, T. Yu. Peresleni, and G. A. Dvorkin, *Biochemistry* (*U.S.S.R.*), 1978, **43**, 626.
369 P. Krkoška, L. Ebringer, M. Ondrišová, and M. Remenár, *Cellulose Chem. Technol.*, 1976, **10**, 155.
370 S. Okamura, F. Crane, H. A. Messner, and T. W. Mak, *J. Biol. Chem.*, 1978, **253**, 3765.
371 T. J. Painter, *Carbohydrate Res.*, 1977, **55**, 95.
372 Y. Horiuti and S. Imamura, *J. Biochem.* (*Japan*), 1977, **81**, 1639.
373 I. Takagahara, Y. Suzuki, T. Fujita, J. Yamauti, K. Fujii, J. Yamashita, and T. Horio, *J. Biochem.* (*Japan*), 1978, **83**, 585.
374 K. Harada, S. Saheki, K. Wada, and T. Tanaka, *Biochim. Biophys. Acta*, 1978, **524**, 327.
375 H. Yamada, *J. Biochem.* (*Japan*), 1978, **83**, 1583.
376 H. Yamada, *J. Biochem.* (*Japan*), 1978, **83**, 1577.
377 N. Simeonov and K. Dimov, *Cellulose Chem. Technol.*, 1976, **10**, 433.
378 N. I. Ilieva, L. S. Galbraikh, and Z. A. Rogovin, *Cellulose Chem. Technol.*, 1976, **10**, 547.

prepare active immobilized β-amylase.[379] Activation of cellulose with *sym*-triazine trichloride followed by reaction with glutathione yielded an affinity support suitable for chromatographic purification of glutathione reductase.[380] *sym*-Triazine-activated cellulose has been used for the preparation of immobilized DNA.[381]

6-*O*-Tritylcellulose has been oxidized with a dimethylsulphoxide–acetic acid anhydride mixture to yield products containing aldehyde and carboxy-groups.[382]

The use of azobisisobutyrylonitrile and benzoyl peroxide as initiators for the grafting of vinyl acetate on to cellulose has been compared.[383] The chain-transfer mechanism occurring in these systems has been verified by carrying out grafting experiments in the presence of the chain-transfer agent propane-1-thiol.

Cross-linked cellulose has been vinylsulphonated and derivatives with a high degree of substitution obtained, particularly when the reaction was performed in glycerine or D-glucose.[384]

Chitin and Chitosans.—The ultrastructure of the solid-phase of chitin gel has been examined by scanning electron microscopy.[385] The gel is considered to be a polyphasic gel with small droplets of water held up in micropores.

A series of partially and heterogeneously *N*-acylated chitosans have been prepared and isolated in 50—100% yield after treatment of chitosan with carboxylic anhydrides.[386] The structure of the *N*-acyl group influenced the gelation: minimum requirements for gelation were found to be about 0.4 *N*-lauroyl (C_{12}), 0.6 *N*-fatty acyl (C_3—C_{10}), or 0.7 *N*-benzoyl groups per 2-amino-2-deoxy-D-glucosyl residue, and gelation did not occur with higher fatty acyl (C_{14}—C_{18}) groups.

N-Acetylchitosan gel and partially *O*-acetylated *N*-acetylchitosan gel have been prepared by reaction of chitosan with acetic anhydride in 10% acetic acid and in aqueous acetic acid–methanol at room temperature.[387] Physical properties, including specific rotation, of the gels were investigated.

Chitosan reacts readily with aldehydes (formaldehyde, acetaldehyde, acrolein, benzaldehyde, cinnamaldehyde, propionaldehyde, salicylaldehyde, and 4-tolualdehyde) to form novel chitosan-aldehyde gels.[388] The gels displayed a variety of physical properties depending on the nature of the aldehyde.

Chitosan activated with a water-soluble carbodi-imide has been used to prepare active immobilized derivatives of glucoamylase, D-glucose isomerase,

379 H. Maeda, G. T. Tsao, and L. F. Chen, *Biotechnol. and Bioeng.*, 1978, **20**, 383.
380 J. Danner, H. M. Lenhoff, and W. Heagy, *Analyt. Biochem.*, 1977, **82**, 586.
381 S. Biagioni, R. Sisto, A. Ferraro, P. Caiafa, and C. Turano, *Analyt. Biochem.*, 1978, **89**, 616.
382 G. A. Vikhoreva, L. S. Galbraikh, and Z. A. Rogovin, *Cellulose Chem. Technol.*, 1974, **8**, 115.
383 B. N. Misra, J. K. Jassal, and C. S. Pande, *J. Polymer Sci., Polymer Chemistry*, 1978, **16**, 295.
384 N. Simeonov and K. Dimov, *Cellulose Chem. Technol.*, 1976, **10**, 419.
385 S. Hirano, R. Yamaguchi, and N. Matsuda, *Biopolymers*, 1977, **16**, 1987.
386 S. Hirano, O. Miura, and R. Yamaguchi, *Agric. and Biol. Chem. (Japan)*, 1977, **41**, 1755.
387 S. Hirano and R. Yamaguchi, *Biopolymers*, 1976, **15**, 1685.
388 S. Hirano, R. Yamaguchi, N. Matsuda, O. Miura, and Y. Kondo, *Agric. and Biol. Chem. (Japan)*, 1977, **41**, 1547.

D-glucose oxidase, trypsin, and urease.[389] β-D-Glucosidase has been immobilized on chitosan activated with glutaraldehyde.[390]

Chitosan and sodium dextran sulphate solutions have co-reacted under various experimental conditions to form a water-insoluble hydrous precipitate.[391]

Cycloamyloses.—The chemistry of cycloamyloses has been reviewed.[392]

The diffusion of cyclohexa-amylose and cyclohepta-amylose (α- and β-cyclodextrin) has been studied in aqueous solutions of poly(methacrylic acid), sodium poly(styrene sulphonate) and co-poly(styrene–methacrylic acid).[393] A decrease in the diffusion coefficients of the cycloamyloses in these polymer solutions was found to be dependent on the polymer concentration, degree of sulphonation, styrene content, and degree of neutralization. The results were interpreted assuming a 1 : 1 complex between the cycloamylose and an appropriate residue in the polymer.

The direction of sodium benzoate and benzoic acid penetration of the cyclohexa-amylose cavity in aqueous solution has been determined by ^{1}H n.m.r. spectroscopy of the respective cycloamylose complexes.[394] Results suggested that benzoic acid has a strong preference for binding in the cycloamylose cavity, while sodium benzoate binding is more random. Both compounds form 1 : 1 complexes with cyclohexa-amylose.

Data from a fluorescence-probe study of the interaction of cyclohepta-amylose with arenes and amphiphilic molecules have been presented.[395] The arene (pyrene) appears to interact with cyclohepta-amylose in an initial, fast inclusion process which is followed by a slower aggregation of the resulting complexes. Addition of surfactant molecules leads to complexing of the three components and thence to an increase in the hydrophobicity of the arene environment.

Conductometric and ^{13}C n.m.r. spectrometric data have allowed calculation of formation constants (at 30 °C) of cyclohexa-amylose–4-methyl cinnamate anion complexes with 1 : 1 and 2 : 1 stoicheiometries.[396] The data suggest direct binding of the carboxylate terminal of 4-methyl cinnamate by cyclohexa-amylose in the 1 : 1 complex, while the structure of the 2 : 1 complex involves occlusion of the anion between two cyclohexa-amylose molecules in a head-to-head orientation.

The asymmetric reduction of trifluoromethyl ketones using 1-propyl-1,4-dihydronicotinamide has been reported.[397] This achiral NADH model compound is oxidized in an aqueous solution of cyclohepta-amylose, which has a

389 T. Kasumi, M. Tsuji, K. Hayashi, and N. Tsumura, *Agric. and Biol. Chem.* (*Japan*), 1977, **41**, 1865.
390 F. Bissett and D. Sternberg, *Appl. Envir. Microbiol.*, 1978, **35**, 750.
391 H. Fukuda and Y. Kikuchi, *Makromol. Chem.*, 1977, **178**, 2895.
392 M. L. Bender and M. Komiyama, 'Cyclodextrin Chemistry', Springer-Verlag, New York, 1977.
393 T. Iijima, T. Uemura, S. Tsuzuku, and J. Komiyama, *J. Polymer Sci. Polymer Physics*, 1978, **16**, 793.
394 R. J. Bergerson, M. A. Channing, and K. A. McGovern, *J. Amer. Chem. Soc.*, 1978, **100**, 2878.
395 H. E. Edwards and J. K. Thomas, *Carbohydrate Res.*, 1978, **65**, 173.
396 R. I. Gelb, H. M. Schwartz, and A. Laufer, *J. Amer. Chem. Soc.*, 1978, **100**, 5875.
397 N. Baba, Y. Matsumura, and T. Sugimoto, *Tetrahedron Letters*, 1978, 4281.

chiral hydrophobic binding site capable of including water-insoluble 'guest' ketones. The reaction is regarded as a new biomimetic model for NADH-dependent reduction in biological systems.

The cyclic imidocarbonate derivative of cyclohexa-amylose has been treated with a 6-aminohexyl derivative of agarose and the support used for affinity chromatography of pullulanase.[398]

Hydrolysis of 4-nitrophenyl ferrocinnamate in the presence of cyclohepta-amylose is accelerated approximately 50 000-fold compared with hydrolysis of the ester by buffer alone.[399] It is suggested that cycloamylose-based catalysts have the potential to act as artificial enzymes, providing accelerations of enzymatic magnitudes.

Crosslinked cyclohexa- and cyclohepta-amylose) gels have been used for the chromatographic resolution of racemic mandelic acid and its derivatives.[400] Crosslinked cyclohepta-amylose bound L-isomers preferentially and resolved DL-methyl mandelate to give the D-isomer in 100% purity. Mandelic acid, ethyl mandelate, and *O*-methylated mandelic acid were partially resolved in initial fractions. Binding was studied quantitatively by an equilibrium method. Cross-linked cyclohepta-amylose bound D-isomers more strongly.

An investigation of the roles of the active species in the γ-radiolytic decomposition of cyclohepta-amylose has been reported.[401] The products were examined by gel filtration and the effects of various additives investigated.

Dextran.—^{13}C and ^{1}H n.m.r. spectroscopic studies of a bacterial dextran and its acetate derivatives and of a synthetic tri-*O*-benzyldextran obtained by polycondensation have been reported.[402] All three polymers showed regular linearity of 1,6-linked D-glucose residues. Proton signals were assigned by homonuclear double irradiation and the ^{13}C signals by selective proton irradiation.

Chromatography of wheat enzymes (β-D-glucosidase, β-D-galactosidase, and β-D-fructofuranosidase) on a macroporous crosslinked dextran (Sephadex® G-200) column displays delayed elution of the proteins.[403] Possible explanations are discussed.

Using a method applicable to other biological systems, the binding constants of dextrans and isomaltose oligosaccharides to dextran-specific myeloma proteins have been determined by affinity electrophoresis.[404] The retardation of electrophoretic mobility of myeloma proteins on addition of dextran to the separating gel may be reversed on addition of suitable oligosaccharides to the gel.

Affinity electrophoresis of proteins interacting with Blue Dextran (dextran treated with the *sym*-triazine group of Cibacron Blue F3GA) has been reported; the potential of the method as a purification technique was discussed.[405]

398 B. S. Enevoldsen, L. Reimann, and N. L. Hansen, *F.E.B.S. Letters*, 1977, **79**, 121.
399 M. F. Czarniecki and R. Breslow, *J. Amer. Chem. Soc.*, 1978, **100**, 7771.
400 A. Harada, M. Furue, and S.-I. Nozakura, *J. Polymer Sci., Polymer Chemistry*, 1978, **16**, 189.
401 T. Komiya, T. Yamada, and S. Nara, *Agric. and Biol. Chem. (Japan)*, 1976, **40**, 2095.
402 D. Gagnaire and M. Vignon, *Makromol. Chem.*, 1977, **178**, 2321.
403 D. W. A. Roberts, *J. Chromatog.*, 1978, **153**, 532.
404 K. Takeo and E. A. Kabat, *J. Immunol.*, 1978, **121**, 2305.
405 M. Tichá, V. Hořejši, and J. Barthova, *Biochim. Biophys. Acta*, 1978, **534**, 58.

A spectrophotometric method of determining *via* acidic hydrolysis the extent of substitution of Cibacron Blue F3GA in Blue Sephadex® and Blue Dextran–Sepharose® preparations has been developed.[406]

Carbonated cross-linked dextrans have been used for the covalent immobilization of trypsin.[407]

A cross-linked lipophilic derivative of dextran (Sephadex® LH-20) has been reacted with veratridine *via* oxirane coupling and used to purify a veratridine-binding proteoglycolipid from rat.[408]

Diethylaminoethyl-Sephadex® has provided a support for the immobilization of β-D-galactosidase.[409]

Turbidity and electrophoretic measurements of the co-acervate formed between diethylaminoethyldextran and bovine plasma albumin have been made.[410] The interaction, which appears to be mainly electrostatic with hydrophobic contributions, was sensitively affected by pH, ionic strength, temperature, and the weight ratio of both compounds.

Dextran cyclic imidocarbonate derivatives have been prepared and used to immobilize phloretin,[411] fetuin,[257] and co-enzyme A.[412] The co-enzyme was immobilized both by reaction with dextran cyclic imidocarbonate, and also *via* carbodi-imide-mediated coupling to 6-aminohexanoyl-dextran. γ-Irradiation of dextran in the presence of trypsin, pepsin, or ribonuclease leads to production of active immobilized forms of these enzymes.[45]

The phase transformations and phase compositions during interaction of dextran sulphate and proteins precipitating in the vicinity of their isoelectric points have been investigated using nephelometry and chemical analysis.[413] Sodium dextran sulphate and chitosan solutions have been co-reacted under various experimental conditions to form a water-insoluble hydrous precipitate.[391]

Glucoamylase has been immobilized on sulphopropyl derivatives of cross-linked dextran (SP-Sephadex C-25 and C-50) by ionic linkage.[414]

D-Galactans.—L-Arabino-D-galactan has been cross-linked with epichlorohydrin and used for affinity chromatographic purification of the lectins from *Arachis hypogaea*,[415] and *Ricinus communis*.[416] Larch arabinogalactan copolymerized with L-leucine *N*-carboxyanhydride and then mixed (5 : 2) with agarose has been used for the purification of the lectin from *Tridacna maxima*.[47]

Guaran gum crosslinked with divinylsulphone yielded gels with good flow properties and chemical stability.[417] The support was used to purify lectins

406 G. K. Chambers, *Analyt. Biochem.*, 1977, **83**, 551.
407 M. Rudel, A. Gabert, and G. Mobius, *Z. Chem.*, 1978, **18**, 178.
408 R. F. Taylor, *J. Neurochem.*, 1978, **31**, 1199.
409 V. S. Baeva, L. V. Kozlov, V. K. Antonov, and A. S. Tikhomirova, *Biochemistry (U.S.S.R.)*, 1977, **42**, 544.
410 K. Gekko, H. Harada, and H. Noguchi, *Agric. and Biol. Chem. (Japan)*, 1978, **42**, 1385.
411 S. J. H. Ashcroft and S. Nino, *Biochim. Biophys. Acta*, 1978, **538**, 334.
412 F. Le Goffic, S. Sicsic, and C. Vincent, *Biochimie*, 1978, **60**, 421.
413 A. N. Gurov, E. S. Wajnerman, and V. B. Tolstoguzov, *Stärke*, 1977, **29**, 186.
414 S. Adachi, Y. Kawamura, K. Nakanishi, R. Matsuno, and T. Kamikubo, *Agric. and Biol. Chem. (Japan)*, 1978, **42**, 1707.
415 T. Majumdar and A. Surolia, *Prep. Biochem.*, 1978, **8**, 119.
416 T. Majumdar and A. Surolia, *Experientia*, 1978, **34**, 979.
417 N. M. Young and M. A. Leon, *Carbohydrate Res.*, 1978, **66**, 299.

from *Bandeiraea simplicifolia* and *Arachis hypogaea*. Crosslinking of guar gum with epichlorohydrin yielded an affinity support for the isolation of D-galactose-binding proteins from *Ricinus communis*.[418]

D-Glucans.—Curdlan, a (1→3)-β-D-glucan from *Alcaligenes faecalis*, has been activated with cyanogen bromide and reacted with 1,6-diaminohexane.[419] Coupling of the product, *via* a carbodi-imide reaction mediator, with L-tyrosine provided an affinity chromatographic support for purification of *Klebsiella* sp. arylsulphatase.

Extracellular polysaccharides from *Rhizobium trifolii*, *R. phaseoli*, *R. leguminosarum*, and *R. meliloti* have been reduced and deacetylated.[420] The resistance of the native polysaccharide to attack by *Streptomyces* cellulase (E.C. 3.2.1.4) or *Rhizopus arrhizus* laminarinase (E.C. 3.2.1.6) is lost on chemical modification. Results were discussed in relation to the available data on the structure of the polysaccharides and the specificity of the enzymes.

The crystal structures of two polymorphic forms of pachyman acetate, the fully acetylated derivative of a naturally-occurring (1 → 3)-β-D-glucan, have been determined by a combination of stereochemical and *X*-ray diffraction analyses.[421]

Glycogen.—The dissolution of precipitates of glycogen and dimethyldodecylbenzylammonium chloride complexes in an excess of a solution of the salt, and the dependence of the process on pH, have been studied.[422] The possible utilization of these properties for the separation of a mixture of α-D-glucans was evaluated.

Oyster glycogen has been treated with cyanogen bromide to form a soluble cyclic imidocarbonate derivative which was covalently linked to phosphorylase *b*.[423] The enzyme conjugate had enhanced thermal stability and decreased stability to low temperature, while other of its properties were not altered. Glycogen cyclic imidocarbonate yielded on reaction with 1,6-diaminohexane an insoluble product which provided an effective affinity support for adsorption at 0 °C of *Helix pomatia* α-amylase.[424] Desorption of the enzyme was effected by increasing the temperature.

Heparin.—*N*-Desulphated heparin has been treated in formamide with acid anhydrides to yield a series of *N*-acyl derivatives.[425]

Hog mucosal heparin and its partially *N*-desulphated derivatives retained biological (anticoagulant) activity after reaction with 5-isothiocyanatofluorescein to yield the corresponding *N*-fluoresceinylcarbamoylheparins.[426]

[418] P. S. Appukuttan, A. Surolia, and B. K. Bachhawat, *Indian J. Biochem. Biophys.*, 1977, **14**, 382.
[419] Y. Murooka, M.-H. Yim, T. Yamada, and T. Harada, *Biochim. Biophys. Acta*, 1977, **485**, 134.
[420] M. A. Anderson and B. A. Stone, *Carbohydrate Res.*, 1978, **61**, 479.
[421] T. L. Bluhm and A. Sarko, *Biopolymers*, 1977, **16**, 2067.
[422] B. N. Stepanenko and T. S. Ulitina, *Biochemistry* (*U.S.S.R.*), 1977, **42**, 1125.
[423] T. G. Sotiroudis, N. G. Dikonomakos, and A. E. Evangelopoulos, *European J. Biochem.*, 1978, **88**, 573.
[424] J. J. Marshall and W. Woloszczuk, *Carbohydrate Res.*, 1978, **61**, 407.
[425] S. Hirano and W. Ohashi, *Carbohydrate Res.*, 1977, **59**, 285.
[426] K. Nagasawa and H. Uchiyama, *Biochim. Biophys. Acta*, 1978, **544**, 430.

Chymotrypsin has been immobilized on heparin after activation of the support with a water-soluble carbodi-imide.[427]

Miscellaneous Polysaccharides and Oligosaccharides.—The reaction of acidic polysaccharides with cetyltrimethylammonium bromide has been studied by recording the turbidity with an automatic titration device.[428] The optimal conditions for the quantitative determination of various acidic polysaccharides were found from a correlation between ionic strength and turbidity.

A crude D-glucose- and 2-acetamido-2-deoxy-D-glucose-containing polysaccharide from *Aspergillus niger* has been adsorbed (1 : 1) on diatomaceous earth and used for affinity chromatographic purification of the lectins from *Datura stramonium*.[429]

A number of lectins have been purified by affinity chromatography on *O*-glycosyl polyacrylamide gels prepared by radical co-polymerization of acrylamide, *N*,*N'*-methylene bis-acrylamide, and the appropriate allyl glycoside.[430, 431] *O*-α-D-Galactosyl-derivatives have been used for purification of lectins from a variety of plant, animal, and microbial sources, such as *Viscum album*,[432] *Fomes fomentarius* (fungus), *Ononis spinosa* (plant), *Marasmius oreades*[433] (fungus), *Coregonus lavaretus mavaena*,[434] *Rutilus rutilus*, *Vimba vimba*, *Scardinus erythrophtalmus*, and *Abramis brama*[435] (fish). *O*-α-L-Fucosyl-gels have been used to purify lectins from *Anguilla anguilla*[433] (animal) and *Perca fluviatilis*[435] (fish); *O*-β-lactosyl-gels for lectins from *Ononis hirara*,[435] *Viscum album*,[432] and *Ricinus communis*[433] (plants); *O*-α-D-glucosyl gel for *Canavalia ensiformis* lectin (Concanavalin A);[433] and 2-acetamido-2-deoxy-α-D-galactosyl-polyacrylamide gels to purify lectins from *Vicia cracca*, *Phaseolus lunatus*, *Glycine soja*, *Dolichos biflorus*, *Maclura pomifera*, *Sarothamnus scoparius* (plants), *Helix pomatia* (animal), and *Clitocybe nebularis*[433] (fungus).

Free monosaccharides can be used for direct glycosylation of Spheron®, a spherical macroporous hydroxyalkyl methacrylate–ethylene dimethacrylate co-polymer.[437] Derivatives of L-fucose, D-galactose, D-glucose, D-mannose, 2-acetamido-2-deoxy-D-galactose, and 2-acetamido-2-deoxy-D-glucose have been prepared in this way and shown to be efficient affinity carriers for the isolation of a number of lectins.

Hydrophilic, water-insoluble gels suitable for affinity chromatography of lectins have been prepared by co-polymerization of acrylamide, *N*,*N'*-methylene-bis-acrylamide, and alkenyl-1-thioglycosides, and used for purification of

[427] V. P. Torchilin, E. V. Il'ina, Z. A. Streltsova, V. N. Smirnov, and E. I. Chazov, *J. Biomed. Materials Res.*, 1978, **12**, 585.
[428] J. Szejtli and E. Dósa, *Acta Chim. Acad. Sci. Hung.*, 1976, **91**, 433.
[429] V. Hořejší and J. Kocourek, *Biochim. Biophys. Acta*, 1978, **532**, 92.
[430] V. Hořejší, P. Smolek, and J. Kocourek, *Biochim. Biophys. Acta*, 1978, **538**, 293.
[431] G. Uhlenbruck, B. A. Baldo, G. Steinhausen, H. G. Schwick, B. P. Chatterjee, V. Hořejší, A. Krajhanzl, and J. Kocourek, *J. Clin. Chem. Clin. Biochem.*, 1978, **16**, 19.
[432] P. Ziska, H. Franz, and A. Kindt, *Experientia*, 1978, **34**, 123.
[433] V. Hořejší and J. Kocourek, *Biochim. Biophys. Acta*, 1978, **538**, 299.
[434] A. Krajhanzl, V. Hořejší, and J. Kocourek, *Biochim. Biophys. Acta*, 1978, **532**, 209.
[435] A. Krajhanzl, V. Hořejší, and J. Kocourek, *Biochim. Biophys. Acta*, 1978, **532**, 215.
[436] V. Hořejší, O. Chaloupecká, and J. Kocourek, *Biochim. Biophys. Acta*, 1978, **539**, 287.
[437] K. Filka, J. Coupek, and J. Kocourek, *Biochim. Biophys. Acta*, 1978, **539**, 518.

Ricinus communis and *Glycine soja* lectins.[438] A partial tentative structure of the derivative obtained with allyl 1-thio-β-D-galactopyranoside was proposed (37).

Highly porous and hydrophilic supports have been prepared by co-polymerization of *N*-tris(hydroxymethyl)methyl acrylamide or *N*-tris(hydroxymethyl)-methyl methacrylamide with acrylic monomers bearing carboxyl functions, and then treated with 4-aminophenyl α-L-fucoside using a water-soluble carbodiimide mediator.[439] The support was used for purification of the L-fucose-binding lectin from *Ulex europeus*.

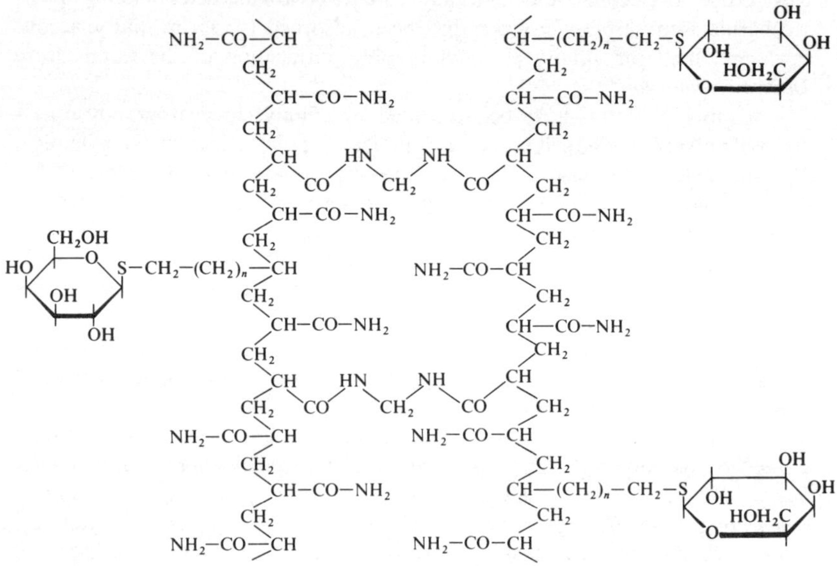

(37)

A number of polyacrylamide derivatives of oligosaccharides have been prepared for studies with intact cells, particularly chicken hepatocytes.[440] Ligands coupled included 6-aminohexyl glycosides of β-D-glucopyranose, 2-acetamido-2-deoxy-β-D-glucopyranose, α-D-mannopyranose, β-D-galactopyranose, 6-*O*-α-D-galactopyranosyl-β-D-glucopyranose, 4-*O*-β-D-glucopyranosyl-β-D-glucopyranose, 4-*O*-α-D-glucopyranosyl-β-D-glucopyranose, 4-*O*-β-D-galactopyranosyl-(α or β)-L-glucopyranose, 1-thio-β-D-galactopyranose, 1-thio-β-D-glucopyranose, and 2-acetamido-2-deoxy-1-thio-β-D-glucopyranose.

Starch.—The molecular structures of unmodified and modified (*O*-2-hydroxypropyl di-starch phosphate) manioc starches have been investigated after fractionation of their pullulanase, β-amylase, and/or glucoamylase hydrolyzates.[441] A model depicting the possible sites of 2-hydroxypropyl and/or phosphate groups on the modified manioc amylopectin molecule was presented.

[438] J. Pipková, V. Hořejší, and J. Kocourek, *Biochim. Biophys. Acta*, 1978, **541**, 515.
[439] A. Racois and E. Boschetti, *Biochimie*, 1978, **60**, 193.
[440] R. L. Schnaar, P. H. Weigel, M. S. Kuhlenschmidt, Y. C. Lee, and S. Roseman, *J. Biol. Chem.*, 1978, **253**, 7940.
[441] L. F. Hood and C. Mercier, *Carbohydrate Res.*, 1978, **61**, 53.

Utilizing resonance Raman and iodine-129 Mossbauer spectroscopy, the structure of the blue-black starch-iodine adduct has been investigated, and unambiguous evidence presented to show that the predominant polyiodide species within the amylose helix is I_5^-.[442]

'High' and 'low' viscosity carboxymethylstarches and their benzylated derivatives have been rheologically characterized.[443]

Colominic acid, a homopolymer of *N*-acetylneuraminic acid, has been cross-linked with soluble starch using epichlorohydrin to yield a gel which is useful for the affinity chromatographic isolation of neuraminidase.[444]

Mixed esters of hydrolysed starch have been prepared in pyridine using anhydrides as acylating agents.[445] The mixed esters are hydrophobic and organic-soluble, but may be readily solubilized in aqueous base through the half-ester groups. The solutions show surface activity which varies according to the type and extent of substitution.

Cationic dialdehyde-starch dispersions have been prepared in water at 15% concentration by a new cooking procedure which has potential commercial application.[446]

Starch derivatized by reaction with ethene oxide has been hydrolysed and silylated, and the distribution of the hydroxyethyl groups in the modified D-glucose units determined by gas chromatography.[447] Six monosubstituted and twelve disubstituted forms were completely separated and identified from their mass spectra.

The use of hydroxyethyl starch as a cryoprotective agent for human erythrocytes has been examined in detail.[448] The pertinent molecular properties of hydroxyethylstarch are described and conditions for its optimal use in cryoprotection summarized.

Properties of absorbent starch co-polymers which exhibit interesting characteristics have been reviewed particularly with relevance to their industrial application.[449]

Styrene has been graft-co-polymerized with acrylonitrile on to gelatinized starch at 50 °C with cerium(IV) initiation.[450] The isolated sidechain grafts are soluble in dimethylformamide and may be characterized by gel permeation chromatography.

The i.r. spectra of graft-co-polymers of poly(acrylic acid) and poly(methacrylic acid) on starch have been investigated.[332] The poly(acrylic acid) graft forms strong interchain hydrogen bonds with the polysaccharide backbone but poly(methacrylic acid) grafts are less likely to do so.

The location of grafted polymer within the starch granule matrix has been

442 R. C. Teitelbaum, S. L. Ruby, and T. J. Marks, *J. Amer. Chem. Soc.*, 1978, **100**, 3215.
443 A. El Thalouth, L. Colombe, and R. A. Schutz, *Stärke*, 1977, **29**, 160.
444 Y. Uchida, Y. Tsukada, and T. Sugimori, *J. Biochem. (Japan)*, 1977, **82**, 1425.
445 S. E. Rudolph and R. C. Glowaky, *J. Polymer Sci., Polymer Chemistry*, 1978, **16**, 2129.
446 H. D. Heath, B. T. Hofreiter, P. J. Borchert, M. I. Schulte, J. L. Neff, and C. R. Russell, *Stärke*, 1976, **28**, 303.
447 J. W. Mourits, H. G. Merkus, and L. D. Galan, *Analyt. Chem.*, 1976, **48**, 1557.
448 C. T. Greenwood, D. D. Muir, and H. W. Whitcher, *Stärke*, 1977, **29**, 343.
449 W. F. Lindsay, *Tappi*, 1976, **59**, 126.
450 L. A. Gugliemelli, C. L. Swanson, W. M. Doane, and C. R. Russell, *J. Polymer Sci., Polymer Letters*, 1977, **15**, 739.

determined by scanning electron microscopy for two wheat starch graft co-polymers prepared by ^{60}Co-initiated graft polymerization of a mixture of acrylamide and *N,N,N*-trimethylaminoethylmethacrylate methyl sulphate.[451] A series of starch–polyacrylonitrile graft co-polymers has been prepared by cerium (IV) ammonium nitrate initiation, the only differences being in the degree of swelling and gelatinization of the starch granules.[452] After saponification to convert the nitrile substituents to sodium carboxylate and carboxamide, scanning electron micrographs were obtained for each co-polymer and compared with those of the unsaponified derivative. The granules were not totally disrupted or dissolved on saponification, suggesting that a crosslinking reaction occurs either during graft polymerization or during alkaline saponification.

Optimal conditions for the precipitation of starch xanthate from sodium starch xanthate as the zinc salt have been determined as a xanthate degree of substitution greater than 0.02 and a zinc ion concentration in solution between 0.005 and 0.015 M.[453] Other properties of the starch xanthates were also reported. Aqueous solutions of starch xanthate, butane-2,3-diol dixanthate and alkali-gelatinized starch have been treated with butane-2,3-diol diglycidyl ether and other diepoxides to produce viscous gels.[454] Epoxy resins also reacted with xanthate to form viscous gels. Solids were precipitated from the gels and their compositions deduced from their elemental contents and infrared spectra.

3 Modification of Glycoproteins and Uses of the Modified Glycoproteins

Albumins.—Albumin has been treated with neutralized 2,4,6-trinitrobenzene-sulphonic acid to yield a 2,4,6-trinitrobenzenesulphonic acid–albumin derivative which when immobilized is useful for affinity chromatography.[126] 2,4-Dinitrophenyl derivatives of albumin have been prepared for immobilization and use as immunoadsorbents.[103]

Albumin activated with glutaraldehyde has been coupled with catalase and the product used in a thermometric enzyme-linked immunosorbent assay.[81] Folic acid has been used to modify albumin in a carbodi-imide-mediated reaction.[52] The folate-derivative was then coupled to agarose and used for affinity chromatography. Albumin alkylated with 1-[^{14}C]iodoacetamide has been used in the development of a new method for preparing affinity chromatography matrices.[309]

Albumin and diethylaminoethyldextran interact to form coacervates, the behaviour of which has been investigated.[410]

Radioiodinated D-glucose–albumin conjugates have been used in an inhibition binding assay for alveolar macrophages,[455] while D-glucosyl- and D-galactosyl-

[451] G. F. Fanta, F. L. Baker, J. A. Stolp, R. C. Burr, and W. M. Doane, *Stärke*, 1976, **28**, 382.
[452] G. F. Fanta, F. L. Baker, R. C. Burr, W. M. Doane, and C. R. Russell, *Stärke*, 1977, **29**, 386.
[453] E. I. Stout, D. Trimnell, W. M. Doane, and C. R. Russell, *Stärke*, 1977, **29**, 299.
[454] G. G. Maher, *Stärke*, 1977, **29**, 335.
[455] P. D. Stahl, J. S. Rodman, M. J. Miller, and P. H. Schlesinger, *Proc. Nat. Acad. Sci.* (*U.S.A.*), 1978, **75**, 1399.

derivatives of albumin have been characterized with respect to their binding to the hepatic protein receptor.[456]

Albumin has been modified, by *sym*-triazine-mediated coupling of 2-aminoethyl glycosides and by mixed anhydride coupling of 8-carboxyoctyl glycosides, to yield conjugates which function as artificial antigens and give rise to precipitating antibodies in rabbits.[457]

Diethylenetriamine penta-acetic acid (DTPA) may be covalently coupled to proteins to provide hexadentate chelating sites for metal ions.[458] The metal–macromolecule complexes are useful as probes in biological systems, and a variety of metals with useful physical properties can be specifically bound. The potential of the system was demonstrated by the preparation of [^{113}In]DTPA-human serum albumin, which was shown to have an *in vivo* distribution in mice nearly identical to that of [^{125}I]human serum albumin.

Antibodies.—Purified parasite antigens have been complexed with precipitating antibodies from rabbits, and boosted into rabbits to allow selective production of precipitins to the complexed antigens on boosting with whole parasite extracts.[99]

2,4-Dinitrophenyl derivatives of immunoglobulin E have been treated with the IgE receptors on rat basophilic leukaemia cells.[87] After solubilizing the cells, the IgE–receptor complexes could be isolated by affinity chromatography.

Sulphonation of human IgG, by treatment with sodium sulphate and sodium tetrathionate, results in selective cleavage of interchain S–S bonds to yield *S*-sulphonated groups.[459] The associated conformational changes reduced the activity of the first component of complement (C1) in normal serum without any reduction of the activity of other components of complement.

Carcinoembryonic Antigen.—The biliary excretion rates of intravenously-injected native and desialylised carcinoembryonic antigen have been determined.[460]

Fetuin.—Fetuin has been immobilized and then desialylized for use in affinity chromatography,[135] and for use in kinetic studies of lectin–cell surface receptor interactions, for which studies soluble iodinated desialylized fetuin has also been used.[257] Desialylized fetuin has been shown to be an antagonist of ligand binding to alveolar macrophages, while agalacto-fetuin is an effective inhibitor of ligand binding.[455] Desialylized fetuin has also been used in an investigation of the biliary excretion of circulating asialoglycoproteins in the rat.[460]

Orosomucoid.—Agalacto-orosomucoid has been shown to be an effective inhibitor of ligand binding to alveolar macrophages, and has been used in an assay method for their determination.[455]

Asialo-orosomucoid has been used in a comparative study of the binding of

456 C. P. Stowell and Y. C. Lee, *J. Biol. Chem.*, 1978, **253**, 6107.
457 R. R. King, F. P. Cooper, and C. T. Bishop, *Carbohydrate Res.*, 1977, **55**, 83.
458 G. E. Krejcarek and K. L. Tucker, *Biochem. Biophys. Res. Comm.*, 1977, **77**, 581.
459 Y. Fukumoto, S. Inai, K. Nagaki, K. Iida, E. Yamagami, Y. Masuho, and T. Watanabe, *J. Biochem.* (*Japan*), 1977, **82**, 955.
460 P. Thomas and J. W. Summers, *Biochem. Biophys. Res. Comm.*, 1978, **80**, 335.

D-glucosyl-albumin to the hepatic asialoglycoprotein receptor.[456] Studies of the biliary excretion rate of asialo-orosomucoid following intravenous injection into rats have led to the proposal of a novel pathway for glycoprotein uptake by the liver.[460] Desialylized orosomucoid has been activated with glutaraldehyde and coupled to *Pseudomonas 7A* glutaminase-asparaginase.[461] After iodination the conjugate was injected into mice and the kinetics of its hepatic uptake investigated. The enzyme was rapidly inactivated by the liver, but liver uptake could be blocked by excess desialylized orosomucoid.

The carboxy-group of the terminal *N*-acetylneuraminic acid residue of a glycopeptide prepared from orosomucoid by protease digestion has been esterified using a carbodi-imide mediator and then reduced.[462] The reduced glycopeptide should be a useful substrate for the assay of *endo*-glycanases in the presence of *exo*-glycanases.

Ovalbumin.—Structural studies of ovalbumin glycopeptides I and II have been performed using β-D-2-acetamido-2-deoxyglucosidase.[264]

Changes in the conformation and immunological activity of ovalbumin during its modification with different acid anhydrides have been investigated.[463] Derivatives prepared included acetylated (21% and 95%), butyrated (90%) and succinylated (32% and 92%) forms of the glycoprotein. Definite conformational changes were demonstrated with u.v. difference spectra, solvent perturbation, gel filtration, and viscosity data, while decreased immunological activity towards rabbit anti-ovalbumin was observed.

Phytohaemagglutinins.—Abrin, ricin, and their isolated A and B chains have been subjected to various chemical treatments and the effect of these on the biological activity of the lectins assessed.[464] Reductive methylation, periodate oxidation, or succinylation of abrin or ricin reduced the toxicity and haemagglutinating ability of the former but not the latter. Treatment with *N*-acetylimidazole or *N*-bromosuccinimide resulted in strongly reduced toxicity of abrin and ricin, while 2-hydroxy-5-nitrobenzyl bromide had much less effect.

Binding of concanavalin A with 4-methylumbelliferyl α-D-mannopyranoside, α-mannobioside, and α-mannotrioside has been studied using difference absorption spectrometry and titrations of ligand fluorescence quenching.[465] The changes observed in the optical properties of the 4-methylumbelliferyl group are specific for the binding of carbohydrates and depend upon the length of the carbohydrate moiety.

Study of periodate oxidation or α-D-mannosidase treatment of the phytohaemagglutinin from *Dolichos biflorus* suggests that complete integrity of the carbohydrate unit is not essential for biological activity.[466] A 16% loss of D-mannose, on periodate treatment, did not affect the lectin's ability to agglutin-

[461] G. Schmer, J. S. Hollenberg, and J. Roberts, *Biochim. Biophys. Acta*, 1978, **538**, 397.
[462] A. Amemura, R. H. Shah, O. P. Bahl, and J. M. Merrick, *Carbohydrate Res.*, 1978, **63**, 241.
[463] M. A. Qasim and A. Salshuddin, *Biochim. Biophys. Acta*, 1978, **536**, 50.
[464] K. Sandvig, S. Olsnes, and A. Pihl, *European J. Biochem.*, 1978, **84**, 323.
[465] A. van Landschoot, F. G. Loontiens, and C. K. de Bruyne, *European J. Biochem.*, 1978, **83**, 277.
[466] S. L. Biroc and M. E. Etzler, *Biochim. Biophys. Acta*, 1978, **544**, 85.

ate type A erythrocytes, precipitate blood group A and H substances or be precipitated by concanavalin A. After enzymic removal of 40% of the D-mannose residues the lectin was not precipitated by concanavalin A but retained its agglutinating ability. However, most of the agglutinating activity was lost on removal of 40% of the D-mannose with periodate, suggesting that periodate treatment alters the protein moiety.

The toxic activity of modeccin is lost on treatment with 2-mercaptoethanol but can be recovered after dialysis.[135]

When conjugated to ricin *via* glutaraldehyde, a non-toxic mutant protein (CRM 45) of diphtheria toxin has been shown to have greater toxicity in mice than either the ricin or toxin alone.[467]

Miscellaneous Glycoproteins.—Cell-free enzyme preparations from cultured fibroblasts infected with Semliki forest virus or fowl plague virus have been used to incorporate D-[^{14}C]mannose from GDP-D-[^{14}C]mannose into endogenous virus-specific glycoproteins. This work formed part of an *in vitro* study of glycosylation reactions.[468]

Immobilized Glycoproteins.—The preparation of many immobilized, insolubilized, and coupled glycoproteins has been described; these derivatives, together with their applications, are summarized in Table 6.[469–477]

Immobilized glycoproteins continue to be widely used in preparative affinity chromatographic procedures. The pH-dependent tetramerization of the major seed protein globulin of *Phaseolus vulgaris* has been exploited: at pH's above 6.5 the monomeric form of the protein predominates, while below pH 6.5 it exists as a polymer.[170] By immobilizing a small amount of highly purified globulin a support is obtained which allows production of gram quantities of the globulin.

Co-immobilization of albumin during immobilization of xanthine oxidase has improved the operational stability of the enzyme.[56]

Soybean trypsin inhibitor immobilized on a photosensitive spiropyran matrix has been used to investigate the photocontrolled binding and release of trypsin.[476]

Immobilized phytohaemagglutinins, particularly concanavalin A, are becoming increasingly important in the purification of carbohydrate-containing molecules. The elution of collagen D-glucosyltransferase from concanavalin A–agarose requires very high concentrations of methyl α-D-mannopyranoside, and the enzyme elutes in a retarded position relative to the protein peak.[128]

[467] T. Uchida, M. Yamaizumi, and Y. Okada, *Biochem. Biophys. Res. Comm.*, 1978, **81**, 268.
[468] R. T. Schwarz, M. F. G. Schmidt, and L. Lehle, *European J. Biochem.*, 1978, **85**, 163.
[469] V. I. Lozinskii, Y. A. Davidovich, B. Y. Zaslavskii, T. A. Egorov, and S. V. Rogozhin, *Biochemistry* (*U.S.S.R.*), 1978, **43**, 207.
[470] D. B. Johnson, *Biotechnol. and Bioeng.*, 1978, **20**, 1117.
[471] J. L. Boitieux, G. Desmet, and D. Thomas, *F.E.B.S. Letters*, 1978, **93**, 133.
[472] M. Paulssen, H. J. Schuurman, B. G. Dekker, and C. L. de Ligny, *Immunochemistry*, 1978, **15**, 275.
[473] Y. Hamaguchi, K. Kato, H. Fukui, I. Shirakawa, E. Ishikawa, K. Kobayashi, and N. Katunuma, *J. Biochem.* (*Japan*), 1976, **80**, 895.
[474] Y. Ikehara, K. Oda, and K. Kato, *J. Biochem.* (*Japan*), 1977, **81**, 1293.
[475] R. K. Genung and H. W. Hsu, *Biotechnol. and Bioeng.*, 1978, **20**, 1129.
[476] I. Karube, *Analyt. Biochem.*, 1978, **86**, 100.
[477] M. Saito, S. Toyoshima, and T. Osawa, *Biochem. J.*, 1978, **175**, 823.

Table 6 *Modification of glycoproteins by coupling to insoluble matrices and other macromolecules*

Glycoprotein	*Macromolecule or matrix coupled and mode of coupling*	*Use of product*	*Ref.*
Abrin and abrus phytohaemagglutinins	Reaction with agarose cyclic imidocarbonate	Study of lectin interaction with cell-surface receptors	257
Agalactoorosomucoid	Reaction with agarose cyclic imidocarbonate provided with linkage extension facilities [see (19), Scheme 6a, p. 525]	Isolation of a hepatic receptor for non-reducing 2-acetamido-2-deoxy-D-glucosyl residues.	203
Albumin	Reaction with agarose cyclic imidocarbonate	Stabilization of co-immobilized enzymes	56
	Covalent reaction by thiol-disulphide exchange with a macroporous silica support bearing an activated thiol group	Reversible immobilization of albumin to allow removal of contaminating substances not covalently bound	469
	Reaction with an amino-propyl derivative of glass or hornblende with glutaraldehyde cross-linking; reaction with titanium(IV) oxide coated with polymerized, diazotized 1,3-diaminobenzene	Stabilization of co-immobilized enzymes	470
Albumin, folate-derivative	Reaction with agarose activated with 4-benzoquinone	Affinity chromatography of folate-binding protein	52
Anti-(β-D-2-acetamido-2-deoxyglucosidase) antibody	Reaction with agarose cyclic imidocarbonate	Purification of β-D-2-acetamido-2-deoxyglucosidase by immunoadsorption	79
Anti-(serum albumin) antibody		Development of an enzyme thermistor for measurement of a variety of substrates using different immobilized enzymes	80
		Removal of serum albumin during purification of α-fetoprotein	82
		Removal of albumin during purification of erythropoietin	83
Anti-(bitter peptide from tryptic hydrolysate of casein) antibody	Reaction with agarose cyclic imidocarbonate	Removal of bitter peptide from tryptic casein hydrolysates	84

Anti-cardiotoxin antibody	Reaction with agarose cyclic imidocarbonate	Purification of synthetic cardiotoxin by immunoadsorption	85
Anti-(pig cobalophilin) antibody	Reaction with agarose cyclic imidocarbonate	Purification of vitamin B-12 binding proteins	86
Anti-(2,4-dinitrophenyl) antibody	Reaction with agarose cyclic imidocarbonate	Isolation of immunoglobulin E receptor complexes	87
Anti-(*Pseudomonas aeruginosa* exotoxin) antibody	Reaction with agarose cyclic imidocarbonate	Purification of *Pseudomonas aeruginosa* exotoxin by immunoadsorption	88
Anti-ferritin antibody		Purification of ferritin by electrophoretic desorption from the immobilized antibody	89
Anti-(human α-fetoprotein) antibody	Reaction with agarose cyclic imidocarbonate	Purification of bovine α-fetoprotein by immunoadsorption	90
Anti-fibrinogen antibody	Reaction with agarose cyclic imidocarbonate	Isolation of fibrin(ogen) degradation products from normal plasma	91
Anti-(D-fructose-1,6-diphosphatase) antibody	Reaction with agarose cyclic imidocarbonate	Quantitative immunochemical determination of radioactive proteins synthesized by chick embryo hepatocyte cultures	92
Anti-(bovine glial fibrillary acidic protein) antibody	Reaction with agarose cyclic imidocarbonate	Purification of bovine glial fibrillary acidic protein from brain extracts by immunoadsorption	93
Anti-(hepatitis B surface antigen) antibody	Reaction with collagen mediated by glutaraldehyde	Preparation of immobilized forms of antibodies with retention of biological activity	471
Anti-(Ia antigen from guinea-pig lymphocytes) antibody	Reaction with agarose cyclic imidocarbonate	Purification of detergent solubilized Ia antigens by immunoadsorption	94
Anti-(human immunoglobulin A) antibody	Reaction with 4-aminobenzyloxy-methylcellulose	Study of the interaction between IgA and immobilized anti-IgA using radioimmunoassay	472
Anti-(rabbit immunoglobulin G) antibody	Reaction with agarose cyclic imidocarbonate	Immunoadsorptive purification of goat anti-rabbit Fab′ fragments	95
Anti-(human leucocyte interferon) antibody	Reaction with agarose cyclic imidocarbonate	One-step purification of human leucocyte, lymphoblastoid, and fibroblast interferon by immunoadsorption	96
Anti-(human intrinsic factor) antibody	Reaction with agarose cyclic imidocarbonate	Purification of vitamin B-12 binding proteins	86
Anti-(α-2-macroglobulin) antibody	Reaction with agarose cyclic imidocarbonate	Purification of α-2-macroglobulin by immunoadsorption	97
Anti-(oestriol 16 α-D-glucopyranosyl uronic acid) antibody	Reaction with agarose cyclic imidocarbonate	Purification of oestriol 16 α-D-glucopyranosyl-uronic acid by electrophoretic desorption from the immobilized antibody	89

Table 6 (*cont.*)

Glycoprotein	*Macromolecule or matrix coupled and mode of coupling*	*Use of product*	*Ref.*
Anti-(ornithine–oxo-acid transferase) antibody	Reaction with an aminoalkyl derivative of porous glass treated with glutaraldehyde	Development of an enzyme-linked sandwich immunoassay of ornithine–oxo-acid transferase	473
Anti-[parasite antigen (*Dirofilaria immitis*, *Toxocara canis*, *Angiostrongyeus cantonensis*, and *Ascaris lumbricoides*)] antibodies	Reaction with agarose cyclic imidocarbonate	Purification of parasite antigens for immunodiagnosis of parasitic zoonoses	98
Anti-[parasite antigen (*Fasciola hepatica* trematode)] antibody		Purification of *Fasciola hepatica* antigen by immunoadsorption	99
Anti-(phospholipase A_2) antibody	Reaction with agarose cyclic imidocarbonate	Purification of phospholipase A_2 from *Vipera ammodytes* by immunoadsorption	68
Anti-(pregnancy zone protein) antibody	Reaction with agarose cyclic imidocarbonate	Purification of pregnancy zone protein	100, 101
Anti-renin antibody	Reaction with agarose cyclic imidocarbonate	Purification of canine plasma renin	102
Anti-transferrin antibody	Reaction with agarose cyclic imidocarbonate	Removal of contaminant transferrin during purification of α-fetoprotein	82
Avidin	Reaction with agarose cyclic imidocarbonate	Purification of inactive biotin enzymes from crude cellular extracts by affinity chromatography	114
Bandeiraea simplicifolia lectin I	Reaction with agarose cyclic imidocarbonate	Study of specificity of the phytohaemagglutinin by competitive binding assays with oligosaccharides	260
Bovine submaxillary mucin	Reaction with agarose cyclic imidocarbonate	Affinity chromatography of lectins of *Tachypleus tridentatus*	119
Casein	Reaction with agarose cyclic imidocarbonate	Affinity chromatography of rat liver nuclear protein kinases	120
Casein hydrolysate	Reaction with agarose cyclic imidocarbonate	Affinity chromatographic purification of 'acid' protease from *Aspergillus niger*	121
Collagen	Reaction with agarose cyclic imidocarbonate	Purification of D-galactosyl-L-hydroxylysyl-D-glucosyltransferase by affinity chromatography	122

Concanavalin A	Reaction with agarose cyclic imidocarbonate	Affinity chromatography of proteases of germinating peanut seeds	123
		Removal of albumin and immunoglobulins from human serum	124
		Affinity chromatography of allergens from cow's hair and dander	125
		Purification of anti-picryl T suppressor factor	126
		Purification of arylsulphatase A by affinity chromatography	127
		Affinity chromatography of collagen glucosyltransferase	128
		Purification of α-fetoprotein by affinity chromatography	78
		Affinity chromatography and purification of D-galactosyl-L-hydroxylysyl-D-glucosyltransferase	122
		Affinity chromatography of glycopeptides obtained by proteolytic digestion of rat-brain glycoproteins	129
		Affinity chromatography of phytotoxic glycopeptide mixtures from *Ceratocystis ulmi*	130
		Purification of pituitary lutrotropin and thyrotropin by affinity chromatography	131
		Purification of human fibroblast interferon by affinity chromatography	132
		Affinity chromatography of lymphocyte plasma membrane subfractions	133
		Affinity chromatography of sialyltransferases	134
		Affinity chromatography and purification of *Adenia digitata* phytohaemagglutinin (modeccin)	135
		Identification of a breast-tumour-associated orosomucoid	136
		Purification of an oviducal enzyme from *Bufo arenarum* by affinity chromatography	137
		Purification of tartrate-sensitive acid phosphatase by affinity chromatography	138
		Removal of unreacted enzyme from synthetic enzyme–dextran conjugates after glycosylation	139

Table 6 (*cont.*)

Glycoprotein	*Macromolecule or matrix coupled and mode of coupling*	*Use of product*	*Ref.*
Concanavalin A (*cont.*)		Electrophoretic desorption of bound immunoglobulins from the affinity matrix	263
		Structural studies of oligosaccharides from *endo*-β-D-2-acetamido-2-deoxy-glucanase digests of ovalbumin	264
		Binding of malaria schizonts which on maturation release readily recoverable merozoites	265
		Concentration of 'acid' glycosidase and arylsulfatase enzymes	266
		Immobilization of β-D-fructofuranosidase and investigation of the involvement of salt links in the stabilization of the enzyme	267
		Comparison of binding affinities of bovine γ-glutamyltransferases from different organs	268
		Affinity chromatographic removal of minor contaminants from conventionally purified rat serum albumin	474
2,4-Dinitrophenyl$_{33}$-human serum albumin	Reaction with agarose cyclic imidocarbonate	Purification of specific anti-(2,4-dinitrophenyl) antibodies by immunoadsorption	103
Fetuin	Reaction with agarose cyclic imidocarbonate	Purification of *Datura stramonium* phytohaemagglutinin by affinity chromatography	149
		Affinity chromatography of *Phaseolus vulgaris* phytohaemaglutinin	150
	Reaction with dextran cyclic imidocarbonate	Kinetic studies of interactions between fetuin and abrin and abrus agglutinins	257
Fetuin, neuraminidase-treated	Reaction with agarose cyclic imidocarbonate	Purification of *Adenia digitata* phytohaemagglutinin ('modeccin') by affinity chromatography	135
D-Glucan–protein complex (*Saccharomyces cerevisiae*)	Reaction with agarose cyclic imidocarbonate	Isolation of (1 → 3)-β-D-glucanase by affinity chromatography	154
Haptoglobin	Reaction with agarose cyclic imidocarbonate	Isolation of anti-haptoglobin antibodies by immunoadsorption	105
		Investigation of binding of haptoglobin with α- and β-chain tryptic peptides of haemoglobin	271

Heparin	Reaction with agarose cyclic imidocarbonate	Affinity chromatographic separation of ribosomes from reticulocyte postmitochondrial supernatants	159, 160
		Purification of hepatitis B surface antigen by affinity chromatography	142
		Affinity chromatography of rat high density lipoproteins	161
		Purification of oestradiol receptor of calf uterus by affinity chromatography	162
		Purification of collagenase by affinity chromatography; isolation of anti-haptoglobin antibodies by immunoadsorption	60
Heterologous immunoglobulin G (rabbit or human)	Reaction with agarose cyclic imidocarbonate	Removal of interspecies reactivity from anti-(immunoglobulin G) sera (human or rabbit)	106
Immunoglobulin G	Adsorption on to nonporous polystyrene latex beads	Investigation of the properties of the immobilized IgG	475
Kunitz bovine trypsin inhibitor	Reaction with agarose cyclic imidocarbonate	Purification of trypsin by affinity chromatography	163
Lens culinaris lectin	Reaction with agarose cyclic imidocarbonate	Purification of detergent-solubilized Ia antigens	94
D-Mannan–protein complex		Isolation of α-D-mannosidase from jack bean by affinity chromatography	154
unc 54-Myosin	Reaction with agarose cyclic imidocarbonate	Preparation of specific anti-(unc 54-myosin) antibodies by immunoadsorption	107
Ovalbumin	Reaction with agarose cyclic imidocarbonate	Affinity chromatography of glycosidases	169
Phaseolus vulgaris globulin	Reaction with agarose cyclic imidocarbonate	Isolation of large amounts of the *P. vulgaris* globulin	170
Phytohaemagglutinin	Reaction with agarose cyclic imidocarbonate	Purification of erythropoietin by affinity chromatography	83
Phytolacca americana mitogens Pa1 and Pa2	Reaction with agarose cyclic imidocarbonate	Purification by affinity chromatography of membrane receptors for Pa1 and Pa2	172
Picrylalbumin	Reaction with agarose cyclic imidocarbonate	Purification of anti-(picryl T suppressor factor) antibody by immunoadsorption	126
Prealbumin	Reaction with agarose cyclic imidocarbonate	Purification of retinol-binding protein by affinity chromatography	175
Ricinus communis lectin I	Reaction with agarose cyclic imidocarbonate	Study of microheterogeneity of rat α-fetoprotein	280

Table 6 (*cont.*)

Glycoprotein	*Macromolecule or matrix coupled and mode of coupling*	*Use of product*	*Ref.*
Soybean trypsin inhibitor	Reaction with agarose cyclic imidocarbonate provided with linkage extension facilities see (19), Scheme 6a, (p. 525)	Affinity chromatographic purification of *Clostridium* collagenase and clostripapain	112
	Reaction with agarose cyclic imidocarbonate	Affinity chromatographic purification of trypsin	177
		Purification of chymotrypsin by affinity chromatography	178
		Affinity chromatography of proteases of germinating peanut seeds	123
	Reaction with periodate-oxidized agarose previously reacted with spiropyran anhydride and then activated with glutaraldehyde prior to coupling	Investigation of photocontrolled affinity chromatography of trypsin	476
	Reaction with agarose–polyacrylamide gel ('Ultrogel ACA 34') after treatment of the gel with glutaraldehyde	Purification of trypsin by affinity chromatography	313
α-Thrombin	Reaction with agarose cyclic imidocarbonate	Purification of heparin by affinity chromatography	180
Transferrin	Reaction with agarose cyclic imidocarbonate	Affinity chromatographic separation of reticulocytes from erythrocytes and leucocytes	181
Wheat germ phytohaemagglutinin	Reaction with agarose cyclic imidocarbonate	Affinity chromatography of glycopeptides obtained by proteolytic digestion of rat-brain glycoproteins	129
		Affinity chromatographic separation of mouse bone marrow cells bearing different amounts of surface immunoglobulin	183
		Purification of erythropoietin by affinity chromatography	83
		Isolation of a major sialoglycoprotein of human T lymphoblastoid cells of a MOLT-4B cell line	477

Elution can be greatly enhanced by adding ethylene glycol to the eluant, suggesting that hydrophobic interactions may be of importance in the binding.

A novel use of immobilized concanavalin A for the large-scale production of malaria merozoites has been described.[265] Concanavalin A binds erythrocytes but not malarial parasites: cells from infected blood were therefore specifically bound. On maturation of the schizonts they released merozoites which could be collected conveniently.

Immobilized glycoproteins have been applied to kinetic investigations of various systems including trypsin–soybean trypsin inhibitor interactions[177] and interactions between abrus agglutinins and fetuin.[257]

Immunoadsorbents continue to find considerable application in preparative procedures[39,40] and in analytical techniques such as enzyme-linked immunoadsorbent assays.[81,473] A simple alternative to direct immunoprecipitation in solution has been reported for the quantitative estimation of newly synthesized radioactive proteins.[92]

Immobilized anti-(Fab′ fragment) antibodies have been used to isolate rabbit lymph node cell populations bearing different proportions of surface immunoglobulin G: those cells with readily detectable surface IgG (*i.e.* available for binding) were retained on the column.[95] Immunoadsorption using immobilized antibodies of very low affinity has been applied successfully to the purification of bovine α-fetoprotein.[90] By using immobilized goat anti-(human α-fetoprotein) serum, which is cross-reactive with bovine α-fetoprotein, the purified protein could be obtained without resort to chaotropic elution buffers.

Immobilized Cells.—The immobilization of cellular material, both for use in affinity chromatography and for the production of enzymatically active preparations, has received considerable attention.

Achromobacter aceris cells have been immobilized by entrapment in polyacrylamide and applied to the continuous production of NADP.[478] Cells of *Arthrobacter simplex* entrapped in polyacrylamide show steroid-Δ^1-dehydrogenase activity and catalyse the transformation of cortisol to prednisolone.[479] Use of the immobilized cells in semicontinuous batchwise operations was investigated using various nutrient media.

Immobilized whole cells of *Bacillus subtilis* have been prepared in acrylamide and used for the continuous production of α-amylase.[480] *Candida lipolytica* yeast cells co-polymerized with acrylamide provided a useful affinity support for the purification of lectins of *Ricinus communis*.[481] Although the yields were low the support proved superior to columns of agarose or erythrocyte stroma.

Clostridium butyricum cells embedded in polyacrylamide have been used in a biochemical fuel cell.[482] The immobilized cells produce hydrogen gas from D-glucose and this is used as a reactant at the cell anode.

The properties of bacteria (*Escherichia coli*) during immobilization have been

478 T. Uchida, T. Watanabe, J. Kato, and I. Chibata, *Biotechnol. and Bioeng.*, 1978, **20**, 255.
479 S. Ohlson, P. O. Larsson, and K. Mosbach, *Biotechnol. and Bioeng.*, 1978, **20**, 1267.
480 T. Kokubu, I. Karube, and S. Suzuki, *European J. Appl. Microbiol. Biotechnol.*, 1978, **5**, 233.
481 L. Genaud, J. Guillot, M. Damez, and M. Coulet, *J. Immunol. Methods*, 1978, **22**, 339.
482 I. Karube, T. Matsunaga, S. Tsuru, and S. Suzuki, *Biotechnol. and Bioeng.*, 1977, **19**, 1727.

investigated using a new immobilization technique.[483] Cells were mixed with bovine serum albumin and glutaraldehyde and dried on glass to form a film. By freezing the film and then raising its temperature an insoluble sponge-like structure was obtained. After grinding and lyophilization the living immobilized cells were used in a packed bed reactor. Polyacrylamide-entrapped β-lactamase-producing *E. coli* cells have been used to study the inactivation of β-lactamase antibiotics.[484]

The yeast glycolysis system of *Hansenula jadinii* has been co-polymerized with polyethylene glycol hydroxyethylacrylate and used for the production of ATP and the subsequent phosphorylation of cytidine monophosphate.[485] The CTP thus formed was effectively converted to CDP-choline in the same system, which operated as a kind of bioreactor.

Yeast microbodies from a *Kloechera* sp. have been immobilized by entrapment in a photocrosslinkable resin composed of hydroxyethylacrylate and polyethylene glycol, and by intra-crosslinking with glutaraldehyde in the presence of albumin.[486] The activities and properties of gel-entrapped and free microbodies were compared.

Lactobacillus arabinosus cells, entrapped in acrylamide, agar, or collagen supports, have been used in combination with a glass electrode for rapid analytical determination of nicotinic acid.[487]

Cells of *Micrococcus lysodeikticus*, hydrolysed with lysozyme and then immobilized by reaction with agarose cyclic imidocarbonate, have been used for the affinity chromatographic isolation of lysozyme.[165] *Pseudomonas testosteroni* cells have been used, after entrapment in polyacrylamide, to catalyse the continuous Δ^1-dehydrogenation of Reichstein's substance S under various conditions.[488]

Continuous production of glutathione has been achieved using *Saccharomyces cerevisiae* cells embedded in polyacrylamide.[489] *Staphylococcus aureus* cells covalently coupled to agarose cyclic imidocarbonate have been used to isolate antibody-bearing cells, whose immunoglobulin G Fc regions bind to the protein A present in *S. aureus* cells.[110] *Streptococcus lactis* cells immobilized by entrapment in polyacrylamide have been studied and optimal conditions for the production of the antibiotic 'nizin' by the cells determined.[490] Characteristics of *Streptomyces phaeochromogenes* cells immobilized by radiation-induced polymerization of 2-hydroxyethylmethacrylate at low temperatures have been studied.[491] The D-glucose isomerase activity of the immobilized cells increased with increasing cell concentration.

483 D. Petre, C. Noel, and D. Thomas, *Biotechnol. and Bioeng.*, 1978, **20**, 127.
484 W. Pache, *European J. Appl. Microbiol. Biotechnol.*, 1978, **5**, 171.
485 A. Kimura, Y. Tatsutomi, N. Mizushima, A. Tanaka, R. Matsuno, and H. Fukuda, *European J. Appl. Microbiol. Biotechnol.*, 1978, **5**, 13.
486 A. Tanaka, S. Yasuhara, G. Gellf, M. Osumi, and S. Fukui, *European J. Appl. Microbiol. Biotechnol.*, 1978, **5**, 17.
487 T. Matsunaga, I. Karube, and S. Suzuki, *Analyt. Chim. Acta*, 1978, **99**, 233.
488 H. S. Yang and J. F. Studebaker, *Biotechnol. and Bioeng.*, 1978, **20**, 17.
489 K. Murata, K. Tani, J. Kato, and I. Chibata, *European J. Applied Microbiol. Biotechnol.*, 1978, **6**, 23.
490 N. S. Egorov, I. P. Baranova, and Y. I. Kozlova, *Antibiotiki*, 1978, **23**, 872.
491 M. Kumakura, M. Yoshida, and I. Kaetsu, *European J. Applied Microbiol. Biotechnol.*, 1978, **6**, 13.

Stroma, erythrocyte membrane residues from haemolysed red cells, have been intra-crosslinked with glutaraldehyde and used for the affinity chromatographic purification of two nonspecific and noninhibitable lectins from *Phaseolus vulgaris* and *P. coccineus*.[492] Formaldehyde-fixed human erythrocytes have provided a specific adsorbent for the purification of the lectin from *Robinia pseudocacia*.[493]

4 Modification of Enzymes and Uses of Modified Enzymes

α-Amylase.—*Bacillus amyloliquefaciens* α-amylase has been modified by covalent coupling with soluble dextran cyclic imidocarbonate and the stability of the resulting complex investigated.[139]

D-Glucose Oxidase.—D-Glucose oxidase has been treated with glutaraldehyde to yield water-soluble derivatives which are intramolecularly crosslinked.[494] The modified enzyme has high activity and good stability towards denaturing agents. Results of comparative circular dichroism studies, performed with the native and crosslinked enzymes and their corresponding apoenzymes, suggest that flavin adenine dinucleotide is not a gross structural determinant of D-glucose oxidase.

In a study of the periodate oxidation of D-glucose oxidase from *Aspergillus niger*, 50% of the carbohydrate was removed with little effect on enzymic activity.[495] Gel permeation chromatography results suggested that the oxidized enzyme exists as a complex oligomerizing system.

Lysozymes.—Three derivatives of hen egg-white lysozyme have been prepared: 2-chlorobenzyloxycarbonyl-lysozyme, 3-(4-methylbenzyl)-(reduced lysozyme) and 2-chlorobenzyloxycarbonyl-*S*-(4-methylbenzyl)-(reduced lysozyme).[496] On treatment of the derivatives with anhydrous hydrogen fluoride, fully active lysozyme could be recovered, with yields of 23, 12, and 7.4% respectively. The properties of the recovered lysozyme were identical to those of the native enzyme, and crystals of the two materials were indistinguishable.

The chemical modification of hen egg-white lysozyme by *N*-bromosuccinimide has been studied kinetically using the stopped-flow method and previous results have been confirmed.[497] One most rapidly reacting L-tryptophan residue (probably Trp-62) was clearly distinguished, by its rate of modification, from four other residues. The residue could be protected by ethylene glycol, chitin, 2-acetamido-2-deoxy-D-glucose, or chitotriose, but not by D-glucono-1,4-lactone.

α-Chymotrypsin-catalysed hydrolysis of specific 4-nitroanilide substrates is enhanced in the presence of a 4-phenylbutylamine derivative of hen egg-white lysozyme.[498] The apparent dissociation constant of the phenylbutylamine-

492 J. L. Ochoa and T. Kristiansen, *F.E.B.S. Letters*, 1978, **90**, 145.
493 V. Hořejší, C. Haškovec, and J. Kocourek, *Biochim. Biophys. Acta*, 1978, **532**, 98.
494 B. Solomon, N. Lotan, and E. Katchalski-Katzir, *Biopolymers*, 1977, **16**, 1837.
495 M. Gerber, O. Bodmann, and G. V. Schulz, *F.E.B.S. Letters*, 1977, **80**, 294.
496 S. Aimoto and Y. Shimonishi, *Bull. Chem. Soc. (Japan)*, 1978, **51**, 205.
497 K. Hiromi, T. Kawagishi, and M. Ohnishi, *J. Biochem. (Japan)*, 1977, **81**, 1583.
498 Y. Shechter and A. Gertler, *Biochim. Biophys. Acta*, 1978, **527**, 42.

modified lysozyme–chymotrypsin complex was found to be 0.03 M and was independent of substrate concentration.

^{1}H n.m.r. relaxation times have been measured for lysozyme crystals cross-linked with glutaraldehyde.[499]

The properties and composition of the complex formed between purified lysozyme and flavoprotein have been investigated.[500]

α-D-Mannosidase.—Modification of the carboxy-groups of α-D-mannosidase with glycine ethyl ester in the presence of carbodi-imide, or with Woodward's reagent K, abolished the enzymic activity.[501] Methyl-α-D-mannoside, a substrate analogue, protected the enzyme against inactivation. Incorporation of a ^{14}C-labelled nucleophile in the presence or absence of the analogue suggested that 2—4 carboxy groups were protected.

Phosphatase-Human Prostatic Acid.—Incubation of human prostatic acid phosphatase with glutaraldehyde resulted in formation of soluble enzymatically active species of both higher and lower molecular weight than the native enzyme.[502] The modified enzyme shows essentially unaltered affinity towards different substrates and inhibitors, but is more resistant toward heat denaturation than the native enzyme.

Phosphorylase *b*.—Glycogen cyclic imidocarbonate has been treated with glycogen phosphorylase *b* to yield a soluble modified enzyme more resistant to heat inactivation and less stable to cold than the native enzyme.[423] Other properties were found to be little altered.

Ribonuclease A.—Lactose has been coupled to L-lysine residues in the cross-linked dimer of bovine pancreatic ribonuclease A by reductive amination with cyanoborohydride.[503] Derivatives of the dimer containing up to ten N^6-(1-deoxy-lactitoyl)-L-lysine residues per molecule retained more than 75% of the activity of the native enzyme. The effect of modification of the enzyme on hepatic uptake was investigated.

Immobilized Derivatives of Enzymes.—Numerous reviews have appeared surveying the field of enzyme immobilization.[24–30, 504] Diffusional influences on the parametric sensitivity of immobilized enzyme catalysts have been mathematically analysed.[505] Particular account was taken of external mass transfer and intraparticle diffusion effects which can significantly influence the dependence of the observed reaction rate on operating parameters such as temperature and pH.

Performance characteristics of multi-enzyme reactions in a continuous stirred tank reactor,[506] in fixed-bed reactors,[507] in fluid-bed reactors,[507, 508]

[499] E. Hsi and R. G. Bryant, *Arch. Biochem. Biophys.*, 1977, **183**, 588.
[500] Y. Hidaka and Y. Matsuoka, *J. Agric. Chem. Soc.* (*Japan*), 1977, **5**, 159.
[501] E. Paus, *Biochim. Biophys. Acta*, 1978, **526**, 507.
[502] J. Gryszkiewicz, E. Dziembor, and W. Ostrowski, *Bull. Acad. polon. Sci.*, 1978, **26**, 657.
[503] G. Wilson, *J. Biol. Chem.*, 1978, **253**, 2070.
[504] D. I. C. Wang, C. L. Cooney, A. L. Demain, P. Dunnill, A. E. Humphrey, and M. D. Lilley, 'Fermentation and Enzyme Technology', John Wiley & Sons Ltd., 1978.
[505] N. G. Karanth and J. E. Bailey, *Biotechnol. and Bioeng.*, 1978, **20**, 1817.
[506] S. H. Lin, *Biophysical Chem.*, 1978, **8**, 105.
[507] A. Sadana, *Biotechnol. and Bioeng.*, 1978, **20**, 781.
[508] A. N. Emery and J. P. Cardoso, *Biotechnol. and Bioeng.*, 1978, **20**, 1903.

in a hollow fibre reactor,[509] and in mixed reactor systems[510] have been investigated and theoretical mathematical models developed for the immobilized enzyme systems.

The determination of the proportion of the original enzyme activity retained on immobilization has been discussed with regard to both physical (diffusional) and chemical effects.[511] A mathematical formulation of the problem is presented and a method for the experimental determination of the 'true' fractional retention of enzyme activity on immobilization proposed.

Many preparations of immobilized enzymes and of enzymes to give water-soluble and -insoluble forms have been reported during the past year. These derivatives and their applications are summarized in Table 7. Immobilized enzymes are referred to in this Table whether or not they are enzymically active.

509 I. A. Webster and M. L. Shuler, *Biotechnol. and Bioeng.*, 1978, **20**, 1541.
510 H. N. Chang and P. J. Reilly, *Biotechnol. and Bioeng.*, 1978, **20**, 243.
511 T. Yamane, *Biotechnol. and Bioeng.*, 1977, **19**, 749.
512 A. Pollak, R. L. Baughn, O. Adalsteinsson, and G. M. Whitesides, *J. Amer. Chem. Soc.*, 1978, **100**, 302.
513 T. T. Ngo and K. J. Laidler, *Biochim. Biophys. Acta*, 1978, **525**, 93.
514 D. B. Johnson, *Biotechnol. and Bioeng.*, 1978, **20**, 1125.
515 S. Barry, T. Griffin, and D. B. Johnson, *Internat. J. Biochem.*, 1978, **9**, 289.
516 H. N. Chang and P. J. Reilly, *Biotechnol. and Bioeng.*, 1977, **19**, 923.
517 Y. Morikawa, T. Tezuka, M. Teranishi, K. Kimura, Y. Fujimoto, and H. Samejima, *Agric. Biol. Chem. (Japan)*, 1976, **40**, 1137.
518 A. M. del C. Battle, G. A. W. de Xifra, and A. M. Stella, *Internat. J. Biochem.*, 1978, **9**, 407.
519 J. F. Kennedy, S. A. Barker, and C. A. White, *Stärke*, 1977, **29**, 240.
520 N. Carvajal, J. Martinez, F. M. de Oca, J. Rodriguez, and M. Fernandez, *Biochim. Biophys. Acta*, 1978, **527**, 1.
521 Y. Morikawa, I. Karube, S. Suzuke, Y. Nakano, and T. Taguchi, *Biotechnol. and Bioeng.*, 1978, **20**, 1143.
522 J. R. Bertino, S. Condos, C. Horvath, K. Kalghatgi, and H. Pedersen, *Cancer Res.*, 1978, **38**, 1936.
523 R. C. Barabino, D. N. Gray, and M. H. Keyes, *Clinical Chem.*, 1978, **24**, 1393.
524 K. Buchholz and B. Godelmann, *Biotechnol. and Bioeng.*, 1978, **20**, 1201.
525 H. D. Schell, A. Turcu, M. Pele, and D. Popescu, *Cellulose Chem. Technol.*, 1977, **11**, 331.
526 N. Peitersen, J. Medeiros, and M. Mandels, *Biotechnol. and Bioeng.*, 1977, **19**, 1091.
527 W. Halwachs, C. Wandrey, and K. Schügerl, *Biotechnol. and Bioeng.*, 1978, **20**, 541.
528 T. H. Finlay, V. Troll, M. Levy, A. J. Johnson, and L. T. Hodgins, *Analyt. Biochem.*, 1978, **87**, 77.
529 T. S. Lai and P. S. Cheng, *Biotechnol. and Bioeng.*, 1978, **20**, 773.
530 V. P. Torchilin, V. S. Golmacher, and V. N. Smirnov, *Biochem. Biophys. Res. Comm.*, 1978, **85**, 983.
531 P. Cremonesi, R. Bovara, and G. Mazzola, *Cellulose Chem. Technol.*, 1976, **10**, 567.
532 W. Halwachs, C. Wandrey, and K. Schürgerl, *Biotechnol. and Bioeng.*, 1977, **19**, 1667.
533 E. Brynda and M. Bleha, *Biopolymers*, 1977, **16**, 1945.
534 H. E. Swaisgood, V. G. Janolino, and H. R. Horton, *Arch. Biochem. Biophys.*, 1978, **191**, 259.
535 B. Mattiasson, K. Mosbach, and A. Svenson, *Biotechnol. and Bioeng.*, 1977, **19**, 1643.
536 J. F. Kennedy and I. M. Kay, *Carbohydrate Res.*, 1977, **56**, 211.
537 J. F. Kennedy and I. M. Kay, *Carbohydrate Res.*, 1977, **59**, 553.
538 Y. V. Rodionov, T. V. Avilova, and V. O. Popov, *Biochemistry (U.S.S.R.)*, 1977, **42**, 1594.
539 S. Fukushima, T. Nagai, K. Fujita, A. Tanaka, and S. Fukui, *Biotechnol. and Bioeng.*, 1978, **20**, 1465.
540 L. F. Chen and G. T. Tsao, *Biotechnol. and Bioeng.*, 1977, **19**, 1463.
541 K. Ohmiya, S. Tanimura, T. Kobayashi, and S. Shimizu, *Biotechnol. and Bioeng.*, 1978, **20**, 1.
542 E. M. Beier and G. Y. Vidershain, *Biochemistry (U.S.S.R.)*, 1977, **42**, 684.

Table 7 *Modification of enzymes by coupling to insoluble matrices and other macromolecules*

Enzyme	*E.C. No.*	*Matrix or macromolecule coupled*	*Use of product*	*Ref.*
Acetate kinase	2.7.2.1	Reaction with poly(acrylamide–*co-N*-acryloxysuccinimide) with *N,N'*-di-(2-aminoethyl)-1,2-diaminoethane as mediator	Active immobilized enzyme used to test versatility of matrix	512
Acetylcholinesterase	3.1.1.7	Entrapped in polyacrylamide gel. Glutaraldehyde-crosslinked to inner surface of nylon tubing	Active immobilized enzyme; kinetic and stability studies of the immobilized enzyme	513
Achromobacter aceris cells	—	Entrapment in polyacrylamide gel	Living immobilized cells for enzymic conversions; continuous production of NADP	478
Adenosine deaminase	3.5.4.4	Reaction with poly(acrylamide–*co-N*-acryloxysuccinimide) with *N,N'*-di-(2-aminoethyl)-1,2-diaminoethane as mediator	Active immobilized enzyme used to demonstrate versatility of matrix	512
Adenosine kinase	2.7.1.20	Reaction with poly(acrylamide–*co-N*-acryloxysuccinimide) with *N,N'*-di-(2-aminoethyl)-1,2-diaminoethane as mediator		512
Adenylate kinase	2.7.4.3	Reaction with poly(acrylamide–*co-N*-acryloxysuccinimide) with *N,N'*-di-(2-aminoethyl)-1,2-diaminoethane as mediator		512
Alcohol dehydrogenase	1.1.1.1	Reaction with poly(acrylamide–*co-N*-acryloxysuccinimide) with *N,N'*-di-(2-aminoethyl)-1,2-diaminoethane as mediator		512
		Reaction with cellulose cyclic imidocarbonate	Active immobilized enzyme used for the fluorimetric determination of ethanol	364
		Reaction with porous glass and hornblende supports previously derivatized with 3-aminopropyltriethoxysilane and glutaraldehyde Reaction with porous titanium oxide coated with polymerized 1,3-diaminobenzene	Study of the stability of the active immobilized enzyme preparations	470, 514

		Adsorption on to agarose cyclic imidocarbonate previously reacted with an *N*-hydroxysuccinimide ester	Active immobilized enzyme; study of the kinetics and stability of the immobilized enzyme	515
Aldolase	4.1.2.13	Reaction with poly(acrylamide–*co*-*N*-acryloxysuccinimide) with *N*,*N*′-di-(2-aminoethyl)-1,2-diaminoethane as mediator	Active immobilized enzyme used to demonstrate versatility of matrix	512
Aldose-1-epimerase (mutarotase)	5.1.3.3	Glutaraldehyde-crosslinked to porous alkylamine silica	Active immobilized enzyme	516
D-Amino-acid oxidase	1.4.3.3	Reaction with agarose cyclic imidocarbonate	Active immobilized enzyme; investigation of optimal number of groups in spacer arm and kinetic studies	54
Aminoacylase	3.5.1.14	Reaction with 4,6-dichloro-*sym*-triazinyl Duolite A7	Investigation of suitability of the matrix for enzyme immobilization	517
Aminoacyl-tRNA synthetases (baker's yeast)	—	Adsorption on to cellulose phosphate	Isolation of the enzymes by specific affinity elution	376
Aminolaevulinate dehydratase	4.2.1.24	Reaction with agarose cyclic imidocarbonate	Active immobilized enzyme; study of porphyrin biosynthesis	518
Aminolaevulinate synthetase	2.3.1.37	Reaction with agarose cyclic imidocarbonate		
α-Amylase	3.2.1.1	Reaction with diazotized 1,3-diaminobenzene coated on magnetic iron oxide	Active immobilized enzyme; study of the effects of immobilization on enzyme stability	519
		Reaction with hydrous titanium(IV) oxide coated on magnetic iron oxide		
β-Amylase	3.2.1.2	Adsorption on to a thiol derivative of agarose previously treated with 2-hydroxy-5-nitrobenzyl bromide to form a stable thioether	Demonstration of the applicability of the new support to enzyme immobilization	297
		Reaction with cellulose activated with tolylene-2,4-di-isocyanate	Active immobilized enzyme used for the continuous conversion of starch to maltose	379
Arginase	3.5.3.1	Glutaraldehyde-crosslinking on to nylon followed by dissociation into active immobilized dimers using 4-hydroxymercuribenzoate	Study of the subunit interactions of the active immobilized enzyme dimers	520

Table 7 (*cont.*)

Enzyme	*E.C. No.*	*Matrix or macromolecule coupled*	*Use of product*	*Ref.*
Arthrobacter simplex cells	—	Entrapment in polyacrylamide gel	Living immobilized cells for enzymic conversions; conversion of cortisol to prednisolone in a semicontinuous batchwise operation	479
L-Asparaginase	3.5.1.1	Co-casting into a thin film with collagen and subsequent cross-linking with glutaraldehyde	Study of kinetic and storage properties of the immobilized enzyme; continuous conversion of L-asparagine to L-aspartate and ammonia; analytical determination of L-asparaginase levels in dogs before and after treatment and using an extracorporeal shunt	521
Bacillus subtilis cells	—	Entrapment in polyacrylamide gel	Batch production of α-amylase by the living immobilized cells	480
Carbonic anhydrase	4.2.1.1	Reaction with agarose cyclic imidocarbonate	Removal of soluble enzymes, prederivatized with an arylsulphonamide, from reaction solutions	55
Carboxypeptidase G_1 (*Pseudomonas* sp.)	—	Glutaraldehyde-crosslinked on to nylon tubes and hollow fibres	Extracorporeal enzyme reactors for the removal of methotrexate from canine blood	522
Catalase	1.11.1.6	Reaction with glutaraldehyde-activated alkylamine silica	Development of immobilized multi-enzyme systems	510
		Reaction with agarose cyclic imidocarbonate	Improvement of the stability of co-immobilized xanthine oxidase	56
		Immobilized on a particulate porous alumina	Used as part of a coupled system of immobilized enzymes for the rapid and continuous estimation of α-amylase	523
		Co-polymerization with D-glucose oxidase and acrylamide	Operational studies on the batch-wise and continuous oxidation of D-glucose; investigation of the associated macrokinetics	524
		Reaction with cellulose cyclic imidocarbonate	Active immobilized enzyme; study of changes in kinetic and stability properties on immobilization	525

Cellulase (*Trichoderma viride* culture broth)	3.2.1.4	Adsorption on to cellulose	Active immobilized enzyme	526
Chymotrypsin	3.4.21.1	Reaction with agarose cyclic imidocarbonate	Isolation of an inhibitor from *Culex pipiens* by affinity chromatography	58
		Carbodi-imide-mediated coupling to heparin	Study of the properties of the active immobilized enzyme	427
		Adsorption on to a thiol derivative of agarose previously treated with 2-hydroxy-5-nitrobenzyl bromide to form a stable thioether	Demonstration of the applicability of the new support to enzyme immobilization	297
		Enzyme acrylamide matrix system entrapped in porous aluminium silicate particles	Active immobilized enzyme. Pore diffusion model for study of various kinetic parameters	527
		Reactive with *sym*-triazine derivative of crosslinked agarose	Active immobilized enzyme	528
		Reaction with activated alkylamino or acylamino derivatives of a styrene–maleic anhydride co-polymer crosslinked with 1,6-diaminohexane	Active immobilized enzyme. Investigation of properties and stabilities of the immobilized enzyme and the coupling capacities of the supports	529
		Reaction with agarose cyclic imidocarbonate	Active immobilized enzyme. Study of the accessibility to proteolysis of apolipoproteins at the surface of human plasma high density lipoproteins	57
		Adsorption on liposome surfaces Dimethyl suberimidate- or glutaraldehyde-mediated coupling to liposome surfaces Enzyme incorporation into liposomes during formation	Demonstration that enzyme and liposome preserve biological activity and integrity respectively during immobilization; comparison of methods of attachment of protein to liposome surfaces	530
		Reaction with cellulose cyclic imidocarbonate	Active immobilized enzyme	531
		Entrapment in acrylamide and then in porous aluminium silicate particles	Active immobilized enzyme for stereospecific hydrolysis of D,L-phenylalanine methyl ester, as a model reaction for racemate resolution of aromatic amino-acids in a 5-stage fluidized bed reactor	532

Table 7 (*cont.*)

Enzyme	*E.C. No.*	*Matrix or macromolecule coupled*	*Use of product*	*Ref.*
Chymotrypsinogen	—	Reaction with hydroxyethyl methacrylate beads ('Spheron 300') pre-activated with cyanogen bromide	Study of conformational transitions of insoluble proteins using fluorescence measurements	533
		Carbodi-imide-mediated reaction with succinylated aminoalkyl derivatives of porous glass beads	Study of structural transitions of the immobilized protein in urea and guanidinium chloride	534
Clostridium butyricum cells	—	Entrapment in polyacrylamide gel mounted on to a rolled platinum electrode	Immobilized living cells for the production of hydrogen from D-glucose in a biochemical fuel cell	482
Colipase	—	Reaction with agarose cyclic imidocarbonate	Active immobilized enzyme. Study of binding of lipase and colipase	59
Collagenase	3.4.4.19	Reaction with agarose cyclic imidocarbonate	Purification of anti-collagenase antibodies by immunoadsorption	60
Creatine kinase	2.7.3.2	Reaction with poly(acrylamide–*co*-*N*-acryloxysuccinimide) with *N*,*N*′-di-(2-aminoethyl)-1,2-diaminoethane as mediator	Active immobilized enzyme used to demonstrate versatility of matrix	512
3-Cyanoalanine synthase (injectase)	4.4.1.9	Reaction with glutaraldehyde-treated amino-substituted glass	Active immobilized enzyme; assay of cyanide in an enzyme thermistor	535
Deoxyribonuclease I	3.1.4.5	Reaction with agarose cyclic imidocarbonate	Purification of human erythrocyte actin free from spectrin	61
Dextranase	3.2.1.11	Adsorption on to porous titanium(IV) oxide	Active immobilized enzyme	536
		Reaction with porous titanium(IV) oxide particles coated with diazotized 1,3-diaminobenzene	Active immobilized enzyme; investigation of parameters affecting its use in the continuous conversion of dextran	537
Escherichia coli cells	—	Entrapment in polyacrylamide gel	Living immobilized cells for enzymic conversions; degradation of β-lactam antibiotics	484
		Glutaraldehyde-crosslinked with albumin	Active immobilized β-D-galactosidase and its application in a packed bed reactor	483
Formate dehydrogenase	1.2.1.2	Reaction with controlled pore glass activated with glutaraldehyde	Comparative study of the properties of the soluble and immobilized enzymes	538

β-D-Fructofuranosidase	3.2.1.26	Reaction with poly(acrylamide–*co-N*-acryloxysuccinimide) with *N,N′*-di-(2-aminoethyl)-1,2-diaminoethane as mediator	Active immobilized enzyme used to demonstrate versatility of matrix	512
		Adsorption on to diethylaminoethyl- and carboxymethyl- derivatives of crosslinked dextran		
		Reaction with cellulose preactivated with titanium(IV) chloride	Demonstration of the importance of salt-links in maintaining the conformational stability of the enzyme	267
		Reaction with immobilized concanavalin A		
		Entrapment in polyurethane gel	Active immobilized enzyme	539
		Reaction with cellulose cyclic imidocarbonate		
		Reaction with cellulose activated with tolylene- or hexamethylene-2,4-di-isocyanate or glutaraldehyde	Active immobilized enzyme; comparison of different immobilization procedures	540
		Reaction with cellulose treated with 3-aminopropyltriethoxy-silane followed by glutaraldehyde treatment		
		Adsorption on to diethylaminoethyl- or guanidino-cellulose		
		Glutaraldehyde-crosslinked to Dowex MWA-1	Active immobilized enzyme	541
α-L-Fucosidase	3.2.1.51	Adsorption on to immobilized concanavalin A	Study of substrate specificity of the active immobilized enzyme	542
α-D-Galactosidase	3.2.1.22	Reaction with agarose cyclic imidocarbonate	Active immobilized enzyme; investigation of potential therapeutic use of the immobilized enzyme in Fabry's disease	62
β-D-Galactosidase	3.2.1.23	Reaction with agarose cyclic imidocarbonate	Active immobilized enzyme; model system for investigation of the use of immobilized lysozyme as a bactericide in enzyme columns	68
		Adsorption on to diethylaminoethyl-crosslinked dextran after modification of the enzyme with Brilliant Orange KKh	Active immobilized enzyme; continuous conversion of lactose	409

Table 7 (*cont.*)

Enzyme	*E.C. No.*	*Matrix or macromolecule coupled*	*Use of product*	*Ref.*
β-D-Galactosidase (*cont.*)		Entrapment in polyacrylamide gel	Active immobilized enzyme; investigation of effect of presence of protective agents during immobilization	543
		Glutaraldehyde-coupling to Dowex® MWA-1	Active immobilized enzyme	541
Glucoamylase	3.2.1.3	Adsorption on to sulphopropyl-derivatives of crosslinked dextran	Active immobilized enzyme	414
		Reaction with porous ceramic or lead glass beads coated with titanium(IV) oxide	Parameter evaluation and performance studies in a fluidized-bed immobilized enzyme reactor	508
		Reaction with glutaraldehyde-treated porous silica beads	Model immobilized enzyme for study of multi-enzyme systems	510
		Immobilized on a particulate porous alumina	Used as part of a coupled system of immobilized enzymes for the rapid and continuous estimation of α-amylase	523
		Reaction with cellulose cyclic imidocarbonate Reaction with cellulose activated with tolylene-2,4-di-isocyanate, hexyl-1,6-di-isocyanate, or glutaraldehyde Reaction with 3-aminopropyltriethoxy-silane-treated cellulose followed by glutaraldehyde Adsorption on to diethylaminoethyl- or guanidino-cellulose	Active immobilized enzyme; comparison of different preparation procedures	540
		Glutaraldehyde-coupling to Dowex® MWA-1	Active immobilized enzyme	541
		Reaction with porous titanium(IV) oxide and hornblende using a metal-link method Reaction with hornblende treated with 3-aminopropyltriethoxysilane and glutaraldehyde	Study of the stabilities of the active immobilized enzymes	544

		Carbodi-imide-mediated reaction with activated carbon	Active immobilized enzyme	545
		Co-casting with tropocollagen followed by glutaraldehyde crosslinking	Active immobilized enzyme	546
		Reaction with poly[(glycidyl methacrylate)–co-(ethylene dimethacrylate)] carrier or its amine derivatives	Active immobilized enzyme	547
		Reaction with activated carbon mediated by a carbodi-imide or followed by glutaraldehyde crosslinking	Active immobilized enzyme; investigation of changes in enzyme stability on immobilization	548
		Glutaraldehyde-crosslinked to chitin	Active immobilized enzyme; use in column form for continuous conversion of partially hydrolysed starch to D-glucose	549
		Carbodi-imide-mediated reaction with chitosan	Active immobilized enzyme	389
(1 → 4)-α-D-Glucan phosphorylase	2.4.1.1	Crosslinking on to porous glass with glutaraldehyde; Reaction with diazotized alkylamine glass; Carbodi-imide-mediated reaction with alkylamine glass	Investigation of properties of active immobilized enzyme	550
Glucocerebroside β-D-glucosidase		Entrapped in liposomes	Active immobilized enzyme for possible therapeutic application in Gaucher's disease	551
D-Gluconolactonase	3.1.1.17	Carbodi-imide-mediated reaction with activated carbon	Active immobilized enzyme	545
D-Glucose dehydrogenase	1.1.1.47	Glutaraldehyde-crosslinked to cellulose cyclic imidocarbonate pre-treated with 1,2-diaminoethane	Immobilized enzyme stirrer for analytical determination of D-glucose in plasma	552
		Reaction with nylon tubes treated with titanium(IV) chloride	Active immobilized enzyme	553
D-Glucose isomerase	5.3.1.18	Adsorption on to diethylaminoethyl-derivatives of cellulose or crosslinked dextran	Active immobilized enzyme; continuous isomerization of D-glucose in a packed-bed reactor	349

Table 7 (*cont.*)

Enzyme	*E.C. No.*	*Matrix or macromolecule coupled*	*Use of product*	*Ref.*
D-Glucose isomerase (*cont.*)		Carbodi-imide-mediated reaction with chitosan	Active immobilized enzyme	389
		Reaction with cellulose cyclic imidocarbonate Reaction with cellulose activated with tolylene-2,4-di-isocyanate, hexyl-1,6-di-isocyanate or glutaraldehyde Reaction with cellulose treated with 3-aminopropyltriethoxysilane followed by glutaraldehyde Adsorption on to diethylaminoethyl- or guanidino-cellulose	Active immobilized enzyme; comparison of different preparation procedures	540
		Reaction with glutaraldehyde-activated alkylamine silica	Development of immobilized multienzyme systems	510
		Not given	Active immobilized enzyme for continuous isomerization of D-glucose in fixed-bed and expanded-bed systems	554
		Occlusion in a gelatin mixture followed by glutaraldehyde-crosslinking	Determination of optimal performance conditions for large-scale isomerization of of D-glucose	555
		Adsorption on to the internal surface of controlled-pore alumina	Active immobilized enzyme; continuous production of D-fructose from D-glucose	556
D-Glucose oxidase	1.1.3.4	Carbodi-imide-mediated reaction with chitosan	Active immobilized enzyme	389
		Reaction with glutaraldehyde-activated alkylamine silica	Development of immobilized multienzyme systems	510
		Immobilized on a particulate porous alumina	Used as part of a coupled system of immobilized enzymes for the rapid and continuous estimation of α-amylase	523
		Co-polymerized with catalase and acrylamide	Operation studies on the batch-wise and continuous oxidation of D-glucose; investigation of the associated macrokinetics	524

		Reaction with cellulose cyclic imidocarbonate	Active immobilized enzyme for the determination of D-glucose	531
		Carbodi-imide-mediated reaction with activated carbon	Active immobilized enzyme	545
		Reaction with activated carbon mediated by a carbodi-imide or followed by glutaraldehyde crosslinking	Active immobilized enzyme; investigation of changes in enzyme stability on immobilization	548
		Glutaraldehyde-crosslinked on to a carbon paste electrode	Active immobilized enzyme; determination of D-glucose with a rotating disc enzyme electrode	557
		Reaction with diazotized polyaminostyrene	Active immobilized enzyme for determination of D-glucose	558
		Adsorption on to Amberlite® IRA-938 ion-exchange resin	Active immobilized enzyme	559
		Glutaraldehyde-crosslinked onto a Kieselguhr support	Active immobilized enzyme; characteristics of the deactivation of the immobilized enzyme investigated in a continuous tubular-flow reactor	560
D-Glucose-6-phosphate dehydrogenase	1.1.1.49	Reaction with poly(acrylamide–*co-N*-acryloxysuccinimide) with *N*,*N*′-di-(2-aminoethyl)-1,2-diaminoethane as mediator	Active immobilized enzyme for demonstration of versatility of matrix	512
		Co-immobilization with hexokinase by reaction with agarose cyclic imidocarbonate	Immobilized enzyme microreactor for the continuously-referenced monitoring of isoenzyme activities of creatine kinase after liquid-chromatographic separation	561
		Co-immobilization with hexokinase on nylon tubing	Evaluation of the analytical performance of the enzyme-containing coils, adapted for use with the Technicon SMAC analyser in the determination of D-glucose	562
β-D-Glucosidase	3.2.1.21	Reaction with cellulose cyclic imidocarbonate	Kinetic studies of active immobilized enzyme	363
		Co-casting into a thin film with fibroin followed by treatment with ethanol	Active immobilized enzyme	563
		Glutaraldehyde-crosslinked on to chitosan	Active immobilized enzyme	390

Table 7 (*cont.*)

Enzyme	*E.C. No.*	*Matrix or macromolecule coupled*	*Use of product*	*Ref.*
L-Glutamate dehydrogenase	1.4.1.3	Glutaraldehyde-crosslinked with albumin	Active immobilized enzyme for kinetic investigations	564
Glyceraldehyde 3-phosphate dehydrogenase	1.2.1.12	Reaction with agarose cyclic imidocarbonate	Active immobilized enzyme; dissociation of the enzyme into active immobilized dimers for study	63
			Demonstration of a new method for the estimation of agarose-bound protein	64
			Active immobilized enzyme	65
Glycerol kinase	2.7.1.30	Reaction with poly(acrylamide–*co-N*-acryloxysuccinimide) with *N,N'*-di-(2-aminoethyl)-1,2-diaminoethane as mediator	Active immobilized enzyme used to demonstrate versatility of matrix	512
Guanylate cyclase	4.5.1.2	Adsorption onto hexyl-, octyl-, or decyl-derivatives of agarose	Comparative study of the properties of the soluble and immobilized form of the enzymes	565
Hansenula jadinii cells		Entrapment in poly(ethylene glycol hydroxyethylacrylate) gel	Living immobilized cells for enzymic conversions of the glycolysis system of yeasts	485
Hexokinase	2.7.1.1	Reaction with poly(acrylamide–*co-N*-acryloxysuccinimide) with *N,N'*-di-(2-aminoethyl)-1,2-diaminoethane as mediator	Active immobilized enzyme used to demonstrate versatility of matrix	512
		Co-immobilization with D-glucose-6-phosphate dehydrogenase by reaction with agarose cyclic imidocarbonate	Immobilized enzyme microreactor for the continuously-referenced monitoring of isoenzyme activities of creatine kinase after liquid-chromatographic separation	561
		Co-immobilization with D-glucose-6-phosphate dehydrogenase on nylon tubing	Evaluation of the analytical performance of the enzyme-containing coils, adapted for use with the Technicon SMAC analyser in the determination of D-glucose	562

L-Iditol dehydrogenase	1.1.1.14	Reaction with collagen membranes pre-modified by esterification, hydrazone treatment and diazotization	Study of the kinetics of the active immobilized enzyme as a model for 2-substrate or cofactor-requiring enzyme reactions	566
Lactate dehydrogenase	1.1.1.27	Reaction with agarose cyclic imidocarbonate	Active immobilized enzyme; fluorimetric determination of lactate in clinical specimens	66
		Reaction with *sym*-triazine derivative of crosslinked agarose	Active immobilized enzyme	309
		Reaction with poly(acrylamide–*co-N*-acryloxysuccinimide) with *N,N′*-di-(2-aminoethyl)-1,2-diaminoethane as mediator	Active immobilized enzyme used to demonstrate versatility of matrix	512
Lactobacillus arabinosus cells	—	Entrapment in agar or polyacrylamide gel	Active immobilized enzyme for the determination of nicotinic acid with a glass electrode	487
		Co-casting with collagen followed by glutaraldehyde crosslinking		
Lactoperoxidase	1.11.1.17	Reaction with poly(acrylamide–*co-N*-acryloxysuccinimide) with *N,N′*-di-(2-aminoethyl)-1,2-diaminoethane as mediator	Active immobilized enzyme used to demonstrate versatility of matrix	512
Lactose synthase	2.4.1.22	Reaction with glutathione derivative of agarose pre-activated with 2,2′-dithiobis-5-nitropyridine	Active immobilized enzyme; development of a method of topographical analysis of the enzyme active site	567
Lipase	3.1.1.3	Reaction with agarose cyclic imidocarbonate	Active immobilized enzyme; study of binding of lipase and colipase	59
			Active immobilized enzyme	67
Lipoamide dehydrogenase	1.6.4.3	Reaction with diazotized aminoalkyl derivatives of controlled pore glass	Active immobilized enzyme; investigation of effect of presence of NAD^+ during immobilization	568
Lysozyme	3.2.1.17	Reaction with agarose cyclic imidocarbonate	Active immobilized enzyme; investigation of potential use as a bactericide in enzyme columns	68
			Affinity chromatographic purification of ovomucin	69

Table 7 (*cont.*)

Enzyme	*E.C. No.*	*Matrix or macromolecule coupled*	*Use of product*	*Ref.*
Lysozyme (*cont.*)		Reaction with 4,6-dichloro-*sym*-triazinyl Duolite® A7	Investigation of suitability of the matrix for enzyme immobilization	517
'Maltase'	3.2.1.20	Immobilized on a particulate porous alumina	Used as part of a coupled system of immobilized enzymes for the rapid and continuous estimation of α-amylase	523
Methanol oxidase (*Hansenula polymorpha*)	—	Adsorption on to diethylaminoethylcellulose	Active immobilized enzyme; study of the kinetics of formaldehyde formation in batch and continuous reactors	569
'Naringinase' (*Aspergillus niger*)	—	Adsorption on to 6-aminohexyl-derivative of cellulose pre-treated with cyanogen bromide-activated gallotannin	Investigation of the suitability of the active immobilized enzyme for industrial removal of the bitter substance in Japanese summer orange juice	570
Neuraminidase	3.2.1.18	Reaction with agarose cyclic imidocarbonate	Active immobilized enzyme	70
			Active immobilized enzyme applied to desialylization of glycoconjugates using fetuin as a model	71
		Reaction with agarose cyclic imidocarbonate Reaction with azide derivative of agarose Reaction with periodate-oxidized glass ('Glycophase')	Active immobilized enzyme; study of the stability and desorption of the immobilized enzyme	571
Papain	3.4.22.2	Reaction with agarose cyclic imidocarbonate	Active immobilized enzyme	67
Peroxidase	1.11.1.7	Reaction with poly(acrylamide–*co*-*N*-acryloxysuccinimide) with *N,N'*-di-(2-aminoethyl)-1,2-diaminoethane as mediator	Active immobilized enzyme used to demonstrate versatility of matrix	512
		Reaction with cellulose cyclic imidocarbonate	Active immobilized enzyme for the determination of D-glucose	531
		Reaction with diazotized polyaminostyrene	Active immobilized enzyme for the determination of D-glucose	558

L-Phenylalanine ammonia lyase	4.3.1.5	Adsorption on to walls of hollow fibres	Active immobilized enzyme; investigation of continuous degradation of phenylalanine at physiological concentrations	572
		Reaction with inner wall of glutaraldehyde-activated nylon tubing		
Phosphoglucomutase	2.7.5.1	Reaction with poly(acrylamide–*co-N*-acryloxysuccinimide) with *N,N′*-di-(2-aminoethyl)-1,2-diaminoethane as mediator	Active immobilized enzyme used to demonstrate versatility of matrix	512
Phospho-D-glucose isomerase	5.3.1.9	Reaction with poly(acrylamide–*co-N*-acryloxysuccinimide) with *N,N′*-di-(2-aminoethyl)-1,2-diaminoethane as mediator	Active immobilized enzyme used to demonstrate versatility of matrix	512
Phospholipase	3.1.1.4	Reaction with agarose cyclic imidocarbonate	Active immobilized enzyme; isolation of specific anti-phospholipase antibodies	72
		Reaction with diazotized arylamine derivative of porous glass beads	Study of the kinetics and stability of the immobilized enzyme and its potential use as a probe for membrane surface phospholipids	573
Phosphorylase *b*	—	Reaction with diazotized ester derivative of agar	Study of the properties of the active immobilized enzyme; isolation of specific anti-(phosphorylase *b*) antibodies	314
Polygalacturonase	3.2.1.15	Reaction with agarose cyclic imidocarbonate	Active immobilized enzyme	73
Poly(methoxygalacturonide) lyase	4.2.2.10	Reaction with titanium-treated diethylaminoethylcellulose or glass	Active immobilized enzyme	574
Pseudomonas testosteroni cells	—	Entrapped in polyacrylamide gel	Living immobilized cells for enzymic conversions; continuous Δ^1-dehydrogenation of Reichstein's substance S	488
Pullulanase (*Streptomyces flavochromogenes*)	3.2.1.41	Tannic acid treatment of the *S. flavochromogenes* culture filtrate to yield a tannin–pullulanase complex which is adsorbed to triethylaminoethyl-cellulose	Active immobilized enzyme	575
Pyridoxamine phosphate oxidase	1.4.3.5	Covalent reaction with a commercially available organomercurial agarose derivative	Active immobilized enzyme	282

Table 7 (*cont.*)

Enzyme	*E.C. No.*	*Matrix or macromolecule coupled*	*Use of product*	*Ref.*
Pyruvate kinase	2.7.1.40	Reaction with poly(acrylamide–*co*-*N*-acryloxysuccinimide) with *N*,*N*′-di-(2-aminoethyl)-1,2-diaminoethane as mediator	Active immobilized enzyme used to demonstrate versatility of matrix	512
Rennin	3.4.23.4	Glutaraldehyde-coupling to Dowex® MWA-1 resin	Active immobilized enzyme	541
Saccharomyces cerevisiae cells	—	Entrapped in polyacrylamide gel	Continuous production of glutathione by immobilized living cells	489
Streptococcus lactis cells	—	Entrapped in polyacrylamide gel	Living immobilized cells for enzymic conversions; production of the antibiotic 'nizin'	490
Streptomyces phaeochromogenes cells	—	Radiation-induced entrapment in 2-hydroxyethyl methacrylate gel	Active immobilized D-glucose isomerase for the conversion of D-glucose to D-fructose	491
Superoxide dismutase	1.15.1.1	Reaction with agarose cyclic imidocarbonate	Improvement of the stability of co-immobilized xanthine oxidase	56
Thiosulphate sulphur transferase	2.8.1.1	Reaction with glutaraldehyde-treated amino-substituted glass	Active immobilized enzyme; assay of cyanide in an enzyme thermistor	535
Thrombin	3.4.21.5	Reaction with agarose cyclic imidocarbonate or Affigel-10®	Comparison of the action of the soluble and immobilized enzyme on fibrinogen, factor VIII and a synthetic tripeptide	74
		Reaction with agarose cyclic imidocarbonate	Affinity chromatographic demonstration of a thrombin binding protein from the platelet plasma membrane	75
Transketolase	2.2.1.1	Reaction with agarose cyclic imidocarbonate	Study of the kinetics and stability of the immobilized enzyme; analytical determination of thiamine pyrophosphate	76
Trypsin	3.4.21.4	Reaction with agarose cyclic imidocarbonate	Active immobilized enzyme. Study of the accessibility to proteolysis of apolipoproteins at the surface of human plasma high density lipoproteins	57
		Reaction with *sym*-triazine derivative of crosslinked agarose	Active immobilized enzyme	309

		Reaction with poly(acrylamide–*co*-*N*-acryloxysuccinimide) with *N,N′*-di-(2-aminoethyl)-1,2-diaminoethane as mediator	Active immobilized enzyme used to demonstrate versatility of the matrix	512
		Reaction with epoxide groups on glycidyl methacrylate and ethylene dimethacrylate co-polymer	Active immobilized enzyme for high pressure chromatography	576
		Glutaraldehyde-crosslinked on to alkylamine derivative of porous silica	Study of the kinetics and stability of the immobilized enzyme; investigation of the effect of the number of available reactive aldehyde groups on the silica matrix on the efficiency of coupling	577
		Reaction with agarose cyclic imidocarbonate	Quantitative study of the interaction of the immobilized enzyme with competitive inhibitors using frontal affinity chromatography	578
		Glutaraldehyde-crosslinked to Spherosil®	Active immobilized enzyme for the continuous optical resolution of D,L-lysine methyl ester	579
		Adsorption on to molecular sieves	Active immobilized enzyme; process engineering studies in a packed column and a continuous flow stirred reactor	580
Urate oxidase	1.7.3.3	Reaction with porous titanium(IV) oxide	Study of the stability of the active immobilized enzyme	77
		Glutaraldehyde-crosslinked on to aminoalkyl derivative of hornblende		
		Adsorption on to dialysis membrane	Automated microassay of plasma uric acid in a membrane reactor	581
		Reaction with *O*-alkylated nylon	Immobilized enzyme reactor for routine clinical determination of serum uric acid	582
		Glutaraldehyde-crosslinked on to a nylon–polyethyleneimine co-polymer		
Urease	3.5.1.5	Adsorption on to the inner surface of nylon tubing followed by glutaraldehyde crosslinking	Analytical determination of plasma urea using the active immobilized enzyme	583
		Glutaraldehyde-crosslinked on to aminoalkyl derivatives of glass, 'Glycophase-G', non-ionic cellulose, or nylon	Analysis of urea in clinical specimens using a flow coulometric immobilized enzyme analyser	584
		Glutaraldehyde-crosslinked on to porous glass	Investigation of urea conversion in an immobilized enzyme analyser	585

Table 7 (*cont.*)

Enzyme	*E.C. No.*	*Matrix or macromolecule coupled*	*Use of product*	*Ref.*
UDP-Glucuronyltransferase	2.4.1.17	Reaction with agarose cyclic imidocarbonate	Synthesis and characterization of tetrahydrocannabinol type glycosides of D-glucopyranosyluronic acid	78, 586
			Synthesis of 4-nitrophenyl glycosides of D-glucopyranosyluronic acid	587
			Enzymic synthesis of D-glucuronic acid-conjugates of meprobamate, diethylstilboestrol, bilirubin, borneol, benzoic acid, and 4-nitrothiophenol	588
Urokinase	3.4.99.26	Carbodi-imide-mediated reaction with nylon Carbodi-imide-mediated reaction with a co-polymer of maleic anhydride and methyl vinyl ether in the presence of polyethyleneimine	Active immobilized enzyme for thrombotic coagulation studies	589
Xanthine oxidase	1.2.3.2	Reaction with agarose cyclic imidocarbonate	Study of the kinetic properties of the active immobilized enzyme; investigation of the effect of co-immobilization with catalase, superoxide dismutase, or albumin on the enzyme's stability	56
		Reaction with agarose or cellulose cyclic imidocarbonates Reaction with glutaraldehyde-activated hornblende Reaction with porous titanium(IV) oxide	Comparative study of the stabilities of the immobilized enzymes	77
		Reaction with poly(acrylamide–*co-N*-acryloxysuccinimide) with *N,N'*-di-(2-aminoethyl)-1,2-diaminoethane as mediator	Active immobilized enzyme used to demonstrate versatility of matrix	512
Yeast microbodies (*Kloechera* sp. cells)	—	Entrapped by copolymerization with hydroxyethylacrylate and polyethylene glycol Glutaraldehyde-crosslinked with albumin	Activities and properties of the enzymes in the entrapped microbodies compared with those of the free microbodies	486

543 K. Ohmiya, C. Terao, S. Shimizu, and T. Kobayashi, *Agric. and Biol. Chem.* (*Japan*), 1975, **39**, 491.
544 A. Flynn and D. B. Johnson, *Biotechnol. and Bioeng.*, 1978, **20**, 1445.
545 Y. K. Cho and J. E. Bailey, *Biotechnol. and Bioeng.*, 1978, **20**, 1651.
546 S. Gondo and H. Koya, *Biotechnol. and Bioeng.*, 1978, **20**, 2007.
547 F. Svec, J. Kalal, I. I. Menyailova, and L. A. Nakhapetyan, *Biotechnol. and Bioeng.*, 1978, **20**, 1319.
548 Y. K. Cho and J. E. Bailey, *Biotechnol. and Bioeng.*, 1977, **19**, 769.
549 W. L. Stanley, G. G. Watters, S. H. Kelly, and A. C. Olson, *Biotechnol. and Bioeng.*, 1978, **20**, 135.
550 F. Wengenmayer, D. Linder, and K. Wallenfels, *Biotechnol. and Bioeng.*, 1977, **19**, 1387.
551 I. P. Braidman and G. Gregoriadis, *Biochem. J.*, 1977, **164**, 439.
552 J. W. Kuan, S. S. Kuan, and G. G. Guilbault, *Clinical Chem.*, 1977, **23**, 1058.
553 E. Bisse and D. J. Vonderschmitt, *F.E.B.S. Letters*, 1978, **93**, 102.
554 J. Oestergaard and S. L. Knudsen, *Stärke*, 1976, **28**, 350.
555 J. V. Hupkes and R. van Tilburg, *Stärke*, 1976, **28**, 356.
556 R. A. Messing and A. M. Filbert, *J. Agric. Food Chem.*, 1975, **23**, 920.
557 F. R. Shu and G. S. Wilson, *Analyt. Chem.*, 1976, **48**, 1679.
558 J. N. Miller, B. F. Rocks, and D. T. Burns, *Analyt. Chim. Acta*, 1977, **93**, 353.
559 H. E. Klei, D. W. Sundstrom, and R. Gargano, *Biotechnol. and Bioeng.*, 1978, **20**, 611.
560 S. Krishnaswamy and J. R. Kittrell, *Biotechnol. and Bioeng.*, 1978, **20**, 821.
561 M. S. Denton, W. D. Bostick, S. R. Dinsmore, and J. E. Mrochek, *Clinical Chem.*, 1978, **24**, 1408.
562 C. C. Garber, D. Feldbruegge, R. C. Miller, and R. N. Carey, *Clinical Chem.*, 1978, **24**, 1186.
563 S. Miyairi, M. Sugiura, and S. Fukui, *Agric. and Biol. Chem.* (*Japan*), 1978, **42**, 1661.
564 J. N. Barbotin and M. Breuil, *Biochim. Biophys. Acta*, 1978, **525**, 18.
565 D. L. Garbers, *J. Cyclic Nucleotide Res.*, 1978, **4**, 271.
566 F. Paul, P. R. Coulet, D. C. Gautheron, and J. M. Engasser, *Biotechnol. and Bioeng.*, 1978, **20**, 1785.
567 J. Grunwald and L. J. Berliner, *Biochim. Biophys. Acta*, 1978, **523**, 53.
568 W. H. Scouten, H. Knowles, L. C. Freitag, and W. Iobst, *Biochim. Biophys. Acta*, 1977, **482**, 11.
569 J. Baratti, R. Couderc, C. L. Cooney, and D. I. C. Wang, *Biotechnol. and Bioeng.*, 1978, **20**, 333.
570 M. Ono, T. Tosa, and I. Chibata, *Agric. and Biol. Chem.* (*Japan*), 1978, **42**, 1847.
571 T. L. Parker, A. P. Corfield, R. W. Veh, and R. Schauer, *Z. physiol. Chem.*, 1977, **358**, 789.
572 H. Pedersen, C. Horvath, and C. M. Ambrus, *Res. Comm. Chem. Path. Pharm.*, 1978, **20**, 559.
573 M. Adamich, H. F. Voss, and E. A. Dennis, *Arch. Biochem. Biophys.*, 1978, **189**, 417.
574 W. H. Handsch, P. A. D. Rickard, and S. Nyo, *Biotechnol. and Bioeng.*, 1978, **20**, 95.
575 R. Ohba, H. Chaen, S. Hayashi, and S. Ueda, *Biotechnol. and Bioeng.*, 1978, **20**, 665.
576 J. Turková, K. Bláha, M. Malaníková, D. Vančurová, F. Švec, and J. Kálal, *Biochim. Biophys. Acta*, 1978, **524**, 162.
577 P. Monsan, *European J. Appl. Microbiol. Biotechnol.*, 1978, **5**, 1.
578 K. Kasai and S. Ishii, *J. Biochem.* (*Japan*), 1978, **84**, 1061.
579 P. Monsan and G. Durand, *Biochim. Biophys. Acta*, 1978, **523**, 477.
580 R. N. Mukherjea, P. Bhattacharya, B. K. Ghosh, and D. K. Taraphdar, *Biotechnol. and Bioeng.*, 1977, **19**, 1259.
581 P. Kamoun and O. Douay, *Clinical Chem.*, 1978, **24**, 2033.
582 P. V. Sundaram, M. P. Igloi, R. Wassermann, and W. Hinsch, *Clinical Chem.*, 1978, **24**, 1813.
583 J. S. Colliss and J. M. Knox, *Med. Lab. Sciences*, 1978, **35**, 275.
584 R. E. Adams and P. W. Carr, *Analyt. Chem.*, 1978, **50**, 944.
585 R. S. Schilfreen, D. A. Hanna, L. D. Bowers, and P. W. Carr, *Analyt. Chem.*, 1977, **49**, 1929.
586 M. A. Lyle, S. Pallante, K. Head, and C. Fenselau, *Biomed. Mass Spectrometry*, 1977, **4**, 190.
587 I. Parikh, D. W. MacGlashan, and C. Fenselau, *J. Medicinal Chem.*, 1976, **19**, 296.
588 C. Fenselau, S. Pallante, and I. Parikh, *J. Medicinal Chem.*, 1976, **19**, 679.
589 A. Sugitachi, K. Takagi, S. Imaoka, and G. Kosaki, *Thrombosis and Haemostasis*, 1978, **39**, 426.

The following information on immobilized glycoside and polysaccharide hydrolases and carbohydrate isomerases and oxidases is worthy of note. Supports incorporating magnetic properties have been used for the preparation of immobilized α-amylase.[519] The use of diazotized 1,3-diaminobenzene coated on magnetic iron oxide, itself uncoated or coated with hydrous titanium oxide, was found to result in higher bound activities of the immobilized enzyme. Thermostability was improved by immobilizing the enzyme on magnetic iron oxide coated with hydrous titanium(IV) oxide while preparations of diazotized 1,3-diaminobenzene coated iron oxide had excellent stability towards freezedrying. The properties of dextranase immobilized on porous titanium(IV) oxide,[536] or on titanium(IV) oxide coated with diazotized 1,3-diaminobenzene,[537] have been investigated and optimal conditions for preparation of the immobilized enzymes determined.[538]

Hydrophilic urethane prepolymers are convenient materials for entrapment of enzymes; β-D-fructofuranosidase has been successfully immobilized in such a matrix.[539] Immobilization of the same enzyme on a number of support matrices (cellulose, ionically charged dextrans, concanavalin A) has provided evidence for the involvement of salt links in its stabilization, as shown from pH activity curves, K_m values, and stability measurements.[267]

α-L-Fucosidase has been immobilized on concanavalin A-derivatized agarose and the effects of immobilization on pH dependence and thermal stability investigated.[542] The enzyme was very strongly adsorbed to the matrix and its complete removal even under harsh elution conditions was not possible.

The feasibility of using α-D-galactosidase, immobilized by reaction with agarose cyclic imidocarbonate, in the treatment of patients with Fabry's disease has been investigated.[62] In an artificial system, where sodium taurocholate was used to solubilize the substrate, *O*-α-D-galactopyranosyl-(1 → 4)-*O*-β-D-galactopyranosyl-(1 → 4)-*O*-β-D-glucopyranosyl-(1 → 1)-ceramide was successfully hydrolysed by the immobilized enzyme, but when incubated with plasma (normal or Fabry's) no hydrolysis of the glycosphingolipid was observed. Techniques for the immobilization of enzymes in liposomes for enzyme therapy continue to arouse interest. Glucocerebroside β-D-glucosidase has been entrapped in liposomes for possible therapeutic application in Gaucher's disease.[551]

β-D-Glucosidase has been immobilized on cellulose and the effect of immobilization on the enzyme's pH activity response investigated.[363] The evidence suggests that two local pH maxima previously reported for β-D-glucosidase covalently bound to cellulose (V. R. Srinivasan and M. W. Bumm, *Biotechnol. and Bioeng.*, 1974, **16**, 1413) (Vol. 10, p. 460, ref. 240) may have been due to the use of tris buffer which is known to inhibit β-D-glucosidase. Properties of β-D-glucosidase immobilized on chitosan have been studied with a view to the eventual use of the immobilized enzyme, together with cellulases, for the saccharification of cellulose.[390] The stabilities of β-D-glucosidase towards heating, electrodialysis, and protease treatment have been considerably enhanced by immobilization in a fibroin membrane.[563]

Cells of *Escherichia coli*, immobilized by glutaraldehyde-crosslinking with albumin, retain β-D-galactosidase activity, which has been measured as a function of a number of parameters during the immobilization process.[483] The immobilized cells were initially prepared as a film which after being frozen slowly to −30 °C and then warmed to 4 °C yielded an insoluble sponge-like structure; this was ground and lyophilized before use. The behaviour of the immobilized cells in a packed-bed reactor was investigated and the stability of the β-D-galactosidase activity found to be good. The continuous-flow hydrolysis of lactose has been investigated using β-D-galactosidase modified by attachment of active Brilliant Orange KKh and then adsorbed on to a diethylaminoethyl derivative of crosslinked dextran.[409] Activity yields and enzyme activities of β-D-galactosidase immobilized by entrapment in crosslinked polyacrylamide have been improved by including an enzyme inhibitor, D-glucono-1,4-lactone or D-galactono-1,5-lactone, in the polymerization mixture.[543] The effects of adding bovine serum albumin, dithiothreitol, or glutathione were also investigated. Temperature and pH optima and K_M were not affected by immobilization, but stability to lyophilization was improved.

Factors affecting the stability of glucoamylase immobilized on a number of supports have been investigated and kinetic parameters determined.[544,546,547] Glucoamylase immobilized on chitin has been used in column form for the conversion of a partially hydrolyzed α-amylolytic digest of starch to D-glucose.[549] The half-life of the column was estimated to be about 240 days. Using immobilized glucoamylase, the conversion of starch to D-glucose has been compared for a fluidized-bed and a fixed-bed reactor.[508] The former was found to be superior and was used to determine the suitability of a dispersed plug-flow model to predict the conversion obtained.

D-Glucose oxidase, glucoamylase, and D-glucose isomerase have been immobilized on activated carbon with a view to alleviating deactivation of the enzyme by hydrogen peroxide formed during the enzyme reaction.[545,548] Batchwise and continuous oxidation of D-glucose have been investigated using immobilized D-glucose oxidase and catalase, particularly with respect to the macrokinetics and operational stability of the system.[524] A biochemical fuel cell has been developed which exploits the production of hydrogen from D-glucose by immobilized whole cells of *Clostridium butyricum*.[482]

Mixed immobilized enzyme systems of glucoamylase–D-glucose isomerase and D-glucose isomerase–D-glucose oxidase–catalase have been prepared and operated.[510] Experimental results verified theoretical predictions that optimal bifunctional catalyst configurations could be superior to those where the catalysts were mixed uniformly or arranged sequentially in a tubular reactor.

Coupled reactions of immobilized enzymes continue to be developed for analytical use, particularly for the determination of D-glucose. Oligosaccharides obtained by α-amylolytic hydrolysis of starch have been passed through a column or columns of immobilized D-glucose oxidase, catalase, and glucoamylase or maltase (β-D-glucosidase); the hydrogen peroxide released was detected with a three-electrode amperometric cell.[523] Peroxidase and D-glucose oxidase have been immobilized on activated cellulose[531] or diazotized polyaminostyrene[558] and used for the analysis of D-glucose. D-Glucose oxidase immobilized at the surface of a carbon paste electrode has been used for surface catalysis studies.[557] Hydrogen peroxide formed on oxidation of D-glucose was treated with iodide and then with molybdate to form free iodine which was determined amperometrically. The deactivation of immobilized D-glucose oxidase has been investigated in a continuous tubular flow reactor with particular respect to the effects of oxygen and hydrogen peroxide concentrations.[560]

The analytical performance of co-immobilized preparations of hexokinase and D-glucose 6-phosphate dehydrogenase has been investigated in a microreactor column[561] and in a nylon-tubing coil.[562] The first system proved to be applicable to the automated and continuously referenced analysis of creatine kinase isoenzymes in a clinical laboratory. The analytical performance of the second, adapted for use with the Technicon SMAC analyser, was usable for one month (12 000 D-glucose estimations), was not affected by plasma constituents, and was subject only to errors within the clinically acceptable range. Plasma D-glucose has also been determined using a stirrer containing immobilized D-glucose dehydrogenase.[552] Within- and between-assay precision was good and the method is well-suited to routine use in small-volume clinical laboratories.

The conversion of D-glucose to D-fructose using immobilized preparations of D-glucose isomerase continues to attract much interest.[349,556] Cells of *Streptomyces phaeochromogenes* immobilized at a high concentration by radiation-induced polymerization of glass-forming monomers have also been successfully used for the enzymic conversion.[491] The properties of D-glucose isomerase immobilized on chitosan have been thoroughly investigated: the pH optimum of reaction and the pH stability range were somewhat lowered and the immobilized enzyme was considerably more sensitive to sodium dodecyl sulphate, L-cysteine, or mercaptoethanol than the soluble enzyme, although other properties were little affected.[389] Commercially available preparations of immobilized D-glucose isomerase have been described. In one immobilization procedure an enzyme–gelatine mixture is sprayed into a cold water-immiscible organic solvent and the particles so formed are separated and thoroughly washed before crosslinking with glutaraldehyde and final washing.[555] The particles, which can be stored in solution or dried, are usable in column, batch, continuous feed, stirred tank, and fluid-bed reactors; optimal performance conditions were reported. The use of another commercially available immobilized D-glucose isomerase in large-scale fixed-bed and expanded-bed continuous isomerization processes has also been reported.[554]

The use of co-immobilized lysozyme as a bactericide in columns of immobilized enzymes has been studied.[68] Three experimental columns were prepared; β-D-galactosidase immobilized alone, immobilized β-D-galactosidase with a separate layer of immobilized lysozyme above it, and co-immobilized β-D-galactosidase and lysozyme. Media infected with *Bacillus subtilis* were then passed through the columns. Bacteria appeared in effluent from the first column after a few hours, but their appearance in effluent from the second column was only delayed for a short time. Co-immobilization of lysozyme with β-D-galactosidase appeared to enhance resistance towards bacteria, suggesting that 'self-sterilizing' columns of this design may be useful industrially.

Neuraminidase has been immobilized on a number of matrices and the kinetic and stability properties of the preparations reported.[70,571] Optimal conditions for the use of neuraminidase, immobilized on agarose cyclic imidocarbonate, for the desialylisation of glycoconjugates have been established with fetuin as model compound.[71] The immobilized enzyme, which can also be used for the desialylization of cells, binds large molecular weight substrates differently from the soluble enzyme.

The covalent coupling of polygalacturonase, a pectinolytic enzyme of practical interest for fruit juice clarification, has been investigated using a number of matrices and detailed studies have been made of the efficiency of the coupling reactions and the characteristics of the resulting preparations.[73]

Treatment of the culture filtrate of *Streptomyces flavochromogenes* with tannic acid yields a tannin–pullulanase complex which is readily adsorbed on triethylaminoethylcellulose.[575] The immobilized enzyme shows quantitative retention of enzymic activity, is stable for more than six months, and provides a simple and inexpensive immobilization method which may have industrial potential.

Immobilized whole cells of *Bacillus subtilis* have been used for the production of α-amylase.[480]

Author Index